Principles of Data Transfer Through Communications Networks, the Internet, and Autonomous Mobiles

IEEE Press
445 Hoes Lane
Piscataway, NJ 08854

IEEE Press Editorial Board
Sarah Spurgeon, *Editor-in-Chief*

Moeness Amin
Jón Atli Benediktsson
Adam Drobot
James Duncan

Ekram Hossain
Brian Johnson
Hai Li
James Lyke
Joydeep Mitra

Desineni Subbaram Naidu
Tony Q. S. Quek
Behzad Razavi
Thomas Robertazzi
Diomidis Spinellis

Principles of Data Transfer Through Communications Networks, the Internet, and Autonomous Mobiles

Izhak Rubin
Distinguished Professor Emeritus
Electrical and Computer Engineering Department
University of California, Los Angeles (UCLA)
Los Angeles,
CA, USA

Copyright © 2025 by The Institute of Electrical and Electronics Engineers, Inc. All rights reserved.

Published by John Wiley & Sons, Inc., Hoboken, New Jersey.
Published simultaneously in Canada.

No part of this publication may be reproduced, stored in a retrieval system, or transmitted in any form or by any means, electronic, mechanical, photocopying, recording, scanning, or otherwise, except as permitted under Section 107 or 108 of the 1976 United States Copyright Act, without either the prior written permission of the Publisher, or authorization through payment of the appropriate per-copy fee to the Copyright Clearance Center, Inc., 222 Rosewood Drive, Danvers, MA 01923, (978) 750-8400, fax (978) 750-4470, or on the web at www.copyright.com. Requests to the Publisher for permission should be addressed to the Permissions Department, John Wiley & Sons, Inc., 111 River Street, Hoboken, NJ 07030, (201) 748-6011, fax (201) 748-6008, or online at http://www.wiley.com/go/permission.

Trademarks: Wiley and the Wiley logo are trademarks or registered trademarks of John Wiley & Sons, Inc. and/or its affiliates in the United States and other countries and may not be used without written permission. All other trademarks are the property of their respective owners. John Wiley & Sons, Inc. is not associated with any product or vendor mentioned in this book.

Limit of Liability/Disclaimer of Warranty: While the publisher and author have used their best efforts in preparing this book, they make no representations or warranties with respect to the accuracy or completeness of the contents of this book and specifically disclaim any implied warranties of merchantability or fitness for a particular purpose. No warranty may be created or extended by sales representatives or written sales materials. The advice and strategies contained herein may not be suitable for your situation. You should consult with a professional where appropriate. Neither the publisher nor author shall be liable for any loss of profit or any other commercial damages, including but not limited to special, incidental, consequential, or other damages. Further, readers should be aware that websites listed in this work may have changed or disappeared between when this work was written and when it is read. Neither the publisher nor authors shall be liable for any loss of profit or any other commercial damages, including but not limited to special, incidental, consequential, or other damages.

For general information on our other products and services or for technical support, please contact our Customer Care Department within the United States at (800) 762-2974, outside the United States at (317) 572-3993 or fax (317) 572-4002.

Wiley also publishes its books in a variety of electronic formats. Some content that appears in print may not be available in electronic formats. For more information about Wiley products, visit our web site at www.wiley.com.

Library of Congress Cataloging-in-Publication Data applied for:

Hardback: 9781394267750

Cover Design: Wiley
Cover Image: © chombosan/Alamy Stock Photo

Set in 9.5/12.5pt STIXTwoText by Straive, Chennai, India

To Nira

to our Children:
Orly, Amir and Michael

and to our Grandchildren:
Sophie, Tess, Chloe, Naomi, Ben, Jonathan, Samantha and Jacob

Contents

List of Figures *xv*
About the Author *xxv*
Preface *xxvi*

1   **Introduction: Networking in a Nutshell** *1*
1.1   Purpose *1*
1.2   Networking Terms and Network Elements *2*
1.3   Network Transport Processes *6*
1.4   An Illustrative Transport Process: Sending Packages Across a Shipping Network *9*
1.5   A Layered Communications Networking Architecture *14*
1.6   Communications Network Architecture: User, Control, and Management Planes *27*
1.6.1   Network Architectural Planes *27*
1.6.2   The Data (User) Plane *29*
1.6.3   The Control Plane *30*
1.6.4   The Management Plane *33*
1.7   Illustrative Network Systems *34*
1.7.1   Highway Transportation *34*
1.7.2   Inter-regional Road System *35*
1.7.3   Train Transportation Network *35*
1.7.4   Enterprise Computer Communications Network *35*
1.7.5   Packet-Switching Network and the Internet *36*
1.7.6   Cellular Wireless Networks *38*
1.7.7   WiFi: Wireless Local Area Networks (WLANs) *40*
1.7.8   Satellite Communications Networks *40*
1.7.9   Autonomous Vehicular Networks *43*
1.7.10   Sensor Networks and Internet of Things (IoT) *43*
           Problems *44*

2   **Information Sources, Communications Signals, and Multimedia Flows** *47*
2.1   End Users *47*
2.2   Message Flows *48*
2.3   Service Classes *51*
2.4   Analog and Digital Signals *53*
2.4.1   Analog and Digital Sources *53*

viii | *Contents*

2.4.2	Analog Signals	*54*
2.4.3	Digital Signals	*54*
2.4.4	Discretization: Analog-to-Digital Signal Conversion	*55*
2.5	Frequency Spectrum and Bandwidth	*56*
2.5.1	Time Domain and Frequency Domain	*56*
2.5.2	Frequency Spectrum of Periodic Signals	*58*
2.5.3	Frequency Spectrum of Nonperiodic Signals	*59*
2.5.4	Nyquist Sampling Rate	*61*
2.6	Audio Streaming	*62*
2.6.1	Audio Encoding and Streaming Across a Communications Circuit	*62*
2.6.1.1	Audio Encoding	*62*
2.6.1.2	Replay and Reconstruction of a Transported Stream	*64*
2.6.1.3	Transport of a Stream Across a Circuit-Switched Communications Network	*66*
2.6.2	Audio Streaming across a Packet-Switching Communications Network: Voice Over IP (VoIP)	*68*
2.6.2.1	Voice Over IP (VoIP)	*68*
2.6.2.2	The VoIP Streaming Process and the Realtime Transport Protocol (RTP)	*70*
2.6.2.3	Other CODECs and VOCODERs	*73*
2.6.2.4	Quality Metrics	*75*
2.7	Video Flows and Streams	*77*
2.7.1	Conversion of Light Waves to Electrical Signals	*77*
2.7.2	Digital Still Images	*78*
2.7.3	Full Motion Video	*81*
2.7.4	Video Compression	*81*
2.7.5	Transporting IP Video Streams over Communications Networks	*83*
2.7.6	Dynamic Adaptive Streaming over HTTP (DASH)	*86*
2.7.7	Performance Measures	*87*
2.8	Data Flows	*88*
	Problems	*90*

3	**Transmissions over Communications Channels**	*93*
3.1	Communications Media	*93*
3.2	Wireline Communications Media	*94*
3.3	Wireless Communications Media	*95*
3.4	Message Transmission Over a Communications Channel	*97*
3.5	Noisy Communications Channels	*98*
3.6	Illustrative Calculation of Signal-to-Noise-plus-Interference Ratio (SINR)	*102*
3.7	Channel Capacity	*104*
3.8	Modulation/Coding Schemes (MCSs)	*107*
3.8.1	The Modulation Concept	*107*
3.8.2	Analog Modulation Techniques	*108*
3.8.3	Digital Modulation Techniques	*110*
3.8.4	Illustrative Digital Modulation/Coding Schemes	*113*
3.8.4.1	Modulation/Coding Schemes Used by a Wi-Fi Version	*113*
3.8.4.2	MCS Configurations for an LTE Cellular Wireless Radio Access Network	*115*
	Problems	*117*

Contents | ix

4   **Traffic Processes**   *119*
4.1   A Multilevel Traffic Model   *119*
4.2   Message Traffic Processes   *122*
4.3   Modeling a Traffic Flow as a Stochastic Point Process   *123*
4.4   Renewal Point Processes and the Poisson Process   *125*
4.5   Discrete-Time Renewal Point Processes and the Geometric Point Process   *129*
4.6   Traffic Rates and Service Demand Loads   *131*
4.6.1   Client–Server Traffic Association   *131*
4.6.2   Call Level Traffic Rates   *132*
4.6.3   Burst Level Traffic Rates   *134*
4.6.4   Message Level Traffic Rates   *134*
4.7   Traffic Matrix: Who Communicates with Whom   *136*
    Problems   *139*

5   **Performance Metrics**   *143*
5.1   Quality of Service (QoS) and Quality of Experience (QoE) Metrics   *143*
5.2   Quality of Service (QoS) Metrics for Communications Networking   *144*
5.2.1   Throughput Metrics   *144*
5.2.2   Message Delay Metrics   *148*
5.2.3   Error Rate Metrics   *150*
5.2.4   Availability and Reliability Metrics   *151*
5.2.5   Cyber Security   *153*
5.2.6   Illustration: QoS Metrics for a Cellular Wireless Network   *155*
5.3   Quality of Experience (QoE)   *157*
    Problems   *159*

6   **Multiplexing: Local Resource Sharing and Scheduling**   *161*
6.1   Sharing Resources Through Multiplexing   *162*
6.2   Fixed Multiplexing Methods   *164*
6.2.1   Time Division Multiplexing (TDM)   *166*
6.2.2   Frequency Division Multiplexing (FDM)   *169*
6.2.3   Wavelength Division Multiplexing (WDM)   *170*
6.2.4   Code Division Multiplexing (CDM)   *170*
6.2.5   Space Division Multiplexing (SDM)   *171*
6.3   Statistical Multiplexing Methods   *171*
6.4   Scheduling Algorithms and Protocols   *173*
6.5   Statistical Multiplexing Over One-to-Many Media   *183*
    Problems   *186*

7   **Queueing Systems**   *189*
7.1   A Basic Queueing System Model   *189*
7.2   Queueing Processes and Performance Metrics   *192*
7.3   Queueing Systems: Properties   *196*
7.3.1   Busy Cycle Properties   *196*
7.3.2   Little's Formula   *197*
7.4   Markovian Queueing Systems   *199*

x | *Contents*

7.5	Performance Behavior of Markovian Queueing Systems	*201*
7.5.1	$M/M/1$: A Single Service-Channel Queueing System	*201*
7.5.2	$M/M/1/N$: A Finite Capacity Single Server Queueing System	*207*
7.5.3	A Multi-server Queueing System	*208*
7.6	A Queueing System with General Service Times	*213*
7.7	Priority Queueing	*218*
7.8	Queueing Networks	*222*
7.9	Simulation of Communications Networks	*227*
7.9.1	Monte Carlo Simulations of Communications Networks	*227*
7.9.2	Illustrative Discrete-Event Monte Carlo Simulation of a Queueing System	*230*
	Problems	*235*

8	**Multiple Access: Sharing from Afar**	*241*
8.1	Multiple Access: Sharing from Afar	*241*
8.2	Fixed Multiple Access Schemes	*244*
8.2.1	Time Division Multiple Access (TDMA)	*244*
8.2.2	Frequency Division Multiple Access (FDMA)	*246*
8.2.3	Space Division Multiple Access (SDMA)	*249*
8.2.4	Code Division Multiple Access (CDMA)	*253*
8.3	Demand-Assigned Multiple Access (DAMA) Schemes	*258*
8.3.1	Demand- Assigned Schemes	*258*
8.3.2	Demand-Assigned Reservation Schemes	*258*
8.3.3	Polling Schemes	*262*
8.3.3.1	Polling Methods and Procedures	*262*
8.3.3.2	Performance Behavior of Polling Systems	*269*
8.4	Random Access: Try and Try Again	*272*
8.4.1	Uncoordinated Transmissions Using Random Access	*272*
8.4.2	Pure Random Access: The ALOHA Protocol	*275*
8.4.3	Carrier Sense Multiple Access (CSMA): A Listen Before Talk Protocol	*284*
8.4.4	Carrier Sense Multiple Access with Collision Detection (CSMA/CD) and Ethernet Local Area Network (LAN)	*291*
8.4.5	The Carrier Sense Multiple Access with Collision Avoidance (CSMA/CA) Protocol and the Wi-Fi Wireless Local Area Network (WLAN)	*297*
8.4.5.1	WLAN Layout and Shared Wireless Medium Resources	*297*
8.4.5.2	Frame Types	*299*
8.4.5.3	Distributed Coordination Function (DCF): The Basic CSMA/CA Medium Access Control Scheme	*300*
8.4.5.4	Point Coordination Function (PCF): A Polling-Based Contention-less Access Scheme	*302*
8.4.5.5	Alleviating the Hidden Terminal Problem: An Optional RTS/CTS Scheme	*302*
8.4.5.6	Hybrid Coordination Function (HCF): Providing QoS to Designated Traffic Categories (TC)	*304*
	Problems	*306*

9	**Switching, Relaying, and Local Networking**	*309*
9.1	Switching	*309*
9.2	Extending the Coverage Span: Repeaters and Relays	*317*

Contents | **xi**

9.3	Local Networking Across a Switching Fabric: Bridging of MAC Frames	*321*
9.3.1	Local Internetting Using Bridges and Layer 2 Switches	*321*
9.3.2	Building a Frame Forwarding Table via a Flooding Protocol	*324*
9.3.3	Spanning Tree Protocol (STP) Methods for Constructing a Forwarding Table	*325*
9.3.4	Multipath Networking Across Local Switch Fabrics: Shortest Path Bridging (SPB)	*330*
9.3.4.1	Shortest Path Bridging (SPB)	*330*
9.3.4.2	Illustrative SPB Network	*334*
9.3.4.3	The Control Plane: Link State Dissemination and SPT Constructions	*334*
9.3.4.4	Multitier Overlay: Data Transport Across Multiple Equal-Cost Paths	*337*
9.3.4.5	The Forwarding Data Base (FDB)	*339*
	Problems	*341*

10	**Circuit Switching**	*345*
10.1	Circuit Switching: The Method	*345*
10.2	The Circuit Switching Network System Architecture	*346*
10.3	The Switching Fabric	*351*
10.4	The Signaling System	*354*
10.5	Performance Characteristics of a Circuit Switching Network	*356*
10.6	Cross-Connect Switching and Wavelength Switched Optical Networks	*359*
	Problems	*368*

11	**Connection-Oriented Packet Switching**	*371*
11.1	Connection-Oriented Packet Switching: The Method	*372*
11.2	The Virtual Circuit Switching and Networking Processes	*373*
11.3	Technologies That Use a Connection-Oriented Packet-Switching Method	*376*
11.4	Performance Characteristics of a Virtual Circuit Switching Network	*379*
	Problems	*384*

12	**Datagram Networking: Connectionless Packet Switching**	*387*
12.1	Connectionless Packet Switching: The Method	*388*
12.2	Packet Flows and the Packet Router	*390*
12.3	Performance Characteristics	*392*
	Problems	*395*

13	**Error Control: Please Send It Again**	*397*
13.1	Error Control Methods	*397*
13.2	Error Control Using Forward Error Correction (FEC)	*400*
13.3	Automatic Repeat Request (ARQ)	*404*
13.3.1	Error Detection Coding	*404*
13.3.2	The ARQ Process	*406*
13.3.3	Stop-and-Wait ARQ	*407*
13.3.4	Go-Back-N ARQ: A Sliding Window Protocol	*415*
13.3.5	Selective-Repeat ARQ: Resend Only Uncorrectable Received Blocks	*420*
13.4	Hybrid ARQ (HARQ) Error Control	*423*
	Problems	*429*

xii | *Contents*

14 Flow and Congestion Control: Avoiding Overuse of User and Network Resources *431*

14.1 Flow and Congestion Controls: Objectives and Configurations *431*
14.2 Feedback-Based Closed-Loop Flow Control *434*
14.3 Open-Loop Input-Rate Flow and Congestion Controls *436*
14.4 Congestion Control: Relieving Bottlenecks *444*
14.4.1 Reactive Congestion Control *444*
14.4.2 Proactive Congestion Control *448*
Problems *451*

15 Routing: Quo Vadis? *453*

15.1 Routing: Selecting a Preferred Path *453*
15.2 Route Metrics *455*
15.3 Routing Domains and Autonomous Systems *457*
15.4 Route Selection Methods *461*
15.5 Shortest Path Tree (SPT): Mapping the Best Path to Each Node *464*
15.6 Distance Vector Routing: Consult Your Neighbors *465*
15.7 Link-State Routing: Obtain the Full Domain Graph *470*
Problems *473*

16 The Internet *475*

16.1 The Internet Networking Architecture *476*
16.2 HTTP: Facilitating Client–Server Interaction Over the Internet *482*
16.3 Internet Protocol (IP) Addresses *485*
16.3.1 Internet Protocol Version 4 (IPv4) Addresses *485*
16.3.2 Internet Protocol Version 6 (IPv6) Addresses *491*
16.4 Internet Protocol (IP) Packets *492*
16.4.1 Internet Protocol Version 4 (IPv4) Packets *492*
16.4.2 Internet Protocol Version 6 (IPv6) Packets *494*
16.5 Transport Layer Protocols *496*
16.5.1 Transmission Control Protocol (TCP) *496*
16.5.2 User Datagram Protocol (UDP) *501*
16.5.3 QUIC: A Fast and Secure Transport Protocol *503*
16.6 Routing Over the Internet *508*
16.6.1 Autonomous Systems as Routing Domains *508*
16.6.2 Intra-domain Routing: OSPF *509*
16.6.3 Inter-domain Routing: Border Gateway Protocol (BGP) *511*
Problems *518*

17 Local and Personal Area Wireless Networks *521*

17.1 Illustrative Personal Area and Local Area Wireless Networks *522*
17.2 WiFi: A Wireless Local Area Network (WLAN) *523*
17.3 Personal Area Networks (PANs) for Short-Range Wireless Communications *528*
17.3.1 Personal Area networks (PANs) *528*
17.3.2 Short-Range Wireless Communications Using Bluetooth *528*
17.3.3 Short-Range Low Data Rate Wireless Communications Using Zigbee *533*
Problems *539*

Contents | **xiii**

18   **Mobile Cellular Wireless Networks** *541*
18.1   Configurations of Mobile Wireless Networks *541*
18.2   Architectural Elements of a Cellular Wireless Network *544*
18.2.1   The Cellular Coverage *544*
18.2.2   Cellular Networking Generations *546*
18.2.3   Key Components of a Cellular Network Architecture *549*
18.3   Cellular Network Communications: The Process *551*
18.4   The 4G-LTE Protocol Architecture *554*
18.4.1   Allocation of Wireless Access Resources *558*
18.5   Next-Generation 5G, 6G, and Millimeter-Wave Cellular Networks *560*
    Problems *562*

19   **Mobile Ad Hoc Wireless Networks** *567*
19.1   The Mobile Ad Hoc Wireless Networking Concept *567*
19.2   Ad Hoc On-Demand Distance Vector (AODV) Routing *569*
19.3   Dynamic Source Routing (DSR) *573*
19.4   Optimized Link State Routing (OLSR): A Proactive Routing Algorithm *576*
19.5   Mobile Backbone Networks (MBNs): Hierarchical Routing for Wireless Ad Hoc Networks *582*
    Problems *592*

20   **Next-Generation Networks: Enhancing Flexibility, Performance, and Scalability** *595*
20.1   Network Virtualization *595*
20.2   Software-Defined Networking (SDN) *597*
20.3   Network Functions Virtualization (NFV) *599*
20.4   Network Slicing *601*
20.5   Edge Computing, Open Interfaces, Technology Convergence, Autonomous Operations *602*
    Problems *604*

21   **Communications and Traffic Management for the Autonomous Highway** *607*
21.1   Data Communications Services for Vehicular Wireless Networks *608*
21.2   Configurations of Vehicular Data Communication Networks *610*
21.3   Vehicular Wireless Networking Methods *613*
21.3.1   VANET-Based Vehicle-to-Vehicle (V2V) Networking Protocols *613*
21.3.2   Selection of Relay Nodes *616*
21.3.3   Flow and Congestion Controls *625*
21.3.4   Vehicular Backbone Networks (VBNs): Hierarchical Networking Using Cluster Formations *628*
21.3.5   Vehicular Backbone Networks (VBNs): Backbone Network Synthesis *633*
21.3.6   Infrastructure-Aided Vehicle-to-Vehicle (V2V) Networking *638*
21.3.7   Cellular Vehicle-to-Everything (CV2X) Networking *641*
21.3.8   Networking Automated and Autonomous Vehicles *648*
21.3.9   Traffic Management of Autonomous Highway Systems *650*

xiv | *Contents*

21.3.9.1 Achieving the Highest Vehicle Flow Rate *650*
21.3.9.2 Traffic Management Under Queueing and Transit Delay Limits *656*
Problems *665*

22 **Networking Security** *671*
22.1 Network Security Architecture and Cybersecurity Frameworks *671*
22.2 Message Confidentiality: Symmetric Encryption *675*
22.3 Public Key Encryption (PKE) *677*
22.4 Digital Signature *679*
22.5 Secure Exchange of Cryptographic Keys *680*
22.6 Secure Client–Server Message Transport Over the Network *682*
Problems *683*

References *685*
Index *689*

List of Figures

Figure 1	Coverage of Topics as Divided into Parts I–III. *xxviii*
Figure 1.1	Illustrative Communications Network *3*
Figure 1.2	Message Fields and Wrapping Headers *11*
Figure 1.3	Vertical and Horizontal Message Communications Between Protocol Layer Entities at Layer-$(i+1)$ and Layer-i *16*
Figure 1.4	Services Provided by Open Systems Interconnection (OSI) Layers *18*
Figure 1.5	The Open Systems Interconnection (OSI) Layered Reference Model *19*
Figure 1.6	The TCP/IP Internet Model *27*
Figure 1.7	Vehicles on a Highway *29*
Figure 1.8	Three Level Hierarchical Network *31*
Figure 1.9	Hierarchical Organization of Conducting Links in a Leaf *32*
Figure 1.10	Multilane Highway *35*
Figure 1.11	Inter-regional Road System in California (Truncated) *36*
Figure 1.12	Union Pacific Railroad System *37*
Figure 1.13	Enterprise Computer Communications Network *37*
Figure 1.14	ARPANET 1980 Network Layout *38*
Figure 1.15	A Cellular Network *39*
Figure 1.16	LTE Cellular Network Radio Access System Architecture *39*
Figure 1.17	Wi-Fi-Aided Network *40*
Figure 1.18	Satellite Communications *41*
Figure 1.19	Vehicular Network *43*
Figure 1.20	IoT Architecture *44*
Figure 2.1	Realtime and Store-and-Forward Message Flows. (a) Realtime Transmission of a Stream (b) Store & Forward Message Stream *49*
Figure 2.2	Analog Signal *54*
Figure 2.3	Digital Signal *55*
Figure 2.4	Digitized Signal *56*
Figure 2.5	Sine Signals: (a) a Sine Signal; (b) A Two-Tone Composite Signal; The spectrum of the Two-Tone Signal *57*

xvi | *List of Figures*

Figure 2.6 Signal Representation via Fourier Series *58*

Figure 2.7 (a) A Rectangular Pulse of Width $T = 1$; (b) Spectrum of a Rectangular Pulse with Width $T = 1$ *60*

Figure 2.8 Build-up and Replay of a Data Stream *64*

Figure 2.9 A Bursty Flow Alternating Between Spurt (Active) and Pause (Inactive) Periods *68*

Figure 2.10 The Voice over IP (VoIP) Streaming Process *70*

Figure 2.11 (a) RTP Header; (b) Voice over IP (VoIP) Packet *71*

Figure 2.12 Illustrative Packet Replay Scenario *72*

Figure 2.13 Illustrative Array of Image Pixels *78*

Figure 3.1 Message Transport across a Digital Communications System *97*

Figure 3.2 Signal Perturbed by Noise *99*

Figure 3.3 Analog Signal Modulation Techniques *108*

Figure 3.4 Illustrative Spectral Spans of Modulated Signals *109*

Figure 3.5 Illustrative Digital Signal Modulations *112*

Figure 3.6 Modulation/Coding Schemes Used by IEEE 802.11n Wi-Fi *114*

Figure 3.7 CQI, MCS, and Spectral Efficiency for LTE *116*

Figure 4.1 Traffic of Vehicles Moving Along a Highway *120*

Figure 4.2 Multilevel Traffic Model *121*

Figure 4.3 Realizations of (a) Continuous-Time and (b) Discrete-Time Arrival Point Processes *125*

Figure 4.4 A Realization of a Counting Process Associated with the Displayed Point Process *125*

Figure 4.5 Flows Across a Network Graph *137*

Figure 5.1 Offered, Carried, and Throughput Traffic (Load) Rates *146*

Figure 5.2 Output Flow Rate vs. Input Flow Rate *147*

Figure 5.3 Average Message Delay vs. Normalized Throughput *150*

Figure 5.4 QoS Class Identifier (QCI)-Based Measures as Specified by 3GPP Standard TS23.203/with Permission of 3GPP *156*

Figure 5.5 QoE Factors *157*

Figure 5.6 Illustrative Message Delay Requirements for Applications That Are: (a) Error Sensitive and (b) Error Tolerant *158*

Figure 5.7 Minimum Transport Layer QoE Performance Requirements for IP-HDTV *159*

Figure 6.1 Vehicle Merging Demonstrating on Ramp Multiplexing *162*

Figure 6.2 Input and Output Service Modules in a Packet Router *163*

Figure 6.3 Multiplexer—DeMultiplexer System Arrangement *164*

Figure 6.4 Sharing a Downlink Wireless Communications Channel on a TDM Basis *167*

Figure 6.5 (a) Joint Time–Frequency Plane, (b) TDM Schemes, and (c) FDM Scheme *167*

Figure 6.6 A TDM Circuit Consisting of a Single Time Slot per Frame in Frequency Band F1 *168*

Figure 6.7	Resource Allocation and Scheduling at a Multiplexing Node	*175*
Figure 6.8	Scheduling Parameters	*176*
Figure 7.1	A Basic Queueing System Model	*190*
Figure 7.2	Realization of a System Size Process	*195*
Figure 7.3	Equality of Areas Used to Derive Little's Formula	*197*
Figure 7.4	The $M/M/1$ Single Server Queueing System	*201*
Figure 7.5	Mean System Size vs. Traffic Intensity for the $M/M/1$ Queueing System	*203*
Figure 7.6	Statistical Multiplexing Gain: (a) Each Flow Assigned a Dedicated Channel; (b) Statistical Multiplexing of all Flows Across a Shared Channel	*205*
Figure 7.7	Blocking Probability vs. Message Capacity N for the $M/M/1/N$ System	*208*
Figure 7.8	The $M/M/m$ Multi-server Queueing System	*208*
Figure 7.9	A Service System That Contains No Queueing facility	*211*
Figure 7.10	A Jackson-Type Queueing Network	*222*
Figure 7.11	Illustrative Queueing Network	*224*
Figure 7.12	Illustrative Tandem Queueing Network	*225*
Figure 7.13	Illustrative Discrete Event Simulation of a Queueing System: Program Routines	*230*
Figure 7.14	Global Parameters of the $M/M/1$ Simulation Program	*231*
Figure 7.15	Initialization of the Simulation Program	*232*
Figure 7.16	The Main Program Routine	*232*
Figure 7.17	The Simulation's Timing( ) Routine	*233*
Figure 7.18	Simulation Performance Updating	*233*
Figure 7.19	The Simulation's Arrive( ) Routine	*234*
Figure 7.20	The Simulation's Departure( ) Routine	*234*
Figure 7.21	The Simulation's Report( ) Routine	*235*
Figure 8.1	A Multiple Access Network	*242*
Figure 8.2	An Illustrative Time Division Multiple Access (TDMA) Network Whereby a Medium is Time-Shared by Two Stations	*245*
Figure 8.3	An Illustrative Frequency Division Multiple Access (FDMA) Network Whereby Each Station Is Dedicated a Frequency Band	*247*
Figure 8.4	A 3-Color Cellular Space Division Multiple Access Network	*251*
Figure 8.5	Directional Communications: (a) Peer-to-Peer Directional Communications; (b) Simultaneous Uplink Communications in Four Quadrants of a Cell	*251*
Figure 8.6	Illustrative Baseband (Non-spread) Spectrum and Spread Signal Spectrum in a CDMA System	*254*
Figure 8.7	Illustrative Message and Chip Symbols in a CDMA System	*255*
Figure 8.8	Uplink and Downlink Signaling and Traffic Channels in a DAMA System	*260*
Figure 8.9	(a) Hub Polling in a Ring Network; (b) Hub Polling in a Multi-dropped Tree Network	*263*

xviii | *List of Figures*

Figure 8.10 (a) Token-Passing Ring with Early Token Release; (b) Dual Counter Rotations Ring Network Layout (as for FDDI) *266*

Figure 8.11 (a) Wireless Net Whose Subscriber Stations Communicate with Their Managing Access Point (AP) Station. (b) Wireless Net with Peer-to-Peer Communicating Stations *273*

Figure 8.12 Illustrative Packet Transmission Dynamics Across (a) an Unslotted ALOHA Channel; (b) a Slotted ALOHA Channel *278*

Figure 8.13 Throughput (S) vs. Channel Load (G) Performance Curves Under Slotted and Unslotted ALOHA Schemes *280*

Figure 8.14 Throughput (S) Performance Dynamics Under the Slotted ALOHA Scheme: (a) Without the Use of Flow Admission Control; (b) When Flow Admission Control Is Applied *281*

Figure 8.15 Average Number of Packet Transmissions vs. Throughput (S) Under the Slotted ALOHA Scheme *283*

Figure 8.16 Average Packet Delay (D [slots]) vs. Throughput (S) Under the Slotted ALOHA Scheme: (a) Under Unrestricted Load; (b) Under Flow Control *284*

Figure 8.17 Performance Behavior of a CSMA Scheme: (a) Throughput Performance Under Acquisition Factor $a = 0.2$ (b) Throughput as Function of the Acquisition Factor (a): Throughput Capacity (Series 1) and Throughput Under $E(N_T) = 1.3$ (Series 2) *289*

Figure 8.18 Ethernet Local Area Network (LAN) Single Broadcast Domain Segment Configurations: (a) Broadcast Bus Segment; (b) Repeater Hub Base Segment *295*

Figure 8.19 Switched Ethernet Local Area Network (LAN) *296*

Figure 8.20 A Station Hears Two AP's and Associates with a Selected One *299*

Figure 8.21 Illustrative Frame Transmission Process Under the Contention-Based Access Mode of the CSMA/CA DCF and EDCA Protocols *302*

Figure 8.22 Alternating Point Coordination Function (PCF) Contention Free (CFP) and DCF Contention-Based Periods *302*

Figure 8.23 Illustrative Hidden Terminal Scenario and the RTS/CTS Dialog: Stations A and B Do Not Hear Each Other but Each Can Hear the AP *303*

Figure 8.24 Illustrative RTS/CTS and Data/ACK Transmission Dialog in a Wi-Fi WLAN *303*

Figure 8.25 Default Access Parameters Under 802.11e per Access Class(AC): Backoff Window Ranges and Illustrative Access Parameter Settings *304*

Figure 8.26 LANs in Close Proximity *306*

Figure 9.1 Switching Highways at an Intersection *310*

Figure 9.2 A Fully Connected Network Requiring the Use of No Switches *310*

Figure 9.3 An Illustrative Switch Module *311*

Figure 9.4 An Illustrative Switching Network *312*

Figure 9.5 A Radio Relay Node Placed on a Hill to Enable Communications *317*

Figure 9.6 Dissemination of Data Messages by Relay Nodes Across a Vehicle Highway *320*

Figure 9.7 Format of Ethernet II Frame (a) Without a VLAN Tag and (b) With a VLAN Tag *322*

List of Figures | **xix**

Figure 9.8	Bridges B1 and B2 Are Used for Internetting Frames Between LAN1, LAN2, and LAN3 *324*
Figure 9.9	(a) Topological Layout of a Network of LANs Interconnected by Bridges; (b) an Embedded Spanning Tree Layout *326*
Figure 9.10	(a) A Network of Interconnected Layer-2 Switches; (b) A Spanning Tree for the Network in (a); (c) A Shortest Path Tree (SPT) Rooted at Switch 8 Which May Be Employed by a Shortest Path Bridging (SPB) Protocol such as IEEE 802.1aq *329*
Figure 9.11	A Multicast Flow in a SPB Network Using VSN I-SID Involving Group Members S1, S2, S3, S4 *332*
Figure 9.12	Illustrative Layout of a Shortest Path Bridging (SPB) Network *333*
Figure 9.13	Illustrative Shortest Path Trees in a SPB Network: (a) Network Layout, (b) SPT1/B-VID=0300 Rooted at Node 1, (c) SPT2/B-VID=0100 Rooted at Node 1, (d) SPT1/B-VID-0300 Rooted at Node 2, (e) SPT1/B-VID-0300 Rooted at Node 3, and (f) SPT2/B-VID-0100 Rooted at Node 4 *338*
Figure 9.14	Forwarding Table/Forwarding Data Base (FDB) at Node 4 Based on Multicast Shortest Path Tree (MSPT) Rooted at Node 4 at Tier B-VID=0100 *340*
Figure 10.1	Architectural Depiction of a Circuit-Switched Network System *347*
Figure 10.2	Illustrative Circuit Capacity Allocation in a Circuit Switched Network: (a) For a Circuit Connecting Edge Switches N1 and N5 Through Tandem Switch N3; (b) For a Circuit Connecting Edge Switches N1 and N4 Through Tandem Switch N3 *348*
Figure 10.3	Illustrative Circuit-Switching Table *350*
Figure 10.4	Time–Space–Time (TST) Circuit-Switching Module *351*
Figure 10.5	A Network System Using Cross-Connect Modules *363*
Figure 10.6	All-Optical Cross-Connect Networks Using WDM Lines: (a) Ring Network Using Optical ADM (OADM) Nodes; (b) Optical Mesh Network Using Optical Cross-Connect (OXC) Nodes *364*
Figure 10.7	A Highway Transportation Path Analogous to an All-Optical Cross-Connect WDM-Based Lightpath *366*
Figure 11.1	Message Flows in an Illustrative Connection-Oriented Packet-Switching Network *375*
Figure 11.2	An Illustrative VCS Switching Table at Node N3 *375*
Figure 11.3	Formats of Asynchronous Transfer Mode (ATM) Cells: (a) Across the UNI; (b) Across the NNI *377*
Figure 11.4	Functional Modules in a VCS-Switching Node *381*
Figure 12.1	Packet Flows in a Datagram Packet-Switching Network *389*
Figure 12.2	Functional Modules of a Packet Router *391*
Figure 12.3	Network Layer Pipelining in a Store-and-Forward Communication Network: (a) Message Switching Transport; (b) Packet Switching Transport Illustrating the Impact of Packet Pipelining Across the Network Layer *393*
Figure 13.1	Encoding and Decoding Error Control Blocks Transported Across a Communications Channel *401*

xx | *List of Figures*

Figure 13.2   Information and Code Fields in a Systematic Error Control Block for: (a) Forward Error Control (FEC) Block; (b) ARQ Error Control Block   *402*

Figure 13.3   Communications in a Classroom   *405*

Figure 13.4   Illustrative Exchange of Error-Control Blocks and ACK Messages Between Two Stations Under Stop-and-Wait ARQ Scheme   *408*

Figure 13.5   Message and ACK Flows Across a Half-Duplex Channel Connection Under a Stop-and-Wait ARQ Scheme   *410*

Figure 13.6   Block and ACK Flows Under a Go-Back-N ARQ Scheme: (a) Timeout Triggered Retransmission of Block F2; (b) Time Delay Incurred in the Transport of Block 1 That Is Retransmitted Three Times   *418*

Figure 13.7   Block and ACK Flows Under a Selective-Repeat ARQ Scheme: (a) Timeout Triggered Selective Retransmission of Block F2; (b) channel Occupancy and Time Delay in the Transport of Block M1 That Is Retransmitted Two Times   *421*

Figure 13.8   ARQ Configuration Consisting of Eight Parallel Stop-and-Wait ARQ HARQ Processes   *428*

Figure 14.1   Flow and Congestion Control Methods: (a) Closed-Loop Feedback Based; (b) Open-Loop Input Rate Control   *432*

Figure 14.2   Regulated Access   *434*

Figure 14.3   Traffic Shaping Using a Token Bucket Algorithm   *440*

Figure 14.4   Impact of Traffic Shaping by Using a Token Bucket Algorithm. (a) Steady Flow, (b) Bursty Flow, and (c) Quasi-bursty Flow   *441*

Figure 14.5   Illustration of a Leaky Bucket Algorithm   *443*

Figure 14.6   TCP Congestion Control   *446*

Figure 15.1   Network Layout and Routes   *454*

Figure 15.2   Driving Map to Las Vegas   *458*

Figure 15.3   Routing Across Multiple Domains: Intra-domain Routing Schemes Employ Protocols such as RIP, OSPF, EIGRP. Inter-domain Routing schemes Employ Exterior Gateway Protocols (EGPs) such as BGP. Boundary Routers (BRs) Are Used to Interconnect Domains   *460*

Figure 15.4   Shortest Paths for an Illustrative Network Using Distance Vector Routing Algorithm: (a) The Weighted Network Graph Showing the Link Cost Values; (b) Routing Table Showing Shortest Paths That Are No Longer Than 1 Hop per Path; (c) Routing Table Showing Shortest Paths That Are No Longer Than 2 Hops per Path; (d) Routing Table Showing Shortest Paths That Are No Longer Than 3 Hops per Path; (e) Shortest Path Tree (SPT) Rooted at Router A; (e) SPT Rooted at Router E   *464*

Figure 15.5   Distance Vector Routing   *466*

Figure 15.6   A Tandem Path That Connects Routers A, B, and C   *469*

Figure 15.7   Calculation of the Shortest Path Tree (SPT) by Using a Link State Routing Algorithm: (a) The Weighted Graph Representing the Network's Domain Layout; (b) Step-by-Step Calculation Performed at Source Node A by Using Dijkstra's Algorithm; (c) The Resulting SPT Rooted at Router A   *471*

List of Figures | **xxi**

Figure 15.8 Dijkstra's Algorithm for Calculating the Shortest Path Tree (SPT) *472*

Figure 16.1 The Internet's Protocol Architecture and Commonly Employed Protocols *476*

Figure 16.2 Illustrative Configuration of an IP Internetwork System *479*

Figure 16.3 Hierarchical Organization of Internet Service and Backbone Networking Providers *480*

Figure 16.4 Illustrative Internet Access and Backbone Networks *481*

Figure 16.5 Application, Transport and Network Layer Protocol Services for HTTP under (a) HTTP 1.1 and HTTP/2; (b) HTTP/3 *484*

Figure 16.6 Illustrative Internet Protocol Version 4 (IPv4) Address Representation in Binary and Dot-Decimal Formats *486*

Figure 16.7 IPv4 Public and Private Address Ranges *487*

Figure 16.8 Illustrative Source Network Address Translation (S-NAT) Under IPv4 *488*

Figure 16.9 IP Network Configuration and Illustrative IPv4 Addresses *489*

Figure 16.10 Home Network Configuration Showing DNS and Web Servers *490*

Figure 16.11 IP Version 6 (IPv6) Address Fields *492*

Figure 16.12 IP Version 4 (IPv4) Address Fields *493*

Figure 16.13 IP Version 6 (IPv6) Packet's Fixed Header Fields *495*

Figure 16.14 IP Version 6 (IPv6) Packet's Hop-by-Hop Options Extension Header *496*

Figure 16.15 IP Version 6 (IPv6) Packet's Routing Extension Header *496*

Figure 16.16 A Transmission Control Protocol (TCP) Message *498*

Figure 16.17 User Datagram Protocol (UDP) Datagram Header *502*

Figure 16.18 Handshake Transactions under TCP and QUIC: (a) TCP and TLS; (b) QUIC *503*

Figure 16.19 OSPF Routing Areas and Router Types *511*

Figure 16.20 A Multi-AS Network Layout *513*

Figure 16.21 Contents of a BGP Update Message *514*

Figure 16.22 Route Update at a BGP Border Router *515*

Figure 17.1 Comparison of Power Consumption and Complexity vs. Data Rate for Versions of WiFi, Bluetooth, and Zigbee Systems *522*

Figure 17.2 Illustrative Network Configuration Involving a Wireless Personal Area network (WPAN) and a Wireless Local Area Network (WLAN) *523*

Figure 17.3 Format of a WiFi MAC Frame *524*

Figure 17.4 Attributes of WiFi Versions *526*

Figure 17.5 Bluetooth Net Configurations: (a) A Piconet Consisting of a Master (M) Device and a Slave (S) Device; (b) A single Master Device Communicating with Multiple Slave Devices; (c) A Scatternet Consisting of Several Interconnected Piconets *529*

Figure 17.6 Bluetooth-Layered Protocol Stacks: (a) Bluetooth BR/EDR Classic; (b) Bluetooth Low Energy (BLE) *531*

Figure 17.7 Zigbee Net Layouts: (a) Star Topology. (b) Tree Topology. (c) Mesh Topology *535*

Figure 17.8 Protocol Layer Stack for a Zigbee System *536*

xxii | *List of Figures*

Figure 18.1   An Infrastructure-Based Wireless Network   *542*

Figure 18.2   An Ad Hoc Wireless Network Employing Multi-hop Peer-to-Peer Communications   *543*

Figure 18.3   A Hybrid Infrastructure of a Wireless Network   *543*

Figure 18.4   Spatial Reuse in a Cellular Network   *545*

Figure 18.5   High-Level LTE Network Architecture   *549*

Figure 18.6   The LTE Radio Access Network Architecture   *550*

Figure 18.7   The LTE Evolved Packet Core (EPC)   *550*

Figure 18.8   The LTE Protocol Architecture: (a) The User Plane; and (b) The Control Plane   *554*

Figure 18.9   LTE Bearers   *556*

Figure 18.10   The LTE Protocol Stack   *557*

Figure 18.11   A Data Flow Across Radio Access Network (RAN) Layers   *558*

Figure 18.12   A Resource Block   *558*

Figure 19.1   Illustrative AODV Protocol Messaging: (a) Topological Layout of an Ad Hoc Network; (b) Selective Route Request (RREQ) Packets; (c) Route Reply (RREP) Control Packet Flows and Data Packet Flows Across Selected Path   *571*

Figure 19.2   Illustrative DSR Protocol Packet Flows and Nodal Route Caches   *575*

Figure 19.3   Illustrative Selection of Multi-Port Routers (MPRs) Under Optimized Link-State Routing (OLSR) Protocol in MANET: (a) 2-Hop Neighborhood of Node A and Its Selected MPR(A) Routers; (b) MPRs Selected by Node A and by Node D for the Underlying MANET Layout   *579*

Figure 19.4   Illustrative Network Layout of a Mobile Backbone Network (MBN) Under the MBNP Protocol   *583*

Figure 19.5   Instances of Mobile Backbone Network (MBN) Layouts   *587*

Figure 19.6   Two Instances of Unmanned Ground Vehicle (UGV) Aided Mobile Backbone Network (UGV-MBN) Synthesized by Using MBNP   *587*

Figure 19.7   Unmanned Aerial Vehicle (UAV) Aided Mobile Backbone Network (UAV-MBN) Synthesized by Using MBNP   *588*

Figure 19.8   Comparative Performance Behavior of the MBNR, MBNR-FC and AODV Ad Hoc Routing Schemes: (a) Throughput vs. Offered Flow Rates; (b) Average Delay vs. Throughput Rate Performance; (c) Delay Jitter vs. Offered Traffic Load Rate   *591*

Figure 19.9   Connectivity Graphs for Multi-radio MBN System: (a) When Employing Higher-Power Radio Modules for Bnet and Anet Communications; (b) When Using Lower Power Radio Modules for Intra-Anet Communications   *592*

Figure 20.1   High-Level SDN Architecture   *598*

Figure 20.2   High-Level NFV Framework   *600*

Figure 20.3   NFV-Based Network Service Represented as an NF Graph   *601*

Figure 21.1   Safety-Oriented Key Services and Message Types for Vehicular-Networked Systems   *610*

List of Figures | **xxiii**

Figure 21.2 Non-safety-Oriented Key Services and Message Types for Vehicular-Networked Systems *611*

Figure 21.3 V2X Performance Requirements for Wireless Communications Systems Employed by Autonomous Vehicular Networked *611*

Figure 21.4 Vehicular Networking Configurations: (a) Vehicular Ad Hoc Network (VANET) Communications Using Vehicle-to-Vehicle (V2V) and Vehicle-to-Infrastructure (V2I) Links; (b) Backbone-Aided Vehicular Network Using Vehicle-to-Network (V2N) Communications; (c) Hybrid VANET and Backbone-Aided Vehicular Communications Network *612*

Figure 21.5 US Wireless Access in Vehicular Environments (WAVE) Protocol Stack and ETSI GeoNetworking layers *614*

Figure 21.6 Vehicular Configurations and Relay Selections: (a) a Platoon of Vehicles Configured to follow each other at Fixed inter-vehicle distances; (b) Vehicles assuming randomly varying inter-vehicle distances *615*

Figure 21.7 GeoNetworking (GN) Protocol Layers *618*

Figure 21.8 Structure of a GeoNetworking (GN) Packet *619*

Figure 21.9 Cross-Layer Distributed Congestion Control (DCC) *626*

Figure 21.10 Architecture of a Vehicular Backbone Network (VBN) *629*

Figure 21.11 A Vehicular Backbone Network (VBN) Configuration in Support of Broadcast of a Data Packet Flow Originating at RSU Node RN1 *633*

Figure 21.12 Performance Behavior of a Backbone Network Synthesized for a Vehicular Backbone Network (VBN): (a) Broadcast Throughput Capacity (C_B) and Transport Throughput Capacity (C_T) Performance Curves vs. the Inter-BN Distance (D) Under Several Reuse-M Levels; (b) Broadcast Throughput Capacity Curves vs. the Product of the Vehicular Traffic Density Rate and the Inter-BN Distance (λD) *636*

Figure 21.13 A VBN-Based V2V Network That Is Aided by the Use of an Infrastructure Core Network *639*

Figure 21.14 Infrastructure-Aided Vehicular Network Using Platoon Formations *640*

Figure 21.15 Cellular V2X Configurations: (a) Uplink, Downlink, and Sidelink; (b) A CV2X Network Configuration Showing Infrastructure Access (Using the Uu Interface) I2N Links and Sidelink Based (Using the PC5 Interface) V2V, V2I, and V2P Links *642*

Figure 21.16 Vehicular Network System That Employs V2V, V2I and C-V2X Communications Links Supplemented by the Use of UAVs and/or Satellites *650*

Figure 21.17 Flow of Vehicle Platoons Across a Highway Link *650*

Figure 21.18 Platoon-Oriented Parameters *651*

Figure 21.19 Illustrative Performance Curves for Vehicular Platoon Traffic Moving Along an Autonomous Highway: (a) Flow vs. Speed and (b) Speed vs. Flow *655*

Figure 21.20 Illustrative Performance Curves for Vehicular Platoon Traffic Moving Along an Autonomous Highway: (a) Flow vs. Density and (b) Speed vs. Density *655*

Figure 21.21 A Link Span of a Highway *656*

Figure 21.22 Delay Constrained Performance of Platoon Flows Along a Lane: (a) Optimal Speed vs. Link Length; (b) Optimal Vehicle Flow Rate vs. Link Length *659*

xxiv | *List of Figures*

Figure 21.23 Single-Link Multiple Lane Configuration *660*

Figure 21.24 Illustrative Transportation Network *662*

Figure 22.1 National Institute of Standards and Technology (NIST) Cybersecurity Framework *673*

Figure 22.2 Symmetric Key Encryption *675*

Figure 22.3 An Illustrative Message Transfer Under Public Key Cryptography *677*

Figure 22.4 A Digital Signature Scheme Employing Public Key Cryptography *679*

Figure 22.5 The Diffie–Hellman (DH) Secure Key Exchange Algorithm *681*

About the Author

Izhak Rubin, PhD, is Distinguished Professor Emeritus of Electrical and Computer Engineering at UCLA, California, USA. He has decades of experience in research and development studies of the Internet, and has published very widely on networking methods, performance modeling, and analysis techniques. He has served as the editor of leading professional journals, and has been elected as an IEEE Life Member Fellow.

Preface

Communication networks have transformed our lives. From the first telegraph networks in the 1840s to today's ubiquitous wireline, wireless, and satellite networks, our world has been connected in ways that would have been unimaginable a few decades ago. The Internet, cellphones, Wi-Fi, telemedicine, video conferencing videoconferencing, chat programs, social networking, company marketing and user support, remote education and training, and more have changed how we work, where we work, and how we communicate with friends, family, and service providers worldwide.

In addition to teaching the fundamentals of communications networks to our students, as I have doing for many years in the electrical and computer engineering department at UCLA, and to members of commercial and governmental organizations, I have been finding out that many other persons are highly interested in gaining a conceptual understanding of the methods used to transport messages across communications networks, including the Internet and wireless-based Wi-Fi, cellular, and satellite networks. There is also much recent interest in learning about connected self-driving vehicles and in understanding the principles of message communications between mobiles that use an autonomous highway system.

Interested persons have been noted to have educational and professional background and experience in diverse areas, including engineering, physical sciences, business, management, law, finance, political science, biological sciences, medical sciences, information technology (IT), education, law enforcement, public safety, defense, military, humanities, art, psychology, sports, accounting, artificial intelligence (AI), and other. There is wide interest in learning about the approaches and techniques used by communications networks to transfer messages among persons, websites, computers, embedded sensing systems, and intelligent devices. It is essential today for students and practitioners in diverse areas and disciplines to learn the principles of message transfer in communications networks, as taught in this book.

I have been teaching, carrying out research investigations, and engaged in network development projects since 1970. Serving as a UCLA distinguished professor in the electrical and computer engineering department, I have been leading studies and engineering teams in contributing to projects that relate to the development of wireline and wireless telecommunications and computer communications networked systems, such as the Internet, wireless cellular, Wi-Fi, and autonomous networked mobile wireless systems.

I have been teaching courses on computer communication networks and telecommunications systems, and related engineering subjects to BS, MS, and PhD students. I have also been lecturing at continuing education and engineering training programs and have received an excellence in teaching award from UCLA Engineering. My classes have been exposing students to the principles of wireline and wireless computer communications network systems. Course attendees include persons that work in diverse fields, engaging in jobs that involve industrial, commercial, governmental, academic, and research and development institutions in the United States and throughout the world.

I have been contributing to commercial, governmental, and military organizations in the design, analysis, implementation, and testing of communications network systems. I have been serving as a technical leader in contributing to network design and analysis projects. I have been innovating methods, algorithms, and protocols, aimed for the design and implementation of wireline and wireless communications networks. I have received an IEEE Life Fellow membership award for my contributions to education and research in the area of computer communication networks.

The networking material, systems, and methods presented in this book are derived in large part from my corresponding experience over more than 50 years in basic and applied research developments, in the teaching of fundamental and advanced concepts of communications networking to persons in academia and in commercial and governmental organizations, and in my practical experience in contributing to the design and evaluation of network systems. My continuing education and in-house courses have also included nonengineering students. In many situations, I have been called upon to explain concepts and methods used for the transfer of messages across network systems in a manner that would promote understanding by persons that have no expertise or background in the underlying disciplines.

My objectives in this book are to provide a reader with conceptual understanding of the principles of message transfer in computer communications networks. I explain and illustrate fundamental networking concepts, and describe the working of communications network systems, including the Internet and autonomous mobile systems, in a manner that is accessible to many readers. The book serves the educational needs of undergraduate and postgraduate students, as well as a valuable resource to persons at large, including researchers and practicing engineers, that are interested to learn the principles of data transfer through and operations of communications network systems. In my presentations in this book, the following subject matters are highlighted:

- Architectures of networking systems and protocols.
- Characteristics of video, audio, and data flows and streams.
- Methods for transmission of signals across communications channels.
- Techniques for sharing communications media and service entities among message flows. Scheduling schemes that provide for the effective allocation of shared communications resources, including multiplexing mechanisms that are used for sharing a communications link by local flows and multiple access algorithms that are employed for sharing communications media among geographically distributed information sources.
- Switching methods; concepts of operation and service provided by circuit and packet switching networks.
- Schemes and algorithms for performing error control, flow control, congestion control, and routing of message flows across a network.
- Protocols, communications networking algorithms, and performance behavior features of specific classes of network system technologies, including the Internet, mobile wireless networks, Wi-Fi wireless local area networks, cellular wireless networks, satellite networks, personal area networks, and connected vehicle systems. The latter make use of data message dissemination among self-driving vehicles that use the autonomous highway.
- Secure message transfer across computer communications networks.

xxviii | *Preface*

To make the material accessible to students, including advanced (junior-senior level) undergraduate and graduate students, and to a wide community of readers, I have structured my presentations to feature the following ingredients:

- My descriptions of basic concepts and methods assume no prior knowledge of the subject matter.
- It is not my purpose to present the numerous details that are involved in the implementation of protocols, algorithms, systems, and management schemes that are used in the operation of specific networking technologies. My aim is to foster understanding of the fundamental concepts of networking architectures and schemes that are used for the transfer of message flows across a network.
- I have included illustrative numerical examples that demonstrate the functioning and performance behavior of networking protocols and algorithms under various operational scenarios.
- The description of networking algorithms, protocols, and processes is presented in a manner that promotes the reader's understanding of the underlying principles without requiring prior knowledge of the subject matter and mostly without having to rely on the use of intricate mathematical tools.
- To accommodate readers that are interested in learning the performance characteristics of networking schemes and systems, I have included, in certain sections, analytical (including Probability based) models, and used them to present mathematical expressions that are employed for carrying out performance analysis and system design. I include discussions that explain and illustrate the essence and use of displayed mathematical expressions. A reader may skip reviewing the detailed descriptions of these analytical models and analyses while still keep reading to learn their applicability and implications. By reviewing the associated performance evaluation discussions and by examining the provided examples, the reader would gain valuable understanding of the merits and performance characteristics of involved networking algorithms, schemes, and protocols.
- To highlight the essence of the presented material, I have included at the start of each chapter a brief overview of the underlying objectives of the networking schemes and algorithms that are described in the chapter, and presented a brief executive summary of key methods that are employed to achieve these objectives.

As noted in Figure 1, the material presented in the book is divided into three parts. In Part I, network users, architectures, and protocols in a nutshell are described. Networking schemes and algorithms are presented in Part II. It is followed by Part III, which includes descriptions of key network systems. A brief description of each part follows.

As shown in Table 1, Part I consists of Chapters 1–5. The essence of *networking in a nutshell* is presented in Chapter 1. Fundamental concepts, terms, and notations are introduced. A network layout is modeled as consisting of nodes and links. Nodes act as switches, routers, and end users. Links represent communications channels that provide for the transmission of messages between attached nodes. Communications networking protocols enable entities that populate nodes to intelligently interact with each other in assuring the consistent, reliable, and timely dissemination of messages across a network. A commonly used networking protocol architecture is described. It is compatible with a reference model whose universal use allows different network systems to intercommunicate, forming an Internet architecture. Widely used network systems are illustrated, including analogous transportation networks. Reviewing this chapter provides the reader with an overview and a basic level of understanding of the networking principles that are used to execute message transfer through communications networks such as the Internet, satellite, and mobile wireless networks.

Figure 1 Coverage of Topics as Divided into Parts I–III.

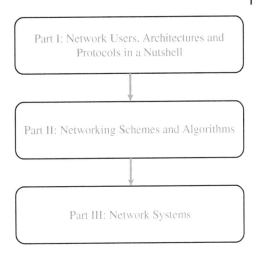

Table 1 Part I: Network Users, Architectures, and Protocols in a Nutshell.

Chapter	Topic
Chapter 1	Introduction: Networking in a Nutshell
Chapter 2	Information Sources, Communications Signals, and Multimedia Flows
Chapter 3	Transmissions over Communications Channels
Chapter 4	Traffic Processes
Chapter 5	Performance Metrics

Messages that are transported by end users across a communications network carry information that is created by variety of media sources and represented in multitude of forms. Audio, data, and video streams and message flows are commonly transported. In Chapter 2, we describe the characteristics of *analog and digital electronic signal flows*. Frequency characteristics of such signals and their ensuing spectral bandwidth levels are discussed.

To transfer a message from one node to a neighboring node, it is transmitted across a wireline or wireless communications link. For this purpose, the sending module employs a transmitter while the receiving device uses a receiver. A single device is commonly used for both transmission and reception of electronic signals; it is identified as a transceiver. As a transmitted message signal propagates across a communications channel, its power strength deteriorates. As a link's distance increases and as the power levels caused by environmental noise and interference signals increase, it becomes more challenging for a receiver to correctly detect and process a received signal in aiming to extract the intended message that it carries. A minimal Signal to Interference plus Noise Ratio (SINR) level is required for a receiving entity to be able to successfully extract a transported message and deliver it, at an acceptable quality, to its intended user. In Chapter 3, we address such issues, as we describe methods used for *reliable transmission over a communications channel*. The use by a transmission module of a modulation / coding scheme (MCS) to enhance the quality of a received message signal is discussed.

Message flows that traverse a communications network are handled by network nodes and end-user devices. To determine the loading levels imposed by flows on network elements, and to assess the quality of service that a network system provides in its transfer of messages, it is essential to characterize the traffic processes that load the system and its nodes and links. In Chapter 4, we employ probabilistic tools to *mathematically model traffic processes* and to calculate the message loading rates that traffic flows induce across a network.

Performance measures that are used to characterize the quality of data transfer across a communications network are presented in Chapter 5. Such *Quality-of-Service (QoS)* performance measures include message throughput, delay, error, and blocking rates. A *Quality-of-Experience (QoE)* metric is used to express the level of satisfaction derived by a destination user upon reception of a designated message, flow, or stream.

Networking schemes and algorithms that are used by a network system for the effective transfer of message flows and streams are described in Part II. As shown in Table 2, this part consists of Chapters 6–15, 22. A network system generally supports a variety of flow types, which are widely distributed across the network. Communications and processing resources, provided by the system's links and nodes, are shared among flows produced by diverse source and application types. Local sharing of a communications link for the transmission by a node of its messages and flows is managed by the use of a *multiplexing* scheme. Multiplexing methods are described in Chapter 6. In turn, the sharing of the resources of a communications medium among geographically distributed message sources is regulated by the employment of a *multiple access* scheme. A wide diversity of multiple access methods have been devised and employed. They are described in Chapter 8.

Performance analysis and design of operations carried out and services rendered by diverse user and network hardware and software entities are performed by using *queueing system models*. Mathematical tools are used to analyze and synthesize a service system. End-user devices, network nodes, and network systems are modeled as service systems. At a node, messages are served by a multitude of processors whose operations and resources are shared by multiple messages and flows. As they wait for their service, messages experience queueing (i.e., waiting in line) delays. Fundamental elements of basic queueing system models are presented in Chapter 7. Background in Probability is required to follow the presented mathematical models and formulas. A reader who does not have the underlying mathematical background, or is not interested in the ensuing mathematical analyses, can skip the corresponding sections. It is however still beneficial for this reader to review the material that describes the ingredients of the models and follow the discussions and

Table 2 Part II: Networking Schemes, and Algorithms.

Chapter	Topic
Chapter 6	Multiplexing: Local Resource Sharing and Scheduling
Chapter 7	Queueing Systems
Chapter 8	Multiple Access: Sharing from Afar
Chapter 9	Switching, Relaying, and Local Networking
Chapter 10	Circuit Switching
Chapter 11	Connection-Oriented Packet Switching
Chapter 12	Datagram Networking: Connectionless Packet Switching
Chapter 13	Error Control: Please Send It Again
Chapter 14	Flow and Congestion Control: Avoiding Overuse of User and Network Resources
Chapter 15	Routing: Quo Vadis?
Chapter 22	Networking Security

displayed graphs that outline and illustrate key patterns that govern the performance behavior of service systems. Queueing models are used to analyze networking schemes and algorithms, including those that are employed to perform multiplexing, multiple access, switching, error control, flow control, congestion control, and resource allocations, as well as those used for the analysis and design of protocol processors.

For learning the principles of operation and the features of networking schemes and algorithms that are presented in Chapters 6 and 8–15, the reader is not required to delve into the underlying mathematical models.

Switching methods are presented in Chapters 9–12. Configurations and algorithms that are employed by using bridging devices for switching messages that are distributed among stations that are located in proximity to each other, such as those that are situated in a single facility, a service center, a web server farm, or a campus setup, are presented in Chapter 9. The principles and features of a *circuit switching (CS)* system, as used by the Public Switched Telephone Network (PSTN) are presented in Chapter 10. A signaling system is used to configure a circuit across a selected route for the support of a connection, across which call messages are transported. A circuit's communications link resources are dedicated for use by its call's data units. Once the circuit is set up, the communications message transport process is initiated. When a call is terminated, the circuit is torn down.

The concept and operation of a *connection-oriented packet switching network system* is described in Chapter 11. A signaling system is instituted and employed to configure an end-to-end route that is used for the distribution of messages that belong to a connection, forming a virtual circuit (VC). A VC is not allocated communications resources on a dedicated basis. Data units that are transported across a VC share the resources of the communications links that they traverse on a statistical multiplexing basis. Frame Relay, Asynchronous Transfer Mode (ATM), and various mobile wireless systems have been employing networking technologies that make use of connection-oriented packet switching methods.

The principles of operation of a *connection-less packet switching network*, also known as a *datagram packet switching network*, are described in Chapter 12. No signaling system is used to manage the core networking process. A message flow is not pre-allocated communications capacity resources across a network route. Rather, a route is dynamically determined by intermediate routers as data units (identified as packets) are forwarded across the network to their destination. Link communications resources occupied by the flows that traverse the link are shared by packets on a statistical multiplexing basis. A packet flow occupies communications resources across the links that it traverses only during its periods of activity. Networking across the *Internet* employs a datagram packet switching method.

Schemes and algorithms that are used in a communications network for performing error control, flow control, congestion control, and routing are presented in Chapters 13–15. *Error control* methods are used to regulate, and react to, the occurrence of message errors. Forward Error Correction (FEC) coding schemes are employed for correcting (when feasible) errors that are detected at a receiving entity. Automatic Repeat Request (ARQ) methods employ error detection and retransmission algorithms. A received message that contains uncorrectable errors may have to be retransmitted.

Flow control schemes are employed to throttle end-to-end message flows. Under a closed-loop transport layer protocol, such as *Transmission Control Protocol (TCP)*, a receiving user regulates the aggregate amount of data that it is willing to receive and handle, over a time period, across a transport connection that is initiated by a sending user. The authorized quota is refreshed when the destination determines that it can allocate the underlying flow modified level of resources.

Under an open-loop flow control scheme, a user is obligated to regulate the data flow that it feeds into a network in accordance with a negotiated traffic flow resource allowance. A *congestion control* scheme is used to control the occurrence of network congestion. Under a reactive scheme, resolution steps are undertaken upon the detection or reporting of congestion. Under a proactive scheme, actions are taken to prevent the occurrence of congestion.

Message flows often choose a route from among several available candidate routes that lead to their intended destination(s). A *routing algorithm* is used to select a preferred route based on monitored or reported conditions of network nodes and links. Distance-vector and link-state routing algorithms are commonly employed for routing messages inside a routing domain. Under a *distance-vector routing algorithm*, each router in a routing domain periodically calculates the performance metric value (identified as its distance) of the best known path that leads from itself to each destination node in its routing domain, forming a distance vector. It passes its distance vector to its neighboring routers. Each node updates its route evaluation as it receives updated distance vectors from its neighboring routers. It passes the updated distance vector to its neighbors. Under a *link-state routing algorithm*, each router in a routing domain broadcasts messages that advertise the states of its attached links to each other router in its domain. Using these announcements, each router configures a weighted graph representation of its domain. It uses this graph to calculate the best path from itself to every other router in the domain.

Border routers are used for routing messages across domains, noting that distinct domains may be managed by different IT administrations. A border router is able to access several routing domains. An inter-domain route is selected by making use of reachability advertisements that are periodically disseminated among border routers. Across the Internet and other networks, intra-domain algorithms, such as Open Shortest Path First (OSPF), often employ link-state routing methods. The *Border Gateway Protocol (BGP)* is commonly used as an inter-domain routing protocol.

Approaches to *secure message transport* across a communications network, including the Internet and other wireline and wireless networks, are described in Chapter 22. Security schemes are employed to provide message confidentiality, integrity and authenticity.

Under a *symmetric-key encryption* system, confidentiality is achieved by having communicating users exchange an encryption key, which they use as their shared secret. A secure key exchange process is often implemented by using a *public-key encryption (PKE)* system. Each user configures a public-key and a secret private-key. The public-key is distributed openly to other users. The private-key is kept as a secret by its user.

Methods are employed to guarantee the *integrity of a received message*, assuring a destination entity that a received message was not modified during its network transport. The secure transport of a *digital signature* is used to provide a destination entity with an authenticated confirmation of the identity of the author of a received message.

As shown in Table 3, in Part III, key network systems are described in Chapters 16–21. Elements of the operation of the *Internet* are presented in Chapter 16. The Internet's protocol architecture and its key components are described, including its use of the *TCP/IP and UDP/IP protocol stacks*. The *Internet Protocol (IP)* is employed for addressing and inter-networking. Transport layer protocols, including *Transmission Control Protocol (TCP)* and *User Datagram Protocol (UDP)* provide end user to end-user transport services. Connection-oriented TCP services include end-to-end multiplexing, error control, flow and congestion control. Fast, extensible and secure transport layer connections can be formed by using the QUIC protocol, which operates on top of UDP. At the network layer, intra-domain and inter-domain routing schemes include OSPF and BGP, respectively. Network and transport layer algorithms employed by the Internet and other networks are also described in Part II.

Table 3 Part III: Network Systems.

Chapter	Topic
Chapter 16	The Internet
Chapter 17	Local and Personal Area Wireless Networks
Chapter 18	Mobile Cellular Wireless Networks
Chapter 19	Mobile Ad Hoc Wireless Networks
Chapter 20	Next-Generation Networks: Enhancing Flexibility, Performance, and Scalability
Chapter 21	Communications and Traffic Management for the Autonomous Highway

Wireless Personal Area Networks (WPANs) provide low power short distance communications over a wireless medium. A *Bluetooth network* is a WPAN that is used to replace cable connections between devices. A *Zigbee network* is used as short distance, low power and low data rate wireless network that is employed for monitoring and control purposes. Such systems are commonly used by *wireless sensor network (WSN)* systems. A *wireless local area network (WLAN)* that uses protocols based on *IEEE 802.11 WiFi* specifications provides local connectivity and wireless access to the Internet for local devices. Message transport methods used by such networks are described in Chapter 17.

Mobile wireless networks enable local and long distance communications that accommodate mobile users. Multiple access, routing, transmission and transport services provided by such key network technologies are described in Chapters 18–19. As described in Chapter 18, under a *cellular wireless network* architecture, an area of operations is divided into cells, where each cell is managed by a base station (BS). Mobile initiated and terminated data flows access a network-core across a shared wireless medium, commonly implemented as a Radio Access Network (RAN). A mobile communicates with a base station that manages the cell in which it is located. Under an infrastructure based architecture, a backbone network such as the Internet or the public switched telephone network is used to connect base stations.

As described in Chapter 19, a different approach is employed by a *mobile ad hoc network (MANET)*. This system does not make use of a backbone network. Rather, peer-to-peer communications techniques are employed to connect mobiles across paths that consist of multihop wireless links, using intermediate mobiles to relay messages.

A network system that consists of network nodes that employ proprietary networking means is difficult to dynamically manage. It is not readily re-configured to accommodate new services. It is more difficult to dynamically manage and control when it needs to rapidly adapt to support variations in traffic flows and service profiles. Several methods and technologies that are being employed for broader support by next generation network systems are delineated in Chapter 20. They include Network Virtualization; Network Slicing; Cloud and Edge Computing; Open Network systems and interfaces. Such operations benefit from the use of machine learning (ML) and artificial intelligence (AI) techniques.

Vehicular wireless networks provide communications between vehicles. They enable a vehicle to disseminate messages to other vehicles as they are moving along a highway. Principles of operation of connected vehicles that move along highway roads are described in Chapter 21. As a primary objective, message transmissions are used to enhance driving safety. They also enable the implementation of vehicular traffic regulation schemes aided often by the synthesis of vehicular platoon formations. Such controls also serve to enhance the vehicular throughput rates attained across a system of highways. Vehicular wireless networks are employed to transfer messages that are used for the regulation and management of *autonomous vehicle highway systems*. Self-driving vehicles make use of such data networks to interact with each other and to communicate with Road Side Units (RSUs).

Methods and schemes employed in the operation of vehicular wireless networks are described. They include peer-to-peer mobile ad hoc networking systems, also known as *Vehicular Ad hoc Networks (VANETs)*, and infrastructure aided networking schemes. VANET networking methods provide for message transport along vehicle-to-vehicle (V2V) multihop paths. A standards based method for V2V networking that is based on using a geo-routing approach is described. A wireless cellular network can be employed for the operation of backbone aided vehicular communications. Invoking vehicle-to-infrastructure (V2I) or vehicle-to-network (V2N) operations, a vehicle is able to transmit and receive messages to and from an associated RSU or base station. Long distance dissemination of such messages is performed across a backbone network.

Under a hybrid networking scheme, both peer-to-peer (V2V) and backbone-based (V2I) transport methods are employed. Based on our developments and studies, we describe the design and performance analysis of *Vehicular Backbone Networks (VBNs)*. Specific vehicles are dynamically elected to serve as backbone nodes (BNs), forming Access Nets (Anets). Vehicles that are members of distinct Anets communicate through a dynamically synthesized backbone network (Bnet). These systems can be aided by the use of an infrastructure system such as that provided by a wireless cellular network system, making use of both V2V and V2I message communications. Such systems may also employ 3-dimensional multi-tier arrays of unmanned aerial vehicles (UAVs), satellite systems, and Internet-based core network systems.

We describe a mathematical model that is used for the design of schemes that provide *traffic management for the autonomous highway*. We present models and analyses that are based on developments that we have been carrying out jointly with collaborating researchers. Advantageous vehicular formations, coupled with the setting of proper vehicle speeds and the application of vehicle flow controls, are used to sustain high vehicle flow rates. The employed traffic management scheme also accounts for the regulation of the time delay levels that are incurred by autonomous vehicles as they wait to access the highway and as they transit the highway to reach their destinations.

A reader that is not interested in the presented mathematical analyses of VBN-networks and of traffic management schemes for autonomous highway vehicles, would still benefit by reviewing the underlying concepts and the performance characteristics that are displayed and discussed.

To aid the reader, student, or teacher, the following is noted. In Part I, it is beneficial to review the material presented in Chapter 1, as it overviews basic concepts, the principles of message transfer processes, network architectures, and networking protocols. Chapters 2 and 5 are of interest for learning about information sources and performance metrics, respectively. Chapter 3 could be skipped by those who do not wish to delve into methods that are used for signal transmission across communications channels. Chapter 4 presents mathematical models of traffic processes. It is useful to review its contents. A reader that does not have the corresponding Probability background will still benefit by reviewing the presented concepts and methods, and by learning about Poisson traffic processes. Undergraduate students that have been attending my classes have previously taken an introductory Probability course.

In Part II, the reader learns about schemes that are used for sharing network resources among message flows. Local and remote sharing techniques are presented as multiplexing and multiple access schemes in Chapters 6 and 8, respectively. Chapter 7 covers queueing system models.

It is useful to understand the structure of such models and to learn about the performance behavior of basic service modules, even if the presented mathematical details are skipped. The remaining chapters in Part II provide important descriptions of networking and network control schemes, protocols, and algorithms, as they describe the elements of network switching, error control, flow and congestion control, routing, and message transport security. Operations of such schemes are related to the algorithms and protocols used by commonly employed network systems, including the Internet, wireless local area, and wireless cellular networks. The coverage of schemes and algorithms in each chapter is largely self contained. I have typically been covering Part II material in my basic networking courses.

In Part III, message transfer procedures and protocols that are used by commonly employed networking technologies are presented in Chapters 16–21. The material covered in each chapter can be reviewed independently of other Part III chapters. The Internet, is presented in Chapter 16, and wireless local area networks (WLANs), such as Wi-Fi, are covered in Chapter 17. Infrastructure based mobile wireless networks, such as cellular wireless networks, are discussed in Chapter 18, while the principles of operation of mobile ad hoc wireless networks are described in Chapter 19. Methods used to enhance the flexibility and performance scalability of networks are highlighted in Chapter 20. Chapter 21 covers a topic of significant recent and ongoing interest. It presents approaches to communications networking for, and traffic management of, autonomous vehicular highway systems. A reader may skip the parts that involve mathematical analyses, focusing on the associated descriptions, performance behavior illustrations and discussions. I have been covering selected Part III topics in my basic networking courses, while including a more detailed coverage of Part III material in my advanced and applied networking courses.

Material included in the book relates in part to papers, presentations, and notes that I have been authoring and using over many years in my lectures and in courses that I have been presenting at academic institutions, R&D organizations, commercial companies, and governmental organizations. Information relating to specific network systems and associated networking methods was also derived by selectively reviewing publicly available documents, such as those published in protocol and standards documents, technical reports, presentations, and overviews that are available on the public Internet, including such that have been presented in technical magazines and journals, Wikipedia, the free encyclopedia, and other sources.

I have been aiming to provide the reader a clear understanding of the principles of message transfer as used by networking schemes and protocols. My descriptions, illustrations, numerical examples, and discussions serve to foster understanding of the elements that drive the operation of networking systems, while illuminating the characteristics of their performance behavior. To motivate the understanding of the functioning and behavioral patterns of basic networking processes, I have compared at times their operations with those provided by analogous systems that are familiar to many readers, such as those used in vehicle transportation and other systems. I have included material that is based on research investigations that I have been carrying out in collaboration with colleagues and members of my research group, using it to demonstrate the application of evaluation and synthesis techniques to the analysis and design of networked systems.

In learning about telecommunications and computer communications network systems, I have benefited from the many interactions that I have had over the years with students, colleagues, and practitioners. Many innovators, engineers, and scientists have contributed to the development and implementation of telecommunications and computer communications network systems. I have gained valuable knowledge about these systems by participating and lecturing at conference, symposium, and workshop meetings, as well as by organizing and leading such meetings, and by reviewing excellent related papers and technical manuscripts. I have also obtained firsthand knowledge of network systems as a consultant and as a system development and implementation collaborator, in serving commercial, governmental and legal organizations in the United States and around the world. These works have been providing me opportunities to communicate with and learn from highly talented engineers, computer scientists, managers, marketers, quality assurance experts, and many other. I am greatly indebted to them.

I wish to thank the reviewers of my manuscript for their comments and suggestions. I also wish to thank several of my past PhD students that have graciously agreed to review book chapters and to provide me with their comments. I am especially thankful to Dr. Joe Baker, Dr. Zhensheng Zhang, Dr. Yu-Yu Lin, and Dr. Yulia Sunyoto. I am grateful to the publisher's administrative staff at Wiley/IEEE Press for their assistance during the book production process, particularly to Ms. Michelle Dunckley, Ms. Kavipriya Ramachandran, Mr. Sindhu Raj Kuttappan, and Ms. Veena S. Rajendran.

23rd November 2024

Izhak Rubin
Santa Monica, CA, USA

1

Introduction: Networking in a Nutshell

Objectives: *End users communicate with each other by sending their messages across a communications network. A source end user employs an application program to produce messages and message streams that it wishes to send across a network to a destination end user. The quality of the transport service that a communications network is requested to provide depends on the involved application and message types. A message is transported across the network by making use of the switching and routing services provided by network nodes that are located along its end-to-end route.*

Methods: The transport of a message across a communications network makes use of processes performed at user access devices and at network nodes. To enable dialogs between processes that reside at different nodes, a layered protocol architecture is employed. Nodal and internodal processing operations are classified into distinct protocol layers. The Transmission Control Protocol/Internet Protocol (TCP/IP) model is a layered protocol architecture that is employed by the Internet and other packet switching network systems. When sending data to the network, a protocol layer entity (PLE) requests a service from a lower PLE and provides service to a higher protocol entity. Higher layer protocols employ application and transport layer protocol entities to provide end-user to end-user (end-to-end [ETE]) services. Lower layer protocols provide services at the network, link, and physical layers. A PLE forms a message for transport across a network by encapsulating a higher layer message and by appending to it control data. Such control data aids in performing the processing, networking and transmission services carried out at network and user nodes. A network layer protocol is used to select an advantageous network route, which is used for the transport of a message to its destination node. A message is forwarded along its route from one node to the next one, whereby each network node acts as a switching and routing device. Control data that is embedded in a message enables a receiving peer-layer protocol entity to properly process the received message. Upon receiving a transported message from the network, a PLE processes the received message, extracts its embedded payload data unit and transfers it to a higher layer entity. To assure orderly, reliable and timely message transport, a network system employs a wide range of networking algorithms such as those that provide multiplexing, multiple-access, error control, flow control, congestion control, routing, security, and network management services.

1.1 Purpose

Key advances have been enabled by the invention of communication networks – the Internet, cellular phones, videoconferencing, Global Positioning System (GPS), telemedicine, the Internet of Things (IoT). Everyone's lives are touched and transformed in profound ways by communication networks.

Principles of Data Transfer Through Communications Networks, the Internet, and Autonomous Mobiles, First Edition. Izhak Rubin.
© 2025 The Institute of Electrical and Electronics Engineers, Inc. Published 2025 by John Wiley & Sons, Inc.

2 | *1 Introduction: Networking in a Nutshell*

How do persons or computers communicate with each other across an electronic communications network such as the Internet, a wireless cellular system, a Wireless Fidelity (Wi-Fi) network or a satellite network? The purpose of a communications network is to provide for communications among end users. Users are provided means to transfer information, which can take the form of data, voice, imaging or video messages, or their hybrids. In this book, I describe the principles and key concepts and methods that enable the effective operation of a system that serves to transfer message flows among local or remote users.

My descriptions are of interest to students and persons that have technical or scientific background, though not necessarily in the area of communication networks, as well as to curious readers that may not have had scientific background. To illustrate key concepts, I use at times corresponding (though not fully analogous) processes or methods taken from transport and other every-day systems that a reader may be personally familiar with.

In this chapter, I describe a communications networking protocol architecture that is widely used for the purpose of enabling the timely and reliable transfer of message flows between end users across a communications network.

In Chapters 16–21, I describe the formats of messages produced by user sources that they wish to disseminate across various network technologies to other users. A user may wish to transport across a network a single message or a group of messages, or engage in the transfer of a message flow, as is the case when streaming voice and video messages.

Users, also identified as end users, access a network by employing the services provided by stationary or mobile devices, aided by the programs embedded in such devices. These devices may be operated directly by persons or by automated or autonomous processes. The following questions are addressed. What are the characteristics of information messages and data flows that users wish to communicate to other users? Typically, we are interested in communicating information bearing messages whose contents assume the format of audio (including voice or music), data, or video, alone or in combination. What are the performance requirements that are imposed by each data flow on the communications network system to assure its reliable and timely transport across the network and to guarantee its effective reception and use at an intended destination entity?

I then discuss mechanisms that are employed to transmit information signals over physical communications channels.

Subsequently, I present the principles of key methods, processes, and mechanisms involved in transporting messages over a communications network. You will learn about networking architectures, communications protocols and algorithms that are used to regulate network flows and to navigate messages across network paths. Also examined are methods that are used to allocate sufficient data transport capacity resources for meeting a message flow's performance objectives.

In the following sections of this chapter, I overview key aspects of the communication networking process.

1.2 Networking Terms and Network Elements

End users, network elements, and message transport processes are defined and illustrated in the following with reference to the network layout as shown in Fig. 1.1. We start by presenting definitions for key terms and elements. Further explanations and illustrations of these and other terms are presented in subsequent chapters.

1) **End users**, also identified as **users**, are entities that produce and consume messages (or flows, streams) that are transported across a communications network, also identified in brief as a

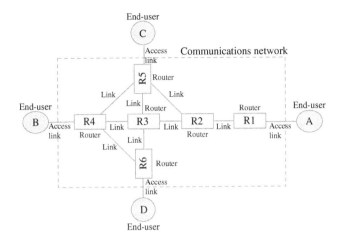

Figure 1.1 Illustrative Communications Network.

network. A user's computer, terminal, or intelligent phone is a device that is used by an end user entity to produce and consume messages. For example, end user A, which uses a computer employed by person A, sends a message to end user B, which may be a web server that is operated by company B. User A may use this message to request the download of data from organization B.

2) An **end-to-end (ETE) flow** between end users consists of a train or stream of messages that are transported between these end users. For example, in response to a request received from user A, web server B would send across the network a flow of messages that contain the requested information. User A would process the received messages and may subsequently display the received data.

3) A **communications link**, also identified in the following as a **link**, provides for the transfer of message transmissions between attached devices. Communications means are employed by attached transmitting and receiving devices, enabling them to respectively send and receive messages. A link may make use of embedded communications channels to conduct the communications process. It is also identified at times as a *communications channel*. Figure 1.1 shows several links.

A link that connects a network node with an end user node is often identified as an **access link**, as it provides the end user access to a *core network*.

A **wireline communications link** employs a wired medium, such as one that uses copper wires, a coaxial cable or a fiberoptic line. A **wireless communications link** makes use of a wireless medium, operating at a radio frequency (RF) or at other frequency bands. A link may be employed for data transfer by using a multitude of means, such as terrestrial communications, satellite communications, space communications, or underwater communications.

Other link attributes include:

1) A **one-way link** provides for the transmission of messages in a single direction; it is also identified as a *simplex* or *directional* link.
2) A **two-way link** provides for the transmission of messages in both directions; it is also identified as a *duplex* or unidirectional link.
3) A **composite link**, also identified henceforth as a link, consists of several (physical and/or virtual) links, which may be of different types. For example, a composite link may consist of

several physical communications wireline and/or wireless media. A two-way link may be composed of several counter-directional one-way links.

4) A **point-to-point link** is used for (simplex or duplex) direct communications between two devices, not requiring the use of an intermediate switch or router.

5) A **point-to-multipoint link** is used for direct communications from a single device to multiple devices. A message transmitted across such a link may be targeted for reception by all destination entities attached to the link. In the latter case, such a message dissemination scope is identified as **broadcasting**.

6) A **multipoint-to-point link** is used for directing communications from multiple devices to a single device. The link's transmission medium is shared among multiple devices, also known as stations.

   For example, in a wireless local area network (WLAN) such as a Wi-Fi network, user devices share a wireless medium for the transmission of their messages to the network's access point (AP). Similarly, in a wireless cellular network, user devices that are located in the same cell share the wireless medium, such as a radio access network (RAN), for the transmission of their messages to a base station (BS) node.

7) Resources of a **multiple access link** are shared by multiple devices. Each device, when scheduled to do so, would use the link for the transmission of its messages to either a single device (serving then as a multipoint-to-point link) or to multiple devices. If a transmission by a device across the link is intended for direct reception by all medium sharing devices, it is then identified as a **broadcast link**.

The following end-user and network nodes are identified:

1) **A node** in a communications network is an entity that is used to transport data units, or communicate messages, across a **link** to another node. Both network nodes and end-user nodes are identified as nodes.

2) An **end-user source node** sends messages (or other data units) that it receives from an attached end user across a network to another node. An **end-user destination node** provides messages that it receives across a network to an attached end user. Typically, an end-user node serves as both a source and a destination node, receiving messages to send across the network from an attached end user device as well as sending messages that it receives from the network to an attached end user device.

   For example, end user A uses node A (such as an intelligent phone) to produce a message (aided by a selected application), format the message for transport across the network and then to send it across an access link that connects it to a network node, such as network node R1.

3) A **network node** is used to send and/or receive messages (or other data units) traveling across a network. Such messages may originate or terminate at other nodes. A network node is connected via links to other nodes and possibly to end users. To illustrate, consider the network shown in Fig. 1.1. Network node R1 is able to communicate with end-user node A across an attached access link that connects these devices. It is also able to communicate with neighboring router R2 across the internodal network link that connects them.

The functionality of a network node may also be embedded in an end-user node.

Network nodes may perform switching and routing functions:

1) A **switch** is a network node that directs a message (or a data unit) that arrives at an attached incoming link, interfacing one of its terminations, known as an incoming **port**, to an outgoing port, which is attached to an outgoing link. This redirection action is identified as a **switching** operation. Switching entries maintained by a switch identify the output port to which an

arriving message should be switched. They are stored in a **switching table**. The calculation of switching entries may be performed by the switch or by another entity.

2) A **routing algorithm** is used to determine the entries stored in a *routing table* (RT). A routing entry identifies the outgoing port to which a message that arrives at a specific incoming port should be switched. A routing algorithm is employed to determine a preferential route that leads to a specific destination node.

3) A **router** is a switching node that performs routing calculations, often by interacting with other routers, and accordingly configure its RT and perform the switching action. A node that engages in only switching processes, while its TR entries are calculated and configured by another entity, is identified as a **switch**.

A message flow is transported across a network route or path to reach its destination node(s):

1) A **network topology** identifies the layout of a network by describing it as a *graph that consists of points (or nodes, or vertices) and edges (or lines)*. In the graph, a vertex represents a communications node, including a network or user node, a switch, or a router. An edge represents a communications link.

   Considering a network layout that is modeled by a graph, the following terms are defined. A node that connects to another node across a direct point-to-point link is said to be its **neighbor**. The number of a node's neighbors is defined as its *degree*. When a graph contains directed links, the corresponding in-degree and out-degree terms are defined. Unless stated otherwise, assume henceforth that we consider the links to be bidirectional (or undirected).

   The graph model of an illustrative network is shown in Fig. 1.1. For this network, the nodal neighbors of node R3 are nodes R2, R4, R5, and R6. The degree of node R3 is therefore equal to 4. The *graph is connected* in that every node can use the network to communicate with any other node.

   End nodes are noted to access the *core network* across access links. The core network is identified in the figure by a dashed line.

2) A **network path**, also identified as a **route**, consists of a sequence of successive distinct neighboring **nodes**. The nodes at the two ends of a path are identified as its terminating nodes. The path provides a route across which its terminating nodes can communicate.

   For the network shown in Fig. 1.1, end-user nodes A and B can communicate across the path that consists of nodes {A, R1, R2, R3, R4, B}, and their interconnecting links, identified as path $P1(A, B)$. These end-user nodes are also able to communicate across other paths, such as the path that consists of nodes {A, R1, R2, R5, R4, B} and their interconnecting links, identified as path $P2(A, B)$.

3) A **flow dissemination scope** characterizes certain attributes of a flow's destination entities. Messages of a **unicast** flow are sent to a single destination entity. **Broadcast** flow messages are sent to all destination entities that are members of a specified network domain. **Anycast** flow messages are sent to a single entity that is a member of a destination group of entities. **Multicast** flow messages are destined for reception by only entities that have joined as authorized members of an identified multicast group. **Geocast** flow messages are targeted for reception by only those destination entities that are located in a specified geographical region.

In the following, we note several measurement units, terms, and notations that are used in subsequent chapters:

1) **Unit multipliers**: 1 K = 1 kilo = 1000. 1 M = 1 Mega = 1 million = 10^6. 1 Giga = 1 G = 10^9. 1 milli = 1 m = 1 thousandth = 0.001 = 10^{-3}. 1 micro = 1 μ = 1 millionth = 10^{-6}.

2) **Time** measures are typically expressed in *seconds*, also denoted in brief as sec or s. 1 millisecond = 1 [ms] = 1 [ms] = 0.001 sec. 1 *microsecond* = 1 μ sec = 1 millionth of a second = 10^{-6} sec.

3) **Distance** measures are typically expressed in meters, also denoted [m].

4) A **message content** is typically expressed in **bits**. A binary symbol, which may assume one of two possible values with equal probability, such as 0 or 1, is said to represent 1 bit of information. Information contents of characters and other data symbols are often expressed in terms of 8-bit **Bytes**.

5) A **message rate** represents the average (over a specified period of time) number of messages per unit time that are received at or traverse a point, an interface, a node, a processor, or a link. It is typically denoted as λ and is measured in units such as [messages/sec], [mess/sec], [mess/s], [packets/sec], [calls/sec], and [calls/hour]. When other entities, such as vehicles, replace message units, the corresponding entity rate (as well as its flow rate) is expressed as: [entities/sec], such as [vehicles/sec]. For example, when a message's arrival rate to a system is said to be equal to 1000 [mess/sec], an average of 1000 messages per second arrive to the system. Hence, over 10 seconds, an average of 10,000 messages would arrive.

6) A **message (or entity) flow rate** represents the average (over a specified period of time) number of messages (or other entities) per unit time that flow across a system, such as a node, a processor or a link. It is typically denoted as f and is measured in units that include [bits/sec] = [bps], [packets/sec] = [pps], [calls/sec], [calls/hour], [vehicles/sec]. For example, a flow rate across a link that is equal to 5000 [packets/sec] indicates that an average of 5000 packets flow across this link per second.

7) **Service rate** represents the average number of entities that can be served by a fully engaged service module, such as a processor, a switch, or a protocol engine. It is often denoted as μ [entities/sec], or μ [messages/sec] when the served entities are messages.

8) **Utilization** or **traffic intensity** represents the average fraction of time (or of resources) that are occupied during a specified period of time. For example, a system peak-busy-hour utilization of 0.75, or 75%, indicates that the system is utilized, or kept busy, during 75% of the time over a peak busy hour. If this system represents a communications line whose data rate capacity (or service rate) is equal to 100 Mbps (Megabits/sec), a peak-busy-hour utilization level of 75% indicates that the flow rate across this link (averaged over a peak busy hour) is equal to 75% of its capacity, which is equal to 75 Mbps.

9) A system's **traffic intensity** metric represents the ratio of the message arrival rate relative to the system's peak message service rate. It is often denoted as ρ, so that $0 \leq \rho \leq 1$. For example, when a system whose service capacity is equal to 10 [messages/sec] is loaded by a traffic process whose total message arrival rate is equal to 7 [message/sec], the ensuing traffic intensity level is equal to $\rho = 0.7 = 70\%$.

1.3 Network Transport Processes

How is a message or a message flow transported across a network from a source node to its destination node? In the following, we overview several key principles of networking methods. Detailed descriptions and illustrations of network transport operations, protocols and algorithms are presented in subsequent sections and chapters.

In the following, we introduce key networking terms and concepts.

1) A **routing algorithm** is employed to determine the best network route to be used by a message flow that is sent from a source node to a destination node. For example, when considering

Fig. 1.1, using a routing algorithm, it may be determined that (at the present time) a specific flow that consists of messages that are sent between end-user nodes A and B (in both directions, assuming bi-directional links) should use route $P1(A, B)$. In this case, when router R3 receives such a flow's message from preceding node R2, it would examine its RT and use a corresponding routing entry stored in this table to determine the outgoing port, which connects it to router R4. Accordingly, it switches this incoming message to the latter outgoing port. It then requests its transmission module to transmit the message across an outgoing link that connects it to router R4.

2) Under a **connection oriented networking process**, a preliminary **signaling** phase is run before the actual data transport process can be initiated. Two connection-oriented networking mechanisms are noted: virtual circuit switching (VCS) and circuit switching (CS).

3) Under a **VCS** method, also known as *Connection-Oriented Packet Switching*, the following process is used. Networking technologies that are based on the VCS approach include X.25, Frame Relay, and asynchronous transfer node (ATM). Also related are techniques used by label switching and mobile ad hoc wireless packet-switching network systems.

   a) During the **signaling phase**, a physical or a virtual (logical) connection is made. A management entity employs a routing algorithm to determine a preferred end-to-end route to be used by the messages that are sent within the connection that serves the underlying end-to-end call. For example, for the illustrative figure, the signaling management system may determine the preferred route between end user nodes A and B to be route $P1(A, B)$. The signaling systems (SSs) inform the routers located along the selected path, leading to the specification of routing entries at involved routes. No link capacity resources are pre-allocated for the support of the flow associated with this call. This configured end-to-end network connection setup is known as a **virtual circuit (VC)**.

   b) Once the VC setup is complete, the **data transport phase** is initiated. Messages are then transported across the network along the path that has been selected during the signaling phase. Each intermediate network node that is located along the selected path acts to switch messages that belong to this flow from an incoming link to an outgoing link, as specified by the routing entries stored in its RT. Each such node operates as a **switch**.

   c) When end users decide to terminate their call, another signaling phase is initiated. It is used to notify the involved network nodes and tear down the associated VC. The corresponding routing entries for the networking of this flow at each involved switch are deleted.

4) Under a **CS** method, the following networking process is employed. Networking technologies that are based on the CS approach include the public switched telephone networks (PSTNs) and other network systems that allocate fixed resources for the transfer of data units that belong to critical data flows that require dedicated communications capacity resources. It is also used at times for the transport of ultrahigh-speed multimedia flows that are transported across optical switching modules.

   a) During the **signaling phase**, a physical connection is made. A management entity uses a routing algorithm to determine a preferred end-to-end route to be used by the messages that are sent within the connection that serves the end-to-end call. For example, for the illustrative figure, assume that the SS has determined the preferred route between end-user nodes A and B to be route $P1(A, B)$. The SSs inform the routers located along the selected path, leading to the specification of the associated routing entries.
   In contrast with the operation undertaken by the VCS scheme, link communications capacity resources are now pre-allocated. They are reserved for the internodal transmission of the connection's messages. Such a reserved end-to-end configuration is known as a **circuit**.

b) Once the signaling setup phase is complete, the **message transport phase** is initiated. Messages are transported across the network along the path that has been reserved during the signaling phase. Each intermediate network node along the selected path switches messages that belong to this flow from an incoming link to an outgoing link, as specified by the routing entries stored in its RT. Switches forward messages across their outgoing links, making use of their pre-allocated communications capacity resources. In this manner, messages that are generated by a call's source are transported across a pre-set path and are guaranteed to have been pre-allocated communications capacity resources across each outgoing link of each network node that they traverse. The corresponding network nodes now operate as circuit switches. In this manner, messages are guaranteed to be supported at a prescribed performance quality level.

c) When end users decide to terminate their call, another signaling phase is initiated. It is used to tear down the associated circuit. The corresponding routing entries at each involved switch are deleted and the link capacity resources that were allocated to the terminated circuit become now available for allocation for the establishment of new circuits that would support new calls.

5) Under a **Connectionless packet switching (PS)** method, also known as **datagram packet switching**, the networking operation within the core network system is characterized by the following features:

a) **No SS is required** as no configuration and the planning phase is performed prior to the start of message transport across the core network system. Consequently, a network signaling process is not implemented. The operation within the communication core network is based on a single data transport phase. The Internet employs a datagram packet switching networking technique.

b) **No communication capacity resources are reserved for message flows across the core network's internodal links**. A network node uses outgoing link communications capacity resources for the transmission of messages only when it actually transmits a message across the link. In this manner, the communications capacity resources of an internodal link are occupied on a statistical basis. Such a method of sharing a communications link is known as **statistical multiplexing**. The message units that are transported across a packet switching network are identified as **packets**.

c) A **dynamic routing** algorithm is often employed by packet switching networks. The outgoing link that is used to forward a packet that arrives at a network node is selected around the time that the packet arrives at the node. The selection is based on current network congestion conditions.

For example, for the network shown in Fig. 1.1, consider a packet that belongs to a flow that produces messages that flow from end user A to end user B. Consider the flow's packets to have arrived at router R2. At certain times, the routing algorithm employed by R2 would determine the best route to consist of nodes {R2, R3, R4} and the corresponding interconnecting links. It would then forward the packet across the outgoing link that connects it with node R3. At other times, the routing algorithm may detect router R3 to be highly congested, or the link that connects it to this router to be unreliable or to have failed. It may then determine the best route to consist of nodes {R2, R5, R4} and subsequently forward the packet across the outgoing link that connects it to router R5.

Such an alternate routing decision is made promptly at (or around) the time that a packet arrives at the packet switching node (identified as a packet switch or as a router). In comparison, when a connection-oriented networking operation is employed, a failure of a node or a link that belongs to a preconfigured route would require the resetting of the end-to-end connection and route, delaying (or interrupting) the ensuing message transport process.

d) A connection-oriented access process may be undertaken by an end user across an access link for the purpose of sending (or receiving) messages to (or from) a terminal network node. Thus, a user may engage in a log-on process to configure a connection across an access link for a session based activity. No connections are configured for moving packets across the core network when the latter operates by using a connectionless (datagram) packet switching process.

The following terms are used in the descriptions of communications networking elements and operations.

As illustrated in Fig. 1.1, a communications network is represented as a **system** that consists of interconnected **entities**. In our context, *such an entity represents an element that is capable of processing data and may also communicate with other elements.* Entities include nodes and processing elements within nodes.

The *topological layout of a network is represented by its nodes and the sets of neighboring nodes of each node.* A node is connected to a neighboring node by a direct link.

A **communications network architecture** represents the organization of network system elements and the rules that guide their functions, interactions and inter communications.

Communications network elements send messages to each other. **A communications networking protocol defines the formats and meanings (syntax and semantics, respectively) of such messages, serving to enable their purposeful interactions.** Protocols may be implemented by hardware, software, or combination of both.

A networking **process** consists of a progression of activities over time, or joint time and space, aiming to achieve targeted transfer objectives. A **networking procedure** is used to carry out a process networking activity and may involve the execution of several **tasks**.

An **algorithm** defines a step-by-step procedure that may be employed to perform a specific task or solve a particular problem.

A method used to perform a networking function is also identified henceforth as a **scheme**. A multitude of **techniques** may be employed for implementing a scheme or for carrying out a particular task.

1.4 An Illustrative Transport Process: Sending Packages Across a Shipping Network

In this section, we illustrate several of the above described networking concepts and methods by considering a package shipping process. Chloe and her colleagues are engaged in a scientific experiment that they carry out in a remote village in Thailand, involving their study of environmental science phenomena. They have no access to the Internet. To aid their experimentation and research works, additional scientific testing machines and analysis kits are required. Izhak, who is located in Los Angeles, has contacted an organization that specializes in the selection and purchasing of the equipment that is required, including test machines, associated components, and computer-aided analysis programs. These items will be wrapped as multiple packages and submitted to a shipping agent.

For illustrative purposes, assume that the transportation network under consideration is represented by the layout shown in Fig. 1.1. End-user node A represents the organization that

Izhak has engaged, which purchases and supplies the required items. End-user node B represents the laboratory office used by Chloe and her colleagues. The core network consists of network nodes that represent regional shipping offices that the engaged shipping company could employ for the shipping of the packages. Each internodal link represents a transportation means that can be used to move items between its neighboring offices. A specific link may employ ground-, air-, or sea-based mode of transportation.

The number of packages that user A is able to purchase and ship varies over time. The package-sending process initiated by user A is highly *bursty*. A burst consists of a variable number of packages that are individually ready to ship within brief time intervals. During a period of burst activity, each ready package is individually wrapped, marked, and made available for shipping. A longer and randomly variable interval of time elapses between the start of different burst activities.

When end user A has a ready to ship package, denoted as an application level original package *PO*, it attaches to it several labels. An application-oriented user-to-user label (*LA*) is attached to the application package. It identifies the contents of the package, its source address A, and its destination address B. It may also include other information that is helpful to the destination user. The original package *PO* with its attached application level label *LA* is identified as an application level package *PA*. Upon its reception at B, the application label will be reviewed and the package's contents will then be examined by an application level expert. If determined to be in good order, it will be delivered for use by Chloe and her colleagues.

Still at user node A, a user-to-user transport-oriented label (*LT*) is attached to the application package to form a transport package *PT*. This package contains the original package (*PO*), accompanied by an attached application label (*LA*) and an attached transport label (*LT*).

The transport label includes information that relates to the end-to-end transport of the package. It will be used by a transport expert at destination user B. It identifies the use type of the application package. It may also include information that enables a destination user to check whether the received package is damaged and needs to be resent. It may also be used by end users to pace each other, as they adjust the rate at which packages should be sent and returned. User B may instruct user A to reduce the rate at which the packages are sent, or even request it to temporarily stop sending packages, until requested to resume the shipping flow.

A transport expert at user A may determine a specific package to be too large, or too heavy, and therefore break it down into multiple smaller package segments. Each segment is separately identified as a transport level package. It contains a segment of the corresponding application item, accompanied by a corresponding transport level label. The latter label also includes a *sequence number* that is used by a receiving end user to identify the order at which the segments should be assembled to produce the original item.

At this stage, a person or a computerized process at node A that has networking expertise, known as a network level entity, would prepare a network level label (*LN*). A network level package (*PN*) would then be assembled. It contains the transport level package (*PT*) to which a network level label (*LN*) is attached. The network label contains information that will be used by network nodes, such as network routers, to ship the package across a route that leads to the destination node. It will include the network addresses of destination user B and of source user node A. It may also contain other information that relates to the networking of the package, such as specifying a desired means of transport and preferred transfer performance levels. For example, it may specify a desired time delay objective, identifying the maximum time that the package should be delayed in transit across the network.

The process of *progressive encapsulation through the addition of labels* is illustrated in Fig. 1.2. In this figure, the transported item is assumed to be a data message, identified as a data payload. The

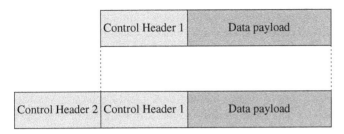

Figure 1.2 Message Fields and Wrapping Headers.

label included by a specialist such as a transport level expert is shown as Control Header 1. The ensuing labeled message may represent a transport level package (*PT*) which contains the original item (i.e., the data payload) accompanied by a transport label (*LT*), whereby the latter is represented as Control Header 1. It is then processed by a network level expert that attaches to it a network level label (*LN*), identified as Control Header 2. Control Header 2 contains network level information that has been produced by a networking expert. The resulting data unit is a network level message, analogous to a network level package (*PN*).

Back to the package shipping process carried out in node A, the networking expert passes the network level package to a link level expert, requesting it to plan the shipping of the network level package across the first link of its selected route. In this illustrative scenario, an access link is used to move the network level package to transport agent R1. The latter is located at office node R1. A link level label (*LL*) is formed. It is attached to the network level package to form a link level package (*PL*). The link level label provides details relating to the transport of the network level package across a link. A link level expert in node A may select a specific road to use to move the package across this link. It would then identify this road, as well as specify the link level addresses of the source and destination entities that form the road's termination points, which represent in this case the road addresses of nodes A and R1, respectively.

The *PL* package is next picked up by a physical level transport expert, which in this case may be a company A driver that uses a company vehicle to drive across an identified road to deliver the package to transport office R1. For this delivery, a physical layer label (*LP*) is formed and attached to form a physical layer package (*PP*).

Office R1 provides a network shipping service. Its physical level expert unloads the arrived package, examining and then removing the physical level label, exposing the link level package. A local link level expert then examines the received link level package *PL* and its attached link label (*LL*) to confirm the proper reception of the package following its move across the access link. If damaged, it may request a replacement package to be re-delivered . Assuming its reception to be positively confirmed, it removes the link label to extract the network level label (*LN*). A networking expert that resides in routing office R1 then examines the network level label. It uses its control data to determine the address of destination user node B. Node B is determined to be connected through an access link to destination network node R4. Studying the current loading levels of involved nodes and links, it uses its routing expertise to find a good route that leads to R4.

Under a connectionless data packet-switching approach, no network connection is established and no arrangements are made with network nodes across a selected path to reserve resources for handling the transport of packets that belong to an underlying flow. In an analogous manner when considering the package transport scenario, each package is handled independently of other packages, following a route that is selected by the network nodes that it traverses.

12 | *1 Introduction: Networking in a Nutshell*

Each intermediate router independently selects the best route to the destination node and forwards the package across a link that connects to the next node that is located across this route. Such a packet-switching routing operation is known as **next hop routing**, where a **hop** identifies a path that consists of a link (or of multiple links) that connects two neighboring routers. It is represented here as a link.

Assume that the network level expert at R1 determines the best route to destination node R4 as the path {R1, R2, R3, R4, B}. It may have to change some data included in the network level label (LN) attached to the received network package. The identities of the destination and source nodes stay unchanged.

The contents of the transport package PT remain unchanged. The new network level package is then passed at R1 to a local link layer expert that produces a link level package PL to which a new link label is attached.

This link label is used to guide the package across the link that connects node R1 with node R2. A physical level entity at router R1 is then used to physically transport the package (e.g., by loading it unto a cargo plane) for transfer to office R2.

A similar process is followed by the entities located at node R2 at the network, link, and physical levels. Continuing in this fashion, the routing experts at subsequently visited nodes may, for example, determine to route the package from R2 to R3, then from R3 to R4 and finally across an access link from R4 to user node B.

A subsequent package sent from user A to user B may be sent across a different path. For example, when router R2 receives a subsequent package, it may find router R3 at that time to be overloaded and thus unable to accept new packages. The network routing expert at router R2 may then select route {R2, R5, R4, B}, as an alternate route. Accordingly, the package will be forwarded next to router R5. Router R5 may then decide to forward the package to router R4.

In this manner, if all goes well, a sent package will be eventually received at user node B. The package will then be progressively examined by node B's experts. A physical layer entity will take care of the physical reception of the package. A link layer expert will examine the link level label and package. It will determine whether the package has been properly transported over the access link (e.g., incurring no link transport induced damages). If successfully received, it will remove the link level label and extract the network level package.

Recognizing itself as the terminating destination node, it would then remove the network level label and extract the transport level package PT. Examining this package and label, it follows the ensuing end-to-end transport level processes, as specified by the employed transport level protocol, to determine whether the received package can serve its useful aim, not incurring malfunctions during its end-to-end transport across the network, which may cause it to contain damaged parts or to miss certain components.

It may have to assemble, in proper order, distinct transport level segments, using the sequence numbers included in the attached labels, to produce the application level package. The latter contains the originally shipped item. The attached application level label would identify the source and destination users and may also include links to use guidelines.

Consider next an illustrative package transport operation that follows a connection-oriented method for shipping packages within the core network. In analogy with a **CS** network operation, the shipping process proceeds as follows. A planning phase precedes the package transport phase. Planning is carried out via a SS. During the planning phase, end user A will inform its access router R1 of the details of its pending package transport process. In doing so, it identifies the end-user entities, users A and B, and also specifies the characteristics of its expected package traffic flow. This can be performed for each direction. Based on this data, an end-to-end (simplex or duplex) connection between the end users will be configured.

The SS is used to select an end-to-end route and inform the involved network-switching nodes about this selection. A connection's flow consists of packages that will be sent across the route and that are allocated sufficient resources across each one of the route's links. The combined specification of the route and the associated link resources is known as a **circuit**.

To illustrate, considering the network in Fig. 1.1. Assume the configured circuit to use the route that consists of nodes {A, R1, R2, R3, R4, B} and to reserve specific transport capacity and means for use by this flow across the route's internodal links and switching connectivity and capacity at each involved switch. If such a circuit can be successfully configured, the SS notifies end users A and B of the impending packaging flow. User A can now start its shipping process. The packages that it sends will be delivered to circuit switch R1 across the access link. R1 will direct received packages to circuit switch R2, using the transportation means and capacity assets that have been reserved across the link that connects itself to switch R2. The process continues in a similar manner for the transport of packages across switches R2, R3, and R4. Destination user B will be receiving in this way this flow's packages, as they arrive across its access link using the resources that were reserved for this purpose.

Packages may be scheduled to be shipped via trucks that are scheduled to arrive every day between 8 a.m. and 9 a.m. At each truck, sufficient cargo capacity is set aside for this flow. The signaling process would reserve, at each truck, a cargo space that serves its needs.

To assure a package transport process that incurs low latency and proceeds at a high flow rate, the configured circuit will be assigned sufficient transport resources. At times, the flow process will produce a high-intensity burst of package arrivals, occupying most allocated reserved cargo spaces. If sufficient resources are assigned, a flow's packages will be reaching their destination in a timely and reliable manner. At other times, the package production rate at the source could be rather low, leading to poor utilization of allocated capacity levels.

In analogy with the operation of a **connection-oriented packet-switching network**, the following package transport process is observed. As described for the CS operation, a SS is used to select an end-to-end route and inform the involved network-switching nodes upon this selection. However, in contrast with the operation undertaken by a CS system, no capacity resources are reserved across a prearranged circuit that supports a flow's connection. The nailed-down route is now identified as a **VC**. Considering the network in Fig. 1.1, assume that the selected route consists of nodes {A, R1, R2, R3, R4, B}. The packages that are sent by user A to node R1 will be automatically switched by R1 to an outgoing link that connects to node R2, in accordance with the setting of the VC, so that no routing calculations need to be performed by R1 during the package transit phase. The switching operation can therefore be performed at a higher speed. Packages are however not guaranteed to have reserved cargo spaces in the trucks that flow across an outgoing link. As a package arrives to a node, it will join a temporary waiting line, known as a **queue**, and will be loaded into a forthcoming truck that has sufficient cargo space for its accommodation.

Thus, under a connection-oriented packet-switching transport operation, cargo transport capacity is better utilized, as it is used to carry only packages that are ready for immediate transport, rather than be a priori reserved to packages that belong to specific flows that may or may not be currently active. In turn, during high loading conditions, packages may experience high queueing delays.

1.5 A Layered Communications Networking Architecture

In the previous section, I have illustrated a process employed for the transport of physical or electronic message units across a communications network. It is noted that the transport process requires nodes to properly interact with each other for the purpose of effectively moving messages (or packages) that are transported from a source entity to a destination entity across a communications or transportation network. For this purpose, multiple message processing operations must be executed at and by multiple nodes. We discuss a framework that enables such interactive operations by considering the transport of messages, flows, and streams across electronic communications networks.

Processing operations are performed by end-user devices and by network nodes. The user's message is carried in a *payload field* of a message and is identified as the message payload data. A networked message also includes control data that is embedded within one or several *control fields*. Control data enables the destination end-user node and the network nodes that are located across the traversed path to properly process and forward the message, aiming for a reliable and timely delivery of the message to its targeted destination node.

The transport of a message between end-user nodes across a path that may include multiple network nodes requires the involved nodes to interact with each other, as they perform the processes that they need to carry out. Though different nodes that a flow traverses may employ different hardware engines and software tools, these nodes must be able to interact among themselves in a manner that enables them to properly execute the algorithms and functions needed to produce their intended outcomes.

To this aim, interface rules and formats are defined. They enable processes that reside at different nodes to speak a common language in corresponding with each other. A set of rules and formats that relate to performing coordinated message transport functions across a network is identified as a **network protocol suite**. These rules enable processes that reside at different network nodes to intelligently communicate with each other in performing message transfer related functions.

Each networking service provided by a process that is embedded at an end-user or at a network node is governed by an associated service-specific protocol. The module within a node that performs such a service in cognizance with an underlying service-specific protocol, using software and/or hardware means, is identified as a service-specific **protocol entity**. The service performed by such an entity follows rules and formats prescribed by a service-specific protocol. Such rules relate to both the involved meanings, or **semantics**, including those that are attached to the information embedded in the control field of a message, and to the underlying message structure, or **syntax**.

To implement such a complex system that requires the coordinated execution of multiservice operations across multiple network nodes, a manageable, scalable, flexible, and robust network protocol architecture is employed. The multitude of networking services that must be performed at individual nodes and network-wide across nodes is classified into distinct layers. A protocol entity that performs services that are associated with a specific layer is identified as a *protocol layer entity* (PLE).

To effectively implement, control and manage the multitude of services that must be performed by network nodes, a *layered protocol architecture* is used. Internodal interfaces and dialog rules are defined. Specific networking-related processes and algorithms that are used by different protocol entities are described in subsequent chapters.

The services that are used to support the message transport process across a network are divided into several functional groups. To make this division universally workable, a standardized *networking reference model* has been developed. It classifies communication networking services into separate categories, identified as **layers**. This division enables a network designer to readily evaluate and modify specific services provided by each layer's protocol entity, the interactions that take place between protocol layer entities that reside in distinct nodes, and the ensuing interface protocols.

A **PLE** (also identified in brief as layer entity) represents a computer-aided process that is implemented by using software and/or hardware means that carries out the services provided by the underlying layer. Multiple protocol entities may be employed at a node in providing layer services. An entity uses the associated message formats and algorithms to perform its services. A collection of protocols that are employed across layers by protocol entities that reside at a node form its **protocol stack**.

Traditionally, seven layers (supplemented by several sublayers) have been recognized. Using an Application ("app") protocol, an end user produces a message that it wishes to transport across a network. This message is processed in a downstream chained fashion by layer entities that reside at an end-user source node. A layer-7 (L7) protocol entity processes the data submitted to it by a user's application by forming an Application layer (L7) message. It does not modify the data that it receives from the user. Rather, it may add to it control data. Control information is used by an application layer entity residing in a destination end-user device to properly process the received information for transfer to the application program employed by the destination end-user. In this manner, **the application layer provides a human–computer interface, enabling applications to access networking services**. An L7 protocol entity hands its product down the protocol stack to a layer-6 (L6) protocol entity. The latter performs its own processing and hands down its produced message to a layer-5 (L5) entity; and so on down to layer-1 (L1). The latter provides for the physical transmission of messages across a communications medium.

The product of a protocol entity residing in layer-i (denoted hereby as L-i, or Li, or L_i) is a Layer-i message. Layer services are ordered by following the following principle: **a layer entity requests a service from a lower layer entity while providing service to a higher layer entity.**

Consider the message build-up process that takes place as we move down the protocol stack. Consider a layer-$(i + 1)$ protocol entity, denoted as E_{i+1}. In providing its service, it forms a layer-$(i + 1)$ message, denoted as M_{i+1}. It then submits this message to a layer-i entity, E_i. M_{i+1} consists of a **payload field** and one or several **control fields**. Its payload field contains the data that it has received from its user, E_{i+2}, for $0 \leq i \leq 5$. Layer-$(i + 1)$ entity passes M_{i+1} to a lower layer entity, a layer-i entity. In passing, it indicates the service that it requests E_i to provide. The message that E_{i+1} passes to E_i is known as a layer-i **service data unit (SDU)**, denoted as SDU_i. Entity E_i processes this message in accordance with its embedded processes and algorithms as it performs its service functions. Its outcome is a layer-i message, M_i, which contains a control field and a message payload. The message payload is derived from the message that it has received from a higher layer entity, as included in SDU_i. Its control field conveys information that enables a layer-i entity that is resident at another node, known as a **peer-layer entity**, to properly process the received layer-i message as it performs layer-i services.

A layer-i message, M_i, that is intended for processing by another layer-i entity, is identified as a layer-i **PDU** and is denoted as PDU_i. A layer-i entity E_i forms a layer-i PDU by appending a layer-i control field (or fields) to its payload, which is equal to (or derived from) SDU_i.

Thus, using an **encapsulation** process, a layer-i protocol entity E_i produces its message product, PDU_i, by including as its payload the (SDU_i) message that it has received from a layer-$(i + 1)$ entity

16 | *1 Introduction: Networking in a Nutshell*

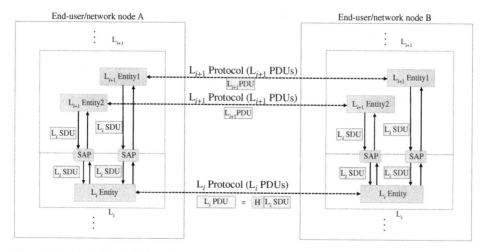

Figure 1.3 Vertical and Horizontal Message Communications Between Protocol Layer Entities at Layer-(i + 1) and Layer-i.

and encapsulating it by a control field. This is equivalent to the labeling process described in the previous Section.

The message communications flows that take place between protocol layer entities associated with layer-i and layer(i + 1) and their ensuing interfaces are illustrated in Fig. 1.3. Two communicating end-user or network nodes, node A and node B are shown. At layer-(i + 1), two protocol entities are noted. They execute different layer-(i + 1)-based processes. For example, for a layer-7 operation, different entities may interface different user applications, such as email and video streaming. The product of a layer-(i + 1) entity is moved as SDU_i for processing by a layer-i entity.

Service access points (SAPs) are used to identify cross-layer interfaces. Different interface points are identified by different SAP addresses. This serves to provide a multiplexing service. It identifies to a layer-i entity located at another node the specific layer-(i + 1) entity to which the payload of a received layer-i PDU should be passed. As noted, the data units that are exchanged and processed by peer-layer entities are identified as PDUs.

How does a layered system architecture work and what are the typical ways that are used to classify networking services into distinct layers?

Consider a message that is prepared by an end-user by using an Application program. The source end-user wishes to communicate the message to another end-user across a communications network. For example, the source and destination end users may employ copies of a program that enables a communications dialog between a client and a remote website server. Realized as a computer-based process, an application layer entity maintains a proper interface with the user's Application program. It collects the data contents provided by the source user and adds to it control information to produce an application layer PDU. In doing so, it follows the semantic and syntactic rules imposed by the employed application layer protocol.

The produced PDU is transferred as the payload of a corresponding SDU down the protocol stack to a local transport layer entity, requesting it to transport the application layer PDU to a specified remote end-user. The transfer of this SDU from an application layer entity to a transport layer entity is carried out (within a node) across a specific interlayer SAP interface. This SAP identifies the specific application layer entity that is being used. The application layer PDU becomes the payload of a transport layer PDU. Its control header includes information that is used to provide services that

1.5 A Layered Communications Networking Architecture | **17**

involve the source and destination transport layer protocol entities employed at the corresponding source and destination end user nodes.

Upon reception at the destination end user, the destination user's transport layer protocol entity processes the received transport layer PDU. If checked valid, it extracts its payload, deriving the application layer PDU. This message is then transferred to the associated application layer entity. The destination SAP specified in the control field of the received transport layer PDU enables the transport layer protocol entity used at the destination end user to determine the specific application layer protocol entity that should be used. Following this transfer, the application layer PDU is processed by the identified application layer entity.

To intelligently process a received application layer message, the communicating source and destination end-user nodes employ compatible copies of the application program. In this manner, an application layer end-user entity is able to properly read, play, display, and process application layer (PDU) messages produced by a peer-layer entity. The mechanics and dynamics of such an interaction, or **handshaking** procedure, is defined by the underlying **application layer protocol**. An end-user may be able to access several locally resident application programs. This is, for example, the case when engaging a multimedia message flow, which makes use of associated voice, data, and video applications.

Network nodes are tasked to transport the messages produced by communicating end-to-end users. Network nodes act to receive, switch, forward, and transmit messages across a network route, aiming them to reach an identified destination end-user node. They employ for this purpose communication networking protocols. Such protocols employ routing algorithms that enable the selection of a desirable network route. Link layer protocols enable the formatting of data for transmission across a communications link, as well as for scheduling transmissions across a shared-medium link. A physical layer protocol is used for carrying out the physical transmission of information-bearing signals across a link's communications channel.

A source end-user may not be directly connected to the intended destination end-user by a dedicated point-to-point communications link. It would then require its messages to be transported to the destination by using the networking services provided by a communications network. A network tends to employ multiple geographically distributed *nodes*. A network node is connected by communications *links* to a limited number of "neighboring" nodes. The topological layout of a communications network is described by a *graph* model. A graph consists of nodes and links. Its layout is prescribed by specifying the end nodes at which each link is terminated. This description identifies those pairs of nodes that are neighbors to each other and that are thus connected to each other by direct links. A graph layout of a network is illustrated in Fig. 1.1.

Several intranodal and internodal interactions need to take place to assure the reliable and timely transport of messages across a network between source and destination nodes. **Layered reference models** are used to enable the effective implementation of multiple interacting protocols, in providing communications networking services that span multiple protocol layers. Such models contribute to the process of standardization of the interfaces between nodes and between networks. Subscribing to a standardized reference model, have made it possible to seamlessly connect network systems, making the construction of the Internet, as a network that interconnects a large number of distinct networks, a reality. The most popular such universally employed layered architectural models are the **Open Systems Interconnection (OSI) reference model** and the **Transmission Control Protocol/Internet Protocol (TCP/IP) model**. While similar, the Internet system employs the TCP/IP protocol layering architecture, which has become the one that is most frequently employed.

Under a layered architecture, distinct service functions are embedded in different layers. Layer protocol entities residing in a node interact with each other in an orderly manner. *Processing operations carried out by a protocol entity that resides at a layer provide service to a protocol entity that resides at a higher layer and requests service from a layer entity that resides at a lower layer.*

In the following, we first discuss the architecture of the **OSI reference model**. The OSI model uses seven basic layers, as well as several sublayers. The higher four layers provide **end-to-end services**, each one involving interactions between a protocol entity residing at a source end user and a protocol entity residing at a destination end user.

The lower three layers provide services that involve the transport of messages across a communications network. They enable message communications within a network and between networks. Messages are transported across a network by following a network path, or route. An end-to-end route may include several intermediate switching or routing nodes.

Key services provided by each OSI layer are identified in Fig. 1.4. They are further described below.

An end-user device makes use of protocol entities that may provide higher and lower layer services. In turn, a network node that does not need to provide end-user interface functionality, such as a routing or switching node, employs only protocol entities that operate at layers 1–3.

An OSI-based layered architecture is depicted in Fig. 1.5. We note that the physical movement (i.e., transmission) of messages is performed only by a physical layer entity. Interactions between peer layer entities at layers above the physical layer are **logical** in character. They are guided by

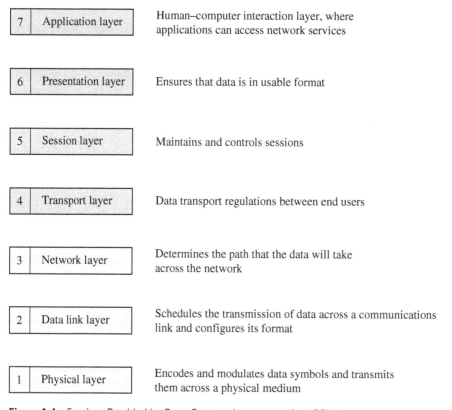

Figure 1.4 Services Provided by Open Systems Interconnection (OSI) Layers.

1.5 A Layered Communications Networking Architecture

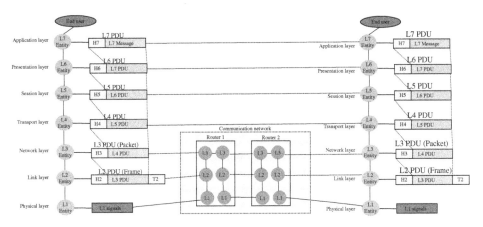

Figure 1.5 The Open Systems Interconnection (OSI) Layered Reference Model.

the rules and formats imposed by the ensuing protocols. The corresponding logical interactions are depicted in the figure as dashed lines.

Adherence to the specifications imposed by a layered protocol model makes it possible for distinct network systems to interface each other. It enables intelligent interconnections between nodes and between networks that may be operated by different branches of a single company or that belong to different organizations. The information contents embedded in message control fields, at the different layers, enable the proper use of a multitude of networking algorithms by protocol layer entities, resulting in the orderly transport of message flows across a communications network. Mechanisms and algorithms that are employed for this purpose are presented, explained, and illustrated in subsequent chapters.

In accordance with the OSI reference model, the following higher layer protocols provide end-to-end services:

1) **Application Layer Protocol (L7):** Serves to interface end-user's application programs, enabling them to communicate application contents across a network, in interacting with remote end users. End-user messages are formatted and processed in following the interaction and format rules defined by an L7 protocol. A source protocol entity is used to form an application layer message, which includes application data received from an application program and an added control header. A destination application layer entity uses the control information included in a received message to properly process the received message, as it follows the rules defined by the employed application layer protocol. The data included in the message payload field is then extracted and transferred to the application program employed by the destination end-user. Many different types of application layer protocols have been employed. Examples include:
    a) *Simple Mail Transfer Protocol (SMTP)* is used for the transfer of email messages.
    b) *Telnet* is used for terminal logging into a remote host computer and for terminal–host and terminal-to-terminal communications.
    c) *File Transfer Protocol (FTP)* is employed for transferring files.
    d) *Hypertext Transfer Protocol (HTTP)* is used for web browsing.
    e) *Domain Name System (DNS)* is employed over the Internet for associating human readable domain names with computer readable nodal (IP) addresses.
    f) *Simple Network Management Protocol (SNMP)* is used for managing network devices and nodes and for monitoring their performance.

A layer 7 (L7) application layer entity prepares a PDU that, with directives that are associated with its request for service, is submitted as a layer 6 (L6) SDU to a layer 6 entity, which is a presentation layer entity. An L6 PDU is submitted as a layer 5 (L5) SDU to a layer 5 entity, which is a session layer entity. When presentation and session layer services are not provided by using separately employed protocol layer entities, or are performed at a higher layer, as is the case under the TCP/IP model, an L7 PDU is submitted as a layer 4 (L4) SDU to a layer 4 entity, requesting its service. The L4 entity provides end-to-end transport services in accordance with the rules specified by a transport layer protocol.

2) **Presentation Layer Protocol (L6):** Serves to change, translate, or map the format of data included in the message produced by an application layer entity at a source user. At the destination user, the L6 entity performs a corresponding inverse-mapping translation. The source presentation layer entity may map the payload data so that it assumes a desirable format, such as one that is acceptable by a targeted application. The control information included in an L6 PDU informs the destination presentation layer entity as to the mapping function or scheme that has been applied, as well as identify associated parameter values, enabling it to perform its processing in accordance with the rules imposed by the employed presentation layer protocol. Examples include:

a) *Data encoding/formatting* for compatibility with a code or format employed by a destination entity.

b) *Data encryption* for secure transfer.

c) *Data compression* for reducing the communications bandwidth required for the transfer of a message or stream of messages across a communications network.

Corresponding inverse-mapping operations make use of decoding, decryption, or decompression processes, respectively.

The presentation layer entity prepares a PDU that is submitted as a session layer (L5) SDU to a session layer entity. Alternatively, when a session layer service is not used or is performed by another layer, it prepares a transport layer (L4) SDU that it submits to a local transport layer entity.

3) **Session Layer Protocol (L5):** Serves to establish, manage, and terminate communications sessions between end-user devices. A session is opened to initiate the transfer of data. It is closed when data transfer ends. Protocol services can be used to regulate the timing and duration of data transfer, setting it to proceed in a half-duplex (HDX; one way at a time) or full-duplex (FDX; simultaneous two-way) transfer mode. It may specify conditions under which data transfer is halted and subsequently resumed, as triggered by checkpoint markings. Services may include log-on, password validation, and rules for dialog discipline. Its functionality is often absorbed by software developers into an application program, which may then implement it by using remote procedure calls.

4) **Transport Layer Protocol (L4):** Provides end-to-end transfer of data between end-user devices. A source transport layer entity receives a message or a data stream from a source application, which requests its transport across a network to a destination end-user. The source L4 entity includes this data in a payload field of one or several transport layer PDUs. Each transport layer PDU contains a header field that carries L4 control data.

At the source user, a transport layer L4 PDU is passed to a network layer (L3) protocol entity for transport across a network to the destination user. The network layer (L3) entity forms a network layer (L3) PDU, known as a **packet**, which carries in its payload field the transport layer message received (as an L3 SDU) from the transport layer entity.

The L3 packet's header contains control information that enables the routing of the packet across a network. In networking across the Internet, a packet's header includes, among other control data, the network-based address of the destination node. It is identified as its IP address, as IP rules are used to identify nodal addresses over the Internet. To obtain this IP address, the source user's application layer entity makes use of the **DNS**. A source user connects to a regional DNS server that maps a destination name descriptor (such as its Internet address, or URL, received from the end-user's application program) to an IP address that identifies the network address of the destination user.

A connectionless or a connection-oriented transport layer (L4) protocol can be employed. Transport layer protocols provide the following services:

a) A **connectionless transport layer protocol**, such as the **User Datagram Protocol (UDP)**, is used for the rapid (one-way) end-to-end transport of an application message. The message is encapsulated by a control header, forming an L4 PDU that is identified as an **L4 datagram**. This datagram can be targeted for reception by multiple end-users under broadcast or multicast dissemination services. The UDP protocol does not provide error-control and flow-control services and does not aim to identify the order at which received datagrams should are arranged for transfer to the destination end-user application entity.

b) Under a **connection-oriented transport layer protocol**, such as **TCP**, or its secure version (as employed when using HTTPS), under which an encryption layer is inserted between HTTP (an application layer protocol entity) and TCP, end-to-end connection is configured between a transport layer entity at the source end-user's device and a corresponding entity that resides at the destination end-user's device. A TCP connection is opened at the start of a data transport session and is closed when data transport terminates, or when it is sensed to be idle for a period of time that is longer than a specified interval. During the period that it is open, it serves as a *data pipe* that carries data flows (as streams of bytes) between a pair of end users.

c) **QUIC is a connection-oriented transport layer protocol that operates over UDP**. It is used for supporting a stateful interaction between a client and a server. It provides transport layer services to an application such as HTTP, performing error, flow and congestion control regulations of data streams that are multiplexed across an Internet connection between a client and a web server. QUIC transport connections include data encryption and offer lower transport latency levels. Several application streams can be multiplexed across a single QUIC connection, while separately managed. QUIC packets are transported across the network within UDP datagrams. It is further described in Section 16.5.3.

TCP can serve to transport multiple two-way data transactions that take place between end users. For example, under a HTTP-based web browsing application, multiple request messages are produced by a user client across a TCP transport layer connection in the client-to-server direction. A TCP connection also provides for the transport of data streams from a web server to the requesting user client. Data flow activity may last for as long as the connection remains open.

A transport layer entity that resides at a receiving end-user processes the received transport layer data by following the rules specified by the underlying transport layer protocol. It provides the key transport layer services noted below. It then transfers the received message (or flow) to the application process identified by the associated SAP. The application layer entity at the receiving user then produces an outcome that meets targeted format and semantics.

1 Introduction: Networking in a Nutshell

In the following, we describe key features of UDP and TCP transport layer services. Key features of the QUIC transport layer protocol are described in Section 16.5.3.

a) **End-to-End Multiplexing**: A source transport layer entity is requested by a higher layer entity to transfer data that is produced by a specific application program, such as an email program or a HTTP application that provides for web browsing. A higher layer entity, such as an application layer entity under the TCP/IP model, specifies the network address (such as an IP address) of the destination user or node. A source SAP number is used to identify the source application program that has produced the data message. A destination SAP number is used to identify the application program at the destination user to which the destination transport layer entity should transfer the received message.

Under a connectionless transport layer protocol, the use of SAP numbers at the header of each L4 datagram makes it possible for the transport layer entities to support datagram flows that are associated with distinct application types. In this manner, flows originated by an end user that are associated with different applications can be differentiated from each other, a process known as *multiplexing*. For this purpose, SAP numbers, known as *port numbers* under the TCP/IP model, are included in the header of each transport layer PDU.

A network flow is identified by the following attributes: {source-IP-address, source-port, destination-IP-address, destination-port}. Applications use software structures known as *sockets* to send and receive data flows. A socket address consists of an IP (L3) address and a port number (as well as the specification of the employed L3 protocol, such as IP). An IP address identifies an interface of a device, while an application port (or SAP) number identifies an application residing on this device.

This is similar to identifying the address of an apartment unit by specifying it as a combination of the address of a multi-apartment building and the number of the unit that is located inside this building. In exchanging data across the Internet, user devices use *Internet sockets* to send data to a specific application that resides in a specific node, and to deliver incoming data messages to the appropriate application process that resides in a specified destination node. Under a connectionless transport layer protocol such as UDP, each L4 message is sent to an individually addressed datagram socket. Under a connection-oriented transport layer protocol, such as TCP, stream sockets are employed. They include mechanisms for creating and terminating TCP connections, as they support the delivery of error-free in-sequence data streams. A TCP connection is associated with a pair of sockets, identified by the source and destination IP and port (SAP) addresses.

b) **End-to-End Segmentation**: Messages received at a source transport layer entity from a higher layer entity may at times have to be fragmented into several transport layer *segments* and then be reassembled into the corresponding higher layer message, such as an application layer message, for proper processing by a destination user's higher layer entity. Such a segmentation is impacted by constraints imposed by lower layer protocols, such as those used at the network and link layers, which may restrict the maximum length of each message that they accommodate. In performing this fragmentation, the segmenting entity may use information about the maximum PDU lengths supported by network and link layer operations across a network path. The lowest (over the path) maximum length among the latter values induces the underlying *maximum transfer unit (MTU)*, when accounting for the inclusion of overhead fields. The MTU specifies the maximum length of a segment that can be accommodated across a path. Using such a segmentation process prevents the need for network layer or link layer entities employed by nodes that are located across a network path from having to engage in fragmentation and reassembly operations, enhancing the message delivery process.

For a connectionless transport layer protocol such as UDP, the maximum length of a network layer (L3) packet imposes a constraint on the length of the transport layer datagram, as the latter is carried in the payload field of a single L3 packet. A UDP transport service does not guarantee an orderly, reliable, and complete delivery of datagrams to the destination transport layer entity. UDP Datagrams may be delivered to the destination user's UDP entity in random order, may contain errors, and certain datagrams may be lost and not delivered at all. Such issues may not be critical for some transport operations. These and other operations may rely on a higher layer protocol for handling such issues.

Due to its lower complexity in processing and in the transport process itself, a connectionless transport layer protocol service is often employed when performance focuses on low latency, while a certain extent of message losses is deemed acceptable. Such protocols are used in providing transport services for certain voice and video messages, as well as for data transport services provided by DNS, SNMP and various query–response applications. Repetitive sending of messages may be used by some applications, such as telemetry and monitoring, as well as in situations under which a message is likely to be lost or delayed. Due to its simplicity and protocol modification flexibility, UDP is also used by QUIC, which is a connection oriented transport layer protocol that is used for secure client–server interactions across the Internet.

c) **Sequencing**: Under a connection-oriented protocol, such as TCP, each segment transported across a TCP connection is associated with a sequence number. This allows the destination transport layer entity to reorder received segments, as they need to be reassembled in the correct order for transfer to the targeted destination application layer entity. A sequencing break indicates to the destination entity that certain segments have been lost. Retransmission of missing segments may be requested from the source transport layer entity. Such a service, coupled with end-to-end error control, flow control, and congestion control TCP services that are noted below, is important for preserving the transport integrity of data flows and for the regulation of network traffic flows. Examples of applications that use TCP include: HTTP, which is employed for web browsing, and has also been used by QUIC; FTP, which is used to transfer files; SMTP, which is employed for the transfer of email messages; Telnet, which is used for client–server interactions and data transfers.

d) **End-to-End Error Control**: For certain flow types, a source end user wishes to be informed by the destination end user as to whether a message has been received and whether it does not contain data errors. A message that is detected to contain such errors may be resent.

e) **End-to-End Flow Control**: A destination end user may wish to dynamically regulate the rate at which it receives messages from a source end user. Such a flow regulation process protects (and regulates the use of) the destination end-user's processing and memory storage resources, preventing them from becoming overly congested.

f) **End-to-End Congestion Control**: By observing the transport flow of TCP segments, a transport layer protocol entity may conclude that the network is experiencing congestion. Employing a congestion control algorithm, it acts to reduce the rate at which segments are sent across a transport layer connection. When it determines normal segment flow conditions to have resumed, the rate at which segments are sent into the network is gradually increased. Such flow rate regulation actions, dynamically and automatically undertaken by message traffic sources as they monitor network congestion conditions, contribute to easing performance degradation effects that are caused by the occurrence of congestion hot spots across a network system.

24 | *1 Introduction: Networking in a Nutshell*

The following services are provided by lower layer protocols:

1) **Network Layer Protocol (L3):** Serves to navigate messages across a network from their source node to their destination node(s).

   a) Once a message has been processed by its source end-user device and is ready for transport across a communications network, the source transport layer entity transfers the message to its local network layer entity, which constructs a network layer message. This message is identified as a **packet** under the TCP/IP based Internet model. Henceforth, L3 PDUs will be often identified as packets. A packet contains a transport layer segment as its payload. This L4 segment is encapsulated by a packet header. The control information included in the packet header enables the routing of the packet across its network path.

   A packet is targeted for transport from its source network node, which is attached to (or embedded in) the end-user device or node, to a destination network node that connects to the destination end-user device. A packet's header identifies the network addresses of the packet's destination and the source nodes. Under the TCP/IP model, these addresses are represented as destination and source IP addresses. The header may also include other information, such as data that specifies desired performance metrics for the L3 transport of the underlying packet. Also included is a SAP number that identifies the transport layer entity (such as one that uses a TCP or UDP protocol) that should handle the payload of the received packet at the end-user node.

   b) Intermediate network nodes, identified as routers, or switches, are employed for moving a packet along its route. To route an incoming packet, a network node determines the route to be used by the packet to reach its desired destination node. It then switches the packet to an outgoing port that uses an outgoing link that connects to the next node along the route. In this manner, a packet will be switched by intermediate network nodes, eventually reaching its destination end user.

   c) Networks, such as the Internet, consist of nodes that are often arranged in a meshed topological layout. Several routes may then be available to connect source and destination nodes. A *routing algorithm* is employed to provide for the selection of an effective route. A chosen route may have to guarantee a packet transport experience that meets desired **Quality of Service (QoS)** performance metric values. Such metrics often specify reliability, correctness, integrity, and timeliness objectives.

   d) In the Internet, packets are handled by a router in following a **store and forward** process. A router's L3 protocol entity instructs the link layer entity to switch an incoming packet to an outgoing link that connects to the next node that resides along a selected route. Incoming packets are *stored* in the router as they wait for their turn to be *forwarded* across the selected outgoing link to the next node.

   Under a **connectionless packet switching** operation, as performed by the Internet, each networking node performs as a network layer (L3) packet switch. A router employs an L3 protocol entity that configures the route to be used by a packet by using a routing algorithm. The network layer PDU is extracted from an incoming link-layer frame. It is then processed by a network layer entity. The latter selects the route to be used to reach the destination node. Link Layer and physical layer entities are then employed for forming the link layer frame and for the transmission of this frame across the outgoing link that connects the node to the next node.

   e) Networking messages across multiple network systems, known as **Internetworking**, is readily feasible when the different networks implement a compatible networking interface, which is based on a commonly employed protocol layering structure.

Consider a packet that is sent from a source node that is located in network A and is destined to a node that is located in network B. Assume these networks to be connected to each other via a gateway router. Assume the traversed networks to employ the TCP/IP layering architecture. As a packet arrives at the gateway router, following its transit along a network A's route, the gateway router examines the packet's header to identify its destination IP address. It then proceeds to determine the best route to use to navigate the packet to its destination node across network B.

Network nodes that are located in such distinct networks can conveniently examine and process packets that travel across networks when they employ the same network layer protocol structure. Each network can internally employ its own preferred routing algorithm, as well as its own resource allocation methods. The use of a packet format and semantics that the gateway router understands enables inter-networking, as performed across the Internet. Principles and illustrations of inter-networking operations that are performed across the Internet and other networks are presented in Chapters 9–12, 15–21.

f) As noted earlier, and further explained in subsequent chapters, the route to be used to navigate a packet that belongs to an identified end-to-end flow may be configured also through the execution of a connection-oriented process. A SS is used to select and configure an end-to-end connection and a network route that will be used by the connection's data flows. Once a route has been set, network nodes that are located along the route are informed as to the identity of the underlying incoming and outgoing links. This information is used by a network node that is located along the route to switch an incoming message that belongs to an identified flow and that arrives across a specific incoming link to a specified outgoing link. The identity of the latter is recorded in a switching table during the configuration phase that is carried out by the SS. During the signaling phase, a network layer protocol entity is involved in the selection of the route. During the subsequent data transport phase, link layer and physical layer protocol entities are employed for switching arriving messages to an outgoing port and transmitting them across the corresponding outgoing link to the next network node. Such an operation is identified as a **connection-oriented switching** process. As will be described in later chapters, we differentiate between two connection-oriented networking methods: a **CS** method, which has been employed by the legacy PSTN, and a **connection-oriented packet switching** method, as used by ATM and Frame Relay networks.

2) A **Link Layer Protocol (L2):** includes a Link Layer Control protocol sublayer and a medium access control (MAC) sublayer:

a) **Link Layer Control (LLC) Protocol Sublayer (L2):** Serves to transport the network layer PDU (such as an Internet packet) across a communications link that connects a node to a neighboring node. It may offer, if requested by a higher layer entity, a link layer error-control service, assuring the reliable transfer of data across the link, and a link layer flow-control service that is used to regulate the flow of data across the link. For this transfer, header and trailer control fields are appended to the layer-3 packet payload, resulting in a link layer message that is identified as a **link frame**, or just a **frame**. A link layer protocol is often subdivided into a **LLC** upper sublayer protocol, which can provide the link with services such as those mentioned above, and an MAC lower sublayer protocol.

b) **Medium Access Control (MAC) Sublayer Protocol**: Serves to control the transport of frames across a shared-medium link. Resources for the link are drawn from a communications medium that is shared among multiple nodes. Such a communication link is identified as a **multiple access link**, or as a link that employs a **multiple-access communications medium**. A MAC protocol employs a *multiple access algorithm* to control, regulate and schedule the sharing of the link's medium communications resources among its nodal users. To transport a link layer frame across such a link, the frame is encapsulated by MAC layer fields that contain control data, forming a **MAC frame**. A MAC layer protocol functions as a sub-layer of a link layer protocol. Ethernet local area networks (LANs), Wi-Fi WLANs, RANs that a mobile uses to access a BS in a cellular wireless network are systems whose users share the resources of a communications medium. A multiple-access algorithm is used to schedule the transmission of MAC frames produced by medium sharing members.

3) A physical layer protocol provides for the physical transmission of data across a communications medium:

a) **Physical Layer Protocol (L1)**: The physical transmission of a data frame across a communications link is executed by using a physical layer (PHY or L1) protocol. A sending node employs a transmission device, identified as a transmitter, such as a *radio transmitter*, to transmit the data symbols (carrying its data bits) included in a message (such as a MAC frame) across the link to a targeted node that is attached to (or has direct access to) the link. The latter node employs a receiving device, known as a receiver, such as a *radio receiver*, to detect, process, and reproduce the transmitted data symbols.

A physical layer device employs mechanisms that enable it to perform its transmission and reception operations. It is identified as a **transceiver**. Certain such devices are able to simultaneously perform reception and transmission operations. This is known as a **FDX** operation. Other devices are designed to be able to transmit and receive but not in a time simultaneously manner. They are said to operate in a **HDX** mode. Other devices may operate in a **simplex** mode, acting as sole transmitters or as sole receivers.

b) The physical layer protocol determines the electrical and mechanical characteristics of the interface of a physical layer device with a link's communications medium. It also characterizes the mechanisms used for the transmission of physical data units across a communications link. The basic communications unit that is transferred across a communications link is known (in the jargon of Communications Systems) as a **Signal**. To reliably transmit a Signal across a communications link, the physical layer entity would typically employ a *Modulation/Coding Scheme (MCS)*. Using this scheme, a Signal is formed by modulating a carrier (such as a RF tone) in accordance with the data bits included in the frame that is supplied to the physical layer protocol entity by the link layer (such as its MAC sublayer) protocol entity. These bits may be encoded through the use of error-detection or error-correction codes to enhance the reliability of the reception of the signals, and thus also of its corresponding data symbols.

The layered architectural model used by the Internet is based on the **TCP/IP model**. A version of this model is illustrated in Fig. 1.6. In comparing the layering functionality presented by the TCP/IP model with that employed by the OSI model, we note the following. The services provided by OSI layers that are located above the transport layer are provided under the TCP/IP model by a single application layer. The essential services provided by a transport layer protocol as offered by the two models are similar. Network layer services are provided by the TCP/IP model through the *Internet* protocol layer, supporting networking and inter-networking operations.

The corresponding link layer and physical layer protocols and their ensuing services that are provided under the two models are generally similar. Earlier versions of the TCP/IP model identified the link layer functionality as providing network-access or network-interface services. The physical layer entity was also identified as a hardware entity.

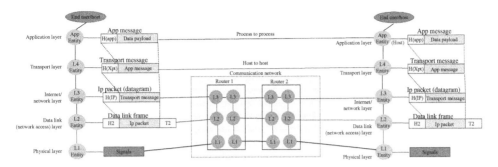

Figure 1.6 The TCP/IP Internet Model.

1.6 Communications Network Architecture: User, Control, and Management Planes

1.6.1 Network Architectural Planes

End users generate message flows that they wish to transport to other end users. They expect, and at times request, the message transport process to be carried out in accordance with specified performance objectives. How does a network system go about accomplishing it?

Communication networks experience wide variability in their network traffic loading conditions, topological layouts, and transport service profiles. It is therefore essential to dynamically adapt network operations and resource allocations to underlying conditions. Network transport regulations require the use of mechanisms that are classified into several categories. Their specifications and implementations refer to a network's data, control, and management planes.

Data plane elements provide for the transport of end-user messages and data flows. Each network node or end-user node that engages in the message transfer process employ a *data plane element*. It is used to execute its targeted data transfer tasks by invoking computing, processing, storing, switching, routing, and transmission-related operations. A component of a data element that performs operations at a specific layer is identified as a *layer data entity (LDE)*.

A network node also makes use of **control plane** elements. A node monitors and processes control information embedded in messages and also makes use of information and directives provided by system control stations. It adjusts and adapts its processing and service operations based on received or monitored control data. A component of a control element that controls operations at a specific layer is identified as a *layer control entity (LCE)*. Nodal control elements interact with the network system and subsystem control stations that maintain ongoing views of system and subsystem conditions and performance behavior. The data derived from such observations and interactions are used by network control algorithms to adjust, over a relatively short time scale, resource allocations, and networking operations, in aiming to enhance system operations. System, nodal, and layer control entities use local and/or system-driven calculations to drive the reconfiguration of impacted parameters and processes to best meet user demands under current system conditions. Control operations are designed to be *highly agile, responding in a timely manner* to variations in system layout and message traffic conditions, as they adjust to best meet message transport requirements.

To illustrate, consider a routing node. It examines the headers of packets that it receives, while temporarily holding the packets in its input memory, known as a buffering module, or as a **buffer**. These packets form a waiting line, known as a **queue**. A nodal processor examines the header of a

1 *Introduction: Networking in a Nutshell*

packet that has reached the head (i.e., front) of the queue, reading its embedded destination address. The router's processor then inspects a RT, searching for an entry that contains a destination nodal address that matches the destination address cited at the header of the incoming packet. This entry identifies the router's output port to which this packet should be switched. Following the switching transaction (occurring within the router's module), the packet is stored at an output buffer, joining an output queue, as it is waiting for its transmission across an output communications link that connects it to the next node along a selected route.

While a router would typically be designed to have the capability to perform these operations, the intelligence of its switching operation is largely expressed through the setting of the routing entries stored in its RT. To determine these routing entries, a sophisticated routing algorithm is employed. For this purpose, the network's topological layout is routinely rediscovered, and the performance states of its nodes and links are regularly inspected. Network-wide interactions among multiple nodes are carried out for the purpose of learning the operational and performance conditions of the communications network system and of its nodal and link elements.

These interactions induce network elements to update their processes and algorithms and their associated parameters. For network systems such as the Internet, rapid fluctuations in traffic processes are regularly observed, inducing the undertaking of dynamic update and adaptation processes. For example, if certain links have just been determined to fail, or if certain nodes have just been observed to be overly congested, control mechanisms may then react by inducing proper reconfigurations. Routes that avoid failed links or bottleneck zones may then be selected.

Monitoring and control operations of network elements and systems that are conducted *over a longer time scale* are also performed. They aim to assure the system's ongoing robust behavior and to maintain high-performance message transport operations. This is the role of functions embedded in the *management plane*. A network managing element is embedded in network management stations and in managed network nodes. A component of a managing or managed element that pertains to operations at a specific layer is identified as a *layer management entity (LME)*.

A managing station can be configured to regularly poll several managed routers and end-user access devices, as it collects system and element performance data. Using such collected status data, and using its embedded system management analysis and synthesis algorithms, a network management station would decide at certain times to modify the service profile that it is able to provide to its network users and accordingly change the operational processes or process parameters that are configured at certain network elements. Adjustments are also triggered in response to change events that include nodal element or system-oriented failures, traffic process changes, modifications in service demand profiles, performance degradation states, and security alarms. A managing station would then react by advising affected layer management entities to properly modify operations.

To illustrate the use of related concepts as they apply to vehicular flows, consider an autonomous vehicle metropolitan highway system. A management station monitors the operation of the system, collecting data about road conditions, traffic flow patterns, speeds of vehicles, time delays experienced by vehicles in moving from entry to exit points, vehicle energy resources, as well as other system states and performance metrics. Based on such observations, and on the use of predictive traffic flow statistics, management algorithms are used to update mobility and access regulation rules. It may also compute new parameters or methods to use to regulate the flow of vehicles across each lane of a managed road. The management station may also calculate the availability status of communications resources that are used to accommodate data communications networking between vehicles and with an infrastructure network that consists of interconnected road side units (RSUs). Resources are dynamically assigned for enabling effective vehicle-to-infrastructure

Figure 1.7 Vehicles on a Highway.

(V2I) communications and for accommodating vehicle-to-vehicle (V2V) data communications. For example, considering a highway system as that shown in Fig. 1.7, RSU access, management, and control modules may be attached to the posts that are placed along the road.

RSUs may be employed to exert shorter-term control of mobility parameters and of vehicle V2I and V2V communications means. Such controls may be used to reorganize the formations of vehicle platoons, to adjust the spacing distances between cars in each platoon and the speed levels at which vehicles should move, modify the processes used for traveling through an intersection, reselect the route to follow by a vehicle to reach its destination exit point, set which lane crossings to initiate, and prescribe the allocation of frequency bands and time slots to use by vehicles for transmitting status, safety, and background data messages.

Autonomous vehicle platforms that move along such a highway system employ embedded communications devices. They initiate at proper times message transmissions that are critical for traffic safety operations. Networking operations support low delay message dissemination between a vehicle and other vehicles and RSUs. In-vehicle intelligent traffic control modules are kept informed of road safety events. A control and management system tracks system and vehicle status states. Status data messages convey information about vehicles' intended destinations, route selections, resource limitations, and monitored critical safety events and traffic congestion conditions. In responding to such system status observations, control and management stations employ algorithms that are used to adjust resource allocations and to modify the operational modes of vehicles and of the communications networking systems that they employ. Networking methods for the autonomous highway are presented and discussed in Chapter 21.

1.6.2 The Data (User) Plane

The Data (or User) Plane contains functions performed by the network system in handling, processing, switching, and transporting data messages. The system's network elements (or nodes) are structured to perform their data networking missions. A switch or a router inspects a message that arrives at one of its input ports and then switches it to a properly selected output port. The switching operation depends on the destination address (or a related label) carried in the header of an incoming packet and on related entries stored in the router's RT. A selected routing entry informs the switch as to which output port should a message be switched to.

Once the RT has been configured, the actual switching operation is readily executed by transferring an incoming packet from its incoming port, which is connected to its incoming link, to an output port, which is connected to the outgoing link pointed out by a routing entry. The switching process may be implemented so that it is performed in hardware, and consequently accomplished at an ultrahigh-speed rate. In contrast, dynamic computations that are carried out in updating routing entries that populate the RT are typically performed by using software-based algorithms.

1.6.3 The Control Plane

Processes performed by the control plane are used to enable the effective and efficient operation of message transport processes that are carried out over the data plane. They entail the use of algorithms that determine the best way to allocate available system resources to network elements and to active users as well as to inform nodes and users of current traffic and loading conditions and of presently recommended system operational parameters and networking processes.

Such determinations are based on collecting system status data, and on determining traffic loading and performance indicators by monitoring message flows, and network system, link, and element states. Control information that is used to dynamically update the calculation of routing entries is provided to switching and routing nodes by using an intelligent (generally software based) control plane utility. Control messages may be generated and disseminated by control stations to network elements and at times also to end-user entities. They can also be autonomously produced by network elements themselves through interactions with other network elements and monitoring systems.

Due to the higher intelligence embedded in calculating and carrying out control plane functions, their operations tend to be performed at lower speeds than those executed across the data plane. To be able to efficiently fuse observations of distributed network states to effectively impact high performance control processes, control points are shared by multiple network elements.

A centralized or hierarchical control architecture is often employed. Under the concept of **Software Defined Networking (SDN)**, physical separation is maintained between entities that perform data plane functions and those that execute control plane processes. As noted in Section 20.2, a central control management operation is then employed.

As noted when discussing the format of PDUs that are formed at distinct layers, control data that is embedded in network messages is typically placed in the message header and/or its trailer fields. Control data included in a PDU that is formed by a layer-*i* protocol entity at a sending node enables proper processing by a layer-i protocol entity that, when applicable, is employed at another node that receives this PDU.

For example, a layer-3 PDU (i.e., a packet) transported across a TCP/IP network, such as the Internet, contains a header field that includes destination and source nodal (IP) addresses, which are used by a packet router to switch and forward an incoming packet. A control field that is included in a packet may also include information that is used by routers to perform packet filtering, error control, flow and congestion control, and carry out at times also other processes, and adapt its behavior as it strives to meet targeted performance objectives.

The *Internet Control Message Protocol (ICMP)* is employed by an Internet router. It is used, as it processes received packets, to produce error messages, which are disseminated to other

1.6 Communications Network Architecture: User, Control, and Management Planes

network elements. This protocol also provides mechanisms for users and routers to carry out network system measurements and tests, such as performing *pinging* operations that are used to examine the integrity of a network path and to determine message round-trip transit time across a network path.

A control plane operation can involve interactions among multiple network elements. To implement a scalable control function, it can be advantageous to subdivide the network control system into control subsystems or domains. A centralized or distributed control operation may be enacted in each subsystem.

To enhance scalability, multiple hierarchical levels may be established. For example, under a 3-level hierarchical control architecture, as depicted in Fig. 1.8, multiple subsystems at hierarchical level 3 are configured so that they are controlled by the same level 2 control module. A central control point (resident at hierarchical level 1) may be set to regulate control functions performed by multiple controllers nested in hierarchical level 2.

Hierarchical networking architectures are highly effective and are prevalent in nature. Figure 1.9 shows the hierarchical composition of conducting links and paths distributed across a leaf. We notice the manner in which multiple narrow conductors connect to a wider one, while the latter connects to a wider core trunk that spans the leaf.

A **SS** is employed by a connection-oriented network to set up connections that are used to transport message flows. In a circuit-switched communications network, such as the *PSTN*, a connection is configured across the network to enable a source user to communicate with a destination user. The signaling systems is used in configuring a *circuit*, whose resources are used to transport the messages that are produced by the connection's user.

Circuit assets entail the dedication of capacity resources across each link that belongs to a selected path, for the duration of a connection. When wishing to set up a connection, an end-user node generates a signaling message, identified as a call set-up request message. This request message is sent

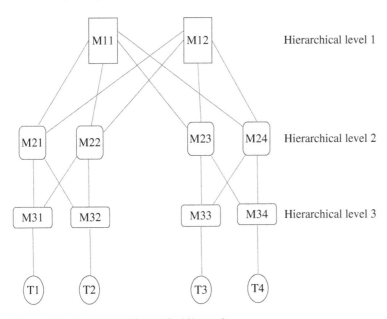

Figure 1.8 Three Level Hierarchical Network.

Figure 1.9 Hierarchical Organization of Conducting Links in a Leaf.

across a signaling network to signaling nodes, also known as signaling transfer points (STPs). In a modern PSTN system, STP nodes communicate across a packet switched signaling network system. It is used to set up and tear down call connection circuits, inform underlying end users and configure the switching tables of involved switching nodes. Signaling System 7 (SS7) is a commonly employed out-of-band SS. For the transport of signaling messages, it employs communications channels that are separate from those used for data (including voice) transport. Such a system is referred to as a *common-channel signaling (CCS)* system. It is also identified as a *common channel interoffice signaling (CCIS)* system.

In a connection-oriented packet switched systems, a SS is used for the setting of an end-to-end path, also known as a VC.

In a cellular wireless network, uplink signaling channels are established across the RAN. They are used by mobile users for the purpose of sending reservation messages to the BS. Such a reservation message carries a call setup request that a mobile user sends to its associated BS node, requesting the allocation of communications resources. The BS node uses radio resource control (RRC) algorithms to determine the sharing of its available communication capacity resources, across a RAN, among requesting mobiles. The BS uses downlink signaling channels to send radio resource assignments to requesting mobiles.

As SSs are used to configure message transport processes across the data plane, we often regard the signaling plane to offer functionality and services that are embedded in the control plane. The boundary between the later mechanisms can depend on the networking technology that is employed.

In a Wi-Fi WLAN system, the data plane accounts for data frame transmissions. An acknowledgment (ACK) frame that is produced by a targeted station that correctly received a data frame is regarded as a control frame. Management frames include messages that are processed by a Wi-Fi net's (AP). An AP station manages the Wi-Fi net. It interacts with a user station while carrying out authentication, association, and disassociation functions. An AP can also be configured to manage the start and end times of specially configured access modes that provide exclusive use of the shared wireless medium to selectively identified user stations.

1.6.4 The Management Plane

A network management system is used by management entities to *monitor and control* the operation of a network system and its elements. A managing entity can consist of a network administrator, a network management station, an automated network management application, or their composite. A network client is associated with a networking service profile that is specified by a **service level agreement (SLA)**.

Management entities *monitor* the performance and operational behavior of the network system. Monitored data can include the status, dynamic behavior, and performance features of network flows and of network elements and subsystems. Status and behavioral data are monitored on a continuous or periodic basis, and/or when exceptional events occur. Alarm messages may be triggered by network elements at times that they are impacted by exceptional conditions. A management system monitors alarm notifications and responds by taking corrective actions.

According to the *International Standards Organization (ISO)*, network management processes are divided into five main functional areas:

1) **Fault Management**: Handling of network system faults involving subsystem failures, performance degradation, and discovery of unexpected processes and message flows. Management functions include isolation and identification of the faults and their location and causes, and the activation of resolution and correction processes. Parameters are set at selected network elements to produce alarm signals and send them to network management system entities when exceptional events take place. Network elements may keep track of the fraction of messages that are received incorrectly and produce alarm signals when the message error rate exceeds a specified threshold. A threshold level is configured by the network management system. It is identified as a *trap* setting.
2) **Configuration Management**: Handling the configuration of the network system, including naming, hardware, and software updates and update of connectivity layouts.
3) **Accounting Management**: Monitoring and regulation of resource allocation and asset utilization levels and their associations with client accounts.
4) **Performance Management**: Monitoring, measurement, and regulation of the performance behavior of system, subsystem, and message flows. Performance metrics include throughput rates, message delays, error rates, and subsystem utilization levels.
5) **Security Management**: Control, regulation, and protection of user and system resources, assuring their use by only authorized entities. Employed schemes include network access control, user authentication, system security, data confidentiality, data integrity, operational robustness, and efficiency. Protection of sensitive data and of system and device elements across the data, control, signaling, and management planes.

For TCP/IP networks, such as the Internet, the **SNMP**, which is an application layer protocol, has been widely used as a network management tool. The underlying operational paradigm is that of a *manager-agent architecture*. A system information technology (IT) person, or a managing program, centrally manages devices that are members of an associated management domain. The TCP/IP networking architecture is employed for the transport of management system messages.

Each managed device contains a protocol entity that is identified as an *SNMP agent*, which interacts with the manager. The managed device keeps a **management information base (MIB)**, which is a database that stores key device parameters and settings. It may also store ongoing measurements of traffic statistics and selective performance records.

The manager polls a managed agent when it wishes to activate, set, or modify operational device parameters, or when requesting to obtain data collected by the managed device. Such data helps

34 | *1 Introduction: Networking in a Nutshell*

the manager to characterize the performance behavior of the underlying device and of message flows that traverse it, and consequently deduce the performance behavior of an underlying system. Observed performance functions can include statistical traffic flow parameters, error rates, and message delay distributions, per flow, and per interface type. A manager can also set *traps*, which are used to induce the managed agent to automatically send performance reports or alarms to the manager upon the occurrence of prescribed events or when statistical values of set performance metrics fall within specified ranges.

In WLAN systems that use a Wi-Fi technology, a user device listens to special announcement messages issued by the AP. The latter acts as the manager of the WLAN. A user station uses the disseminated management data to derive essential system data and parameters. To join and become a member of a WLAN system, a device exchanges association management frames with the AP. When moving away, disassociation management frames are issued by the device and used to remove its membership. Management frames are also used to execute a member authentication process.

Network management functions are used to **monitor and control** devices and system operations over a time scale that tends to be longer than that covered by control plane functions. For example, considering a wireless access network, a management function may perform device authentication and allocate certain communications bandwidth resources (such as specifying the frequency channels that a device is authorized to use) at the time that the device (or session) is activated. Updates are performed relatively infrequently thereafter. In turn, as the need arises, a control plane function may assign to a user bandwidth resources across a wireless access medium within a very short time latency from the time that the corresponding request was issued.

1.7 Illustrative Network Systems

In the following, we present several examples of network systems. Included are vehicular and train transportation networks, as well as electronically driven computer communications and telecommunications networks. Each network provides for the transfer of user entities. Users are often identified as network clients. Transporting entities involve vehicles or trains on the one hand, and on the other hand, data, voice, or video message flows. A network layout consists of access and core network nodes and links. Users access and exit the core network at *network-edge nodes*. Within the core network, a flow traverses core network nodes, such as intersection junctures of a vehicle transportation network. Nodes are interconnected by network links. A core network node redirects or switches a flow from an incoming link to an outgoing link. In a vehicle transportation network, highway roads serve as internodal links. In data networks, electrical communications channels are used as the corresponding links.

A network edge or core node is often identified simply as a node. It may function as both an edge node, as it provides certain users with access to the network, and a switching node, as it directs an incoming traffic flow to a proper outgoing link.

1.7.1 Highway Transportation

As shown in Fig. 1.10, a multilane highway provides the vehicular transportation across multiple lanes. Its nodal elements consist of network-edge nodes that are used as entry and exit points, and core network nodes that reside at road intersections, at which vehicles switch from an incoming road to a proper outgoing road. Highway roads thus serve as links that connect nodes. Vehicles transport users' *payloads*, which can include cargo items and passengers.

Figure 1.10 Multilane Highway.

1.7.2 Inter-regional Road System

A truncated map of an inter-regional road network system spanning the state of California is shown in Fig. 1.11. We note the intricate topological layout of the road system and its nodal and link elements. Typically, a driver has a choice of several routes to use to reach a specific location. Urban regions, such as the Los Angeles metropolitan area, offer an intricate network of intra-regional highways and local roads (which are not depicted in the figure). Lower capacity local roads connect to a backbone (core) network of higher capacity highways.

1.7.3 Train Transportation Network

A layout of a train transportation network operated by Union Pacific is depicted in Fig. 1.12. Train stations serve as the network's nodal elements. Rail lines serve as the network's links. Neighboring rail lines are connected to each other by railroad ties. Rails are typically bolted to the ties. The ties are set into loose gravel or ballast units. A link that serves to connect two stations may consist of several linearly interconnected rails.

In observing the topological layout of this train system, we note that certain areas are densely covered while longer distance lines are laid out in a more sparse manner. Densely used locations make use of hub junctions at which multiple train lines meet and interconnect.

1.7.4 Enterprise Computer Communications Network

In Fig. 1.13, we show an illustrative layout of an enterprise computer communications network. *LANs* connect devices that are attached directly to them. Such devices are located in close proximity to each other, such as being situated in a single office or at close-by offices that are situated in a single facility. A *Corporate Backbone* network consists of interconnected router nodes, which attach to local LANs. The backbone network's routing devices switch data flows to LAN lines that lead to their destination. The backbone network is often structured to assume a mesh layout, so that multiple routes exist between pairs of backbone routers. Such a layout, known as a *mesh* topology, offers higher reliability, as alternate routes can be used when reacting to element failures or to congestion events.

36 | *1 Introduction: Networking in a Nutshell*

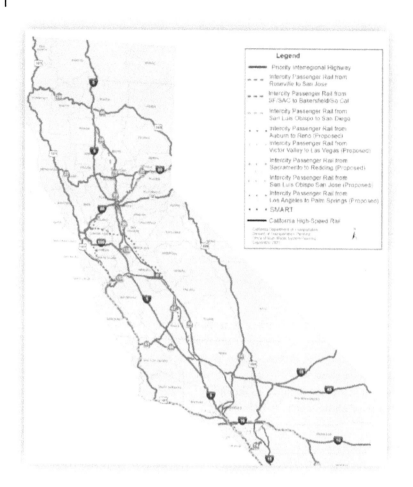

Figure 1.11 Inter-regional Road System in California (Truncated). Source: Caltrans, Division of Transportation Planning, Interregional Transportation Strategic Plan, Fig. 4, October 2021, https://dot.ca.gov/programs/transportation-planning/multi-modal-system-planning/interregional-transportation-strategic-plan.

1.7.5 Packet-Switching Network and the Internet

In Fig. 1.14, we show an early topological layout of a computer communications network that was a precursor to the Internet, representing a topological model of the Advanced Research Projects Agency Network (ARPANET) layout around the year 1980. The first transmission of a packet over an early 4-node network layout took place in 1969, originating from a computer located at UCLA. This network was funded by the US ARPA (Advanced Research Projects Agency) office. It served to interconnect laboratory and research centers to each other and to computing facilities. The ARPANET, which was implemented as a pioneering packet switching computer communications network, made use of the TCP/IP protocol architecture. It incorporated dynamic routing mechanisms, which served to increase the system's reliability and robustness in adapting to element failures and to overloading occurrences at communications links and switching devices.

This network system employed packet-switching nodes using processors that were identified as interface packet processors (IMPs) and as terminal interface processors (TIPs). End-user terminals

1.7 Illustrative Network Systems

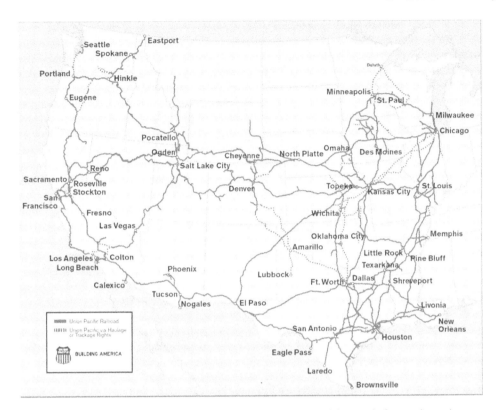

Figure 1.12 Union Pacific Railroad System. Source: www.up.com/aboutup/reference/maps/system_map/index.htm.

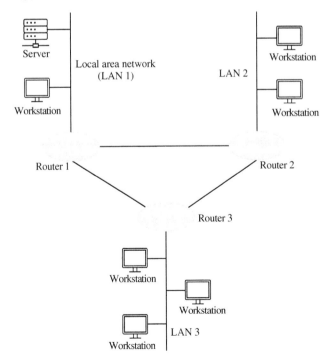

Figure 1.13 Enterprise Computer Communications Network.

38 | *1 Introduction: Networking in a Nutshell*

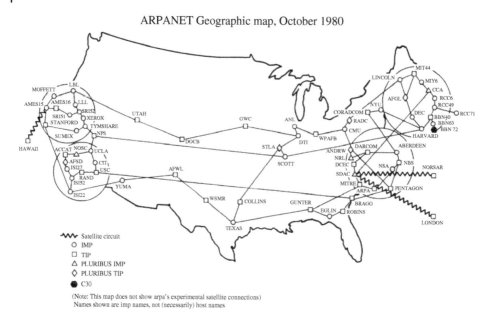

Figure 1.14 ARPANET 1980 Network Layout. Source: An Atlas of Cyberspaces – Historical Maps, ARPA Geographic Map, October 1980, The University of Rochester; https://personalpages.manchester.ac.uk/staff/m.dodge/cybergeography/atlas/arpanet4.gif.

were connected to TIPs. A TIP could also be used by a remote terminal to log-in to the ARPANET. An IMP acts as a packet-switching node. Today, this role is performed by Internet Routers. Originally, these devices were implemented through the use of minicomputers. Today, such devices are often implemented on a single microchip, with processing functions embedded into the operating systems of involved devices. Nodes were interconnected initially through the use of 56 kbps leased telephone lines. Satellite links were employed to provide longer distance connections.

The ARPANET served as a network of networks, providing for intercommunications and message transfer across networks. The ARPANET was decommissioned in 1990. Subsequently, industrial computer and telecommunications organizations formed partnerships, contributing to the construction of a commercialized public packet-switching network, enhancing its operation and capacity through the introduction of many technological developments, evolving it into the today Internet. The layout of the latter is significantly more involved as it enables communications networking among a large number of domain networks, serving a very large number of end users.

1.7.6 Cellular Wireless Networks

Elements of a cellular wireless network system are shown in Fig. 1.15. A region served by a cellular wireless network is divided into cells. Each cell is managed by a BS. BSs are interconnected by backbone networks, which may include the Internet, the PSTN, private networks, satellite links, fiberoptic links, or a heterogeneous combination of such and other network systems.

A client, identified as a mobile station (MS), accesses the network by communicating with a BS that manages the cell in which the mobile is currently located. Uplink (or upstream) communications is directed from an MS to its BS, while downlink (or downstream) communications is directed from a BS to an MS that resides in its cell. A communications medium such as a RAN provides uplink and downlink wireless communications channels that are used to connect a BS and

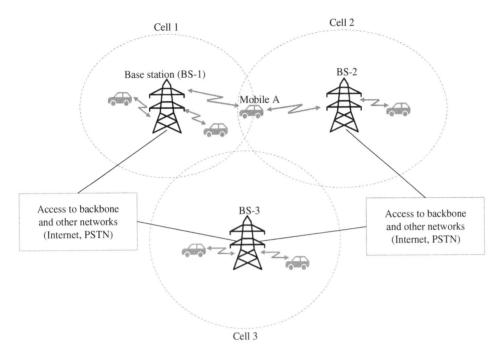

Figure 1.15 A Cellular Network.

Figure 1.16 LTE Cellular Network Radio Access System Architecture.

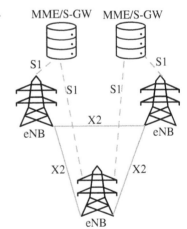

its users (including its mobiles) that reside in its managed cell. As it moves, an MS may cross into an area that is served and managed by a different BS. It would associate with the BS of this new cell and its ongoing communications would then be handoffed (or handovered) to the new BS node.

In Fig. 1.16, we present a diagram that shows several components of a cellular wireless network. The indicated interfaces follow the specification of the 4G-long term evolution (LTE) cellular network architecture. BS nodes are identified as eNB nodes. BS nodes communicate with each other across the a standardized X2 interface. Such communications is also used by BSs to coordinate the scheduling of their uplink and downlink transmissions, aiming to mitigate intercell signal interference effects. A BS node connects with a mobility management entity (MME), which enables the handling of handoff operations for its mobile users. Serving gateway (S-GW) stations provide connections of BS nodes with backbone networks, including connections to a packet gateway (P-GW) that connects the cellular access system with a packet-switching backbone data network such as the Internet.

1.7.7 WiFi: Wireless Local Area Networks (WLANs)

A WLAN uses a shared wireless medium for interconnecting stations that are located in close proximity to each other. Such a network is illustrated in Fig. 1.17. The sharing of the wireless medium in this system is governed by physical, MAC, and link layer protocol specifications, as cited in IEEE 802.11 Standard documents. Such a WLAN system is also known as a Wi-Fi system.

A Wi-Fi cell is typically managed by a central node that is identified as an AP. The AP is often attached to a router that connects to a backbone network such as the Internet. A home gateway node may be used to implement Wi-Fi access control functions. Under a commonly used configuration, all communications to and from a user station proceeds through the AP node. Wi-Fi communications between user nodes and their AP make use of a shared radio channel. A home gateway consists of an AP node that provides wireless communications with local devices and a router that provides access to external networks such as the Internet.

1.7.8 Satellite Communications Networks

In Fig. 1.18, we depict a satellite platform that provides a communications link between a sending earth station and a receiving earth station. A satellite that uses a geosynchronous orbit, which is

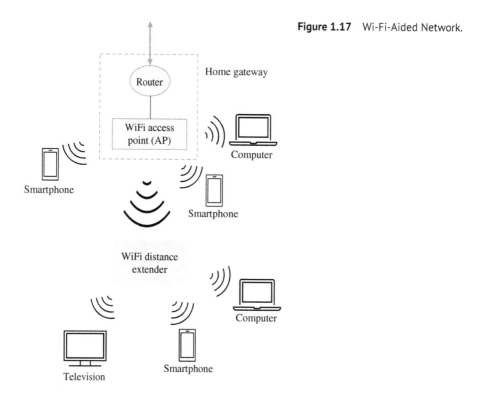

Figure 1.17 Wi-Fi-Aided Network.

Figure 1.18 Satellite Communications.

located 22,236 miles (35,786 km) above the Earth's equator, rotates around the earth at a speed that is equal to the earth's rotation speed. A point on earth will observe such a satellite to stay above it at a fixed position relative to its own position. Connectivity between earth stations is then simplified, as each earth station is associated with a specific satellite node.

Satellite systems provide excellent connectivity means when used for one-way communications. As such, they enable the transmission of messages from a central station Hub through a broadcast satellite to multiple widely distributed client earth stations. Such use is made possible by employing a satellite based broadcast system. It is used to distribute television signals directly to the home or to earth station hubs. Other applications include the use of satellites for wide distribution of navigation signals, as implemented by the satellite-based GPS. For data networking purposes, satellite systems are designed to support two-way communications. They serve to transport messages originating as well as terminating at earth stations.

To transport a message across an earth–satellite link, a sending earth station encodes and modulates its message and then transmits it across an **uplink** communications channel to the satellite. For this purpose, the sending station is allocated capacity resources (such as frequency bands and time slots) across an uplink. The onboard satellite system processes the signal that it receives across the uplink, in aiming to improve its quality. It then proceeds to re-encode and re-modulate it and transmit the ensuing signal across a **downlink** channel, which employs assigned downlink resources (such as allocated frequency bands and time slots). Under such a configuration, a satellite platform is said to act as a *mirror in the sky*. In this manner, a satellite system is designed and managed to enable communications between stations that are located at wide distances from each other. Using inter-satellite links, multiple satellite platforms that are located at geosynchronous or at lower earth orbits are employed to provide communications among earth stations that are located around the globe.

A satellite employs a wide beam antenna to directly reach any station that is located in the wide scope *footprint* of its beam. Such use is beneficial when disseminating broadcast (or multicast) transmissions that are destined to all, or to a targeted group of, earth stations located within a wide earth region.

To increase communications networking efficiency and conserve spectral (i.e., frequency) resources, satellite systems often use uplink and downlink narrow beam antennas. In this manner, uplink and downlink communications resources are spatially separated. Also, the higher directionality of employed antenna beams enables the use of higher transmission data rates and/or reduced transmit power levels. By using spatially nonoverlapping antenna beams, multiple stations are scheduled to transmit uplink at the same time, sharing the same frequency band. Such a frequency-reuse operation saves spectral resources.

Different beams are used to cover different earth regions. A switching mechanism is then used onboard the satellite platform to enable a message that is received from uplink beam A (issued, e.g., by a station located in Los Angeles) to be switched to downlink beam B. The later would cover the location of the destination earth station. Such satellite systems are identified as onboard switching satellites.

Due to its high altitude, a signal transmitted via a satellite that is placed in a geosynchronous orbit experiences a one-way propagation delay (up and down the satellite) of about 250–270 ms. This is a rather high latency level when considering the use of a communications link for the transport of data messages that are produced by interactive applications. Such applications require fast round-trip response times.

Signal propagation latency levels are reduced when using systems that employ satellites that are placed at medium earth orbits (MEOs). They are even much lower when using satellites that are placed at low earth orbits (LEOs). Due to the shorter distances traveled by signals between earth stations, and these satellites, higher data rates, lower transmit power levels, smaller and lower-cost satellite platforms and earth station terminal modules are implemented. MEO satellites are positioned at orbit altitudes that range from 2000 km to below the geosynchronous orbit (with orbital periods that range from 2 hours to 24 hours). MEO systems are often used for navigation (such as GPS-based systems), communications, and space science. GPS satellites are located at an altitude of about 20,200 km (12,552 miles), leading to an orbital period of 12 hours (vs. the 24 hours orbital period realized by a geosynchronous satellite). Communications satellites that cover the North and South Pole are also often placed in MEO orbit.

LEO satellites are placed in earth orbits at altitudes of 2000 km (1200 miles) or lower (at approximately one-third of the radius of Earth). The duration of such an orbital period is 128 minutes or less. Placing satellites at lower altitudes enables the LEO system to provide higher communications capacity and much lower message propagation latency. LEO-based satellites and space stations are more accessible for servicing. Such systems are employed to support Internet access and communications for mobile satellite network systems. They accommodate communications that require short response times, including interactive data and multimedia communications.

Since the latter systems use satellites that are not placed in a geosynchronous orbit, the network topological layout becomes more complex and is fluctuating over time. A much higher number of satellites is now required to cover stations that are located in a given earth region, inducing layout-aware management, monitoring, control, and replacement costs. An earth station is in view of a specific LEO satellite only for a limited period of time. Messages that belong to a given flow need therefore, over time, to be handed off to different satellites. To achieve the required networking connectivity, use is made of a satellite network mesh that employs inter-satellite links.

Examples of LEO systems include earth observation satellites, the International Space Station (which is placed at an altitude of 400 to 420 km above earth's surface), remote sensing satellites, the Hubble Space Telescope (at an altitude of about 540 km), and LEO-based networked systems of communications satellites that form an *Internet in the Sky*. The latter network systems make use of inter-satellite links. Such systems include the Iridium system, whose satellites orbit at an altitude of about 780 km, and LEO-based satellite networks such as Starlink, which is designed and operated by SpaceX and makes use of thousands of small satellite platforms that are placed at altitudes that range from 540–570 km to 1100–1300 km above earth.

1.7.9 Autonomous Vehicular Networks

In Fig. 1.19, we illustrate a layout of an autonomous vehicular network system. Communications links and networks are established to enable message transport among vehicles as well as between vehicles and RSUs. The latter include cellular wireless BSs, Wi-Fi AP stations or other gateways. A network that provides for peer-to-peer communications among its entities without resorting to the use of an infrastructure system (such as the backbone network provided by a cellular wireless network system) is known as an **ad hoc network**.

A vehicular network that provides *peer-to-peer* wireless communications between vehicles, so that it does not make use of a backbone network, is known as a **VANET**. V2V, V2I, and vehicle-to-anything (V2X) message flows are essential ingredients in the implementation of autonomous highway systems. Such systems support the deployment of self-driving vehicles.

Autonomous systems that make use of communications between their elements are also employed in robotic, biological, sensor, human-body, medical, educational, and gaming network systems.

1.7.10 Sensor Networks and Internet of Things (IoT)

Sensor devices are embedded today in many systems and are planned for inclusion in many future ones. Sensors collect information about the surrounding environment. They monitor the behavioral characteristics of the underlying system and of associated entities. An IoT system may employ, or connect to, actuators that are dynamically driven in accordance with the observations collected by the sensors. Systems, subsystems, nodes, or other entities, identified as *things*, may contain sensors that are often microscopic in size. Such systems may be highly restricted in terms of availability of energy, computing, storage, and processing resources.

A control or management station is commonly used to collect data from such system elements. Such data collection often requires message transport via a communications network. A management station carries out processing, computations, and evaluations that can involve the fusion of data collected from multiple sources. In consultation with other management and control stations, a managing station transports resolution messages to managed devices and actuators, inducing them to adjust operations and, when applicable, trigger designated actions.

An IoT system is generally designed to autonomously execute its operations. Involved devices are programmed to carry out sensing, actuation, and data transport functions over a communications network without requiring human-to-human or human-to-computer interactions. To illustrate such a system, consider a *smart home* that uses devices such as lighting fixtures, thermostats, home security systems, and appliances that can be regulated by an intelligent control and management system whose elements can be embedded in smartphones.

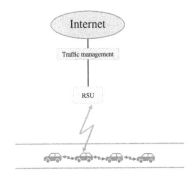

Figure 1.19 Vehicular Network.

1 Introduction: Networking in a Nutshell

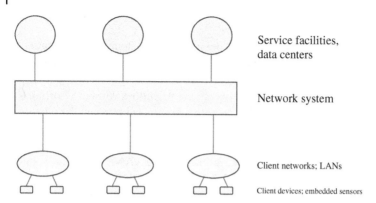

Figure 1.20 IoT Architecture.

As noted in Fig. 1.20, the IoT system's network architecture includes the following entities: client devices (which include embedded sensors, actuators, management, and control agents); client networks (including LANs; a network that serves to transfer data from and to client systems; service facilities that include data service centers and communications networking service providers. Client and core networks include entities that process and manage data and provide for the secure transport of data to and from service facilities that may be embedded in a cloud. A cloud system can be used to provide support for management and control of applications, including the management of data storage, the collection of sensor data, and the triggering of actuation processes.

Capitalizing on the structure and operation of an IoT system, the Web of Things is an architecture for the application layer of the IoT. It examines and fuses data supplied by IoT devices, transporting it to Web application entities that offer innovative use cases.

As it involves a large number of entities, including embedded sensors and actuators, an IoT network system must be designed to scale in its support of a very high number of client devices. For this purpose, the system tasks certain IoT devices (which are often power limited) to focus on supplying data only when a targeted critical set of events takes place. Collected data is transported across a communications network, such as the Internet, aiming to reach cloud-based servers. The latter perform fast computations to produce timely responses that are transported across a network to targeted IoT and/or actuation devices.

Problems

1.1 Consider the network shown in Fig. 1.1. Assume that the network carries the following message flows: (1) Message flows between node A and node B across the route R4–R3–R2–R1, at a total message rate of 20 [message/sec], at equal rate in each direction, and an average message length of 500 bits. (2) Message flows between node C and node D across the route R5–R3–R6 at a total message rate of 15 [message/sec], at equal rate in each direction, and an average message length of 750 bits. (3) Message flows between node A and node C across the route R5–R3–R2–R1 at a total message rate of 15 [message/sec], at equal rate in each direction, and an average message length that is equal to 1200 bits. Calculate the following:
  a) The total flow rate across each network link, expressed in [message/sec] and in [bits/sec].
  b) The loading rate of each node, expressed in [message/sec] and in [bits/sec].

c) The internal loading bit rate of all network links by each flow and by all flows.

d) Assume that the capacity levels of link R2–R3 and of node R3 are reduced by 20%. Assume that impacted flows will consequently reduce their rate in a proportional manner (relative to their flow rates). Calculate the modified bit rates of impacted flows.

1.2 For the network described in the preceding problem, considering the unmodified message flow rates, describe the operation of the network, and compose the corresponding switching/RTs at each node, under each one of the following methods:

a) Circuit switching.

b) Virtual circuit switching.

c) Datagram packet switching.

1.3 Describe the functions performed by a SS in aiding the operation of a network that uses the following switching methods:

a) Circuit switching.

b) Virtual circuit switching.

1.4 Describe the structure of a layered protocol architecture and identify the services provided by each layer, under the OSI and the TCP/IP protocol reference models.

1.5 Explain the function of the following entities in a layered protocol system:

a) Service access point (SAP).

b) Service data unit (SDU).

c) Protocol data unit (PDU).

d) Protocol layer entity.

1.6 Consider the transport of a data message across a network, starting at source end-user node A and terminating at destination end-user node B. Assume the route used by messages flowing between the end user nodes to consist of sequentially ordered nodes (acting as switches or routers) R1, R2, and R3. The nodes are connected by communications links (A,R1), (R1,R2), (R2,R3), and (R3,B). Draw the flows of message units across the route, using time–space flow diagrams, as they traverse each node, and as they are processed by involved protocol layer entities, under the following switching methods. When a SS is employed, show separately the corresponding flows undertaken by signaling units across involved layer entities.

a) Circuit switching.

b) Virtual circuit switching.

c) Datagram packet switching.

1.7 Considering the TCP/IP and OSI reference models, discuss the differences between the functions provided by protocol layer entities operating at layers 1–3 and those operating at higher layers.

1.8 Considering the TCP/IP and OSI reference models, discuss the functions of each PLE and provide examples of services rendered by each entity.

1.9 Describe the error control services provided at the link layer and at the transport layer. When would it be beneficial to provide such services at both layers and when would it be sufficient to provide these services just at the transport layer?

1.10 Describe and contrast the flow control services provided at the link layer and at the transport layer. When would it be beneficial to provide such services at both layers? Just at the transport layer?

1.11 Describe the key services provided by the following transport layer protocols and identify situations under which each would be preferred:
a) User datagram protocol (UDP).
b) Transmission control protocol (TCP).
c) QUIC protocol.

1.12 Describe the role served by SAPs, or TCP/IP ports, in multiplexing application layer streams across the transport layer.

1.13 Describe the principle of store-and-forward operations as performed by switching nodes in handling a packet flow across a network.

1.14 Describe the working of a medium access control (MAC) protocol in enabling the sharing of a multiple-access communications medium.

1.15 Describe the function of a network management system and identify different services that are rendered by a network management system across a management domain.

1.16 Discuss the networking elements and links used by the following technologies for the transport of payload units and identify aspects of similarity and dissimilarity:
a) Train transportation network system.
b) Autonomous vehicular highway network system.
c) Embedded sensor-based IoT system.
d) Cellular wireless network.
e) Wi-Fi WLAN.
f) The Internet.
g) Satellite communications network.

2

Information Sources, Communications Signals, and Multimedia Flows

Objectives: *End users produce message flows that they communicate over a communications network to other end users. Included are audio, video, and data messages and their multimedia combinations. End-users wish their destinations to enjoy high-quality message reception, as expressed by a targeted high measure of experience (MoE). To this aim, the network transport process is tasked with providing a quality of service (QoS) level that meets the requirements of communicating users. Accordingly, a communications network offers several classes of transport services. Networking services are configured to accommodate the transport of real-time audio and video streams as well as to support the reliable and timely transport of non-realtime data, audio and video flows.*

Methods: *Messages that are transported by a modern communications network generally carry information in a digital format, represented by information bits. Information signals that are originally produced by audio and video sources as analog waveforms undergo a digitization conversion process. They are sampled in time and quantized in magnitude. To enable effective real-time interactions between communicating users, the network system's mechanisms are configured to carry out message transport in a reliable and timely manner. For this purpose, destination users often inform senders as to the current status of the message reception process. Senders use this information to adapt message formats and adjust network transport directives.*

2.1 End Users

End users send and/or receive messages that are transported across a communications network. An end-user entity can assume different forms. It can be a person, a computer process, an autonomous vehicle, a robot, a sensor device, an implanted medical chip, or a smartphone device.

Information units generated by a source tend to be converted to a computer readable format, if not yet already in such a form. A source node uses communications protocols to format the messages that it wishes to transport across a network to a targeted *destination node*. The latter processes the information that it receives and may convert it to a format that is preferred by the destination end-user entity. In the following, several illustrative information sources are noted.

1) **Example 1**: A person records a voice message using her smartphone. The latter acts as a source node. It uses a microphone as a *transducer* that serves to convert analog audio signals into electrical signals, which are then digitized, encoded, and represented as digital symbols. Several such symbols are gathered to form a message whose structure is governed by the associated application layer protocol. Following processing by transport, network, and link layer protocols, the produced message is submitted to a physical layer protocol entity (such as one that makes

Principles of Data Transfer Through Communications Networks, the Internet, and Autonomous Mobiles, First Edition. Izhak Rubin.
© 2025 The Institute of Electrical and Electronics Engineers, Inc. Published 2025 by John Wiley & Sons, Inc.

2 *Information Sources, Communications Signals, and Multimedia Flows*

use of a radio module). The latter encodes and formats the message to enable its reliable transmission across a communications link to a communications network's edge node.

2) **Example 2**: A person types a document by employing a word-processing application, using her computer or smartphone, which acts as the source node. In this case, the source message is produced by its source in electronic format. Upon completion, the person initiates the transport of the document across a communications network to a destination user. Invoking the underlying networking protocol, the document may be fragmented into segments. Each fragment, such as a TCP segment, is formatted as a transport layer message. Following its transport across the network, it is processed by a transport layer protocol entity embedded at a destination end-user device. The latter assembles received segments into the original application message, which it delivers to the destination end-user entity.

3) **Example 3**: A person records a video session using a smartphone device. The latter acts as the video's source node. If not yet represented in digital form, the video signals are digitized. Ensuing digital symbols are gathered to form video streams. This flow is divided into video segments, each of which is known as a *chunk*, which is set to contain a prescribed number of video frames. Each chunk is carried as a payload of a message that is properly formatted to enable its transport across a communications network.

2.2 Message Flows

Classification of message flows relates to the application type and the source user's intended mode of interaction with the destination user. The following key considerations are used:

1) **Message Production at the Source User**:
   a) **Real-time Stream**: A *real-time stream* (such as an audio or video flow) as produced at the source user tends to assume a temporally **synchronous** structure. During an activity burst, messages are often generated at a fixed rate, so that they are produced at fixed time intervals. A message content may consist of a fixed or a variable number of information units.
   b) **Non-real-time Message Flow**: A *non-real-time message flow*, as is often the case for a messaging data flow, as produced at the source user, follows a temporally **asynchronous** structure. Messages may be generated at random times. A message may carry a variable number of information bits.
2) **Message Replay at the Destination User**:
   a) **Realtime Replay of Real-Time Streams**: A destination user is interested in replaying a real-time message stream in real-time. For instance, consider a user that wishes to view a video flow, or listen to a audio stream, as it is in the process of being produced and received. For real-time experience that accounts for the timely mutual production and replay of a stream, it is desirable that the network transport process induces low message time delays and low message time delay variations. This is the case when transporting in real-time across a communications network ongoing lecture messages, whether it involves one-way real-time transport of a video/voice stream from a teacher to her students, or whether a two-way interaction in real-time is involved.
   b) **Non-realtime Replay of Streams**: The stream produced by the source user is transported to the destination user for non-realtime replay. For example, it may be recorded, transported and then stored locally at a destination entity. It is used for replay by the user at a later time.
   c) **Non-realtime Interactive Message Flows**: The source and destination users engage in a dialog that induces non-realtime two-way message interactions. For example, under a

query–response application, a source user initiates a dialog with a destination server across a communications network. The destination entity may be a web server managed by a commercial organization. A client may send request (or query) messages asynchronously (at random times). These requests would trigger responses by the server, producing documents that are sent to the client across a communications network. To enhance the timeliness of such an interaction process, it is desirable to configure the network transport process to produce rapid response times.

  d) **Non-realtime Non-interactive Message Flows**: A source user produces messages at fixed or random times. They are targeted for reception by a destination user in a time asynchronous manner. It is generally not essential that end-to-end (ETE) network message time delay levels are strictly as low as they should be when accommodating interactive message flows. As an exception, timeliness is essential when disseminating such messages that are produced upon the occurrence of designated critical events.

3) **Message Transport Performance Requirements**:
  a) **Message Transport Performance Requirements for Realtime Flows**: For real-time replay of a real-time stream, such as an audio or video stream, the communications network is required to provide low ETE message delays, in measuring both the average message delay and the message time delay variation levels. Such a requirement is generally not essential for non-realtime replay purposes. The required settings of the transport process depends on the targeted fidelity of the reproduced stream. To achieve higher fidelity reproductions, the source user would have to produce the stream at an acceptably high-quality level and the network would have to provide an effective connection that supports the real-time transmission pattern of produced flow data units at an acceptably high throughput rate. Depending on the application, the required message error rate (representing the rate at which received messages contain errors) is generally low but may not be as low as that required for the reception of critical data flows.
  b) **Message Transport Performance Requirements for Non-realtime Message Flows**: Interactive non-realtime message flows generally require the network to induce low ETE message transport delays. Noninteractive data transport processes may incur higher message transport delay levels. The required transport throughput rate depends on the flow's data rate. The error-rate is required to be very low for critical data flows, as is the case for many safety, security, financial, and medical data transactions.

In Fig. 2.1, we illustrate two message flow modes. The figure shows a recurring time-frame duration of 125 ms which consists of 4 time slots. Transmissions of message frames across a

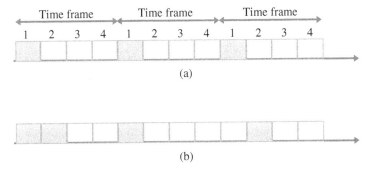

Figure 2.1 Realtime and Store-and-Forward Message Flows. (a) Realtime Transmission of a Stream (b) Store & Forward Message Stream.

communications channel are performed in a time slotted manner. Under this illustration, the information source is set to transmit (at most) a single frame at the start of each time slot. During an activity burst, messages are generated and transported across this channel in a time synchronous manner. In contrast, messages that are transported across the communications channel in a *store-and-forward* fashion are transmitted at time slots that correspond to the times at which the source is ready (and permitted) to transmit its frames across the channel, when these time slots are available. Such transmissions take place at random times. A message is *stored* at the source until it is ready to *forward* it by transmitting it across an assigned communications channel. Such an operation is said to proceed by following a *store-and-forward* process.

In connection with the underlying message application type and reproduction aims, we observe the following message flow types:

1) **Real-Time Streaming**: The information segments generated by the source are temporally structured. A segment represents a fragment of information that is linked to a specific time window. The segments of an audio flow (including speech and music), or of a video flow, form a stream in that each segment represents a portion of the information that corresponds to a specific time span. For example, a chunk of data included in a video stream may represent a 20 seconds duration of a video message that is slated for transport across the network.

    Interactive communications in real-time is also identified as *realtime two-way* communications. It is essential that the communications network system transports a stream's segments so that they incur low ETE delays and low delay-variation levels. It is a challenging task to implement interactive voice communications in real-time through a geosynchronous satellite communications link as the round-trip propagation delay is of the order of 0.5 seconds. A one-way ETE message delay that is lower than 100 ms (i.e., 0.1 seconds) is desirable for interactive voice communications, as normally realized via transport over a terrestrial network path.

    One-way real-time streaming includes broadcasting of information by a source user to single or multiple receive-only end-user entities. This is the case for real-time broadcasting of television signals, movies, or radio audio streams, destined for reception by multiple subscribers. For playback of received messages in real-time, a limited amount of message storage (i.e., buffering) may be made available at the receiving end user.

2) **Non-Real-Time Streaming**: A message flow consists of message segments that are temporally linked. The transport of the message segments is however not limited by strict time delay requirements as those imposed for real-time streaming. Illustrative uses include the archival transport of voice, video, and structured data streams. Messages received at a destination node can be stored (buffered) for a longer period of time. Yet, due to their structured temporal linking, each segment must be identified by an associated time mark, so that when the playback process is executed, message segments can be reassembled in proper order and timing. Hence, when subjected to network transport that induces time delay variations, segments are marked by temporal signatures. Corresponding markings can also include time and 3D-space joint (time, space) identifiers.

3) **Interactive Store-and-Forward Data Flows**: Many applications involve interactive communications between end users. For example, an end user may initiate a session involving two-way communications between her and an application program residing on a host server. The user may request data to be downloaded from the server and then after receiving the data, following a think-time pause, may proceed to request another record of data. For many such interactive applications, it is essential that the **response time** latency, which accounts for the time delay elapsed from the instant that a user submits its request message to the network to the instant

of time that she receives a corresponding response message, will be sufficiently short so that a timely conduct of the interactive process is not impeded. Messages that belong to such a non-streaming flow, may be stored for a brief period of time at the receiving node, or at intermediate nodes, before being forwarded to the destination end-user entity. The segments of a downloaded document may reach the receiving node at random times and are then stored at its buffer for a period of time that is sufficiently long to allow segments that belong to a single document to accumulate at the buffer before they are assembled into the original document and forwarded to the targeted application entity for viewing by a destination end user. The corresponding data flow may be transported across the network in a *store-and-forward* manner.

4) **Non-interactive Store-and-Forward Data Flows**: Certain (typically noninteractive) applications do not impose strict ETE message transport delay requirements. Message segments may be transported in a store-and-forward fashion. Segments can be delayed at the source, destination, or intermediate network nodes as they are routed across the network. At the destination node, received messages may be buffered and delayed until the receiving system is ready to assemble them, as it proceeds with the reproduction of the original message.

2.3 Service Classes

The following categorization of classes of service that are offered by telecommunications and computer communications network systems is noted. Each category of service is of particular applicability for the transport of messages that belong to identified message flow types.

1) **Continuous Bit Rate (CBR)**: A transport service that preserves the temporally synchronous features of a message flow. It offers an admitted flow quality of service (QoS) that guarantees throughput and message delay performance levels. The temporal pattern of a flow is largely preserved as its messages are transported across a network. It is highly useful when employed for the transport of a real-time stream. It provides effective transport for flows whose messages are generated at a prescribed rate in a synchronous manner. In many cases, messages are produced at essentially constant time intervals, and they should be faithfully reproduced at such constant intervals at a destination receiver. Flows that desire such service include uncompressed video streams. It is also useful for the transport of high-speed critical data flows that require network transport at specified sustained throughput rates while incurring very low message delay levels. This service class is also known as *"emulated leased line service"*, as it is viewed by a client to be equivalent to being assigned a communications line that is leased from a telecommunications service provider. Such a line is dedicated for use by a source node in transporting its messages to a destination node.

2) **Real-Time Variable Bit Rate (RT-VBR)**: A service that provides for near-synchronous real-time transport of a message flow, whereby the message information rate is variable. Such a service is typically required for the network transport of real-time streaming flows and for supporting the network transfer of interactive delay-sensitive data flows. Network transport is configured to induce ETE message delays that assume low average and low variation levels. This service is of essential value for accommodating the transport of real-time flows whose messages may experience wide variations in their information contents, exhibiting therefore variability over time in their bit rates. This is, for example, the case for the real-time transport of compressed video streams. In this case, the information contents embedded in a video-frame may vary widely. Frames that exhibit a video scenario that is highly dynamic (e.g., an active

scene in a basketball game) would contain a larger number of information bits. Such a service is also applicable for the real-time transport of speech flows. Typically, such a flow consists of periods of activity, identified as activity bursts (or simply bursts), spaced by inactivity periods (or pauses). Each activity and inactivity period may last for a random period of time.

RT-VBR message flows (including real-time streaming flows) may also be transported across the network by using CBR service. However, such a service will have to be sized so that it offers the client an emulated leased line data rate level that is equal to the *peak rate* that is exhibited by the flow. For example, a compressed video flow whose average bit rate is equal to 1.5 Mbps and whose peak bit rate is equal to 6 Mbps would require, under a CBR service, an ETE network connection that offers a data rate that is equal to 6 Mbps. In turn, by using RT-VBR service, it may be feasible to accommodate such a flow by configuring the network to allocate to it such a peak data rate level only during the flow's burst periods. Communications capacity that is not used by the stream during its inactivity periods could be allocated to the support of other flows.

3) **Non-Real-Time Variable Bit Rate (NRT-VBR)**: A service that accommodates a variable bit rate flow but does not require the network transport process to be executed in real time. This service is not required to offer an admitted flow strict ETE message latency performance. For example, consider a non-real-time download of a data file that is fragmented into multiple message segments, whereby different segments may contain distinct amounts of data. This service sustains a message throughput rate that varies in accordance with the statistics of the VBR source. However, due to the NRT character of the transport process, it is acceptable to have messages experience looser average and variable ETE delay levels. This service is often employed for the support of non-RT streaming flows and non-interactive store-and-forward data flows.

4) **Available Bit Rate (ABR)**: A service that provides for the transport of a message flow for which data rate and message delay support requirements are not imposed. To provide a satisfactory service, a flow may be offered a minimal throughput rate level. Such a service is often applied for the support of certain interactive and noninteractive store-and-forward data flows.

5) **Best Effort (BE) service**: A service that accommodates the transport of a message flow while not targeting throughput and message delay performance objectives. The service is designed to allow a network to make its BE in accommodating the transport of messages. Sources do not impose transport performance requirements. Also, they may not be required to specify the features and associated parameters of the traffic flows that they drive into the network. The original implementation of the Internet (in its early ARPANET version) was set to provide BE service for the transport of data packets. Such a service is often employed to accommodate the transport of certain noninteractive store-and-forward data flows.

CBR and RT-VBR services are employed for the transport of streaming and interactive data flows. NRT-VBR services are used to support the transport of non-real-time streaming flows. The latter flows can also be supported by ABR or BE services when lower (or no) performance expectations are involved. Noninteractive store-and-forward data flows may be offered ABR or BE-type services when no QoS provisions are required; otherwise, an NRT-VBR service is used.

Services that offer a user QoS performance guarantees in the transport of a flow tend to employ a *flow admission control* mechanism. To determine admissibility, a user is required to provide the network a **traffic descriptor** that characterizes the statistical behavior of its flow's traffic process. Alternatively, based on a characterization of the underlying application type or the flow's traffic class, the network system may associate QoS values that a network transport service should offer for the proper support of the flow.

The Public Switched Telephone Network (PSTN) was originally designed as a telecommunications network system that offers users message transport by setting dedicated circuit-switched connections. They were designed to carry primarily voice streams. As such, it provides a CBR service. A voice flow is allocated a network circuit whose capacity is reserved for the exclusive use by the flow's voice connection. The circuit's data rate level is set to support high-quality transmission of voice signals. Typically, a voice circuit data rate of 64 Kbps has been allocated for the provision of high quality (known as toll quality) voice connection. Leased line services that have been provided by telecommunications networks, offer a user a dedicated high speed line. For example, $T1$ and $T3$ leased lines provide a user with data rates that are equal to 1.544 Mbps and around 45 Mbps, respectively.

Cellular wireless networks have originally been designed to provide voice connection services to mobile users via the use of circuit-switched CBR service. As they have evolved, these networks have been designed to employ packet switching-based networking techniques, offering mobiles a wide range of transport services, in the support of data, voice, and video flows.

The Internet was originally designed as a datagram packet-switching network that offers BE service for data transport. As it has evolved, network layouts and communications networking mechanisms and protocols have been expanded to enable it to provide network services that accommodate the transport of data, voice, and video flows. BE and NRT-VBR type services have been traditionally accommodated. Advanced mechanisms have been incorporated and enhanced communications capacity allocation schemes have been employed to enable the network to offer RT-VBR-oriented services, including such that are used to accommodate the transport of audio and video streaming flows.

As communications transport resources that are available across an ETE route in a packet switching network are statistically shared among multiple flows, it is more demanding in these systems to provide a flow with dedicated capacity resources. This is especially the case for flows that require such capacity resources during randomly occurring burst periods. In Chapters, 8,11,12,16,18–21, we present techniques that are used for the support of message flows across packet switching networks.

When considering networks that possess transport capacity resources that are more than sufficient for the support of the aggregate message traffic rate, it could be feasible to offer high-performance transport service levels to many or even all active flows, independently of the underlying flow type and its service requirements. This is at times the case when considering network system domains that employ high-capacity fiberoptic-based communications links. In contrast, in networks that employ wireless links, communications capacity resources are often highly limited. Resources must be shared among multiple flows, whose activity levels may fluctuate in a random manner. Flows, whose service requirements cannot be met, may not then be admitted for network transport.

2.4 Analog and Digital Signals

2.4.1 Analog and Digital Sources

Information produced by end users is categorized as being of data, voice, or video types. In the following, we discuss the key features of each as well as the performance requirements imposed for effective communications transport of messages that carry such information. We first discuss the key concepts of analog and digital signals.

An **analog** variable assumes values from a set of infinite (or relatively high) dimensionality, which contains a very large number of entries. For example, a variable X whose assumed value can be any real number in the set [0, 1], which of course contains infinite such entries, is said to be an analog variable. In turn, a variable, say K, which assumes a value in a set of finite dimensionality M, such as $\{0, 1, 2, \ldots, M-1\}$, is said to be a **discrete** or a **digital** variable.

For $M = 2$, so that the discrete set in which the variable assumes its values is of dimension 2, containing (say) the two values $\{0, 1\}$, the variable is said to be a **binary** variable.

An information source produces data that assumes either an analog or digital format. The output of a source over a time period of activity, say $0 \leq t \leq T_A$, is represented as a signal $S = \{S(t), 0 \leq t \leq T_A\}$.

2.4.2 Analog Signals

As you speak into a microphone, the sound waves produced by you are converted by the microphone into electrical signals. Upon producing sounds over an ongoing activity period of duration T_A [s], the corresponding electrical signals are noted to display variable intensity levels. The signal's intensity (also expressed as its amplitude, strength, or power) at time t, denoted as $S(t)$, is observed to assume values that vary over an analog range. The variation of the signal intensity over this period of time is represented as the signal $S = \{S(t), 0 \leq t \leq T_A\}$. This signal assume values that are analog in amplitude over an analog period of time. In Fig. 2.2, we illustrate the format of a waveform of such an analog signal.

2.4.3 Digital Signals

Modern communications channels are designed to function in an efficient and reliable manner by providing the transport of digital signals. A digital signal is discrete in amplitude and discrete in time. Over time, the amplitude (intensity) assumed by the signal is observed at only certain points in time, represented by a discrete set of K sampling times, such as $\{0 \leq t_0, t_1, \ldots, t_{K-1}\}$. The intensity of a discrete signal at a sampling time, identified as a sample's amplitude, assumes a value that is equal to one of a finite number of values, say M, forming a set of possible values such as $S_M = \{A_0, A_1, A_2, \ldots, A_{M-1}\}$. The resulting discrete-space and discrete-time signal is represented as a sequence of discrete symbols $S_D = \{S_{t_0}, S_{t_1}, \ldots, S_{t_{K-1}}\}$, so that the i_{th} sample, S_{t_i}, assumes a value

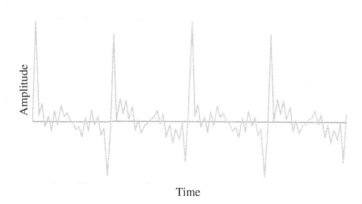

Figure 2.2 Analog Signal.

Figure 2.3 Digital Signal.

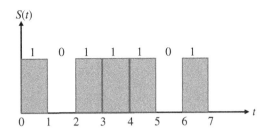

that belongs to the set S_M. An illustrative digital signal is shown in Fig. 2.3. In this case, $M = 2$ so that each sample assumes one of the two possible amplitude values, representing a binary signal.

Digital computers and other devices carry out processing and computation tasks by using data that is in the digital form. They produce output signals that also tend to assume digital format. Computations, processing, and signal transmissions over communications channels are generally carried out by using digital information symbols.

2.4.4 Discretization: Analog-to-Digital Signal Conversion

It is more effective to send signals over a communications channel when they are presented in the digital form. If the original signal produced by a user is analog in amplitude and analog over time, it is converted to a digital signal by performing analog-to-digital (A/D) conversions in amplitude and in time. The process of discretizing the amplitude of a signal is called **quantization**. The process of selecting a discrete sequence of times at which to convey the signal's amplitude is called **sampling**.

By discretizing the analog signal in amplitude and time, we produce a digital signal that would often represent a *distorted version* of the original analog signal. A weighted averaged (over amplitude and time) of the difference between the original signal and its digitized version is often used to express the *level of signal distortion* that is caused by the ensuing digitization process. It combines the effects of quantization and sampling errors. When a digitized signal is conveyed to a receiving entity, the latter will act to reconstruct a version of the original signal. A targeted *signal's reconstruction fidelity* level serves to limit the magnitude of the difference between the original signal and the reconstructed signal. Various mathematical metrics have been employed to express the ensuing distortion level.

The error caused by a quantization process is identified as a *quantization error*. It can be reduced by using a larger number (M) of quantization levels. Such a selection would however increase the signal's digital information content that would need to be transported over a communications network system.

The error caused by a sampling process is identified as a **sampling error**. It can be reduced by employing a higher sampling rate ($R(sampling)$), representing the number of samples taken per unit time; that is, $R(sampling) = \frac{K}{T_A}$. The sampling rate is often set as equal to the *Nyquist rate*, which is equal to twice the *signal's bandwidth* value.

An illustrative digitized signal is shown in Fig. 2.4. We note that the original analog signal is sampled at discrete times and that the amplitude of each sample is digitized by using a 3-bit quantization process. The amplitude variation range is divided into 8 levels (noting that by using a 3-bit range, one is able to identify $2^3 = 8$ distinct levels), so that the discrete amplitude level of each sample is represented by a 3-bit digital value. The quantization process may be carried out by setting the amplitude of a sample to be equal to the discrete value that is closest (based on a specified distance measure that could change in a nonuniform manner) to its analog amplitude level.

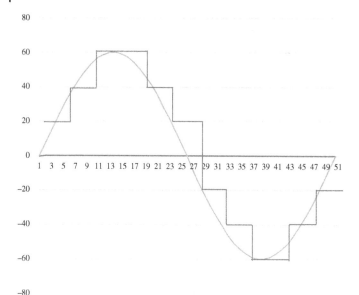

Figure 2.4 Digitized Signal.

2.5 Frequency Spectrum and Bandwidth

2.5.1 Time Domain and Frequency Domain

Communication channels that are used to transport messages have limited frequency bandwidth that they can make available for the transmission of signals. For example, in the United States, an FM radio station is allocated a frequency band (whose size is identified as its bandwidth) in the frequency spectrum range of 88 to 108 MHz. Each station requires a channel bandwidth that is lower than 200 KHz = 0.2 MHz for the transmission of its broadcast signals. Hence, a total of 100 FM stations can be locally accommodated.

The frequency range that is made available for signal transmissions across a communications channel by different stations is limited. Higher-frequency bandwidths are available for allocation to stations when transmitting signals at higher carrier frequencies. However, the power of the signal received at a destination receiver experiences higher attenuation over distance when a higher carrier frequency is used.

A signal generated by an information source is characterized in terms of its **time-domain** features (i.e., its properties related to its variation over time) as well as in terms of its **frequency-domain**-oriented features. The key range of frequencies that needs to be used to characterize (or construct) a signal is identified as the signal's **frequency bandwidth**, also identified in brief as its **bandwidth**.

Consider a source that generates information bearing signals that it wishes to transmit over a communications channel. What is the minimal frequency bandwidth level that a communication channel should allocate for the transmission of a signal such as one that represents a message produced by an end user or a nodal station (identified henceforth as simply a station)? If the allocated bandwidth level is too low, the received version of a transmitted signal will be distorted. In this case, the received signal waveform may widely deviate from its transmitted version, which may cause the transmitted data symbols to be incorrectly decoded by an intended receiver.

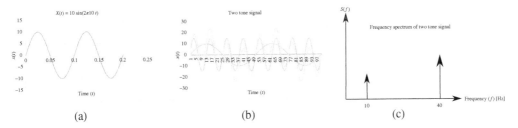

Figure 2.5 Sine Signals: (a) a Sine Signal; (b) A Two-Tone Composite Signal; The spectrum of the Two-Tone Signal.

To effectively transmit a signal across a communications channel, the signal is encoded and modulated by using a modulation/coding scheme (MCS). The **baseband** signal represents the signal's waveform that is generated by the information source prior to its processing by the MCS. The frequency bandwidth required for the transmission of a modulated signal is impacted by the *bandwidth of the baseband signal*, denoted as B_m, and by the bandwidth index of utilization associated with the employed MCS.

What is a frequency spectrum and how do we measure the frequency bandwidth of a signal?

We start by noting that a signal waveform can be expressed as the weighted sum (or integral) of pure sine and cosine signals. A sine signal assumes the shape of a sinusoidal waveform, which is expressed as follows (see Fig. 2.5(a)):

$$S(t) = A\sin(2\pi f t + \phi), t \geq 0. \tag{2.1}$$

The sine's signal level at time t is denoted as $S(t)$, and often also as $s(t)$ or $X(t)$, while its maximal level is represented by its *amplitude A*. Its magnitude level at time t is also identified as its amplitude $A(t)$ level at time t. Sine is a periodic function, repeating its waveform every period of duration T_p [s], also identified as a **cycle**. The frequency of the sine signal is denoted as f. It represents the rate at which the signal's cycles are repeated and is expressed in units of Hertz = cycles/s, or 1/second. Hence:

$$f = 1/T_p. \tag{2.2}$$

An illustrative sine signal is shown in Fig. 2.5(a). It is represented as $X(t) = 10\sin(2\pi 10\, t)$, so that its amplitude is of magnitude 10, and its frequency is equal to $f = 10$ [Hz] so that its cycle period is of duration $1/10 = 0.1$ [s]. The figure illustrates the evolution of a sine waveform over two cycles.

A sine signal which represents a single tone at frequency f has a frequency spectrum $S(f)$ whose single component is a spike at frequency f.

As another example, consider a signal that consists of two tones. The signal waveforms corresponding to these two tones and the composite signal are shown in Fig. 2.5(b). The magnitude level of the composite signal at any instant of time is equal to the sum of the amplitudes of its tone components. The first tone is as noted in Fig. 2.5(a), produced at frequency $f = 10$ [Hz] at amplitude $A = 10$. The second tone is produced at frequency $f = 40$ [Hz] at amplitude $A = 15$, so that it is expressed as $X_2(t) = 15\sin(2\pi 40\, t)$. Its cycle period is of duration $1/40 = 0.025$ [s]. The figure illustrates the evolution of the three signals over a period of 0.2 [s], which spans two cycles of the first signal's component and 4 cycles of the second signal's component. As shown in Fig. 2.5(c), the frequency spectrum $\{S(f), f \geq 0\}$ of the composite two-tone signal consists of two spectral spikes, located at $f = 10$ [Hz] and at $f = 40$ [Hz].

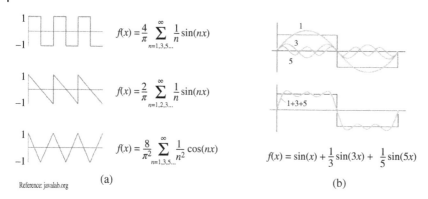

Figure 2.6 Signal Representation via Fourier Series. Source: https://javalab.org/en/fourier_series_en/, Javalab, Public domain, Last accessed 8 August 24.

2.5.2 Frequency Spectrum of Periodic Signals

How about the frequency spectrum of a signal that does not assume the shape of a sinusoidal waveform?

Consider first a periodic signal. The waveform of such a signal repeats every cycle (or signal period) time of duration T_p. The interesting property to note here is that such a signal can be represented as a weighted sum of sine and cosine signals. Such a presentation is known as **Fourier Series**. It is named in honor of Jean-Baptiste Joseph Fourier (1768–1830). Fourier series representations of several periodic signals are shown in Fig. 2.6(a). It is noted that each representation consists of a fundamental spectral component associated with a sine wave whose frequency is equal to the signal's cycle rate, $f = 1/T_p$ [Hz], known as the **fundamental frequency**, and additional sinusoidal components whose frequency values are equal to integral multiples of the fundamental frequency, known as **harmonic frequencies**. It is also observed that a large number of sinusoidal components (and/or cosine signal components, noting that a sine signal with appropriate phase translation becomes a cosine signal) is often required to provide a precise representation of the periodic signal as a weighted sum of sine components.

Consequently, the frequency spectrum of a periodic signal consists of a spike at the fundamental frequency and additional spikes at frequency harmonics. The magnitude of an harmonic component tends typically to decrease as the harmonic frequency increases. Hence, as shown in Fig. 2.6(b), a good approximate representation of a periodic signal can be attained by including the fundamental sine wave component and a small number of sine harmonics. We note that a close reconstruction of the original illustrative signal is obtained by using just two harmonics. The frequency spectrum of the latter approximation consists of just three spikes, located at the fundamental frequency and at the third and fifth harmonics.

The **bandwidth** of a signal whose spectral representation $S(f)$ consists of a finite number of frequency components measures the difference between the values of the highest and lowest frequency components. For example, for the two tone signal whose spectrum, as shown in Fig. 2.5(c), consists of frequency components at 10 [Hz] and at 40 [Hz], the signal bandwidth is equal to 30 [Hz]. It is often the case that a precise representation of a signal would demand the use of a frequency band that is wider than the one that can be made available. As the range of frequency components is truncated, the ensuing signal's temporal waveform changes, as it becomes an approximation of the original signal.

2.5 Frequency Spectrum and Bandwidth | 59

For the signal representation shown in Fig. 2.6(b), the signal bandwidth level is calculated as the difference between the frequency level of the fifth harmonic and that of the first harmonic.

2.5.3 Frequency Spectrum of Nonperiodic Signals

What is the frequency spectrum of a nonperiodic signal?

In this case, we use a **Fourier Transform** representation. It represents the signal not as a sum of a discrete number of sine and cosine waveform components whose frequency levels are equal to integral multiples of the fundamental frequency but rather as a weighted sum of a continuum of waveforms that are mathematically modeled as complex exponential (CE) components. The Fourier Transform formula that is used to calculate the frequency spectrum $\{S(f)\}$ of a time-domain signal $\{s(t)\}$, and the *Inverse Fourier Transform* formula used to calculate the time signal waveform $\{s(t)\}$ from its spectrum $\{S(f)\}$, are presented in Eqs. 2.3. It is noted that $\{S(f)\}$ assumes values that are represented as complex numbers that have real and imaginative parts. They can be presented by their corresponding magnitude and phase components.

$$S(f) = \int_{-\infty}^{+\infty} s(t)e^{-j2\pi ft}\, dt, \quad -\infty < f < +\infty;$$

$$s(t) = \int_{-\infty}^{+\infty} S(f)e^{j2\pi ft}\, df, \quad -\infty < t < +\infty. \tag{2.3}$$

Using this transform, we show in Fig. 2.7 the spectrum $\{S(f)\}$ of a rectangular pulse of time duration width that is equal to $T = 1$ and whose amplitude is equal to 1. It is noted that $\{S(f)\}$ spans an infinite range of the parameter f, $-\infty < f < +\infty$. The reconstruction of the time function $\{s(t)\}$ from its spectral function $\{S(f)\}$ makes use of the values of the later for both positive and negative values of the parameter f. For a time signal that assume real values (having no imaginary components), the magnitude of the Fourier Transform function is even, being symmetric around the y-axis, so that $\|S(f)\| = \|S(-f)\|$, for all values of f. (It can also be readily noted that the Fourier Transform of real and even, or real and odd, time-domain signals are spectral functions that are real and even, or pure imaginary and odd, respectively.) We note in Fig. 2.7 that for a real and even time domain function, such as the rectangular signal waveform that is shown in the figure that spans the period $(-T/2, T/2)$, where we have set $T = 1$, the corresponding Fourier transform is real and even, and it can be calculated by just integrating over the positive values of f. It is expressed as the integrated aggregate of weighted cosine components. (A similar statement applies to the Fourier Transform of a real and odd signal function in relation to the computation of its absolute value as the aggregate integral of sine components.) We use the positive values of the f parameter to assess the effective frequency bandwidth of the signal.

How do we calculate the frequency bandwidth of a nonperiodic signal?

As noted above, the frequency spectrum of a nonperiodic signal may include an infinitely wide frequency span. Communications media however allocate a limited (and finite) frequency band for the transmission of a particular signal. Hence, we employ a different definition for a signal's effective frequency bandwidth. As illustrated in Fig. 2.6(b), a good approximate representation of the bandwidth of a signal can be attained by accounting for only a limited portion of its spanned frequency spectrum. If the communications channel is set to transport only a limited portion of a signal spectrum's frequency components, the signal waveform that is reconstructed by the receiver could then represent a **distorted version** of the original signal's waveform.

Hence, a possible definition of a signal's bandwidth would be to set it equal to the magnitude of the range of frequencies that would be required to induce a reconstructed version of the signal whose distortion level is not higher than a prescribed value. For this purpose, one would have to define a relevant distortion measure.

An alternate definition of the bandwidth level of a signal is to set it equal to the magnitude of the range of frequency components whose absolute spectral amplitude (or power) value is not lower by a specified level relative to a specified level, such as the value assumed at $f = 0$.

Alternatively, the frequency bandwidth is often defined as the size of the range of frequencies over which the aggregate area under the magnitude level (i.e., absolute value) of the spectral function is fractionally sufficiently high. Alternatively, we require the corresponding energy of the accommodated (i.e., conducted) signal to be equal to a high fraction of the energy level of the original signal. For this purpose, we note that the square of the amplitude value of the signal, $s(t)^2$, is representative of the signal's power level, so that its integration over time yields its energy level. It is also noted that the latter integral is equal to the integral of the signal's Fourier transform over the frequency domain.

We often simplify the computation of the signal's frequency bandwidth by examining the specific functional behavior of the signal's spectral function $\|S(f)\|$. For example, consider the spectrum $\{S(f)\}$ of a rectangular pulse of time duration T over $(-T/2, T/2)$. Its spectrum is calculated to be given by the following expression:

$$\text{Spectrum of a rectangular pulse of width } T: S(f) = T\frac{\sin(\pi T f)}{(\pi T f)}. \tag{2.4}$$

A rectangular pulse of width $T = 1$ is shown in Fig. 2.7. Its spectrum is calculated by using Eq. 2.4 and is shown in the same figure. A high fraction of the signal's energy level is embedded in its *main lobe*, spanning the frequency range $(-1/T, +1/T)$. By considering the positive part of this interval, recalling that the spectral function is a real and even function and thus symmetric around the y-axis, so that $S(f) = S(-f)$, we set the bandwidth $B(T)$ of a baseband rectangular pulse of time-width T to be proportional to $1/T$:

$$\text{Bandwidth of a rectangular pulse of width } T: B(T) = 1/T. \tag{2.5}$$

By examining the frequency spectrum and the bandwidth level calculated for different signals, we note the following. Consider the frequency components of the three periodic signals shown in

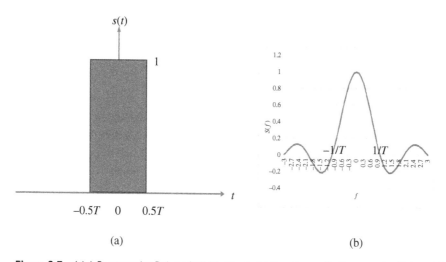

Figure 2.7 (a) A Rectangular Pulse of Width $T = 1$; (b) Spectrum of a Rectangular Pulse with Width $T = 1$.

Fig. 2.6(a): the rectangular, sawtooth, and triangular signals. The first two signals have sharp edges. In turn, signal level variations exhibited by the triangular signal are relatively smoother. We note that the magnitude of the n_{th} harmonic, which represents the spectral level incurred at the frequency $f_n = nf_0$, where $f_0 = 1/T$, where T is the signal's time period duration, for the rectangular and sawtooth signals, is proportional to $1/n$. In turn, it is proportional to $1/n^2$ for the triangular signal. Thus, a signal whose time waveform exhibits more abrupt variations over time produces higher-intensity frequency components over a wider frequency range. Such higher-frequency components need to be receive at an intended receiver for it to be able to reproduce the signal waveform at an acceptable fidelity level.

Similarly, when observing the spectral span of a rectangular pulse. As the pulse's width narrows, so that its duration T is reduced, more abrupt variations in the magnitude of the signal are incurred. The frequency bandwidth of this signal has accordingly been noted to be proportional to $1/T$. **Hence, to transmit or reproduce a signal that is represented as a narrower pulse would require a wider range of frequency components and thus a communications medium that offers a higher bandwidth level**.

For example, the transmission of a rectangular pulse of duration $T = 1$ ms would require a communication channel whose bandwidth is of the order of $B = 1/T = 1000$ Hz $= 1$ KHz. If the signal's duration is 1000 times narrower, so that $T = 1\,\mu\text{s}$, the required bandwidth would be 1000 times wider, $B = 1/T = 1,000,000$ Hz $= 1$ MHz. If each pulse carries information content that is equal to 100 bits, and the corresponding bandwidth levels are assigned, the attained data rate transmitted over the channel (and denoted as R) for these two cases would be equal to $R = 100$ bits/1 ms $= 100$ Kbps and $R = 100\,bits/1\,\mu\text{s} = 100$ Mbps, respectively.

2.5.4 Nyquist Sampling Rate

Consider an analog signal whose frequency spectral span is limited to a frequency band of $B\,[\text{Hz}]$. Such a limitation may also be imposed when a source is allocated limited frequency bandwidth for the transmission of its signal across a communications channel. In 1924, Harry Nyquist showed that by sampling such a band limited analog signal at a rate that is equal to at least $2B\,[\text{samples/s}]$, the information contents of the analog signal can be fully recovered by a receiving entity by using the produced (discrete time) samples. **The sampling rate of $2B\,[\text{samples/s}]$ is identified as a Nyquist Sampling Rate**.

This result is employed when performing the **sampling** over time of an analog signal $\{S = S(t)\}$ during the process of converting it to a digital signal. If the frequency bandwidth of the signal is equal to $B\,[\text{Hz}]$, the sampling rate is then set to a value that is not lower than the Nyquist rate. Each $1/2B\,[\text{s}]$, a sample is taken. The latter period is known as the sampling interval.

This result guides the selection of a sampling rate also when a high fraction of its spectral energy is limited to finite span of frequencies even though the signal's spectral distribution is not completely band limited.

A toll quality voice signal is generally assumed to have a bandwidth of about $4\,[\text{KHz}]$. It represents a version of the signal that preserves frequency components that are not higher than $4000\,[\text{Hz}]$. A high quality discrete-time version is obtained by sampling this signal at a rate of $8000\,[\text{samples/s}]$. The original continuous-time voice signal is then sampled every $125\,\mu\text{s}$.

To fully digitize a sampled sequence of pulses, we digitize the magnitude (amplitude) of each sample. This operation is identified as a **quantization** process. For example, assume that 256 possible discrete amplitude levels are used. Each sample's amplitude level is approximated by setting it equal to the discrete value that is closest to it. A quantization process is shown in

Fig. 2.4. When using such a quantization span, the value of the digitized sample's amplitude is expressed by a sequence of $\log_2 256 = 8$ bits. As each binary digit assumes a value in $\{0, 1\}$ (i.e., it is equal to either 1 or 0), it is noted that a sequence of 8 binary digits (or bits) represents a total of $2^8 = 256$ numbers, whose values vary from 0, which is represented by a sequence of 8 consecutive 0s, 00000000, to 255, which is represented by the sequence 11111111, which consists of 8 consecutive 1s. Under such a quantization level, a toll quality digital voice sequence produces information at a rate of 8000 [samples/s], whereby each sample carries 8 bits of information. Hence, such a toll quality digital voice signal loads the system at a data rate that is equal to $R_d = 64$ [Kbits/s].

In Sections 2.6–2.7, to demonstrate the use of the discussed voice and video encoding and processing techniques, we briefly outline networking-related approaches employed by a selective set of commonly used products. These outlines are based on descriptions that are provided in manufacturer sites, including such that appear in product specifications sheets, as well as in outlines provided online by various organizations, including in Wikipedia and other sites. As existing products are continuously updated and new ones emerge, the interested reader should examine manufacturer based and other online sites to learn about the latest implemented methods and associated equipment offerings.

2.6 Audio Streaming

2.6.1 Audio Encoding and Streaming Across a Communications Circuit

2.6.1.1 Audio Encoding

As you talk into a microphone, an electrical analog signal is produced. This signal fluctuates in a continuous fashion over time and amplitude. We describe in the following a mechanism that is used to perform analog-to-digital conversion (also denoted as A/D or ATD) of an analog signal. The described mechanism is known as **Pulse Code Modulation (PCM)**.

In telephony, the usable voice frequency band ranges from approximately 300 Hz to about 3500 Hz. We assume the effective frequency bandwidth of high-quality telephone voice signal, also known as **toll quality voice**, to not exceed 4000 Hz. To transmit this signal over a digital communications system, it is converted to a discrete signal in time and amplitude. To discretize it in time, the signal is *sampled at the Nyquist rate*, which is equal to twice the bandwidth level. Hence, this signal is sampled at a *sampling rate of 8000 [samples/s]*. The sampling interval is thus equal to $T_{sampling\ interval} = 1/8000 = 125\,\mu s$.

The amplitude of each sample is discretized by using a **quantization scheme**. The number of employed quantization levels is selected. As this number is presented in a binary form, it is also known as the *bit depth*, or *sample depth*. The higher the number of quantization levels, the higher the reproduction precision of the replica of the original voice signal that would be produced at a receiving entity. In telephony, to produce toll quality voice, the system often uses a sampling depth of 8 [bits/sample]. Thus, a total of $2^8 = 256$ quantization levels is used. The A/D conversion of the amplitude of a sample is performed by selecting the quantization level whose value is closest to the signal's original analog amplitude level.

In this manner, a toll quality digital voice signal is produced by sampling the voice signal at a rate of 8000 [samples/s] and by setting the bit depth as equal to 8 [bits/sample]. Consequently, such a PCM-encoded digital voice signal produces data at a data rate that is equal to $R_d = 8000 \times 8 = 64$ Kbps.

In some systems, a 64 Kbps voice circuit uses 8 Kbps for the transport of control data. Control data has been used to maintain time synchronization between communicating entities. In this case, a 7-bit quantization process is employed, so that the ensuing net voice flow data rate is equal to 56 Kbps.

Under a basic quantization process, identified as **Linear Pulse Code Modulation (LPCM)**, quantization levels are uniformly configured over an amplitude range of interest. The output of the quantizer is then linearly dependent on its input. In this case, within the range of operation, an amplitude increase by a factor k of the input signal produces an output discrete signal whose intensity is larger by a factor of about k.

Due to the intricate character of the functional behavior of human produced voice signals, the dynamic range of these signals is higher than that which is accommodated by ordinary network communications media. Noise processes that impact communications transport processes tend to compromise the ability of a receiver to determine the detailed behavior of lower volume signal levels. Hence, it is customary to compress the dynamic range of a quantized audio signal (and then decompress it at the receiver), contributing to an increase in the *Signal-to-Interference-plus-Noise (SINR)* level measured at a destined receiver. Such a compression–expansion process is identified as *companding* (short for compression and expansion). The target values configured for the employed quantization levels vary as a function of the amplitude of the analog sample. A commonly employed companding scheme uses a logarithmic function, so that when the input signal is lower than a specified value, the output of the quantizer is linearly related to its input, while otherwise a logarithmic relation is used. In this manner, when an analog signal at the input assumes a higher amplitude level, it is quantized over a narrower span. Commonly employed companding mechanisms use an A-law or a μ-law algorithm. The A-law algorithm is used in European 8-bit PCM digital communications systems while the μ-law algorithm is used in North America and Japan. Under the latter algorithm, for a given input sample amplitude value x, the compander output value is proportional to $\ln(1 + \mu|x|)$.

Under a Differential PCM (DPCM) encoding scheme, rather than encoding the magnitude of a sample, the difference between the magnitude of the current sample and the predicted magnitude of the next sample is quantized and encoded. An algorithmic process is used to predict the value of the next sample based on values assumed by previous samples. Such an encoding mechanism tends to reduce the number of bits used in the quantization process of audio signals by a factor that can be as high as 25%–50%. Further improvement, as well as an equivalent improvement in the received SINR value or in the required bandwidth level of the communications channel used to transport the digital signal, is attainable by using adaptive variations in the setting of the quantization steps. Such a scheme is known as Adaptive DPCM (ADPCM) encoding. An ADPCM algorithm has been used to map a series of 8-bit μ-law or A-law PCM samples into a series of 4-bit ADPCM samples. The ensuing data rate of a voice flow is then equal to 32 Kbps. A modified more simple DPCM operation is offered by Delta Modulation. One bit per sample is used to indicate whether the sample's amplitude has increased or decreased when compared to the amplitude of the previous sample. When used, over-sampling (i.e., sampling at a rate that is higher than the Nyquist rate) is often employed.

PCM encoding mechanisms and their various adaptations are widely used in encoding audio and video signals. Common sample depths for LPCM are 8, 16, 20 or 24 [bits per sample]. LPCM is commonly used to encode a single sound channel. Multiple LPCM streams are also supported, as is the case for surround sound systems, which may support as many as 8 audio channels. Common DVD-based videos use a sampling frequency of 48 KHz. A sampling frequency of 44.1 KHz has been used in Compact Discs.

2.6.1.2 Replay and Reconstruction of a Transported Stream

The **reconstruction of a voice signal** at a receiving entity proceeds as follows. During a voice activity burst, the sender is continuously active. A continuous train of digital samples is then produced and included as payloads within message units that are transported across a communications network. These transported messages reach the intended user station after network induced time delays. Following demodulation and decoding, the receiving module of the destination user collects the received samples and uses them to reconstruct the voice signal. This procedure is identified as a **replay** process.

Different messages transported across a communications network may incur different ETE delays. Hence, voice samples may arrive at the replay module at variable times. Received samples are stored in a **replay buffer** prior to the start of the voice reconstruction process.

To reconstruct the voice signal, the replay module executes a D/A conversion process that is complementary to the A/D process used at the sending station. Sample units that belong to an activity burst must be picked up by the replay processor at time intervals that are equal to the sampling interval. For example, for a 64 Kbps PCM voice stream, the replay processor will need to pick up from the replay buffer a single sample every 125 μs. If the next sample to be picked up does not reside in the replay buffer at that time, possibly because it was delayed while transiting the network, the replay module will not have a sample to include in its reconstruction process and it would then experience a replay gap (i.e., an interruption in the reproduction of the voice signal). To avoid the occurrence of such gaps, the D/A process is delayed so that it is not initiated before a sufficient number of samples have been received and stored in the replay buffer. This operation causes the start of the reconstruction process to be time delayed. This delay is identified as **replay latency**. The duration of such replay latency should be set to be sufficiently long so that it yields a low probability of incurring voice reconstruction gaps.

Key stages involved in the production and replay of a stream are illustrated in Fig. 2.8:

1) **Analog-to-Digital Conversion**: At the source user, the A/D module converts the user's original analog signal to a digital sequence of bits.

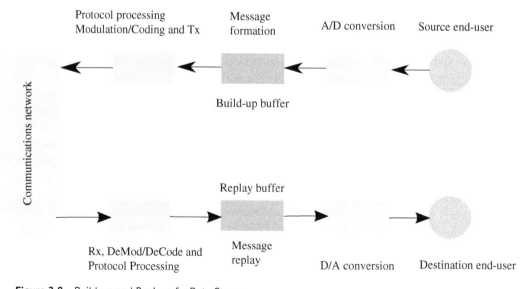

Figure 2.8 Build-up and Replay of a Data Stream.

2) **Buildup Buffering and Message Formation**: The bits produced by the A/D converter are queued (i.e., temporarily stored) in a build-up buffer. When a designated number of bits accumulate (or when there are no further bits available for transmission within a specified period of time), the queued bits are included in a message unit as payload bits.

The message build-up delay time, designated as $D_{build-up}$, is equal to the time used in accumulating the designated number of samples, denoted as $N_{no.\ samples\ per\ message}$. We thus have:

$$D_{build-up} = N_{no.\ samples\ per\ message}\ T_{sampling\ interval}. \tag{2.6}$$

For example, in the case of a PCM-encoded voice signal, if a message is designated to include 100 samples, whereby each sample consists of 8 bits, the build-up buffer will store up to 100 samples (or 800 bits) and the system would then proceed to create a message and include these samples as its payload. In this case, the build-up time delay will be equal to $D_{build-up} = 100 \times 125\,\mu s = 12.5\,ms$.

3) **Protocol Processing, Modulation/Coding, and Transmission**: The formed message is processed by protocol layer entities prior to its submission to a physical layer protocol entity. Often, a carrier is used and encoded information bits are used to modulate the channel carrier. Produced message signals are transmitted across a system's communications channel. For real-time streaming purposes, once the first message in an activity burst has been transmitted, the sender strives to continuously transmit subsequent messages that are included in this burst such that intra-burst messages are transmitted every interval of duration $D_{build-up}$ seconds.

4) **End-to-End Message Transport Across the Communications Network**: The message is transported across the communications network by following a network route that leads to a designated destination node. As it traverses the network, a message may have to transit several intermediate links and switching nodes, experiencing delays while being switched by the nodes that it visits and while being transmitted across the communications links that it traverses. An ETE network transport delay is incurred by a message, denoted as $D_{ETE\ Network\ Message\ Delay}$. The loading levels impacting intermediate network nodes and links may stochastically (i.e., randomly) fluctuate, causing ETE message transport delays to vary randomly.

5) **Reception, DeModulation/DeCoding, and Protocol Processing**: At the destination user node, the received message is demodulated and decoded. Protocol processing is performed by the involved protocol layer entities.

6) **Buffering and Message Replay**: Received message bits are stored in a replay buffer at the destination node. After a replay delay time at this buffer, the system initiates the replay process, using a D/A process. The analog signal is constructed and presented to the end-user. Once initiated, the system aims to continue the process without experiencing gaps in the playback of the analog signal. Such gaps are prompted by messages incurring random delay fluctuations as they are transported across the network, leading to the occurrence of message delay levels that are longer than the replay delay threshold level. The latter, denoted as $D_{playback}$, is used to smooth out such variations. Key factors affecting the setting of this threshold level include the average and variation (also known as **jitter**) values characterizing the ETE network delay incurred by a message. While it is easier to include a playback delay that is equal to the average value of the ETE message delay, it is generally not sufficient. To assure (with high probability) the timely arrival of packets at the replay buffer, it is essential to account for the randomly fluctuating components of the ETE network transport delay. The playback delay is typically configured to a threshold level, $D_{playback\ threshold}$, that would allow a high fraction of messages to arrive to the replay buffer before a specified occupancy level is exhausted.

2 Information Sources, Communications Signals, and Multimedia Flows

As an example, consider a PCM voice stream under which an 8-bit sample is produced by the A/D module every 125 µs for as long as a voice burst activity is ongoing. Assume a message to be produced every 10 ms. In this case, each message contains 80 samples (as 8 samples are produced every 1 ms). Each message thus carries a payload of 640 information bits. Assume the transport of messages across the communications network to fluctuate randomly, and the following statistics to have been observed when measuring the latency values incurred by messages from the instant that their first sample is collected to the instant that they are received at the destination user: an average message delay of 50 ms and a 99-percentile message delay of 82 ms. The latter metric means that 99% of the messages experience a delay level that is not higher than 82 ms, so that no more than 1% of the messages incur a delay higher than 82 ms. In this case, if the replay delay threshold value is set to $D_{playback\ threshold} = 82$ ms, the operation assures a relatively smooth playback in the processing of at least 99% of the messages. In turn, 1% of the messages may arrive late at the replay buffer and may then be ignored and discarded. At such a time, the replay process may also use (as a replacement of the missing unit) an intermediate value, which is calculated by interpolating past observed and future predicted values, aiming to produce a smooth replay output. For many audio and video streaming purposes, an acceptable fidelity level, or Quality of Experience (QoE) value, may involve a sufficiently small fraction (such as 0.1%–1%) of late (and possibly discarded) message arrivals.

2.6.1.3 Transport of a Stream Across a Circuit-Switched Communications Network

To accommodate an **interactive conversation** between communicating end users, the cumulative delay of a voice message should be sufficiently low for high percentage of the stream's messages. We are familiar with the corresponding impact of transport delays induced by messages that are transmitted across a satellite communications link. When considering a geosynchronous satellite system, the propagation time delay across the satellite's link is equal to 250–270 ms. Hence, the round trip message time delay is of the order 500–540 ms. Consequently, following the start of a conversation by a source user, the time delay incurred to the receipt of a voice reply signal, accounting for the round trip (RT) time, is equal to 500–540 ms plus additional time delay components that account for queueing and processing delays incurred at visited network nodes. Such a long delay level is not conducive to the maintenance of an effective interactive conversion. A user would need to pause for a noticeable time period to give a chance to its remote party to receive the message, respond to it, and have the response message propagate across the satellite link. To promote an effective interactive dialog process, we prefer the system to induce a relatively short ETE delay. For example, when communicating across a terrestrial communications network, this delay is targeted to generally not exceed a level of about 100 ms for high fraction of the time, and preferentially be much lower.

To secure a high fidelity reproduction of a real-time message stream, such as that involving interactive audio and video streams, it is desirable to reduce the following involved time delay components:

1) **Message Buildup Delay**: The time it takes to collect digitized signal samples. As noted above, this delay is expressed as: $D_{build-up} = N_{no.\ samples\ per\ message} \times T_{sampling\ interval}$.
2) **Average Latency and Delay Jitter for Network Transport**: The average time latency and the variation of the delay incurred by a message in its ETE transport across the communications network.
3) **QoE: Round-Trip Response Time and Message Discard Ratio**: Such sufficiently low values contribute to the achievement of high-quality reproduction of the stream at a destination

user. The system aims to provide a high QoE to a destination user, enabling the support of an interactive dialog and the reception of high-quality data streams. To maintain timely interactions, exchange of messages between communicating end-users must be accomplished in a consistent and timely manner. To assure high quality of the received data, quantization error rates and communications interference levels must be properly limited.

A simple implementation that would meet desired performance levels involves the use of a dedicated point-to-point communications link (whether wireline or wireless). In this case, the source end user would be directly connected to the destination end user across a communication channel whose capacity resources are assigned for the exclusive use of their connection. Assuming the user's peak traffic loading level to not exceed the capacity of its allocated channel capacity, the user's transported stream would not incur unacceptable latency and delay variation levels.

Such a favorable performance behavior is also attained when the source and destination users are connected across the network via a high-capacity path that employs multiple circuit switches. As we will note in Chapter 10, this is the case when a data flow is transported across a properly provisioned connection in a **circuit-switched communications network**. The *Public-Switched Telephone Network (PSTN)* provides dedicated capacity connections. Also known as Plain Old Telephone Service (POTS), this network system assigns a **communications circuit** for the transport of call messages that flow between source and destination nodes. A circuit is assigned on a dedicated basis. It can thus serve for the real-time transport of a data stream. Communications capacity resources that are allocated to the circuit are sufficiently high to enable the ETE transport of a data flow at a specified data rate. For the above-mentioned PCM voice stream, the circuit's data rate is set equal to $R_d = 64$ Kbps. For supporting the transport of PCM voice streams in PSTN telephony systems, the capacity resources of the system's communications channels are frequently utilized in the following manner. Under a **Time Division Multiplexing (TDM)** mechanism, a communications link is shared among multiple flows on a time division basis. Recurring time frames are defined across each link. Each time frame consists of multiple time slots. Consider the resources allocated to a single PCM voice stream under TDM. Assume each time slot to be sufficiently long to accommodate the transmission of a single sample of a voice stream. The sample is transmitted across the communications channel at a prescribed channel data rate R_c [bits/s]. Assume that for the transmission of a single PCM voice stream, the source user is allocated a single time slot every sampling period duration of 125 μs. Also assume an arrangement under which the TDM time frame duration is equal to the voice sampling interval of 125 μs. In this case, a PCM voice stream would be allocated a single slot in every time frame, for as long as it is assigned the underlying circuit, which can last for the duration of its call or connection, whether it is active or inactive throughput the call.

In this manner, transported voice samples incur minimal delays as they flow along a network path and handled by intermediate circuit switches. A sample that is transmitted from one switch to the next one is set to not incur a waiting time at the subsequent switch that is longer than a time frame duration of 125 μs. The cumulative latency incurred for the transport of a sample across a network route, which is upper bounded by the product of the number of the route's links and the frame duration, would thus be lower than several milliseconds.

At a destination user, the received samples will therefore arrive within a brief and predictable period of time. Assuming voice call samples to be transmitted sequentially in time across an allocated circuit, and to not be lost or perturbed by noise (beyond an acceptable level), the samples will arrive at a destination user within a properly bounded time window and in *sequential* fashion. In this case, the replay process can be carried out in an orderly manner, avoiding the occurrence of most latency-driven interruptions.

A circuit-switched-based communications connection process thus secures high-performance stream transport operation, as it provides the following advantageous settings:

1) **Message Buildup Delay**: This delay is lowered by setting a low value of $N_{no.\ samples\ per\ message}$. The lowest such value is attained by setting $N_{no.\ samples\ per\ message} = 1$, so that each message consists of a single sample. The ensuing delay is equal to $T_{sampling\ interval}$, so that under a PCM voice encoding process, we have $D_{build\text{-}up} = T_{sampling\ interval} = 125\,\mu s$.
2) **Average Message Latency and Delay Jitter**: Average and variation values of the incurred message network transport delay are kept low by dedicating a circuit, in each direction, for the sole purpose of transporting a call's messages. Since communications capacity resources are dedicated to the transport of messages that are members of only this call, delay factors that can arise when multiplexing flows generated by distinct users across a connection are avoided.
3) **QoE: Round-Trip Response Time and Message Discard Ratio**: High interactivity is achieved as the round-trip delay time is lowered. This can be accomplished by keeping the message buildup, replay, and network transport delay components at low and steady values. Since a call's messages use dedicated capacity resources across their assigned circuit, they do not need to share resources with messages that belong to other calls. Message transit times across the network assume predictable values. Messages arrive within prescribed times at their destination node. Messages are set to incur, with high probability, low discard ratios and to be received in a correct sequential order.

2.6.2 Audio Streaming across a Packet-Switching Communications Network: Voice Over IP (VoIP)

2.6.2.1 Voice Over IP (VoIP)

A real-time interactive voice (and often audio and video) stream tends to be active in **spurts**. Observing the signal output at the source, we note it to temporally alternate between *busy* and *idle* modes, or equivalently between *active/spurt* and *inactive/pause* periods. An end user may produce a signal for a period of time and then she pauses, waiting for a destination user to receive the information and respond to it. This is illustrated in Fig. 2.9.

For interactive voice flows, it is often assumed that the spurt period, which is also identified as a burst or "on" period, is of random duration. It is at times modeled to follow an Exponential distribution whose average length is $t(on) = 0.35$ seconds. The pause period, also identified as an "idle" or "off" period, also tends to last for a random period and is at times assumed to be governed by an Exponential distribution with mean that is equal to $t(off) = 0.65$ seconds. In this case, the user generates an active voice signal during only 35% of the time, while it generates no signal for 65% of the time.

The average fraction of time that a user (or any data source) is active in producing a signal that it aims to send over a communications network to a destination end user is defined as the **duty-cycle** of its activity. For the above-noted illustrative voice activity model, the user's activity duty-cycle is

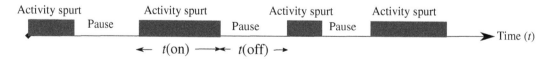

Figure 2.9 A Bursty Flow Alternating Between Spurt (Active) and Pause (Inactive) Periods.

equal to 35%. Denoting $t(cycle)$ as the average duration of a cycle time, which is defined as the sum of the average lengths of successive on-time and off-time periods, we note the following relationship:

$$\text{Duty-Cycle} = \frac{t(on)}{t(cycle)} = \frac{t(on)}{t(on) + t(off)}. \tag{2.7}$$

If a circuit-switched network is used to transport the above noted user's stream, so that communications capacity is dedicated for the transport of the stream across the network, only 35% of the allocated capacity would be actively utilized.

Next consider the network that is used for the transport of a stream to be a datagram **packet switching network**, such as the Internet. In such a network system, transported message units are formatted as **network layer packets**. As no circuit is dedicated for such a transport, each packet includes control fields that inform intermediate network routers as to its destination and possibly also identifies its transport quality needs. Such control information includes source and destination addresses. Consequently, each packet, no matter what is the length of its payload field (which carries the information-bearing data produced by the user), must include control fields, adding packet overhead bits. For packets that carry voice information, the number of control bits per packet can be, for example, of the order of 40 Bytes, which is equal to 320 bits, where a *Byte* consists of 8 bits.

The process used to transmit a voice signal across a packet switching network follows the one described in Section 2.6.2, as depicted in Fig. 2.8. The analog signal is first discretized. Encoding schemes such as PCM and ADPCM are employed. The produced digital streams may represent the voice signal at a lower fidelity level and thus transported at lower data rates, or at higher fidelity level and higher data rates, entailing various levels of implementation complexity.

For illustrative purpose, assume the following description to focus on a PCM-encoded digital stream. A similar process takes place when other encoding schemes are used. Several samples are stored and aggregated to form a packet. It is not throughput effective to include in a packet just a single (8 bit) sample, as the packet must now also contain overhead control bits. Consequently, to decrease a packet's control overhead, it is configured to carry multiple samples. How many samples should be included in each packet?

In considering an interactive real-time streaming session, it is essential to limit the time latency involved in the build-up of a packet. Therefore, the number of samples included in a packet must be limited. For example, for the underlying streaming process, if a packet was set to consist of 800 samples, the time that it will take to collect these samples would be equal to 100 ms. This implies that before a packet can be formed, samples need to be stored in the build-up buffer for a period of 100 ms. During this time delay, the transport of the packet across the network cannot yet be initiated. An excessively high delay reduces the interactive character of the dialog.

Common voice packet build-up times are of the order of 10, 20, or 40 ms. To illustrate, consider a PCM voice encoding scheme that uses a sampling rate of 8000 [samples/s] and a quantization level of 8 [bits/sample]. Over a period of 10 ms, 80 samples are collected, yielding a 640 bits payload. Under a control overhead that is equal to 320 bits, the throughput efficiency of the operation is equal to 67%. If an ADPCM-encoding scheme is used, so that the quantization level is equal to 4 bits/sample, the corresponding payload consists of only 320 bits and the ensuing throughput efficiency is reduced to 50%.

This process, whereby a voice stream is transmitted across a datagram packet switching network, such as the Internet, is identified as **Voice over IP** and is denoted as **VoIP**. Our discussion indicates that to enhance the interactivity of a voice connection, voice packets should be constructed to be relatively short. As control overhead then becomes an issue, overhead compression algorithms are often used.

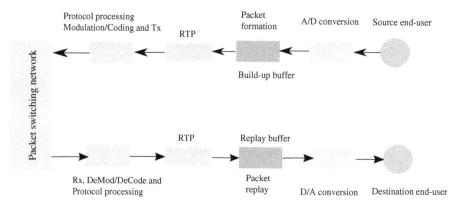

Figure 2.10 The Voice over IP (VoIP) Streaming Process.

2.6.2.2 The VoIP Streaming Process and the Realtime Transport Protocol (RTP)

Key elements of the VoIP streaming process are depicted in Fig. 2.10. The process is distinguished from that shown in Fig. 2.8 for streaming voice over a circuit switched network by the following noted factors:

1) **Packet Buildup Delay**: Several samples are gathered to build the VoIP packet. The ensuing packet buildup delay is equal to $T_{packet\ build\ up} = T_{sampling\ interval} \times N_{no.of\ samples\ per\ packet}$.
2) **Statistical Network Delay Variations**: Packets transported over a packet-switching network such as the Internet tend to experience *wide variations of transit latency*. Different packets that are members of the same stream may also be transported across different routes and arrive at the destination node *out of order*. As they make their way across the network to their destination, packets may be *lost or discarded*. A network layer protocol entity in a datagram packet switching network does not provide for the retransmission of missing packets.
3) **Use of the Real-Time Transport Protocol (RTP) to assist in the Replay Process**: Once the replay of a burst of a stream is initiated at a receiving node, it is necessary to continuously feed the receiver's D/A module with stream packets. It is also essential that packets are presented to the replay module in the right order and time instants; that is, *at the same order and at the same time intervals* that existed when they were produced at the source.

RTP employs a scheme that enables the real-time transfer of streaming media, including audio and video applications, such as VoIP, audio over IP, WebRTC, and Internet Protocol television. Its services include jitter compensation and detection of packet losses and out-of-order receptions. It also enables transfer to multiple destinations through IP multicast. As such, it plays a key role in the transport of audio and video streams in packet-switching networks. Its protocol functions are implemented at the application protocol layer, above the transport protocol layer. The User Datagram Protocol (UDP) is often employed at the transport layer for use by real-time streaming applications. UDP is a connection-less protocol, lacking the ability to identify the correct in-sequence order of received packets and to bind packets to their respective flows. RTP control overhead is of the order of 12 Bytes (i.e., 96 bits). Included are the following control fields, which are essential for the proper execution of the replay process (and which require a total of 6 Bytes out of the total 12 Bytes in the RTP control header):

a) **Sequence Number (16 bits)**: The sequence number is incremented for each RTP data packet that is transported across the network. It is used by the receiver to detect missing packets and to handle out-of-order delivery. The initial value is selected at random.

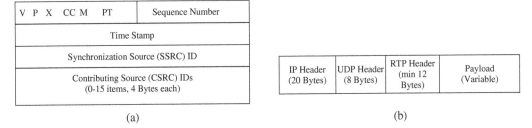

Figure 2.11 (a) RTP Header; (b) Voice over IP (VoIP) Packet.

b) **Time Stamp (32 bits)**: Each packet is attached a time stamp. It is used by the receiver to play back the received packets at correct time intervals. The time stamp identifies the sampling instant of the first octet of an RTP data packet. This time is derived from a clock that is incremented linearly in time to allow synchronization and jitter calculations. For example, a PCM voice stream may use an audio application that produces samples at a rate of 8 KHz, which is used as the time-resolution clock. Video streams often use a 90 KHz clock. The clock granularity for an application is specified in the corresponding RTP profile.

c) **Synchronization Source (SSRC, 32 bits)**: Identifies the synchronization source, such as that used by the sender of a stream of packets; it may be derived from a signal source such as a microphone or a camera, or an RTP mixer.

d) **Contributing Source (CSRC, 32 bits)**: Identifies an array of contributing sources for the payload contained in the packet. It is used, for example, by an audio-conferencing application where a mixer indicates all the talkers whose speech was combined to produce the packet.

The structure of an RTP packet header is shown in Fig. 2.11(a). The header is noted to consist of 4 words, whereby each word consists of 4 Octets (or 4 Bytes), so that it is 32 bits long. In particular, we note the header to include sequence number, time stamp, SSRC and CSRC fields. In Fig. 2.11(b), we show a VoIP packet to include an IP header (20 Bytes), a UDP header (8 Bytes), and an RTP Header (12 Bytes minimum), which are followed by a payload field that contains the streamed data. A total of 40 overhead Bytes is noted.

RTP sessions are typically initiated by using a signaling protocol, such as the Session Initiation Protocol (SIP). A separate RTP session is established for each multimedia stream. Audio and video streams may use separate RTP sessions, enabling a receiver to selectively receive components of a particular stream. The RTP (and the below noted RTCP) design is independent of the employed transport protocol. The Stream Control Transmission Protocol (SCTP) and the Datagram Congestion Control Protocol (DCCP) may be used when a reliable transport protocol is desired. The Secure Real-Time Transport Protocol (SRTP) is used to provide cryptographic services for the transfer of payload data.

Associated with RTP is the **Real-Time Transport Control Protocol (RTCP)**. It is used to provide out-of-band statistics and control information for an RTP session. In particular, it provides feedback on the QoS of the stream's transport process. Performance metric values that are distributed to participants in a multimedia streaming session include packet and byte counts, packet loss data, packet delay variation values, and round-trip delay times. An application may use this information to control QoS values. For example, when the network transport mechanism is indicated to be congested, inducing lower QoS values, an application at the source entity may act to adjust the quality of the stream by changing the encoding mechanism or the transport parameters.

In a **multicast session**, multiple users are involved. Certain users are set to serve as senders of information, which may include audio and video streams and messaging flows. All session

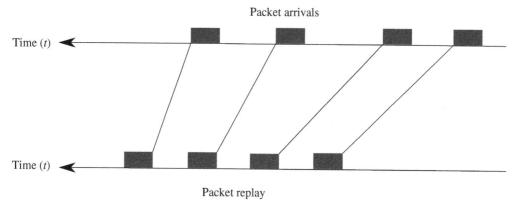

Figure 2.12 Illustrative Packet Replay Scenario.

users may be able to receive transmissions issued by a sending user. In support of multiuser multicast sessions, RTCP services include the following features. RTCP enables reaching all session members, providing unique canonical end-point identifiers (CNAME) to all session participants. This is useful as the instantaneous binding of source identifiers (SSRC) to end-points may change during a session. CNAMEs provide unique identifications of end points across an application instance, which can involve multiple use of media tools.

RTCP reports are sent by all session participants. A sender's report includes transmission and reception statistics for all RTP packets sent during a time period. It employs an absolute time stamp, which allows the receiver to synchronize RTP messages. It is of essential utility when both audio and video streams are transmitted simultaneously, as they can employ independent relative time stamps.

To avoid congestion occurrences when a large number of participants are involved, the rate at which reports are generated is regulated, aiming to set the loading level to not occupy more than about 5% of the session's allocated bandwidth. About 25% of RTCP bandwidth is reserved for media sources, so that all participants can readily receive the CNAME identifiers of all senders. The RTCP reporting interval is randomized to reduce the occurrence of high burst rates. A minimum reporting interval per user (such as 5 seconds) is specified.

4) **Replay Process**: RTP packets are created at the application layer and handed to the transport layer for delivery. Each data unit of RTP media created by an application begins with an RTP packet header. This header is nominally 12 Octets (i.e., 96 bits) long. Each packet contains a sequence number, a time stamp, a source ID, and a contributing source ID that enumerates other sources that may have contributed to the production of the stream. The packet is processed by protocol layer entities, including processing and transmission by the physical layer module. The packet is then transported across a packet switching network (such as the Internet), arriving at the destination user station. The received signal is demodulated and decoded and the packet's payload is processed by the receiving user's higher layer protocol entities. The receiving RTP layer protocol entity detects missing packets and may reorder packets. Packets are decoded in accordance with processes identified by the payload type, stored and delayed at the replay buffer, in aiming to minimize replay pauses. Packets are then submitted in proper order and time intervals to the D/A module for replay and for presentation of the stream to the destination user.

An illustrative Packet Replay Scenario at a destination user station is shown in Fig. 2.12. As packets arrive at a destination node at random times, they are delayed in the replay buffer so that when the replay process is initiated, stored packets are continuously fed to and processed by the D/A module.

As will be discussed in Section 2.7, when considering video streaming over IP networks, which is also applicable to integrated video/audio streaming, an effective alternative to using RTP over UDP, which is a connection-less transport layer protocol, is to use a connection-oriented transport layer, such as TCP, in conjunction with a higher layer dynamically adaptive streaming utility such as **MPEG-DASH (Dynamic Adaptive Streaming over HTTP)**. Such a mechanism enables the targeted destination user to inform the sender about the streaming parameters that it wishes the sender to use. The use bases its requests through monitoring of the features of its ongoing reception and replay processes. This is particularly an advantageous operation when considering the transport of a stream across a communications network that uses wireless access links for which congestion levels tend to randomly fluctuate. During degraded transport conditions, when high packet delay levels are incurred, this adaptation mechanism is used to induce the source user to adjust its encoding (including compression) parameters or schemes to enable the destination user to experience acceptable replay performance, while possibly incurring lower, yet acceptable, video and audio reproduction quality levels.

2.6.2.3 Other CODECs and VOCODERs

Two major groups of codecs (encoding/decoding) schemes are noted when considering Packet Telephony, including VoIP technologies. In the first group, the encoding/decoding principle of the codec does not depend on the type of the underlying signals. Such Group I codecs, including G.711 PCM codecs, are called waveform codecs.

For group II codecs, the encoding/decoding mechanism is based on using a model that describes the process used to generate human speech. Speech modeling codecs do not operate exceptionally well on non-speech signals (such as music). An example of such a codec is G.729A, a Linear Prediction type. A major advantage of these codecs is that they require a much lower transport data rate. They are therefore of utility in cellular and other wireless communications networks, as well as for transport across media that are highly bandwidth limited (such as submarine communications channels). Group II CODECS include: G.729 (ITU-T), with the sampling rate is 8 KHz, the bit rate is 8 Kbps, the frame size is equal to 10 ms; GSM 06.10 (ETSI), with the sampling rate is 8 KHz, the bit rate is 13 Kbps, frame size that is equal to 22.5 ms.; SILK (Skype), LPC (linear Predictive Coding based), for which the sampling rate is variable, the bit rate is of the order 6–40 Kbps, and the frame size is equal to 20 ms.

In relation to Group II mechanisms, we note that an associated **vocoder** is a voice codec that analyzes and synthesizes the human voice signal for performing audio data compression, multiplexing, voice encryption, or voice transformation. The vocoder models speech process by characterizing its spectral (frequency) components and their dynamic variations over time. Control signals are produced for representing these frequency components over time as the speech process progresses. The sender needs to transport across the network just the parameters of these control signals. To reproduce the speech waveform, a vocoder reverses the process. Thus, the voice is synthesized by using the control signals to determine the intensity of ensuing spectral components and then to combine them. Since transported parameters change slowly compared to the original speech waveform, the bandwidth that is required to transmit such a speech signal is reduced.

A wide range of other audio (and audio/video) encoding algorithms have been developed. For example, we note the *Opus audio coding* algorithm, which extends the use of SILK and CELT algorithms. It was developed by the Xiph.Org Foundation and standardized by the Internet Engineering Task Force (IETF). It provides high-quality speech and general audio coding at low latency (accommodating interactive communications) and at relatively low complexity. Delay is of the order of 26.5 ms, which can be reduced to 5 ms when a lower-quality signal is produced. It supports constant (CBR) and variable (VBR) bit-rate encoding from 6 to 510 Kbps (and up to 256 Kbps per channel for multichannel tracks), frame sizes from 2.5 to 60 ms, and five sampling rates from 8 K samples/s (with 4 KHz bandwidth) to 48 K samples/s (at 20 KHz bandwidth, relating to a human's hearing range). An Opus stream can support up to 255 audio channels. Under the default 20 ms frame length and application settings, a very short latency is incurred (26.5 ms), accommodating real-time telephony, VoIP, and videoconferencing applications. In a stream, the bit-rate, bandwidth, and delay values can be continually varied without introducing distortion.

Several transformation algorithms have been developed and used to compress the bandwidth required to represent a signal. For this purpose, key use has been made of the **discrete cosine transform (DCT)**. As discussed in considering the Fourier series-based representation of a periodic function, which consists of a weighted sum of sine and cosine functions, DCT expresses a finite sequence of data points in terms of a weighted sum of cosine functions oscillating at different frequencies. The corresponding transformation maps a finite set of n input values (representing n input signal sample values) to a finite set of n output points, characterizing n output weights of cosine components operating a various frequency levels. As a *block compression mechanism*, DCT compresses data in sets of discrete DCT blocks. DCT blocks can be configured to use several integer sizes, such as those that vary between (4×4) and (32×32) pixels. Such a transformation yields a highly significant compression ratio. It is therefore widely employed for compressing digital signals for storage and for transport over communications media. It has been used for compressing digital images (such as JPEG and HEIF), digital video (such as MPEG and H.26x), digital audio (such as Dolby Digital, MP3 and AAC), digital television (such as SDTV, HDTV and VOD), digital radio (such as AAC+ and DAB+), and speech coding (such as AAC-LD, Siren and Opus).

The 1-D **discrete cosine transform (DCT)** is a linear, invertible function that is represented as an $N \times N$ square matrix, say $C = \{C(i,j) = (x_i, y_j), i, j = 0, 1, \ldots, N - 1\}$. A variant of this transformation is presented as follows. An input set of N real numbers, denoted as $\{x_0, x_1, \ldots, x_{N-1}\}$ is transformed to produce an output (i.e., the DCT transform) that is represented by the N real numbers $\{y_0, y_1, \ldots, y_{N-1}\}$ through the application of the following formula:

$$y_j = \sum_{i=0}^{N-1} x_i a(j) \cos \left[\frac{\pi(2i+1)j}{2N} \right] \qquad j = 0, \ldots, N-1$$

where

$$a(0) = \sqrt{1/N}$$
$$a(j) = \sqrt{2/N} \qquad j = 1, \ldots, N-1. \tag{2.8}$$

The inverse matrix, yielding the input sequence from the output sequence, is readily calculated, as it is functionally identical to the transformation matrix.

To compress a sampled voice signal and other audio signals using DCT, a group of N audio samples is transformed by using one of the variants of the DCT mapping. The outcome identifies the spectral components of the selected signal segment. As the fidelity of the voice segment often depends on certain frequency bands in a more pronounced manner than on other, the magnitude levels of the former frequency components are quantized in a finer fashion, while the other components undergo a more crude quantization process. The information content of the ensuing encoded sequence is often further compressed by using lossless compression techniques. Such a process leads to a significant reduction in the ensuing rate of the audio data transported across a packet switching network, while providing for high quality signal reproduction. To produce the audio signal at a receiving node, the corresponding decoding and inverse DCT transformation processes are employed, in addition to the above-noted packet buffering and replay processes.

The mentioned techniques, including the one that employs DCT and related transformations, employ *lossy compression* mechanisms. Thus, a reduction in the data contents is produced at an expense of lowering the *fidelity* of the reconstructed signal. Aggregate segment-oriented transformations such as that performed under DCT lead to high-quality reproductions at highly reduced data rates. Yet, further reduction in the data rate may result in noticeable reduction in the quality of the reproduced signal. When a compression process yields a signal that is reproduced at a sufficiently high quality level, it is often the case that an increase in the data rate level may not much enhance the fidelity of the reconstructed signal.

As further discussed in the Video section, a two-dimensional DCT transformation scheme is performed on spatial segments of video pixels within each video frame, yielding highly compressed representations of picture, image, and video signals. Compression schemes that make use of the fact that lower-frequency components of video signals, tend to often make a more pronounced contribution to the quality of the signal, have been shown to yield higher-quality signal reproductions at highly compressed data rates.

The **modified discrete cosine transform (MDCT)** is based on performing DCT of overlapping data. It is performed on consecutive blocks of a larger dataset, where subsequent overlapping blocks are used, so that the last half of one block coincides with the first half of the next block. This mechanism, in addition to the energy-compaction qualities of DCT, makes MDCT especially attractive for signal compression applications, as it helps to avoid artifacts stemming from block boundaries. Hence, MDCT is a widely used lossy compression technique that has been used in audio data compression. It is employed by many audio coding standards, including MP3, Dolby Digital (AC-3), Vorbis (Ogg), Windows Media Audio (WMA), ATRAC, Cook, Advanced Audio Coding (AAC), High-Definition Coding (HDC), LDAC, Dolby AC-4, and MPEG-H 3D Audio, and in speech coding standards such as AAC-LD (LD-MDCT), G.722.1, G.729.1, CELT, and Opus.

While human voice signals occupy a frequency spectrum that ranges from 80 Hz to 14 KHz, voiceband or narrowband telephone calls tend to limit audio frequencies to a 300 Hz–3.4 KHz range. **Wideband audio** relaxes the bandwidth limitation, forming transmissions that occupy an audio frequency range of 50 Hz–7 KHz, and can also be reaching 22 KHz. Several wideband codecs use a higher audio quantization bit depth in encoding samples, such as 16-bits per sample, resulting in improved voice quality. Mobile wireless systems use typically low data rate audio streams, as they strive to attain higher spectral use efficiency. Yet, wideband audio has also been employed. For example, the 3GPP Standards Group has designated G.722.2 as its wideband codec, identifying it as Adaptive Multirate – Wideband (AMR-WB). Wideband audio coding standards such as G.718, G729.1, and G711.1, are based on the above-noted MDCT compression algorithm.

2.6.2.4 Quality Metrics

The International Telecommunications Union (ITU) specifies network delay for voice applications in Recommendation G.114. This recommendation defines three bands of one-way delay. The range of 0–150 ms is deemed acceptable for most applications. The range of 150–400 ms is stated to be acceptable provided that administrators are aware of the transmission time and its impact on the quality of the underlying user application. Values above 400 ms are stated to be generally unacceptable.

These recommendations do not account for the round-trip delay. To assess the quality of a transported audio stream, it is essential to account for multiple elements. We have discussed the impact of fixed and variable delay parameters on voice quality and session interactivity. Other key factors include such that affect the desired fidelity of the reproduced signal. To achieve a targeted fidelity measure, sufficient frequency bandwidth must be allocated. Interference signals, including side-tones and background noise processes, must be kept at acceptable levels. Packet drops must be limited. Upon a dropping of a packet, a period of the order of 10–40 ms of voice may be lost. When considering interactive voice sessions, no voice packet retransmissions are performed.

For a voice signal transported across a communications network, how do we assess the quality of its reproduction at a destination user?

The following quality metrics have often been used: Mean Opinion Score (MOS), the Perceptual Speech Quality Measurement (PSQM), and the Perceptual Evaluation of Speech Quality (PESQ):

1) **Mean Opinion Score (MOS)**: An MOS value is generated when multiple users listen and evaluate prerecorded sentences that have been impacted in various ways, such as through the application of compression algorithms. Listeners rate the quality of the sound by assigning a rating level on a scale from 1 to 5, where 1 is the worst and 5 is the best. An illustrative sentence that has been used for English language MOS testing is: "Nowadays, a chicken leg is a rare dish." This sentence is used because it contains a wide range of sounds found in human speech, such as long vowels, short vowels, hard sounds, and soft sounds.

   Test scores are averaged to produce an average composite score. Test results are subjective as they are based on the opinions of listeners. Results are also relative, as a score of 3.8 from one test cannot be directly compared to a score of 3.8 from another test. Therefore, a baseline needs to be established for all tests, so that the scores can be normalized and compared directly.

2) **Perceptual Speech Quality Measurement (PSQM)**: This is an automated method of measuring speech quality "in service," or as the speech happens. PSQM software is often embedded in IP call management systems. It is then integrated into the network management system (such as by using SNMP). A comparison is made between the original transmitted speech signal and the corresponding signal received at the destination. Measurements are made during the progression of a conversation across the network. It was determined that such an automated testing mechanism can have over 90% accuracy, as compared to subjective listening tests, such as MOS tests. Scoring is based on a scale from 0 to 6.5, where 0 is the best and 6.5 is the worst. This method was originally designed for circuit-switched voice, so that it does not take into account average and jitter delay problems that are experienced when communicating VoIP streams across a packet-switched system.

3) **Perceptual Evaluation of Speech Quality (PESQ)**: MOS and PSQM metrics are not recommended for use in VoIP networks. They do not measure typical VoIP problems such as those that are impacted by fixed latency and delay jitter. It is possible to obtain an MOS score of 3.8 for an VoIP network when the one-way delay is high as an MOS evaluator only rates audio quality and does not assess the two-way efficiency of a conversation. PESQ takes into account CODEC errors, filtering errors, jitter, and fixed delay problems that are typical in a VoIP network. PESQ combines elements of PSQM while using a method that is called Perceptual Analysis Measurement System (PAMS). PESQ scores range from 1 (worst) to 4.5 (best), with 3.8 considered "toll quality" (i.e., acceptable quality when used in a traditional telephony network). PESQ measures the effect of ETE network conditions, including CODEC processing, delay jitter, and packet loss. Effects of two-way communications, such as loudness loss, delay, echo, and sidetones, are however not reflected in PESQ scores.

As will be discussed in Chapters 5,6,8,11,12,16–21, and illustrated for various technologies, traffic management, and processing mechanisms are employed in packet-switching networks to provide VoIP, audio, and video-streaming sessions with high-quality performance. The network transport mechanism is configured to grant a stream its desired levels of fixed and variable transport delays and acceptable packet loss ratio values. Class-Based Queueing methods are employed at packet routers in aiming to provide stream packets their desired packet delay and packet loss performance levels.

In transporting audio or video streams over a packet switching network such as the Internet, a user whose communications connection lacks sufficient bandwidth may experience stops, lags, or slow buffering of its content. It is therefore useful to have the receiving user send experienced performance (QoS and QoE) values to the sender. The sending entity uses reported performance metric values to dynamically adapt stream parameters to provide the receiving user with the best quality level that is currently feasible.

For example, when the receiving user indicates her detection of degradation in the received data rate and signal quality, which may be induced by higher congestion or resource availability issues that occur along the communications medium, a source may act to decrease the spectral resources that it requests. The source may reduce the data rate of the stream; for example, by lowering the number of quantization bits per sample, or by switching to a different codec. This may reduce the quality of the transported stream. When the corresponding network state degradation conditions are resolved, the quality of the stream would be restored. In a mobile communications system, such degradation conditions often occur when a destination mobile enters a location at which it is subjected to signal blockage or higher interference conditions.

When considering delay insensitive applications, such as one-way or noninteractive audio flows, non-live video streaming, or archival-oriented downloads, the strict requirements that are imposed on fixed and variable ETE latency values, and at times on packet discard ratios, may not be as critical. The destination user can proceed to download a flow, or a specified portion of a stream, at an advanced (or preferred) point in time. For example, a recorded audio/video lecture stored on a remote server can be downloaded by a user in advance. A destination user's replay process can start when a sufficient amount of data has already been stored in its buffer. The replay module can then be continuously driven by downloaded data without incurring buffer starvation and replay gap conditions. System requirements can still be of interest in relation to the targeted fidelity of the reproduced signal, the communications bandwidth assets that should be allocated, and the buffering and processing resources that are required by a user system.

2.7   Video Flows and Streams

2.7.1   Conversion of Light Waves to Electrical Signals

Image sensors are used to convert light waves to electrical signals, which are then formatted and stored, producing a photographic electrical record of a scenery. To characterize light image attributes over a 2D area, the area is divided into cells or pixels. A **pixel** identifies a **picture element or picture cell** of the image. A pixel is the smallest component of a digital image. Its definition can be context dependent. Pixels used in a digital camera can correspond to photo-sensor elements. By using a higher number of pixels, a more accurate representation of the image is obtained. The intensity of each pixel is variable. A color is typically represented by three or four components such as red, green, and blue (RGB), or cyan, magenta, yellow, and black (CMYB).

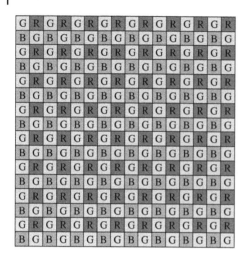

Figure 2.13 Illustrative Array of Image Pixels.

Common types of electronic image sensors are charge-coupled device (CCD) and active-pixel CMOS sensors. Both are based on metal–oxide–semiconductor (MOS) technology, with CCDs based on MOS capacitors and CMOS sensors based on MOSFET (MOS field-effect transistor) amplifiers.

CMOS sensors are cheaper and consume less power. They have therefore been used in many commercial devices. CCD sensors are used by high-end broadcast quality video cameras. Each CCD image sensor is an analog device. When light strikes the chip it is held as a small electrical charge in each photo sensor. The charge is amplified and the output is linked to its underlying spatial location and is stored. A CMOS image sensor uses an amplifier for each pixel, while few amplifiers are used by a CCD image.

In typical digital cameras, the sensor array is covered with a patterned color filter mosaic. Each sensor element (also identified as a pixel) can record the intensity of a single primary color of light. The camera interpolates the color information of neighboring sensor elements to create the final image.

An illustration of a pixel array is depicted in Fig. 2.13. The shown configuration is based on a 1976 development by Bryce Bayer of Eastman Kodak, known also as a Bayer Filter. This pixel mosaic consists of 2 × 2 squares. Each square contains 2 green pixels, 1 blue pixel and 1 red pixel. An arrangement is identified as RGBG. During daylight, a human retina is more sensitive to green light. Each pixel sensor is sensitive to a single color. A de-mosaicking algorithm is used to associate an image state value for a pixel. The algorithm interpolates the color outputs of pixel sensors that surround an underlying pixel to derive an image value for that pixel.

To assess the resources required for the transport of images and video streams over a packet switching network such as the Internet or a packet switched wireless network, we differentiate between flows that carry still images, freeze-motion or slow-motion presentations, and full-motion and conferencing video streaming sessions involving live events or stored video files.

2.7.2 Digital Still Images

A **digital image** is a picture or photo that consists of *picture elements*, or **pixels**. The pixels cover a 2D spatial frame. Each pixel is characterized by its location in the frame and by its intensity or gray level.

A **raster image**, or *bitmap image*, is a digital image that consists of a fixed number of rows and columns of pixels. It is described as an array of bits within a rectangular grid of pixels or dots. The data structure is based on a (usually rectangular) tessellation of the 2D plane into cells.

Raster images can be created by various devices, such as digital cameras, scanners, and airborne radar. Such multidimensional images can be synthesized by using *computer graphic* techniques.

A **vector image** consists of graphic image components, such as lines and other geometric shapes. The geometrical shape assumed by a vector image is defined through the use of a mathematical formula. Each vector is characterized by magnitude (or length) and direction attributes. Raster and vector elements can be combined in a single image. The raster's image represents a rectangular grid of pixels, with each pixel's gray level and color specified by a digital parameter whose value is represented by a sequence of bits. As a bitmap image, the raster's image contents are stored as a dot matrix data structure in an image file. A raster is characterized by the width and height of the image, measured in pixels, and by the number of bits per pixel, or color depth, which determines the number of represented colors.

A high-quality digital image, also identified as a **still image**, consists of a large number of pixels, whereby each pixel's parameter is encoded and represented by a specified number of bits. Thus, a representation of the contents of an image may require a large number of information bits. To transmit an image across a communications channel in a bandwidth efficient manner, its content is compressed. Several lossy image compression algorithms have been devised, whereby key techniques employ a **discrete cosine transform (DCT)**-based mechanism. This approach is the basis for JPEG image codecs. JPEG was introduced by the Joint Photographic Experts Group in 1992. It has been extensively used for the distribution of digital images, digital photos, and FAX documents across the Internet. Other commonly used formats include GIF and PNG.

The data contents of an image depend on the quality level at which the image is presented. This depends on **image-resolution attributes**, including the following key parameters:

1) **Pixel Count**: The number of recorded pixels is a key parameter that determines image resolution. Often, the effective pixel count is based on the number of associated sensors that contribute to the composition of the image. A pixel resolution of $N \times M$ identifies the image as having a width of N pixel columns and a height of M pixel rows. The total number of image pixels is then equal to $N \times M$. Often, also stated is the number of pixels per unit length, **PPI = pixels per inch**, or per unit area, **PSI = pixels per square-inch**.

   An image that is 2048 pixels in width and 1536 pixels in height has a total of $2048 \times 1536 = 3,145,728$ pixels or 3.1 mega-pixels. Its use will produce a very low-quality image (72 ppi) if printed on a page that is about 28.5 inches wide but will yield a very good quality (300 ppi) image if printed on a page which is about 7 inches wide.

   The number of photo-diodes used in a color-based digital camera image sensor is often equal to a multiple of the number of pixels in the image that it produces, as information from an array of color-based image sensors is used to reconstruct the color of a single pixel. The image has to be processed to produce three color components for each output pixel.

2) **Spatial Resolution**: Identifies the precision at which image objects can be differentiated. It depends on the nature of the source of the image and is generally related to the order of independent pixel values used per unit length. The spatial resolution of consumer displays ranges from 50 to 800 pixel lines per inch.

3) **Spectral Resolution**: Identifies the resolution of the different spectral components of the image, such as those that capture the image components of different colors. Higher than regular spectral resolution involves storage of more than the regular 3 RGB image color components.

The number of distinct colors that can be represented by a pixel depends on the number of bits per pixel (bpp). A 1 bpp image uses 1-bit for each pixel. Then, each pixel can reside in either an on-state or an off-state mode. An 8 bpp image can be represented in 256 colors. For color depths of 15 or more bits per pixel, the depth is normally the sum of the bits allocated to each of the red, green, and blue components. High-color usually refers to 16 bpp, with often setting of five bits each for red and blue, and six bits for green, as the human eye is more sensitive to errors in green. In certain systems, extra bits per pixel are added and used to characterize opacity.

4) **Temporal Resolution**: Relates to time-oriented resolution. This is of particular importance as it relates to full-motion video. The time resolution used for movies is usually set to 24 to 48 frames per second [frames/s], whereas high-speed cameras may use 50–300 [frames/s] or higher resolutions.

5) **Radiometric Resolution**: Identifies the level of intensity resolution. It relates to the number of bits used to describe each intensity level, such as the magnitude of a pixel attribute. For example, computer image files often use a resolution of 8 [bits/pixel], enabling the use of 256 different magnitude (such as shades and colors) levels. It is also identified as *color depth*.

As an image is being transported across a communications channel, its resolution may become limited by the underlying noise and interference levels, as determined by the SINR level measured at the receiving node. As noted in Chapter 3, the reported SINR level determines the spectral efficiency level that should be set across the communications media employed for transporting the image across a communications network, and the corresponding transmission data rate.

6) **Dynamic Range**: Identifies the range of luminosity that can be reproduced accurately. At a lower dynamic range, sensor elements for different colors may saturate in turn, leading to hue shifts or other resolution distortions.

Many camera phones and digital cameras use memory cards that employ flash memory to store image data. Cards may use the Secure Digital (SD) format; many are Compact Flash (CF). The XQD card format has been targeted for use at high-definition camcorders and high-resolution digital photo cameras. Modern digital cameras also use internal memory, which can be employed for pictures that may be transferred to or from the card or through the camera's connections.

To illustrate, we note the following pixel resolution of samples of several commercial system digital images: UDV, HD DVD, Blue-Ray, HDCAM: 1920×1080; 4K UHDTV: 3840×2160; 16K Digital Cinema: 15360×8640. Illustrative Digital Camera: 10380×7816, 81.1 Mega Pixels (MP); Illustrative digital still camera (Canon EOS 5DS): 8688×5792, 51 Mega Pixels (MP). Illustrative mobile phone (Nokia 808 PureView): 7728×5368, 41 Mega Pixels (MP).

An image contents may be represented by a very high number of information bits. To efficiently transport an image across a communications channel at a reasonably low time latency, a high spectral efficiency level (i.e., a high data rate per allocated frequency bandwidth) would be required. Since bandwidth resources are often highly limited or expensive, in particular when considering transport across a wireless medium (such as a cellular wireless system), it is essential to reduce the information contents of a transported image. Accordingly, image and video transport systems employ highly effective image compression techniques. Methods based on the use of DCT have proven to be highly effective, leading to the transfer of high-quality images across communications networks at much reduced data rates.

2.7.3 Full Motion Video

Video streams and flows that are transported across communications networks convey moving visual media. While originally using analog formats, today video signals are generally produced, stored, transported, and processed in digital bit-mapped format. Effective use is made of powerful data compression and processing algorithms, tools, and network transport techniques.

As noted above, a still video image is represented spatially as an array of pixels. The intensity and color of each pixel are described by a digital attribute, requiring a specified numbers of bits per pixel (bpp).

To represent a moving video segment, image frames are produced over time. How many image frames per unit time should be presented to a viewer? For *freeze frame or slide/photo show* purposes, the frame rate would match the targeted time interval between successive image displays. For example, in a slide (or photo) presentation session, if new image frames (such as slides or photos) are displayed at 10 second intervals, the targeted frame rate is set to 0.1 frames/s, or 1 frame every 10 seconds.

A much higher frame rate is required for displaying a **full motion video** segment. In the following, we focus on full-motion video flows as they impose much higher storage, processing, display, and communications transport bandwidth requirements.

For a person to experience an effective full-motion display of a video segment, the minimum frame rate that is required to achieve a comfortable illusion of a moving image is about 16 frames per second. PAL (as used in Europe, Asia, and Australia) and SECAM (France, Russia, and parts of Africa) standards specify a frame rate of 25 [frames/s], while NTSC standards (USA, Canada, and Japan) specify 29.97 [frames/s]. Film is generally shot at the slower frame rate of 24 [frames/s]. Certain professional video display applications employ frame rates of 120 [frames/s] or higher.

Computer display standards specify video display parameters that include a combination of aspect ratio (representing the number of horizontal vs. vertical pixels per image frame), display size, display resolution, color depth, and refresh (or frame) rate. Video can be *interlaced or progressive*. In progressive scan systems, each refresh period updates all scan lines in each frame in sequence. Interlacing was developed in aiming to reduce flicker in early displays without increasing the frame rate. In interlaced video, the horizontal scan lines of each complete frame are captured as two fields: an odd field (upper field) consisting of the odd-numbered lines and an even field (lower field) consisting of the even-numbered lines. NTSC, PAL, and SECAM are interlaced formats.

High-Definition TV (HDTV) systems include the following. A 1920 × 1080 p 25 system uses progressive scanning format at a rate of 25 [frames/s (FPS)], whereby each image frame is 1920 pixels wide and 1080 pixels high, presenting an aspect ratio of 1920:1080 = 16:9. The 1080 *i* 30 system uses interlaced scanning format at a rate of 30 [frames/s] (corresponding to scans of 60 [fields/s] to account for the interlace scanning that is performed at twice the basic scanning speed), each frame being 1920 pixels wide and 1080 pixels high. The latter frame thus contains a total of 2,073,600 (about 2.1 Mega) pixels. Standard definition TV (SDTV) systems produce images that consist of 1280 × 720 pixels. A frame rate of 30 [frames/s] (more precisely set to 29.7 [frames/s], at 99% of the former) is employed. Other standards and schemes have been employed under a wide range of video display and scanning technologies, standards, and applications.

2.7.4 Video Compression

Uncompressed video delivers maximum quality, but its effective transport requires high bandwidth and needs to be transferred at a very high data rate to realize low latency levels. It can also require the use of high-storage capacity at a receiving entity. Consider a video frame that is 1920 pixels wide and 1080 pixels high. This frame contains a total of about 2.1 Mega pixels. If each pixel is represented by a 3-bits wide attribute per color, and if 3 colors per pixel are employed, and a frame rate of 30 [frames/s] is used, the uncompressed data rate is equal to about 1.49 Gbps. Even if implemented at a fraction of this data rate, a significant compression ratio is required to enable bandwidth efficient transport.

Variety of algorithms have been used to compress video streams. Effective schemes have made use of a group of pictures (GOPs) approach to reduce spatial and temporal redundancy. Spatial redundancy is reduced by identifying differences between different parts of a single frame; this process is known as intra-frame compression and its function is closely related to that used for image compression. Temporal redundancy is reduced by registering differences between frames. This task is known as inter-frame compression, and it involves the use of motion compensation techniques.

Commonly used modern compression standards for full motion video include MPEG-2, which has been used for DVD, Blu-ray and satellite television, and MPEG-4, which has been used for Advanced Video Coding High Definition (AVCHD) and mobile phones (3GPP). H.264/MPEG-4 AVC is widely used on the Internet. MPEG-4 Part 14 (MP4) is employed by many platforms, such as YouTube and Apple Computer systems. High-Efficiency Video Coding (HEVC/H.265) is a successor technique that achieves compression upgrade. Flash Video (FLV) employs the Adobe Flash software. It has been used by various video-sharing websites, including YouTube.

A broadband speed of 2 Mbps or higher is often recommended for streaming standard definition video without experiencing buffering or skips, especially for live video. It has been used in TV-based streaming to Roku, Apple TV, Google TV, or in Sony TV Blu-ray Disc Player. A rate of 5 Mbps is recommended for High-Definition content and 9 Mbps for Ultra-High-Definition content.

A receiving user tends to gain higher QoE when receiving video streams that contain more detailed video information contents (within specified bounds). Such streams are produced at higher data rates and will therefore require the communications network system to allocate higher bandwidth levels for their transport. Also, the receiving device may have to maintain higher buffer storage levels. The latter can be set to store a full video segment download for non real-time operation, or a segment of the video stream when interactive video sessions are involved, or when limited buffering capacity is available. For example, a video segment that is one hour long and is streamed at a data rate of 300 Kbps produces a total of 1080 Mbits, or 128 MBytes (noting that 1 MBytes = $8 \times 1024 \times 1024$ bits).

Practical implementation of video streaming has been enabled by the use of Discrete Cosine Transform (DCT). It is a lossy compression algorithm. The DCT algorithm is the basis for the first (dated 1988) practical video-coding format, H.261, in the family of MPEG video formats that have been developed, starting at 1991. The video compression process is a multidimensional extension of the 1D DCT formalism described above for use in audio compression. A 2D version of DCT is derived from the 1D DCT scheme by being performed along the rows and then along the columns (or vice versa). Following the transformation, the contents of an image frame is characterized in terms of the relative magnitude of its frequency components. Since the spectral components of typical video segments that belong to lower-frequency bands tend to be of higher relative significance, it is beneficial to attach higher encoding significance to these components. Hence, after employing a DCT-based mapping, the magnitude values of the ensuing frequency components are examined. Following it with quantization and encoding operations, many higher-frequency components are eliminated. Proceeding in this manner, the process normally yields significant compression gains.

The inverse of a multidimensional DCT is a separable product of the single-dimensional inverses applied along each dimension at a time.

The following formula describes a version of the 2D **Discrete Cosine Transform (DCT)**. An input of $N_1 \times N_2$ real numbers, representing the magnitude level of frame pixels, and denoted as $\{x_{i,j}\}$, where $x_{i,j}$ denotes the video attribute value associated with the cell element at row-i and column-j, is transformed to produce an output (i.e., the corresponding DCT transform) of $N_1 \times N_2$ real numbers $\{y_{m,n}\}$, whereby:

$$y_{m,n} = \sum_{i=0}^{N_1-1} \sum_{j=0}^{N_2-1} x_{i,j} \cos\left[\frac{\pi(2i+1)m}{2N_1}\right] \cos\left[\frac{\pi(2j+1)n}{2N_2}\right]. \tag{2.9}$$

As noted in Eq. 2.9, the 2D DCT transform provides a frequency spectral characterization of the spatial cells of the video frame, spanning its rows and columns. Typically, video contents over a frame tends to not undergo significant variations as one scans the frame's columns and rows, as expressed by the spectral values presented at the output of the DCT module, noting that lower (m,n) values correspond to lower frequency components and slower spatial variations.

Video compression schemes use motion-compensated DCT (MC DCT) coding, also called block motion compensation (BMC). This scheme combines DCT coding in the spatial dimension with predictive motion compensation in the temporal dimension.

The order in which intra- and inter-frames are arranged in a stream is specified by using the concept of **Group of Pictures (GOP)**. GOP identifies a collection of successive frames or pictures within a coded video stream. The stream consists of successive GOPs, whereby a new GOP in a compressed video stream indicates to the decoder that previous frames are not required for the decoding of subsequent ones. The following types of pictures or frames are used: an (Intra frame coded) I-frame is coded independently of other frames; a (Predictive coded) P-frame contains motion-compensated difference information relative to previously decoded frames; a (Bi-predictive coded) B-frame contains motion-compensated difference information relative to previous and following pictures; a (DC direct coded) D-frame serves as a fast-access representation of a frame (and is only used in MPEG-1 video). GOP order has been arranged so that primarily an I-frame indicates the beginning of a GOP, and subsequently several P and B frames follow. However, several later designs employ more flexible structures.

2.7.5 Transporting IP Video Streams over Communications Networks

Communications networks, including IP networks and the Internet, are used for the transport of video flows. An end user may send a message to a video server, requesting a download across the network of a designated video segment. Users may also exchange video segments across a network.

Using a video streaming application, combined video and audio flows are downloaded or exchanged. Non-streaming flows include downloads of stored media, such as video, audio, and written material (books). High-quality network transport, reception, and reproduction of video streams across a communications network can be highly demanding. It usually requires the network to allocate high network bandwidth resources. High-quality reproduction of a video stream in real-time requires the network to transport video packets at high data rates, to incur low latency, and to experience low packet loss rates.

At times, the availability of adequate network bandwidth resources may statistically fluctuate and be lacking at times. It is encumbered by the variability of packet traffic loading conditions. Streaming packet flows may have to be routed across high-congestion hot spots, as often occurs when traversing busy wireless access network segments. Such conditions induce high variability in the QoE levels incurred during the reproduction of video segments at a destination end user.

While audio streams tend to impose upon the network a requirement for a **CBR** type service, compressed video streams require a **Variable Bit Rate (VBR)** type network service. This is the case since the magnitude of compression attained by spatially encoding a video-based Group of Pictures (GOP) depends in a critical manner on the spectral characterization of the underlying video contents. When presenting a dynamically active scene (as when considering a rapidly changing sport event), higher spectral (frequency) components are involved, inducing lower compression ratios. A higher compression ratio is attained under static or slowly changing scenes.

A video-streaming operation may be live or non-live. Under a live **streaming session**, video (and associated audio) streams are presented to the destination in real-time. Under non-live streaming, a user may access a video service provider at any time and then request the download of a video segment. Using a file loading process, an end user requests a video service provider for a download of a video file. The user may run it at a convenient future time. Under a streaming transport, an end user may listen to its video/audio contents on-the-go before the entire file has been transported.

A service provider may store video segments in a central location or in multiple geographically distributed sites. A user may access a *proxy server* that stores a segment of interest and is located in the user's vicinity. Illustrative streaming services include: Netflix, Hulu, Prime Video, YouTube, and other sites that stream films and television shows; Apple Music and Spotify, which stream music; and video game-oriented live streaming sites.

The process used for transporting a video flow across an IP communications network follows the general principles described above (see Section 2.6.2.1) for the transport of audio flows and streams. Key elements of the VoIP streaming process are depicted in Fig. 2.10. The following key elements are involved when transporting a video stream:

1) **Video Stream and Flow Production and Dissemination Modes**: Video frames are aggregated and encoded. Video messages are produced for dissemination to a targeted user, or for multicast to multiple users. We note the following typical modes of dissemination:

   a) **Viewing in Realtime of Live Video Streaming**: For a live video event, the video stream is produced at a designated data rate, R_V [bps], and disseminated in real-time across a communications network to its destination users. To assure real-time viewing, it is necessary to limit the delay level incurred across the network by video packets. For viewing a stream in real-time (as video events are happening) and without playback interruptions, it is necessary to configure a steady network transport process, inducing limited packet delay and delay variations.

   The designated encoding scheme and parameters are adjusted in accordance with the targeted QoE metric level. For higher QoE metric values, streams are produced at lower compression ratios. More data is then included in video frames. Furthermore, to assure viewing of a live event in real-time, this data must be transported across the network within a limited time span, inducing a higher stream data rate.

   The targeted QoE value and ensuing data rate level can be dynamically adjusted in real-time by a video source based on the receipt of feedback control messages that are sent by a destination end user.

   b) **Non-live Video Streaming**: A video stream is produced, encoded and saved at a source video server. It is transported to a user upon the receipt of a user's request for its download.

Its viewing by a destination user does not need to keep up with the progress of an ongoing live event.

When a destination user's device contains sufficient buffering and processing resources, it can store the complete video stream's content. It is then available for viewing by the user at any time. The time delay incurred in the download process is impacted by the level of network capacity resources allocated for the support of this flow.

In turn, under replay buffer size limitations at the destination user, while the initial download of video data may not be time sensitive, network transport delays of subsequent video segments have to be performed in a timely manner to avoid replay interruptions. The stream's encoding and ensuing data rate should be adjusted to assure a stream's timely network transport and a replay at an acceptable QoE level.

The targeted QoE value and associated data rate level are often dynamically adjusted based on feedback received from the destination end user. The video server may keep multiple versions of a video stream, each encoded at a different code rate. Different stored versions represent the video stream at different QoE levels, and ensuing data rates. The server transports across the network the stored version of the stream that best matches the user's requested QoE value while also aiming to transport data in a timely manner under current network congestion conditions and available communications capacity levels.

2) **Video Message Buildup**: A video message is constructed by a video application process. The length of a video message is adjusted by taking into consideration system parameters, such as those that relate to the allocated network bandwidth level and to the storage and processing resources available at the receiving end user.

For video-streaming purposes, the video message is often constructed so that it consists of **video chunks**. The content of a chunk consists of video data that represents a segment of the video that is being played over a *specified period of time*. Often, a *video chunk* is defined as a data unit that carries a quantity of video data that is included in a single video frame. For example, a frame may represent a video run-length of 25 ms. Consider a video stream that is encoded to operate at a data rate of 1 Mbps. Then, each chunk contains a total of 1 Mbps × 25 ms = 25 Kbits video data. At certain times, the source node may determine that it needs to reduce the video data rate by increasing the video compression ratio, induced by limitations in available network bandwidth resources, or by limitations in the receiver's storage capacity, or by impaired reception conditions. The QoE level experienced at the receiver would then be reduced. For example, assume that the data rate level is reduced to 500 Kbps. Assuming again that each chunk conveys video contents that correspond to a single video frame of duration 25 ms, a message chunk now carries only 12.5 Kbits of video data.

3) **Network Delay Variations**: In transporting a video packet across a packet-switching network such as the Internet, the packet's ETE delay level is likely to incur statistical variations. Packets may also be *lost or discarded* as they make their way across a network. Users engaged in an interactive application, such as those participating in a real-time video conference, tend to experience loss of fidelity when replay buffering delays exceed a certain level, such as 200 ms.

Video flows impose significant bandwidth and throughput rate requirements on a network system. A compressed video flow loads the network at a *Variable Bit Rate (VBR)*. Hence, the transport mechanism needs to be able to reliably and expeditiously handle a flow whose data rate may exhibit significant random variations. Such a VBR flow tends to exhibit a high *peak-to-average* data rate ratio.

To reduce transport delays, regional servers are geographically spread across a service area. A central server disseminates copies of stream data to proxy regional servers.

2 Information Sources, Communications Signals, and Multimedia Flows

4) **Protocol Processing at the Source, Including the Use of Real-Time Transport Protocol**: A video message produced at a source station is processed by protocol layer entities resident at the source. For transport across an IP packet switching network, an IP packet is constructed by a network layer entity. For real-time streaming applications, the UDP is typically employed at the transport layer. UDP is a connection-less protocol, lacking the ability to regulate the sequencing of packets, which would identify time attributes of packets and allow their orderly reassembly by a receiving module. For this purpose, the Realtime Transport Protocol (RTP) is used. Key features of RTP have been discussed in Section 2.6.2.2, noting that it provides packets with sequence numbers and time stamps.

Associated with RTP is the **RTCP**. It is used to deliver out-of-band statistics and control information for an RTP session. It provides feedback on the QoS of a stream transport process. Performance metrics that are distributed to participants in a multimedia streaming session include packet and byte counts, packet loss, packet delay variation, and round-trip delay times. An application may use this information to control its configuration of QoS levels. For example, when a network transport mechanism is reported to experience congestion, inducing lower video QoE values, an application at the source entity may act to adjust the quality of the video stream by changing the encoding mechanism or by modifying the requested network transport parameters.

RTCP services also support multiuser multicast sessions. In a **multicast multimedia session**, users that have joined a multicast group participate in the reception of group associated multimedia messages. Certain users may be configured to serve as senders of multimedia streams and messaging flows.

As an alternative, technologies such as Apple's HLS, Microsoft's Smooth Streaming, Adobe's HDS and nonproprietary formats such as MPEG-DASH have been employed to enable adaptive bit-rate streaming over HTTP. Destination users provide performance data to the source entity such as a video server. The latter uses it to adapt its video stream's parameters. As noted in Section 2.7.6, a Standards based DASH (Dynamic Adaptive Streaming over HTTP) protocol makes use of the connection-oriented transport layer services provided by TCP. Often, a streaming transport protocol is used to send video from an event venue to a "cloud"-based transcoding service, which uses HTTP-based transport protocols to distribute the video to users.

5) **Replay Process**: Consider video packets that are transported across a packet switching network. As a packet arrives at a destination station, the received signal is demodulated and decoded. Message data units are processed by ensuing protocol layer entities. An RTP or DASH/TCP layer entity detects missing packets and reorders packets that arrive out of order. Decoded and processed messages are stored in a video replay buffer. To avoid run-time replay gaps, packets are delayed at the replay buffer. Message chunks are assembled and decompressed. Video data is arranged in the proper spatial and temporal order to construct the video frames that drive the video replay process.

2.7.6 Dynamic Adaptive Streaming over HTTP (DASH)

The QoE enjoyed by a destination user in her replay of a video stream depends on several factors, including the following ones: the received video flow's data rate, whose realizable level is impacted by the SINR recorded at the receiver; the frequency bandwidth that the network system makes available to support the transport of the stream; parameters characterizing the reception mechanism, including its storage capacity, processing capability, and energy resources. When considering a mobile user, the data rate at which the user is able to reliably receive a video stream tends to vary as its location changes, as it is impacted by blockage, interference, and fluctuating signal power levels.

Consequently, it is beneficial for the video-streaming system to dynamically adapt the data bit rate level of the produced video stream to underlying signal reception conditions. Based on reports received from the receiving user, an adaptive system would adjust the video stream's parameters, including its bit rate, as it adapts to reception conditions. In this manner, the system can provide the user with the best quality video experience that is feasible under current conditions and resource allocations, as it aims to meet user performance objectives.

The **DASH** scheme, also known as *MPEG-DASH*, implements such an adaptive process. The bit rate settings of video streams that are provided by conventional HTTP web servers are dynamically adapted. Video content is fragmented into subsegments, identified as **chunks**. Each chunk carries the bit content produced over a short interval of playback time. A media presentation description (MPD) field contains parameters that characterize a chunk's segment, including information about: timing, URL, and media features such as video resolution and bit rates. Different video-encoding schemes are used to produce different chunk payloads loads that would be sent across the network at different data rates. A higher QoE level would be enjoyed by the recipient when the stream is produced and transported at a higher data rate.

The destination user employs an Adaptable Bit Rate (ABR) algorithm to automatically select the highest data rate at which it is feasible to replay a segment under current system and network conditions. The specific algorithm that a system uses is not specified by a corresponding Standard, allowing therefore different vendors to make use of their own proprietary or favorite scheme. The server is informed apriori about the prospective parameters that are preferred by a receiving user for use over a forthcoming period of time. In this manner, a destination user would receive a targeted segment at a preferred rate within a prescribe period of time. Such an operation aims to avoid stalls and re-buffering events during the playback process. It strives to realize a network transport process that exhibits a targeted performance behavior, achieving desired information throughput rate, chunk latency, and packet error rate.

MPEG-DASH is an international standard. It employs TCP as a transport protocol. It follows the structure used by web servers in the production of streams and in their transport across the Internet. It is able to deliver multimedia (video, audio, and data) streams to a wide range of devices, such as IP TV platforms, desktop computers, smartphones, and tablets. While various vendors offer proprietary adaptive streaming mechanisms, DASH offers an open Standard-based solution for a wide range of systems. It is also adopted for use by cellular wireless network Standards such as those developed by 3GPP. It is codec-agnostic, so that it can use contents that is encoded with any coding format.

2.7.7 Performance Measures

QoE requirements are set by the client of a video stream in accordance with a level of satisfaction that is acceptable to the end user. The QoE measure is dependent upon the video flow's session type, as well as on a variety of conditions relating to features and parameters associated with the replay and receiving system, the video source production system and the communications network transport system. As the QoE requirement is frequently end-user **subjective**, it is often expressed by a *Mean Opinion Score (MOS)*, which rates the quality of a specific video flow process in terms of an average score awarded by a group of viewing users. Other metrics are noted in Chapter 5. A QoE rating may be linked to the underlying video flow's class. Different video flow classes may employ different compression algorithms and data rates, as well as impose distinct requirements of video production attributes, such as frame rates and quantization levels.

QoS requirements are imposed on the communications network system for the purpose of characterizing the quality of a message flow's network transport. QoS metrics for network communications are **objective** measures. They specify the performance parameters that must be met by the communications network system in the transport of a message flow, including a video stream, under prescribed video production and replay mechanisms and their associated parameters. As outlined in Chapter 5, communications networking QoS specifications include statistical measures of network throughput, packet delay, packet delay-jitter, and packet loss ratio.

2.8 Data Flows

Audio and video streams are often presented as **realtime** flows. To preserve the real-time replay feature of a stream, it is essential that following their transport across a communications network, the stream's packets be received in a timely manner. Packet receptions should be properly ordered for the replay process. Depending on the specified end-user's QoE value of interest, it may be acceptable for a limited fraction of packets to be discarded, as can be caused by transport-induced packet errors or losses. Strict limits are generally imposed on the packet's transit time and delay variation across the network.

Data flows carry messages that are generated by a wide range of data applications. For many applications, it is critically important that sent messages are correctly received at the destination, avoiding the occurrence of message errors. This is generally the case for fund transfers, medical scans, critical documents and files, and a multitude of high priority and safety-oriented messages. Messages that upon reception are determined to contain errors may be discarded and retransmitted by the sending user. Data flows may not be subjected to strict inter-message timing reception requirements.

In contrast, audio and video real-time flows and streams require the use of protocols such as Realtime Transport Protocol (RTP) that attach message time stamps and sequence numbers. This is illustrated in Fig. 2.1, whereby data flows are identified as store-and-forward flows, which carry data messages that are generated at random times. In comparison, real-time streaming flows follow a more temporally regular message flow pattern.

Performance requirements imposed by a data flow relating to its transport across a communications network depend on the flow's application class. For interactive data applications, the network transport process may have to sustain data exchange dialogues that are performed on a timely basis. The following categories of data flows are noted:

1) **Client–Server Interactive Session**: A client user interacts with a server. Typically, the client sends a request message to the server. The latter responds by returning a message or file. A client may also initiate a dialog with a server that induces the server to send to the client successive messages that the client needs to respond to. Telnet is an application layer protocol that has been used across the Internet for the transport of messages that are produced via client–server interactions.

2) **Peer to Peer**: An end user directly interacts with another end user, resulting in messages (directly) transported between the end-uses.

3) **Multicast Peer to Peer**: Message transactions are disseminated by an end user directly to a group of other session end users. Involved users may have joined a multicast session as members.

4) **Email Messaging**: Interchange of email messages among end users. The Simple Mail Transfer Protocol (SMTP) has been widely employed over the Internet such as an application layer protocol.

5) **File Transfer**: Transfer files from an end user or server to another end user or to a server. File Transfer Protocol (FTP) and Trivial File Transfer Protocol (TFTP) have been widely employed over the Internet such as application layer protocols

6) **File Download**: Downloading of a file, or a chain of files.

7) **Network Control**: Dissemination of network control messages to targeted entities, aiming to control the operation of a network or a system.

8) **Network Management**: Transmission of query/command messages from a management station to a managed object embedded in a targeted device, aiming to manage the device. A management message may, for example, request a managed device to send the manager performance data that the device has been collecting, or to adjust its operation. The Simple Network Management Protocol (SNMP) has been widely used over the Internet as a network management application layer protocol.

9) **Network Signaling and Session Control**: Transmission of signaling messages from a control station to designated network nodes for the purpose of establishing or tearing-down network connections or sessions. Session control messages are used for the negotiation of session or flow parameters (such as employed codecs, bandwidth utilization, and message contents). Related Internet-oriented application layer signaling and control protocols include H.323 and SIP. SIP is employed for setting up a call connection, ringing, terminating, and managing VoIP and VoIP calls and for message interactions. Such protocols are also used to modify connection configuration or parameters during a call, such as adding an additional participant. SIP is based on a "request–response" protocol, accepting requests from one computer and returning responses from another. Creation and tearing-down of media connections are performed by using Media Control Protocols.

10) **Browsing the Internet**: Browsing the Internet in search of specified users, information, data, and sites.

We differentiate between data flows based on a multitude of attributes, such as:

1) **Time Sensitivity**: Involving data flows that must be disseminated to their destination entities in a timely manner.
   a) Carry information that needs to reach the destination node in a designated time frame.
   b) Highly time sensitive as to the value (or perishable character) of the information that is being transported.
   c) Response-time sensitive, whereby a request message must trigger the production and reception of a response message within a strict timing bound.

2) **Throughput Rate**: Certain flows involve the transport of a high volume of data and thus require the network to allocate them a sufficiently high level of capacity resources, enabling high throughput rates. This is, for example, the case for high-volume data dumps, large program update downloads, high-volume document dissemination, and non-realtime flows that carry high image-oriented contents. Also, a real-time high-quality video flow can require the provision of high network throughput means.

3) **High Priority/Critical Flows**: Data flows that are of critical importance for the operation of an underlying system or in terms of their essential value to end users. Included are network signaling, control flows, and critical network management flows.

90 | 2 *Information Sources, Communications Signals, and Multimedia Flows*

4) **Burstiness Character**: A highly **bursty** data flow that requires the network to provide it with a high data rate when it is active, while not being active very often, so that it exhibits a *low duty-cycle* ratio (see Eq. 2.7). As illustrated in Fig. 2.9, a bursty source could produce a high data rate activity for a relatively short period of activity. During other periods of time, the source is either idle or produces data at a low data rate. Many data flows, such as those involving inter-active session dialogues, tend to be highly bursty, while requiring fast response and network transit times.

Problems

2.1 Give three examples of information sources that produce real-time streams and three examples of information sources that produce non-realtime streams.

2.2 Draw a space–time diagram that depicts the progress of data units of an illustrative flow across a tandem connection of communications links produced by a real-time stream, assuming each link is shared among flows on a time-division basis.

2.3 Describe the operations performed at a destination end user when receiving and processing the following message units:
a) Data units of a non-realtime flow.
b) Data segments of a real-time stream such as a voice or video stream.

2.4 Describe the network performance requirements that are typically imposed for the success-ful transport of the following flow types:
a) Interactive data.
b) File transfer data.
c) Realtime voice and video streams.
d) Archival voice and video.

2.5 Describe the features of each one of the following service classes and identify for each class an illustrative application and typical QoS requirements.
a) Continuous Bit Rate (CBR).
b) Real-Time Variable Bit Rate (RT-VBR).
c) Non-Real-Time Variable Bit Rate (NRT-VBR).
d) Available Bit Rate (ABR).
e) Best Effort.

2.6 Characterize the features of analog and digital signals and provide three examples of infor-mation sources whose original signals are of each type.

2.7 Describe the processes that are undertaken in performing the following conversions and identify performance measures that are applicable in aiming to execute the conversions at higher precision levels.
a) Analog-to-digital conversion.
b) Digital-to-analog conversion.

2.8 Describe methods that are used to calculate the bandwidth of a signal and of a communications link.

2.9 Consider a rectangular pulse signal of temporal width (baud time) of length $T = 1$ ms. What is the effective bandwidth of this signal? What would be the effective bandwidth occupied by this signal if its temporal width is reduced to $T = 0.1$ ms?

2.10 Consider an analog signal whose bandwidth is equal to 10 MHz. It is converted to a digital signal by sampling it at the Nyquist rate, and by quantizing each sampled signal amplitude to one of the eight different values. Calculate the bit rate of the produced digital signal, expressed in [bits/s].

2.11 Describe common techniques that are employed to perform the following procedures. Also identify functions that are used as performance measures in selecting the associated parameters of each procedure:
a) Signal sampling.
b) Signal quantization.

2.12 Describe the process used by a Linear Pulse-Code-Modulation (LPCM) encoder of an audio stream.

2.13 Describe the factors that affect the quality of production of a digital voice stream.

2.14 Describe the factors that affect the quality of production of a digital full-motion video stream.

2.15 Describe the process used in the production, networking and replay of VoIP streams.

2.16 Identify the services provided by Real-Time Protocol (RTP) when used for the transfer of VoIP audio streams across a communications network.

2.17 Describe the impact of network delay, network delay-variation, and replay delay, on the quality of voice flow reception at an intended destination end user. Identify measures that can be taken to attain high Quality to Experience (QoE) levels. Consider separately each of the following voice flow types:
a) PCM encoded voice connection across a network circuit.
b) VoIP stream over a packet switching network.

2.18 Outline and discuss methods that are used to characterize the quality of produced audio streams.

2.19 Outline and discuss key image resolution attribute parameters.

2.20 Consider a full-motion IP video stream that is transported across a packet switching network to a destination end user. Identify and discuss key factors that impact the quality of experience (QoE) of its reception at a destination user.

3

Transmissions over Communications Channels

Objectives: *A wireline or wireless communications channel is used to carry information signals transmitted by a sending station to a receiving station. A segment of the channel's frequency spectrum of specified bandwidth is allocated for signal transmission. The sending station aims to have its data transmitted across the channel at a bit-error-rate level that does not exceed a prescribed value. The sending station must set its transmission data rate so that it is lower than a value identified as the channel's capacity. To efficiently utilize the allocated channel bandwidth, a sending station strives to synthesize the signals used to transmit its data across the channel in a manner that achieves a high data rate level per utilized bandwidth unit, as it aims to achieve a high spectral efficiency level.*

Methods: *In accordance with Shannon's Noisy Channel Coding Theorem, assuming an additive-white-Gaussian-noise channel model, the highest spectral efficiency level achievable by a sending station is limited by the channel's capacity. The latter increases in a logarithmic manner as the signal-to-noise ratio (SNR) measured at the intended receiver increases. To achieve a reliable transmission of its message at a targeted spectral efficiency level, in limiting the incurred error rate level, a sending station processes its signal by applying a properly configured modulation/coding scheme (MCS). Encoded data is used to modulate a carrier signal. The outcome is a channel signal that conveys the user's data while being configured for effective transmission across a communications channel.*

3.1 Communications Media

Multitude of communications media can be employed for message transmissions. **Wireline** media use wired communications lines while **wireless** media use radio communications or other non-wired media. Wireless media are advantageous in that they can accommodate communications between mobile users. A wireless communications link that connects mobile or stationary users is normally established much faster than a wired connection, which requires the installation of wires. In turn, wireless communications channels are more susceptible to interference and eavesdropping. They generally require signal transmissions to be carried out at lower data rate levels. They require enhanced security and privacy protections. In the following, we overview key wireline and wireless communications media.

Principles of Data Transfer Through Communications Networks, the Internet, and Autonomous Mobiles, First Edition. Izhak Rubin.
© 2025 The Institute of Electrical and Electronics Engineers, Inc. Published 2025 by John Wiley & Sons, Inc.

3.2 Wireline Communications Media

Wireline media make use of the following types of wired conductors.

1) **Copper Cable**: A copper cable carries signals as electrical pulses that travel across its metal strands. Copper cables are sensitive to interference caused by electrical radiations. Interference sources include power lines, lightning, and intentional jamming by electrical signals. Noise and interference signals reduce the effective message data rate that can be conducted across a copper cable.

   Copper wires are less secure than fiber-optic cables. Copper wires are sensitive to eavesdropping and security breaches that can take place by tapping into the cable (such as by draining its propagating energy level by a small amount) and by sensing the electrical fields created across the cable.

   The interference rejection capability of a copper cable is improved through the use of shielding, leading to better interference immunity and thus to increased message throughput capacity. A twisted pair cable consists of conductors of a single circuit that are twisted together. This arrangement reduces electromagnetic radiation from the pair, lowers crosstalk interference between neighboring pairs and lowers external electromagnetic interference. Its invention is attributed to Alexander Graham Bell.

   We observe two types of twisted pair cables: unshielded and shielded.
   An unshielded twisted pair (UTP) cable entity consists of groups of multiple (such as 25) pairs. The cables are typically made with copper wires measured at 22 or 24 American Wire Gauge (AWG), using polyethylene insulator and a polyethylene jacket. Other materials are also employed. The cable's bandwidth is sufficiently wide to allow it to carry TV signals. They are widely used in data networks for short and medium-length connections because of their lower cost, as compared with optical fibers and coaxial cables.

   Shielded twisted pair (STP) cables are able to better protect from electromagnetic interference. Shielding serves as an electrically conductive barrier that is used to attenuate electromagnetic waves that are external to the shield. The shield also provides a conduction path by which induced currents can be circulated and returned to the source via a grounded reference connection. Such shielding can be applied to individual pairs or to a collection of pairs. Shielding may assume foil (F) or braided formations. The following notations are used: U for unshielded, S for braided shielding (in the outer layer only), and F for foil shielding. They are used to indicate the type of screen used for overall cable protection and for protecting individual pairs or quads. A two-part abbreviation in the form of x/xTP is customary. Shielded Cat 5e, Cat 6/6A, and Cat 8/8.1 cables typically have F/UTP construction, while shielded Cat 7/7A and Cat 8.2 cables use S/FTP construction.

   The following data illustrates the variety of bandwidth levels accommodated by different twisted pair cables: a Cat 3 UTP cable provides a bandwidth of 16 MHz; Cat 5 UTP cable provides a bandwidth of 100 MHz; Cat 7 (S/FTP, F/FTP) cable provides a bandwidth of 600 MHz; Cat 8.2 (S/FTP, F/FTP) cable provides a bandwidth of 2000 MHz.

2) **Coaxial Cable**: Like a copper cable, a coaxial cable is used for information transport by using a copper conductor to carry electrical signals. The copper conductor is surrounded by an insulating layer. All is enclosed by a shield that consist of metallic braid and tape, and an outer insulating jacket. The design induces the carried signal's electric and magnetic fields to incur little leak outside the shield. Electric and magnetic fields that exist outside the cable are prevented from inducing much interference of signals that travel inside the cable. Consequently,

the cable is well protected from noise and interference signals and is able to effectively carry even weak signals. It accommodates the transmission of signals across a wide frequency band, offering high data throughput rates. It is used to transmit radio frequency (RF) and microwave signals. It has been used in computer communications, Ethernet Local Area Networks (LANs), and in the dissemination of digital audio and cable-TV (CATV) signals.

To illustrate the features of a coaxial cable, consider the RG-6/U coaxial cable. It uses an 18 AWG (1.024 mm) center conductor and offers 75 ohm characteristic impedance. It is used in residential and commercial applications. It is commonly employed for CATV signal distribution within homes. It typically employs a copper-clad steel (CCS) center conductor and a combined aluminum foil and aluminum braid shield. Signal attenuation increases with the used frequency level. For frequency levels of 1, 10, 100, and 1000 (MHz), the corresponding attenuation levels are 0.2, 0.6, 2.0, and 6.2 (dB/100 ft), respectively.

3) **Fiber Optic Cable**: A fiber optic cable carries signals as pulses of light that travel along flexible glass threads. A light emitting source, such as a Light Emitting Diode (LED), is used as a transmitter. The cable does not conduct electrical signals and is thus not impacted by electrical and electromagnetic interference signals. A fiber-optic cable is capable of carrying information at a data rate that is significantly higher than that carried by a copper cable, offering a very high message throughput rate. Transported data experience very low interference, so that communicated messages are received at extremely low bit error rates (BERs). A fiber-optic cable typically includes multiple fiber-optic threads and is thus able to simultaneously carry multiple message flows.

A fiber-optic cable offers higher data transport security level. Unauthorized tapping into the fiber can be readily detected as it tends to drain a noticeable level of its signal energy, triggering security alarms.

Multiple message flows can be multiplexed for transport across a fiber-optic cable by using *Wavelength Division Multiplexing (WDM)*. Each flow is allocated a separate wavelength channel. A channel is configured for use at a specified wavelength (and associated frequency band) and is identified as a wavelength pathway. Under a *dense WDM operation (DWDM)*, many wavelengths are carried by a fiber-optic glass thread, realizing an ultrahigh-throughput rate. For example, if each wavelength pathway carries information at a data rate of 1 Gbps ($= 10^9$ bits/s), and if 1000 wavelengths are multiplexed across a fiber cable, the total throughput capacity rate offered by this fiber-optic cable is equal to 1000 Gbps $= 1$ Tbps ($= 10^{12}$ bits/s).

3.3 Wireless Communications Media

Frequency spectral range used for communications is divided into frequency bands. A band consists of a contiguous range of frequencies, identified as the corresponding spectral range. Communications performed over a specific band are often used to provide similar services. The International Telecommunications Union (ITU) has divided the radio spectrum into 12 bands. This classification terminates at 300 GHz, as above this frequency level, electromagnetic waves radiated by a transmitter are highly absorbed by the Earth's atmosphere.

In the following (reference: https://en.wikipedia.org/wiki/Radio_spectrum), we point out a classification of these bands. We note for each band the spanned range of frequencies as well as the associated wavelengths. The wavelength of a signal that is transmitted at frequency f is denoted as λ. It is calculated as equal to $\lambda = c/f$, where c is the speed of light. The size of an effective antenna element is of the order of a signal's wavelength λ. Hence, when operating at a higher frequency band, smaller transmit and receive antennas are employed.

1) **Ultra Low Frequency (ULF)**: Spanning the spectral range 0.3–3 kHz (wavelength range of 1000–100 km). Low-frequency bands have been used for underwater communications.
2) **Very Low Frequency (VLF)**: Spanning the spectral range 3–30 kHz (wavelength range of 100–10 km). Illustrative uses: underwater communications, navigation, medical monitoring.
3) **Low Frequency (LF)**: Spanning the spectral range 30–300 kHz (wavelength range of 10–1 km). Illustrative uses: navigation, AM long-wave broadcasting, time dissemination, RFID, amateur radio.
4) **Medium Frequency (MF)**: Spanning the spectral range 300–3000 kHz (wavelength range of 1000–100 m). Illustrative uses: AM medium-wave broadcasting, amateur radio.
5) **High Frequency (HF)**: Spanning the spectral range 3–30 MHz (wavelength range of 100–10 m). Illustrative uses: short-wave broadcasting, citizens band radio, amateur radio, over the horizon communications, sky wave radio communications, marine communications, mobile radio communications and telephony. Reflections from the Earth's ionosphere layer enhance associated long-range communications.
6) **Very High Frequency (VHF)**: Spanning the spectral range 30–300 MHz (wavelength range of 10–1 m). Illustrative uses: television and FM broadcasting, line of sight (LOS) to/from aircraft communications, land mobile communications, maritime communications, amateur, and weather radio.
7) **Ultra High Frequency (UHF)**: Spanning the spectral range 300–3000 MHz (wavelength range of 1–0.1 m). Illustrative uses: television broadcasting, microwave ovens, microwave communications, wireless local area networks (WLANS such as Wi-Fi), Bluetooth, ZigBee, GPS, two way radio communications, LOS to/from aircraft communications, land mobile communications, maritime communications, amateur and satellite radios, remote control systems.
8) **Super High Frequency (SHF)**: Spanning the spectral range 3–30 GHz (wavelength range of 100–10 mm). Illustrative uses: microwave communications, WLANs (wireless LANs, Wi-Fi), radar, satellite communications, cable and satellite TV broadcasts, satellite radios, and amateur radios.
9) **Extremely High Frequency (EHF)**: Spanning the spectral range 30–300 GHz (wavelength range of 10–1 mm). Illustrative uses: microwave communications, WLANs, mmWave communications such as those used for radio access networks (RANs) in mobile cellular and other mobile and autonomous vehicle wireless networks, remote sensing, scanners, and amateur radios.
10) **Terahertz (THz) or Tremendously High Frequency (THF)**: Spanning the spectral range 300–3000 GHz (wavelength range of 1–0.1 mm). Illustrative uses: medical imaging, terahertz communications, and remote sensing.

Other band designations have also been employed. For example, letter band designations (used by the United States, EU, and several military organizations) include the following: Band A: 0–250 MHz; Band B: 250–500 MHz; Band C: 500–1000 MHz; Band D: 1–2 GHz; Band E: 2–3 GHz; Band F: 3–4 GHz; Band G: 4–6 GHz; Band H: 6–8 GHz; Band I: 8–10 GHz; Band J: 10–20 GHz; Band K: 20–40 GHz; Band L: 40–60 GHz; Band M: 60–100 GHz; Band N: 100–200 GHz; Band N: 200–300 GHz.

Another frequency band designation is offered for radar-frequency-oriented bands by an IEEE standard: HF Band: 0.003–0.03 GHz; VHF: 0.03–0.3 GHz; UHF: 0.3–1 GHz; L Band (long wave):

1–2 GHz; S Band (short wave): 2–4 GHz; C Band: 4–8 GHz; X Band: 8–12 GHz; Ku Band: 12–18 GHz; K Band: 18–27 GHz; Ka Band: 27–40 GHz; V Band: 40–75 GHz; W Band: 75–110 GHz; mm or G Band: 110–300 GHz.

It is noted that propagating signals incur high attenuation at higher-frequency bands. At higher frequencies, signal radiations are also more sensitive to blocking by objects that are located along their paths. While at lower-frequency bands, signals can penetrate walls and be well received by user devices located inside buildings, at higher-frequency bands, signals travel effectively only when not blocked by objects that obstruct their direct propagation path. A signal is then able to often reach its destination, rather than be blocked, if there is a **LOS** pathway between its transmitting and receiving entities (so that they can effectively see each other). Electromagnetic wave radiations that use higher-frequency bands would thus generally require a LOS pathway. Signals that use such bands are subjected to blockage by obstructions (including walls and metal objects) as well as by topographical objects (such as hills or mountains).

3.4 Message Transmission Over a Communications Channel

The transport of a message across a digital communications system that consists of a single communications channel which connects a source end-user to a destination end-user is illustrated in Fig. 3.1. In the following, we briefly describe the underlying communications transport process; further details are provided in subsequent sections.

Under the illustrated scenario, the source generates an analog signal. This signal is discretized, undergoing analog-to-digital (A/D) conversion. If the source data is already in discrete form, as is often the case for data communications, no A/D conversion would be required. The digital data is then processed by several protocol layer entities, in accordance with the process imposed by the employed protocol architecture. At each layer, a corresponding protocol data unit (PDU) message segment is created. As the PDU formation reaches (down the protocol stack) the link layer entity, a layer-2 (L2) frame is created. The corresponding bits embedded in L2 frames form a digital message flow (or stream) which is then processed by a modulation/coding scheme (MCS). Applying a *channel coding* scheme, code bits are added. They enable a decoder at the receiver to detect, and at times to also correct, transmission errors. Under a *source coding* scheme, a *data compression* process is used to remove redundant data and thus reduce the amount of information that needs to be transmitted over a communications channel. This is commonly done for the transmission of video signals. *Channel symbols* are then formed. Each symbol represents a bunch of bits. It is used to modulate a carrier, forming a *channel signal* that the sender transmits across the communications channel.

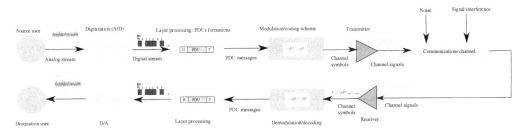

Figure 3.1 Message Transport across a Digital Communications System.

Many communications channels, including wireless ones, require a transmitter to send its information across the channel via the transmission of analog signals. A signal may convey several bits of information. Such a group of bits is used to modulate a carrier signal, producing an information-bearing modulated channel signal.

A channel signal may be formatted to carry information that is represented as an M-**ary symbol**. The latter carries a quantity of information that is measured as $\log_2 M$ bits. For example, a 2-level (or binary) channel signal, for which $M = 2$, carries 1 bit of information. Its reception conveys to the receiver the realization of a single binary unit (i.e., it is either a 0 or a 1). A 4-level signal (for which $M=4$) carries 2 bits of information, so that it conveys to the receiver a sequence of two binary digits (which could be 00, 01, 10 or 11).

The transmitted modulated signal **propagates** across the communications channel, reaching its destination user's station **receiver**. During the propagation process, the signal's power is attenuated. Due to perturbations caused by environmental and system induced stochastic *noise* and *signal interference* processes, the received signal's magnitude level tends to randomly fluctuate. Consequently, following demodulation of the received signal at the receiver, the recovered bit stream may contain errors. If the number of incurred errors is limited, and the employed coding scheme is sufficiently powerful, the decoding scheme may be able to correct these errors and recover in an error-free manner the original message. Segments that contain errors that cannot be corrected may be discarded or possibly retransmitted. For certain applications, such as audio and video, a limited error rate level can be acceptable.

At a receiving entity, proceeding upward the protocol stack, protocol layer entities process their received PDU messages. The transport layer entity reassembles the underlying message segments to reconstruct the original transport layer message, which is then delivered to and processed by an application layer protocol entity. A digital-to-analog (D/A) conversion is performed in case the destination end-user device expects to produce an analog signal (such as for audio playback applications).

3.5   Noisy Communications Channels

As a message signal propagates across a communications channel, its power is attenuated and its shape is distorted. Perturbations are caused by environmental and system-based stochastic noise processes. In addition, other signal transmissions that take place in the vicinity during overlapping times may interfere with the reception of the intended signal at a destination receiver.

A signal that has been perturbed by channel noise is shown in Fig. 3.2. It is received in a distorted format. Following its time sampling at the receiving entity, we note that two of the received data bits are in error, whereby a 0 bit was detected by the receiver to be a 1 bit, and a 1 bit was determined to be a 0 bit.

Among the various types of incurred noise and interference processes, we note the following ones.

Thermal noise and **shot noise** processes are induced by thermal activity and random movement of elementary particles (such as electrons) occurring in electrical (including signal reception) devices and components. A noise process is typically modeled to follow the statistics of a *Gaussian Stochastic Process* and to have a frequency spectrum that is uniformly distributed over a frequency band that is relevant to the reception of the underlying signal. It is referred to as **white noise**, as it induces an equal measure of noise activity at each frequency range, when considering the frequencies included in the involved communications band. We measure the intensity of this noise

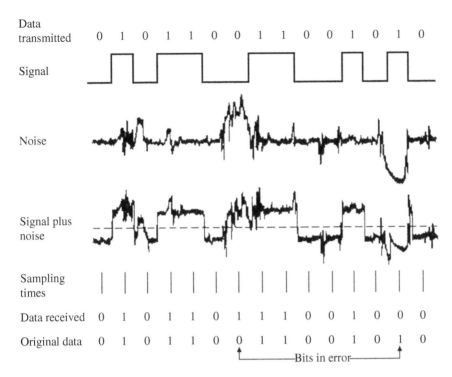

Figure 3.2 Signal Perturbed by Noise. Source: With Permission of TechnologyUK.

process (impacting signal reception at the receiver's system) by specifying its **power spectral density (PSD)**. It expresses a noise power level per unit bandwidth. It is measured in units of [watts/Hz] and is denoted as N_0.

Impulse or burst noise processes consist of random spikes of noise. They tend to occur at random times, last for a random (often short) duration, and assume relatively high-power levels. They can be caused by electromagnetic disturbances such as those generated by heavy-duty electrical machine emissions or other activities. They are often of such high intensity that they cause the receiver system to incur **reception outages**. During a period of high noise burst activity, a receiving device may not be able to correctly process its intended user signals.

To reduce the chance of erasures of consecutive information bits received during a **noise burst activity period**, which would lead to increased rate of message reception errors, the transmission system tends to **interleave** the information bits sent by a user prior to their processing by the transmitting module. The order in which a source user's bits are ordered in the transmitted information stream is rearranged. In this way, when a segment of transmitted bits is wiped by a noise burst, it will not impact a bunch of bits that were positioned in consecutive order within the original stream, increasing the probability of error recovery at the receiver. The receiver system **de-interleaves** the received bits, arranging them back in their original order, and is employing a decoding mechanism that strives to correct incurred errors (a process that can be performed successfully if a decoded block of bits does not contain too many bit-errors). The time delay applied by an interleaving mechanism is designed to be longer than the expected duration of a high fraction of expected noise bursts for the underlying communication channel system. Under such a setting, the interleaving operation would induce error events that impact message symbols in a random, rather than in a batched, manner, enhancing the effectiveness of an employed error correction scheme.

A signal transmitted across a communications channel can often take multiple paths as it propagates toward the intended receiver. Such a scenario can give rise to **multipath fading** phenomena, which may cause reception bit errors. The intended receiver would then detect a transmitted signal to arrive via a direct path as well as via one or multiple indirect paths. The receiver is then said to detect multipath induced signal duplicates. Different signal duplicates may arrive at the receiver at different time delays, causing the received waveform components of the signal to incur distinct phase shifts.

Duplicate signals travel over indirect alternate paths, caused by reflections of the signal waves by pathway objects or by atmospheric layers (such as reflections and defractions caused by the earth ionosphere layer) and reflections from mountains, buildings and water concentrations. The end result is that received signal reflections may cancel or re-enforce each other, depending on the corresponding induced relative time delay (and ensuing signal phase) differences. Consequently, signals that represent the same message, when detected and combined at an intended receiver, would incur power level fluctuations, a phenomena that is identified as **fading**. The received signal intensity level will **fade** when multipath signals combine in a manner that causes destructive interference. At times, a dominating direct path exists, under which the statistics of the corresponding process is shown to follow a Rician probability distribution function. This is known as Rician fading. On other occasions, multiple paths contribute in a significant way to the interference process. The corresponding stochastic process is then set to be problematically governed by a Rayleigh distribution. It is known as Rayleigh fading.

The mobility of users in conjunction with multipath fading can induce signal strength fluctuations and signal waveform perturbations at the receiver due to **Doppler** effects. Signals transmitted by a mobile user over a wireless communications channel to a base station (BS) node, or between two users, whereby one user or both users are mobile, incur frequency changes, a phenomenon that is known as Doppler effect. Such effects become more pronounced as the speed of the mobile user increases.

A *coherence time* characterizes the time duration over which the fading level does not incur a significant change. In a *fast-fading* channel, the latency incurred in message transport across a communications channel is longer than the coherence time. The message transit time may then span multiple coherence periods. The overall interference level incurred in this case would be determined through statistical averaging. Otherwise, the communications channel is characterized as a *slow fading* channel. In this case, a message segment that is transmitted across a channel during a high fade period may be wiped out, leading to signal outage at the receiver.

Adjustments in the configuration of the antenna system, including the use of multiple directional-beam antenna arrays can be helpful. This is of particular interest when employing highly directional antenna beam arrays for communications over mmWave communications media.

The use of an **Orthogonal Frequency Division Multiplexing (OFDM)** system, as undertaken by several advanced communication systems including cellular and Wi-Fi network systems, is often advantageous. A user's stream, which consists of a sequence of symbols that are intended for transmission across a communications channel, is divided into several sub-streams. Symbols that belong to different sub-streams modulate different sub-carriers. Sub-carriers are closely spaced. Signals generated across different sub-carriers produce overlapping frequency spectra. The symbol structure and the associated sub-carrier spacing are set in a manner that induces the orthogonality of these signals so that the peak amplitude of a sub-carrier signal coincides with the null amplitudes of the other sub-carrier signals, avoiding cross-interference effects. Sub-stream symbols are transmitted in parallel. An Inverse Fast Fourier Transform (IFFT) process is used for signal production at the transmitter while a Fast Fourier Transform (FFT) process is used at the receiver. As sub-stream symbols span a narrow frequency band, a sub-stream is transmitted at a lower data rate. The time duration of each sub-carrier symbol is therefore relatively long. Hence, fading induced time shifts incurred by different sub-signals have a reduced impact on reception effectiveness. Such an operation is also effective in combating frequency-selective interference effects. In addition, forward-error-correction (FEC) encoding is often applied, serving to handle signal losses.

To illustrate, consider an OFDM system for which a channel symbol carries 64 bits of data. Assume it to be transmitted at a data rate of 20 [Mbps]. The symbol transmission time is then equal to T = 64 [bits] / 20 [Mbps] = 3.2 [µs]. The spacing between sub-carriers in this system is equal to 1/T = 1/3.2 [µs] = 312.5 [kHz]. A system that allocates a total data transport bandwidth of 312.5 [MHz] for use by such a communications channel, will thus accommodate a total of 1000 sub-carriers.

Inter Symbol Interference (ISI) occurs when successively received channel symbols overlap each other in time. Such a condition can lead to reception errors. To handle such interference, guard-time gaps are set between successive channel signals. In an OFDM system, the guard interval assumes the form of a frame field that is generated as a *Cyclic Prefix*. It is taken into account when calculating the frame overhead ratio.

Signal interference is a major source of degradation in a system that accommodates multiple transmissions that are executed in a time simultaneous fashion over the same, or in neighboring, frequency bands. This is the case for many wireless communications networks, such as **cellular wireless systems**. In such a system, the area of operations is divided into multiple cells, whereby each cell is managed by a BS. A mobile user is associated with a BS that manages the cell in which it is currently located. Messages generated by a mobile are transported across the network by first being transmitted across an *uplink* wireless communications channel to its associated BS node. Messages that are destined to another mobile user reach the destination mobile's associated BS. The latter then transmits them across a *downlink* wireless channel to the destination mobile user.

The **RAN** in a wireless cellular system includes the underlying uplink and downlink communications channels. To increase its spectral efficiency, noting that the transport capacity of a cellular wireless system is limited by the aggregate total RAN bandwidth that is available, a proper bandwidth reuse scheme is used. Selected groups of BSs are set to reuse the same uplink and downlink frequency bands. In this case, the reception of an uplink message at a BS may experience signal interference from transmissions carried out by nearby mobiles or other BSs, whereby these transmissions occur during overlapping periods of time and occupy overlapping frequency bands. Similarly, the reception at a mobile user of a downlink transmission performed by its associated BS may experience interference from signals issued by transmissions that are carried out by nearby base stations or mobiles that take place during overlapping times over overlapping frequency bands.

The following measures are commonly used to express the relative power levels of intended, noise and interference signals.

To illustrate, consider a receiving station that is perturbed by a noise process which is modeled as white noise whose PSD is equal to 10^{-11} [watts/Hz]. Assume that the bandwidth of the received signal is equal to $B = 100$ [MHz] $= 10^8$ [Hz]. The overall noise power perturbing the reception of a desired signal is then equal to $P_N = 10^{-11} \times 10^8 = 0.001$ [watts]. If the un-perturbed intended signal received power level is determined to be equal to 0.01 [watts], then the ratio of the received signal power, denoted as P_S, to the received noise power, P_N, referred to as the **Signal to Noise Ratio (SNR)**, is equal to $SNR = 10$. Measured in *dB* units, it becomes $SNR = 10 \log_{10} 10 = 10 \, dB$.

Thus, a *SNR* measure, when the received signal is subjected to white noise interference, is calculated as follows:

$$SNR = \frac{P_S}{P_N} = \frac{P_S}{N_0 B}. \tag{3.1}$$

The *SNR* metric is often measure in dB units. We express the value of a variable X in dB units, identified as $X[dB]$, by calculating it as $X[dB] = 10 \log_{10} X$. Hence, when the *SNR* is expressed in dB units, it is calculated as follows:

$$SNR[dB] = 10 \log_{10} SNR = 10 \log_{10} P_S - 10 \log_{10}(N_0 B) = 10 \log_{10} P_S - 10 \log_{10} N_0 - 10 \log_{10} B. \tag{3.2}$$

When the reception of a signal is subjected to both white noise perturbation and signal interference, we use the *SINR* measure to express the ratio of the received signal power to the sum of the noise power and the signal interference power, as detected at the receiver. Denoting the interference power detected at the receiver by P_I [watts], the *SINR* measure is calculated as follows (when the background noise process is modeled as white noise, assuming the signal reception process to be performed over a bandwidth level that is equal to B):

$$SINR = \frac{P_S}{P_N + P_I} = \frac{P_S}{N_0 B + P_I}. \tag{3.3}$$

In examining this expression, we note a communications system to exhibit the following key operational modes.

- **Signal Interference-Dominated Mode**: Under this mode, the power measured at the receiver that is produced by interfering signals is higher than that induced by perturbing background noise processes. A receiving mobile user may detect a relatively high signal interference power. Its signal reception is subjected to high interference by signals that are issued by BSs or mobiles that are located in other cells. Such interfering entities may have been scheduled to use the same time and frequency transmission slots as those used by entities communicating with the underlying receiving mobile.
- **Noise-Dominated Mode**: Under this mode, the power measured at a receiver that is produced by interfering signals is lower than the power intensity level that is produced by noise processes. In cellular wireless network systems, this is often the case when interfering signals issued by other mobiles or by BSs are received at power levels that are lower than the received power of an intended signal. This is the case when the distance between a mobile user and its BS is relatively short so that the received intended signal power is relatively higher than the received power of signals produced by transmissions originated by entities located in other cells.

3.6 Illustrative Calculation of Signal-to-Noise-plus-Interference Ratio (SINR)

We illustrate the calculation of a SNR by considering a scenario that involves a transmission between two stations across a wireless communications channel. It may represent a transmission between an end-user node and a BS or an Access Point (AP) station in a Wi-Fi system. The operation is performed at a carrier frequency level f [Hz], and a corresponding wavelength that is equal to $\lambda = c/f$ [meters], where $c = 10^8$ [m/s] is the speed of light. The two nodes are assumed to be at distance d [m] from each other, whereby $d \geq d_0$ [meters]. The power of the transmitted signal (as measured at the transmitting node) is equal to P_t [watts], or equivalently, when expressed in [dB] units, $P_{t,dB} = 10 \log_{10} P_t$ [dB].

As the signal propagates across the channel, its power level is attenuated (i.e., it decreases). The attenuation of the transmitted signal as a function the distance d between the transmitting and receiving nodes is inversely proportional to a power of the distance. The power of the received signal, denoted as P_r [watts], is often calculated as follows:

$$P_r = C \frac{P_t}{d^\alpha}. \tag{3.4}$$

In this formula, the parameter C is a constant whose value depends on the features of the underlying channel and the possible presence of obstacles that may block, or otherwise affect, the propagation of the signal across its path, not including above mentioned noise and interference effects. The parameter α is a distance oriented **attenuation exponent**. It typically assumes values in the range $2 \leq \alpha \leq 4$. For a signal that propagates in free space, we set $\alpha = 2$.

A more involved model accounts for the observation that, for certain channels, the attenuation exponent α assumes higher values at longer distances. This gives rise to the following *two-ray model*:

$$P_0 = P_{r,dB}(d_0) = P_{t,dB} + 10 \, \log \left(\frac{\lambda^2}{(4\pi)^2 d_0^2} \right)$$

$$P_{r,dB}(d) = \begin{cases} P_0 - 10\alpha_1 \log_{10} \frac{d}{d_0}, & \text{if } d_0 \leq d \leq d_c; \\ P_0 - 10\alpha_2 \log_{10} \frac{d}{d_c} - 10\alpha_1 \log_{10} \frac{d_c}{d_0}, & \text{if } d > d_c. \end{cases} \tag{3.5}$$

For an illustrative Wi-Fi system following the 2019 IEEE 802.11p Standard recommendation, and using a commonly employed transceiver parameter values, we assume the following. Transmit power levels $P_{t,dB}$ are set to a value in the range 23–33 dBm corresponding to a transmit power value in the range 100–1000 mWatts (i.e., 1 watt); $\alpha_1 = 1.9$ and $\alpha_2 = 3.8$; $d_0 = 10$ [m], $d_c = 80$ [m].

To assure successful reception and decoding of the signal, the following requirements are imposed.

A **receiver sensitivity parameter**, denoted hereby as r_s, identifies the minimum signal power level that is required at the receiver to enable it to perform its reception function. For the underlying system, assuming an allocated bandwidth level of 10 MHz, the minimum acceptable receiver sensitivity r_s level induces the minimum received signal power level to be equal to -85 dBm, -82 dBm, or -77 dBm when using transmission rate levels that are equal to 6 Mbps, 12 Mbps, or 24 Mbps, respectively.

To assure reliable reception under specified data rate and channel bandwidth levels, while handling the level of noise and interference perturbations that are detected at the receiver, it would be necessary to assure the recording of an acceptable SINR level at the receiver. The minimum required SINR for achieving a desired spectral efficiency can be calculated by using Shannon's Channel Capacity formula.

This system is assumed to employ a specific set of MCSs. A message transmitted across this system's wireless medium is assumed to carry an average of 3024 bits. Assume that we wish to achieve an effective message error rate that is not higher than $10^{-3} = 1\%$, so that more than 99% of received messages do not contain errors. Consider the employed transmission rate level to be set equal to 6, 12 or 24 Mbps. Communications system analysis shows that the MCS settings that

104 | *3 Transmissions over Communications Channels*

should be used would then require the SINR level at the receiver to be no lower than 7 dB, 11 dB, and 20 dB, respectively.

For another illustration, consider the following parameters, relating to a transmission performed across a RAN of an LTE wireless cellular system:

$$C = \text{path loss factor} = -47.86\,\text{dB}$$

$$\alpha = \text{path loss exponent} = 2.75$$

$$d = \text{distance between transmitter and receiver}$$

$$P_t = \text{transmit power} = 33\,\text{dBm} = 2\,\text{Watts}$$

$$W = \text{bandwidth for signal reception [Hz]} = 10\,\text{MHz}$$

$$N_0 = \text{noise power spectral density} = -174\,\text{dBm/Hz}$$

$$P_N = \text{noise power at receiver} = N_0 W\,\text{Watts}$$

$$P_N[\text{dBm}] = -174\,\text{dBm} + 70\,\text{dB} = -104\,\text{dBm}. \tag{3.6}$$

Using these parameter values and Eq. 3.4, we calculate the power of the received signal transmitted across distance ranges $d = 100\,\text{m}$ and $d = 1000\,\text{m}$ to be equal to $-33.36\,\text{dBm}$ and $-36.11\,\text{dBm}$, respectively. Assume the power of interference signal measured at the receiver to be equal to $P_I = -46.11\,\text{dBm}$. In this case, as $P_I = -46.11 \gg P_N = -104$, the receiver operates in *signal interference dominated mode*. We conclude that the ensuing signal-to-interference-plus-noise ratio (SINR) is equal to about 12.75 dB and 10 dB, for sender to receiver distances of 100 m and 1000 m, respectively. Using the specific settings of the MCSs employed for the underlying version of the LTE system, and assuming a targeted message error rate that is not higher than 1%, we find that the corresponding attained spectral efficiency levels are equal to about 3.32 [bps/Hz] and 2.73 [bps/Hz], respectively. Consequently, noting the allocated channel bandwidth to be equal to 10 [MHz], the highest values that can be set for the transmission rates across this radio channel are equal to 33.2 [Mbps] and 27.3 [Mbps], respectively.

3.7 Channel Capacity

We have seen that a rectangular signal of duration T [s] occupies an effective frequency bandwidth of $B_S = 1/T$ [Hz]. As discussed in the next section, to effectively transmit this signal across a communications channel, a MCS is employed. The modulated channel-signal occupies a frequency resource of the communications channel whose bandwidth is equal to about $B = 2/T$ [Hz]. Assuming this channel signal to use an M-ary MCS, so that each information symbol that it transmits across the channel carries $\log_2 M$ data bits. Noting that a data rate of $R = \log_2 M/T$ [bps] is realized over a bandwidth $B = 2/T$ [Hz], we conclude that the attained **spectral efficiency** level of this transport mechanism, denoted as η [bps/Hz] is equal to:

$$\eta = \frac{\log_2 M}{2}\,\text{[bps/Hz]}. \tag{3.7}$$

Thus, one could increase the spectral efficiency level by increasing the modulation order M. To make best use of the spectral resource, it is desirable to configure a design that achieves the highest possible spectral efficiency level.

What prevents us from upgrading the spectral efficiency level by just increasing the modulation parameter M, so that a given channel signal would carry a higher number of information bits?

The answer to this question is provided by the observation that as the channel signal is detected by its receiving station, following its transmission across a communications channel, it is combined with stochastic noise and interference signals. The detected signal is then a distorted version of the intended signal. As we increase the quantity $\log_2 M$ [bits/symbol] of information included in a symbol (and conveyed by the corresponding channel signal) per prescribed bandwidth and signal duration levels, there is a higher likelihood that the received signal will have its information bits scrambled. It would then be more difficult for a receiver to differentiate one symbol from another, leading to reception errors. This will be further discussed in the next section.

The question raised next is the following: What is the highest spectral efficiency that a communications system can achieve in its transmission of information across a noisy communications channel?

Perturbed by randomly fluctuating noise processes, there is a possibility that a signal transmitted across a communications channel will be corrupted to an extent that will cause the receiving mechanism to mistake it for another signal, making errors in attempting to recover the data that is being conveyed. A key metric that measures such an occurrence is the **Bit Error Rate (P_b) (BER)**. It expresses the average ratio of erroneously received data bits vs. sent data bits.

For example, if a transport of data across a communications channel yields $P_b = BER = 10^{-3} = 0.1\%$, then an average of 1 bit out of every 1000 bits would be received in error. To improve the transport efficiency attained under noisy channel conditions, messages (or ensuing segments) are encoded by using a **FEC** code. The performance of a FEC coding scheme relates to the power of the code that it employs, which is characterized by a code rate r_C, where $0 < r_C < 1$. A fraction r_C of the total number of bits carried by an encoded message accounts for information bits. The remainder are code bits. The error correction capability of a FEC increases as the employed code rate decreases, as a higher fraction of the transmitted bits function as code bits. All bits included in a received encoded message are used at the receiver to decode the message. Provided the total number of incurred errors in an encoded message is not higher than a specific code-dependent limit, the receiver's decoding mechanism is able to detect and correct errors. By using an encoding scheme, it is possible to reduce the *raw BER* level to an improved *encoded BER* value.

To illustrate, consider a MCS that employs a FEC code at code-rate 0.25. Assume a transmitted encoded message segment to consist of a total of 4000 bits. Thus, 25% of the message bits are information bits and the remaining 3000 bits are code bits. If this message is allocated a channel bandwidth of $B = 1$ MHz, and if it is transmitted across the channel at a channel rate of $R_C = 1$ Mbps, then the segment's transmission time is equal to 4000 bits/1 Mbps = 4 ms. During this time, the message conveys a net of 1000 information bits and is thus attaining an effective information data rate of $R_C = 1000$ bits/4 ms = 0.25 Mbps = $r_C R_C$. The attained spectral efficiency level is equal to $\eta = 0.25$ Mbps/1 MHz = 0.25 [bps/Hz] = $r_C R_C / B$.

As noted, when no FEC coding is used, and a channel transmission rate of R_C [bps] is used over a channel bandwidth of B [Hz], the attained spectral efficiency is equal to $\eta = R_C / B$. However, such a scheme will generally not lead to reception of the information bits at a sufficiently low error rate. To attain an acceptable (lower) error rate, an encoding mechanism is employed through the use of a selected MCS. To achieve a targeted BER, a properly selected MCS, whose code rate is equal to r_C, will be configured. The ensuing spectral efficiency would be equal to $\eta = r_C R_C / B$.

Incorporating a coding scheme, an issue of interest is to determine the highest feasible spectral efficiency level that can be achieved when targeting the reception of data bits to yield a sufficiently low BER.

A mathematical model for studying this problem was developed by Claude E. Shannon in 1948. His developments provided the foundation for the discipline of *Information Theory*. He defined the concept of **Channel Capacity**, denoted as C [bps], as an upper bound measure on the information data rate that can be reliably transmitted across a communications channel that assigns a bandwidth of B [Hz] for the underlying communications, assuming that an arbitrarily low error rate level is achieved at the targeted receiver.

The following result, known as the *Shannon–Hartley Theorem*, and also as *Shannon's Noisy Channel Coding Theorem*, was derived by assuming an **Additive White Gaussian Noise (AWGN)** channel model. Under this model, the signal received at the intended station is modeled as the outcome of the **addition of two signals**: the intended signal and a noise process that is modeled as White Gaussian Noise (WGN). The noise process is characterized by a PSD that is equal to N_0 [watts/Hz]. The allocated bandwidth level is equal to B [Hz]. The result states that the capacity of this channel, C, is calculated as follows:

Shannon's Noisy Channel Capacity Theorem:

$$C = B \log_2(1 + SNR) = B \log_2 \left(1 + \frac{P_S}{P_N} \right) = B \log_2 \left(1 + \frac{P_S}{N_0 B} \right) \text{ [bps]}. \tag{3.8}$$

This is a mathematical result that is derived through the use of a stochastic model, probabilistic considerations, and a random coding approach that involves the encoding of arbitrarily long messages. The random coding method used to derive the result does not lead to the production of practically implementable codes. Over the years, implementable coding schemes that yield data rates that are quite close to those predicted by this formula have been identified.

It is noted however that for many commercial systems, it is essential that the employed MCS modules will be designed so that they are implementable at relatively low complexity and at low cost levels. Never the less, the expression provided by the channel capacity formula is highly revealing and valuable in guiding the design of a communications channel system.

Examining the channel capacity formula presented by the Theorem, we note the following operational modes.

When the SNR value is high ($SNR \gg 1$), the channel capacity level is approximated as:

$$C \approx B \log_2(SNR) = B \log_2 \frac{P_S}{N_0 B} \text{ [bps]}. \tag{3.9}$$

Consequently, under high SNR, the attained channel capacity level is logarithmic in power and linear in bandwidth. An incremental increase in the SNR level induces a logarithmic increase in the capacity rate. In this case, the received signal power is already quite high when compared to the power embedded in the perturbing noise process. An increase in the power of the transmitted signal produces just a limited improvement in the achievable level of the data rate or spectral efficiency. Under this mode of operation, the system is thus noted to operate in a *bandwidth-limited regime*.

In turn, when the SNR value is low ($SNR \ll 1$), the capacity level is approximated as:

$$C \approx \frac{P_S}{N_0 \ln 2} \text{ [bps]}. \tag{3.10}$$

Consequently, under low SNR levels, the attained channel capacity level grows in linear proportion to the received power level of the intended signal, while it is insensitive to the underlying assigned bandwidth level. The system is then said to operate in a *power-limited regime*.

The formula presented by Shannon's Noisy Channel Capacity Theorem is simple and easy to use when performing system design and conducting analysis and performance evaluations. It is also often used to carry out approximate system performance evaluations when the signal's reception at an intended receiver is perturbed by both noise and signal interference processes. For this purpose, we replace in the above formulas the *SNR* term with an *SINR* term, leading to the following formula:

Shannon's Channel Capacity Formula Under Noise Plus Interference:

$$C = B \log_2(1 + SINR) = B \log_2 \left(1 + \frac{P_S}{P_N + P_I}\right) = B \log_2 \left(1 + \frac{P_S}{N_0 B + P_I}\right) \text{ [bps].} \qquad (3.11)$$

Other channel capacity expressions have been employed for other communications channel models. For example, in a fast-fading channel, the corresponding channel capacity expression is obtained by averaging over per-fade channel capacity components. The scenario model is more complex when considering a slow-fading channel model. Other calculations are performed when considering other system and interference models. In certain wireless systems, the aggregate interference signal is modeled at times as a random entity that is governed by a Log-Normal probability distribution function.

Under certain network system operational scenarios, the interference process is stochastically modeled to have long-range dependence (LRD). Occurrence of interference events that may randomly occur on infrequent occasions may be correlated. Such conditions can impact system performance in a significant manner. Fractal (or multi fractal) type stochastic process models have been used in studying such scenarios.

3.8 Modulation/Coding Schemes (MCSs)

3.8.1 The Modulation Concept

A modulation process is used to condition a signal for effective transmission and propagation of information across a communications channel. The modulated signal conveys to the intended receiver the information that has been prepared by a sending user.

Information prepared by a user is presented as a *baseband signal*. The baseband signal, including audio-, video-, and computer-based digital data, may be targeted for transmission across a communications channel to a receiving station (such as a BS or a Wi-Fi AP). The modulation process enables the effective transmission of a user's baseband signal across the communications channel by embedding the baseband information into a carrier signal that provides for the effective propagation of the signal across a communications channel medium.

The modulation process also makes it possible for multiple users to share the channel's frequency spectrum, so that multiple distinct message flows can be transported across a communications channel over overlapping time periods. Different flows can be carried across a channel by using different **carrier signals**. Each signal may occupy a distinct frequency band. A device known as a modulator is used at a station's transmitter to **modulate** a carrier signal by modifying one or several of its parameters in a manner that enables a **demodulator** residing at the intended receiver to extract the baseband information produced by the sending user.

The frequency spectrum of a modulated signal occupies a frequency band that surrounds a central frequency. The latter is the frequency at which the carrier signal operates. The bandwidth of the modulated signal depends on the bandwidth of the information signal that it carries (i.e., of the baseband signal produced by an end-user) and on the type of modulation scheme that is being used.

A **transceiver** is a device that performs both transmission and reception functions. A **full-duplex transceiver** is able to perform signal transmission and reception at the same time, while a **half-duplex transceiver** is capable of performing signal transmission and reception, though not at the same time. The modulating section of a transmission module performs the modulation process while the demodulating section of a reception module executes the inverse function, performing demodulation. A system entity that performs both operations is identified as a **modem**, a device that contains modulator and demodulator modules.

A carrier is represented as an analog sinusoidal signal. Its intensity (magnitude or amplitude) level at time t, denoted as $c(t)$, is expressed as follows:

$$c(t) = A \sin(2\pi f t + \phi), \quad t \geq 0. \tag{3.12}$$

3.8.2 Analog Modulation Techniques

Under analog modulation, an analog information bearing baseband signal is used to modulate a carrier in a time continuous fashion. Common analog modulation schemes include the following:

1) **Amplitude Modulation (AM)**: The amplitude of the carrier signal is varied in accordance with the instantaneous level (such as amplitude value) of the modulating baseband signal. If the user's baseband signal level at time t is denoted as $s(t)$, we set the carrier's amplitude at time t to a value that depends on the magnitude of $s(t)$, denoting it as $A(s(t))$. The AM modulated signal, denoted as $s_{AM}(t)$, is then represented as:
$s_{AM}(t) = A(s(t)) \sin(2\pi f t + \phi), \quad t \geq 0.$

2) **Frequency Modulation (FM)**: The frequency of the carrier signal is varied in accordance with the instantaneous level of the modulating baseband signal. The FM-modulated signal, denoted as $s_{FM}(t)$, is then represented as:
$s_{FM}(t) = A \sin(2\pi f(s(t))t + \phi), \quad t \geq 0.$

3) **Phase Modulation (PM)**: The phase shift of the carrier signal is varied in accordance with the instantaneous intensity of the modulating baseband signal. The PM-modulated signal, denoted as $s_{PM}(t)$, is then represented as:
$s_{PM}(t) = A \sin(2\pi f t + \phi(s(t))), \quad t \geq 0.$

In Fig. 3.3, we illustrate the process used to modulate a carrier. The process performed under AM is depicted in Fig. 3.3(a). The amplitude of the modulated carrier is varied over time in accordance with the amplitude of the modulating baseband signal. In Fig. 3.3(b), we show an illustrative

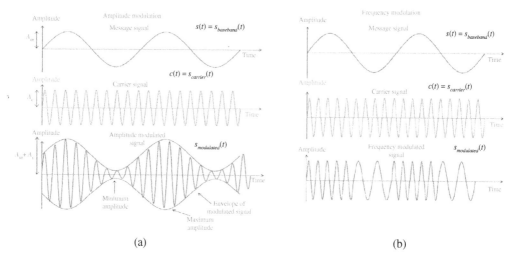

Figure 3.3 Analog Signal Modulation Techniques. Source: https://www.physicsand-radio-electronics.com/blog/amplitude-modulation/; https://www.physics-and-radioelectronics.com/blog/frequency-modulation// with permission of Physics and Radio-Electronics/Last accessed on 08-08-24.

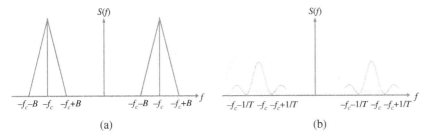

Figure 3.4 Illustrative Spectral Spans of Modulated Signals.

signal that is modulated by using FM. The frequency of the modulated carrier is varied over time in accordance with the amplitude of the modulating baseband signal.

In Fig. 3.4(a), we illustrate the frequency span produced by a modulated carrier signal. While, for illustration purposes, the figure shows the spectral span of a modulated signal to be approximately equal to $2B_b = 2B$, where $B = B_b$ is the bandwidth of the baseband signal, the actual span would depend on the MCS that is employed, as illustrated in Fig. 3.4(b) for a rectangular signal of time span T.

The modulated signal is shaped to contain much of its power in its allocated spectral band. A bandpass filter is used to limit the produced frequency range to this band. This band spans the frequency range $[f_c - B_b, f_c + B_b]$, where f_c is the frequency of the carrier signal. Typically, upper- and lower-frequency guard-bands are included, serving to reduce interference from and to neighboring bands. In this manner, each modulated signal occupies a distinct frequency range, enabling the time-simultaneous transmission of multiple modulated signals. The intended receiver of a signal would filter and process only signals that occupy the corresponding band(s) that it uses. By *correlating* the received modulated signal with its associated carrier frequency, the receiver is able to demodulate the signal, extracting from it the information bearing baseband signal that carries the information that the source user aims to transport to it.

To illustrate the use of such modulation techniques, we next discuss broadcast radio systems. Such systems have generally used AM or FM modulation techniques.

Consider *public radio AM broadcasts*. As noted in Fig. 3.4(a), the modulated signal occupies a frequency band that is equal to $2B_b = 2f_m$, where f_m is the highest frequency of the baseband signal that is designated for transport across the communications channel. Speech signals of good quality, and music signals of medium quality, are assumed to occupy a frequency range that is not higher than $f_m = B_b = 4$ kHz. Hence, the bandwidth that needs to be assigned for AM broadcasts is set to at least $2f_m = 8$ kHz. When including frequency guard bands, we expect different stations to be spaced by about 9–10 kHz. Indeed, channels are spaced at 10 kHz intervals in the Americas, and at 9 kHz intervals in certain other places. We note that AM broadcasts in North America, when using the medium wave frequency range, use the frequency span of 525–1705 kHz, allowing a total frequency span of 1180 kHz, and therefore accommodating up to 118 AM bands.

Across its operational spectral band, an AM signal is susceptible to variety of interference signals, including electrical storms (lightning) and various other electromagnetic interference (EMI).

FM radio broadcasts in the United States (and other countries) use the VHF frequency range of 88–108 MHz. FM broadcasts are designed to carry high-fidelity audio (speech and music) signals, accommodating a baseband signal for which $B_b = f_m = 15$ kHz. The FM operation uses a peak deviation from the carrier frequency that is equal to $\Delta f = 75$ kHz. The frequency bandwidth span B occupied by the FM signal is calculated by using the following computation (known as Carson's Rule): $B = 2\Delta f + 2f_m = 180$ kHz. Incorporating channel guard bands, FM stations are allocated carrier frequencies that are 200 kHz (0.2 MHz) apart. Accordingly, the above-mentioned VHF band accommodates a maximum of 100 FM bands.

The range of frequencies used by AM stations is such that at times, propagating waves may incur reflections from the ionospheric layer of the earth's atmosphere. This occurs mainly during night hours, leading then to longer signal propagation ranges. As a consequence, interference signals may be detected at stations that are located remotely from each other, such that these stations are assigned the same frequency band for *frequency-reuse* purposes. Generally, FM broadcast transmissions propagate along line-of-sight ranges, and are not subjected to ionospheric reflections. Due to their wider bandwidth levels, they are more susceptible to phase dispersion issues.

3.8.3 Digital Modulation Techniques

Under a digital modulation technique, a digital baseband signal is used to modulate an analog carrier. If the original information signal produced by the end-user source is in analog form, as can be the case for various audio and video signals, this signal is digitized, producing a discrete-time and discrete-amplitude signal. The user generated original signal is often already in digital format when generated by a computer based device such as a smartphone. To produce the modulated signal, a Modulation/Coding Scheme (MCS) is employed, proceeding as described in the following. Implementation details would depend on the specific technology being used.

1) **Encoding the Source's Digital Information to Produce Error-Control Blocks**: Due to noise and interference processes that affect the reception of a signal transmitted across a communications channel, information included in a transmitted message may be determined by the receiver to contain errors. For instance, a bit that the receiver has determined to represent a 1 digit may have originally represented (and transmitted as) a 0 digit. To reduce the realized BER, the transmission system employs an **error control** scheme. The **channel encoding** process may employ an *Error Detection Code (EDC)*, which provides the receiver with the ability to determine whether a received message contains errors. A transmission system tends to also use a *FEC code*. Making use of the latter code, the receiver is able to determine whether the received message contains errors as well as to correct those bits that it deems to be in error, provided the received messages contains a limited number of bit errors. The use of a powerful FEC is of particular importance for attaining rapid and reliable data transport for critical delay-sensitive data flows.

   For this purpose, upon receiving a message to transmit across a communications channel, such as a link layer frame or a Medium Access Control (MAC) layer frame, the transmission system proceeds to encode the message data to produce one or several error-control blocks. If the encoded segment is longer than the maximum length set for such a block, the message is segmented into multiple error-control blocks. Each such block is then submitted to the transmission module for modulation and transmission across the communications channel.

   Each error-control block produced by the transmission module consists of two fields: a data field and a code field. The data field contains the message payload bits, or a fragment thereof. Following its transmission, the receiving entity examines the received error-control block. Using the corresponding decoding scheme, it is able to determine if the received block contains errors (in making use of the error-detection capability of the code), and (when a FEC is used), if a limited number of errors occur, it is also able to correct incurred errors (in making use of the error-correction capability of the employed Error Correction Code).

As mentioned, for a rate r_c code, a fraction r_c of the total number of bits included in the error-control block accounts for information (payload) bits, while the remainder bits represent the block's code bits. A code that operates at a lower code rate would generally achieve higher error correction capability. Yet, the correction capability of a code is generally effective only if the total number of incurred errors per block is not higher than a specified limit.

For example, when using a code whose code-rate is equal to $3/4 = 0.75$, 75% of the block's bits are payload bits (so that an average of 3 bits out of every 4 bits are data bits). In turn, when a code-rate $1/10 = 0.1$ is used, only 10% of the block's data represent payload bits. A rate 0.1 code provides a higher error correction capability, so that it can correct a higher number of errors, avoiding more often the need to discard a block or to request for its re-transmission. However, a lower rate code yields a transport operation whose data throughput rate may be lower, as a lower fraction of its transported bits are payload bits. When using a rate 0.1 code, the transmitting entity is using only 10% of the channel capacity level that has been allocated for the transport of its data.

2) **Production of a Channel Symbol**: When using an M-ary modulation scheme, each channel signal (i.e., the signal waveform that is transmitted across the communications channel) carries $\log_2 M$ bits. The transmission of the bits included in an error-control block is performed by segmenting the block into sub-blocks, whereby each sub-block consists of $\log_2 M$ bits. Each such sub-block is represented as a **channel symbol**. For example, when using a 16-level modulation scheme, each symbol consists of $\log_2 M = 4$ bits. In this case, the encoded block is fragmented into consecutive 4-bit symbols. *Each symbol is used to modulate a carrier*. The ensuing modulated signal forms the **channel signal**.

Basic digital modulation schemes commonly employ the following techniques:

1) **PSK (Phase-Shift Keying)**: The modulated channel signal uses M distinct phases to represent the M different values that can be undertaken by the modulating symbol. For example, for $M = 2$, each symbol carried $\log_2 2 = 1$ bit of information, so that it is a binary variable (referred to as a *bit*). The modulated channel signal would then assume the format of a sinusoidal pulse (whose frequency is equal to the carrier frequency) of duration T (its baud time) that assumes one out of two possible phase values, representing the corresponding 0 or 1 information digits, and thus carries 1 bit of information. Under $M = 4$, the modulated signal is a sinusoidal pulse that assumes one out of four possible phase values, corresponding to 2-bit symbols, whereby each symbol represents one of the following four possible groups of information bits: 00, 01, 10, or 11. The modulated signal thus carries $\log_2 4 = 2$ bits of information.

2) **FSK (Frequency-Shift Keying)**: The modulated channel signal uses M distinct frequencies to represent the M different values that can be undertaken by the modulating symbol. For $M = 2$, each symbol carried $\log_2 2 = 1$ bit of information. The modulated channel signal is then represented as a sinusoidal pulse of duration T (its baud time) that operates at one out of two possible frequencies (or tones), say f_1 and f_2, which represent the corresponding 0 and 1 symbol digits. Under $M = 4$, the modulated signal is a sinusoidal pulse that operates at one out of four possible frequencies, such as f_1, f_2, f_3, or f_4, which represent the corresponding 00, 01, 10, or 11 symbols, and thus carries $\log_2 4 = 2$ bits of information.

3) **ASK (Amplitude-Shift Keying)**: The modulated channel signal uses M distinct amplitude levels to represent the M different values that can be undertaken by the modulating symbol. For $M = 2$. The modulated channel signal is represented as a sinusoidal pulse (the carrier signal) of duration T (its baud time) that assumes one out of two possible amplitude values, representing the corresponding 0 and 1 symbol digits.

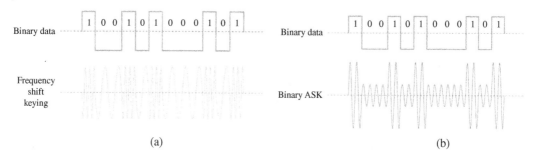

Figure 3.5 Illustrative Digital Signal Modulations. Source:https://www.technologyuk.net/telecommunications/telecom-principles/digital-modulation-partone.shtml#ID04andID06/with pemission of TechnologyUK/Last accessed on 08-08-24.

4) **QAM (Quadrature AM)**: Under QAM, an in-phase signal (referred to as an I—signal, using for example a cosine waveform) and a quadrature-phase signal (referred to as a Q—signal, using a sine wave) are amplitude modulated and then summed. It can be seen as a two-component system, whereby each component is using ASK. The resulting signal is equivalent to a combination of PSK and ASK.

In Fig. 3.5(a), we show illustrative signals involved in the operation of an FSK scheme. A 0 symbol is represented by a sinusoidal waveform operating at one frequency level while the 1 symbol is represented by a sinusoidal waveform operating at a second frequency level. A binary ASK modulated signal is shown in Fig. 3.5(b) where a nonzero signal amplitude is used to represent a binary 0 symbol.

The frequency spectrum of a modulated pulse of duration (baud time) T [s] is shown in Fig. 3.4(b). It is noted that the bulk of the energy of the modulated signal is included in the main lobe of its spectral power distribution curve. Accordingly, the (minimal value of the) frequency bandwidth of such a modulated signal is approximately set equal to $B = 2/T$ [Hz].

Additional factors are often taken into account in determining the bandwidth level that needs to be assigned for sustaining the effective transmission of a modulated signal across a communications channel. It is essential to account for signal interference effects that could be detected at a receiver from transmissions of other signals. To assure the reception of a channel signal at an acceptable BER, a minimal SINR level must be present at the receiver. To reduce interference, systems use frequency guard bands, consequently increasing the actual frequency bandwidth that is allocated for message transmission. To account for such overhead, we denote the effective allocated bandwidth per channel signal as B_a and express its approximate value as $B_a = B(1 + g)$, where g is a factor that represents an excess bandwidth factor.

The **symbol rate**, denoted as R_s [symbols/s], also identified as the **baud rate**, or just as **baud**, represents the rate at which symbols are transmitted over a communications channel. The channel rate, denoted as R_c [bits/s], represents the rate at which message bits, including data and code bits, are transmitted over a communications channel. The channel data rate, denoted as R_d [bits/s], represents the rate at which message data bits, excluding code bits, are transmitted over the communications channel. The spectral efficiency attained in the transmission of channel bits (or symbols), denoted as η_c [bps/Hz], represents the realized channel rate per unit bandwidth that is occupied by the underlying signals. The spectral efficiency attained in the transmission of data bits, denoted as η_d [bps/Hz], represents the realized data rate per unit bandwidth.

For the above-described case whereby an *M*-ary digital modulation is used, the modulated signal is a pulse of duration T [s] that occupies a bandwidth of $B_a = (1 + g)2/T$ [Hz]. In this case, we deduce the following relationships:

$$\text{symbol rate} = R_s = 1/T \ \text{symbols/s}$$
$$\text{channel rate} = R_c = \frac{\log_2 M}{T} \ \text{bits/s}$$
$$\text{data rate} = R_d = r_c \frac{\log_2 M}{T} \ \text{bits/s}$$
$$\text{channel spectral efficiency} = \eta_c = R_c/B_a = \frac{\log_2 M}{2(1 + g)} \ \text{bits/s/Hz}$$
$$\text{channel data spectral efficiency} = \eta_d = R_d/B_a = r_c \frac{\log_2 M}{2(1 + g)} \ \text{bits/s/Hz}. \tag{3.13}$$

Other modulation schemes are often employed. For example, a *Minimum Shift Keying (MSK)* scheme belongs to a family of continuous-phase frequency shift keying (CPFSK) schemes. A symbol is represented by a pulse whose phase is linearly increasing.

Digital baseband modulation (or digital baseband transmission) is also identified as *line coding*. A digital bit stream is transferred over an analog baseband channel as a train of pulses by directly modulating the level of voltage or current used across a medium such as a cable or a serial bus. Common examples include unipolar, non-return-to-zero (NRZ), Manchester and alternate mark inversion (AMI) coding schemes.

Under an *Orthogonal Frequency Division Multiplexing (OFDM)* scheme, the information stream is split into several parallel data sub-streams. Each component sub-stream is transferred over its own sub-carrier, using a common digital modulation scheme. The modulated sub-carriers are summed to form an OFDM signal. The resulting information stream is transmitted as a single flow over the communications channel.

Protocol-based control header and trailer fields included in message segments account for overhead ratios, inducing throughput reductions. Compression algorithms are applied at times to reduce the lengths of selective control fields. This is particularly useful when considering messages that contain short data payloads, such as voice packets.

3.8.4 Illustrative Digital Modulation/Coding Schemes

In the following, we illustrate the use of MCSs employed in certain versions of network system technologies.

3.8.4.1 Modulation/Coding Schemes Used by a Wi-Fi Version

In Fig. 3.6, we list MCSs, and the corresponding attained data rates, employed by a Wi-Fi WLAN system that follows the IEEE 802.11n Standard recommendation, for 20 MHz frequency band and a 0.8 µs guard interval (GI) (see Tables 20-29 and 20-30 in [24]). This system uses an OFDM channel transmission operation. It also provides for the possible use of multiple-beam transmit and receive antennas. Under a Multiple-In Multiple-Out (MIMO) operation, the communications channel uses a space division multiplexing (SDM) approach to carry out simultaneous transmissions and corresponding targeted receptions of multiple streams. Up to four multiple streams can be configured by this technology. When N multiple streams are used, the achieved data rate is increased by a factor of N. Implementation requirements for a MIMO system are more demanding, as a separate radio frequency chain and signal processing module is employed for setting-up and processing each stream.

MCS index	# streams	Modulation	Coding rate	Data rate (Mbps)
0	1	BPSK	1/2	6.50
1	1	QPSK	1/2	13.00
2	1	QPSK	3/4	19.50
3	1	16-QAM	1/2	26.00
4	1	16-QAM	3/4	39.00
5	1	64-QAM	2/3	52.00
6	1	64-QAM	3/4	58.50
7	1	64-QAM	5/6	65.00
8	2	BPSK	1/2	13.00
9	2	QPSK	1/2	26.00
10	2	QPSK	3/4	39.00
11	2	16-QAM	1/2	52.00
12	2	16-QAM	3/4	78.00
13	2	64-QAM	2/3	104.00
14	2	64-QAM	3/4	117.00
15	2	64-QAM	5/6	130.00

Figure 3.6 Modulation/Coding Schemes Used by IEEE 802.11n Wi-Fi.

This system can use a single 20 MHz wide channel as well as a 40 MHz wide channel. The latter doubles the transmission rate. As is the case for other Wi-Fi versions, the system can operate in the 2.4 GHz or at the 5 GHz frequency range.

Each employed MCS, combined with a number of configured MIMO streams, is identified in the table by an MCS index value. For example, Index value 2 identifies a system that uses QPSK modulation (i.e., a 4-level PSK modulation scheme) and a FEC code operating at rate $r_c = 1/2$. Two separate Guard Interval (GI) values are used, providing for proper separation between sub-carriers (recalling that each symbol modulates a separate data sub-carrier).

Considering the 20 MHz allocation, we note the following system parameters. A total of 56 OFDM subcarriers are used, whereby 52 are used for data and 4 are pilot tones, with a carrier separation of 0.3125 MHz (=20/64 MHz). Each of these subcarriers supports the transmission of a symbol, using the noted modulation scheme, including BPSK, QPSK, 16-QAM, or 64-QAM. Out of a total bandwidth of 20 MHz, the occupied bandwidth is equal to 17.8 MHz. Symbol duration is equal to 3.2 µs, to which a guard interval (GI) of duration 0.4 or 0.8 µs is added.

We next focus on the setting that employs a 20 MHz band and a 0.8 µs guard interval. Under MCS index 0, the transmission operation uses a single stream, BPSK modulation (i.e., M = 2, a 2-level modulation) and a rate $r_c = 1/2$ code. Each symbol thus carries $\log_2 M = 1$ channel bit/symbol, and $r_c \log_2 M = 0.5$ data bits/symbol. The data rate is calculated to be equal to 6.5 Mbps. The gross spectral efficiency is calculated to be equal to $\eta_{gross} = 6.5$ Mbps/20 MHz = 0.325 [bps/Hz]. The spectral efficiency attained by accounting only for the bandwidth occupied by 52 subcarriers, considering then a bandwidth level that is equal to $52 \times 0.3125 = 16.25$ MHz, is calculated to be equal to $\eta_{data} = 6.5$ Mbps/16.25 MHz = 0.4 [bps/Hz].

The data rate values attained under other employed MCSs relate to those calculated for the case of modulation index 0. For example, considering the other single stream MIMO schemes, we note that the data rates are related to that attained by the index 0 MCS case by a factor that accounts for the two involved components: the number of $\log_2 M$ bits/symbol and the code rate r_c, through their product $r_c \log_2 M$, which represents the number of data bits per symbol. As the corresponding values of the later for MCS index 0 is equal to 0.5, we conclude that the data rate attained by each one of the other MCS schemes relates to that attained by MCS index 0 setting by being higher than it by a factor that is determined by number of streams $\times 2r_c \log_2 M$, assuming the same GI and bandwidth levels to be employed (i.e., 0.8 μs and 20 MHz, respectively). For example, under MCS index 1, we conclude the rate to be higher by a factor of 2 as each symbol carries 2 bits rather than 1 bit. Under MCS index 7, a 64 QAM modulation scheme is used, so that $\log_2 M = 6$ [bits/symbol], and the code rate is equal to 5/6 leading to $r_c \log_2 M = 5$ [data-bits/symbol], as compared with the level of 0.5 data-bits per symbol achieved under MCS index 0. Hence, the data rate achieved under MCS index 7 is higher than that achieved under MCS index 0 by a factor of 10, leading to a data rate of 65 Mbps. When a MIMO arrangement is used, the data rate is further increased by the MIMO gain factor, which accounts for the number of configured simultaneous streams.

How does a Wi-Fi system determine which MCS should be employed? As we have seen earlier, the effective data rate and spectral efficiency that can be achieved in transmitting messages across a communications channel depends on the SINR value that is measured at the receiving station at the time that a transmission takes place. In a Wi-Fi system, the SINR value at the receiver is generally not available to the transmitting station. Consequently, Wi-Fi systems would estimate it and accordingly adjust the employed MCS configuration. Under such an operation, the MCS is selected dynamically by the transmitting module based on its observation of the behavior of the transmission process. This is referred to as an **Adaptive Modulation/Coding Scheme (AMCS)**.

One version of such an operation proceeds as follows. The sending entity starts by selecting an efficient MCS (aiming to attain high spectral efficiency and data rates levels). It then monitors the flow of positive acknowledgment (PACK) frames that it receives in response to its data frame transmissions. If no PACK is received after repeating the transmission of a data frame for a specified number of times, the transmission module will switch to a lower efficiency MCS. This process is continued until a positive ACK is received (or until a maximum number of attempts is reached). Thereafter, if the transport process is monitored to proceed well for a certain number of data frame transmissions, the sending mechanism switches to a higher efficiency MCS setting. The configured efficiency is progressively increased for as long as no new data frame retransmissions are imposed.

3.8.4.2 MCS Configurations for an LTE Cellular Wireless Radio Access Network

In the following, we illustrate an MCS selection process by considering a version of the RAN of a 4G LTE cellular wireless network system. Downlink wireless channels are used for communications from a BS to mobile users that reside in its cell. Uplink wireless channels are used for communications from mobile users to their associated BS.

The allocation of channel resources and the determination of the MCS to be employed for downlink and uplink transmissions within a cell, are performed by the cell's managing BS, which is identified in 4G LTE documents as eNodeB. In Fig. 3.7 [1], we show the system parameters that are used by the BS for configuring its MCS settings for downlink transmissions. For this purpose, the BS needs to determine the current SINR level recorded at the intended destination mobile user, identified in 4G LTE documents as a UE, a User Equipment entity.

The BS transmits periodically downlink reference signals that mobile users employ to perform signal power measurements. The result is expressed by each user as a **Channel Quality Indicator (CQI)**, which is sent to the managing BS. As demonstrated in Fig. 3.7(a), a higher CQI number indicates to the BS that its signal is received at the user at a higher SINR (also expressed as SNR) level. For example, if the user's receiver determines its SINR level to be equal to about 10 dB, the reported CQI value will be set to 10. If at a later time, the user moves away from the BS node to a location at which a weaker BS signal is received, and/or a higher interference signal has appeared, and the measured SINR level is degraded to about 3 dB, the corresponding CQI value is then reduced to 6.

116 | *3 Transmissions over Communications Channels*

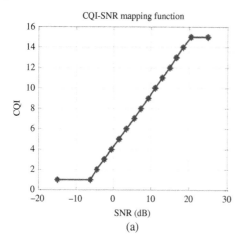

CQI index	Modulation	Code rate × 1024	Efficiency
0	Out of range		
1	QPSK	78	0.1523
2	QPSK	120	0.2344
3	QPSK	193	0.3770
4	QPSK	308	0.6016
5	QPSK	449	0.8770
6	QPSK	602	1.1758
7	16QAM	378	1.4766
8	16QAM	490	1.9141
9	16QAM	616	2.4063
10	64QAM	466	2.7305
11	64QAM	567	3.3223
12	64QAM	666	3.9023
13	64QAM	772	4.5234
14	64QAM	873	5.1152
15	64QAM	948	5.5547

(a)   (b)

Figure 3.7 CQI, MCS, and Spectral Efficiency for LTE. Source: 3GPP TS 36.213, v. 8.3.0/with permission of 3GPP.

The BS uses the reported CQI value to determine the MCS configuration to be used for the transmission of data to the reporting user, depending upon the targeted error rate value that it is aiming to achieve. For the illustrative example, a frame (identified in this system as a "transport block") error rate of 10% is specified. As demonstrated in Fig. 3.7(b), when a higher CQI value is reported by a user, the BS is able to select a more throughput-efficient MCS by setting a higher modulation level (M) and/or a higher code rate.

It is noted that in LTE, OFDM is used as the basic signal format. An OFDM method is used in the downlink. Often, a QAM modulation is then used for each sub-stream leading to OFDM-QAM. The OFDM technique leads to high peak power to average power ratio at the transmission module. Induced by the requirement to attain an efficient power use (and energy consumption) operation at a mobile station while conducting its uplink transmissions, extending battery lifetime, a Single Channel Orthogonal Frequency Division Multiple Access (SC-FDMA) scheme is employed in the uplink.

Example 3.1 To illustrate, we consider the following scenario. When the reported CQI is equal to 10, the corresponding $SINR_{dB}$ value is equal to 10 dB, and the BS employs a 64 QAM modulation scheme, for which $M = 64$, and each channel signal carries $\log_2 64 = 6$ [bits/symbol]. The corresponding coding scheme that is employed operates at a code rate of $r_c = 466/1024 = 0.455$. The ensuing spectral efficiency is determined to be equal to $\eta_d = 2.7305$ [bps/Hz]. Hence, if the BS allocates a bandwidth level of $B = 10$ MHz for this transmission, the attained data rate would be equal to $R_d = \eta B = 27.305$ Mbps.

The upper bound value for the spectral efficiency, when aiming to achieve a very low error rate level, is calculated by using Shannon's formula. It states that $\eta = \log_2(1 + SINR)$. For $SINR_{dB} = 10$ dB, we have $SINR = 10$, so that the power of the received user signal is 10 times higher than the power of the aggregate noise plus interference signal. Shannon's formula then yields the upper bound on the attainable spectral efficiency to be equal to $\eta = \log_2 11 = 3.459$ [bps/Hz]. In comparing this value with that realized by the underlying LTE scheme, we observe that the LTE system achieves a spectral efficiency level that is equal to about 79% of that calculated by the upper bound, noting however that the LTE system's configuration specifies a much higher bound on the message error rate level.

Assume next that the mobile user has moved away from its previous location and is now experiencing higher noise and interference conditions and/or lower data signal power. The user is now reporting to the BS a lower CQI level, which is hereby assumed to be equal to 6. The corresponding SINR level is now reduced to $SINR_{dB} = 3$ dB. Accordingly, the BS now employs a QPSK modulation scheme, for which $M = 4$, so that each channel signal carries $\log_2 4 = 2$ [bits/symbol]. The employed coding scheme operates at a code rate of $r_c = 602/1024 = 0.588$. The attained spectral efficiency level is reduced to $\eta_d = 1.1758$ [bps/Hz]. Hence, if the BS allocates a bandwidth level of $B = 10$ [MHz] for this transmission, the attained data rate is reduced to $R_d = \eta B = 11.758$ Mbps. If the BS wishes to maintain the same transmission data rate value, it would have to increase the frequency bandwidth level that is allocated for this transmission. This increase would be equal to $2.7305/1.1758 \approx 2.32$, so that the allocated bandwidth would increase by a factor of about 232%.

Problems

3.1 Identify several wireline communications media and describe the key features of each one.

3.2 Identify several wireless communications media and describe the key features of each one.

3.3 Consider a station whose receiver detects the following signals: an intended message signal that is received at a power level of 40 [mWatts]; background noise at a power level of 1[mWatts]; an interference signal at a power level of 5 [mWatts]. Calculate the Signal-to-Interference-plus-Noise-Ratio (SINR) [dB] level.

3.4 Consider a wireless communications channel whose parameters are equal to those presented in Fig. 3.6, except for the modifications noted below. Using the signal attenuation model presented by Eq. 3.4, calculate the corresponding SINR [dB] levels detected at the receivers under the following scenarios. Compare and discuss the results.
   a) Scenario 1: The transmit power is set to 0.5 [Watts], and no interference signal is detected at the receiver.
   b) Scenario 2: The transmit power is upgraded to 10 [Watts], and the detected power of an interference signal is equal to $P_I = -46.11$ [dBm].

3.5 Using Shannon's Channel Capacity formula, calculate the attainable spectral efficiency levels for each one of the scenarios presented in the preceding Problem.

3.6 Discuss the operation of the following digital modulation techniques: FSK, PSK, ASK, and QAM. Illustrate the operation of each technique when set to convey 3 [bits/symbol].

3.7 Consider a cellular wireless radio access network whose parameters assume the values presented in Fig. 3.7. Assume the base station to obtain a report that identifies a mobile's receiver to determine its reception channel state to be equal to CQI = 6. Calculate the spectral efficiency level achieved by a message signal transmitted by the base station across this channel to this mobile user. Also, compute the ensuing transmission data bit rate when a bandwidth level of 5 MHz has been allocated for this purpose.

3.8 Consider a mobile whose receiver determines the communications channel used for its message reception to be interference dominated.

a) Describe the implication of this situation, and present an expression that can then be used to provide an approximate calculation of the attained spectral efficiency level by using Shannon's channel capacity formula.

b) Calculate the attained spectral efficiency level when the power of the received intended message signal is higher by 4 dB than the power of the detected interference power.

c) What would be the impact on the attained spectral efficiency level if the power level of each signal (including intended and interference ones) decrease by 20%?

3.9 Consider a mobile whose receiver determines the communications channel used for its message reception to be noise dominated.

a) Describe the implication of this situation, and present an expression that can then be used to provide an approximate calculation of the attained spectral efficiency level by using Shannon's channel capacity formula.

b) Calculate the attained spectral efficiency level when the power of the received intended message signal is higher by 10 dB than the background noise power.

c) What would be the impact of the attained spectral efficiency level if the power level of the intended signal decreases by 20% (while the background noise power does not change)?

4

Traffic Processes

Objectives: *A network system serves to transport message flows and streams that originate at source nodes and are directed for arrival at destination nodes. To size the capacity of resources that are required for such an operation and to synthesize and tune-up the employed transport mechanisms and algorithms, it is essential to statistically characterize the traffic flows that arrive at network nodes and that load communications links, routers, and other networking elements and processors. Stochastic process methods are used to model traffic processes, representing the message loads that are imposed on network entities by each flow class. Such characterizations are used to calculate service demand levels imposed by traffic processes on network links and nodes.*

Methods: *The statistical activity of a message flow is modeled by considering its behavior over distinct time scale levels. A flow's call (or session) level activity is characterized over a longer-term time scale. A stochastic point process is used to model the random times at which sessions are initiated, and session lengths are represented as random variables. Burst events are defined, for each involved traffic class, over a medium-term time scale. Bursts are characterized as in-session processes that alternate between activity and inactivity periods. The random times at which burst activities are initiated are represented as a stochastic point process. The lengths of in-session activity bursts are modeled as random variables. Message flows are statistically characterized over a short-term time scale. A stochastic point process is used to model the random times at which message arrivals occur, and probability distributions specify the quantity of data carried by a message. A traffic process specification is used to calculate the level of service that message flows require from nodal and link entities that they encounter while flowing across a network route. The random times at which messages arrive at a service element are often modeled as a Poisson point process. Traffic matrices are used to specify the spatial distribution of message flows across a network system.*

4.1 A Multilevel Traffic Model

As it is engaged in an underlying application, an end-user generates messages that it sends to a destination party across a communications network. During a period of activity, the user would produce a train of messages that are sent across the network for transport to a destination user. These **message flows** load a communications network.

Consider a *flow of vehicles* across a highway. Each vehicle is analogous to a message, and the processes that represent successive vehicles entering and moving along a highway are analogous to message traffic processes. Vehicle traffic flows are depicted in Fig. 4.1, where we note a cars moving along a highway over multiple lanes. One notes the statistical variations of distances that separate variable size vehicles that are moving along each highway lane. These entities are

Principles of Data Transfer Through Communications Networks, the Internet, and Autonomous Mobiles, First Edition. Izhak Rubin.
© 2025 The Institute of Electrical and Electronics Engineers, Inc. Published 2025 by John Wiley & Sons, Inc.

Figure 4.1 Traffic of Vehicles Moving Along a Highway.

analogous to variable size data messages that arrive at random times following inter-vehicular random spacing distances.

How do we mathematically model a traffic process?

The statistical characterization of a traffic process that represents a flow of messages that are produced by a user depends on the user's employed application mix and its User Activity Profile (UAP). Users that follow a similar activity profile are categorized as members of the same **User Class (UC)**. Illustrative UCs include: Small Office/Home Office (SOHO), corporate, transportation, enterprise, medical, fire, police, first responders, educational, remote learning, autonomous vehicles, robotics, and Internet of Things (IoT) sensor-based entities.

When engaged in an application of a specific category, a user produced traffic process is characterized as an instance member of a corresponding **Traffic Class (TC)**. Statistical features for such a traffic process can be derived and updated by monitoring the corresponding traffic flows. Illustrative TC types include: Packet Voice (VoIP), full motion video streaming, medical imaging, email data, client–server-oriented request–response, interactive dialogues, signaling, network management, network control, file transfer, airline reservation, web download, and multimedia.

A multilevel traffic model is shown in Fig. 4.2. Under this model, which was developed by Professor Izhak Rubin for use in the PLANYST modeling and analysis software program, the temporal characteristics of the traffic process produced by an information source are described in terms of their dynamic evolution over three time granularity levels: Call/Session, Burst, and Message/Packet levels.

1) **Call/Session Level**: A temporally long-term description of the dynamics of the traffic process. Over certain periods of time, the user is engaged in a Call, or is logged-on while in- session.

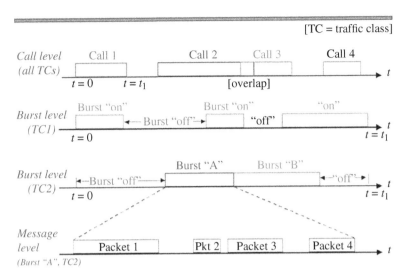

Figure 4.2 Multilevel Traffic Model.

For users that are members of a given UC, call/session rates, which relate to the average time interval incurred between call start times, and the average call time duration, and often other related statistical characterizations, are described.

For example, a user may generate a VoIP call during a Peak Busy Hour (PBH) at a rate of 2 [Calls/Hour], whereby each call is, on the average, 10 minutes long. In this case, the user generates, on the average, a new call every 30 minutes and each call lasts an average of 10 minutes. An activity period of average length of 10 minutes is followed by an idle period of average length of 20 minutes. This user is active at a duty-cycle ratio that is equal to $10/30 = 33.3\%$. The actual activity ratio may be lower as the user may be involved in producing message traffic for only a fraction of the time while in an on-call state, as represented by the user's activity rates during the burst and message levels.

2) **Burst Level**: A temporally medium-term description of the dynamics of the traffic process. During an in-session (call-on) period, messages are produced only during certain periods of time, which are identified as burst periods. The statistical characterization of burst activity, including specifications of the statistics of burst time lengths and of the time intervals incurred between the start of burst activities, depend on the associated TC.

In Fig. 4.2, we show the user activity to involve two TCs, TC1 and TC2. Burst-on and Burst-off periods are noted for TC1 and TC2. The model shown for TC2 burst activity process accounts for two burst types, type-A and type-B bursts. For example, the user may be involved in a joint voice/data multimedia activity. TC1 may represent a voice process. The burst-on period for TC1 may have an average duration of 350 ms, and be governed by an Exponential distribution. The burst-off period for TC1 may have an average duration of 650 ms, and be also governed by an Exponential distribution. A burst level duty-cycle of 35% thus ensues for TC1. In turn, TC2 may model a data activity process. Its burst level dynamics alternate between burst-off, burst-on-A, and burst-on-B periods, whereby the corresponding periods may last for average times of 800 ms, 150 ms, and 50 ms, respectively. In this case, a burst level duty-cycle of 20% is noted for TC2.

3) **Message/Packet Level**: A temporally short-term description of the dynamics of the traffic process. During an in-session (call-on) period, while a burst activity (a burst-on period) is incurred, messages are produced in a manner that follows the statistical features of the underlying TC.

4 Traffic Processes

Example 4.1 Considering the activity shown in Fig. 4.2, we examine the following illustrative characterizations. TC1 may model, as noted above, a VoIP process. During its (burst-on) activity period, voice packets are constructed and transmitted across the network. Assume that a PCM voice encoder is employed, using a sampling rate of 8000 [samples/s], and that each sample is encoded by using 8 [bits/sample] quantization. Every 20 ms a VoIP packet is formed, so that a packet contains 160 samples, yielding a packet length of 1280 bits. Assuming continuous voice activity during a burst-on period, a VoIP packet is generated every 20 ms during this period.

The time that it takes to transmit a VoIP packet over a communications channel, such as one that provides access to a network system, depends on its transmission data rate. Assume that this data rate is equal to 10 Mbps. The transmission time of a VoIP packet is then equal to 1280 [bits/packet]/10 [Mbits/s] = 0.128 ms. In this case, the fraction of time that the access network is utilized by TC1 packet transmissions is equal to 0.128 [ms]/20 [ms] = 0.64%. Since control bits will be added to the voice packet, the utilization ratio at the packet level will be somewhat higher, by say 10%, so that it will be approximately equal to 0.7%. The remaining 93.3% of the time is available for TC2 transmissions by this source as well as for message transmissions by other sources that share this communications channel.

Under TC2, the source generates data type traffic. For the model shown in Fig. 4.2, during a burst-A-on activity, the source generates data packets, whereby each data packet may contain a random number of bits. They will be transmitted across the channel when ready, which will happen at random instants of time. For example, consider an application under which, during the latter burst period, packets are fed into the network at a rate of 200 [packets/s], whereby each packet contains an average of 1000 bits. In this case, during TC2 burst-A activity of average duration of 150 ms, the source feeds into the network 200 [packets/s] × 0.15 [s] = 30 data packets. If the data rate at which packets are transmitted across the access channel is equal to 10 Mbps, the time that its takes to transmit 30 data packets, each containing an average of 1000 bits, is equal to 30 [kbits]/10 [Mbps] = 3 ms. Hence, a burst-A-on duration of 150 ms is utilized for the transmission of data packets for a fraction of 3/150 = 2% of the burst-A time.

4.2 Message Traffic Processes

A message-level traffic process represents a flow of messages over a period of time, taking place at an identified system location or across a specified link, path, or region. Such a location can be selected as an output port of a station that is used by an end-user source. A traffic process of interest can, for example, include the aggregate (superimposed) message traffic produced by a group of end users. A message traffic flow may also be evaluated at an input port(s) of a network element, such as a router. A network router tends to receive message flows at random times from neighboring routers and from attached end-user sources.

A class of message traffic flows is statistically characterized by following a relevant traffic model with specified parameters that are identified as its traffic flow attributes. Illustrative such attributes include: TC, source–destination spatial identifiers, application types, and UC. Characterizations may relate to a flow's statistical activity over various time periods, such as over a PBH or over a Peak Busy Minute (PBM).

The dynamic behavior of a message traffic process is specified in a **statistical** manner. The specification is typically used to characterize a communal category of flows. Statistical data may be collected through the monitoring of past and/or recent flows of the same type. In the following, we describe a method that is used to specify a *stochastic model* for a message traffic flow.

Consider a flow that consists of data packets that arrive at a router during a PBH at a rate of 500 [packets/s]. Each packet contains an average of 1200 bits. The probability distributions of packet transmission times and of packet inter-arrival times are prescribed.

In the following, we describe a statistical characterization of a message flow, assuming the flow to consist of *data messages whose lengths are mutually statistically independent.*

This tends to be the case when observing the superimposed aggregate of traffic flows at an input to a network element, such as a router. These flows may belong to a single or a mix of several TCs.

Communications links that feed message flows into a network router may operate at relatively high data rates. Messages arriving along these links originate at wide ranging sources. The rate of messages produced by a given source is typically much lower than the data rate of an internal network communications link. Hence, during a relatively short period of time, a router may be fed by messages that originate by a wide range of distinct sources. The message production processes at the latter sources tend to be statistically independent.

As noted above and in previous sections, the dynamics of *realtime traffic processes*, such as those produced by VoIP, imaging, and video sources, can be well characterized at the output of a source by using the multi-level traffic model described above. For example, consider the burst level-based on-off process that models the activity of a VoIP source. The average length of the burst-on period is equal to 350 ms and the average duration of the burst-off period is equal to 650 ms. Burst-on and burst-off period lengths are statistically independent and each is assumed to be governed by an Exponential probability distribution. At the packet level, during a burst-on period, we noted VoIP packets to be produced at fixed intervals, so that a packet is produced, for example, every 20 ms. Each packet contains 160 voice samples, yielding a packet length of 1280 data bits, to which control bits are added.

As another example, consider a full-motion realtime video stream. Under a frame rate of 30 [frames/s], a frame is generated about every 33 [ms]. The information embedded in each frame is compressed. The number of bits included in a frame depends on the spectral characterization of the spatial content included in the frame. Similarly for the bit contents of a Group of Pictures. Compressed data is assembled to form segments that are known as message chunks. A chunk may contain a quantity of information that is produced (and compressed) by a video player over a specified period of time. Different message chunks may thus contain different numbers of bits. To transport video chunks across a packet switching communications network, the chunks are segmented into packets. A larger number of packets will be created when a dynamically active video scene is involved. Such a video source produces Variable Bit Rate (VBR)-type packet flows.

The illustrated VoIP-voice and IP-video real-time streams follow the above noted characterizations when observed at the output of their originating sources. Yet, when observed at an input to a router located inside a communications network, each flow's data rate constitutes just a small fraction of the data rate capacity of the internal communications link across which it is transmitted. Voice and video packet flows arriving to a router from a multitude of sources are mixed and multiplexed as they traverse internal links. The packet process that loads a router tends to consist of a mix of randomly arriving packets, whose lengths are often statistically independent. The traffic model described below is frequently used to characterize message flows that may be produced by a single source of a single type or by a statistical mix of traffic source types.

4.3 Modeling a Traffic Flow as a Stochastic Point Process

The basic characterization of a message flow process observed at a specific location involves two key elements: the times at which messages are deemed to have arrived at this location, identified as

message arrivals times and the sizes (or lengths) of messages arriving at this location, identified as **message lengths**.

This characterization is also applicable when considering general traffic process types, whereby other entities than messages are used, such as events, customers, vehicles, general protocol data units (PDUs). To simplify the description of a corresponding traffic process model, we use the following terminology: traffic flow associated *event occurrence times* are referred to as *arrival times* and the arriving entities are identified as *messages*.

Consider messages arriving to a system. The nth message, $n = 1, 2, \ldots$, to arrive is said to arrive at time A_n and its size is equal to L_n. The size of a message can be measured in terms of the total number of bits that it carries. We set $A_0 = 0$ as a time of reference relative to which we describe the times of other message arrivals. The first message arrives at time A_1 [s], and it carries a total of L_1 [bits].

The lengths of messages that are expected to arrive to the system may not be known a priori. They are represented as *random variables*. Under the basic traffic model considered henceforth, these variables, represented by the stochastic sequence $\{L_1, L_2, \ldots\}$, are assume to be **identically distributed and statistically independent**, denoted as **i.i.d.**, random variables. The length of any message is statistically independent of the length of any other message and the lengths of any two messages follow the same probability distribution. Hence, under this model, while their lengths are not necessarily the same, they are governed by the same probability distribution.

Therefore, under this model, the lengths of different messages included in the underlying traffic flow (or process) have the same average length (also termed as their mean) and variance values. The message length distribution is specified by a probability mass function $l(i) = P\{L = i\}, i \geq 1$, whereby $l(i)$ represents the probability that a message contains i bits. Note that the random variable representing the message length has been denoted as L without referring to its order (n) of arrival, as all messages included in this process follow the same length statistics. Given the message length probability distribution function (or its above-noted probability mass function), one readily calculates the message average (or mean) length using the Expectation operator E, which is denoted as $E(L)$, as well as its Variance, denoted as $Var(L)$, and Standard Deviation $\sigma(L) = \sqrt{Var(L)}$.

Consider the following illustrative example. A message flow has been observed to consist of messages whose lengths are specified by a probability mass function (or distribution) that is characterized as follows: 10% of the messages are such that each contains 12 kbits, 40% of the messages are such that each contains 6 kbits, while the remaining 50% of the messages are such that each contains 1 kbits. The average message length (for this flow) is calculated to be equal to $E(L) = 0.1 \times 12K + 0.4 \times 6K + 0.5 \times 1K = 4.1$ [kbits/message].

The standard-deviation of the message length is calculated to be equal to 5.413 [kbits/message]. Note that in this case, there is a noticeable fluctuation in the message length. The ratio of the standard deviation to the mean is identified as the **Coefficient of Variation = COV** of the random variable, so that $COV(L) = \frac{\sigma(L)}{E(L)}$. It is used as a measure of statistical fluctuation in the value assumed by the random variable. When $COV \ll 1$, the fluctuation level is low, while for $COV > 1$, the variation level is higher. When the variable assumes a deterministic (i.e., fixed) value, we have $COV = 0$. For the underlying example, we obtain $COV = 1.32 > 1$. Such a fluctuation level is relatively high, noting that 10% of the messages assume a length that is equal to $\frac{12}{4.1} = 2.9$ times the average length.

The arrival traffic process that represents the times at which messages arrive to a system is represented as the stochastic sequence $A = \{A_0, A_1, A_2, \ldots\}$. As noted above, A_n represents the time at which the nth message arrives to the system, noting that $A_{n+1} > A_n$. We set $A_0 = 0$ and assume, for mathematical convenience, time 0 to be a reference time at which an *uncounted arrival* takes

Figure 4.3 Realizations of (a) Continuous-Time and (b) Discrete-Time Arrival Point Processes.

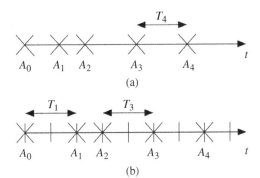

Figure 4.4 A Realization of a Counting Process Associated with the Displayed Point Process.

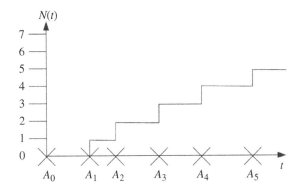

place. Realizations of continuous-time and discrete-time arrival processes are depicted in Fig. 4.3. The times at which messages arrive to the system are represented as points along a time axis. Such a process is called a **stochastic point process**.

Also depicted in Fig. 4.3 are the intervals that represent the times elapsed between successive arrival times. The nth inter-arrival time is denoted as T_n. It represents the time incurred between the arrival times of the $(n-1)$-st message and the nth message. We thus have:

$$T_n = A_n - A_{n-1}, \; n \geq 1. \tag{4.1}$$

Associated with a stochastic point process A is a **stochastic counting process** $N = \{N(t), t \geq 0\}$, where $N(t)$ represents the number of arrivals that take place in the time interval $(0, t]$. Under this definition, we exclude an arrival that occurs at time 0. A realization of a counting process associated with a displayed realization of the point process is shown in Fig. 4.4. A sample function of a counting process is a nondecreasing step function as the total number of arrivals that occur over a period $(0, t + s]$ includes those arrivals that occurred over the sub-interval $(0, t]$, for each $s \geq 0, t \geq 0$.

4.4 Renewal Point Processes and the Poisson Process

The statistical specification of a stochastic point process is simplified by assuming that the inter-arrival times are statistically independent and identically distributed (i.i.d.) random variables (RVs). For such a process, any inter-arrival time interval, denoted as a random variable (RV) T, follows a probability cumulative distribution function (CDF) $A(t), t \geq 0$, with mean value m_T, standard-deviation σ_T, and variance $var_T = (\sigma_T)^2$. Thus:

126 | 4 *Traffic Processes*

Definition 4.1   Renewal Point Process: A stochastic point process $A = \{A_n, n = 0, 1, \dots \}$ is said to be a Renewal Point Process if its intervals $\{T_n, n = 1, 2, \dots \}$ form a sequence of independent identically distributed (i.i.d.) random variables. The inter-arrival time distribution is governed by the cumulative distribution function (CDF)

$$A(t) = P\{T \leq t\}, \qquad t \geq 0,$$

and its probability density function (pdf) is given by

$$a(t) = \frac{dA(t)}{dt}, \qquad t \geq 0.$$

The corresponding counting process $N = \{N(t), t \geq 0\}$ is said to be a **Renewal Counting Process**.

We note that $1 - A(t) = P\{T > t\}$ and that $a(t)dt = P\{t < T \leq t + dt\}$.

The **rate** at which messages occur, also identified as **intensity**, and also as a message arrival rate in case the stochastic point process models message arrivals, is defined as the following.

Definition 4.2   Message Intensity or Arrival Rate: The message occurrence (or arrival) rate (or intensity) is denoted as λ. It expresses the (long-term) average number of messages that occur (or arrive) per unit time:

$$\lambda = \lim_{t \to \infty} \frac{E\{N(t)\}}{t}.$$

For an arrival process that is modeled as a Renewal Point Process, the message arrival rate is calculated as follows:

$$\lambda = \frac{1}{E(T)} = \frac{1}{m},$$
$$\text{where} \quad m = E(T). \tag{4.2}$$

To illustrate, consider a message arrival process that is modeled as a Renewal Point Process, under which the average time between two successive message arrivals is equal to $E(T) = m = 0.1$ s. In this case, on the average, a message arrival occurs every 0.1 second. We conclude the message arrival rate to be equal to $\lambda = 1/E(T) = 1/0.1 = 10$ [messages/s].

A special case of a renewal point process and that is widely used for modeling traffic processes is the **Poisson Process**. It is defined as the following.

Definition 4.3   Poisson Point Process: A Poisson Point Process is a renewal point process $A = \{A_n, n = 0, 1, \dots \}$ whose inter-arrival times are i.i.d. random variables that are exponentially distributed.

Accordingly, for a Poisson point process whose intensity parameter is equal to λ, the inter-arrival time distribution, density, and moments are given by the following:

$$A(t) = 1 - e^{-\lambda t}, \qquad t \geq 0;$$
$$P\{T > t\} = 1 - A(t) = e^{-\lambda t}, \quad t \geq 0;$$
$$a(t) = \lambda e^{-\lambda t}, \qquad t \geq 0;$$
$$E(T) = m = 1/\lambda;$$
$$VAR_T = \sigma_T^2 = \frac{1}{\lambda^2}. \tag{4.3}$$

Consider a continuous-time renewal point process. What is so special about an inter-arrival time whose duration is governed by an Exponential probability distribution function? The following property provides the answer.

Definition 4.4 **Memory-less Inter-arrival Time Distribution**: An inter-arrival time variable T is said to be governed by a **memory-less** probability distribution function if the following property holds:

$$P\{T > t + s | T > t\} = P\{T > s\}, \quad t, s > 0.$$

We note that when the inter-arrival time is exponentially distributed, we obtain:

$$P\{T > t + s | T > t\} = \frac{e^{-\lambda(t+s)}}{e^{-\lambda t}} = e^{-\lambda s} = P\{T > s\}, \quad t, s > 0.$$

Hence, an exponential distribution is memory-less. Consequently, given the event $\{T > t\}$, which states that there was no arrival taking place over the last t units of time, the probability that we will have to wait longer than an additional s units of time for the next arrival to occur is statistically independent of the time passed since the last arrival, as well as on previous past inter-arrival times. Furthermore, it is interesting to note that, when considering renewal point processes, Exponential distribution is unique in that it is the only probability distribution of a continuous RV that is memory-less, as stated by the following:

Property 4.1 **Exponential Distribution is the only Memory-less Probability Distribution Function**: An inter-arrival time variable T is governed by a **memory-less** probability distribution function if and only if it follows an Exponential probability distribution function.

Prompted by the **memory-less property** of the exponential probability distribution function, we note that a Poisson point process statistically models message arrivals (or a general flow of events) that occur **at random** in the sense that a past record of message occurrence times does not impact the occurrence times of future messages.

We note that this property applies also when considering a discrete-time renewal point process, whereby the continuous valued exponential distribution is replaced by a Geometric distribution.

To assess the extent of random fluctuations that characterize inter-arrival times, whereby each is governed by an exponential distribution, we note that the interval's Coefficient of Variation, which is equal to the ratio of its standard-deviation to its mean, is equal to 1.

Property 4.2 **Coefficient of Variation of an Exponential Distribution**: The Coefficient of Variation of an exponential distribution with intensity λ, mean m and standard-variation σ is equal to 1:

$$COV = \frac{\sigma}{m} = \frac{1/\lambda}{1/\lambda} = 1.$$

To further illustrate the statistical fluctuations of message inter-arrival times that follow an exponential distribution, we note the following. The probability of that an interval is longer than its mean ($m = 1/\lambda$) is calculated as $P\{T > m\} = 1/e \approx 0.368$, so that this event will occur about 36.8% of the time. In turn, noting that the tail of the distribution is decreasing exponential fast, the probability that the interval is longer than $2m$ or $3m$ is calculated to be equal to $e^{-2} \approx 0.135$, or $e^{-3} \approx 0.05$, which will occur for only 13.5% or 5%, of the time, respectively. Thus, 95% of inter-arrival times are expected to be no longer than $3m = 3/\lambda$.

128 | 4 Traffic Processes

The counting process $N = \{N(t), t \geq 0\}$ that corresponds to a Poisson Point process is defined as a **Poisson Counting Process**. A realization of a counting process is shown in Fig. 4.4. This process is characterized as follows.

Property 4.3   Poisson Counting Process: A Poisson Counting Process with intensity λ is a counting process $N = \{N(t), t \geq 0\}$, where $N(t)$ represents the number of message arrivals in $(0, t]$, $N(0) = 0$, which has **stationary-independent increments**, for which the number of arrivals over a time period $(s, t], t > s$ of duration $t - s$, denoted as $N(s, t) = N(t) - N(s)$, is governed by a **Poisson distribution**:

$$P\{N(s, t) = n\} = e^{-\lambda(t-s)} \frac{(t - s)^n}{n!}, \quad n = 0, 1, \dots . \tag{4.4}$$

The average and variance values of the number of arrivals occurring over a period of duration $t > 0$ are given by:

$$E\{N(t)\} = VAR\{N(t)\} = \lambda t.$$

The property of a process, such as a Poisson counting process N, of having *stationary independent increments* implies that: (1) The number of events occurring over any two disjoint (i.e., nonoverlapping) time periods are statistically independent (the independent increments property). (2) The distribution of the number of events occurring over any time period depends only the period's duration and not on its position in time (the stationary increments property). Hence, when considering any two disjoint time periods, such as $(0, s]$ and $(s, s + t], s, t > 0$, the numbers of events occurring over these respective intervals, $N(0, s) = N(s) - N(0)$ and $N(s, s + t) = N(s + t) - N(s)$ are statistically independent, and they have the same distributions as those that characterize the counts $N(s)$ and $N(t)$, respectively, noting that $N(0) = 0$.

Induced by the memoryless behavior of the occurrence times of Poisson events, we deduce the following property.

Property 4.4   Conditional Distribution of a Given Number of Poisson Events: For a Poisson counting process N, given that, over an interval $(0, T]$, a sum total of n events (such as message arrivals) has occurred, $N(T) = n$, the joint (unordered) times at which these events take place follow the joint probability distribution of n statistically independent random variables, whereby each random variable is **uniformly distributed** over the interval $(0, T]$. When the corresponding times are ordered, expressing the arrival times $\{A_1, A_2, \dots, A_n\}$ of the associated point process A, their probability law follows the order statistics of these n random-variables.

To illustrate, consider the following case. Given that a single message has arrived in an interval $(0, 10]$, the probability that this message has arrived within an interval of duration 1 that is located, for example, in $(2, 3]$, or in $(5, 6]$, is equal in either case to $1/10 = 10\%$. This is demonstrative of a Poisson process as a model of **at random** arrivals of messages.

Another interesting property of Poisson processes relates to the examination of the statistics of an arrival process that is formed as the aggregate of several arrival processes. The composite process is said to be formed as the **superposition of the component processes**. Consider the following two Poisson arrival point processes $A^{(1)}$ and $A^{(2)}$, with respective Poisson counting processes $N^{(1)} = \{N_t^{(1)}, t \geq 0\}$, $N^{(2)} = \{N_t^{(2)}, t \geq 0\}$. The superimposed counting process is defined as $N = \{N_t = N_t^{(1)} + N_t^{(2)}, t \geq 0\}$. The corresponding superimposed point process A is the point process that is associated with the superimposed counting process N. For example, if the message arrival times in a given period for $A^{(1)}$ are $\{A_1^{(1)} = 2, A_2^{(1)} = 6\}$ and for $A^{(2)}$ are $\{A_1^{(2)} = 4, A_2^{(2)} = 8, A_3^{(2)} = 12\}$, the arrival times for the superimposed process A are $\{A_1 = 2, A_2 = 4, A_3 = 6, A_4 = 8, A_5 = 12\}$.

We note that the superposition of multiple statistically independent Poisson arrival processes is characterized as an arrival process that is also a Poisson process:

Property 4.5 Superposition of Statistically Independent Poisson Processes: Consider a counting process N that is formed as the superposition of n statistically independent Poisson processes $\{N^{(i)}, i = 1, 2, \dots, n\}$, whereby the intensity (rate) of $N^{(i)}$ is equal to $\lambda^{(i)}$. Then, the superimposed process N is also a Poisson counting process and its intensity (rate) λ is equal to the sum of the intensities of the component processes:

$$\lambda = \sum_{i=1}^{n} \lambda^{(i)}.$$

The composite point process associated with N is a Poisson point process A whose intensity is equal to λ.

4.5 Discrete-Time Renewal Point Processes and the Geometric Point Process

Systems tend to operate in a discrete time fashion. The system processor records arrivals and produces output flows at discrete instants of time. The periods of time incurred between two successive discrete times (also known as time marks) are identified as **time slots**, simply referred to as *slots*.

A *master clock* regulates the timing of a system's input–output flows, internal computations, interactions and processing operations. For example, when using a system master clock that performs such regulation at a slot rate of 1 MHz, executions are performed at discrete time marks whose slot duration is equal to 1 μs.

Stations that share access to a synchronous communications medium are time synchronized. They maintain precise knowledge of start times of commonly recognized time slots (relative to an identified reference point such as the location of an Access Point). Such stations are said to be **synchronized** and the corresponding system is said to be (temporally) **slotted**. Every transmission or reception operation is initiated at a time instant that corresponds to a time mark.

In this section, we consider a discrete-time system model. It is noted that often the time slot duration is relatively very short so that the analysis of system operation can be well approximated by modeling it to evolve in continuous time.

A message process arriving to a slotted system is modeled as a *discrete-time point process*. The arrival times are represented by the arrival process $A = \{A_n, n \geq 0\}$, and the corresponding inter-arrival intervals $\{T_n = A_n - A_{n-1}, n \geq 1\}$, are discrete valued random variables. The time axis is divided into **time-slots**. A time-slot duration is denoted as τ. We assume henceforth, unless stated otherwise, with no loss in generality, that $\tau = 1$. Hence, start times of time slots are set to occur at times $\{k = 0, 1, 2, \dots\}$. The first slot starts at time $k = 0$ and ends at time $k = 1$. Message arrivals that occur during the n_{th} slot are recorded at time $n+$.

A realization of a discrete-time point process is illustrated in Fig. 4.3(b). This figure shows messages to arrive at times $\{A_0 = 0, A_1 = 2, A_2 = 3, A_3 = 5, A_4 = 7\}$. The corresponding inter-arrival times are noted to be $\{T_1 = 2, T_2 = 1, T_3 = 2, T_4 = 2\}$.

A discrete-time renewal arrival process is defined as the following.

Definition 4.5 A discrete-time point process is modeled as a **discrete-time renewal point process** by modeling inter-arrival times $\{T_1, T_2, T_3, \ldots\}$ as a sequence of independent identically distributed (i.i.d.) random variables.

The corresponding counting variable $N(t)$ records the total number of message arrivals (or events) that occur over the period $(0, t]$. A **discrete-time counting process** is defined as $N = \{N(k), k = 0, 1, 2, \ldots\}$, where $N(k)$ represents the total number of message arrivals that occur over the first k slots, setting $N(0) = 0$.

Let M_n denote the number of message arrivals that occur in the nth time slot. The corresponding *Message Arrival Sequence M* is defined as $M = \{M_n, n = 1, 2, 3, \ldots\}$.

A frequently employed arrival process that is modeled as a discrete-time renewal point process is the **Geometric Point Process**. It is defined as follows.

Definition 4.6 Geometric Point Process: A Geometric Point Process is a discrete-time renewal point process for which in each time-slot either no messages arrive or a single message arrives. Arrival events in different time slots are modeled to occur in a statistically independent and identically distributed manner. Assuming that a single message arrives in each time slot with probability p, $0 < p \leq 1$, and that no messages arrive in a time slot with probability $1 - p$, the *message arrival sequence M* consists of i.i.d. binary-valued variables $\{M_n, n = 1, 2, 3, \ldots\}$, so that

$$P\{M_n = 1\} = p, \quad P\{M_n = 0\} = 1 - p, \quad n \geq 1. \tag{4.5}$$

The inter-arrival time T follows a **Geometric Distribution**:

$$P\{T_n = k\} = p(1 - p)^{k-1}, \quad k \geq 1. \tag{4.6}$$

The average and variance values of the inter-arrival time T for a Geometric Point process are calculated to be given as:

$$E\{T\} = \frac{1}{p}, \quad Var\{T\} = \frac{1-p}{p^2}. \tag{4.7}$$

To explain the reason that the underlying inter-arrival times follow the Geometric distribution noted above, we note the following. Following a message arrival that has occurred at time $t = i$, the next message will arrive at time $t = i + 1$ with probability p, and then the inter-arrival time would be equal to $T = 1$. Alternatively, if no arrival occurs at time $t = i + 1$, which will happen with probability $1 - p$, the next message will arrive at time $t = i + 2$ with probability p. Putting it together, we conclude that the inter-arrival time will be equal to 2 slots, $T = 2$, with probability $(1 - p)p$. We proceed in a similar manner to calculate $\{P(T = k)\}$.

In simulating an arrival message process so that it follows the statistics of a geometric point process, one essentially performs statistically independent tosses of a biased coin, whereby a successful outcome (i.e., an arrival) occurs with probability p while failure (no arrival) occurs with probability $1 - p$. Such a coin tossing experiment is known in Probability Theory to involve the performance of Bernoulli trials. Hence, a geometric point process is often also identified as a **Bernoulli Point Process**.

The counting variable $N(k)$ for a geometric point process, which represents the number of message arrivals that occur during k slots, represents the number of successful outcomes when performing k Bernoulli tosses, so that it follows a *Binomial Distribution*:

$$P\{N(k) = n\} = \binom{k}{n} p^n (1 - p)^{k-n}, \quad 0 \leq n \leq k, \ k \geq 1. \tag{4.8}$$

Accordingly, a geometric arrival process is also referred to as a **Binomial arrival process**.

The **message arrival rate** for a discrete-time message arrival point process is defined as the average number of messages that arrive to the system per slot. For a geometric point process, the message arrival rate is then equal to:

$$\text{Message Arrival Rate} = E\{M\} = p \quad \text{[messages/slot]}. \tag{4.9}$$

To illustrate, consider a message arrival process that is modeled as a geometric point process with arrival rate parameter $p = 0.1$. Thus, there is a 10% chance that a message will arrive in a slot and a 90% chance of an idle slot. The message arrival rate is equal to $p = 0.1$ [messages/slot]. If each slot is $\tau = 1\,\mu$s long, the arrival rate is equal to $\frac{1}{p\tau} = 100,000$ [messages/s]. The average time between successive message arrivals is equal to $E(T) = 1/p = 10$ slots, and its standard deviation is equal to $\sigma_T = \sqrt{\frac{0.9}{0.1^2}} \approx 9.48$ slots. The corresponding mean and standard deviation values of the message inter-arrival time under the above assumed slot duration become $E(T)\tau = 10\,\mu$s and $\sigma_T\tau = 9.48\,\mu$s, respectively.

4.6 Traffic Rates and Service Demand Loads

4.6.1 Client–Server Traffic Association

Message flows produced by end users for transport across a communications network are handled by a multitude of network nodal, link, and user entities. What are the traffic and service demand loads that message flows impose on network entities? How well can an entity provide its service under an imposed demand level?

To answer this question, we observe that network systems, nodal elements, communication links, and end-user entities are modeled as **service modules**. A service module is a system that provides service to its clients. Depending on the function of the underlying module, its services may include: data processing, transmission, multiplexing, switching, routing, buffering, protocol processing, control, management, and their combinations. A system that employs multiple service modules, such as a communications network system, can also be modeled as a service entity.

The traffic and service demand rates imposed on a service module during a specified period of time depend on the characteristics of the sources that produce the associated traffic flows. Invoking a **client–server traffic** model, traffic sources serve as **clients** of a service system, requiring the latter to provide requested service to identified traffic flows.

Consider the following illustrative service modules:

1) A base station that regulates access of mobiles to a cellular network system across a Radio Access Network (RAN). The resources of a RAN wireless medium are shared by the cell's mobile users.
2) A switch (or a router) node that processes packet flows that traverse the communications network, as it directs them across selected routes to their destinations.
3) A physical layer protocol entity that configures a modulation/coding scheme and executes transmissions across a communications channel.
4) A MAC layer protocol entity that regulates the sharing of a communications channel through the implementation of a scheduling scheme.
5) A resource control service that assigns communications and processing resources for the support of message flows.
6) A control and management scheme that provides network control and management services.

132 | 4 Traffic Processes

In the following, we present key measures of traffic flow and of service demand rates as requested by clients from a service module. We relate these metrics to activity measures that characterize flow loads at the call, burst and packet levels.

4.6.2 Call Level Traffic Rates

Consider traffic flow processes produced at the **call level** that load a **service module**. Assume the aggregate call arrival rate loading the service module, accounting for the sum total of all calls loading this module during a specified period of time (such as PBH) to be equal to $\lambda_{call} = \lambda_c$, measured in units of [calls/s], unless stated otherwise.

Assume the time duration of a call, measured from its initiation time to the instant that it is terminated, to be represented by a random variable H [s/call]. The average call duration, when averaging over all calls (or, otherwise, when considering a selective group of calls) loading this server, is denoted as $E(H) = h$ [s/call].

The level of service requested from a service module is modeled as the following.

Definition 4.7 The **Service Time** that a client entity requests from a service module is modeled as a random variable S, generally expressed in units of seconds. The average service time value is denoted as $E(S) = \beta$.

To illustrate, consider a service module that is loaded by 100 active client devices whereby, during a PBH, each client starts a single call, on the average, every 40 seconds, and each call lasts for an average duration of 10 seconds. In this case, the total call arrival rate at the server is equal to $\lambda_c = 100/40 = 2.5$ [calls/s].

Consider a service module that operates in a **circuit switching** communications system. It operates as follows. A client that wishes to initiate a call proceeds first to send a signaling message to the system's manager, requesting it to assign resources that will be used to transport the call for its full time duration H. More specifically, consider a service node, such as a cellular wireless base station node, that manages the allocation to clients of its m communications channels. The latter are identified as its **service channels** or **servers**. An admitted user is assigned a single communications channel (or server) for the duration of its call. Each server is assigned on a full-time basis for the service of a specific call, serving it at a service rate of 1 [second per each second]. A *service demand rate* is defined as the following.

Definition 4.8 A **Service Demand Rate**, also identified as a **load**, requested from a service module over a prescribed period of time, is defined as the average aggregate service time requested by its clients per unit time.

Assume that an active client requests a service module to dedicate to it the use of a single service channel for a period of time that is equal to S, which represents the service time requested by this client in support of its activity. When needed to identify explicitly the activity level for which a service request is made, we use the following notation. The time duration of service requested by a call, burst or message (or packet) entities is denoted as S_c, S_b, and S_m (or S_p), respectively. The corresponding average service time values are denoted respectively as β_c, β_b, and β_m (or β_p). When considering an activity to impose a service request that is represented in reference to such a level, we denote the corresponding service time as S_a, the corresponding average requested service time as $E(S_a) = \beta_a$, and the activity arrival rate as λ_a. We introduce the following definition.

4.6 Traffic Rates and Service Demand Loads | **133**

Definition 4.9   Erlang Loading: Consider a service module that employs a single or multiple service channels (also known as servers). To obtain service, a client requests the system to allocate to it a single service channel for each one of its activities, occupying it for a period of time S_a while an activity is being served. The overall imposed service demand rate, also identified as the system's **load** or **Erlang Loading**, represents the average aggregate service time per unit time that the service system is requested to support. This loading metric is measured in units of **Erlangs**, whereby a service load rate of 1 [s/s], or 1 [min/min], or 1 [h/h], is identified as a load rate of 1 Erlang.

Under an aggregate activity loading rate of λ_a [activities/s], whereby each activity requests an average service time of $E(S_a) = \beta_a$ seconds, the Erlang loading of the service system, denoted as f_E [Erlangs], is calculated as:

$$f_E = \lambda_a \times E(S_a) = \lambda_a \times \beta_a \quad Erlangs. \tag{4.10}$$

To illustrate the calculation of a system's Erlang loading, consider the above described example. Call level activities are examined. We have $\lambda_c = 2.5$ [calls/s], $\beta_c = h = 10$ s. The Erlang loading is calculated to be equal to $f_E = 25$ Erlangs. Since each resource (i.e., server) can provide a service of at most 1 second during each second, offering a service rate of at most 1 Erlang, we conclude that an average of 25 servers would be required to meet the *average* system's load. Note, however that sizing the system to that level will not be generally satisfactory since the loading level of 25 Erlangs is an **average** value. The actual loading level could be at times much higher than 25. In the latter situation, the system will be able to actively serve only a fraction of ongoing activities and will have then to block or delay its provision of service to excess activities.

When considering a *call level based* **circuit-switching** communications network system, also identified as a **call switching** system, such as that provided by the Public Switched Telephone Network (PSTN) and by various all-optical networks, each call (as a client) is allocated a **circuit** for the duration of its call activity. A circuit provides resources for the support of a single nominal call. It is assigned dedicated communications capacity resources over each one of the links that is included in its end-to-end route. A circuit group may be allocated sufficient capacity resources to enable it to support multiple circuits. It is sized to sustain a service demand rate that is equal to f_E (circuits) [Erlangs], so that it is able to accommodate a maximum of f_E (circuit) simultaneous call activities.

Not all flows, or flow activities that are produced by clients are selected for admission to a service system. When a request to serve an activity takes place while the system is saturated with in-service activities, so that it has no residual capacity resources that are available to support a new activity, this activity is either blocked or is accepted but its admission into service is delayed. Under a **blocking service**, such an activity (or flow) would be blocked. At times, a blocked activity is assumed to be lost, as far as its current request for service is concerned. Such a service operation is also known as **Blocked Calls Lost (BCL)** service. Under a **delayed or queueing service**, an activity (or flow) that arrives when the service module is fully occupied, is delayed (or queued) and may be considered for admission into service at a later time when (and if) capacity then becomes available. Such a service operation is also known as a **Camp-On or queueing** service.

Accordingly, as noted in the following, we differentiate between **offered load** and **carried load** rates.

Definition 4.10   Offered and Carried Load and Message Rates: The **offered load or offered message rate** metric measures the corresponding rates of traffic activities that request service from a service system. It includes activities that are admitted into service as well as such that are blocked from entering the system. The **carried load or carried message rate** metric measures the corresponding rates of traffic activities that have been admitted into service.

To illustrate, consider the above-described example of a service system and assume that 15% of the calls that request service from the system are blocked and lost, yielding a call blocking probability that is equal to $P_B = 0.15$. The call admittance probability is then equal to 0.85, so that only 85% of arriving calls are admitted for service. In this case, noting that λ_c(offered) = 2.5 [calls/s], and f_E(offered) = 25 [Erlangs], we conclude the carried rates to be given as λ_c(carried) = 2.5 × 0.85 = 2.0 [calls/s], and f_E(carried) = 25 × 0.8 = 20 [Erlangs].

4.6.3 Burst Level Traffic Rates

Consider a system that provides service to flows on a burst activity basis. The definitions and calculations of traffic flow metrics for the system are analogous to those defined above for call-level oriented service systems. Traffic bursts are now used as the underlying activity periods. The following description illustrates the operation of such a system. Consider a service module in a **burst switching** system that employs m service channels (or servers) in providing service to message flows during burst periods. It serves multiple clients during their bursting activities. A client may send a signaling message to a system manager (such as a base station or an access point) upon the start of a burst activity, requesting service. For its support during a burst activity, the client may request the system to assign it a single server (such as a communications channel that it will use to transmit its messages across a wireless access network). This assignment, if granted, is maintained intact for the duration of the burst activity period. The service time that the server provides is now represented as a random variable S_b, whose value is equal to the burst's duration. The average length of this period is equal to $E(S_b) = \beta_b$ [s/burst].

The corresponding activity loading rate of the service system is equal to λ_b [bursts/s], and the Erlang service demand loading rate is equal to

$$f_E = \lambda_b \times E(S_b) = \lambda_b \times \beta_b \quad Erlangs.$$

A network system, such as a cellular wireless RAN, that uses a **burst switching**-based *resource allocation method* operates in this fashion. A mobile user that initiates a burst activity sends a signaling message to its associated base station manager, requesting the allocation of a communications channel (stating its required performance parameters). If resources are then available, the manager allocates the requesting client the requested resources (such as a communications channel that operates at the requested service rate) for use for the duration of the burst activity period. The corresponding period duration may be prespecified, or the client may inform the manager as to when its burst has terminated so that the manager would then cancel is resource allocation, making it available for the support of newly formed activities.

4.6.4 Message Level Traffic Rates

Consider a service module that models a system that provides service to messages. This is for example the case when we consider protocol engine entities that process the messages that they receive from other layer entities. A Medium Access Control (MAC) protocol layer entity receives a packet from a network layer protocol entity, which it transfers to a physical layer protocol entity for transmission across a communications channel such as one that is configured over a Wi-Fi or a cellular radio access link.

The MAC layer entity provides service to a higher layer entity. It forms a MAC layer frame and schedules its transmission across a shared communications medium. Acting as a service module, it is loaded with packets that arrive from a network layer entity at rate $\lambda_m = \lambda_p$ [packets/s]. The service time of each packet, S_p, with average value $E(S_p) = \beta_p$ [s/packet], depends on the rate at which the service module performs its processing, scheduling, and transmission tasks. The Erlang service demand loading rate is measured as:

$$f_E = \lambda_p \times E(S_p) = \lambda_p \times \beta_p \quad Erlangs.$$

As another example, consider a packet switching router located in a packet switching network such as the Internet. We first examine its header processor. It is modeled as a service module that represents the switching operation performed by the packet router. This operation entails reading the header of an incoming packet to determine its destination address, and reviewing a routing data base to determine the proper output port to which this packet should be switched. This packet is then switched to an output processor and processed by it. This processor manages an output queueing system that buffers its received packets and feeds them to a radio transmitter. The latter transmits the fed packets across a communications channel to next packet router, as packets follow a selected route.

The rate at which the header processor performs its operation is specified in units of r_S [packets/s]. Often, the processing performed by a header processor does not depend on the packet's length. It just examines the packet's header. Assume that the operation performed by the header processor in examining a packet takes it a fixed period of time, which is set equal to $S_p = \beta_p$ seconds. Its service rate is hence calculated as $r_S = 1/\beta_p$ [packets/s]. The rate at which packets arrive to the router is measured as λ_p [packets/s]. The Erlang loading of the header processor is calculated by using the formula noted above. For instance, assume that the arriving packet rate is equal to $\lambda_p = 8000$ [packets/s] and that the time that it takes the header processor to examine and switch a packet is equal to $\beta_p = 0.1$ ms. Then, the Erlang loading rate of the header processor is equal to

$$f_E = \lambda_p \times \beta_p = 8000 \times 10^{-4} = 0.8 \quad Erlangs.$$

Assume the router to use a single header processor. It can then process no more than a single packet at a time so that its processing service capacity is limited to 1 Erlangs. Thus, in 1 second, it can dedicate no more than 1 second of its time to performing its switching service. Hence, under the noted loading level, the header processor is loaded to 80% of its processing capacity.

As another example, consider the transmission service module located at an interface to an output port of a packet router (say Router A) in a packet switching network. The output port is attached to a communications channel that connects the underlying router to a neighboring router (say Router B). Packets that are routed by the switching processor to this output port require this service module to forward them to Router B. Packets switched to this transmission service module are stored in an output buffer controlled by this module and are transmitted, one after the other, by this module across an outgoing communications channel.

In modeling this output transmission system as a service module, we note the following parameters. Assume packets to arrive to this module at a rate that is equal to λ_p [packets/s]. Assume the transmission data rate across the output communications channel to be equal to R [bps]. Each packet is assumed to contain a random number of L_p [bits/packet]. The average packet length is equal to $E(L_p) = l_p$ [bits/packet]. The packet transmission time is equal to $T_p = L_p/R$ [s/packet]. Assuming that a single radio module is employed in transmitting packets across an output channel, we conclude that only a single packet can be served at a time. While the server is engaged in serving (i.e., transmitting) a packet, it cannot provide service to other packets. Hence,

136 | *4 Traffic Processes*

the service time provided to a packet by this service module is represented as $S = S_p = T_p = L_p/R$ [s/packet]. The average packet service time is calculated as $\beta_p = E(S_p) = E(T_p) = l_p/R$ [s/packet]. Hence, the offered Erlang loading at this module is expressed as:

$$f_E = \lambda_p \times \beta_p = \lambda_p \times l_p/R \quad Erlangs. \tag{4.11}$$

The offered data rate, measured in units of [bits/s], denoted as f_b, is given by:

$$f_b = \lambda_p \times l_p \quad [bits/s]. \tag{4.12}$$

To illustrate the calculation of the underlying offered traffic rate, assume the transmission data rate across the communications channel to be equal to $R = 10$ Mbps. The average packet length is equal to 4000 [bits/packet]. The packet arrival rate to the module is equal to 1500 [packets/s]. Using the formulas given above, we obtain the bit rate offered to the output module to be equal to $f_b = 1500 \times 4000 = 6$ Mbps. The average packet service time is equal to $\beta_p = l_p/R = 0.4$ ms. The offered Erlang loading rate is then calculated to be $f_E = \lambda_p \times \beta_p = 0.6$ Erlangs. The offered traffic load would thus occupy, on the average, 60% of the module's service capacity (which is equal to 1 Erlang). Equivalently, the loading bit rate, which is equal to 6 Mbps, makes use of 60% of the service rate capacity of the communications channel (which is equal to 10 Mbps).

4.7 Traffic Matrix: Who Communicates with Whom

In previous sections, we have presented mathematical models for the temporal characterization of traffic flow processes. Flow rates have been measured at an underlying service module that may represent a network element, a transport path, an end-user station, or a network system that employs a composite of such modules.

A communications network system consists of a multitude of end-user stations and network nodes, distributed over geographically diverse positions. Nodes may be operated by various organizations and may provide a mix of services.

Different traffic flows may be associated with the same or with different TC. Class association can be based on the underlying TC (involving audio, video and data based classifications), Service Class, organizational membership, functional aim, and other attributes. Traffic flows originating at groups of users or network elements that are linked to specific TCs, may be targeted for distribution to particular groups of users or network elements.

We characterize the **spatial distribution of traffic flows** through the specification of **traffic matrices**. In the following, we consider the spatial distribution of traffic belonging to a specific TC. The *topological layout* of a network is described by using a *graph model*. A graph consists of a set of *vertices and edges*, also often identified respectively as *nodes and lines*. *A graph vertex represents an end-user entity (such as a user terminal, or a computer server) or a network node (such as a network switch, network router, or a network management and control station), or both. An edge represents a communications link or an attachment link.*

End users attach themselves to certain nodes, identified as *edge nodes*. An end-user device may also include the functionality of a network node and would then be modeled as both an end user and a network node. Other nodes may not provide access to end users but may serve to switch and route internal network traffic flows. Nodes may also provide access to network management and control entities.

Terminal traffic flows relate to flows that originate and terminate at vertices, representing end-users and/or network nodes. *Nodal traffic flows* characterize flows at vertices, including

Figure 4.5 Flows Across a Network Graph.

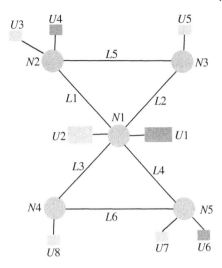

network and end-user nodes. *Traffic flows across lines or edges* characterize flows that traverse communications links that connect network nodes or are used to attach end-users to edge nodes. In this section, we discuss *traffic matrices* that related to *terminal traffic flow distributions*.

A graph model of a communications network is illustrated in Fig. 4.5. End users are represented by the nodes that are identified as $U = \{U1, U2, \ldots, U8\}$. Assume that end users are classified as members of two organizations. The first organization is associated with Group-1 (*G1*) of end-users. It includes the following users: $U_{G1} = \{U1, U4, U6\}$, whereby *U1* is identified as the group's server computer. The second organization is associated with Group-2 (*G2*) of end-users, and it includes the following users: $U_{G2} = \{U2, U3, U5, U7, U8\}$, whereby *U2* is identified as the group's server computer. Network nodes form the nodal set $N = \{N1, N2, N3, N4, N5\}$. These nodes, representing switches or routers. They may be shared among traffic flows that involve the two groups. Network links form the line set $L = \{L1, L2, L3, L4, L5, L6\}$. We assume that these lines represent communications links that connect network nodes and that are shared among traffic flows that belong to the two groups.

In characterizing the distribution of traffic flows between end users, we examine which end-user is communicating with which other end user (or users). Associated traffic flow metrics, such as message rates, are stated for an operation that takes place over a specified time period (such as a PBH, PBM, or over a 24 hour period). The corresponding traffic matrix is also identified as an **End-User Terminal Traffic Matrix**.

Similarly, a traffic matrix that is identified as a **Network Nodal Terminal Traffic Matrix** is used for characterizing the distribution of traffic flows between network nodes.

We thus define the following terminal flow traffic matrices.

Definition 4.11 A **End-User Terminal Traffic Matrix** Λ_u characterizes message traffic rates for flows that are distributed between a designated set of end-users $U = \{U1, U2, U3, \ldots, UN_u\}$, when considering a specific TC and an underlying activity period, is defined as follows: $\Lambda_u = \{\lambda_u(i,j), i,j = 1, 2, \ldots K_u\}$, where $\lambda_u(i,j)$ represents the rate of traffic flows that originate at end-user *Ui* and are destined to end-user *Uj*.

In a similar manner, an end-user-based terminal traffic matrix is defined when using a set of traffic flow metrics, denoted as $F_u = \{f_u(i,j), i,j = 1, 2, \ldots K_u\}$, where $f_u(i,j)$ represents a vector of traffic metrics for traffic flows that originate at end-user *Ui* and are destined to end-user *Uj*. Traffic metrics

138 | 4 *Traffic Processes*

can include bit and Erlang loading rates and a myriad of statistical attributes that characterize the underlying traffic processes.

Definition 4.12 A **Network Nodal Terminal Traffic Matrix** Λ_n characterizes message traffic rates for flows that are distributed between a designated set of network (or sub-network) nodes $N = \{N1, N2, N3, \ldots, NK_n\}$, when considering a specific TC and an underlying activity period. It is defined as follows: $\Lambda_n = \{\lambda_n(i,j), i,j = 1, 2, \ldots K_n\}$, where $\lambda_n(i,j)$ represents the rate of traffic flows that originate at network node Ni and are destined to network node Nj.

In a similar manner, a network nodal-based terminal traffic matrix is defined when using a set of traffic flow metrics, denoting it as $F_n = \{f_n(i,j), i,j = 1, 2, \ldots K_n\}$, where $f_n(i,j)$ represents a vector of traffic metrics for traffic flows that originate at network node Ni and are destined to network node Nj.

It is noted that under a given traffic matrix and a routing protocol, one can calculate the traffic rates that load the network's links and nodes. Routing algorithms and protocols are presented and discussed in a later section.

To illustrate, consider the following example involving the system shown in Fig. 4.5:

1) For user group G1:
   a) User U4 generates traffic destined to its server at U1 at a rate of $f(U4, U1) = f_b = 2$ Mbps. The routing scheme directs the traffic flow from U4 to U1 over the route that consists of the (bi-directional) link (N2,N1) (i.e., along link L1).
   b) User U6 generates traffic destined to its server at U1 at a rate of $f(U6, U1) = 2f_b = 4$ Mbps. The routing scheme directs the traffic flow from U6 to U1 over the route that consists of link (N5,N4) followed by link (N4,N1) (i.e., along link L6 followed by link L3).
2) For user group G2:
   a) Each one of the G2 non-server users, involving users $\{U3, U5, U7, U8\}$, generates traffic destined to its server at U2 at a rate of $5f_b = 10$ Mbps. The routing scheme directs the traffic flow from each user to server U2 over the route that consists of a single direct link.

Based on these traffic flow distributions, we conclude the following traffic matrices.

1) The **end-user terminal traffic matrices** are given as follows:
   a) For group G1 users, the end-user terminal traffic matrix $F_u(G1)$ consists of the following entries:

   $$f(U4, U1) = f_b = 2\,\text{Mbps}; \ f(U6, U1) = 2f_b = 4\,\text{Mbps}; \text{reminder terminal flows} = 0.$$

   b) For group G2 users, the end-user terminal traffic matrix $F_u(G2)$ consists of the following entries:

   $$f(U3, U1) = 10\,\text{Mbps}; \ f(U5, U1) = 10\,\text{Mbps}; \ f(U7, U1) = 10\,\text{Mbps};$$

   $$f(U8, U1) = 10\,\text{Mbps}; \text{remainder terminal flows} = 0.$$

   c) For combined user groups G1 and G2, the end-user terminal traffic matrix F_u consists of entries that are calculated by combining the entries included in the per-group matrices.
   d) Considering all network flows, the **network nodal terminal traffic matrix** F_n consists of the following entries:

   $$f(N2, N1) = 12\,\text{Mbps}; \ f(N3, N1) = 10\,\text{Mbps}; \ f(N4, N1) = 10\,\text{Mbps};$$

   $$f(N5, N1) = 14\,\text{Mbps}; \text{remainder terminal flows} = 0.$$

Based on the routing scheme described above, the traffic flow rates directed across network inter-nodal links are calculated to be given as follows:

$$f(L1) = 12 \text{ Mbps}; \ f(L2) = 10 \text{ Mbps}; \ f(L3) = 14 \text{ Mbps};$$

$$f(L4) = 10 \text{ Mbps}; \ f(L6) = 4 \text{ Mbps}; \ \text{remainder link flows} = 0.$$

Problems

4.1 Describe the structure of the Multilevel Traffic Model presented in the chapter. Draw a multilevel illustration of activities of a traffic flow process that is characterized by the following parameters. Also calculate the duty-cycle values incurred at each level as well as the overall packet-level activity duty-cycle (which expresses the fraction of time that a packet transmission activity takes place).
   a) At the call level: a call duration is exponentially distributed with average call length that is equal to 2 minutes. The inter-call period is exponentially distributed with average time that is equal to 20 minutes.
   b) At the burst level: a burst duration is exponentially distributed with average burst length that is equal to 1 second. The inter-burst period is exponentially distributed with average time that is equal to 4 seconds.
   c) At the packet level: a packet's transmission time duration is exponentially distributed with average that is equal to 2 ms. The time elapsed between the starting instants of two packet transmissions is exponentially distributed with average time that is equal to 40 ms.

4.2 Consider a traffic flow that consists of messages that arrive to the system at an average inter-arrival time of 5 seconds. Each message contains an average of 750 bits. Calculate the flow's arrival bit rate, expressed in [bits/s] (bps), and arrival message rate, expressed in [messages/s].

4.3 Consider a message flow process that is modeled as a Poisson process. Assume messages to arrive at a rate of $\lambda = 4$ [mess/s], starting at time $t = 0$, at which time the system contains no messages. Calculate the following:
   a) The average and variance values of the number of arrivals that occur over the interval $(0,2)$.
   b) The probability that during a period of 2 seconds, there will be no message arrivals; $P\{N_2 = 0\}$.
   c) The probability that during a period of 4 seconds there will be more than two message arrivals; $P\{N_4 > 2\}$.
   d) The conditional probability that, given that there was a single message arrival during the period $(0,4)$ there will be a total of two message arrivals during the period $(0,8)$.
   e) The probability that the first message arrival occurs before $t = 5$ s.

4.4 A processing node is loaded by two (statistically independent) flows, each of which is modeled as a Poisson process. The first flow (identified as flow-1) consists of messages that arrive at a rate of $\lambda_1 = 2$ [message/s]. The second flow (identified as flow-2) consists of messages that arrive at a rate of $\lambda_2 = 4$ [message/s]. Assuming both processes to starts

140 | *4 Traffic Processes*

at time t=0. Consider in the following the superimposed process that consists of messages that belong to either one of the flows.

a) State whether the superimposed process is a Poisson process. Determine its message arrival rate, denoted as λ, and the average message inter-arrival time.

b) Calculate the probability that the first message arrival will be a flow-1 message.

c) Calculate the probability that the first message arrival will occur at a time that is later than 1 second.

d) Given that the first message arrival is one that belongs to flow-1, calculate the probability that it arrives at a time that is later than 1 second.

4.5 Consider a message flow that is modeled as a geometric point process. Its message arrival rate is equal to $p = 0.2$ [messages/slot]. Calculate the following:

a) The average message inter-arrival time (measured in [slots]).

b) The probability that the time to the next message arrival is longer than five slots.

c) The probability that there will be two message arrivals during a period of five slots.

d) The average and standard-deviation values of the number of slots elapsed between two successive message arrivals (i.e., the inter-arrival time).

4.6 Consider a message flow that is loading a service system. Messages arrive at a message arrival rate of $\lambda = 10$ [mess/s]. The average service time requested and provided to each message is equal to 2 [s]. Calculate the following:

a) The Erlang loading of the system.

b) The minimal number of required simultaneously employed service channels.

4.7 Consider message arrivals to a system that are classified as voice and data streams, identified respectively as streams 1 and 2. Each stream ($i = 1, 2$) follows the statistics of a Poisson counting process, with the corresponding message arrival rates equal to $\lambda_1 = 1$ [mess/s], $\lambda_2 = 2$ [mess/s].

a) Given that there is a total of one message arrival occurring during a 1 second long period (0, 1], calculate the probability that there will be no messages of class-1 arriving during this period of time.

b) Calculate the probability that two class-2 message arrivals would occur during the period (0, 1], given that a total of five message arrivals (considering both classes) have been observed to occur during the period (0, 2].

4.8 System 1 is loaded by message flow-1 that is modeled as a Poisson process with message arrival rate $\lambda_1 = 20$ [mess/s]. The average time that the system takes to serve a flow-1 message is equal to 0.04 [s].

System 2 is loaded by message flow-2 that is modeled as a renewal point process, for which the time elapsed between two successively arriving messages is modeled as follows: with probability 0.8, it is governed by an exponential distribution with average value 10 [ms]; with probability 0.2, it follows an exponential distribution with average value 210 [ms]. The average time that the system takes to serve a flow-2 message is equal to 0.1 [second].

a) Considering flow-1, calculate the average (m), standard deviation (σ), and coefficient of variation (COV) values for the message inter-arrival time. Also calculate the Erlang loading level.

b) Considering flow-2, calculate the message arrival rate (λ_2), and the standard deviation (σ) and coefficient of variation (COV) values for the message inter-arrival time. Also calculate the Erlang loading level.

c) Which one of these flow processes produces message arrivals that display wider stochastic variations? Explain the reason for it.

4.9 Vehicles arrive to a highway at an entry point at instants that are observed over a period of 100 minutes to occur at the following times (expressed in units of minutes): {1, 6, 12, 16, 23, 29, 34, 42, 44, 54, 62, 74, 77, 81, 90, 95, 96, 100}. Calculate the average (sample mean) value of the inter-arrival time and the message arrival rate during the observation period.

5

Performance Metrics

Objectives: *Rating of the experience realized by a user upon receipt of a message flow or stream transported across a communications network is expressed by a **Quality of Experience (QoE)** metric. To achieve a desired QoE level, a source end user needs to properly format the message or stream and have it effectively transported across the network, while the destination end-user device is required to properly process the received data. For the latter processing to yield a satisfactory result, the transport of the message or flow across the communications network must be carried out in a reliable and timely basis, following specified **Quality of Service (QoS)** network performance requirements.*

Methods: *QoS metrics specify performance objectives for network transport. Such performance metrics account for end-to-end message delay, message and flow throughput, and message error rates. Key metrics also specify requirements for system's availability and reliability levels. In addition, it is critically important that the transport process employs methods that assure message transport security. To account for varying network loading and resource availability conditions, dynamically adaptive algorithms are employed. They are used to adjust operations undertaken at end-user nodes and across the network. As realized QoS values are being monitored, networking adjustments are triggered in aiming to achieve best feasible QoE levels.*

5.1 Quality of Service (QoS) and Quality of Experience (QoE) Metrics

How well does a communications network system perform in its transport of video, audio, and data messages and streams to their destinations? Network transport-oriented performance metrics are identified as **Quality of Service (QoS)** measures.

How effective is the end-to-end transport system in providing source and destination end users the experience that they desire? Metrics that are used to represent the quality of such an experience are presented as **QoE** measures.

Consider source and destination end users that are engaged in a video-streaming session. To achieve a specific QoE measure, the users may request the network system to meet identified QoS metric levels. For example, users may request the network to allocate an adequate level of communications capacity and processing rates so that their flows incur sufficiently high-throughput rates and low message delays.

Would a network transport operation that conforms to requested QoS levels provide users with their desired QoE levels? Would a transport operation that yields higher (or lower) packet loss rates while providing lower (or higher) packet delays yield an acceptable QoE level?

Principles of Data Transfer Through Communications Networks, the Internet, and Autonomous Mobiles, First Edition. Izhak Rubin.
© 2025 The Institute of Electrical and Electronics Engineers, Inc. Published 2025 by John Wiley & Sons, Inc.

When higher packet transit delay levels and variability are incurred, a user's viewing of a received streaming flow may be interrupted by the frequent occurrence of start and stop events, degrading the QoE realized by the user.

It is of interest to design the network's transport mechanisms in a manner that accounts for both types of quality metrics. As the loading of a network would typically fluctuate in a random manner, the networking system should react by dynamically adapting its networking operations and resource allocations. Such adaptation mechanisms are based on using ongoing monitoring of network system states and its performance dynamics.

In the following sections, we present and discuss key QoS and QoE metrics.

5.2 Quality of Service (QoS) Metrics for Communications Networking

QoS metrics are measures that specify the performance behavior that a message or a flow of messages experience, or wish to experience, while being served by a communications network system.

Message throughput and delay metrics are key network performance measures. To illustrate, consider a flow of vehicles across a highway. The traffic management administration is highly interested in assessing the number of vehicles that traverse a highway during a busy hour. The *flow rate* of vehicles accommodated across a highway represents the average number of vehicles per hour (or other unit of time) that transit it. The individual passenger, in turn, is highly interested in assessing the delay that she would incur in reaching her destination. Included is the *queueing delay* that is incurred in waiting on-ramp to access the highway as well as the *transit time* delay experienced while moving along the highway.

5.2.1 Throughput Metrics

We define a system's throughput measure as a metric that represents the rate of traffic that is delivered to targeted users.

Definition 5.1 The **Network System Throughput** metric measures the quantity of information units that are delivered to targeted destination entities per-unit-time. It is measured over a specified period of time and may be restricted to specified regions. It can also be specialized to include only identified traffic flow classes and prescribed user groups. It is commonly used to express the average quantity of information units per-unit-time that are delivered to targeted destinations, or that depart a network system. For a more detailed statistical description, the throughput measure is expressed also in terms of its standard deviation and tail probability, or by a probability distribution function.

The underlying information units can be associated with specific protocol layers. For example, throughput rate measures can relate to information rates exchanged across a transport layer connection, MAC frames that are sent between an Access Point (AP) and its Wi-Fi client devices, or MAC frames that are transmitted across a Radio Access Network (RAN) of a cellular wireless cell. Throughput rate levels can be displayed for distinct traffic classes, such as for the transport of various types of voice calls, video streams, and data flows.

An analogous measure used in a vehicular highway system, often identified as a *vehicle flow rate*, expresses the average number of vehicles per peak busy hour (PBH) that travel on a highway between specified entry and exit points.

5.2 Quality of Service (QoS) Metrics for Communications Networking | 145

To illustrate, consider a private network that transports data flows among end-users. User devices may include employee and customer computers and corporate servers. Consider the corresponding network flows that are monitored over a PBH. Assume that the *throughput rate* that accounts for data reception at terminals, computers, and servers was measured to have an average of $\lambda_{TH} = 100,000$ [packets/s]. The average packet length was measured to be equal to $L = 4000$ bits. The data throughput rate is thus equal to $f_{TH} = \lambda_{TH} \times L = 100K \times 4000 = 400$ Mbps. It was determined that 50% of the throughput rate is associated with packet receptions at the server. We conclude that the server receives an average of $50,000$ [packets/s], amounting to 200 Mbps.

How about measuring the flow rate of traffic that is carried inside a network? The corresponding metric is identified as the **network's carried traffic rate**, f_C, or as the **network's internal traffic rate**, f_I. When measured in units of [packets/s] (or [messages/s]), we denote the corresponding rates as λ_C or λ_I, respectively. It accounts for the sum total of the traffic rates loading all network links.

Example 5.1 Consider the following scenario. A message flow is routed along a path that consists of three links. The admitted flow rate at the input to the network (assuming that all flow's arriving packets are admitted into the network so that it is also equal to its offered rate) is equal to $\lambda_A = 1000$ [packets/s]. The flow rate across each one of the path's links, denoted as λ_L, is equal to $\lambda_L = 1000$ [packets/s]. Upon exiting the network, the (output) flow rate, identifying its throughput rate, is equal to $\lambda_{TH} = 1000$ [packets/s]. Since each one of the flow's packets traverses each one of the path's three links, we conclude that its contribution to the internal packet flow rate is equal to $\lambda_I = 3 \times \lambda_L = 3000$ [packets/s], or 12 Mbps (assuming each packet to contain an average of $L = 4000$ bits), accounting for the sum total traffic rate that loads the path's three links.

What portion of message traffic that arrives at network edge nodes and wishes to be transported across the network actually gets admitted into the network?

The rate of messages that arrive at a network's edge nodes is identified as the **network's offered rate** and is denoted as f_O [bits/s], or λ_O [messages/s]. If for the above illustrative network system, we determine the fraction of packets that are **blocked** from entering the network to be equal to $P_B = 10\%$, where P_B denotes the **message blocking probability**, then the **admittance probability**, denoted as P_A, and expressing the fraction of offered traffic that is admitted into the network, is equal to $P_A = 1 - P_B = 90\%$. The admitted message rate, which is denoted as λ_A [packets/s], is calculated as:

$$\lambda_A = \lambda_O \times P_A = \lambda_O \times (1 - P_B) \quad [\text{messages/s}]. \tag{5.1}$$

Similarly, the offered traffic (load) rate f_O and the admitted traffic (load) rate f_A are related as:

$$f_A = f_O \times P_A = f_O \times (1 - P_B) \quad [\text{bits/s}]. \tag{5.2}$$

The relationships between offered, carried, and departing (throughput) traffic (load) rates are illustrated in Fig. 5.1. As noted above, the blocking probability (P_B) represents the fraction of data units (such as messages, bursts, or calls) that are blocked from entering the network. The message loss ratio (P_L) is used to represent the fraction of internal messages that are lost and do not successfully depart the network. Internal message losses may be caused by various events, including message discards that are induced by switches or other processors whose buffers or other resources are saturated. Messages that contain bit errors when visiting nodal elements or reaching end-user devices are often also discarded.

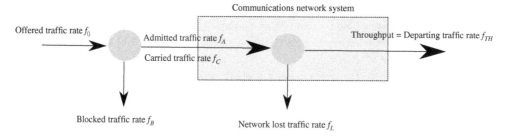

Figure 5.1 Offered, Carried, and Throughput Traffic (Load) Rates.

Example 5.2 Consider a cell in a cellular wireless network system that is managed by a base-station (BS) that is located around the center of the cell's service region. The **cell's downlink throughput rate** expresses the rate of message transmissions sent by the BS and successfully received by its client mobiles. The latter are mobile units that are associated with and managed by the cell's BS.

The **cell's uplink throughput rate** expresses the rate of message transmissions sent from mobiles and successfully received by their associated BS.

In expressing these rates, one can discriminate between mobiles that reside close to the cell's boundary vs. those that are located inside the internal region of the cell, closer to the BS. When considering reception of downlink transmissions, the former mobiles are more susceptible to signal interference effects caused by transmissions that are carried out in neighboring cells. Uplink transmissions by the former mobiles are received at lower power levels at the BS than those executed by internal mobiles that are closer to the BS. Hence, while the downlink (or uplink) **average throughput rate** is an average value that is calculated for transmissions to (or from) all cell mobiles, a certain fraction of the mobiles may experience much lower rates, as can be indicated by the following percentile throughput measures.

A 95-percentile throughput rate, denoted as $f_{TH,95\%}$, measures the throughput value achieved by 95% of the mobiles. It is calculated by discarding the 5% lowest rates while maintaining the rest. To express the throughput rate level attained in the communications system by the bottom 5% users, the 5-percentile throughput rate, denoted as $f_{TH,5\%}$ is used. The $f_{TH,5\%}$ value is used to characterize the throughput rate incurred by mobiles that reside closer to the cell's boundary, while the $f_{TH,95\%}$ level is used to characterize the throughput rate incurred by mobiles that reside closer to the BS. When subjected to limited availability of bandwidth resources, the system may not allocate the bandwidth resources that are required for providing high quality support to boundary users. Consequently, such mobiles would then experience lower uplink and downlink data throughput rates.

We differentiate between **gross throughput** and **net throughput**. In calculating a gross throughput rate, we include protocol overhead and control data. Net throughput rate excludes overhead and control data units.

When considering data flows transported across a network, we note that certain data units may not be useful to involved network elements or to destination end-users. For example, messages received at a destination user that are determined by the user to contain errors, may be discarded by the user. To account for such events, the concept of **goodput** is sometimes employed. It expresses the rate of information units received per unit time at an intended destination entity (or entities) when considering only those information units that have reached their destination(s) in good order or that are deemed useful to their destination nodes.

Another metric of interest when measuring throughput rates is defined as follows. Consider a network that supports peer-to-peer message communications among mobile nodes, also known as an *ad hoc network*. Messages are sent from one mobile to another use multi-hop routes that employ multiple intermediate mobile nodes. The later are said to act as *relay nodes*. Due to the varying topological layout of the network, induced by the dynamics of nodal mobility, the lifetime of a route (during which its links are not broken, so that it can continuously sustain a message flow) is limited and variable. Consequently, the transport of a message flow across such a multi-hop route may be interrupted before the complete quantity of data included in an activity burst (such as that embedded in the download of a file used for software update) has been successfully transported. To account for such occurrences, we have defined the concept of **Robust Throughput** [46]. It expresses the throughput rate achieved when the transport of a file is regarded to have been successfully completed only if all of the file's messages that are required by the destination are successfully received at the destination node within a specified transit time. No throughput credit is given to partially transported data units.

In Fig. 5.2, we demonstrate a typical relationship between the flow rate measured at the output of a communications network system, $f_{Output} = f_{TH}$ vs. the flow rate measured at the input to the system, $f_{Input} = f_{Offered}$. For a system that is well managed and properly flow controlled, we note the desired (ideal) behavior to yield $f_{Output} = f_{Input}$ as we increase the offered flow rate until the output flow rate reaches the maximum flow rate level that the system can deliver. The latter is identified as the **system's flow rate capacity**, or as its throughput capacity rate, or just as its **throughput capacity**, and is denoted in the figure as $f_{Capacity}$. We also show in the figure that under an actual implementation, a realized performance curve does not typically yield as good a performance behavior. For many systems, which tend to not be properly flow controlled, when the input flow rate reaches a certain level, a further increase in the input flow rate would result in a decrease in the throughput rate, as demonstrated in the figure. The latter behavior is often caused by admitting into the system a too high traffic load, which leads to excess contentions for the use of shared resources, resulting in throughput rate (often non-linear) degradation.

To demonstrate the latter behavior when considering a vehicle highway, we observe the following. When the highway is lightly loaded, the flow throughput rate would increase as we admit

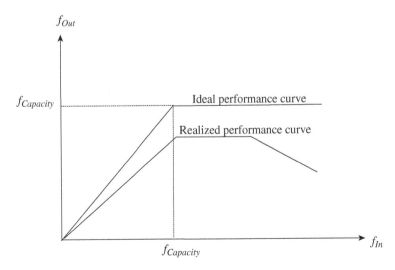

Figure 5.2 Output Flow Rate vs. Input Flow Rate.

more vehicles into the highway. As the highway becomes congested beyond a certain point, as we admit more vehicles into the highway, we find that the higher density of vehicles flowing across the highway causes vehicles to have to decrease their speed (noting that at a lower speed, vehicles can keep shorter inter-vehicular spacing) leading subsequently to a lower flow throughput rate. Such a behavior is modeled and discussed in in Chapter 21, where we describe traffic management and networking schemes for autonomous vehicle highway systems.

We state the following in defining the throughput capacity of a system.

Definition 5.2 The **Throughput Capacity** of a communications network system represents the maximum throughput rate, $f_{CAP} = f_{Capacity}$, that is achievable by the system. The **normalized throughput** attained by the system, also identified as its **throughput efficiency**, represents the ratio of the system's realized throughput rate relative to its throughput capacity, $s = f_{TH}/f_{CAP}$.

Example 5.3 As an example, consider a communications network system that accommodates a maximum throughput rate of $f_{CAP} = 100$ [Mpackets/s]. Assume that currently the system is loaded at a packet offered rate of $f_O = 70$ [Mpackets/s]. Assume that all arriving packets are admitted into the system, $P_B = 0$, and that 80% of admitted packets are successfully departing the system, so that $P_L = 20\%$. The packet throughput rate is equal to $f_{TH} = f_O \times (1 - P_L) = 56$ [Mpackets/s]. The normalized throughput rate is thus equal to $s = f_{TH}/f_{CAP} = 0.56$. Hence, the system operates at a throughput efficiency factor of 56%.

5.2.2 Message Delay Metrics

Message delay is one of the most important QoS metrics that is prescribed and measured when planning and evaluating the transfer of messages across a communications network. It measures the time incurred by a message as it is transported from a source entity to a destination entity. As delays incurred by messages that are transported across a network vary in a random manner, they are characterized in a statistical manner through the use of statistical measures that include the delay's average, standard deviation, jitter, percentile (probability tail), and probability distribution function.

We note the following attributes and parameters that are employed for characterizing the time delay incurred by a message transported across a communications network:

1) **Source and destination demarcation boundaries** across which source and destination entities send and receive the messages for whom the underlying delay time is measured.
2) **End-to-end (ETE) message delay** measures the time delay incurred by a message as it is transported from a source end user to a destination end user. When excluding the access time delays that are incurred, it is often measured as the time period measured from the instant that the last bit of the message is transmitted by a source end-user terminal to the instant that the first bit of the message is received at a destination end-user terminal.
3) **Message backbone Delay**: Represents the message delay incurred across a network's backbone system, measuring it from the instant that the message is fully receive at a backbone's edge node to the instant that the message is fully delivered to the network's backbone destination edge node. Alternatively, when considering associated backbone network edge nodes, it may be measured from the instant that the last bit of the message is obtained at a source edge node to the instant that the first bit of the message is transmitted by a destination edge node in aiming it to reach a destination end-user.

5.2 Quality of Service (QoS) Metrics for Communications Networking | 149

4) **Message access Delay**: Measures the time delay incurred in transporting a message across an access network from a source end user to a corresponding backbone network's source edge node. Similarly for the destination access network, it measures the time incurred in transporting a message from a backbone network's destination edge node to a destination end user. In a wireless cellular network, the RAN that connects a BS to mobile users is such an access network.

5) **Response Time**: Represents a round-trip time delay, often identified as **Round-Trip Time (RTT)**, incurred from the instant that a message is submitted to the network by a source node to that instant that a response message is received by the same node. When accounting for only network-induced delay components, the RTT is adjusted to not include delay components that involve reaction and processing delays that are incurred at the destination node prior to its transmission of a response message.

6) **Queueing and processing Message Delays at a Service Module**: Represent the time delays incurred by messages when processed by a network node, such as a router, or by any service module, such as a protocol engine or a protocol layer entity. A message time delay at a service module consists of its waiting time in a queue and its service time while being processed.

Above described message delays are calculated for message units under consideration. Call, burst, or message (or packet) level data units, exchanged between various protocol layer entities, can be of interest. For example, when considering message transmissions between a client end user and an AP in a Wi-Fi wireless LAN system, the underlying PDUs are message frames (i.e., layer-2 PDUs or MAC frames). The time elapsed in the transport of a message frame is then of interest. When considering the network system latency incurred by messages transported across a packet switching network (such as the Internet), the message unit of interest is a packet. When measuring the end-to-end message delay between end-user terminals, as recorded at the underlying transport or application layer entities, the involved messages are transport layer or application layer PDUs, respectively.

In Fig. 5.3, we show a characteristic relationship between the average delay $E(D)$ experienced by a message that is processed by a service module, and a normalized traffic throughput rate $s = f_{TH}/f_C$. For example, the service module can represent a statistical multiplexing entity, as used in a router when serving incoming packets that are processed by the router's protocol engine. As another example, the service module can represent a router's output radio entity that provides for the transmission of outgoing packets across an output communications channel.

As noted above, a message delay metric represents the cumulative sum of queueing (i.e., waiting-time) and processing time components. A message delay time is measured from the instant that the message arrives at the module to the instant that it completes its service. In Fig. 5.3, the mean (i.e., average) delay value is normalized by the required average message service time, which is set equal to $E(S) = \beta = 1$.

For this illustration, the service system is modeled as an $M/M/1$ queueing system that imposes no storage capacity constraints (see Chapter 7). Messages are assumed to arrive to the service module in accordance with a Poisson process at a rate of λ [messages/s]. Each message requires a random service time that is assumed to be Exponentially distributed with an average message service time that is equal to $E(S) = \beta$ [s]. The loading of the system is equal to $\rho = \lambda \times \beta$ [Erlangs]. If $\rho < 1$, as assumed here, the normalized throughput rate of the system is calculated to be given as $s = \rho$. We note that for this queueing system model, the mean message delay is given by the following formula (which was used to plot the displayed performance curve):

$$E(D) = \frac{E(S)}{1 - \rho}, \tag{5.3}$$

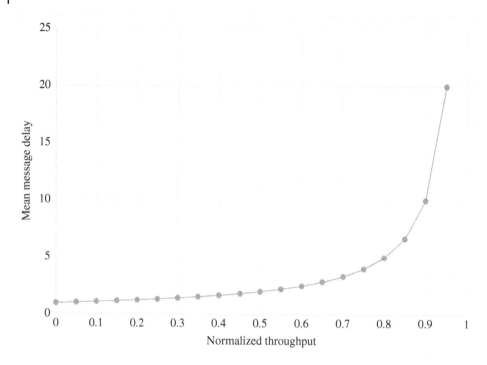

Figure 5.3 Average Message Delay vs. Normalized Throughput.

where $s = \rho < 1$ is the system's Erlang loading level as well as the system's normalized throughput rate.

It is noted in Fig. 5.3 that the incurred average message delay increases with the loading rate in a highly *nonlinear* manner. As long as the loading traffic rate is lower than about 50–75% of the module's capacity, messages incur (on the average) rather low delays, which for the demonstrated system are respectively lower than a factor of 2–4 of the message average service time. As the traffic loading rate (λ or ρ) increases, we note messages to incur rapidly increasing average delay levels. For the illustrated system, when the system is loaded at 95% of its throughput capacity, $s = \rho = 0.95$, we note the average message delay to be equal to a value that is higher than the message service time by a factor of 20, $E(D) = 20\beta$.

5.2.3 Error Rate Metrics

For many applications, it is essential that a message received by a destination device is a reliable replica of the originally produced message. As messages are transported across a communications network, they are liable to be corrupted by noise and interference signals. Errors can also occur while messages are processed by network nodes and interface devices. To reduce the frequency at which messages are received in error at a destination end user, error control schemes are implemented. Error detection and correction codes are used to reduce the error occurrence rate. The following error rate metrics are noted:

- The **Error Rate** incurred in the transport of message units expresses the average rate at which, upon reception, messages contain unresolved errors.
- For example, when considering peer-to-peer Physical Layer operations, the **Raw Bit Error Rate (BER)** metric represents the fraction of message information bits that are received in error when

5.2 Quality of Service (QoS) Metrics for Communications Networking | **151**

transmitted across a communications channel or across a communications path, when no error control scheme is employed by one or several protocol layer entities.

- When an error-control scheme is employed, the ensuing **Bit Error Rate (BER)** metric represents the fraction of message information bits whose errors cannot be resolved (i.e., corrected).
- Such measures are similarly defined for peer-to-peer transfers between protocol entities. Accordingly, a **Frame Error Rate** metric is defined when considering message transfers between Link/MAC layer peers; a **Packet Error Rate** metric is calculated for packet transfers across a network (such as that implemented between Internet-based IP peer layer entities embedded in IP routers); **Transport Layer PDU Error Rate** metric is used for transfers of transport layer segments; and **Application Message Error Rate** metric is used for message transfers between application layer peers.

Example 5.4 Consider a mobile user that transmits to a BS, across a radio access channel, message segments that are encoded for error-control purposes, identified as error-control blocks. If no such an encoding scheme is utilized, it has been determined that this system's raw bit error rate (raw BER) is equal to 10^{-3}. Thus, on the average, 1 out of a 1000 bits is received in error. To reduce this error rate, the system implements a modulation/coding scheme (MCS) that yields a reduced bit error rate level. Assume it to be equal to 10^{-5}. Now only an average of 1 bit out of every transmitted 100,000 bits is received in error. If each error-control block represents a single packet whose average length is equal to 1000 bits, the ensuing packet error rate (PER) would be equal to about $PER = 1\%$, so that an average of about 1 packet out of every 100 packets would contain unresolved errors.

In network operations, messages are often lost or discarded during their transport. Such events can be caused by reception of such messages by entities that cannot resolve incurred message errors. This can, for example, occur when a received message contains too many corrupted symbols. Messages can also be lost, as they may be discarded by network processors, switches, or routers, when the latter are excessively overloaded, so that messages may be discarded due to buffer overflows, or lack of sufficient energy, spectral, or other resources. Messages can be lost while being transported across the network for variety of other causes, never reaching their destination entity. A **Message Discard or Loss Ratio** metric expresses, at the corresponding layer, the fraction of messages that are discarded or deemed lost.

For many data transport applications, it is essential to have a data message accepted by a destination entity only if the message can be confirmed to contain no uncorrectable errors. In turn, voice and video messages are often accepted even if they contain a limited number of uncorrectable errors.

5.2.4 Availability and Reliability Metrics

Availability and reliability metrics are defined for a communications network system in a manner that is similar to that employed for other systems.

The availability of a network is measured as follows.

Definition 5.3 The **availability** of a communications network system is measured over an identified time period as the fraction of time (or, equivalently, the probability) that the system is operational in a manner that makes it available for providing its intended service.

The downtime of a system can be expressed in terms of the time, as measured, for example, by the number of minutes or hours, that the system is down, and thus nonoperational to a desired level, over a specified period of time such as per year. Typically, the *availability* of a system is expressed as the percentage of time that the system is expected to be operational. For example, a system whose *availability is at the five nines level*, is operational for 99.999% of the time.

A system that experiences failures of certain components or subsystems can still continue to be available for performing certain identified intended service functions. To assure a system's robust behavior, failure recovery and reconstitution schemes and mechanisms are implemented. Repair processes or alternate service mechanisms may be rapidly activated. Use may be made of stand-by resources, assuring continuation of the operation of the system at prescribed (though possibly reduced) performance level.

An available communications network system may not however always offer its clients a reliable transport service.

Definition 5.4 The **reliability** of a communications network system is measured over an identified time period as the fraction of time (or, equivalently, the probability) that the system is offering its clients (and flows) properly performing transport service. Subsystem or component failures can degrade the system's performance behavior and therefore not provide its clients the service level that they desire.

A communications network system may continue to be available for transporting its clients' flows even when certain subsystems have failed to operate at their regular capacity level. Malfunction or degraded performance events can be caused by hard failures of key elements or components. They can also be induced by soft failure events, such as malfunctioning of software modules, including such that lead to degradation of switching and processing rates.

Under a **reactive** operational mode, failure detection modules are used to detect the occurrence of failures and to then activate repair mechanisms. To illustrate, consider a system whose operation alternates between operational and down (or breakdown) periods. Assume the system to be repairable, so that once it fails, a repair process is invoked, restoring it to operational mode after an average repair time that is denoted as *Mean Time To Repair (MTTR)*. The average time that the system stays operational, as measured between two consecutive failure events, is identified as its *Mean (operational) Time Between Failures (MTBFs)*. It measures the average length of time that the system is continuously operational until the occurrence time of the next failure event. The fraction of time that the system is operational, denoted as P_{OP} is thus calculated as:

$$P_{OP} = \frac{MTBF}{MTBF + MTTR}.$$

The speed at which a system's repair process is activated and completed is used to characterize a system's efficiency in terms of its Serviceability or Maintainability. When considering failures of elements that are not repairable, the Mean Time To Failure (MTTF) metric is used. It measures the average time to total breakdown.

Example 5.5 Consider a system that is monitored to stay continuously in operational mode for a working period of 10 days, operating 24 hours per-day. Assume that once the system is detected to have failed, the system's repair mechanism is able to repair the system (or activate a stand-by system) within a period of 1 minute. For this system, we thus have: $MTBF = 14,400$ minutes, and $MTTR = 1$ minute. Hence the fraction of time that the system is available (at a proper performance level) is calculated to be equal to:

$$P_{OP} = \frac{14,400}{14,401} \approx 0.9999306 = 99.99306\%.$$

To enhance a system's reliability level, fault detection and isolation mechanisms are implemented. They are tasked to rapidly and reliably identify and isolate failures and activate resolution mechanisms. When available, **failover** processes are activated. Through the institution of redundant resources and mechanisms, *automatic switch-over* processes are triggered, inducing the rapid activation of replacement modules.

Under a **proactive** operation, artificial intelligence (AI)-oriented machine-learning mechanisms may be employed. They are used to execute in the background processes that learn the system's dynamic behavior loading characteristics and failure profiles. Such processes act to proactively alert the management and control system about predictions of forthcoming failure or degradation threats. The system reacts by activating resolution mechanisms that mitigate the impact of expected threats and failure events.

5.2.5 Cyber Security

Cybersecurity methods protect systems and networks from digital attacks. A wide range of mechanisms have been developed to detect, assess, and resolve digital threats that aim to interfere with the secure transport of information across a communications network. It is essential for a communications network system to provide its clients a secure message transport process, while also meeting their requested QoS measures.

The following measures relate to synthesizing a network system to effectively provide for confidentiality, integrity, and access (CIA) security protections:

1) **Access Security** procedures secure access to network system assets, processes, and mechanisms so that only authorized users and processes are admitted and accommodated. Procedures are also used to prevent the blocking of legitimate users from accessing their destination nodes or users and secure acceptable availability levels of networking resources for authorized users.
   a) It is essential that only authorized entities and processes are able to make adjustments in the functioning of critical modules, at each layer, across hardware and software platforms, associated with the network system's data, control, and management planes.
   b) Strong authentication mechanisms are employed to prevent access to essential resources to unauthorized users.
   c) *Virtual Private Networks (VPNs)* are established to provide privacy and security for interactions and ensuing message flows that are preserved for the exclusive private use by users that are members of an established private network. Members may be located across multiple domains and networks. Access control and protection techniques include the use of *firewall gate keeping* mechanisms.
   d) *Distributed Denial of Service (DDoS)* attacks flood the resources of a network device, such as an end user or a network node, preventing it from providing service to legitimate users. The perpetrator may cause a large number of devices that are located across distributed network domains to produce such traffic overloading attacks. For example, in disrupting communications networking services, such attacks have been used to flood processing and memory resources used by UDP and TCP operations, as well as disrupt the proper functioning of a multitude of application layer entities. To illustrate, we note that by executing a SYN flooding attack, a perpetrator causes a large number of computers to send TCP connection setup request packets (identified as SYN packets) to a network device that controls network access to a targeted web server, overflowing its resources and thus preventing it from establishing connections with legitimate users. Such attacks may also involve the use of forged IP sender addresses (known as IP address spoofing).

2) **Confidentiality** measures are enacted to guarantee the privacy and secrecy of user and network system data that requires such measures, allowing only designated entities to have access to the associated information content.

   a) Messages transported across a communications network between end users, such as those that contain financial data that is uploaded by a user to a web server across a network, are kept confidential by the use of encryption methods.

   b) *Private key encryption* methods make use of the distribution of private keys between connected clients, allowing only these clients to decrypt the secret messages that they exchange.

   c) *Public key encryption* methods allow a user to make its key available publicly. An entity that wishes to send confidential data to this user, encrypts its message data with the public key advertised by the user. Only the intended user is able to decrypt these messages.

   d) *Hybrid private and public key encryption* methods are utilized. For example, a public key encryption method is used to exchange private encryption keys, which are then used to exchange encrypted messages in a confidential manner.

3) **Message Integrity** procedures are used to assure the integrity and authenticity of a message.

   a) Message integrity can, for example, be assured through the use of methods that send a secret integrity code that is calculated by the sending user by using data that was originally included in the message. Upon receipt of the message, the recipient uses the received data to calculate a corresponding code. The result must agree with the decrypted integrity code that was received from the sender. A match would guarantee the authenticity of the data, assuring the recipient that the data was not modified during its transport process.

   b) User authenticity can be affirmed through the use of encryption-based methods. For example, a user may encrypt its ID by making its decryption possible only when its public key is utilized. A communications network system can then rely on such a secure process to confirm the identity of the sender. It can then provide access to this user to resources that the user is authorized to access.

Of prime interest to network managers and users are the effectiveness levels of its employed security methods in serving to detect, isolate, and resolve security risks. The following measures are often used to assess the fitness of a system in performing such functions:

1) **Threat Detection** time measures the time that it takes the system to detect security threats. A Mean Time To Identify (MTTI) metric is used.

2) **Threat Containment** time measures the time that it takes the system to contain security threats. A Mean Time To Contain (MTTC) metric is used. Metrics are also employed to identify the time taken to resolve and eliminate threats.

3) **Mean Time to Resolve (MTTR)** measures the average time that it takes the system to react to and resolve a threat.

4) **Window of exposure** measures the relative time that identified network elements and transport processes remain vulnerable.

5) **Dwell Time** measures the relative time that the attacker resides in the system.

6) **Security System Testing and Validation** metrics evaluate ongoing testing and verification processes that confirm the readiness and effectiveness of the security system.

Relating specifically to the operation of and services provided by a communications network, methods should be instituted to rapidly detect and resolve security threats that prevent the system from executing its mission of transporting client messages and flows at desired QoS performance levels. It is essential to prevent threats that: induce packets to be delivered to unintended destinations; provide access of unauthorized users and flows to segregated resources and VPNs; cause the failure of network elements, signaling and control points, and of admission and management sub-systems; used to track user flows and flow statistics in a manner that is not authorized by the system.

Consider autonomous vehicles that move across a highway, whose management depends on the effective operation of an employed inter-vehicle communications network. It is essential to assure the safety of the operation when the performance of the network degrades. One approach is to synthesize a multiple-layer functionality that when detecting a layer failure, fall-over transition to an alternate functionality layer is triggered. In this manner, when vulnerable software-heavy management and control processes embedded at higher layers are infected, the operation automatically falls back into a more robust mode, which may be more hardware focused. The system may then instantly transition to acting in a more conservative manner, closing safety vulnerabilities; for example, acting then to prevent imminent vehicle collisions.

The segregation of the network system into physically or logically disjoint sub-networks, whereby individual sub-networks are designed to offer their own unique security and QoS measures, can reduce security risks and improve the detection and resolution of security threats. However, each subsystem may then be designated as available to only a limited group of users or functionalities.

Commonly employed transport security methods are described in Chapter 22.

5.2.6 Illustration: QoS Metrics for a Cellular Wireless Network

In the following, we illustrate the specifications of QoS metric values in a communications network system by displaying certain quality measures that were specified in a released version of the 4G LTE cellular wireless network standard.

Mobile users are connected to the cellular system through their associated BS. A BS then connects a user's flow to a packet gateway, which in turn provides access to a packet switching backbone network. Such a connection tunnel is identified in the 4G LTE system as a *bearer*. To support the transport of a user's messages at a desired QoS level, a proper set of resources are allocated across a bearer. Included are resources that are used to transmit messages across the RAN segment of the bearer that connects a user with its associated BS.

To perform this allocation in a manner that is satisfactory to the user, the system needs to know the user's performance requirements. For this purpose, the system classifies user flows into a specified set of flow classes. Each class is identified by a QoS class identifier (QCI) and is associated with a specific set of QoS parameters. These parameters are used by a network management and control entity in determining the service treatment to be provided, and the resource allocations to be made.

Guaranteed bit rate (GBR) bearers offer transmission resources that are permanently allocated (e.g., by an admission control function managed by the BS) at bearer establishment or modification times. Bit rates higher than those offered by the GBR level may be allocated to a GBR bearer if resources are available. A maximum bit rate (MBR) parameter that is associated with a GBR bearer, sets an upper limit on the bit rate level that can be expected from a GBR bearer.

Non-GBR bearers do not guarantee a particular bit rate. They can be used for applications such as certain web browsing or File Transfer Protocol (FTP). For these bearers, no bandwidth resources are allocated on a dedicated basis.

In Fig. 5.4 [6], the performance parameters that have been specified for each QoS class (as identified by its QCI) for a noted version of the LTE specification are presented. Typical traffic classes that are associated with the corresponding QoS classes are noted. Performance parameters include

QCI	Resource type	Priority	Packet delay budget	Packet error loss rate	Example services
1	GBR	2	100 ms	10^{-2}	Conversational voice
2	GBR	4	150 ms	10^{-3}	Conversational video (live streaming)
3	GBR	3	50 ms	10^{-3}	Real time gaming, V2X messages
4	GBR	5	300 ms	10^{-6}	Non-conversational video (buffered streaming)
65	GBR	0.7	75 ms	10^{-2}	Mission Critical user plane Push To Talk voice (e.g., MCPTT)
66	GBR	2	100 ms	10^{-2}	Non-mission-Critical user plane Push To Talk voice
75	GBR	2.5	50 ms	10^{-2}	V2X messages
5	Non-GBR	1	100 ms	10^{-6}	IMS signaling
6	Non-GBR	6	300 ms	10^{-6}	Video (buffered streaming) TCP-based (for example, www, email, chat, ftp, p2p and the like)
7	Non-GBR	7	100 ms	10^{-3}	Voice, video (live streaming), interactive gaming
8	Non-GBR	8	300 ms	10^{-6}	Video (buffered streaming) TCP-based (for example, www, email, chat, ftp, p2p and the like)
9	Non-GBR	9	300 ms	10^{-6}	Video (buffered streaming) TCP-based (for example, www, email, chat, ftp, p2p and the like). Typically used as default bearer
69	Non-GBR	0.5	60 ms	10^{-6}	Mission critical delay sensitive signaling (e.g., MC-PTT signaling)
70	Non-GBR	5.5	200 ms	10^{-6}	Mission critical data (e.g. example services are the same as QCI 6/8/9)
79	Non-GBR	6.5	50 ms	10^{-2}	V2X messages
80	Non-GBR	6.8	10 ms	10^{-6}	Low latency eMBB applications (TCP/UDP-based); augmented reality
82	GBR	1.9	10 ms	10^{-4}	Discrete automation (small packets)
83	GBR	2.2	10 ms	10^{-4}	Discrete automation (big packets)
84	GBR	2.4	30 ms	10^{-5}	Intelligent transport systems
85	GBR	2.1	5 ms	10^{-5}	Electricity distribution- high voltage

Figure 5.4 QoS Class Identifier (QCI)-Based Measures as Specified by 3GPP Standard TS23.203/with Permission of 3GPP.

priority handling, acceptable delay budget, and packet error loss rate, as well as **Resource Type**, which identifies the service provided to this class as being either GBR or non-GBR.

The **Priority Handling** index identifies the priority that a flow is granted by the system as it executes its flow admission and resource allocation processes. A QCI that is attached a lower priority handling index is granted a higher priority treatment and given higher precedence while allocating resources. Mission-critical delay-sensitive signaling messages are granted the highest priority.

The **Packet Delay Budget (PDB)** parameter expresses the maximum delay value that a packet belonging to the ensuing class should incur in its transport across the RAN that connects a mobile unit and the associated BS. As part of this measure, the specification includes also delay components that account for message transit between the BS and the system's Policy and Charging Enforcement Function (PCEF) entity. It is often assumed that 98% of admitted packets are expected to experience delay values that do not exceed the specified PDB level.

The **Packet Error Loss Ratio (PELR)** parameter specifies an upper bound on the fraction of packets that are lost due to the occurrence of packet errors.

In examining the QoS performance metrics provided to various traffic classes, we note the following:

1) **Conversational voice** flows are attached a relatively high priority index (=2) and are guaranteed a message delay (PDB) value that is not higher than 100 ms and a message loss ratio that should be below 1%.
2) **Live streaming video** flows are assigned a relatively high priority index (=4) and are guaranteed a (PDB) message delay level that should not exceed 150 ms and a message loss ratio that should be below 0.1%.
3) **Realtime gaming and V2X** flows are assigned a relatively high priority index (=3) and are guaranteed a low (PDB) message delay level that should not exceed 50 ms and a message loss ratio that should also be low, below 0.1%. V2X data flows involve communications between a moving vehicle and other entities, such as other vehicles or Road Side Units (RSUs), Aps, or BSs. Due to the variable nature of data flows to/from RSUs and between highway vehicles, they

are associated with QCI indices that provide them a strictly low delay value (=50 ms), while allowing a somewhat higher packet loss ratio of 1%; a higher priority level (= 2.5) is set for certain V2X traffic classes while a lower priority level (=6.5) and non-GBR resource-type assignment is applied to another class of V2X data flows.

4) **Intelligent Transportation Systems (ITSs)** require means for effective data communications, including vehicle-based communications networking service that supports an autonomous vehicle highway. Such data message flows are assigned a relative high-priority index (=2.4) and are guaranteed very low message delay (PDB), following an upper bound of 30 ms. For safety reasons, a very low message loss ratio level is prescribed, setting it to not exceed 0.001%. A vehicle that is forced to make a sudden stop could broadcast safety packets of this class to alert nearby vehicles.

5) **Buffered Video Streaming** flows (that are TCP based) are provided a non-GBR resource-type assignment, a lower priority index (=5), and a higher message delay (PDB) upper bound, setting it to not exceed 300 ms. It is however granted a very low message loss ratio, which is required to be no higher than 10^{-6}, so that on the average, no more than one out of a million streaming messages should be lost.

5.3 Quality of Experience (QoE)

The QoE measure has been defined by ITU-T [27] to express "the degree of delight or annoyance of the user of an application or service." It also notes that a QoE measure accounts for end-to-end system effects, including impacts by client, terminal, network, services infrastructure, and other sub-systems and processes.

As described in [27] and shown in Fig. 5.5, **QoE** is determined by **QoS-based factors** and by **human factors**. The QoS factors involve technical performance elements that are determined in an **objective** manner. They include QoS metrics associated with the network system, including

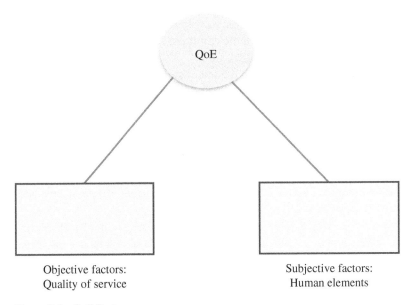

Figure 5.5 QoE Factors.

158 | *5 Performance Metrics*

Application	Sample delay
Management and control	< 0.1 s
Transactional	< 1.5 s
Messaging download	< 8 s
Non-critical	> 8 s

(a)

Application	Sample delay
Realtime voice and video	< 0.1 s
Voice/video messaging	< 1.5 s
Audio/video streaming	< 10 s
Fax	> 10 s

(b)

Figure 5.6 Illustrative Message Delay Requirements for Applications That Are: (a) Error Sensitive and (b) Error Tolerant.

service, transport and application level factors. Considering applications such as those discussed in [26], illustrative ranges of message delay objectives for error tolerant and error sensitive applications are shown in Fig. 5.6. As noted in previous sections, key QoS metrics include throughput, delay, delay jitter, availability, and packet loss ratio.

In turn, human factors that impact a QoE evaluation are **subjective** in nature and may involve elements of experience, scenario type and conditions, emotions, and service billing. A QoE rating specified by an end user tends to be based on the user's experience as part of her engagement in the transport, creation, reception, and adaptation of the underlying information flow. It can be dependent on context, user expectations, the underlying system, environment, the end-user's device and on employed networking platforms and their features.

Consider the reception of a video stream. Generally, higher QoE systems use streams that employ higher-quality video encoding schemes (using a higher density of pixels per frame, a higher frame rate, a higher encoding rate that is associated with a lower compression ratio) and network connections that allocate higher bandwidth levels and lower error rates. Reception at a higher data rate and a lower error rate yields a higher-quality video reception experience. Yet, the ensuing QoE rating will depend in a nonlinear fashion on the values configured for such parameters. Increasing the compression ratio by a certain factor may degrade the reception quality of a video stream by a minimal or by a noticeable magnitude, depending on the employed factor and on the underlying scenario, human elements, and operational conditions.

Often, user and network-oriented resources may be limited, requiring the system to reduce the user's measure of satisfaction. It is then important to determine which transport, user and system processing and signal production parameters should be modified, and when it is feasible to do so.

For the purpose of designing an integrated end-user and communications network transport system, it is convenient to make use of analytical methods that use mathematical expressions to relate the attained QoE rating level to the underlying QoS metrics that characterize the employed flow production and transport system. Such employed formulas, though yielding approximate estimates for evaluating QoE measures, are highly useful. They often incorporate parameters that are based on models of the underlying environment and scenery.

QoE for video is often measured via controlled subjective tests, where video samples are played to viewers, who are asked to rate them on a scale that typically consists of five points. Averaging the results yields a rating that is expressed as a **mean opinion score (MOS)**. The relation between QoE and QoS performance metrics is typically estimated empirically. Using a QoE objective prescribed by a user, one then determines the QoS performance measures that the transport network system should support. Inversely, when a network system offers a user certain QoS transport levels, one

Stream bit rate (Mbps)	Jitter (ms)	Duration of single error event (ms)	Packets lost per error event	Stream packet loss rate
10	< 50	≤ 16	17	≤ 1.24 E−06
12	< 50	≤ 16	20	≤ 1.22 E−06

Figure 5.7 Minimum Transport Layer QoE Performance Requirements for IP-HDTV.

would use the noted empirical estimation model to determine the QoE rating that the user should expect to experience.

Such a relationship is discussed in [27] and demonstrated in Fig. 5.7. It shows the minimum transport layer performance metric levels required to provide satisfactory QoE for the transport of HDTV streams over an IP network, using an encoding scheme such as H.264. We note the maximum-targeted values for message loss, message error rate and message delay jitter, under several prescribed bit-rate levels. It is observed that, at each data rate level, to reach a satisfactory QoE value, it is essential to provide sufficiently low message delay jitter (i.e., message delay variation) and packet loss levels. It is also essential to limit the number of errors occurring during short time periods.

Due to the complexity of the relationship between a user's desired QoE rating and the QoS values that should be accordingly imposed on the transport mechanism, it is advantageous to implement *dynamically adaptive* schemes to control the QoE/QoS relationship. This is accomplished by the DASH (Dynamically Adaptive Streaming over HTTP) system, as outlined in the video modeling section. In addition to using ongoing measurements to plan the transport process, it is also essential to track the ongoing performance fluctuations that impact the behavior of the communications network and the user system and adapt to them. While engaged in the reception of a video stream, a DASH user continuously analyzes the performance of the reception process, informing the sender in realtime as to adjustments that should be invoked in aiming to maintain a high QoE level.

Problems

5.1 Provide examples of Quality of Service (QoS) and Quality of Experience (QoE) metrics and discuss the differences between these performance measures.

5.2 Define network throughput metrics and provide several examples of such measures.

5.3 Define offered traffic and departing traffic message rates and discuss methods for their measurement.

5.4 A network system is loaded by packet flows whose aggregate offered traffic rate is equal to 25,000 [packets/s]. On average, 10% of the packets are blocked from entering the system while 5% of admitted packets are lost or discarded within the network. The remaining packets depart the network successfully. Calculate the network's throughput rate (accounting for successfully departing packets).

5.5 Consider the system's data described in the previous problem and assume that the average packet length is equal to 2500 [bits/packet] including 450 overhead [bits/packet]. Calculate the net throughput rate in units of [bits/s] and in [Erlangs].

5.6 Consider the system's data cited in the previous problem and assume that the attained gross throughput rate is equal to 65% of the system's throughput capacity. Calculate the system's gross and net throughput capacity rate.

5.7 Define and discuss several message delay measures.

5.8 Consider a network processor whose service operation is modeled as an M/M/1 queueing system. Messages arrive at a rate of 280 [messages/s]. The processor's service rate is equal to 300 [messages/s]. The average message service time is equal to 40 ms. Calculate:
a) The message throughput rate, expressed in [mess/s], in [bps] and in [Erlangs].
b) The normalized message throughput rate.
c) The average message delay time in the system.

5.9 Define and discuss several error-rate measures.

5.10 Discuss the definition of availability and reliability measures and describe methods for their measurement.

5.11 Consider a network system that stays operational for an average period of 300 hours prior to failing and subsequently being repaired within a period of 5 minutes. Calculate the fraction of time that this system is operationally available.

5.12 Identify and discuss several malicious scenarios that degrade the operational integrity and security of network operations and describe approaches that can be used to promote the cyber security of the system.

5.13 Consider a Radio Access Network (RAN) of a cellular wireless network, such as that implemented by an LTE system. Point out and discuss the various performance metric values that are associated with the following service types:
a) Guaranteed Bit Rate (GBR).
b) Non-guaranteed Bit Rate (non-GBR).

5.14 Identify message delay performance metric values that are associated with various error-intolerant and error-tolerant applications that are also classified as time-delay sensitive or time-delay insensitive.

5.15 Provide illustrative transport layer performance values that are often associated with satisfactory QoE for IP-HDTV streams.

6

Multiplexing: Local Resource Sharing and Scheduling

Objectives: *A network or user node shares its processing and communications resources among multiple message flows that it handles. A **multiplexing algorithm** is employed for regulating the sharing process and for managing the order at which messages are scheduled for service. Illustrative multiplexing schemes include the following. A network switch's header processor shares its processing power in examining the headers of incoming packets. The packets are subsequently switched to an output buffer, waiting there for transmission across a shared outgoing communications channel link. A base station node in a cellular network shares the communications resources of its downlink communication channels in its message transmissions to targeted mobile users.*

Methods: *A multiplexing scheme is configured to accommodate message flows that may belong to different traffic and service classes, imposing different quality-of-service requirements, while efficiently utilizing shared resources. A multiplexing scheme may employ a procedure that allocates resources to message flows on a fixed multiplexing basis. Resources are then allocated to each flow on a relatively static basis. In turn, a statistical multiplexing scheme allocates resources to a flow only when the flow is active. Resources are then shared in a demand-assigned manner. Proportionally fair multiplexing algorithms are used to allocate shared resources to message flows in a priority-oriented, fair and equitable manner.*

Networking algorithms and resource allocation schemes are configured and dynamically adapted to provide network flows the transfer services that best meet their performance objectives. The operation of each protocol layer entity is accordingly configured.

A network system may offer a network layer scheme that a employs a circuit switching mechanism, as offered by traditional public-switched telephone networks (PSTNs). In turn, it can be based on using a packet-switching operation, as offered by the Internet. A hybrid of these methods may also be implemented. Across the medium access control (MAC) sublayer, a network system may offer a MAC scheme that is similar to that offered by a cellular wireless network in sharing its radio access network (RAN) communications resources, or follow an operation whose protocols and algorithms are similar to those offered by a Wi-Fi wireless-LAN (WLAN) system.

To streamline the description of the wide variety of networking schemes and algorithms that network systems employ, we use the following approach. We identify principles, strategies, and methods that are fundamental to the operation of a multitude of communications networking systems. The corresponding networking schemes are employed at network interfaces, at network access systems, and within the core network. Variations of the basic schemes that are described in the following are often formed for use at distinct layers and when applied to different networking layouts and services. In the following chapters, we describe the principles of operation of key networking schemes, as well as illustrate their application and operations by presenting and discussing specific system implementations.

Principles of Data Transfer Through Communications Networks, the Internet, and Autonomous Mobiles, First Edition. Izhak Rubin.
© 2025 The Institute of Electrical and Electronics Engineers, Inc. Published 2025 by John Wiley & Sons, Inc.

6.1 Sharing Resources Through Multiplexing

A communications network system employs a multitude of processing, storage, communications, and switching modules that are essential to its operation and performance behavior. The number of such network elements is low when compared with the vast number of messages and flows that make use of system resources as they are transported across a communications network. Hence, network elements must be **shared** among the many flows, messages and processes that simultaneously employ them. Processes that act to manage the efficient sharing of network resources are highly critical to the effective functioning of the message transport mission.

To illustrate, consider a transportation highway system. Vehicles share access and exit ramps, roads and intersections. Vehicles that wait on an incoming ramp are regulated for admission to the highway, as they share the spatial resources of entry and exit ramps, and as they share the use of highway lanes. The sharing by vehicles of entry ramps to a highway is illustrated in Fig. 6.1. The flow rate of on-ramp vehicles as they enter the highway may be throttled to meet system loading objectives.

Of key importance is the method used for **sharing a resource** among clients that are *assembling at a single location*. These clients gain access to a shared resource by following the rules prescribed by a local **scheduler**, or by following a *localized self-organization* method. Such clients are said to be **multiplexed** over the shared resource. The regulating module is identified as a *multiplexer (MUX)*, or a *multiplexing scheduler*. It schedules the admission of clients to the shared system.

Definition 6.1 Under a **multiplexing** process, a resource is **shared** by multiple **co-located** entities, such as data sources, messages, flows, clients, stations. These entities are said to be *co-located* in the sense that the underlying module, identified as the **multiplexer**, has direct, instantaneous and essentially low cost means to determine the status and service requirement of each entity. Illustrative shared resources include processors, communications media, and storage assets.

Figure 6.1 Vehicle Merging Demonstrating on Ramp Multiplexing.

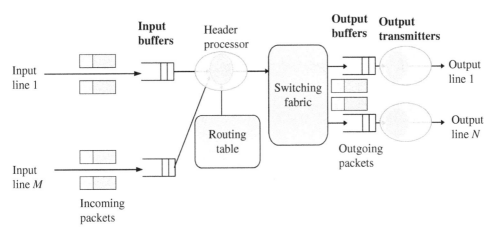

Figure 6.2 Input and Output Service Modules in a Packet Router.

As another example, consider messages that are temporarily stored in an end-user computer or at a network node as they wait to be processed by a local service module. Waiting messages are ordered for service as they form a temporary waiting line identified as a *queue*.

As another illustration, we examine the service modules embedded in a packet router. As shown in Fig. 6.2, a packet router consists of input and output service modules. The header processor serves as the input service module. Packets that arrive at the router are stored in input buffers (or in a single shared buffer), forming input queues (or a single queue) as they wait for service by the header processor. Once it becomes available, the header processor would select a waiting packet to process. It reads the destination address of the examined packet, and based on information stored in its routing table, the packet is switched to an appropriate output port, such as output port 1 that is connected to output link or line 1. The packet then joins a queue of packets that are stored in output buffer 1 as they wait for their turn to be transmitted across output link 1.

If it takes the header processor an average of β_1 seconds to service (examine and switch) a single packet, then the service rate of the header processor, also characterized as the router's packet switching rate, is identified as being equal to $1/\beta_1$ [packets/s]. For example, if the processor's service time is equal to $\beta_1 = 1$ [ms/packet], then the router's service rate (also identified as its switching rate) is equal to $1/\beta_1 = 1000$ [packets/s].

Packets queued in an output buffer are served by a transmission entity such as a radio module. When available, the radio service processor selects a waiting packet, such as one that is positioned at the head-of-the-line, and serves it. The service consists of transmitting the packet across an outgoing communications channel as it is transferred to the next packet switch. The packet's service time by the output module is equal to the time that it takes for the radio module to transmit it. This time is equal to the packet's length divided by the transmission data rate. If a packet is $L = 1000$ bits long, and the transmission data rate is equal to $R = 2$ Mbps, then the time that it takes to transmit this packet is equal to $S = L/R = 0.5$ ms. After a packet is transmitted by a radio module, the next packet, if any, is fetched from the associated output buffer and is transmitted.

As illustrated, a packet router manages two types of **multiplexing** operations. Arriving packets are queued in input buffers as they wait for selection and processing by the header processor. The header processor is shared among all packets that arrive to the router, in providing each packet a switching service. In turn, each output radio module is shared among the packets that require its transmission service. Outgoing message transmissions served by a transmission module are **multiplexed** across the outgoing communications link attached to this module.

Figure 6.3 Multiplexer–DeMultiplexer System Arrangement.

While, as noted above, different multiplexing configurations can be implemented, a generic multiplexing operation is demonstrated by considering the Multiplexer–DeMultiplexer arrangement presented in Fig. 6.3, whereby messages are multiplexed across a single shared communications link. As a single link is being shared, no switching function needs to be employed by the multiplexer. Stations $S1, S2, \ldots, SM$ represent information sources that become active at random times. When active, a station will drive its messages into the multiplexer. These messages are stored at the input buffer (i.e., memory/storage module) of the multiplexer. Managed by an input scheduler, they form a temporary waiting line, a **queue**. When the proper transmission module becomes available, after completing the transmission of a previously handled message (if any), the next scheduled message is transferred to this module. The module then transmits this message across its attached communications channel. The latter is the resource that is being shared by the messages that are waiting in the associated output queue.

When destined for reception by different entities, messages received across a communications channel are stored and queued at the input buffer of a de-multiplexer. These stored messages are examined by the de-multiplexer's controller, which determines the local entity to which each newly arriving packet should be directed, thus switching it to the proper destination entity. In the illustrative system, incoming packets aim to reach one (or several) destination stations $D1, D2, \ldots, DN$. In the shown configuration, a destination station is attached to an output port. It may represent a radio transmitter, a designated processor, an end-user node, or another module.

As a special case, consider a system configuration that consists of a single destination station, $DN = D1$. In this case, multiple message flows are multiplexed across a shared communications channel as they all aim to reach the same entity.

Under a system configuration that is known as **inverse multiplexing**, a single flow is decomposed into several sub-flow components. The superposition of these sub-flows yields the original flow. The sub-flows are transmitted in parallel across distinct communications channels. At the output of the communications channels, the sub-flows are combined to yield the original flow.

What are the methods, schemes and protocols that are being used to multiplex messages and flows across a communications channel? We discuss such mechanisms in this chapter.

6.2 Fixed Multiplexing Methods

A fixed multiplexing technique is defined as the following.

Definition 6.2 Under a **fixed multiplexing** method, also identified as a **static multiplexing** scheme, a client that is admitted for service, such as a message flow, is allocated a service resource on a dedicated basis.

6.2 Fixed Multiplexing Methods | 165

For example, a highway lane may be dedicated for use by a designated class of vehicles, such as those that are equipped with specified autonomous driving capability.

Consider the multiplexing configuration shown in Fig. 6.3. The multiplexer's control module monitors an incoming flow, observing its service requirements. It accordingly determines the level of resources that it is currently able to allocate for servicing the flow's messages. This allocation is maintained, at the same resource level, for the duration of the flow's activity or session. A scheduling module is assigned for implementing the resource allocation scheme.

The multiplexed message flows are received at the de-multiplexer module following their transmission across the shared communications channel. These messages are stored in the DeMux's input buffer and are then processed by a control module. The latter may, for example, forward all received messages to a designated processor, as would be the case in Fig. 6.3 when $N = 1$.

In turn, when $N > 1$, the de-multiplexer also acts as a **switch** as it directs each received message to its designated outgoing port.

Messages that are switched to the same output port are stored at an associated buffer and are then sent to the attached destination entity.

The data transmission phase is preceded by a control and management phase. Prior to engaging in the data multiplexing and transmission operation, a signaling process is used to configure an Si-to-Dj connection, which serves a message flow that originates at source Si and is destined to output port Dj. Since each flow is assigned a fixed and *dedicated resource across the shared communications channel*, the de-multiplexing (and switching) operation is readily accomplished. Upon monitoring the channel resource that has been used by a message that is received at the DeMux, and by then applying the connectivity information conveyed by the signaling process, the DeMux controller is able to identify the input–output connection with which an incoming message is associated. In this way, the DeMux determines the identity of the output port to which the received message should be transferred, proceeding to readily perform its switching function.

A signaling channel is configured and is jointly used by the Mux and DeMux modules to communicate as they exchange signaling, control and management messages. For a combined Switching-Multiplexing operation, the specification of the joint switching and resource allocation process is informed by performing the underlying **signaling** process.

The signaling channel is also employed to modify the specifications of the resource allocation and switching patterns. Yet, under a static multiplexing scheme, resource assignment modifications are not performed on a highly dynamic basis, so that the sharing schedule is nailed down (i.e., kept fixed) over a relatively long period of time. A flow is assigned a resource for a period of time that is sufficiently long to accommodate the provision of service to a large number of messages or bursts, as performed when employing a circuit-switching method. Connection modification, also identified as **programmability**, may be carried out by using a *network management* utility.

A multiplexing module may be connected to several outgoing communications links that connect to distinct nodes. In this case, the module acts as a switching node in that it switches incoming packets that belong to a specific flow to a designated outgoing port. It then multiplexes the messages that have been switched for service by the same port across the outgoing communications channel that is attached to this port. Switching operations will be discussed in Chapters 9–12.

As described in the following, we identify several different fixed multiplexing schemes in accordance with the type of resource that is being allocated across the shared communications channel.

6.2.1 Time Division Multiplexing (TDM)

Under a **time division multiplexing (TDM)** scheme, the system's resources are *time shared* on a fixed basis. To illustrate the functioning of such a system, consider a single server processing system. The service time of each customer takes 5 minutes. Customers are admitted into service at start times of 5 minute long time slots. The 12 slots that make every hour are designated as slot numbers 1 through 12. Upon arrival to the system, a customer is classified as a member of one out of 12 user classes. Under a nominal service arrangement, class-i customers are admitted to service only in slot-i of each hour-long period, $i = 1, 2, \ldots, 12$. As they arrive to the system, customers stay in a waiting room awaiting their turn. The system's manager checks the waiting room at the start of every 5 minute slot. If at the start of slot-i, it finds (at least one) class-i customer to be present in the waiting room, it admits a single customer of this class for service. Otherwise, no action is taken so that no customers are served during this slot and the slot remains idle.

Under this scheme, class-i customers are guaranteed an admission rate that is not lower than 1 customer/h. Their service rate is not interrupted by variations in the traffic loading rates imposed by the arrivals of customers that belong to other classes. If the system manager wishes to provide a higher service rate to a specific class of customers, possibly one that commands higher priority or one that operates at a higher loading rate, it can assign several time slots per hour for the service of customers of this class, while at the same time reducing the service rate (even downgrading it to 0) that is provided for the support of customers that belong to other classes.

As another example that illustrates the operation of a TDM scheme, consider a control station, such as a base station (BS), that receives messages from a core network for distribution to its client devices. Clients may include embedded sensor platforms, intelligent phones, or mobiles, which are locally managed by the BS node through a wireless communications channel.

Messages that are received at the BS are stored in its buffer as they await their transmission across a downlink wireless channel. Assume that currently N client devices are configured to share the capacity of a downlink wireless communications channel. Under a uniform traffic loading configuration, the BS node would configure an N-slot frame, dedicating its ith slot for the transmission of a buffered message that is destined to reach the ith device, $i = 1, 2, \ldots, N$. If at a time, no such message resides in the buffer, the ith slot remains idle. Such a configuration is illustrated in Fig. 6.4, where $N = 4$ devices are shown to time share the downlink channel for the reception of messages that are transmitted to them by the BS.

Such a time-sharing procedure is further illustrated and discussed in the following.

Consider a system for which the resources of a communications channel are segmented across the joint time–frequency plane as shown in Fig. 6.5(a). Across the time axis, time is divided into time slots, which are further organized into time frames. Each time frame is shown to consist of $N_{SF} = M$ equal size time slots, identified as $\{\tau_1, \tau_2, \ldots, \tau_M\}$. Assuming that each time slot is set to be of duration τ sec. The time-frame duration is then equal to $T_F = N_{SF} \times \tau = M\tau$ seconds.

The frequency bandwidth assigned to the underlying communications link is of size B (Hertz). Assume it to be segmented into $N_{FB} = N$ frequency bands, represented as $\{F_1, F_2, \ldots, F_N\}$. In the figure, the channel's bandwidth has been shown to be partitioned into $K = N$ equal size frequency bands, so that the bandwidth of each band is set equal to $B/N = F$ [Hz].

Under a TDM scheme, the communications channel is **time shared** among accommodated flows, following a predetermined time-sharing pattern. In each time slot, a single segment (at the underlying protocol layer, whether a message, packet, frame, bit or Byte) is scheduled for transmission across the shared communications channel. A source station, or flow, is allocated on a fixed basis an identified group of time slots in each time frame. When an accommodated flow

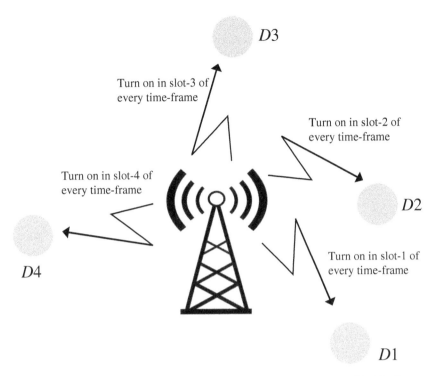

Figure 6.4 Sharing a Downlink Wireless Communications Channel on a TDM Basis.

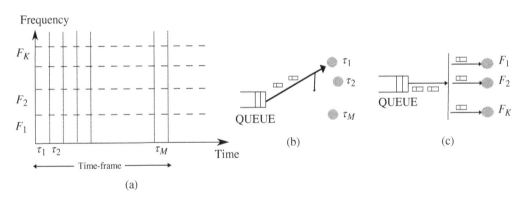

Figure 6.5 (a) Joint Time–Frequency Plane, (b) TDM Schemes, and (c) FDM Scheme.

is active so that it has message segments queued in the multiplexer's buffer, waiting for transmission, the multiplexer's scheduler will transfer these segments to its transmission module for transmission across the channel during the time slots that have been assigned by the multiplexer (MUX) module to this flow for this purpose.

The operation of a TDM multiplexer is further illustrated in Fig. 6.5(b), showing the MUX to transmit messages that belong to identified flows during their respective time slots, within periodically occurring successive time frames. The operation is described to be governed by the following process. At the start of a time slot, the scheduler examines the MUX's buffer occupancy to check whether it holds a segment that belongs to a flow to which this time slot has been allocated. If so, the time slot is used for the transmission of this segment. The scheduler then proceeds to examine the availability of a segment for transmission in the next time slot of the same time frame, or (when all time slots in the current time frame have already been examined) in the first slot of the next time frame. If the scheduler determines at a start of a time slot that there is no queued segment which has been assigned for service in this time slot, *this time slot remains idle and unused.* Subsequently, the operation moves to the next time slot.

Example 6.1 For example, $slot_1$ in each time frame can be reserved by the multiplexer for transmission of a segment (if any) that belongs to flow f_1, $slot_2$ is assigned for transmission of a flow f_2 segment (if any), while $slot_3$, $slot_4$, and $slot_5$ in each time frame are reserved for use for the transmission of flow f_5 segments. Based on such a schedule, the multiplexer is able to determine the segment transmission list that would take place during each time frame. Under this illustrative allocation scenario, the effective average data rates allocated to the three respective flows are proportionally set in accordance with the ratio 1:1:3.

Example 6.2 While transmitting a segment across a communications channel, a TDM multiplexer occupies the full magnitude of the channel bandwidth that has been allocated for this operation, say $BW = B$ [Hz]. If it employs a Modulation/Coding Scheme (MCS) whose spectral efficiency is equal to η [bps/Hz], its transmission data rate across this channel would be equal to $R_T = \eta B$ bps.

For example, if the bandwidth of the channel is equal to 10 MHz, and the spectral efficiency is equal to 2 [bps/Hz], then the transmission data rate is equal to $R_T = \eta B = 20$ Mbps. If each slot is of duration $\tau = 1$ ms, and if an average of 98% of a time slot carries payload data (while the rest is used for control overhead), each data segment is noted to accommodate a payload size of $0.98 \tau R_T = 19.6$ Kbits.

A **TDM circuit** consists of a prescribed number of dedicated time slots per frame for an ongoing number of frames, over a prescribed channel frequency band. (It is noted that slots may also be assigned in a nonperiodic manner for use by a circuit.) For a time frame that consists of M slots, if a flow is allocated a circuit that consists of a single time slot per time frame, this flow is assigned $1/M$th of the capacity of the shared channel. For example, assuming $M = 3$ time slots per frame, the latter circuit occupies 33.3% of the capacity of the channel. Such a TDM circuit is shown in Fig. 6.6, whereby frequency band $F1$ is shared on a TDM basis. A multiplexer can adjust the number of time slots (and frequency bandwidth) that it allocates to each flow in accordance with its capacity needs.

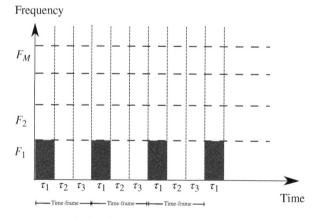

Figure 6.6 A TDM Circuit Consisting of a Single Time Slot per Frame in Frequency Band F1.

6.2 *Fixed Multiplexing Methods* | **169**

A multiplexer should not over-allocate its communications capacity resources. For example, consider a channel whose data communications capacity rate is equal to 20 Mbps. Assume each time frame to consist of 10 TDM slots. Capacity resources are then assigned in integral multiples of a 2 Mbps rate. If each flow is assigned a single time slot per frame, a maximum of 10 flows can be simultaneously multiplexed.

The communications circuit assigned by a MUX for serving a specific flow is dedicated for the exclusive use by this flow. Hence, over periods of time during which the corresponding information source is idle, the slots allocated to this flow remain unused and are therefore wasted. When a flow is active only 5% of the time, its allocated slots remain unused for 95% of the time, resulting in poor utilization of the channel's shared communications resources. In turn, the operation is efficient if multiplexed flows occupy their dedicated communications resources over a high fraction of time.

6.2.2 Frequency Division Multiplexing (FDM)

Under a **FDM** scheme, each flow, or station (or group of stations), is allocated a specific frequency band of the communications channel for its exclusive use.

Example 6.3   To illustrate the operation of an **FDM** system, consider a bank facility that serves its clients in the following fashion. The bank employs M tellers. Each teller specializes in performing a specific type of service or transaction. Customers that arrive to the bank wait in its waiting room until a teller that specializes in the execution of their transaction type of interest becomes available to serve them. Once this teller completes serving a client, she accepts for service the next waiting client that requires the specialized service that she provides. The traffic flow of arriving clients is classified into M traffic flow types. Members of a specific flow type require a specifically identified service type. In comparison, under a TDM scheme, a single teller that is capable to provide any requested service type may be employed. Her services are shared by all traffic flow types.

As shown in Fig. 6.5(a), the frequency bandwidth of a shared communications channel is divided into disjoint frequency bands. The operation of an FDM multiplexer is illustrated in Fig. 6.5(c). A frequency band is dedicated to the transmission of messages that belong to a specified flow. Messages that belong to several active flows can be transmitted during overlapping time periods, as long as each is formatted to occupy its distinct frequency band. For this purpose, the information presented in a message that belongs to a flow that is allocated frequency band $F_i, i = 1, 2, \ldots, K$, modulates a carrier that is filtered in a manner that restricts its frequency spectrum to selectively occupy its allocated frequency band. At an FDM de-multiplexer, a filtering system, corresponding to a bank of K filters, is used to capture the signals received in each band, de-modulate them and extract the embedded data. If a switching operation is then invoked, the switching configuration that was defined during the signaling phase is employed. In this manner, the identity of the band in which an incoming data flow has been detected to be received serves to indicate the output port (or ports) to which this flow should be switched.

In a period of time during which a flow remains inactive, the frequency band allocated to this flow remains unused, inducing lower utilization of communications capacity resources.

Example 6.4   In comparison with the above illustrated TDM operation, a corresponding FDM operation is described as follows. Assume the bandwidth of the shared channel to be equal to $B = 10$ MHz. It is divided into 10 distinct bands, so that each band spans a frequency spectrum that is 1 MHz width.

Assume each band to include guard-bands at its edges, serving to reduce cross-band signal interference effects, so that the useful bandwidth of each band is equal to 98% of its spectral span.

The employed MCS in each band is assumed to have a spectral efficiency that is equal to $\eta = 2\,\text{bps/Hz}$. The data rate used to transmit messages in each frequency band is thus equal to $R_F = \eta\ 0.98\ B/10 = 1.96$ Mbps. Hence, in comparison with the above described TDM operation, we note that, when fully loaded, both systems support 10 flows at an average data rate of 1.96 Mbps per flow. Under the FDM operation, the messages of each flow are transmitted continuously in time, occupying their own band. Each message, which is 19.6 Kbits long, is transmitted at a data rate of 1.96 Mbps, so that its transmission time is equal to 10 ms. In turn, under TDM, a 19.6 Kbits message is transmitted once every 10 slots, at a burst rate of 20 Mbps, so that the message transmission time is equal to only 0.98 ms.

Messages that belong to a specific traffic flow and that share a medium on an FDM basis are transmitted at a fixed data rate. Transmissions progress at a *consistent data rate level over time*, as long as a flow's activity persists.

In comparison, messages that belong to a traffic flow that shares the medium on a TDM basis are transmitted in a highly **bursty** fashion, as they occupy a wider system spectral band over a much shorter time duration (as short as a single time slot). Under a symmetrically loaded and provisioned system scenario, a TDM flow occupies a frequency band that may be about M times wider and is thus transmitted at a data rate that is about M times higher, yielding a message transmission time (such as a slot time) that is about M times shorter.

6.2.3 Wavelength Division Multiplexing (WDM)

When considering an optical communications medium such as an optical fiber link, transmitter and receiver modules are able to communicate simultaneously in time across distinct wavelengths, spanning disjoint spectral bands. In a manner that is similar to that used for FDM, we define **wavelength division multiplexing (WDM)** as a multiplexing scheme under which an optical channel is shared by allocating distinct wavelengths to distinct flows.

Optical links that accommodate the shared use of a large number of wavelengths are said to provide **dense wavelength division multiplexing (DWDM)**. Under this method, the system is able to make effective use of the ultrawide frequency bandwidth that is available when communicating across an optical medium.

By limiting the data rate used by each flow to a level dictated by the bandwidth of its employed wavelength channel, it is possible to process messages that belong to a flow through the use of mechanisms that operate at lower rate (say, at Gbps range), in comparison with the ultrahigh data rate that is attainable when communicating across an optical link, which can be of an order of Tbps.

A fiber-optic link that provides a bandwidth as wide as 1 Tbps can accommodate a DWDM operation that multiplexes data flows across 1000 light circuits, whereby each circuit enables a transmission at a data rate level of 1 Gbps. Using the latter circuit rate is compatible with the operation of software-centric mechanisms that are used to produce and consume many data flows.

6.2.4 Code Division Multiplexing (CDM)

Under a **CDM** technique, several flows are multiplexed across a communications channel in such a way that the flows share the temporal and spectral resources of the channel. For sharing the channel among M flows, a collection of M orthogonal CDM codes is defined. Each flow is encoded by using one of these codes. Code-i is used to encode flow-i. Multiple encoded message signals can be transmitted at the same time, jointly occupying a wide frequency band of the shared channel (which is expanded by an order M in relation to the bandwidth occupied by a single message flow). At a receiver that is a destination of each one of the flows (such as a BS that is associated with the stations that issue these flows), a bank of M decoders is used. The ith decoder is used to decode the superimposed signal received from the communications channel by using Code-i, $i = 1, 2, \ldots, M$. Decoder-i correlates well with the signal that represents flow-i. It thus produces an output signal that yields the original flow-i signal, decorrelating (and thus extracting out) other signals.

Since each flow makes use of a signal that occupies the full bandwidth of the shared communications channel, such an operation is also identified as a *spread spectrum* multiplexing scheme. Further exposition of the underlying method is presented at a later section where the principle of operation of a CDMA scheme is outlined.

6.2.5 Space Division Multiplexing (SDM)

Under a **SDM** scheme, different flows share a communications channel by using *spatially distinct sectors*. BS nodes in a cellular wireless system use multi-sector **directional antenna arrays**. Messages transmitted by a BS node by using a specific sector of the antenna are directed to mobile users that reside in a region that is covered by this sector.

By using a directional antenna to communicate signals across a communications channel, a higher antenna gain factor is realized. As the power emitted by the transmitting antenna is now concentrated into a narrower spatial volume, a receiver is able to receive the intended signal at a higher power level. This can result in a higher *signal-to-noise-plus interference ratio (SINR)* which leads to a better signal reception process, enabling the implementation of a higher data rate operation and lower error rate performance. Similar advantages are gained if the receiving device employs a directional antenna array.

When K directional antenna beams are employed by a multiplexer across a shared communications channel, the multiplexer is able to transmit in a time simultaneous manner K different flows (targeted to destination nodes that reside in distinct sectors) while *occupying the same frequency band* across each space segment. This increases the spectral efficiency of the channel sharing system. It induces a K-level **frequency reuse** increase factor. For such a reuse operation, the sending node employs multiple transmission modules. It is beneficial to avoid signal interference across sectors. To reduce interference effects, it is possible to share distinct (particularly neighboring) sectors on a time division basis (so that such sectors are activated during different time slots). Alternatively, close-by sectors can be configured to operate at different frequency bands.

A **hybrid multiplexing mechanism** can also be invoked. For example, in employing a *joint SDM/FDM/TDM* scheme, a BS node can allocate different frequency bands for communicating across its downlink communications channel to different client groups. The use of each such frequency band can be shared on a TDM basis among flows sent by the BS to a group of client members. In addition, disjoint spatial segments can be configured by employing directional antenna arrays.

6.3 Statistical Multiplexing Methods

Fixed multiplexing methods are advantageous in that the performance behavior of messages belonging to a flow to which resources are dedicated is predictable, as its resources are reserved for its support for a specified duration. Its major disadvantage is that when a flow's activity is

6 Multiplexing: Local Resource Sharing and Scheduling

randomly fluctuating, resources dedicated to the flow over a longer term are not utilized efficiently. Resources may then remain unused during relatively long periods of time, leading to poor channel resource utilization.

Consider a communications system whose resources are shared among multiple flows, whereby each flow, when active, produces data at a burst rate of 1 Mbps. Assume however that each flow is active for an average of only 1% of the time. Also assume that the transmission data rate across the channel is equal to 20 Mbps. If this system is shared (over a long period of time) on a fixed multiplexing basis, only 20 flows would be accommodated. However, if the channel is used to accommodate each flow only during its burst activity, the number of flows that can be accommodated would increase substantially. Under this scenario, since each flow requires channel resources for only 1% of the time, the number of flows that can be sharing channel resources may be increased by an average factor of 100. Hence, rather than supporting just 20 flows, the average number of served flows could be as high as 2000. The corresponding increase is referred to as a **statistical multiplexing gain**. As will be noted later, while a high gain is attainable, the realized gain level would be somewhat lower. This reduction is enacted to attain sufficiently low message delay performance.

To illustrate, consider the operation of a banking branch. Assume now that each one of its tellers can handle any client transactions. Using a statistical multiplexing approach, the following operation is invoked. Arriving clients stay in a waiting room facility until one of the tellers becomes available. Then, a waiting client is admitted into service. In comparison with an above noted fixed multiplexing operation, an available teller does not remain idle when at least a single client is waiting for service.

Under a statistical multiplexing scheme, messages are allocated channel resources on a dynamic basis, based on demand and on server availability. A waiting message (or flow) is admitted into service as soon as the shared service system becomes available.

A commonly implemented statistical multiplexing scheme employs a **first-in first-Out (FIFO) scheduling policy**, also known as *first-come first-served (FCFS)*. Under this discipline, messages are served in *order of arrival*. Arriving messages are stored in a buffer as they await admission into service, forming a FIFO queue. The latter designates the order used in selecting waiting messages for service. Such a module is called a **FIFO buffer**. When the service system is ready to admit the next message, the **head of the line (HOL)** message, which is the message to arrive at the earliest time when considering currently waiting messages, is admitted into service. A FIFO-based statistical multiplexer is also identified in brief as a **FIFO Stat-MUX**.

A FIFO Statistical-MUX that orders messages for transmission across a shared communications link transmits its queued messages in a FIFO order, one after the other. The service sharing process can be dynamically performed across the joint time/frequency/space dimensions.

The operation of a **Statistical-TDM multiplexer** proceeds as follows. As illustrated in Fig. 6.5(b), arriving messages form a queue as they are waiting their turn to be transmitted across an outgoing communications channel. In the shown configuration, time slots are maintained, whereby a time slot is used for the transmission of a message segment. Time frames, which consist of M time slots, are defined. Once a slot becomes available, the MUX's scheduler selects a message segment from the input queue and instructs the transmission module to send it across the outgoing channel. The scheduler selects messages for service in following the order dictated by its configured scheduling algorithm, such as a FIFO policy. In contrast with the operation used by a fixed TDM scheme, time slots are not reserved for specific flows or flow types. Queued messages are continuously transmitted, occupying the channel's time slots for as long as the input buffer is occupied.

The operation of a **statistical-FDM multiplexer** follows in a similar manner, except that messages selected for transmission across the outgoing channel are now sent across a selected frequency band. The underlying configuration is depicted in Fig. 6.5(c), whereby a statistical (rather than static) scheduling policy is invoked. Frequency bands are utilized in a dynamic fashion, as they are used to serve queued messages. Under a FIFO scheduling policy, once one of the frequency bands becomes available, the HOL message is selected for transmission across this band. In contrast with a fixed FDM scheme, a frequency band is not pre-dedicated for use by a specific flow or flow type.

Under a **statistical-FDM-TDM MUX** operation, a group of joint frequency bands and time slots (and jointly possibly spatial sectors) is selected, as a subset of the multidimensional space shown in Fig. 6.5(a). Multidimensional joint resources are statistically shared in serving queued messages.

6.4 Scheduling Algorithms and Protocols

To provide flows with desired quality of service (QoS) levels, a control module of a multiplexer system monitors the traffic flows that desire to be served and those that are being served. It assesses the performance requirements that should be granted to such flows. It determines the resources that are required for the satisfactory support of an incoming flow that should be granted a QoS level and consequently decides whether this flow should be admitted into the system. It accordingly activates scheduling algorithms that manage the sharing of its service resources.

Shared resources include the temporal, spectral, and spatial assets that are available to the multiplexing system for serving its clients. For example, a BS node in a cellular wireless network uses a multiplexing algorithm to allocate resources across its downlink communications media. An allocation may consist of a specified set of time slots that are included in specified time frames, occupying an assigned frequency band and possibly also guided to use a designated spatial sector.

Incoming messages are stored in a local buffer, which serves as a waiting room facility. They are held in identified storage locations. The multiplexer's controller keeps track of each waiting message, as its designates each entry in a wait list by noting the message ID, traffic class, service class, performance and activity attributes, and the location in memory of the associated data. Messages may also be grouped as members of distinct queues. An entry in a *queue of pointers* identifies the location in memory of a specific message and may also identify its attributes.

A multiplexer's queueing management and scheduling policy may involve a simple implementation, such as one that is based on serving all arriving messages by using a **FIFO scheduling policy**. Under a basic version of such an implementation, all messages that arrive to the multiplexer are admitted into the system as long as there is available buffering space. Waiting messages are ordered for service by forming a FIFO queue. When a service resource becomes available, the **HOL** message is admitted into service.

Consider the output section of a packet router, where packets wait in an output queue to be transmitted by a radio module across an outgoing communications channel. Assume packets to be served on a FIFO basis. When the radio module completes the service of a message, consisting of its transmission across an outgoing communications channel, the packet at the HOL position in the queue is selected for service and is then scheduled for transmission across the outgoing channel.

Similarly, when employing a FIFO scheduling policy for selecting packets that arrive to a packet router for service by the header processor. Once the processor completes the service of a packet, examining its header to identify its destination address, searching a routing table, and determining the router's output port to which the packet should be switched, the service of the next HOL packet residing in an input queue is initiated.

As another example, consider a FIFO-based multiplexing operation by a station that transmits messages across a wireless channel to close-by client devices. As a wireless resource becomes available, the HOL packet residing in the transmission queue is examined. It is selected for service once a resource of sufficient magnitude becomes available. Otherwise, the scheduling controller may decide to wait until sufficient service capacity is available. Alternatively, it can select for service the next message in line whose service demand level can be accommodated.

In many situations, multiplexing systems need to offer a more complex and dynamically adaptable service. They need to accommodate multiple flow types with dynamically varying parameters and offer each flow type its desired QoS. A key performance metric is **message delay**, which includes the sum of its queueing delay and effective service time. The message queueing delay (or waiting time) represents the time spent by a message in the queue, waiting until it is selected for service. The message effective service time represents the time that it takes the service system to deliver the message across the outgoing channel so that it is properly received at its designated device and so that it then becomes available to serve the next queued message, if any. The statistical behavior of **queueing processes** incurred by messages of each flow type is of the essence. *For messages of a prescribed flow type, the dynamics of message queueing, scheduling and delivery processes are dependent on the ensuing system traffic loading conditions, service policy, service class affiliations, and the availability of service resources.* As described in Chapter 5, messages that belong to different flow types may impose distinct performance requirements. Related metrics include message delay, error rate, throughput rate, and message loss ratio.

Performance metrics associated with different flow types of an LTE cellular wireless network systems are illustrated in Section 5.2.6. A **QoS Class Identifier (QCI)** is used to identify the performance treatment required by a flow. In Fig. 5.4, performance levels required for each QCI-associated flow type are cited. In this system, a BS node that multiplexes flows across a downlink wireless channel, as it aims to reach mobile users, allocates shared downlink resources. It controls message queueing and scheduling operations in a manner that aims to best satisfy performance requirements imposed by such flows.

In a cellular wireless network system, such as LTE, certain flows are granted **guaranteed bit rate (GBR)** service. A minimum throughput rate level is guaranteed for the support of such a flow. Such a flow is also granted specified performance levels. No bandwidth resources are reserved for flows that are designated to receive **non-GBR** service treatment.

To perform such service functions in a high-speed multiple-flow multiplexing environment, the ensuing processing chain consists of the following stages: flow categorization, classification and characterization; flow admission control; formation of resource pools; scheduling within service pools. A schematic displaying such a multitier functional architecture is presented in Fig. 6.7 and is described in the following.

We identify each client of the multiplexer system as a **flow**. In general, a flow consists of a stream of messages that have common source(s) and common destination(s), and whose messages are characterized by common attributes such as traffic class and requested service class (or a QoS Class Identifier such as a QCI).

As shown in Fig. 6.7, the first stage examines incoming messages, evaluating flow identifiers, and categorizes them into flow classes. A flow class consists of flow members whose attributes assume similar parameter values. Such attributes can include the following parameters:

1) **Application Type**: Such as voice (VOIP), audio, imaging, video, interactive data, batch data, signaling, vehicle control, and management data.

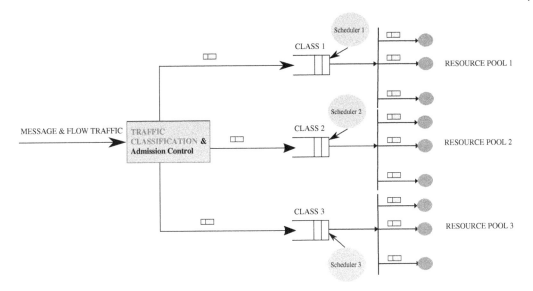

Figure 6.7 Resource Allocation and Scheduling at a Multiplexing Node.

2) **Traffic Descriptor**: Statistical characterization of the traffic process for each flow type. Traffic parameters that are specified at the call/session, burst, and packet levels can be included. For a simplified descriptor, a flow activity model describes the flow's message arrival process to follow a recurring on/off burst process, fluctuating between high-activity and low-activity (or idle) periods, each lasting for a possibly random period of time. Other stochastic models are employed as well for specific applications and traffic types.
3) **QoS**: QoS requirements that prescribe performance and service objectives, such as message priority, average delay, delay jitter, throughput rate, error rate, packet loss ratio, and performance class (such as GBR or non-GBR).
4) **Special Handling**: Requirements for special service features.

This multiplexer system would proceed to classify arriving messages as members of specific flow classes and identify their required service features and parameters. Following classification, each message may be attached a flow type identifier (FTI), such as a QCI.

A pool of service resources, denoted as RP_{f_i}, is set aside for use in providing service to group G_{f_i} of class-f_i flows. The maximum number of flows that the system is willing to admit for sharing a resource pool, denoted as NRP_{f_i}, is calculated through statistical analysis, and may depend on the system's current loading rate and profile. Admission calculations make use of modeling and analysis techniques that follow Queueing System-based modeling methods. Such models are outlined in a subsequent section. An employed adaptive multiplexing process is expected to produce **statistical multiplexing gains**, as a data flow is often active in a statistically fluctuating manner. This is illustrated by the following example.

Example 6.5 Consider a module that multiplexes flows across a shared communications link. The underlying resource pool consists of frequency bands that are shared on a statistical TDM basis. In each band, recurring time frames are defined. Each time frame consists of time slots.

Assume that a pool of time/frequency resources provides an overall service data rate of 20 Mbps to a group of type f_1 flows. Assume that each type f_1 flow requires a service data rate that is equal to 1 Mbps. The multiplexing system is required to provide service at this data rate during a burst activity period of a flow. Clearly, a maximum of 20 such simultaneously active flows can be served. However, if the pool is loaded by a larger group of such flows, depending upon the duty cycle of each flow (i.e., its activity ratio), messages will have at times to be delayed in the system's buffer as they wait for their transmission to take place since more than 20 messages may be queued for service at those times.

For instance, if each flow is active an average of 5% of the time, an average of 400 flows could be sustained if no message queueing delay limits are imposed. However, under a high loading rate, it would be likely that more than 20 flows will be simultaneously active, so that multiple messages could be delayed in the multiplexer's queue for relatively long periods of time.

To meet message queueing requirements imposed by a flow class, it is essential to implement a **flow admission control** scheme. It prevents the number of flows that are admitted into the system from exceeding a threshold level.

In Fig. 6.8, we identify key parameters that can be used in the operation and synthesis of scheduling mechanisms. Three groups of parameter categories, per flow and per flow messages, are noted:

1) **Message Input Time/Frequency/Space Parameters**: The synthesis and operation of a scheduling scheme is impacted by message traffic and requested service parameters, including, for example, the following:
   a) *Message arrival times* are used to order messages in the queue as they await admission into the service facility. Under a FIFO policy, messages are served in order of their arrival.
   b) *Message appointment times* are used to provide preference for the service of messages at times that are close to their desired appointment times.
   c) *Due-date scheduling* schemes provide preference for the admission of messages at times that are close to their due-dates, as may be specified in message headers.
   d) *Frequency band and bandwidth* attribute parameters for the service of messages belonging to the underlying flow class are taken into consideration in determining the scheduling of messages.
   e) *Spatial sectors* and the geographical coverage scopes that they require are taken into account in determining message scheduling.

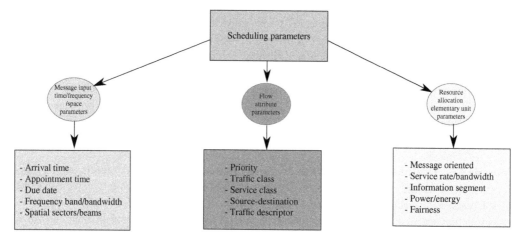

Figure 6.8 Scheduling Parameters.

2) **Flow Attribute Parameters**: Flow parameters are used to determine the structure of a scheduling scheme and the order in which messages are selected for service. For example:

a) *Priority-based ordering* is used to provide higher preference in the allocation of service resources and in system admission.

b) *Traffic Class* association considerations are used to impact the order at which messages are served. For example, conversational voice and realtime streaming video messages are given higher service priority than routine email and file messages.

c) *Service class* identifiers impose QoS requirements that are used to impact the order at which messages are served. For example, high volume data transfer flows may require the scheduling of high-throughput capable resources, while audio flows require the allocation of resources that accommodate lower-throughput rates but provide low message delay-jitter levels.

d) *Source–destination* identities may be highlighted by the system manager as requiring specialized resource allocation and scheduling treatment. For example, communications that involve identified source and/or destination entities of highly critical nature may impose requirements for special QoS and security measures that would lead to preferred resource allocations and scheduling treatments.

e) *Traffic descriptor* specifications of input flows identify the statistical dynamics of message arrivals for each underlying flow class. For example, a flow may produce messages in a highly bursty manner, exhibiting a low activity duty-cycle factor. Burst activities may not occur at a high rate but bursts could last for a relatively long periods of time, during which messages may arrive as a regular stream, or in a stochastic manner, at a high data rate. The scheduler must take into consideration this behavior in providing in a timely fashion sufficient resources for the support of such burst activities, as they are detected to occur.

   A scheduling controller may employ *machine learning (ML)* algorithms and artificial intelligence (AI) managed *traffic activity sensing* methods. Their outcomes are used to dynamically construct models that describe the statistical behavior of detected or prescribed classes of message flows and accordingly allocate resources and adapt scheduling processes in a manner that anticipates the statistical activities of tracked message processes.

3) **Resource Allocation Elementary Unit Service Parameters**: Different specified elementary service parameters are used to determine the basic resource elements that the system uses in allocating its resources and in fairly scheduling the service of its client flows.

a) *Message-Oriented* scheduling schemes make resource allocations on a message by message basis. The message unit used relates to the native operation of the serving entity. For example, the messages handled by a protocol layer entity are its associated protocol data units (PDUs). Such message units may be application layer messages, transport layer messages, network layer packets, link layer frames, MAC layer frames, and Physical layer PDUs.

b) *Service Rate/Bandwidth* features can be used to determine the elementary unit that is scheduled for service. For example, messages of a given flow are allocated prescribed time slot and/or frequency band resources. As further illustrated below, when considering multiplexing operations at a BS node that schedules messages for downlink transmission to client devices, resource availability and ensuing realizable data rates levels may depend on the features of the flow's destination device. The size of the elementary unit that is used for resource allocation purposes may then depend on the service rate level that can be realized.

c) *Information Segment* of a specified size may be used as the elementary scheduling unit. Several factors that impact the selection of a segment size have been noted above. Other impacting factors stem from the specific structure enforced by the service mechanism, the underlying network system and communications channel settings, or the flow's attributes. For example, assume an outgoing channel to be shared on a statistical-TDM basis. A time frame that consists of M equal time slots is defined. Messages that belong to a given flow, and are queued at a buffer as they await service, are transmitted in time slots that were allocated to them. When transmitted in an allocated slot, the transmission data rate of a flow's message may depend on the flow's transmission path or communications channel current parameters. As such, it can be affected by the location of the destination device, the interference environment in the latter's surrounding area, and the directionality of the antenna beam that is used to reach the destination device. The size of the information segment that would be set for transmission within a time slot will be accordingly determined. Flow messages would be fragmented into such segments. For sharing a wireless channel such as a downlink channel in a cellular radio access net, each segment is embedded as an information payload of a MAC frame. This segment is then the elementary message unit that this multiplexing system schedules and serves.

d) *Power/Energy* considerations may constitute key factors in determining the size of an elementary resource sharing unit. For example, when considering energy or power limited multiplexing modules, such as sensor-based systems that are battery powered, it is often necessary to limit segment sizes, transmit power levels, and activity period lengths to reduce energy consumption rates, in aiming to extend the operational lifetime of the module.

e) *Fairness* considerations are important in configuring, scheduling and synthesizing the resource allocation scheme when resources are allocated to multiple message flows and flow types. What should be considered to be a **fair allocation of resources**?

The service unit that is used as the basis for determining a fair allocation scheme can be based on any of the entities noted above, as well as on a composite thereof. When messages are used as service units, queued messages belonging to active flows are often served at an equal per-active-flow message rate. However, messages tend to assume varying lengths, which may depend on their flow class type. Even messages that are members of a single flow, or that belong to the same flow class, may assume variable lengths.

In these cases, when we allocate service resources on an equal per-message basis, the aggregate resources consumed by message activities could fluctuate widely. For example, consider a service channel that is operated at a given data rate. The service time incurred for the transmission of a 10 Kbits long message is 10 times longer than that used for the transmission of a 1 Kbits long message. When it is of interest to allocate the channel in a manner that is equitable in terms of the *cumulative transfer of information units*, a data segment that consists of a specified fixed number of channel symbols can be used as the elementary service unit. A message would then be fragmented into such data segments.

Active flows may be served in a manner that allocates equal time to the service of queued messages of different flows. Consider a period of time during which M active flows produce waiting messages in the multiplexer's buffer. Under a statistical-TDM scheduling mechanism, a cycle time, or time frame, of duration of M slots is defined. Each flow is allocated a single time slot for the transmission of its segment during each cycle in the underlying period.

As will be noted later, under certain conditions, messages that belong to different flows may require service at different data rates. Data segment transmission times will be longer when a lower data rate is prescribed. In this case, under a certain fairness measure, an equitable allocation of shared resources may be based on using an equitable time-sharing schedule. For example, each flow may be assigned a similar number of time slots per frame. In this case, a time slot is used as the elementary service unit, and resources are then allocated based on an equitable time sharing operation. A flow whose messages are transmitted at a lower data rate across the shared medium will transmit a lower number of bits during each one of its allocated time slots.

A variety of scheduling schemes can be employed for the sharing of a resource pool among admitted waiting messages. In the following, we describe several commonly employed scheduling policies.

1) **FIFO Queueing**: Messages are served in order of arrival, on a FIFO basis, also identified as a FCFS service discipline. This is the most frequently employed service policy.

2) **LIFO Queueing**: Messages are served on a Last-In First-Out (LIFO) basis. When a service resource becomes available, the last message to arrive is selected for service. The corresponding service module is often identified as a *stack buffer*. When you enter a bus and pay the driver for your ticket, your coins are inserted on top of a stack register and your change comes out from the top of the stack as well. When serving data, such a scheduling method is advantageous when the value of information decays with time, so that when a user is ready to draw information from the buffer, the latest message to arrive is regarded as most valuable. Such a queueing discipline is frequently invoked in radar and space exploration systems.

3) **Priority-Based Queueing** schemes are based on associating each message with a priority level. Higher-priority messages are served before lower-priority messages. Separate resource pools may be allocated for the service of messages of different priority levels.

    Different queues are formed and maintained for messages that are associated with different priority levels. When selecting messages for service, the higher-priority queue is scanned first. If it is not empty, messages are selected for service from this queue. Its service continues until the residual service capacity allocated to this pool is exhausted. If the buffer becomes empty when residual service capacity exists, or if empty when scanned, a lower-priority queue is instantly scanned and considered for service.

    Messages of the same priority class are ordered for service on a FIFO basis. Other scheduling algorithms can also be employed to serve messages of the same priority class, including the ones noted in the following.

Example 6.6 Consider a resource pool that is allocated 6 time slots in each time frame, over an assigned frequency band that is shared on a statistical TDM basis. Assume that a message (of any priority level) requires the use of a single time slot in a time frame for its transmission service. The pool is shared among multiple flows, whereby a flow can be designated to be of high priority (identified as priority-1) or of a lower priority (identified as priority-2).

Assume that at the start of time-frame-1, no messages are being served and that the pool's priority-1 buffer contains 4 messages and the priority-2 queue holds 6 messages. Consequently, 4 priority-1 and 2 priority-2 messages are selected for service during time frame-1. At the start of time frame 2, 4 priority-2 messages remain. Assume that at that time, 3 new priority-1 and 4 new priority-2 messages arrive. The newly arriving 3 priority-1 messages are served in time frame 2. In addition, 3 priority-2 messages are also selected for service during time frame 2. The remaining 5 priority-2 messages are kept at the priority-2 queue, as they wait their turn to enter service in forthcoming time frames.

Priority-based service policies are highly effective and commonly employed as they are simple to implement and can be used to provide fast service to messages flows that are of critical importance, when feasible. High-priority designations are often applied to signaling, control, and network management flows. Such flows tend to contain short messages that are frequently used to rapidly resolve network malfunctions or failures, adjust the allocation of resources to address overloading conditions, or act to prevent or suppress cyber-security attacks.

Under high overall traffic loading conditions, high-priority message arrivals can be kept isolated from resource contentions imposed by lower-priority traffic flows. The loading rate imposed by high-priority message flows is regulated to assure that such critical messages incur acceptably low message delay levels.

4) **Round Robin (RR)** schemes have been used in time-sharing systems to provide shorter waiting times to messages that require shorter service times. Upon admission to service of a head-of-the-line (HOL) message, it is provided service for only a fixed time period, identified as a **service quantum**, denoted as Δ. Following its service, if its requested overall service is not yet complete, the message is placed at the end of the FIFO queue. When it reaches again the HOL, the message is served for a period of time that does not exceed the service quantum period (Δ). If its service is not yet complete, the message returns again to the back of the queue. In turn, if its service is then complete, the messages departs from the system. In this manner, a message whose required service time is equal to, say, 5Δ, would circulate around in visiting the queue five-times before departing the system. Hence, messages that require longer service times would experience longer delay times.

To achieve such system delay behavior, whereby shorter messages would incur lower delays, a **Shortest Message First (SMF)** policy can also be invoked. To this end, the scheduler must be able to determine the service time requirements imposed by messages that are stored in its queue (e.g., by examining message control fields, or by estimating their lengths based on past behavior and/or associated application type). The message whose required service time is the shortest is then selected as the next one to be served.

5) **Round Robin** and other **Fair Queueing** disciplines are used to share a pool of resources among multiple flows in a **fair** manner.

Multitude of criteria and elementary service units can be employed for realizing a *fair* allocation of shared resources. One approach is to base the scheduling process on the *current state of the message occupancy process* in the system's buffer, whereby each message is identified by its flow. Active flows may be assigned equal service times.

Alternatively, equal number of message service units can be allocated per flow, whereby the message service unit can assume a variable length or be based on an integral multiple of a fixed length segment. In the following, we discuss such a scheduling method.

Consider the following *segment oriented round robin fair scheduling algorithm*. Service resources are shared on an equal basis in selecting for service queued messages that belong to different flows. Longer messages may be divided into multiple fixed length segments.

To illustrate, consider a scheduler that uses a time frame that consists of M time slots for scheduling messages on a time sharing basis. Assume that the scheduler allocates k time slots in each time frame for use by a specific group of flows. Assume each time slot to be of duration τ [s], and the transmission data rate to be equal to R [bps]. A segment transmitted in a time slot contains $L = R\tau$ [bits/segment]. The duration of a time frame is equal to $T_F = M\tau$ [s].

The scheduler keeps track of waiting packets by organizing them into multiple flow-queues. Packets that belong to a specific flow are placed in a FIFO flow-specific-queue that is associated with this flow. If the system has currently admitted N flows, N corresponding flow-queues are formed. Queue-i, denoted as Q_i, holds messages that belong to flow f_i. Assume that the scheduler scans the flow-queues in a cyclic order, and selects for service a single HOL packet from each non-empty flow-queue. Empty queues are instantly skipped.

For example, assume that currently three flows are active. Assume that the corresponding numbers of waiting segment-size packets for active flows f_1, f_2, and f_3 are equal to 2, 3, and 5, respectively. Upon its first scan of the queues, the number of packets selected per flow type, ordered per flow ID, are: $(1, 1, 1)$. At the second scan, the selected numbers are similarly equal to $(1, 1, 1)$. In subsequent scans, assuming no further message arrivals to take place, the selections are equal to $(0, 1, 1), (0, 0, 1), (0, 0, 1)$.

Noting that if the current resource pool provides service to its client flows at a rate of k slots per frame, the pool can accommodate a maximum throughput rate, also known as its throughput capacity, that is equal to $R_p = k/T_F$ [packets/s], or equivalently, $R_b = kR/M$ bps. When the system is fully loaded, so that a total of, say, N flows have been admitted and are simultaneously active, each flow is served at a rate that is equal to $1/N$th of the pool's service rate, or R_p/N [packets/s/flow]. In turn, if currently the system is loaded by a number of active flows that is lower by a factor of 50%, each active flow can at this time be allocated twice as many service resources. In this manner, messages driven by active flows share available resources in a fair manner.

Such a scheduling scheme is often identified as a **processor sharing (PS)** policy as it yields a fair sharing of a processor, a channel, or server resources. For example, if a processor is capable to serve packets at a rate of $R_p = 1000$ [packets/s], then over an underlying period of time during which 4 active flows have been admitted, each flow is effectively served at a rate of $R_p/4 = 250$ [packets/s].

Under the above-noted configuration, (equal length) message segments are used as the service quanta. This corresponds to a statistical TDM service operation under which message segments are transmitted across a shared communications channel at a fixed data rate. The service quantum can also be set as a message of variable length. Under such an operation, the service time of a message would vary accordingly.

6) **Weighted round robin (WRR)** and **weighted fair queueing (WFQ)** algorithms are used to share a pool of resources by multiple flows by assigning resources to flows on a **weighted** basis. When messages are used as the elementary sharing units, each flow is assigned a number, identified as its weight, that specifies the maximum number of message units that can be admitted into service relative to the corresponding maximum number of messages that are allocated for service to other flows during the same service time period, when sufficient service capacity is available. Service ratio weights are specified in a similar manner when other elementary service units are used.

To illustrate, consider messages as the elementary service units and assume the multiplexer to share the resources of an outgoing communications channel on a WRR basis among two active flows which are assigned weights at a ratio of 2:1. Assume the system to be able to serve a maximum of 3 messages per service cycle. When feasible, two flow-1 messages would be served per each single flow-2 served message. Assume that currently the flow-1 queue holds 5 messages and that the flow-2 queue holds 6 messages. Assuming no further arrivals to occur, the number of (flow-1, flow-2) messages that are served under this scenario in the next 4 service cycles is given as: $(2, 1), (2, 1), (1, 2)$, and $(0, 2)$.

Such a weight-based service algorithm can similarly be applied for the scheduling of message segments. When time slots are used as the elementary sharing units, the corresponding weights

6 Multiplexing: Local Resource Sharing and Scheduling

specify the ratios between the relative maximal time duration values allocated for the service of active flows. For example, under a weight ratio of 2:1 allocated to flow-1 vs. flow-2, two time slots will be assigned to serve class-1 segments per each 1 time slot that is allocated to serve a flow-2 segment, for as long as a sufficient number of such segments reside in their queues as they wait for service.

Similarly, weight ratios may be employed when considering elementary service units that are based on a count of information units.

Other scheduling schemes, such as those that are identified as *generalized processor sharing (GPS)*, and at times also as WFQ algorithms, are generally based on achieving an ideal fair allocation of resources when using a small elementary service unit, such as a 1-bit segment. Active flows are then supported in bit oriented fair manner. Such a mechanism is generally not practically feasible as actual data segments include control and other overhead fields. Such algorithms are often used to provide a reference for comparing measures of fairness rendered by various practical scheduling schemes, or as a basis for the dynamic adjustment of the operation of certain scheduling algorithms.

7) **Sliding Window-Based Fair Service** scheduling schemes are employed when it is of interest to allocate resources in a manner that takes into consideration the aggregate allocation made to each admitted flow over a recent period of time. Under such a scheme, flows that have been allocated a high quantity of resources over a recent period of time are assigned lower precedence for current allocations; otherwise, they would be granted higher allocation weights. The sliding window duration T_{SW} specifies the preceding period of time that is used in calculating the aggregate amount of resources allocated to a flow. The magnitude of such allocations over the time interval $(t, t - T_{SW})$ is taken into consideration when determining resource assignments made at time t.

8) **Tiered Resource Sharing** schemes employ multiple resource pools that are shared in a hierarchically tiered manner. Different scheduling algorithms can be employed in sharing the resources of different resource pools. Service resources that remain unused following the service of messages associated with a higher tier can be used to temporarily supplement resources that are assigned to a lower tier.

To illustrate, consider the hierarchical resource pool configuration shown in Fig. 6.7. Consider a statistical-TDM-based operation, though the concept is similarly applicable when using a joint temporal and spectral (as well as spatial) allocation process. Class-1 flows are served by the resources allocated to Resource Pool 1 (RP-1). Messages that are produced by these flows form class-1 flow-queues and are served (in sharing their pool's resources) by using the scheduling scheme that is specified for this pool. If during a service cycle, RP-1 is observed to have unused resources, these resources are added to the resources allocated to RP-2 and used within this period for the service of queued messages produced by class-2 flows, in accordance with the scheduling policy employed for the service of class-2 flows. This process continues in this manner down the tiers within each service cycle. When the service cycle period terminates, the resource allocation process repeats itself, restarting with the allocation of RP-1 resources.

A **Best Effort Resource Pool** contains service resources that are made available, when feasible, on a best effort basis, to messages that have not been provided QoS guarantees. Access of corresponding flows may not be subjected to regulation performed by the system's admission control mechanism. Such a resource pool is placed at a lower (or lowest) tier, so that resources that remain unused at higher tiers become available for use by messages that are served on a best effort basis.

9) **Due-Date Priority**-based scheduling methods make use of due-date specifications cited in the control field of received messages, or deduced by analysis of flow attributes. The priority level of a message that receives such a service may be upgraded as a targeted due-date approaches.

10) **Cost–Benefit**-based scheduling policies employ a **utility function** to determine the scheduling of active flows and queued messages. The system's manager aims to maximize this utility function. A utility function may be configured to account for factors that provide quantified benefits to the system and to its clients, while taking into consideration cost factors such as those that account for resource expenditures.

Typical factors employed by a composite utility function include measures and attributes such as throughput rate, message delay, error rate, priority level, energy consumption, transmission power levels, pollution effects, physical size and weight, cost, revenue, and operational expenditures.

Illustrative composite objectives include flow throughput to message delay ratio, TH/D, which expresses the attained throughput rate per unit message delay; flow throughput to incurred power ratio, TH/P, which expresses the attained throughput level per used unit of power, expressed in units of (bits/sec/watt); or, equivalently, expressing the number of transported information bits per consumed unit of energy, measured in information units per joule, such as (bits/joule).

6.5 Statistical Multiplexing Over One-to-Many Media

Consider a multiplexing operation under which the multiplexing station holds in its buffer messages that it receives from one or multiple sources. It is tasked to transmit these messages over a shared communications channel to multiple receiving devices or stations, whereby different messages may be targeted for reception by different devices. This is, for example, the case when considering the multiplexing function performed by a BS that transmits messages to mobile clients that reside in its cell within a targeted cell sector. A single radio transmitter and a sectorized (directional) antenna array are employed. The receiving devices are located in geographically diverse positions.

The wireless communications channels used for transporting data from the central station to different receiving devices exhibit different physical layer characteristics. Different MCSs may be employed to produce communications signals that are effectively transmitted across each channel, realizing different spectral-efficiency factors. Such a factor, denoted η, is expressed in [bps/Hz] units. The spectral efficiency level that is attained depends on the signal-to-interference-plus-noise ratio (SINR) measured at the receiver of the corresponding destination device. Higher spectral efficiency ratio levels are realized when the associated SINR value is higher, leading to higher transmission data rates per allocated frequency bandwidth level. These data rates are used to calculate the corresponding service rates realized by the multiplexer's scheme.

Assume the bandwidth level that is allocated for the sharing of the communications medium to be equal to B [Hz]. The sending station transmits flow-i data to a client's device-i, at present time t, at data rate $R_i(t)$. Under a realized spectral efficiency level for flow-i communications that is equal to η_i, the ensuing data rate level is equal to $R_i(t) = \eta_i B$ [bps]. For example, given $B = 10$ MHz, $\eta_1 = 2$ (bps/Hz), and $\eta_2 = 0.25$ [bps/Hz], we conclude that $R_1(t) = 20$ Mbps, while $R_2(t) = 2.5$ Mbps. User device-2 may be currently situated at a poor signal reception location (impacted, e.g., by blocking caused by local buildings and/or affected by stronger interference signals originating by nearby

184 | *6 Multiplexing: Local Resource Sharing and Scheduling*

transmission sources). If the system were to require the message transmission data rate to device-2 to be carried out at the same high data rate that is used for communicating to device-1, the bandwidth level that would be required for this purpose would have to be increased from 10 to 80 MHz.

Such a scenario could also occur under variety of other network multiplexing configurations. A common scenario under consideration involves different message flows which are sent by a single source and are destined to different user devices across a shared (wireline or wireless) communications medium. The destination users can be stationary or mobile. A wireline medium to which multiple terminals are attached, and which is managed by a central multiplexing computer, is known as a *multi-dropped network* configuration.

What are key considerations to apply in devising scheduling methods for such a multiplexing network configuration?

We are faced with the following dilemma. Should we provide the same transmission data rate for communicating with each user device, including those that are situated at this time at low-quality communications reception locations? If we do so, assuming a prescribed level of spectral resources, we would end realizing an overall degraded average throughput rate. Alternately, should we compromise by reducing the transmission data rate levels used for reaching destination devices over periods of time during which these devices are situated in low-quality reception areas? When operating this way, different flows will have to be assigned different service rates (such as data rates) at different times.

Users that are located at low-quality reception areas will then incur lower-throughput rates for as long as their lower-quality reception conditions persist. When considering audio, imaging or video messages, message lengths can be reduced by adjusting the compression ratios provided by employed compression algorithms. Such adjustments would produce lower-quality representations of the original signals while improving bandwidth utilization. In a cellular wireless system, such users tend to be situated further away from the BS, closer to the edge of the cell. As they move closer to the center of the cell, the quality of their signal reception improves. Such a selective data compression, or transmission data rate adjustment operation, tends to upgrade the overall averaged throughout rate incurred over the cell area.

In considering such systems, whereby a single multiplexing source is sending messages to different destination devices, identified as a **one-to-many shared medium configuration**, we observe the following scheduling approaches:

1) **Round Robin**-oriented scheduling mechanisms may be employed to assure fair allocation of resources to admitted flows. When messages serve as the elementary sharing units, such an approach is often identified as one that assures **absolute fairness**, as it provides different flows the same message service rate. Under such an operation, assuming a fixed bandwidth per-message allocation, certain messages will require longer transmission times. This operation may reduce the attained system-wide average throughput rate level.

2) **Maximum Throughput** scheduling mechanisms are designed to achieve the highest feasible average-throughput rate. Transmissions of messages that are destined for reception at user devices that are situated at higher-quality reception areas would be preferred. Flows are served at a higher-throughput rate during the time that their destination devices are situated at a higher reception quality location, while they will incur lower-throughput rates during other periods of time. This operation would lead to enhanced overall average throughput performance while producing higher variability in service rates, possibly inducing service fairness issues.

3) **Proportional Fair Scheduling (PFS)** algorithms are designed to strike a balance between the scheduling dynamic offered by a Round Robin scheme and that achieved by using a Maximum

Throughput scheduling approach. They aim to attain a high overall throughput rate level while offering each flow a non-negligible service rate level. In the following, we describe two versions of such an operation.

Under the first version, flow-i messages that are waiting for service at the multiplexer's buffer at time t are assigned service resources (relative to those assigned to queued messages of other flows) in proportion to a weight factor $w_i(t)$ which is computed as follows:

$$w(i, t) = \frac{R_i(t)^a}{AR_i(t - T, t)^b}, \tag{6.1}$$

where $R_i(t)$ denotes the throughput rate achievable at time t when providing service to a flow-i message; $AR_i(t, t - T)$ represents the average throughput rate allocated for serving flow-i messages over the sliding window period $(t - T, t)$ that stretches over a period of time that terminates at time t and that is of window duration T. The exponents a and b assume values in $[0,1]$, so that $0 \leq a \leq 1, 0 \leq b \leq 1$.

When setting $a = 1, b = 0$, the operation follows that of a highest-throughput rate scheduling scheme. When we set $a = 0, b = 1$, a Round Robin-oriented fair allocation scheme is instituted, as the scheduler aims to equalize the average data rates granted to different flows. In turn, when configuring $a = 1, b = 1$, the scheduling scheme would tend to yield a high-throughput rate schedule that is not too unfair.

A second version of a PFS algorithm is represented as follows. Assume the system to currently share resources among N active flows. Under the PFS scheme, the controller aims to schedule active flows so that it maximizes the sum over the flows of the logarithm of the throughput rate per flow, as induced by the service allocations made to the queued messages of each flow. Thus, under this scheduling scheme, the allocation to each active flow is calculated by aiming to maximize the following metric:

$$\text{Maximize} \sum_{i=1}^{N} \log(R_i), \tag{6.2}$$

where R_i denotes the service rate allocation made to serve flow-i messages, $i = 1, 2, \ldots, N$. It is noted that the logarithmic function accentuates the lower-throughput rate values while damping the benefit levels derived from higher-throughput rate values.

To demonstrate the implication of using such an objective measure, consider the following configuration. Assume resources are shared on a time division basis so that an active flow-i is allocated in each time-frame T_F a service time of duration T_i. Assume N flows to be active so that the time frame is set to be of duration $T_F = \sum_{i=1}^{N} T_i$. Thus, flow-i is allocated a fraction T_i/T_F of the shared service resources. Assume that when transmitting flow-i messages to their destination device, the communications channel's characteristics, and the ensuing MCS that is implemented, yield a spectral efficiency level that is equal to η_i [bps/Hz]. Consequently, assuming that a total bandwidth level of B [Hz] is allocated, the data rate calculated for use by flow-i is equal to $R_i = \eta_i B$ [bps]. Noting that flow-i is allocated service for only a fraction T_i/T_F of the time, its realized data throughput rate would be equal to $R_i T_i/T_F$. Hence, the objective function employed by this PFS controller is expressed as:

$$\sum_{i=1}^{N} \log(\eta_i B T_i/T_F) = \log\left(\left(\prod_{i=1}^{N} T_i/T_F\right)\left(\prod_{i=1}^{N} B\eta_i\right)\right). \tag{6.3}$$

In the above-stated equation, the term that expresses the second component ($\prod_{i=1}^{N} B\eta_i$) does not depend on the scheduling allocation. The first component is maximized by setting $T_i/T_F =$

6 Multiplexing: Local Resource Sharing and Scheduling

$1/N, i = 1, 2, \ldots, N$. Hence, the corresponding best scheduling scheme would configure an *equal time allocation for the service of each active flow* [9]. Flows whose messages are transmitted across higher-quality communications channel would then carry more data during their allocated time periods. In turn, flows that at the time communicate with devices that are located at lower communications quality positions, would carry a lower quantity of data.

Using such a PFS objective function, a similar result is concluded when frequency-based allocations are performed.

Problems

6.1 Describe the multiplexing operation that is performed by the header processor of a packet switch. Identify the following elements:
   a) The input flow and its associated traffic parameters.
   b) The queueing facility.
   c) The service module and its associated service rate and service-time parameters.
   d) The outgoing entity.

6.2 Consider the multiplexing operation that is performed by the header processor of a packet switch. Assume that the processor takes an average time of 0.01 [ms] to process each packet that it admits into service. Calculate the switching rate realized by this processor and the ensuing packet switch, measured in units of [packets/s].

6.3 Describe the multiplexing operation that is performed by a radio transmission module located at an output stage of a packet switch. Identify the following elements:
   a) The input flow and its associated traffic parameters.
   b) The queueing facility.
   c) The service module and its associated service rate and service-time parameters.
   d) The outgoing link.

6.4 Consider the multiplexing operation that is performed by a radio transmission module at an output stage of a packet switch. Assume that packets arrive at this stage at a rate of 1200 [packets/s] and that the average packet length is equal to 2400 [bits]. The data rate of the communications link that serves the radio module is equal to 4.8 [Mbps].
   a) Calculate the normalized loading of the transmission module.
   b) Calculate the average service time (i.e., transmission time) of a packet.
   c) Calculate the throughput capacity rate (i.e., maximal throughput rate) of the transmission module in units of [packets/s].

6.5 Describe the operation of each one of the following fixed multiplexing schemes and identify their key parameters and structural elements.
   a) Time division multiplexing (TDM).
   b) Frequency division multiplexing (FDM).
   c) Wavelength division multiplexing (WDM).
   d) Code division multiplexing (CDM).
   e) Space division multiplexing (SDM).

6.6 Describe and discuss the differences in the operation and performance behavior of fixed multiplexing schemes vs. statistical multiplexing procedures.

6.7 Describe the operation of several statistical multiplexing schemes and identify their key parameters and structural elements.

6.8 Describe the operation of statistical multiplexing schemes that employ one of the following scheduling methods and provide examples that illustrate in a numerical manner the operation of each:
a) FIFO.
b) LIFO.
c) Priority oriented.
d) Round robin (RR) and weighted round robin (WRR).
e) Fair queueing (FQ) and weighted fair queueing (WFQ).
f) Sliding window fair service.
g) Tiered resource sharing.

6.9 Describe the operation of statistical multiplexing schemes over a one-to-many shared communications medium under the use of the following scheduling schemes and under selected hybrids and provide numerically illustrations for each:
a) Round robin.
b) Maximum throughput.
c) Proportional fair scheduling (PFS).
d) Round robin and weighted round robin.
e) Priority oriented.

6.10 Consider the statistical multiplexing operation that takes place at an output stage of a packet switch. Assume the radio module, which acts as the server, to be attached to an output channel that operates at a data rate of 100 [Mbps]. Assume packets to arrive at this radio module at a rate of 0.1 [Mpackets/s] = 100,000 [packets/s]. The average packet length is equal to 800 [bits/packet]. This service system is modeled as an M/M/1 queueing system. A FIFO scheduling policy is employed.
a) Calculate the throughput rate delivered by the multiplexing radio module, expressed in [bits/s].
b) Determine the normalized loading and the normalized throughput of the multiplexing module.
c) Calculate the average time that a packet is delayed while being served at this output stage in the system (in waiting in the output queue and in being served).

7

Queueing Systems

The Process: *An entity that serves messages or customers is modeled as a **queueing system**. It may represent the operations undertaken by a wide range of systems, such as a communications networking protocol layer entity, a packet switch, a mobile user, a base station, or an access point that transmits messages across a wireless channel, a satellite communications station, a web server, an autonomous highway system, a hospital, and a business enterprise. Each such entity is loaded by message flows. It provides service to arriving messages in accordance with their requirements. Arriving messages (or clients) are held in a wait facility such as a message buffer, as they wait for their admission to a service module. Waiting messages form a wait line, identified as a queue. A scheduler determines the order used for admitting waiting messages into the service module. Upon completion of service, a message departs from the system.*

Methods: *A basic queueing system module is defined by its message arrival process, the attributes of the storage facility, the queues that waiting messages form, its scheduling policy, and its service operation. Queueing processes are modeled by using stochastic process techniques. Probabilistic analysis methods are used to derive mathematical formulas that provide for the calculation of the probability distributions of system performance metrics. Such measures characterize the statistical behavior of the system and of its served messages. Measures account for message waiting times, message throughput and blocking rates, system occupancy, resource utilization, and busy-period lengths. The ensuing formulas are used as tools for the analysis, design, and management of a queueing system and of its associated service modules. A queueing network model is used for the analysis and design of a network of queueing nodes.*

7.1 A Basic Queueing System Model

Many of the operations involved in transferring, transporting, processing, switching and storing messages in a communications network system involve sharing of service resources. The analysis and design of such service modules is carried out by employing modeling and analysis methods derived through the use of tools from **Queueing Theory**. In this section, we review key basic queueing models, without delving into the involved mathematical analyses. We describe the models and outline results that characterize their performance. Queueing models have been applied for the analysis of many systems, for example [11, 21, 33, 34].

A basic queueing system model is presented in Fig. 7.1. We note the system to use a wait facility and a service facility. Messages that arrive to the queueing system are stored in a wait facility, also identified as a buffer, which is used to temporarily hold messages that are waiting for admission into the service facility. Waiting messages form a queue, or a waiting line, as they are lined up in a

Principles of Data Transfer Through Communications Networks, the Internet, and Autonomous Mobiles, First Edition. Izhak Rubin.
© 2025 The Institute of Electrical and Electronics Engineers, Inc. Published 2025 by John Wiley & Sons, Inc.

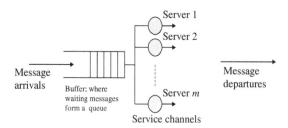

Figure 7.1 A Basic Queueing System Model.

manner that may determine the order of their admission into service. Effectively, messages may be assigned labels that identify their position in the queue. Such labels may be stored in a computer system as a separate list of ordered message pointers that indicate the location in memory where the actual data has been placed. The service facility is shown to consist of $m, m \geq 1$, service channels, whereby each channel is also identified as a server. When one of the servers becomes available, a selected queued message is admitted to service by this server. When its service has been completed, the message departs the system.

To characterize the elements of this basic queueing system model, the following processes and parameters are defined:

1) **Message Arrival Process**: The message arrival process is characterized as a Stochastic Point Process, denoted as $A = \{A_n, n = 1, 2, \ldots\}$. Its corresponding Counting Process is denoted as $N = \{N(t), t \geq 0\}$. The random variable A_n denotes the time of arrival of the n_{th} message. We set $A_0 = 0$. The counting variable $N(t)$ represents the number of messages that have arrived in the interval $(0, t]$, where we set $N(0) = 0$.

It is commonly assumed for this model that the arrival process is a *renewal point process*, so that the inter-arrival times $\{T_n = A_n - A_{n-1}, n = 1, 2, \ldots\}$ are statistically independent and identically distributed (i.i.d.) random variables. The inter-arrival time distribution function is denoted as $A(t) = P(T_n \leq t), t \geq 0, n \geq 1$. The average inter-arrival time is denoted as $E(T) = \alpha$. The message arrival rate is equal to $\lambda = 1/\alpha$ [messages/unit time].

The arrival process is often assumed to be a **Poisson point process**. This process is a renewal point process whose inter-arrivals are governed by an Exponential distribution: $A(t) = 1 - e^{-\lambda t}, t \geq 0$. The probability law of a Poisson process is characterized by a single parameter, its message arrival intensity, which is equal to λ [messages/unit time]. The average inter-arrival time is equal to $E(T) = \alpha = 1/\lambda$.

The message arrival process used for this basic model may represent message arrivals that are associated with a single flow or that are produced by a mix of multiple flows. For example, when the arrival process represents the superposition of two Poisson traffic flows, of class-1 and class-2, whose rates are equal to λ_1 and λ_2, respectively, then the superimposed process follows the statistics of a Poisson arrival process whose arrival rate is equal to the sum of the component arrival rates, $\lambda = \lambda_1 + \lambda_2$. A message that is selected at random from the composite flow of arrival messages would be a class-1 message with probability $\frac{\lambda_1}{\lambda_1 + \lambda_2}$.

2) **Requested Message Service Times**: The service time requested by the n_{th} arriving message is denoted as S_n. We assume that message service times $\{S_n, n = 1, 2, \ldots\}$ form a sequence of i.i.d. random variables. Service time levels are assumed to be statistically independent of the arrival process. The service time distribution is denoted as $B(t) = P(S_n \leq t), t \geq 0$. The average service time requested by a message is denoted as $E(S) = \beta$ [time units]. Its variance is denoted as $Var(S)$.

By assuming each server to serve one message at a time, we note that the server spends an average of β [time units] to serve a single message. The **service rate** of a server is observed when it is continuously occupied in serving messages. It is thus calculated as equal to $\mu = 1/\beta$ [messages/unit-time]. This parameter also represents the throughput rate capacity of a server, as it expresses a server's message output rate when it is kept continuously busy in serving messages.

A server that is loaded at a rate of λ [messages/s] and spends an average period of $\beta = 1/\mu$ [s/message] to serve each message handles an **Erlang loading** level of $f_s = \lambda/\mu$ [Erlangs/server]. This load value also expresses the fraction of time that the server is kept busy in serving messages.

For example, assume that a server is loaded at a rate of $\lambda = 100$ [messages/s] and that each message requires an average service time of $\beta = 8$ [ms] $= 0.008$ [s/message]. The server's service rate is equal to $\mu = 1/\beta = 1/0.008 = 125$ [messages/s]. The Erlang loading of the server is equal to $f_s = 100/125 = 0.8$ [Erlangs/server]. Consequently, this server is kept busy in serving messages for an average of 80% of the time.

The **traffic intensity** parameter, denoted as ρ, expresses the ratio between the rate of arriving messages that wish to enter the queueing system, also identified as the message offered rate, denoted as λ_o, and the overall service rate of the queueing system. For a system that uses m service channels (or servers), the system's overall service rate would be equal to $m\mu$. Assuming the offered message arrival rate to be equal to $\lambda_o = \lambda$, the traffic intensity parameter for this system is given by:

$$\rho = \frac{\lambda}{m\mu}. \tag{7.1}$$

3) **System Capacity/Admission Threshold**: The maximum number of messages that the system is willing to hold at any time may be specified as a finite level N, $N < \infty$. In this case, once the system is saturated, holding then a total of (in-queue and in-service) N messages, no additional messages are admitted into the system. Such a specification can be induced by buffer capacity, weight, power, energy, or processing power limitations. It is also used to represent the impact of a flow admission control function. A flow control throttling process is then used to prevent the system from becoming overloaded, protecting the quality of service (QoS) levels provided to messages that have already been admitted into the system.

4) **Service Discipline**: The service policy employed by the queueing system determines the order used to admit waiting (or arriving) messages into service. The **scheduling schemes** used by a multiplexing system discussed in Chapter 6 represent a wide variety of service disciplines.

The most commonly employed service policy is First-In First-Out (FIFO), also identified as First-Come First-Served (FCFS). When comparing other policies with a FIFO scheme, we note the following definition.

Definition 7.1 Work Conserving Discipline (WCD): A WCD is defined as a service policy that does not cause the service facility to perform more work or less work than requested by its customers. It provides each one of its served messages a service time that is no longer or shorter than the quantity of service (or work) that it requests. The service facility is assumed to employ non-lazy servers, so that a server would not refuse to render service when it is available and selected to serve.

Examples of work conserving policies include: First In First Out (FIFO), Last In First Out (LIFO), Round Robin (RR), Weighted RR (WRR), Shortest Message first (SMF), non-preemptive priority, and preemptive-resume priority. A preemptive-repeat priority service policy is not work conserving, as when the service provided to a message is interrupted in its midst and is thereafter resumed, repeated service times are incurred. In this case, the preempted message is inducing the system to perform excess work, as it occupies the service facility for longer time than its requested service time.

The following notation is commonly used to designate different versions of the basic queueing system model: $A/B/m/N$, where:

1) A = Designates the inter-arrival time distribution ($A(t)$). Unless stated otherwise, we assume the message arrival process to be a renewal point process.
2) B = Designates the message service time distribution ($B(t)$).
3) m = Number of servers or service channels.
4) N = System capacity or admission threshold. $N < \infty$ specifies the maximum number of messages that the system can hold at a single instant of time. When a value for N is not prescribed, it is assumed that $N = \infty$, so that the system is assumed to have no practical message storage or admission threshold limitations.
5) The following symbols are used to designate certain distributions:
   - M = Exponential distribution; M is used as when specifying exponential distributions, queueing processes are often modeled as Markovian stochastic processes.
   - D = Deterministic or Fixed (nonrandom) value.
   - G (or GI) = General distribution function can be incorporated into the model.
   - E_k = Erlang of order k distribution, which is a special case of a Gamma distribution. A variable which is equal to the sum of k statistically independent exponentially distributed random variables is governed by a Gamma (or Erlang) distribution.
   - Geom = Geometric distribution.

7.2 Queueing Processes and Performance Metrics

In the following, we present stochastic processes and performance metrics that are of prime importance in evaluating the behavior of queueing systems.

We aim to calculate the probability distributions of the underlying metrics and queueing process states when the system is assumed to have reached a stable *steady-state* behavior. A queueing system that does not limit its message occupancy size will reach steady-state after a typically short *transient* period of time, provided the system is not overloaded. The traffic loading rate of the system must be lower than the effective service rate of the system. The latter also indicates the *throughput capacity* of the system.

A queueing system that accommodates (or restricts admission to) only a finite number of messages at a time would reach its steady-state behavioral mode after a typically short transient period of time, independently of its traffic loading rate.

It is mathematically easier to derive analytical expressions that exhibit the system's performance behavior and use them to carry out analysis and design evaluations, by assuming the system to have reached steady state. It is the system's preferred operational mode, as otherwise the system is overloaded and its operation would then tend to exhibit degraded performance behavior.

For the basic queueing system model under consideration, the following are the conditions under which the system is assured to be able to reach a stable steady-state operational mode. Under this mode, system queues would hold (with high probability) a limited number of waiting messages, and an admitted message would then experience a finite average delay level.

Property 7.1 Stable Behavior of a Basic Queueing System: A basic queueing system that employs a FIFO service policy operates in a stable manner, reaching steady-state behavior, when it applies message admission controls so that the number of messages that it holds at any time is not higher than a finite occupancy limit level. Such an operation is induced when the system employs a finite message capacity buffering module. When no such message admission control operation is enacted, the system would reach steady-state operation if its message arrival rate is lower than its message service rate, so that:

$$\rho = \frac{\lambda}{m\mu} < 1.$$

It is noted that while the above-mentioned condition for stability was stated for a FIFO service policy, it is widely applicable for many other (work conserving) service policies, such as LIFO, non-preemptive priority and preemptive-resume priority policies, as well as other scheduling schemes. When considering the underlying basic queueing model and a FIFO-based service policy, we note that the above-stated inequality, requiring $\rho < 1$, serves as a necessary and sufficient condition of system stability. Thus, a corresponding (unlimited storage) system for which $\rho \geq 1$ is overloaded and would not be able to reach a stable steady-state operation.

Under a flow admission control scheme, the number of messages waiting in the system for service, at any instant of time, is limited to a finite number, denoted as N. Under a FIFO service policy, a newly admitted message will then have to wait for the service of at most $N-1$ other messages that have arrived earlier, experiencing a limited average wait time, assuming that each message requires limited (finite) average and standard-deviation service time levels.

In turn, if the number of messages admitted to the queueing system is not regulated, as may be the case when no flow admission control mechanism is employed, then for the system to reach steady-state conditions, it is generally necessary to require the message arrival rate to be lower than the system's message service rate.

To illustrate, consider a system that applies no flow admission control regulation and for which $\rho = 2$, so that $\lambda = 2\,m\mu$. In this case, we note that the average time elapsed between two message service completions, assuming all m servers to be busy, is equal to $\frac{1}{m\mu}$. During this period of time, the number of new messages arriving to the system is equal to $\frac{\lambda}{m\mu} = \rho = 2$. Hence, for each single message departure from service, 2 new messages arrive. Consequently, over time, the average number of messages residing in the system will keep constantly growing.

When flow admission control limitations are imposed, under high message offered rate conditions, so that $\rho \geq 1$, many arriving messages will be blocked from entering the system. To reduce the message blocking rate, it is advisable to lower the offered message arrival rate to the system, or increase the service capacity rate of the system. The traffic intensity factor would then be reduced to a lower value, preferably yielding $\rho < 1$. The maximum acceptable traffic intensity level that a system should support depends on the message delay value that is deemed acceptable as will be illustrated in Section 7.5.

The following are key stochastic processes and performance metrics that are employed for evaluating the performance of a basic queueing system model. For performance calculations, conditions are assumed to be such that the processes are stable and have reached a steady-state operational mode. For the stable queueing system models under consideration here, the underlying queueing processes are noted to behave in an *ergodic* manner. The statistical features of such a process can be deduced by observing its evolution over a sufficiently long period of time, as will be revealed by evaluating its behavior through the execution of Monte-Carlo-type simulation runs.

1) The **system size** process is denoted as $\mathbf{X} = \{X_t, t \geq 0\}$, where X_t (also denoted at times as $X(t)$) represents the system size at time t, which accounts for the total number of messages residing in the system at time t, including waiting messages that are held in the queueing facility and messages that are currently in service. The steady-state system size variable is denoted as X. The steady-state system size distribution is denoted as $P(i) = P\{X = i\}, i = 0, 1, 2, \ldots$, the average system size is denoted as $E(X)$ and its variance as $VAR(X)$.

2) The **number of messages in service (or service size)** at time t is denoted as $N_S(t)$ (also denoted as N_{S_t}). The corresponding stochastic process is $\mathbf{N_S} = \{N_S(t), t \geq 0\}$. The steady-state system size variable is denoted as N_S. The steady-state distribution is denoted as $P\{N_S = i\}$, $i = 0, 1, 2, \ldots$, the average system size as $E(N_S)$ and its variance as $VAR(N_S)$.

3) The **queue size** process is denoted as $\mathbf{Q} = \{Q_t, t \geq 0\}$, where Q_t (also denoted at times as $Q(t)$) represents the queue size at time t, which accounts for the total number of messages residing in the queue, waiting to be admitted into service. The steady-state system size variable is denoted as Q. The steady-state distribution is denoted as $P\{Q = i\}, i = 0, 1, 2, \ldots$, its average queue size as $E(Q)$ and its variance as $VAR(Q)$.

For the basic queueing system model under consideration, the system size at time t is equal to the sum of the queue size and service size levels at this time:

$$X_t = Q_t + N_S(t), \quad t \geq 0. \tag{7.2}$$

At steady state, we have:

$$X = Q + N_S.$$

4) The **message waiting time** process is denoted as $\mathbf{W} = \{W_n, n = 1, 2, \cdot\}$, where W_n represents the waiting time of the nth served message, which accounts for the time that the message spends waiting in the queue. Under a FIFO (or non-preemptive priority) service policy, it is equal to the time elapsed from the instant that this message arrived to the system to the instant of time that the message is admitted into service. The steady-state message waiting time variable is denoted as W. The steady-state waiting time distribution is denoted as $W(t) = P\{W \leq t\}, t \geq 0$, the average message waiting time as $E(W)$ and its variance as $VAR(W)$.

5) The **message delay time or system time** process is denoted as $\mathbf{D} = \{D_n, n = 1, 2, \cdot\}$, where D_n represents the delay time of the n_{th} served message, which is equal to the time that the message spends in the system, including its waiting time in the queue and its service time at the service facility. It is equal to the time elapsed from the instant that a message has arrived to the system to the instant of time that the same message departs from the system after its service has been completed. The steady-state message delay time variable is denoted as D. The steady-state delay time distribution is denoted as $D(t) = P\{D \leq t\}, t \geq 0$, the average message delay time as $E(D)$ and its variance as $VAR(D)$.

Considering a system that employs a FIFO service policy, or another WCD such as LIFO, a non-preemptive priority or preemptive-resume priority service policy, we observe that a served message spends in the service facility a period of time that is equal to its requested (and granted) service time, denoted as S. We then write:

$$D_n = W_n + S_n$$
$$D = W + S \tag{7.3}$$
$$E(D) = E(W) + E(S).$$

Under FIFO service policy, we have:

$$VAR(D) = VAR(W) + VAR(S).$$

The last equation follows by observing that W_n is statistically independent of S_n since under FIFO, the waiting time incurred by a message depends on the service times requested by messages that are positioned ahead of itself in the queue but not on its own requested service time.

6) The **unfinished load** (or **unfinished work**) process is denoted as $\mathbf{L} = \{L_t, t \geq 0\}$, where L_t (also denoted as $L(t)$) represents the unfinished load at time t, which accounts for the total amount of service that remains to be performed at time t, including service of messages that are currently waiting in the queue and the residual (remaining) service of messages that are currently being served. The steady-state unfinished load variable is denoted as L. The steady-state unfinished load distribution is denoted as $L(t) = P\{L \leq t\}, t \geq 0$, its average as $E(L)$, and its variance as $VAR(L)$.

7) The **virtual waiting time** process is denoted as $\mathbf{V} = \{V_t, t \geq 0\}$, where V_t (also denoted as $V(t)$) represents the virtual waiting time at time t, which is equal to the waiting time that would have been incurred by a hypothetical message that is assumed to arrive to the system at time t. The steady-state virtual waiting time variable is denoted as V. The steady-state virtual waiting time distribution is denoted as $V(t) = P\{V \leq t\}, t \geq 0$, its average as $E(V)$, and its variance as $VAR(V)$.

When observing the dynamic evolution of the system size process $\mathbf{X}$ of a basic queueing system, as illustrated in Fig. 7.2, we note that the system's state oscillates between busy periods and idle periods. A busy cycle consists of an idle period following by a busy period or a busy period followed by an idle period. The system size process thus evolves from one busy cycle to the next one. We define the following busy period-related variables and processes:

1) The **idle period** process is denoted as $\mathbf{I} = \{I_n, n = 1, 2, \cdot\}$, where I_n denotes the duration of the n_{th} idle period. An idle period represents a contiguous period of time during which the system is idle (i.e., it holds no messages). The idle period's duration is denoted as I, its probability distribution function as $I(t) = P\{I \leq t\}, t \geq 0$, its average as $E(I)$ and its variance as $VAR(I)$.

2) The **busy period** process is denoted as $\mathbf{G} = \{G_n, n = 1, 2, \cdot\}$, where G_n denotes the duration of the n_{th} busy period. A busy period represents a contiguous period of time during which the system is busy in serving one or several messages. The busy period's duration is denoted as G, its probability distribution function as $G(t) = P\{G \leq t\}, t \geq 0$, its average as $E(G)$ and its variance as $VAR(G)$.

3) The **busy cycle** process is denoted as $\mathbf{C} = \{C_n, n = 1, 2, \cdot\}$, where C_n denotes the duration of the n_{th} busy cycle. A busy cycle represents a period of time starting at the instant that the

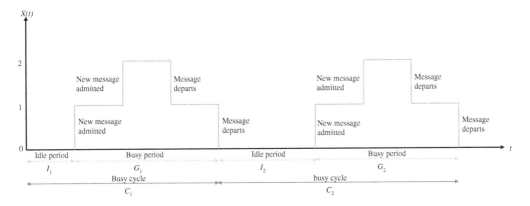

Figure 7.2 Realization of a System Size Process.

system transitions from idle state to busy state and ending at the subsequent instant that the system transitions from idle state to busy state. The busy cycle's duration is denoted as C, its probability distribution function as $C(t) = P\{C \leq t\}, t \geq 0$, its average as $E(C)$, and its variance as $VAR(C)$.

4) The **number of messages served during a busy period** process is denoted as $\mathbf{N} = \{N_n, n = 1, 2, \cdot\}$, where N_n denotes the number of messages served during the n_{th} busy period. The number of messages served during a busy period is denoted as N, its probability mass function as $N(k) = P\{N = k\}, k = 0, 1, 2, \ldots$, its average as $E(N)$, and its variance as $VAR(N)$.

In the following, we note additional performance measures that are often used as performance metrics, assuming steady-state conditions to hold:

1) The **system's utilization**, denoted as $E\{U\}$, is defined as the average fraction of time that the system is busy in the service of messages.
2) The **message blocking probability**, denoted as P_B, is defined as the probability that an arriving message is blocked from entering the system. It also represents the fraction of arriving messages that are blocked.
3) The **system's time-blocking probability**, denoted as P_{tB}, is defined as the probability that at any time the system is in blocking state. It also represents the fraction of time (over the long term) that the system is in blocking state.

Input and output traffic rate parameters are expressed by using the definitions presented in the Traffic section. In the following, several such key measures are noted:

1) The **offered message rate**, denoted as λ_o, represents the rate of messages arriving externally to the system. Not all such messages may be admitted into the system.
2) The **message throughput or message departure rate**, denoted as λ_D, represents the rate of messages departing from the system after their service was successfully completed.

   Assuming steady-state conditions to hold, we observe the following relationship when all admitted messages successfully depart the system, so that the *message carried rate* is equal to the message departure rate, as assumed for the basic queueing model:

$$\lambda_D = (1 - P_B)\lambda_o. \tag{7.4}$$

   When all arriving messages are admitted into the system, we set $P_B = 0$.
3) The **offered and carried Erlang loads** are calculated as $f_o = \lambda_o E(S) = \lambda_o/\mu$ and $f_D = \lambda_D E(S) = \lambda_D/\mu$, respectively.

7.3 Queueing Systems: Properties

In this section, we present several key properties of a basic queueing system model.

7.3.1 Busy Cycle Properties

The duration of a busy cycle (C) is noted to be equal to the sum of the inter-arrival times between admitted messages that are served during the busy cycle (within its associated busy period), yielding:

$$C = \sum_{n=1}^{N} T_{n,ad}, \tag{7.5}$$

where $T_{n,ad}$ denotes the n_{th} inter-arrival time between two successively admitted messages, and N denotes the number of messages served during a busy period. Hence, assuming the system to operate in steady-state mode and the involved average values to be finite, the average duration of the busy cycle, $E(C)$, is calculated as $E(C) = E(N) \times E(T_{ad})$. As $E(T_{ad}) = 1/\lambda_D$, where λ_D is the message departure (or throughput) rate, we have:

$$E(C) = E(N)/\lambda_D. \qquad (7.6)$$

If all externally arriving messages are admitted to the system, assuming it to have sufficiently high storage capacity, we have: $\lambda_D = \lambda = \lambda_o$.

The property expressed by Eq. 7.6 holds for a stable basic queueing system model under general parameter values and any number of employed service channels.

Consider next a queueing system model under which arriving messages are all admitted (so that $P_B = 0$). Also assume that the message arrival process follows the statistics of a Poisson point process with arrival intensity $\lambda_o = \lambda$. In this case, we note that the event $\{I > t\}$ is the same as the event that no messages arrive during a period of length t. Hence, we conclude that:

$$P(I > t) = e^{-\lambda t}, \qquad t \geq 0.$$

We conclude that the idle time period is exponentially distributed:

$$P(I \leq t) = 1 - e^{-\lambda t}, \qquad t \geq 0. \qquad (7.7)$$

The average duration of an idle period is therefore given by:

$$E(I) = 1/\lambda. \qquad (7.8)$$

Consider now an $M/G/1$ queueing system model. The system's message storage capacity is not limited, and message arrivals follow a Poisson process with intensity λ. We assume that $\rho = \lambda/\mu < 1$, where $E(S) = 1/\mu$, and that the system's operation has reached its steady state behavior mode. For this system, we obtain the following results:

$$\begin{aligned} E(N) &= \frac{1}{1-\rho}, \\ E(G) &= E(N)E(S) = E(N)/\mu, \\ P(0) &= 1 - \rho. \end{aligned} \qquad (7.9)$$

7.3.2 Little's Formula

The following property, known as **Little's Formula**, is applicable to a general class of queueing system models. It can also be applied to service systems that are modeled as networks of queueing systems, such as communications network systems. This result is based on the observation that when considering the evolution over time of the system size process **X** and of the corresponding delay time process **D**, the areas under their realized state curves that are calculated over any busy cycle period, are the same. This equality follows by the observation that for each unit of time that a message is delayed in the system, it adds itself (i.e., a count of 1) to the total count (**X**) of messages residing in the system. This is illustrated in Fig. 7.3. Similarly, we observe that for each unit of time

Figure 7.3 Equality of Areas Used to Derive Little's Formula.

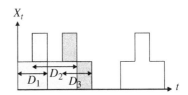

198 | 7 Queueing Systems

that a message is waiting in the queue, it adds itself (a count of 1) to the total count ($\mathbf{Q}$) of messages that are waiting in the queue. Hence, we write:

$$
\begin{aligned}
\int_0^C X_t\, dt &= \sum_{n=0}^N D_n, \\
\int_0^C Q_t\, dt &= \sum_{n=0}^N W_n.
\end{aligned}
\tag{7.10}
$$

Using this property, and the relation noted above stating that $E(N) = \lambda_D E(C)$, we obtain the following property. A stable system operation is assumed, so that it has reached steady-state and the average values of the state variables included in the following expressions are all finite.

Property 7.2   Little's Formula:

$$
\begin{aligned}
E(X) &= \lambda_D E(D), \\
E(Q) &= \lambda_D E(W).
\end{aligned}
\tag{7.11}
$$

By subtracting the latter two formulas, and noting that for a basic queueing system model we have $E(D) = E(Q) + E(S)$, $E(X) = E(Q) + E(N_S)$, and that the Erlang load departing from the system (which, under steady-state condition, is equal to the Erlang load admitted into the system) is expressed as $f_D = \lambda_D E(S)$, we conclude the following result:

$$
E(N_S) = f_D.
\tag{7.12}
$$

Example 7.1   To illustrate the implication of these results, we consider the output buffer of a router where packets are waiting to be transmitted by the radio module across an outgoing communications channel. The transmission data rate across the outgoing communications channel is equal to 100 Mbps. The served message units are identified as packets. The average packet length is equal to 200 [bits/packet]. The transmission rate is equal to 0.5 [Mpackets/s] (denoted also as Mpps).

This output module is modeled as a single server ($m = 1$) queueing system. The radio module is the server and the arriving packets are its customers. It is loaded at a packet arrival rate that is equal to $\lambda = 0.4$ Mpps. The service rate is equal to $\mu = 0.5$ Mpps. The average message service time is thus equal to $E(S) = 1/\mu = 2\,\mu s$. The traffic intensity parameter is equal to $\rho = \lambda/\mu = 0.8 < 1$. Since the arrival rate is lower than the service rate, we conclude that the system's operation is stable, transitioning rapidly in time to its steady-state behavior mode.

Noting that $\lambda_D = \lambda = 0.4$ Mpps, and $f_D = \lambda_D/\mu = 0.8$ [Erlangs], we conclude by using Eq. 7.12 that $E(N_S) = 0.8$. Thus, the average number of messages in service, representing the average number of messages occupying the communications channel while being transmitted, is equal to 0.8. Consequently, 80% of the time, the channel is occupied in transmitting a message, while 20% of the time, the channel is idle, as there are then no messages waiting for transmission.

We also note that for the basic system under consideration, the traffic intensity parameter $\rho = \frac{\lambda}{m\mu} < 1$ represents the fraction of time that the system is busy in serving messages. It also expresses the probability that the system is busy.

The throughput capacity of the system, expressing its maximum output rate, is equal to $m\mu$. Its throughput rate is equal to $\lambda_D = \lambda$. Hence, the normalized throughput rate, expressing the ratio of the system's throughput to its maximum throughput, is equal to $\frac{\lambda}{m\mu} = \rho$. For the example under consideration, $\rho = 0.8$, so that the system is loaded at 80% of its throughput capacity.

Assume now that for this illustrative service system, the average number of messages residing in the system (consisting of a single message in service, if any, plus messages waiting in the queue) is equal to $E(X) = 12$. As $\lambda_D = \lambda = 0.4\,\text{Mpps}$, we concluding by using Little's Formula that $E(D) = \frac{E(X)}{\lambda_D} = 30\,\mu s$. Hence, the average time that a packet spends in the module is equal to $30\,\mu s$, out of which the average transmission (or service) time is equal to $E(S) = 1/\mu = 2\,\mu s$. The average time that a message spends in the queue, waiting for admission into service, is then equal to $E(W) = E(D) - E(S) = 28\,\mu s$.

Using Little's Formula, we have $E(Q) = \lambda_D E(W)$. We then calculate $E(Q) = 11.2$, so that an average of 11.2 packets occupy the waiting queue, as they await their turn to access service via transfer to the radio module. We confirm this outcome by noting that $E(N_S) = \rho = 0.8$, so that $E(Q) = E(X) - E(N_S) = 12 - 0.8 = 11.2$.

Example 7.2 As another example, consider a network system at steady state. The throughput rate from the network, accounting for the aggregate outgoing message rate from all terminal network nodes, is equal to λ_D. The average system size, accounting for the cumulative average number of messages residing in all network nodes at a point in time at steady state, is equal to $E(X)$. Using Little's Formula, we conclude the average delay of a message in the system (averaging over all messages transported across the network) is equal to $E(D) = \frac{E(X)}{\lambda_D}$. For instance, if $E(X) = 120$ and $\lambda_D = 6000$ [messages/s], the average message delay is equal to $20\,\text{ms}$.

7.4 Markovian Queueing Systems

Impacted by the memoryless property of the exponential distribution, when we assume service times requested by messages to be exponentially distributed and message arrivals to follow a Poisson process, the ensuing system size process ($\mathbf{X}$) is characterized as a *Markov process*. For such a process, the Markov Property holds: given a present state and past states, the distribution of a future state depends only the present state. The analysis of **Markovian queueing system** models is carried out by using Markov process-based methods. In this section, we present performance evaluation formulas for several queueing system models whose state processes are Markovian.

A common method employed to analyze these systems is based on modeling the system size process as a continuous-time Birth-and-Death (*B&D*) Markov chain. The system size variables assume non-negative integer values. They form the system's **State Space**, which is denoted as $\mathcal{S}$, so that $X_t \in \mathcal{S}, t \geq 0$. Since only discrete state values are permissible, when a change in system's state takes place, it is represented as a jump in the state level. Such a process is therefore also identified as a *Markov Jump Process*.

As a B&D process, the model allows only jumps that are of unit magnitude. Hence, when the process enters state $k > 0$, it stays there for a random period of time (which is exponentially distributed) and then it jumps to either state $k + 1$, induced by a message arrival to the system, or to state $k - 1$, which is induced by a message departing from the system. Clearly, if $X_t = k = 0$, the next jump to take place must be triggered by an arrival, as no departures can occur from an empty system. A realization of such a process is shown in Fig. 7.2. Upon a message arrival, the system size state is noted to jump up so that a single message is added. When a message departs, the system size is noted to decrease by 1 message.

The parameters that are used to characterize the statistical behavior of such processes are the arrival and departure intensities (or rates), which are defined as follows:

Definition 7.2 **Arrival intensity:** $\lambda_i =$ admitted arrival intensity when the system size is equal to $i, i \in S$.

Definition 7.3 Departure intensity: μ_i = departure intensity when the system size is equal to $i, i \in S$.

The arrival intensity at time t, given the state of the system is $X_t = i$, denoted as λ_i, represents the admitted arrival rate at a time that the system contains i messages. It represents the average number of admitted message arrivals per unit time under the condition that the system contains i messages.

The departure intensity at time t, given the state of the system is $X_t = i$, denoted as μ_i, represents the departure rate of messages from the system at this time. It is equal to the average number message departures per unit time under the condition that the system contains i messages.

Assuming message arrivals to the system to follow a Poisson process with intensity $\lambda_o = \lambda$, the intensity parameter λ_i is calculated as follows. When the system imposes no message holding capacity limitations, so that $S = \{0, 1, 2, \ldots\}$, no message blocking takes place. Hence, $P_B = 0$ and all arriving messages are instantly admitted to the system independently of the current state of the system, so that we have:

$$\lambda_i = \lambda, \quad i = 0, 1, 2, \ldots . \tag{7.13}$$

In the following analyses, we set $\lambda_i > 0$, for each state i. In turn, when the system's capacity or admission threshold level is set at N messages, an externally arriving message is admitted into the system only if the system is not saturated. Otherwise, the message is blocked and assume lost. We then have:

$$\begin{aligned} \lambda_i &= \lambda, \quad \text{if} \quad i = 0, 1, 2, \ldots, N-1 \\ \lambda_i &= 0, \quad \text{if} \quad i = N. \end{aligned} \tag{7.14}$$

In the following analyses, we set $\lambda_i > 0$, for each state $i < N$.

The departure intensity μ_i is calculated as follows. Assume the queueing system to employ m service channels, $m \geq 1$. Each service channel operates at a service rate that is equal to $\mu > 0$. The mean message service time is equal to $E(S) = 1/\mu$. If the total number of messages that currently reside in the system (i) is lower than the number of service channels, $i < m$, then all i messages are in service. From these i in-service messages, the one whose service is completed first would be the next message to depart. Hence, $\mu_i = i\mu$. In turn, when $i > m$, exactly m messages are in service, so that the departure intensity at that time is equal to $\mu_i = m\mu$. Using $min(i, m)$ to denote the minimum value of (i, m), we have:

$$\mu_i = min(i, m)\mu, \quad i = 0, 1, 2, \ldots . \tag{7.15}$$

The system size process for the basic Markovian queueing system model under consideration is noted to be modeled as a continuous-time B&D Markov chain. Its evolution is characterized by the arrival and departure intensities noted above. For such a process, the steady-state system-size probability distribution $\{P(i) = P(X = i), i \in S\}$ is calculated by using the following result.

Property 7.3 Steady-State System Size Distribution For An Unlimited Storage Capacity Queueing System Whose System Size Process Is Modeled As Continuous-Time B&D Markov system:

The system reaches a stable steady-state operation when the following condition is satisfied:

$$\sum_{n=0}^{\infty} a_n < \infty, \tag{7.16}$$

where

$$a_n = \frac{\lambda_0 \lambda_1 \cdots \lambda_{n-1}}{\mu_1 \mu_2 \cdots \mu_n}, \quad n \geq 1,$$

$$a_0 = 1.$$

The steady-state distribution is then calculated as follows:

$$P(i) = \frac{a(i)}{\sum_{n=0}^{\infty} a_n}, \quad i \geq 0. \tag{7.17}$$

For a queueing system whose system size process is modeled as a B&D Markov process, and whose message storage capacity is limited to N, we similarly obtain the following.

Property 7.4 **Steady-State System Size Distribution For A System With Limited Capacity of N messages:**
The steady-state distribution exists and is given as:

$$P(i) = \frac{a(i)}{\sum_{n=0}^{N} a_n}, \quad i = 0, 1, 2, \ldots, N. \tag{7.18}$$

7.5 Performance Behavior of Markovian Queueing Systems

Using the results presented in Section 7.4 for Markovian queueing systems, we consider the performance behavior of several commonly used such system models.

7.5.1 M/M/1: A Single Service-Channel Queueing System

As depicted in Fig. 7.4, the $M/M/1$ queueing system model represents a service system that uses a single service channel (also identified as a server). Messages arrive to the system in accordance with a Poisson process at intensity $\lambda > 0$ [messages/s]. In the following, to simplify the presentation, we use *seconds* as the underlying units of time, unless stated otherwise. The message service time is exponentially distributed with average $E(S) = 1/\mu$ [s], $\mu > 0$. The message storage capacity is deemed to be sufficiently high. The system's (offered and admitted) Erlang loading level is equal to $f = \lambda/\mu$ [Erlangs]. Its traffic intensity parameter is equal to $\rho = \lambda/\mu = f$.

The system size process **X** is a continuous-time B&D Markov chain. Its arrival intensity is equal to:

$$\lambda_i = \lambda, \quad i \geq 0.$$

Figure 7.4 The M/M/1 Single Server Queueing System. Source: Izhak Rubin.

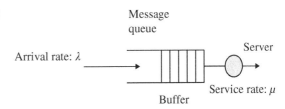

202 | *7 Queueing Systems*

Its departure intensity is given as:

$$\mu_i = \mu, \qquad \text{for} \quad i \geq 1.$$

Hence, the parameter a_i is calculated to be:

$$a_i = \rho^i, \qquad i \geq 0.$$

Consequently,

$$\sum_{i=0}^{\infty} a_i = \sum_{i=0}^{\infty} \rho^i < \infty \quad \text{if and only if} \quad \rho < 1.$$

Therefore, the system reaches steady-state condition if and only if $\rho < 1$.

When $\rho \geq 1$, the system is overloaded, so that the service mechanism is not able to catch up with arrivals. In this case, after the system operates for awhile, in the limit as $t \to \infty$, the number of messages residing in the system will keep growing beyond any finite bound. In this case, the system queues and message waiting times grow in an unbounded manner. Consequently, the average values for the system size, message waiting time, and message delay time would also assume very high values, so that $E(X) \to \infty, E(W) \to \infty, E(D) \to \infty$.

For $\rho < 1$, the system's steady-state system size distribution $P_X = \{P(j), j \geq 0\}$ is calculated by using the formula stated in Property 7.3, yielding the following result.

Property 7.5 Steady-State System Size Distribution for the $M/M/1$ system:

$$P(i) = (1 - \rho)\rho^i, \qquad i \geq 0, \quad \rho < 1. \tag{7.19}$$

Using this result, we obtain the average value of the system size at steady state, expressing the average number of messages that reside in the system to be equal to:

$$E(X) = \frac{\rho}{1 - \rho}, \qquad \rho < 1. \tag{7.20}$$

The probability that there are K or more messages in the system is obtained to be given by:

$$P\{X \geq K\} = \rho^K, \qquad \rho < 1. \tag{7.21}$$

For $\rho < 1$, the above-stated result also implies the following:

$$\begin{aligned} P\{X = 0\} &= 1 - \rho \\ P\{X > 0\} &= \rho. \end{aligned} \tag{7.22}$$

The parameter ρ expresses the ratio between the system's arrival rate and service rate. When $\rho < 1$, it also serves to convey other important performance indications. As indicated by the results presented above, at any time at steady state, the system is idle with probability $1 - \rho$ and it is busy with probability ρ. Also, since the throughput capacity of the system is equal to μ [messages/s], and the throughput rate is equal to $\lambda_D = \lambda$, the ratio of the latter two rates, which expresses the system's normalized throughput, is equal to ρ. Hence, for example, if $\rho = 0.8$, then the system carries and produces a load of 0.8 [Erlangs] and is busy for 80% of the time, and its normalized throughput is equal to 0.8. Its throughput rate is thus equal to 80% of its throughput capacity rate. If the server represents a communications channel that transmits messages at a data rate of 100 Mbps, then its throughput data rate would be equal to 80 Mbps.

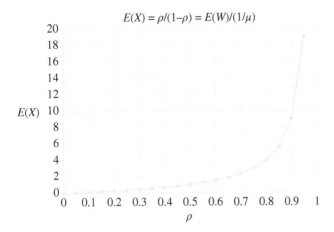

Figure 7.5 Mean System Size vs. Traffic Intensity for the M/M/1 Queueing System.

The average message delay time at steady state, $E(D)$, can be calculated by using Little's formula. Under $\rho < 1$, the system reaches steady-state conditions, so that we have $\lambda_D = \lambda_{Admitted} = \lambda_o = \lambda$. We obtain:

$$E(D) = E(X)/\lambda = \frac{E(S)}{1-\rho} = \frac{1/\mu}{1-\rho}, \quad \rho < 1. \tag{7.23}$$

The average message waiting time at steady state is then equal to

$$E(W) = E(D) - E(S) = \frac{E(S)\rho}{1-\rho} = E(S)E(X) = E(X)/\mu, \quad \rho < 1. \tag{7.24}$$

Using the above-presented formulas, we show in Fig. 7.5 the variation of the mean system size level, $E(X)$, with traffic intensity parameter ρ. We note that for $\rho < 1$, the depicted mean system size curve also represents the mean message waiting time per unit mean service time, $E(W)/E(S)$. As the traffic intensity parameter increases, the mean message waiting time and mean system size levels increase, initially at a slow rate, and then, as the traffic intensity approaches the stability threshold level, $\rho = 1$, at a faster and faster pace. The behavior of the system is thus *highly nonlinear*. It exhibits low message delay levels for values such as $\rho \leq 0.5$. At higher loading rates, message waiting time and system size values rapidly increase as the traffic loading rate increases.

Example 7.3 For a loading level of $\rho = 0.5$, we obtain $E(X) = 1$ and $E(W) = E(S)$, so that the average waiting time of a message prior to admission into service is equal to its average service time. Since $P(X > 0) = \rho$, the system is busy in serving messages during 50% of the time. In turn, under a high loading rate of $\rho = 0.9$, we calculate $E(X) = 9$ and $E(W) = 9E(S)$, so that the average message waiting time is now 900% higher!

When examining the tail probability, we note the wide variability of the number of messages residing in the system. For $\rho = 0.5$, we have $E(X) = 1$, but we also observe that $P(X > 1) = P(X \geq 2) = \rho^2 = 0.25$, so that 25% of the time, the system size is higher than its mean. For $\rho = 0.9$, $E(X) = 9$ and $P(X > 9) = \rho^{10} = 0.9^{10} \approx 0.349$, so that 34.9% of the time, the system size is higher than its mean.

Prompted by such wide variations in system's performance behavior that are induced by changes in the system's traffic loading rate, we often observe network and other service systems to exhibit highly variable performance behavior when subjected to random fluctuations in loading

204 | 7 Queueing Systems

conditions. Therefore, the system provides good response times at certain times while at other times, it takes a long time to get through.

The calculation of the steady-state message waiting time distribution $W(t) = P\{W \leq t\}, t \geq 0$ is carried out in the following manner. For $\rho \geq 1$, we noted that the limiting message waiting time increases in an unbounded fashion. Assume therefore that $\rho < 1$ and that the system has reached steady state.

Assuming henceforth a FIFO service policy, the waiting time of an arriving message depends on the number of messages residing in the system ahead of itself. Due to the memory-less nature of the Poisson arrival process, the number of messages in the system ahead of an arrival is noted to be equal to i with probability $P(i)$. The service time of each message ahead of an arriving message (including the residual service time of a message in service, if any) is exponentially distributed with mean service time that is equal to $E(S) = 1/\mu$. Hence, if there are i messages ahead of an arriving message, the distribution of the waiting time of the arriving message is equal to the sum of the service times of the these i messages. The distribution of the latter is calculated as the i_{th} order convolution of the exponential service time distribution of a single message. Averaging this convolution by applying the weight factor $P(i)$, we obtain the following result for the waiting time distribution.

Property 7.6   Steady-State Message Waiting Time Distribution for the $M/M/1$ system:

$$W(t) = 1 - \rho e^{-\mu(1-\rho)t} \qquad t \geq 0, \quad \rho < 1. \tag{7.25}$$

The message delay time is calculated as the sum of its waiting time and its service time, $D = W + S$. Noting that, under a FIFO service policy, the latter two random variables are statistically independent, we obtain the steady-state message delay distribution $\{D(t), t \geq 0\}$ to be given by the following.

Property 7.7   Steady-State Message Delay Time Distribution for the $M/M/1$ system:

$$D(t) = 1 - e^{-\mu(1-\rho)t} \qquad t \geq 0, \quad \rho < 1. \tag{7.26}$$

Example 7.4   Consider an access ramp to a highway system. Vehicles arrive in accordance with a Poisson process at a rate of $\lambda = 30$ [vehicles/minute]. The highway system is capable of admitting vehicles at a maximum rate of $\mu = 33$ [vehicles/minute]. Thus, the minimal average time elapsed between successive admissions of vehicles into the highway is equal to $E(S) = 1/\mu \approx 1.818$ s. The system is modeled as an $M/M/1$ queueing system, whereby the vehicles are the customers and the highway is the server. The traffic intensity parameter is equal to $\rho = \lambda/\mu = 30/33 \approx 0.909$. This access system is busy in admitting and serving vehicles for about 90.9% of the time. It is loaded at a service load rate of 0.909 Erlangs. The average time spent by a vehicle in waiting to enter the system is equal to:

$$E(W) = E(S)\frac{\rho}{1-\rho} \approx 18.16\,s.$$

The probability that a vehicle will wait on-ramp for longer than 20 seconds is calculated to be:

$$P(W > 20) = \rho e^{-\mu(1-\rho)(20)} = 0.334.$$

Hence, for the case under consideration, 33.4% of the vehicles will wait on ramp for longer than 20 seconds.

7.5 Performance Behavior of Markovian Queueing Systems

Example 7.5 Consider an output module in a router to which packets arrive at a rate of λ [packets/s]. The server is a radio module that transmits packets across an outgoing communications channel at a rate of $R = 10$ [Mbits/s]. Packets are of random length, and the transmission time of a packet is exponentially distributed. The average length of a packet is equal to $L = 2000$ bits. The average time that it takes to transmit (i.e., serve) a packet is equal to $E(S) = 1/\mu = L/R = 200$ μs $= 0.2$ ms. The system is modeled as an $M/M/1$ queueing system with arrival rate λ and service rate $\mu = 5000$ [packets/s].

The system planner aims to guarantee that at least 90% of the packets experience a delay time that is not higher than 1 ms. To this aim, the following condition needs to be satisfied:

$$P(D > t) = 1 - D(t) = e^{-\mu(1-\rho)t} = 0.1,$$

where $\mu = 5000$ [packets/s], $\rho = \lambda/\mu$, $t = 1$ ms, so that only 10% of the packets will experience a delay higher than the specified 1 ms limit value. We use this formula to calculate the highest packet arrival rate λ_M that a flow control mechanism should admit for service by this module.

The resulting calculation indicates that the maximum admitted message arrival rate should be set at $\lambda_M \approx 2697$[packets/s] ≈ 5.394 Mbps, so that the system operates at a traffic intensity level that does not exceed $\rho = 0.5395$, keeping the outgoing channel busy for at most 53.95% of the time. If we admit messages into the system at a higher message rate, the targeted message delay objective will not be satisfied.

Example 7.6 Consider a service system that is used to serve k flows. Each flow consists of messages that arrive to the system in accordance with a Poisson process at intensity λ/k [messages/s.]. The total message arrival rate is thus equal to λ [messages/s]. As shown in Fig. 7.6, we compare two service architectures that are used to allocate resources to these flows. Under the architecture shown in part (a), each flow is dedicated a service channel, whereby each such channel operates at a service rate of μ [messages/s]. Messages of an individual flow wait in their own dedicated buffer and are served by their own dedicated server (e.g., by using dedicated frequency bands and/or time slots).

Under the statistical multiplexing architecture shown in part (b), a combined flow that is formed as the superposition of the k flows is served by a single higher speed channel that operates at a service rate of $k\mu$ [messages/s]. Messages (irrespective of their flow affiliation) are queued in a shared buffer and are served (e.g., by using a FIFO service policy) by a shared high-speed channel.

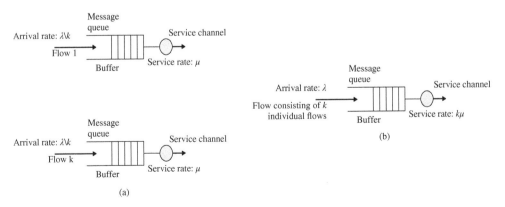

Figure 7.6 Statistical Multiplexing Gain: (a) Each Flow Assigned a Dedicated Channel; (b) Statistical Multiplexing of all Flows Across a Shared Channel.

206 | *7 Queueing Systems*

For each architecture, the overall message loading rate is equal to λ and the overall service rate is equal to $k\mu$. Hence, for each architecture, the overall traffic intensity parameter, which expresses the ratio of the overall input rate and the overall service rate is equal to $\rho_T = \lambda/(k\mu)$. We assume that $\rho_T < 1$ so that the system reaches steady state behavior.

Each service system is modeled as an $M/M/1$ queueing system. The part (a) system, which employs dedicated service channels, is modeled as k separate per-flow queueing systems. Each one of the latter is loaded at a rate λ/k and its server operates at a rate μ, so that the mean message service time is equal to $1/\mu$. The traffic intensity for each per-flow service system is equal to $\rho_a = \frac{\lambda/k}{\mu} = \rho_T < 1$. Hence, using the $M/M/1$ system results presented above, we conclude that the average system size for each per-flow queue is equal to $E(X_a) = \frac{\rho_T}{1-\rho_T}$. The average waiting time experienced by a message (served by its per-flow server) is equal to $E(W_a) = \frac{1}{\mu}\frac{\rho_T}{1-\rho_T}$. The average delay time experienced by a message is equal to $E(D_a) = \frac{1}{\mu}\frac{1}{1-\rho_T}$.

The traffic intensity for the shared queueing system is equal to $\rho_b = \frac{\lambda}{k\mu} = \rho_T < 1$. Hence, using the $M/M/1$ system results presented above, we conclude that the average system size for the shared queue is equal to $E(X_b) = \frac{\rho_T}{1-\rho_T}$. The average waiting time experienced by a message served by the shared server is equal to $E(W_b) = \frac{1}{k\mu}\frac{\rho_T}{1-\rho_T}$. The average delay time experienced by a message served by the shared server is equal to $E(D_b) = \frac{1}{k\mu}\frac{1}{1-\rho_T}$.

We conclude the following:

1) The dedicated per-flow service system and the shared service system are each characterized by the same traffic intensity level $\rho = \rho_T$. Hence, the average system size at each server module is the same, except that the dedicated service system consists of k queues, having thus to store an average total number of messages that is higher by a factor k than that stored by the shared queueing system.

2) The average waiting time experienced by a message that is served by the statistical multiplexing system (b) is lower than that experienced by a message served in the dedicated per-flow service system (a) **by a factor of** k. This significant advantage is characterized as a **statistical multiplexing gain**, as it is induced by statistically sharing a higher-speed channel among multiple flows. This is explained by noting that in the shared system, when certain flows produce temporarily lower message traffic loading of the system, messages that belong to other flows can take advantage of this situation by making use of the excess service rate that then becomes available. The same advantage factor is gained when comparing the incurred average message delay times.

3) If we wish the fixed multiplexing configuration (a) to yield the same average waiting time value as that attained by using statistical multiplexing configuration (b), the following adjustment in the operation of configuration (a) could be applied. The aggregate bandwidth that is made available for either configuration is assumed to be prescribed at a level that induces a total service rate $k\mu$. To meet the targeted waiting time objective, assume that the total message arrival rate that would be accommodated by configuration (a) is reduced to a lower level, denoted as λ_a. To illustrate, assume that we have $k = 10$ flows, and that we set configuration (b) to have a traffic intensity level that is equal to $\rho_b = \frac{\lambda_b}{k\mu} = 0.8$. The shared service channel is thus occupied 80% of the time under configuration (b). The ensuing mean message waiting time is equal to $E(W_b) = \frac{\rho_b}{k\mu(1-\rho_b)} = \frac{0.4}{\mu}$. The average message waiting time under configuration (a) is equal to $E(W_a) = \frac{\rho_a}{\mu(1-\rho_a)}$ where $\rho_a = \frac{\lambda_a/10}{\mu}$, noting that each per-flow queue carries a message rate that is equal to $\lambda_a/10$.

To achieve the same mean waiting time performance under both configurations, we set $E(W_a) = E(W_b)$. The ensuing calculation yields the following result. We should set $\rho_a = 2/7 \approx 0.286$, and subsequently limit the message arrival rate under configuration (a) to $\lambda_a = (20/7)\mu$. In comparing the latter rate with that handled by configuration (b), which is equal to $\lambda_b = 8\mu$, we conclude that $\lambda_b/\lambda_a = 2.8$. Thus, statistical multiplexing arrangement (b) would then accommodate a throughput rate this is higher by a factor of 280% than that supported by the service system that does not share its resources among different message flows.

7.5.2 $M/M/1/N$: A Finite Capacity Single Server Queueing System

Consider a single server queueing system that supports messages that arrive according to a Poisson process at arrival rate λ. Message service times are exponentially distributed with an average message service time that is equal to $1/\mu$. The system can however hold a total of only N messages at a time, including waiting and in-service messages.

The system size process is modeled as a B&D Markov Chain with admitted arrival intensity that is equal to $\lambda_i = \lambda$ for $i = 0, 1, \ldots, N-1$ and $\lambda_N = 0$. The departure intensity is equal to $\mu_i = \mu$, $i = 0, 1, \ldots, N$. Accordingly, using the formula presented by Eq. 7.18, we obtain the following result for the system size distribution at steady state (which exists for any value of ρ).

Property 7.8 **Steady-State System Size Distribution for the** $M/M/1/N$ **system:**

$$P(i) = \frac{(1-\rho)\rho^i}{1-\rho^{N+1}}, \quad i = 0, 1, \ldots, N, \quad \rho \neq 1$$
$$P(i) = \frac{1}{N+1}, i = 0, 1, \ldots, N, \quad \rho = 1. \tag{7.27}$$

The message blocking probability, denoted as P_B, expresses the probability that an externally arriving message would find the system saturated and would therefore be blocked. It is calculated by using the following equation:

$$P_B = P(N) = \frac{(1-\rho)\rho^N}{1-\rho^{N+1}}, \quad \text{if } \rho \neq 1$$
$$P_B = P(N) = \frac{1}{N+1}, \quad \text{if } \rho = 1. \tag{7.28}$$

Example 7.7 Consider a multiplexing or processing module at which packets arrive in accordance with a Poisson process. The packet service time is exponentially distributed. The system's packet holding capacity is limited to N packets, so that no more than N packets can be held in the system (including waiting and in-service packets) at one time. Assume the system's traffic intensity parameter to be equal to $\rho = \lambda/\mu = 0.8$.

Using the formula presented by Eq. 7.28, we calculate the following values for the packet blocking probability. If we set the message capacity (or admission threshold) level to $N = 2$, 5, or 10 messages, we calculate the blocking probability (also representing the fraction of time that the system is saturated, as well as the fraction of arriving packets that are blocked) to be approximately equal to 0.262, 0.089, or 0.0235, respectively. Thus, if we set $N = 2$, 26.2% of arriving packets are blocked, while by increasing the admission threshold to $N = 10$, we conclude that only 2.35% of the packets are blocked. As illustrated in Fig. 7.7, we observe accelerated decrease in the blocking rate as we keep increasing the system's admission threshold or capacity level.

As the system's capacity is finite, the system reaches steady-state behavior at any message loading rate. The message departing rate λ_D is equal to the message admitted rate λ_{AD}. These rates

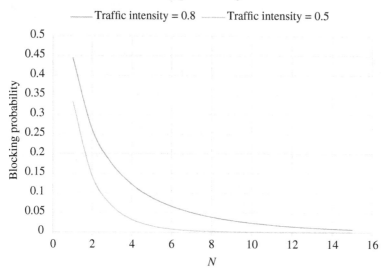

Figure 7.7 Blocking Probability vs. Message Capacity N for the $M/M/1/N$ System.

are calculated as follows in terms of the offered message rate, $\lambda_o = \lambda$, and the message blocking probability P_B whose calculation is given by Eq. 7.28:

$$\lambda_D = \lambda_{AD} = \lambda(1 - P_B). \tag{7.29}$$

Using Little's Formula, the average message delay time, $E(D)$, is obtained by first calculating the average system size, $E(X) = \sum_{i=0}^{N} iP(i)$, where the system size probabilities $\{P(i), i = 0, 1, \ldots, N\}$ are given by Eq. 7.27, yielding:

$$E(D) = \frac{E(X)}{\lambda_D} = \frac{E(X)}{\lambda(1 - P_B)}. \tag{7.30}$$

The mean message waiting time is then calculated as

$$E(W) = E(D) - E(S) = E(D) - \frac{1}{\mu}.$$

7.5.3 A Multi-server Queueing System

Consider a multi-server queueing system with messages arriving in accordance with a Poisson process at arrival rate λ and exponential message service times whose average is equal to $1/\mu$. The number of service channels is equal to $m \geq 1$. Such a configuration is depicted in Fig. 7.8.

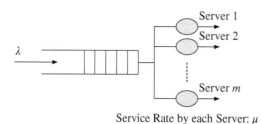

Figure 7.8 The $M/M/m$ Multi-server Queueing System.

The performance behavior of the system size process $\mathbf{X}$ is obtained by using the results presented in Property 7.3. The message arrival intensity is $\lambda_i = \lambda, i \geq 0$, and the departure intensity is equal to $\mu_i = i\mu$ for $i = 0, 1, 2, \ldots, m$, and $\mu_i = m\mu$ for $i \geq m$. The system reaches steady state if and only if $\rho = \frac{\lambda}{m\mu} < 1$. For $\rho \geq 1$, the system size and message waiting time values grow in an unlimited fashion as the system is overloaded.

Under the condition $\rho < 1$, we have $\lambda < m\mu$ so that the arrival rate is lower than the total system service rate. The message Erlang loading of the system is equal to $f = \lambda/\mu$ Erlangs. Under $\rho < 1$, the departing load rate is $f_D = f_o = f$ Erlangs, and the departing message rate, also identified as the message throughput rate, is $\lambda_D = \lambda$. The normalized throughput rate expresses the ratio of the throughput rate and the throughput capacity rate. The latter is equal to $\lambda_D(Max) = m\mu$ [messages/s], as each server can deliver messages at a rate that is equal to at most μ [messages/s] (producing this maximum rate when it is continuously busy). Hence, when $\rho < 1$, the normalized throughput rate is expressed as follows.

Property 7.9 Normalized throughput:

$$\text{Normalized throughput} = \frac{\lambda_D}{\lambda_D(Max)} = \frac{\lambda}{m\mu} = \rho, \quad \text{for} \quad \rho < 1. \tag{7.31}$$

For $\rho \geq 1$, the system is continuously busy in serving its customers, working at its maximum service rate $m\mu$. Hence, the throughput rate (or throughput) approaches its service rate, so that $\lambda_D = m\mu$. The normalized throughput is then equal to 1, or 100%, as the system is kept continuously busy in serving its messages.

The system size distribution is obtained by calculating the limiting probabilities as $P(i) = \frac{a(i)}{\sum_{n=0}^{\infty} a_n}$, obtaining the following expression.

Property 7.10 Steady-State System Size Distribution for the $M/M/m$ system:

$$P(i) = P(0)a_i, \quad i \geq 0, \qquad \rho < 1;$$

where

$$a_j = \frac{f^j}{j!} \quad \text{for} \quad j = 0, 1, 2, \ldots, m;$$

$$a_j = \frac{f^j}{m!m^{n-m}} \quad \text{for} \quad j \geq m;$$

$$f = \lambda/\mu; \tag{7.32}$$

$$P(0) = \left[\sum_{j=0}^{m-1} \frac{f^j}{j!} + \frac{f^m}{m!(1-\rho)} \right]^{-1}.$$

Using Little's Formula, the average message delay time, $E(D)$, is obtained by first calculating the average system size, $E(X) = \sum_{i=0}^{\infty} iP(i)$, where the system size probabilities $\{P(i), i \geq 0\}$ are given by Eq. 7.32, yielding:

$$E(D) = \frac{E(X)}{\lambda_D} = \frac{E(X)}{\lambda}, \quad \text{for} \quad \rho < 1. \tag{7.33}$$

The mean message waiting time is then calculated as:

$$E(W) = E(D) - E(S) = E(D) - \frac{1}{\mu}.$$

210 | 7 Queueing Systems

Example 7.8 In comparing the performance of an $M/M/m$ queueing system with that of an $M/M/1$ queueing system, when assuming both systems to be characterized by the same message arrival rate λ and the same total service rate μ, we note the following. It is statistically more efficient to employ a single higher speed channel, operating at service rate μ, than to use m service channels whereby each operates at a service rate μ/m. The reason being that when using multiple channels, the lower speed per channel operation reduces the efficiency of the statistical multiplexing operation.

To illustrate, consider the performance of configuration-1, a single server $M/M/1$ queueing system, whose channel operates at a service rate μ, as it is compared with the performance of a corresponding $M/M/2$ queueing system, identified as configuration-2. The latter uses $m = 2$ service channels, each of which operates at a rate of $\mu/2$. Each system is loaded at a message arrival rate λ. Assume the traffic intensity at system-1 to be equal to $\rho = \lambda/\mu = 0.8$. Using the $M/M/m$ system formulas presented above and Little's Formula, we obtain the mean message waiting time realized under the single service channel (system-1) to be lower than that incurred at system-2 by a factor of about 11%.

It is however noted that the highest service rate that can be implemented in a specific system may be limited by communications and networking considerations. For example, in a cellular radio access network, the highest data rate that can be employed by the base station to communicate with a mobile depends on the reception conditions detected by the mobile's radio receiver, relating to its current Signal-to-Interference plus Noise Ratio (SINR) level.

When the system is *limited to hold a maximum number N of messages*, the corresponding model becomes that of a $M/M/m/N$ **queueing system**. The system can hold at any time no more than N (waiting plus in-service) messages. We set $N \geq m$. Messages arrive in accordance with a Poisson process at intensity λ. The message service time is exponentially distribution with mean service time equal to $1/\mu$. The number of service channels is equal to m. The service rate of each server is equal to μ. The traffic intensity is equal to $\rho = \frac{\lambda}{m\mu}$. A message that arrives to the system when it is saturated, so that it contains N messages, is *blocked*. The analysis of this system follows the approach described above. In particular, we note the following.

The steady-state system-size distribution $\{P(i), i = 0, 1, 2, \ldots, N\}$ exists for any value of the arrival rate and is calculated in accordance with the results presented in Property 7.3. For this purpose, we note that $\mathbf{X}$ is a Markov Chain whose arrival intensity is given by $\lambda_i = \lambda$ for $i = 0, 1, 2 \cdot, N - 1$ and $\lambda_N = 0$. Its departure intensity is given by $\mu_i = i\mu$ for $i = 0, 1, 2, \ldots, m$ and $\mu_i = m\mu$ for $i \geq m$. The average system size $E(X)$ is calculated by computing $\sum_{i=0}^{N} iP(i)$.

The following formulas are obtained for the calculation of the steady-state system size probabilities.

Property 7.11   Steady-State System Size Distribution for the $M/M/m/N$ system:

$$P(i) = P(0)a_i, \quad i = 0, 1, 2, \ldots, N,$$

where

$$a_i = \frac{f^i}{i!} \quad \text{for} \quad i = 0, 1, 2, \cdot, m$$

$$a_i = \frac{f^i}{m!m^{i-m}} \quad \text{for} \quad m \leq i \leq N$$

$$f = \lambda/\mu, \tag{7.34}$$

$$P(0) = \left[\sum_{i=0}^{m-1} \frac{f^i}{i!} + \frac{f^m}{m!} \frac{1-\rho^{N-m+1}}{1-\rho}\right]^{-1}, \quad \text{for} \quad \rho \neq 1;$$

$$P(0) = \left[\sum_{i=0}^{m-1} \frac{f^i}{i!} + \frac{f^m}{m!} \frac{1}{N+1}\right]^{-1}, \quad \text{for} \quad \rho = 1.$$

The message blocking probability P_B is calculated by using the following expression:

$$P_B = P(N). \tag{7.35}$$

The message departure rate is equal to $\lambda_D = \lambda_{AD} = \lambda(1 - P_B)$. Using Little's formula, we obtain the mean message delay to be calculated as:

$$E(D) = \frac{E(X)}{\lambda_D} = \frac{E(X)}{\lambda(1 - P_B)}. \tag{7.36}$$

What happens when the service module contains no waiting room facility?

In this case, the system is modeled as an $M/M/m/m$ **queueing system**. The configuration of such a system is illustrated in Fig. 7.9. This system employs $m = N$ service channels. Each channel operates at a rate of μ [messages/s]. The message average service time is thus equal to $E(S) = \beta = 1/\mu$ [s]. The offered message arrival rate is equal to λ [messages/s]. The capacity of the system is equal to $N = m$, so that no buffering facility is made available to store messages that wish to wait for an available server. A message that arrives to find all N service channels to be occupied is blocked and assumed to be lost. An admitted message will occupy a service channel for a period of time that is assume to be exponentially distributed with mean of $1/\mu$. The offered Erlang loading rate of the system is equal to $f = \lambda\beta = \frac{\lambda}{\mu}$ [Erlangs].

To illustrate the application of this model, consider a traditional telephone system that provides a **pure blocking service**. The local telephone exchange that handles your call employs a limited number of outgoing trunks. It accommodates a maximum number of simultaneous calls that is equal to N. If, when you call, all the system's trunks are occupied, and the system is unable to provide you with a "camp on" (i.e., waiting) service, you will hear a busy tone and you will be considered **blocked**. An admitted call will occupy a service channel for the duration of the call, freeing the service channel when the call is completed.

Analysis of the $M/M/m/m$ queueing system follows as a special case of the $M/M/m/N$ queueing system, where now we set $N = m$, so that the capacity available to hold waiting messages is set to 0, noting that it is generally equal to $N - m$. Using the formulas presented in Eq. 7.34, we conclude the steady-state system size distribution for the $M/M/m/m$ system to exist for any value of $f = \lambda/\mu$, and to be given as the following.

Figure 7.9 A Service System That Contains No Queueing facility.

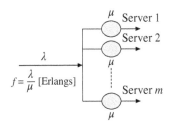

212 | 7 Queueing Systems

Property 7.12   System Size Distribution for the $M/M/m/m$ system:

$$P(j) = \frac{\frac{f^j}{j!}}{\sum_{k=0}^{m} \frac{f^k}{k!}}, \qquad j = 0, 1, 2, \ldots, m. \tag{7.37}$$

Using this result and Eq. 7.35, we obtain the following formula for the calculation of the message blocking probability, denoted as P_B, for the $M/M/m/m$ system. This formula is known as **Erlang's Loss Formula**, noting that it has been used by Erlang at the start of the 20th century to model call traffic behavior in a telephone system. We have the following.

Property 7.13   Erlang's Loss Formula:
The message blocking probability for the $M/M/m/m$ system is equal to:

$$P_B = \frac{\frac{f^m}{m!}}{\sum_{k=0}^{m} \frac{f^k}{k!}}. \tag{7.38}$$

Using a result presented earlier, we note that the average number of messages in service, $E(N_S)$, which is now equal to the number of messages that are in the system, is calculated as:

$$E(N_S) = f = \lambda/\mu. \tag{7.39}$$

It is interesting to note that Erlang's loss formula remains valid if the message service time follows any general distribution, under a prescribed average message service time. It thus applies also to an $M/G/m/m$ queueing system.

As the number of service channels m increases, we observe the formulas presented above for the steady-state system size distribution and for the blocking probability to converge to the corresponding expressions presented in the following. The service system is now modeled as an $M/G/\infty$ **queueing system**. This is a system that employs a sufficiently large number of service channels so that all message arrivals are admitted and assigned a service channel. It is loaded by messages that arrive in accordance with a Poisson process at rate λ. The message service time is governed by an arbitrary service time distribution. The average message service time is equal to $1/\mu$. The offered load is $f = \lambda/\mu$ [Erlangs]. Since there is a high number of service channels, no queueing (waiting time) delays are incurred. The system reaches steady state conditions under any loading rate. The steady-state system size distribution is given as the following.

Property 7.14   System Size Distribution for the $M/G/\infty$ system:

$$P(j) = e^{-f} \frac{f^j}{j!}, \qquad j \geq 0. \tag{7.40}$$

Example 7.9   Consider a telephone switching system (often identified as a Private Branch Exchange or PBX) used by a small business. It is loaded at a rate of 12 [calls/h], or 0.2 [calls/minute]. Arriving signaling messages serve as incoming requests for allocating channels to accommodate call connections. Requests thus arrive to the switch in accordance with a Poisson process at a rate of $\lambda = 0.2$ [calls/minute]. The holding time of each call (i.e., its service time, which is equal to the time that the call holds an outgoing channel) is exponentially distributed with the average call holding time being equal to $\beta = 1/\mu = 4$ [minutes]. The system's service rate (realized while it continuously serves calls) is thus equal to $\mu = 0.25$ [calls/minute]. The offered Erlang loading rate is equal to $f = \lambda/\mu = 0.8$ [Erlangs]. Hence, on the average, requests for a total service of 48 minutes arrive every 60 minutes. It is noted that a single outgoing channel can support a service request rate that is equal to at most 1 [Erlang].

Would it suffice to use a single outgoing channel?

Using the message blocking probability formula presented by Eq. 7.38, we conclude the following performance results. If a single outgoing channel is used, $m = 1$, we obtain $P_B = 0.4444$, so that 44.44% of arriving messages will be blocked. In turn, if we employ $m = 3$ outgoing service channels, the blocking probability is reduced in a significant way to $P_B = 0.0386$, so that only 3.86% of arriving message are blocked.

We note that since $f = 0.8$ [Erlangs], there is an average of 0.8 messages in service in the system at any time. Continuing to assume that $f = 0.8$ [Erlangs], if we further increase the number of service channels, the blocking probability will rapidly decrease. The reason for it is that there is now a higher probability that upon a new message arrival at least a single service channel will be available. As we set the number of service channels to m = 4, 5, or 6, we obtain the message blocking probability to decrease to 0.7%, 0.12% and 0.016%, respectively.

We further note that as the number of channels increases, we can approximate the calculation of the blocking probability by setting the expression for the system size distribution to be approximately equal to the Poisson distribution presented in Property 7.14, yielding:

$$P_B = P(m) \approx e^{-f}\frac{f^m}{m!}.$$

7.6 A Queueing System with General Service Times

The service times requested by arriving messages often follow a non-Exponential probability distribution function. This is, for example, the case when messages require deterministic (fixed) service times, as is the case for header processors. It is also the case for transmission modules that fragment messages into segments, whereby each segment is set for transmission in a time slot of fixed duration, or when considering telemetry and update messages produced by sensors, medical supervision and plant monitoring devices.

A service module can be loaded by message flows whose message lengths and corresponding service requirements may be stochastically variable and characterized by a specific distribution that could be non-Exponential. This is, for example, the case when a flow consists of messages that have been generated by a mix of applications. Such messages can be short (such as Voice over IP [VOIP] packets), assume medium lengths (such as email or modestly sized documents) or quite long (such as imaging, video, and long documents). This is illustrated by the following example.

Example 7.10 Consider a service module that transmits messages across a communications channel at a data rate that is equal to $R = 10$ Mbps. The module serves a flow of messages that arrive in accordance with a Poisson process. It consists of messages that belong to two classes. Class-1 messages are shorter, with an average message length equal to $E(L_1) = 500$ bits. The average transmission (i.e., service) time for such a message is equal to $E(S_1) = E(L_1)/R = 50\,\mu s$. Class-2 messages are longer, with an average message length equal to $E(L_2) = 17.75$ Kbits. The average transmission (i.e., service) time for this message is equal to $E(S_2) = E(L_2)/R = 1775\,\mu s$. Messages of either class are of random length, and the corresponding message transmission times are each modeled as governed by exponential distributions.

214 | *7 Queueing Systems*

A message is a member of class-i with probability p_i. Thus, a fraction p_i of served messages belong to class-i, $i = 1, 2$. Assume that $p_1 = 0.6$ and $p_2 = 1 - p_1 = 0.4$. Thus, 60% of served messages are class-1 members while 40% are class-2 members.

The average service time of a randomly observed message is equal to $E(S) = 1/\mu = p_1 E(S_1) + p_2 E(S_2) = 740\,\mu s$. Noting that for an exponentially distributed variable S with mean $E(S)$, the second moments is equal to $E(S^2) = 2E(S)^2$, we compute the following. The second moment of the service time of a randomly observed message is equal to $\beta_2 = E(S^2) = p_1 E(S_1^2) + p_2 E(S_2^2) = 2p_1 E(S_1)^2 + 2p_2 E(S_2)^2 = 2523500\,(\mu s)^2$. The corresponding standard deviation of the service time is then calculated to be equal to $\sigma(S) = \sqrt{\{E(S^2) - E(S)^2\}} \approx 1404\,\mu s$. The coefficient of variation $COV(S)$ of the service time variable is therefore equal to $COV(S) = \eta_S = \frac{\sigma(S)}{E(S)} \approx 1.9$. In comparison, if we assume the service time variable to be exponentially distributed, we would have $\beta_2 = 2E(S)^2$, $\sigma(S) = E(S)$, so that $COV(S) = 1$.

As noted in the Multiplexing section, a transmission module may handle different flows whose messages are destined to different stations. This is, for example, often the case when a base station sends message flows to distinct mobile client devices. As different receiving devices may be situated at different locations and are thus positioned at different distances from the transmitting station and subjected to different noise and interference signals, the sending station may have to employ different modulation/coding schemes (MCS) and hence different data rates for the transmission of messages to different stations. Therefore, the transmission times of messages destined to different stations will be distinctly different.

Thus, as illustrate by Example 7.10, when the service module is subjected to serving messages that belong to flows produced by different applications, the variations in the values of the required service times can be much different than those that characterize service times that vary in accordance with an exponential distribution. We expect the ensuing message delay time statistics to change accordingly. When the arrival message process contains messages that induce wider random fluctuations in their required service times, higher message delay times are expected. When the required service time variations are lower, we expect message delay times to be reduced.

How do such service time variations impact the resulting performance of the average message delay and waiting time performance? In the following, we present a model that answers this question.

We consider an $M/G/1$ **queueing system**. It is a single server queueing system that is loaded by messages that arrive in accordance with a Poisson process at intensity λ. The system is assumed to have unlimited message storage capacity, so that all arriving messages are admitted into the system and no message blocking is incurred. The message service time is assumed to be described by a general service time distribution, denoted as $B(t)$:

$$B(t) = P\{S \le t\}, \qquad t \ge 0. \tag{7.41}$$

The average service time is denoted as β, the service rate as μ, the second moment of the service time as β_2, the variance of the service time is represented as $VAR(S)$, and the coefficient of variation of the service time is denoted as $COV(S) = \eta_S$:

$$E(S) = \beta; \quad \mu = 1/E(S); \quad E(S^2) = \beta_2; \quad VAR(S) = \beta_2 - \beta^2; \quad \eta_S^2 = \frac{VAR(S)}{\beta^2}.$$

The traffic intensity parameter ρ is equal to

$$\rho = \lambda E(S) = \lambda\beta = \lambda/\mu. \tag{7.42}$$

In the following, we present several key properties of this queueing system. The system reaches steady state conditions, yielding finite mean message waiting time, if and only if the arrival rate is lower than the service rate, $\lambda < \mu$, or $\rho < 1$. Under this condition, at steady state, the system size variable is denoted as X, and its average value as $E(X)$. The steady-state message waiting time variable is represented as W, and its mean message waiting time is denoted as $E(W)$.

As noted in Section 7.3, the system size process alternates between busy and idle periods. The idle period duration I is governed by an exponential distribution, so that

$$I(t) = P\{I \leq t\} = 1 - e^{-\lambda t}, \qquad t \geq 0. \tag{7.43}$$

The mean $E(I)$ and variance $VAR(I)$ of the idle time duration are consequently equal to

$$E(I) = 1/\lambda; \qquad VAR(I) = 1/\lambda^2. \tag{7.44}$$

Assuming henceforth that $\rho < 1$, so that the system reaches steady state, we observe, as noted in a Section 7.3, the number of messages N served during a busy period, the duration G of a busy period, and the duration C of a busy cycle, to assume the following average values:

$$E(N) = \frac{1}{1-\rho}; \qquad E(G) = E(N)E(S) = \frac{\beta}{1-\rho}; \qquad E(C) = E(G) + E(I) = \frac{1}{\lambda(1-\rho)}, \quad \rho < 1. \tag{7.45}$$

The steady-state probability that the system is idle, $P(0)$, is calculated by noting that it is equal to the fraction of time in a busy cycle that the system is idle and therefore is equal to the ratio of the mean duration of an idle period and the mean duration of a busy cycle:

$$P(0) = \frac{E(I)}{E(C)} = 1 - \rho, \qquad \rho < 1. \tag{7.46}$$

The throughput capacity of the system is equal to μ messages/s. If $\rho < 1$, the offered, admitted and departing message rates are all equal, so that $\lambda_o = \lambda_D = \lambda$. The normalized throughput is then equal to $\lambda_D/\mu = \lambda/\mu = \rho$.

We conclude that if $\rho < 1$, the probability that the system is idle is equal to $1 - \rho$, the probability that the system is busy is equal to ρ, normalized throughput rate is equal to ρ.

Example 7.11 Consider a transmission processor that is modeled as an $M/G/1$ queueing system. The transmission data rate (i.e., its service rate) across the outgoing communications channel is equal to $R = 10$ Mbps. The average message length is equal to $E(L) = 7400$ [bits/message]. Hence, the average message service time, or transmission time, is equal to $\beta = E(S) = E(L)/R = 740 \,\mu s$. The service rate is thus equal to $\mu = 1/E(S) \approx 1351$ [messages/s]. Considering a flow that consists of messages that are members of two classes, assume that the message arrival rate is equal to $\lambda = 1081$ messages/s. Consequently, the traffic intensity parameter is $\rho = \lambda/\mu = 0.8$.

Thus, this system is loaded at 80% of its throughput capacity rate. Since $\rho < 1$, the system reaches steady state. At any time in steady state, the system is idle with probability $P(0) = 1 - \rho = 0.2$, so that 20% of the time, the system is idle and 80% of the time it is busy. The busy period, during which the system is continuously busy in transmitting messages, lasts for an average duration that is equal to $E(G) = \frac{\beta}{1-\rho} = 3.7$ ms. The average number of messages served during a busy period is equal to $E(N) = \frac{1}{1-\rho} = 5$ [messages/busy-period]. The mean idle time is equal to $E(I) = 1/\lambda \approx 0.925$ ms. The mean cycle time, which represents the average time elapsed between the start times of two successive busy periods, is equal to $E(C) = E(G) + E(I) \approx 4.625$ ms. Note that we then calculate $P(0) = E(I)/E(C) = 0.2$, as expected.

216 | *7 Queueing Systems*

Next, we determine the average message waiting time in the system at steady state, $E(W)$. It represents the average time that a message waits in the queue before it is admitted into service. Recalling that for $\rho \geq 1$, the system does not reach steady-state behavior, noting that the system is then overloaded and consequently the message waiting time becomes very high, we assume henceforth that $\rho < 1$.

We assume a FIFO service policy. Actually, as we will note later, the stated result will also hold, under certain conditions, for other work-conserving service policies. The steady-state average system size $E(X)$ for the $M/G/1$ queueing system, assuming $\rho < 1$, is given by the following formula.

Property 7.15 **Average Steady-State System size for an** $M/G/1$ **Queueing System:**

$$E(X) = \rho + \frac{\rho^2}{2(1-\rho)} \frac{\beta_2}{\beta^2} = \rho + \frac{\rho^2}{2(1-\rho)}(1 + \eta_S^2), \quad \rho < 1. \tag{7.47}$$

The steady-state average message delay time, $E(D)$, is then calculated by using Little's Formula as $E(D) = \frac{E(X)}{\lambda}$, by noting that $\lambda_D = \lambda$ for $\rho < 1$. The steady-state mean message waiting time is calculated as $E(W) = E(D) - \beta$. We consequently obtain the following formulas.

Property 7.16 **The Pollaczek–Khintchine (PK) Equation: Average Steady-State Message Waiting Time for an** $M/G/1$ **Queueing System:**

$$E(W) = \frac{\rho\beta}{2(1-\rho)} \frac{\beta_2}{\beta^2} = \frac{\rho\beta}{2(1-\rho)}(1 + \eta_S^2), \quad \rho < 1. \tag{7.48}$$

Average Steady-State Message System Delay Time for an $M/G/1$ **Queueing System:**

$$E(D) = E(W) + \beta = \beta + \frac{\rho\beta}{2(1-\rho)} \frac{\beta_2}{\beta^2} = \beta + \frac{\rho\beta}{2(1-\rho)}(1 + \eta_S^2), \quad \rho < 1. \tag{7.49}$$

In the following, we illustrate the computation of the average message waiting time and delay through an example.

Example 7.12 Consider the system configuration presented in Example 7.10 and in Example 7.11. For this system, we have $\beta = E(S) = 740\,\mu s$, $\lambda = 1081$ messages/s, and $\rho = \lambda/\mu = 0.8$. For the above-presented two-class message flow, we have obtained $\beta_2 = 2523500\,(\mu s)^2$; $COV(S) = \eta_S = \frac{\sigma(S)}{E(S)} \approx 1.9$. In turn, when considering a single class message flow for which a message is modeled to require a service time that is Exponentially distributed, we have $\beta_2 = 2\beta^2$, and $COV(S) = \eta_S = 1$.

For the two-class flow scenario, we obtain the mean message waiting time in the queue to be equal to $E(W_1) \approx 9.2\beta = 6812\,\mu s$. In turn, for the single class case, we obtain the average message waiting time to be equal to $E(W_2) = 4\beta = 2960\,\mu s$. We observe that the mean message waiting time for the two-class flow scenario is higher by a factor of $\frac{9.2}{4} = 2.3$. The first flow contains messages whose service time requirements have higher variability, leading to an incurred average message waiting time that is higher by a factor of 130%.

Since $E(D) = E(W) + \beta$, we conclude that the corresponding average message delay times are equal to $E(D_1) = 10.2\beta$ and $E(D_2) = 5\beta$, implying that messages in the higher variability flow experience mean delay times that are higher by about 104%.

In comparing the average waiting times incurred by messages that are exponentially distributed, inducing a corresponding $M/M/1$ queueing system model, and those experienced by messages of

7.6 A Queueing System with General Service Times | **217**

fixed length, inducing a corresponding $M/D/1$ queueing system model, while assuming the same message arrival rate λ and mean message service time β for both systems, and also assuming that $\rho = \lambda\beta < 1$, we conclude the following by using the formulas presented above.

Property 7.17 The mean message waiting time for the $M/M/1$ system is equal to

$$E(W)_{MM1} = \frac{\rho\beta}{1 - \rho}, \qquad \rho < 1. \tag{7.50}$$

In turn, the mean message waiting time for the $M/D/1$ system is equal to

$$E(W)_{MD1} = \frac{\rho\beta}{2(1 - \rho)}, \qquad \rho < 1. \tag{7.51}$$

Hence,

$$E(W)_{MM1} = 2E(W)_{MD1}, \qquad \rho < 1. \tag{7.52}$$

By examining the formula for the mean message waiting time, $E(W)$, we conclude that the **waiting times incurred by messages increase as the variance of the requested message service time increases**. Hence, message waiting times in the queue can be reduced through the reduction of their requested service time variability. By fragmenting longer messages into shorter packet segments, and selecting packets as basic service units, we reduce the variability of the bit contents of served message units. The segmentation process may cause a (typically modest) increase in the traffic intensity parameter $\rho = \lambda\beta$ level due to the addition of segmentation sequence numbers and related overhead. However, if the variability in the requested service times is rather high (as can be the case when the processor serves a mix of message types), a net reduction in the message waiting time would be realized.

It is often of interest to also calculate the ensuing waiting time variability parameters, including $VAR(W)$ and the corresponding standard deviation $\sigma(W)$, at steady state, assuming $\rho < 1$. Denoting $\beta = E(S), \beta_2 = E(S^2), \beta_3 = E(S^3)$, the following result is obtained.

Property 7.18 The second moment of the message waiting time for the $M/G/1$ system is equal to

$$E(W^2) = \frac{\rho}{3(1 - \rho)} \frac{\beta_3}{\beta} + 2E(W)^2, \qquad \rho < 1. \tag{7.53}$$

The variance of the message waiting time for the $M/G/1$ system is then equal to

$$VAR(W) = E(W^2) - E(W)^2 = \frac{\rho}{3(1 - \rho)} \frac{\beta_3}{\beta} + E(W)^2, \qquad \rho < 1. \tag{7.54}$$

The standard deviation of the message waiting time for the $M/G/1$ system is calculated as

$$\sigma(W) = \sqrt{VAR(W)}, \qquad \rho < 1. \tag{7.55}$$

Example 7.13 Consider the system configuration presented in Examples 7.10, 7.11, and 7.12. When messages are modeled, as defined above, to be members of two classes, by using Property 7.18, we obtain the standard deviation of the message delay to be equal to $\sigma_D = 7473\,\mu s$. The average message delay is equal to $7552\,\mu s$.

In comparison, for the single message class flow case, at the same message arrival rate and same average message service time, and thus at the same traffic intensity level of $\rho = 0.8$ so that the system is occupied for 80% of the time as for the 2-message class flow case, we obtain $\sigma_D = E(D) = 5\beta = 3700\,\mu s$.

218 | *7 Queueing Systems*

For the single message class flow case, noting that the message delay is governed by an Exponential distribution, we calculate the following estimate: $D_{0.95} = E(D) + 2\sigma_D = 3E(D) = 11.1$ ms. Thus, while the mean message delay is equal to 3.7 ms, 5% of the messages will incur a delay that is higher than 11.1 ms. In comparison, under the two message class flow, a much higher service time deviation is incurred, and the corresponding mean plus two sigma message delay is obtained to be equal to about 22.5 ms, which is higher than that incurred in the single message class case by 102.7%.

7.7 Priority Queueing

Priority queueing mechanisms are frequently employed to differentiate between the quality of service granted to different messages or message flows, recognizing that certain message classes should be granted precedence over other. As demonstrated in Chapters 5–6, communications networks often guarantee designated flows or flow classes specified quality of service performance metrics, which they tend to enforce through the use of preference based service methods. In the following, we discuss such a priority-oriented operation.

Message flows are divided into priority classes. Class-1 messages are granted highest priority, while class-2 messages are granted lower service priority. If flows are categorized into $r > 1$ priority classes, then priority-r messages are served at the lowest priority, while class-1 messages are provided the highest priority. Messages that belong to the same priority class are served by using a First-In First-Out (FIFO) policy.

Under a **non-preemptive priority policy**, a message in-service is not preempted from service by a higher priority message that arrives in the midst of its service. In turn, under a **preemptive resume priority service policy**, such a preemption is enforced, and the service of a preempted message is resumed at its point of interruption, once its service resumes. In turn, under a **preemtive-repeat priority service policy**, when a preempted message is re-admitted into service, the system performs its service from its starting point, assuming that the portion of its past uncompleted service is lost or is not being used. For example, when a message is transmitted across a communications channel and its transmission is interrupted in the midst of its transmission, a preemptive-resume operation would take advantage of past transmission portions that have been saved by involved entities, while under a preemptive-repeat operation, past transmission fragments, if any, have not been saved or are not taken into consideration.

To illustrate the performance behavior of a queueing system when serving messages on a priority-oriented basis, we assume a **non-preemptive priority** $M/G/1$ **queueing system** model. Messages that are granted priority-i, $i = 1, 2, \ldots, r$, arrive at the system at random in accordance with a Poisson process at rate $\lambda^{(i)}$ [messages/s]. Each class-i message requires a service time whose probability distribution function is given as $B^{(i)}(t)$. The mean service time is equal to $E(S^{(i)}) = \beta^{(i)} = \frac{1}{\mu^{(i)}}$, and the second moment of its service time is denoted as $E\{(S^{(i)})^2\} = \beta_2^{(i)}$. The traffic intensity associated with class-i messages is denoted as $\rho^{(i)} = \lambda^{(i)}\beta^{(i)}$. The traffic intensity associated with the loading of the system by all message classes is denoted as $\rho = \sum_{i=1}^{r} \rho^{(i)}$. We assume that $\rho < 1$ so that the system reaches steady-state behavior and messages incur finite waiting times.

We also set

$$\sigma_j = \sum_{i=1}^{j} \rho^{(i)}, \quad j = 1, 2, \ldots, r,$$

whereby σ_j represents the cumulative traffic intensity associated with message flows whose priority level is higher than, or equal to, priority-j. Thus, $\sigma_1 = \rho^{(1)}$, and $\sigma_r = \rho$. We set $\sigma_0 = 0$.

Analysis of this system yields the following result for the steady-state average waiting time for a class-i message, denoted as $E(W^{(i)})$:

Property 7.19 The average waiting time at steady-state experienced by a class-i message served by an $M/G/1$ system, assuming $\rho < 1$, is equal to:

$$E(W^{(i)}) = \frac{E\{\hat{S}\}}{(1 - \sigma_{i-1})(1 - \sigma_i)}. \tag{7.56}$$

where $E\{\hat{S}\}$, which represents the mean residual service time of a message in service, is calculated as follows:

$$E\{\hat{S}\} = \frac{1}{2}\sum_{k=1}^{r} \lambda^{(k)} \beta_2^{(k)}. \tag{7.57}$$

As a special case, consider the above system when loaded by two classes of flows, so that $r = 2$. Class-1 flows are granted higher priority. Class-1 messages incur a mean message waiting time that is denoted as $E(W^{(1)})$. Class-2 flows are served at a lower priority, and class-2 messages incur a mean message waiting time that is denoted as $E(W^{(2)})$. Assuming $\rho < 1$, we conclude the following mean steady-state average message waiting time formulas:

$$E(W^{(1)}) = \frac{\frac{1}{2}\sum_{k=1}^{2} \lambda^{(k)} \beta_2^{(k)}}{1 - \rho^{(1)}}. \tag{7.58}$$

$$E(W^{(2)}) = \frac{\frac{1}{2}\sum_{k=1}^{2} \lambda^{(k)} \beta_2^{(k)}}{(1 - \rho^{(1)})(1 - \rho)}. \tag{7.59}$$

Using these formulas, we conclude that the average waiting time incurred by a class-2 message is higher than that experienced by a class-1 message by the following factor:

$$\frac{E(W^{(2)})}{E(W^{(1)})} = \frac{1}{1 - \rho}, \qquad \text{for} \quad \rho = \rho^{(1)} + \rho^{(2)} < 1. \tag{7.60}$$

Thus, higher priority messages experience a mean waiting time that is lower by a factor of $\frac{1}{1-\rho}$, where ρ is the total traffic intensity loading the system. To illustrate, consider a system that under a first scenario is loaded at a traffic intensity ρ value that is equal to 50%, while under a second scenario, it is equal to 90%. Higher-priority messages then experience an average waiting time level that is *lower by a factor of 2, under the first loading scenario and by a factor of 10, under the second (higher) loading scenario*. It is interesting to note that the values of these factors are independent of the specific values of the traffic intensity level of each flow class. Rather, it depends only on the total traffic intensity level.

What can we say about the relationship between the average waiting times experienced by messages that belong to different flows when they are served by any *work conserving* service policy? For example, the service discipline can be FIFO, LIFO, non-preemptive priority, or preemptive-resume priority.

The following property, identified often as an $M/G/1$ **conservation law** holds for an $M/G/1$ queueing system which is loaded by multiple message flows, assuming that $\rho < 1$, whereby a general work conserving service discipline is used. The arrival and service models and parameters are the same as those presented above for the multi flow priority system, except that the service policy does not have to be priority class oriented.

We express this result by comparing the performance of the underlying multi-class $M/G/1$ queueing system with that of an $M/G/1$ queueing system that is loaded by a single flow, where messages are served on a FIFO basis. The later system is identified by us as a $\widehat{M/G/1}$ compounded-service system, or in brief a *compounded system*. The overall message arrival rate at this system is equal to $\lambda = \sum_{i=1}^{r} \lambda^{(i)}$, where $\lambda^{(i)}$ is the arrival rate of class-i messages. A message that arrives to the compounded queueing system requests a service time that is noted to have a probability distribution that is the same as that of $S^{(i)}$ with probability $\lambda^{(i)}/\lambda$. This is explained by noting that in the multi-flow system, an arriving message belongs to class-i with probability $\lambda^{(i)}/\lambda$.

For the compounded queueing system, we observe that the system's traffic intensity is equal to $\hat{\rho} = \rho$ and that the message arrival rate is $\hat{\lambda} = \lambda$. Assuming that $\rho < 1$, and using the PK equation 7.48 to calculate the mean message waiting time in an $M/G/1$ queueing system, the average steady-state message waiting time in the compounded system, denoted as $E(\widehat{W})$, is given by the following formula:

$$E(\widehat{W}) = \frac{\sum_{i=1}^{r} \lambda^{(i)} \beta_2^{(i)}}{2(1-\rho)}. \tag{7.61}$$

The $M/G/1$ conservation law states the following.

Property 7.20 $M/G/1$ **Conservation law:**

$$\sum_{i=1}^{r} \frac{\rho^{(i)}}{\rho} E(W^{(i)}) = E(\widehat{W}). \tag{7.62}$$

To interpret this result, we note the following. Given the message arrival rate and message service time distribution, the mean message waiting time in the compounded system is denoted as $E(\widehat{W})$. For its calculation, we assume a FIFO service policy. This value is equal to the summation term presented as the left-hand side of the noted equation, which assumes the employed service policy to be any work conserving policy. In interpreting this summation, we note that $\rho^{(i)}$ is equal to the probability that there is, at any time in steady state, a class-i message in service. Similarly, ρ is equal to the probability that there is a message in service. Hence, the weight $w^{(i)} = \frac{\rho^{(i)}}{\rho}$ is equal to the probability that given that there is a message in service, this message is a class-i message. Thus, the summation term represents the average-weighted message waiting time, $\sum_{i=1}^{r} w^{(i)} E(W^{(i)})$, whereby $w^{(i)}$ represents the fraction of time that a busy server is engaged in the service of a class-i message, and therefore it is noted to also represent the long-term temporal average of the waiting time of a message in service.

As the weighted sum is equal to the calculated value on the right-hand side, the expression states that no matter which work conserving service policy is used, the weighted sum is equal to the cited value. If an employed policy induces reduced average waiting time for messages that belong to a certain class, then the consequent average waiting time incurred by messages that belong to at least one other class has to be higher. For example, when considering a two class system using a priority-oriented service policy, we observe that the lower average waiting time incurred by class-1 messages must result in a higher average waiting time values incurred by class-2 messages by a factor that makes the underlying weighted sum equal to the calculated right hand-side value.

As a special case of this property, we consider next the case whereby the probability distribution function of the service times requested by messages of a given flow class is equal to that requested by messages of any other flow class. Assuming that $\rho < 1$, we note that since $\beta^{(i)}$ and $\beta_2^{(i)}$ are independent of the associated flow class i, the following result is obtained.

Property 7.21 $M/G/1$ **Conservation Law Under A Message Service Time Distribution That Does Not Depend On The Flow's Class:**

$$\sum_{i=1}^{r} \frac{\lambda^{(i)}}{\lambda} E(W^{(i)}) = E(\widehat{W}) = \frac{\lambda \beta_2}{2(1-\rho)}, \quad \rho < 1. \tag{7.63}$$

We recognize the right-hand side term to represent the average message waiting time incurred in an $M/G/1$ queueing system that is loaded by a single message arrival class, and that uses a FIFO service policy, under message arrival rate λ, mean message service time β, and second moment of the service time β_2. Noting that the ratio $\frac{\lambda^{(i)}}{\lambda}$ is equal to the probability that an arriving message is a class-i message, we conclude that the left-hand side term represents the average waiting time incurred by a randomly selected arriving message. The conservation property thus states that the average waiting time incurred by a randomly observed arriving message, under the use of any work conserving service policy, is equal to that incurred when a FIFO service policy is used.

It is also interesting to note the following property. Consider the above noted multi flow $M/G/1$ queueing system that employs any work conserving service policy. The unfinished load at any time t, L_t, represents the total amount of service that remains to be performed when considering all messages in the system at time t. The order at which messages are selected for service does not affect the total service requested by each message, and therefore the statistics of the unfinished load process $L = \{L_t, t \geq 0\}$ do not depend on the specific work conserving policy that is employed.

As a special case, consider a corresponding queueing system model under which the statistics of the system size process $\mathbf{X} = \{X_t, t \geq 0\}$ also do not depend on the specific work conserving policy that is used. For example, such a *system size conserving* scenario takes place when a work conserving service policy is used and when the service time statistics are independent of the class membership of the message being served. In particular, in this case, the mean system size, $E(X)$, does not depend on the policy used, as the order at which messages are selected for service does not affect the statistics of the system size process $\mathbf{X}$.

Consider the steady-state system size performance of the queueing system as we employ different system size conserving service policies. In each case, the system is assumed to be loaded by traffic and service processes that are governed by the same statistics. The $\mathbf{X}$ process statistics are conserved under the queueing system models employed, and in each case the same message departure rate, denoted as λ_D, is incurred. Using Little's Formula, the average message delay is expressed as $E(D) = E(X)/\lambda_D$. We conclude that the average message delay, as well as the average message waiting time, $E(W)$, incurred by messages served under any one of the underlying service policies, assume the same value.

Thus, under the prescribed conditions, assuming all messages, of either class, to follow the same service time statistics, the average message waiting time (and delay time) under policies such as FIFO, LIFO, or (non-preemptive or preemptive-resume) priority oriented are all equal to each other. It is noted, however, as observed above, that while the average waiting time incurred by averaging over all message classes does not vary as we change the service policy, when using a priority oriented policy, higher-priority messages would incur lower average waiting times.

The variance and distribution of a message waiting time do depend on the specific service policy that is being used. While the average waiting times incurred under FIFO and LIFO service policies were noted above to be equal to each other, the variance level of a message waiting time is lower under the FIFO policy.

7.8 Queueing Networks

Many systems operate as networks that consist of interconnected nodes whereby each node functions as a queueing and processing module. We model such systems as **queueing networks**. Messages (or customers, clients, vehicles, and other entities) enter the queueing network at network nodes. Upon completion of service at a node, a message may depart the system or be routed to another node. The following are examples of such networks.

- Computer communications networks such as the Internet or cellular wireless networks, providing message transport services. Network nodes act as switches, multiplexers, and processors that are linked to other nodes by communications links.
- Vehicle transportation networks. Junctions, intersections, and access/departure ramps serve as network nodes. These nodes are linked by highways and roads.
- A manufacturing facility or machine shop. Machine parts or subsystems are transferred among processing stations before departing the facility. Each station may provide a specialized service.
- A cargo shipping network that transports items from a source warehouse to destination stations or depots while traversing on the way intermediate refueling or reconditioning stations.
- A human neural network that serves to transport peripheral sensor signals to the brain traversing interconnected neural nodes.

In the following, we first present a queueing network model that was developed by Professor James R. Jackson. The concept initially appeared in a paper published in 1957 [30]. It is referred to as a **Jackson queueing network model**. The model, as illustrated in Fig. 7.10, is characterized as follows.

The network layout is described by the topology of a graph that consists of K nodes. A node is connected by communications links to its neighboring nodes. Node-i processes the messages that arrive to it at a rate that is equal to μ_i [messages/s], noting that entities other than messages can also be modeled. Network nodes are identified as $n_1, n_2, n_3, \ldots, n_K$, where n_i denotes node-i. The environment outside the network system is identified as node-0, or n_0.

Messages arrive to the system from the outside environment. The external message traffic process arriving to node-i follows the statistics of a Poisson process whose rate is set equal to $\gamma_i, i = 1, 2, 3, \ldots, K$. It is assumed that the service times provided to a message at different nodes that it traverses, or even by the same node upon different visits to the node, are statistically independent. This is at times referred to as the **service independence assumption**.

We model the spatial distribution of messages as they travel across the network by monitoring, over a prescribed period of time, the flows of messages over the network topology. These observations are used to model the transit of a message flow around the network as a **Markovian stochastic routing process**. Upon the completion of a service at node-i, a message is routed with probability r_{ij} to node-j, $i, j, = 1, 2, \ldots, K$, where $0 \le r_{ij} \le 1$. The matrix $R = \{r_{ij}, i, j, = 1, 2, \ldots, K\}$ is the Markovian Routing Matrix. We set r_{i0} to denote the probability that following the completion

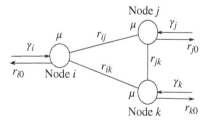

Figure 7.10 A Jackson-Type Queueing Network.

of its service at node-i, a message will depart from the system (and thus exit to the outside environment). It is noted that it is permissible to have $1 > r_{ii} > 0$, in which case a message that completes service at node-i will with probability r_{ii} return to node-i for additional service. Note however that the latter service times are assumed to be statistically independent (though they are performed at the same rate μ_i so that the mean message service time at node-i, per each visit at the node, is equal to $1/\mu_i$).

Assume that each node is modeled as an $M/M/1$ queueing system. We study the statistics of the joint system size vector process $X = \{\underline{X}_t, t \geq 0\}$, where $\underline{X}_t = \{X_1(t), X_2(t), \dots, X_K(t), t \geq 0\}$, and $X_i(t)$ represents the number of messages in node-i at time t. The vector $\underline{X}_t$ denotes the size state of the system at time t. Under our modeling assumptions, this system size process is a continuous-time B&D stochastic process, noting that at any point in time only a single state change event takes place. A change event represents either a single external arrival to one of the nodes, or a single departure out of the network from one of the nodes, or an internal transition involving a message departure from a node followed by an instantaneous arrival of this message to another (or same) node.

The rate of messages traversing node-i over the long-term evolution of the system process is denoted as λ_i. We calculate this rate by noting that two change event types can be involved: 1. External arrivals of messages from outside the network to node-i, which occur at rate γ_i. 2. Internal arrivals of messages to node-i, triggered by having such messages complete their service at another (or same) network node and then routed to node-i. Consequently, the message flow rate through each node-i, λ_i, is calculated by solving the following set of linear equations:

$$\lambda_i = \gamma_i + \sum_{j=1}^{K} \lambda_j r_{ji}, \qquad i = 1, 2, \dots, K. \tag{7.64}$$

The traffic intensity at node-i is then calculated to be equal to:

$$\rho_i = \frac{\lambda_i}{\mu_i}, \qquad i = 1, 2, \dots, K. \tag{7.65}$$

Analysis of this system shows that if none of the employed nodes is overloaded, assuming for this purpose that $\rho_i < 1$, $i = 1, 2, \dots, K$, the system reaches equilibrium. We then set $P\{X_1 = x_1, X_2 = x_2, \dots, X_K = x_K\} = P\{x_1, x_2, \dots, x_K\}$, where X_i represents the steady-state system size at node-i. At steady state, the following significant result for the joint steady-state system size distribution has been obtained by Professor Jackson:

Property 7.22 Joint Steady-State System Size Distribution For A Jackson network of $M/M/1$ queues:

$$P\{x_1, x_2, \dots, x_K\} = \prod_{i=1}^{K} (1 - \rho_i)\rho_i^{x_i}, \qquad \rho = \sum_{i=1}^{K} \rho_i < 1. \tag{7.66}$$

The joint system size distribution is thus noted to be expressed as the product of the distributions of the system sizes at the individual nodes. At node-i, the latter is equal to the steady state system size distribution derived for an individual $M/M/1$ queue with message arrival rate that is equal to λ_i and message service rate μ_i. A solution in this format is known as a *product form solution*. It indicates the following interesting and simple (yet non obvious) result: **the system size processes at individual nodes are mutually statistically independent and each one follows the statistics of an individual $M/M/1$ queueing system.**

Consequently, under stability conditions, at steady state, the mean system size at node-i is equal to

$$E(X_i) = \frac{\rho_i}{1-\rho_i}, \quad \rho_i < 1, \quad i = 1, 2, \ldots, K. \tag{7.67}$$

Using Little's Formula, the mean message delay and waiting time levels at node-i are expressed as follows:

$$E(D_i) = \frac{1}{\mu_i - \lambda_i}, \quad E(W_i) = \frac{\rho_i}{\mu_i(1-\rho_i)}, \quad \rho_i < 1, \quad i = 1, 2, \ldots, K. \tag{7.68}$$

If a message traverses a route that consists of node-1, node-2, and node-3, then its end-to-end average delay time will be equal to $E(D_1) + E(D_2) + E(D_3)$.

Examining the whole network system, we note that, in equilibrium, the average cumulative system size is equal to $E(X_T) = \sum_{i=1}^{K} E(X_i)$. The message system-wide throughout rate, λ_D, which represents the rate at which messages depart the system, is equal to the total rate of external messages arriving to the system (as in equilibrium, the message output rate is equal to message input rate), so that $\lambda_D = \sum_{i=1}^{K} \gamma_i$. Hence, using Little's Formula for the whole system, we conclude that the average (over all messages traversing the network) end-to-end delay time incurred by a randomly selected message flowing in the system, denoted as $E(D_T)$, is given as:

$$E(D_T) = E(X_T)/\lambda_D = \frac{\sum_{i=1}^{K} E(X_i)}{\sum_{i=1}^{K} \gamma_i}, \quad \rho_i < 1, \quad i = 1, 2, \ldots, K. \tag{7.69}$$

The product form result presented as Property 7.22 has been shown to extend to a Jackson Queueing Network where each node is modeled as an $M/M/m_i$ queueing system, whereby node-i employs $m_i \geq 1$ service channels, and each corresponding processor operates at rate μ_i. To express the ensuing joint system size distribution, we now replace in the product form formula the individual nodal $M/M/1$ system size distribution with the system size distribution calculated for the corresponding $M/M/m$ queueing system. In a similar manner, we use Little's Formula to calculate the average message delay time (and waiting time) at each node.

Example 7.14 Consider a network that consists of 3 nodes, node-1, node-2, and node-3, which are directly connected to each other, whose layout and parameters are illustrated in Fig. 7.11. Messages arrive to each node from outside the network in accordance with a Poisson process at rates that are equal to $\gamma_1 = 10$ [messages/s], $\gamma_2 = 0$, $\gamma_3 = 20$ [messages/s]. 80% of the messages that complete service at node-1 are routed to node-2 and 20% of them are routed to node-3. Hence, $r_{12} = 0.8, r_{13} = 0.2$. 90% of the messages that complete service at node-3 are routed to node-2 and 10% to node-1. Hence, $r_{32} = 0.9, r_{31} = 0.1$. Messages that complete service at node-2 depart from the network, so that $r_{20} = 1$. The service rates of the processors at the nodes are specified as $\mu_1 = 15, \mu_2 = 40, \mu_3 = 30$ [messages/s].

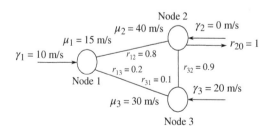

Figure 7.11 Illustrative Queueing Network.

Using these parameters, we solve Eq. (7.64) to obtain the flow rates of messages traversing each node to be equal to: $\lambda_1 = 850/67 \approx 12.686$ [messages/s], $\lambda_2 = 2300/67 \approx 34.328$ [messages/s], $\lambda_3 = 1800/67 \approx 26.866$ [messages/s].

Using these results, and noting that steady state conditions are met at all nodes, we obtain the average message delay at each node to be given as follows: $E(D_1) = \frac{1}{\mu_1 - \lambda_1} = 67/155 \approx 0.432$ [s]; $E(D_2) = \frac{1}{\mu_2 - \lambda_2} = 67/380 \approx 0.845$ [s]; $E(D_3) = \frac{1}{\mu_3 - \lambda_3} = 67/210 \approx 0.319$ [s].

Hence, a message that is routed from node-1 to node-2 (and then departs the network) will experience a mean delay of $E(D_1) + E(D_2) \approx 1.277$ [s] if it is routed directly across the link that connects these two nodes. In turn, if it is routed through node-3, its mean end-to-end delay will be equal to $E(D_1) + E(D_3) + E(D_2) \approx 1.596$ [s]. The total message throughput rate is equal to $\lambda_D = \gamma_1 + \gamma_2 + \gamma_3 = 30$ [mess/s]. Using Little's Formula, we conclude the cumulative (i.e., its sum over all nodes) average system size to be equal to $E(X_T) = \lambda_D \times E(D_T) = 30 \times 1.596 = 47.88$ [messages], where $E(D_T)$ denotes the cumulative sum over all nodes of the average message delay.

The traffic intensity levels at the nodes are calculated to be equal to $\rho_1 = \lambda_1/\mu_1 \approx 0.845$, $\rho_2 = \lambda_2/\mu_2 \approx 0.858$, $\rho_3 = \lambda_3/\mu_3 \approx 0.896$. Hence, the utilization levels at the corresponding nodes (i.e., the fraction of time that each node is busy as well as the normalized loading of each node represented as a fraction of its processing capacity) are equal to 84.5%, 85.8%, and 89.6%, respectively.

While the probability distribution function of the sum of the waiting times experienced by messages that are routed across a prescribed path in the queueing network is not readily calculated, as the ensuing waiting times are not statistically independent, it can be calculated for the following special case.

Consider a **tandem queueing network** which consists of N switching nodes that follow each other in a series configuration. Such a layout is shown in Fig. 7.12 for the special case of $N = 3$. Each node in this network system is modeled as a single server queueing system. It is assumed that the service time of each message is equal to a deterministic (fixed) rather than a stochastic, value.

For example, if each message consists of a single segment that contains a fixed number of L bits, and the nodal processor/switch operates at a rate of μ [bits/s], then the service time of a message at this node (accounting for the combined switching and transmission operation) is equal to L/μ [s/message].

It is often the case that each message consists of a single fixed size segment, as is commonly observed for telemetry, factory control, signaling and management data, as well as for many packets transported across a packet-switching network and/or processed by protocol engines.

This configuration is also applicable when considering a linear path of switching processors whereby the interconnecting communications links operate at high speeds and the bottleneck components involve certain nodal processors that operate at distinctly lower rates. As another example, it is noted that the processing speed of a protocol engine such as a header processor at a packet router is generally independent of the length of the packet's payload. In packet switching networks, longer messages are segmented into packets such that a high fraction of the packets

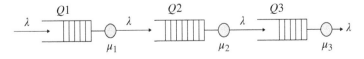

Figure 7.12 Illustrative Tandem Queueing Network

7 Queueing Systems

tend to be of fixed (such as maximum allowable) length. In considering such a packet-switching network application, the following results were derived by Professor Izhak Rubin [42].

Consider such a tandem network that consists of two nodes in series. Assume that $\mu_1 \le \mu_2$. In this case, the (the actual and mean) service times of messages processed (or transmitted, when considering a transmission processor) by node-1, denoted as $\beta_1 = L/\mu_1$, are always longer or equal to the service times of messages processed by node-2, denoted as $\beta_2 = L/\mu_2$. Assume the system to be loaded (starting at an idle state) by an on-going stream of messages that arrive at node 1. Each incoming message is processed by node-1 and is then moved for processing to node-2. Served messages depart node-i at inter-departure times that are equal to (or, in general, not lower than) β_i. Since, in the current case, $\beta_1 \ge \beta_2$, we note that when the second message departs from node-1 and is then instantly moved for service to node-2, the first message has already completed its service at node-2. Hence, under this setting, all messages departing from node-1 incur no waiting times at node-2!

We have shown that when considering such a tandem queueing network configuration, the sum of the waiting times experienced by a message traveling across such a network path does not change if we reorder the path's nodal processors. Consequently, assume that we reorder the processors so that the one that operates at the lowest service rate, denoted as $\mu_{min} = min\{\mu_1, \mu_2, \ldots, \mu_N\}$, is placed as the first processor in the tandem configuration. Note that if several nodes assume a service rate that is equal to the minimal one, then any one of them may be placed as the first processor in the modified network. Messages routed through this modified tandem path would experience positive waiting times only at the first processor while experiencing no waiting times at subsequent processors. We assume the system to have reached steady-state operation. The steady state waiting time of a message at node-i is denoted as W_i. This leads to the following result.

Property 7.23 End-to-end cumulative message waiting time experienced by messages that receive fixed service times across a tandem network of queues. For the tandem queueing configuration under consideration, at steady state, the probability distribution function of the sum of the waiting times experienced by a message in traversing all network nodes, $P(W_T \le t, t \ge 0)$, where $W_T = \sum_{i=1}^{N} W_i$, is equal to the waiting time distribution calculated at a system that consists of a single processing node that operates at a service rate μ_{min}, which is equal to the service rate of the of the slowest processor in the tandem path, and is subjected to the same incoming message stream as that loading the original tandem network.

It is noted that this result is based on combinatorial properties and thus holds for an incoming message flow that is governed by any traffic flow statistics. We have also proven that this result holds for a series of packet transmitters when messages can be of random length, while each message is fragmented into fixed size segments [42].

To illustrate, consider such a tandem queueing network at which messages arrive in accordance to a Poisson process at rate λ. The distribution of the sum of the waiting times experienced by a message traversing the linear path is equal to the steady-state message waiting time distribution incurred by messages served by a single node, which is modeled as an $M/D/1$ queueing system, whereby the service time is equal to $\beta_{max} = L/\mu_{min}$. Steady-state conditions hold under the assumption that $\rho = \lambda/\mu_{min} < 1$.

Consider packets that belong to a certain flow that has been admitted into a packet-switching network. Assume the network to configure the resources allocated along the path used by this flow so that these packets are provided their requested quality of service (QoS) guarantees. For this purpose, packet switches located along the path set aside, or make promptly available, sufficient resources so that they provide each one of this flow's packet a prescribed service rate (or bandwidth) level. In this case, the ensuing packet flow effectively traverses the system across a tandem path in which each node allocates a prescribed service rate for the exclusive use of packets that belong to this flow. The result stated above indicates that the end-to-end cumulative waiting time incurred by packets that are members of such a flow is equal to that incurred at a node whose service rate is equal to that of the slowest (i.e., bottleneck) node that is located along their path!

This result has also been employed for the design of a flow control mechanism that regulates the flow of data messages from a Road Side Unit (RSU) to platoons of vehicles (including autonomous vehicles). As such vehicles move along a highway lane, they communicate with each other, forming a tandem queueing network configuration. Messages received from the RSU by a leading vehicle are transmitted to a downstream neighboring vehicle over a wireless communications channel; the latter relays the received messages to its own downstream vehicle neighbor; and so on. The message flow is disseminated in this way across a designated span of the vehicle network. The inter-vehicle communications channel that offers the lowest data rate constitutes the underlying bottleneck link, as described in the above presented tandem queueing model. The RSU is configured to **pace** the inter-message transmission time intervals of packets to values that are not shorter than the longest inter-vehicular service time level, β_{max}, in aiming to regulate the waiting times incurred by messages disseminated over the path.

It is noted that under a Jackson network model, the average cumulative end-to-end waiting time incurred across the tandem network is equal to the sum of the average waiting times experienced at each network node. In contrast, under the tandem path setting described above, whereby each message of an incoming flow is guaranteed a fixed service time at each node along the path, the average cumulative waiting time is equal to the average waiting time incurred at just a single node, the slowest one! In turn, if the path traveled by messages that belong to a tagged flow consists of nodes that are fed by messages that arrive from other parts of the network, so that these messages would contend for service at intermediate nodes with messages belonging to the underlying linear flow, the conditions prescribed for the system (in providing QoS performance guarantees to packets that are members of a tagged flow) to represent a tandem path in isolation that serves the tagged flow are not applicable and Jackson queueing network type models may then be used to provide an approximate characterization of system performance.

7.9 Simulation of Communications Networks

7.9.1 Monte Carlo Simulations of Communications Networks

Mathematical methods are often used for the analysis and design of communications network systems. For this purpose, a model of the system of interest is created. To reduce its complexity, key events (also identified as the key change events) that drive the system's dynamic behavior are identified and used in the formation of system model processes and the calculation of performance metrics. The characteristics of the system's loading flows, operational scenarios, protocols, and networking schemes are embedded in the composed models.

Consider a network system that is modeled as a queueing network, such as the one presented in Section 7.8. The following elements are defined: the network's topological layout; the network's message loading flows; the message requested service times at the network nodes that they visit; the routing scheme. Performance functions of interest may display message delay and system-size statistics. Key change events identify state transitions, which are induced by message arrivals, message departures, message transitions from a node to another node, message external arrivals to a network node, and message departures from the network. The stochastic mechanisms that drive this model are used for statistical characterizations of message arrival processes, message service times, and when applicable message service disciplines and routing-based message transfers and exits.

As demonstrated in Section 7.8, certain message delay and system-size performance functions can be calculated by using mathematical formulas. The system is assumed to have reached steady-state conditions. For the underlying queueing network model, message arrivals are assumed to follow the statistics of Poisson processes, message service times are assumed to be exponentially distributed and statistically independent (over distinct messages and nodes), and message transfers among nodes are assumed to be governed by a Markovian routing scheme. While assumed system operational models and probability distributions may not faithfully reflect observed operations and characterizations, they are useful as they enable mathematical evaluations of the system's performance behavior.

It is practically often essential to use other characterizations in modeling the system's operations, networking and protocol schemes, and statistical characterizations of involved processes. Message arrival processes may not be governed by Poisson statistics, service times may not follow exponential distributions and may not be statistically independent (over distinct messages and/or nodes), different nodal service policies may be employed, and different routing strategies may be invoked. Specifications of various system and networking processes may vary over time. They may be derived from observations of ongoing network system processes and operations.

It may not then be feasible to derive closed-form mathematical formulas that can be readily used to calculate steady-state performance functions. It is then useful to carry out a computer simulation process. It is still useful, when feasible, to make use of mathematical models that provide approximate analyses. Ensuing analytical calculations can provide guidance to the design of a simulation system and confirmation of simulation results. Mathematical models and corresponding analytical performance expressions also serve as effective tools in carrying out sensitivity analyses, which serve to identify the impact of variations in key system parameters and schemes on the performance behavior of a network system.

Under a **Monte Carlo simulation** approach, the following process is carried out. A model of the system to be evaluated is defined, using analytical, finite state machine, or other system and process characterizations. Key change events are defined and employed in forming the underlying model. Traffic flow processes, nodal service procedures, routing schemes, involved processing, storage, computing, networking algorithms and protocols, and other relevant operations are modeled. The corresponding characterizations may be derived, and dynamically adjusted, by performing ongoing observations of network system processes, operations and outcomes.

A *Monte Carlo computer simulation run* produces **repetitive realizations** of the modeled system. In simulating the above mentioned queueing network system, the computer model generates, over its runtime, message flows that enter the system at network nodes. Computer-based models are used to have arriving messages served by the nodes that they traverse across the network in accordance with their service schemes. When a nodal service of a message is complete, the message is routed to another node, or it departs the system. Message inter-arrival times at system nodes, service times by the nodes that they visit, and the identities of the next nodes that messages are routed to, assume values that are determined as realizations of computer based random number generators, governed by specified probability distributions.

As the computer simulation program dynamically evolves, messages keep flowing across the modeled network, and the simulation program keeps tracking and updating its system states. State data is recorded and used to update the calculation of performance functions, such as those involving system-size and waiting-time processes. At the termination of a simulation run, the

7.9 Simulation of Communications Networks | **229**

calculation of the statistical behavior of targeted performance functions is finalized and the results are displayed. For example, the end-to-end average delay incurred by messages served by the network is calculated as the sample-mean of recorded end-to-end delays experienced by computer generated messages that departed the network model.

As the simulation process progresses, the system's state processes evolve over time, and their assumed values are used to calculate statistical metrics of designated performance measures. For example, let W_n denote the waiting time incurred by the n_{th} message that is served at node A during a simulation run. The waiting time process is denoted as $W = \{W_n, n = 1, 2, 3, \dots\}$. Following a simulation run during which a total of N messages have been recorded to experience waiting times at node A, the average waiting time of a message at this node is calculated as the respective sample mean $AW_{1,N} = \sum_{n=1}^{n=N} W_n/N$. Assume that the aim is to use the outcomes obtained from the underlying simulation-run to estimate the steady-state average of the message waiting time at this node, denoted as $E(W)$, aiming for $lim_{n \to \infty} AW_{1,N} = E(W)$. Assuming steady-state conditions are met for the underlying process, yielding a finite mean message waiting time value, $E(W) < \infty$, the latter equality will be achieved by an *ergodic* stochastic process. For such a stochastic process, time-based averages converge to probabilistic averages. In this case, if the system run is sufficiently long, so that the number of served messages N is sufficiently high, the sample mean value $AW_{1,N}$ that is derived over the simulation-run would serve as a good estimate of $E(W)$.

The queueing systems discussed in this chapter reach steady-state and display ergodic behavior when the number of messages that are allowed to reside in the system is bounded by a finite capacity value, as is the case when a flow control scheme is applied. In general, as noted in the above-described mathematical analyses, when no flow admission controls are imposed, steady-state behavior is generally reached if the system is not overloaded, so that the message arrival rate is lower than the overall system service rate.

A simulation run should be sufficiently long to include a minimum number of replicas of system process dynamics, meeting a targeted accuracy (confidence level) metric. The performance functions that are calculated by using the results of a simulation-run could be set to deviate from the targeted steady-state levels by no more than a specified margin, with probability that is higher than a specified level (such as 95%). For example, the discussion in Chapter 8 of [21] addresses methods that can be used for setting simulation run lengths and for performing analyses by employing of numerical techniques.

In structuring the execution of a Monte Carlo simulation program, we differentiate between a **time-driven computer simulation** program and a **discrete-event computer simulation** program . Under a time-driven computer simulation program, the state of the system is examined and updated every time slot. A time slot duration is set to be sufficiently short so that it properly covers the occurrence of critical events of interest. For example, as the time slot index k is advanced, following the updating and recording of the system size $X(k)$ at time k, the system state level $X(k + 1)$ at time slot $k + 1$ is updated and its new level recorded.

A *time-driven simulation* process may run inefficiently. This is particularly the case when short time slots must be used, such as for the purpose of recording event occurrence times at a high precision. In this case, the probability that a critical event will occur during a time slot may be quite low. It can then be more temporally efficient to conduct a **discrete-event simulation** process. The state of system processes is then updated only at times at which critical events take place. For example, consider a time slot of duration 1 ms, and assume the average time between critical events to be of the order of 1 second. Under a discrete-event simulation program, state updates will occur at an average rate of 1 [update/s]. In contrast, under time-driven simulation, the state of the system will be examined at the high rate of 1000 [updates/s].

230 | *7 Queueing Systems*

To further reduce the complexity and run-time of a simulation program, a hybrid Analytical/ Monte-Carlo simulation scheme may be employed. Under such an arrangement, performance characterizations of certain functions are separately calculated by using mathematical formulas. For example, when focusing on the queueing and delay performance of messages that traverse a network system, we may model the data rate and error rate levels associated with involved communications links by using results obtained from separately conducted mathematical calculations. Simulation program complexity and run-time length reductions can also be realized by segmenting a large simulation program into a multitier collection of smaller component programs.

The $PLANYST^{TM}$ program developed by Professor Dr. Izhak Rubin and implemented by him and his team at IRI Computer Communications Corporation is an innovative hybrid simulation program. Under its burst level simulation mode, burst occurrences are produced through the conduct of Monte Carlo simulations, while the behavior and performance of the system during each burst period, whether governed then by steady-state or transient-state behavior, is conducted by using mathematical analyses. Such a simulation evaluation process has been shown to accelerate the run-length of the simulation run in a significant manner.

To illustrate the operation of a Monte Carlo discrete-event simulation process, we present and discuss in Section 7.9.2 a sample code that can be used to carry out a discrete-event computer simulation of an $M/M/1$ queueing system.

7.9.2 Illustrative Discrete-Event Monte Carlo Simulation of a Queueing System

In the following, we illustrate the structure of a discrete-event Monte Carlo simulation of an $M/M/1$ queueing system. The program commands are partially based on those used by a *MATLAB* code but can be similarly expressed by using other computer program codes. For illustration purposes, we have set below $\lambda = 4$ [mess/s] and $\mu = 5$ [mess/s], so that we have $\rho = 4/5 = 0.8$. These values can be readily changed. Note that the % symbol preceding a sentence in the code identifies it as a comment.

The displayed discrete event simulation program consists of the following routines (Fig. 7.13):

1) Routines that are used to define and calculate system states, variables, and parameters.
2) Initialization: setting the initial values of system states and parameters.

- The program consists of the following parts:
  - Global parameters
    - Definitions and calculation routines of the system states, variables and their parameters
  - Initialize
    - Setting the initial values of system states and parameters
  - Main and called routines
    - Defining the flow of the iterative progress of the simulation program from start to end (when the termination conditions are met).
    - At each step, a cited routine is called; when a result is returned by this routine, the proper next routine, if applicable, is called; or the simulation is terminated, at which time performance metrics are updated.
  - Report
    - Collect simulation data to compute and present performance metrics.

Figure 7.13 Illustrative Discrete Event Simulation of a Queueing System: Program Routines.

7.9 Simulation of Communications Networks | 231

3) Main routine and routines that are called from the Main routine. They are used to define the progress of the simulation program from start to end (when termination conditions are met). At each step, a cited routine is called. When a result is returned by this routine, the proper next routine, when applicable, is called, or the simulation is terminated, at which time the final values assumed by employed performance functions are updated.
4) Simulation data is collected and performance function values are calculated, presented, and included in a Simulation Report.

The simulation system global parameters are shown in Fig. 7.14. The following parameters are noted:

1) Two key events are defined. They consist of message arrivals and message departures. An arrival event is denoted as a type-1 event (setting $k = 1$) while a departure event is denoted as a type-2 event (setting $k = 2$).
2) Message inter-arrival times are calculated as realizations of an Exponential distribution with arrival rate parameter λ [mess/s], and mean inter-arrival time that is equal to $1/\lambda$ [s]. For illustration purposes, we have set $1/\lambda = 2$ [s].
3) Message services times are calculated as realizations of an Exponential distribution with rate parameter μ [s/mess], and mean service time that is equal to $1/\mu$ [s/mess]. For illustration purposes, we have set $1/\mu = 1$ [s].
4) The simulation is set to terminate when the number of message arrivals occurring during the simulation exceeds a specified limit that is denoted as *limit_customers*, unless terminated by first meeting another termination threshold.
5) The simulation terminates when the number of messages waiting in the queue (identified as *queuedmessages*) exceeds a specified limit denoted as *Q_LIMIT*, unless terminated by first meeting another termination threshold.

The initialization of system parameters proceeds as shown in Fig. 7.15.

- time_next_event(1:2) = 0;
    - % Events = {arrivals, departures}
    - % Identifier of arrival event = 1; Identifier of departure event = 2
    - % Refers to class 1 and 2 event times, which are initially set to 0.
- num_events = 2;
    - % The state evolution of a single node queueing system is described through the iterative computation of 2 types of events (arrival and departure times)
    - % Number of events = 2 = Arrival and departure events
- mean_interarrival = 2;
    - % Average interarrival time between message arrivals [sec]
    - % $1/\lambda = 2$ sec (illustrative case)
- mean_service_time = 1;
    - % Average message service time [sec]
    - % $1/\mu = 1$ sec (illustrative case)
- limit_customers = 1e6; (illustrative case)
    - % Maximum number of messages that arrive to the system (whether admitted or blocked) which induce, when met, termination of the simulation run
- Q_LIMIT = 1e30;
    - % The simulation program is terminated if the number of stored (in queue) messages exceeds this level.

Figure 7.14 Global Parameters of the $M/M/1$ Simulation Program.

232 | *7 Queueing Systems*

- sim_time = 0.0; % Initializes simulation time
- server_status = 0;
    - % server is initially idle. Recall that status = 0 implies the system to be idle and status =1 indicates the server to be busy.
- num_in_s = 0;
    - % number of customers in the system (or system size, denoted as X)
- num_in_q = 0;
    - % number of customers in the queue (wait size, denoted as Q)
- time_last_event = 0;
    - % the time of the latest recorded event
- % Initialize the statistical counters:
- num_custs_delay = 0;
    - % cumulative number of customers that have experienced delay up to *this* time
- num_waiting_custs = 0;
    - % cumulative number of customers that have experienced waiting time up to *this* time
- total_of_waits = 0;
    - % cumulative sum of the waiting times experience by all customers served by the system up to *this* time

- total_of_delays = 0;
    - % cumulative sum of the delay times experience by all customers served by the system up to *this* time
- area_num_in_s = 0;
    - % cumulative system size area up to *this* time
- area_num_in_q = 0;
    - % cumulative queue size area up to *this* time
- area_server_status = 0;
    - % cumulative service area up to *this* time
- % Initialize event list. Initially, no customers are present.
- % departure (service completion) events are not yet involved.
- time_next_event(1) = sim_time + exprnd(mean_interarrival);
    - % time of next arrival event
    - % Matlab defines exprnd(mean_interarrival) = exponentially distributed RV with this specified mean
- time_next_event(2) = 1e30;
    - % time of next departure event; 1e30 denotes infinite

Figure 7.15   Initialization of the Simulation Program.

- while total_of_customers-1 < limit_customers
    - % checks that the max number of served messages is below the specified limit and calls the timing( ) routine.
    - % The program should also include termination conditions that involve maximum a simulation run time and queue-size limits to assure timely completion and stable operation.

    - timing(); % Determines the next event type and updates the current simulation time.

    - update_time_avg_stats(); % updates_the values of the computed system state / performance statistical measures.

    - switch next_event_type % iterative progress of the simulation process as it goes through successive arrival or departure events.
    - **case** 1
        - arrive();
    - **case** 2
        - depart();
    - end
- End
- report(); % report subroutine is called to generate performance exhibits.

Figure 7.16   The Main Program Routine.

The simulation flow process follows the Main subroutine whose structure is shown in Fig. 7.16. As long as the program does not reach a termination limit, the following subroutines are called in sequential order: The timing( ) routine is used to determine the next event type, whether it is an arrival or a departure event, identified as the first event that is scheduled to occur next. It calculates the time that this next event is scheduled to take place. It updates the current simulation time by advancing it to the time of the next event. It updates the statistical functions of recorded performance measures. It proceeds with the simulation by calling the subroutine arrive( ) if the next event is an arrival event, or by calling the subroutine depart( ) if the next event is a departure event. When the simulation reaches a termination condition, the report( ) routine is called. The Report document presents the performance outcomes of the simulation run.

- min_time_next_event = 1.0e+29;
- next_event_type = 0;
  - % Initial default value
- for k = 1:num_events
  - % class-k event; k =1 designates an arrival event; k = 2 designates a departure event
  - if time_next_event(k) < min_time_next_event
    - min_time_next_event = time_next_event(k);
    - next_event_type = k;
  - end
  - if next_event_type == 0
    - % i.e. when simulation time is higher than 1.0e+29;
    - sim_time // displays the simulation time
    - **break;**
  - end
- end
- sim_time = min_time_next_event;

Figure 7.17 The Simulation's Timing( ) Routine.

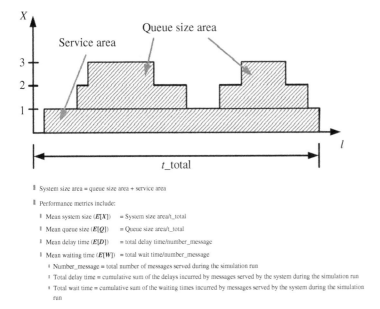

- System size area = queue size area + service area
- Performance metrics include:
  - Mean system size ($E[X]$)   = System size area/t_total
  - Mean queue size ($E[Q]$)   = Queue size area/t_total
  - Mean delay time ($E[D]$)   = total delay time/number_message
  - Mean waiting time ($E[W]$) = total wait time/number_message
    - Number_message = total number of messages served during the simulation run
    - Total delay time = cumulative sum of the delays incurred by messages served by the system during the simulation run
    - Total wait time = cumulative sum of the waiting times incurred by messages served by the system during the simulation run

Figure 7.18 Simulation Performance Updating.

As noted above and as shown in Fig. 7.17, the Timing( ) routine determines the next event type (whether it is a departure or arrival event), updates the time of the next event, and sets the simulation time to be equal to the time of the next event. As shown in Fig. 7.18, the noted routine updates the values of the simulation performance functions.

The updates that take place when the next event is a message arrival are shown in Fig. 7.19, using the Arrive( ) routine. In turn, the updates that take place when the next event is a message departure are shown in Fig. 7.20, using the Departure( ) routine.

234 | 7 Queueing Systems

- time_next_event(1) = sim_time + exprnd(mean_interarrival);
- num_in_s = num_in_s + 1;
- time_arrival_system(num_in_s) = sim_time;
  - % time arrival system (i) = time of arrival of message-i
  - % time arrival system (num_in_s) = time of arrival of the last message to arrive
- total_of_customers = total_of_customers + 1;
- if server_status == 1
  - % busy status is 1 and idle status is 0
  - % Server is busy, so increment number of customers in queue
  - num_in_q = num_in_q + 1;
  - if num_in_q > Q_LIMIT
  - % Checks to see whether termination condition is met: the queue limit has been reached, so stop the simulation.
    - display(['Overflow occurs at simu time = ' num2str(sim_time)]);
    - break;
  - end
  - time_arrival_queue(num_in_q) = sim_time;

- else % i.e., server_status == 0;
  - Server is idle, a new message arrives and is admitted into service.
  - % Arriving customer has wait time of zero.
  - % The following two statements are for program clarity and they do not affect the results of the simulation.
  - wait = 0;
  - total_of_waits = total_of_waits + wait;
  - % Increment the number of waiting customers and identify the server to be in busy status.
  - num_waiting_custs = num_waiting_custs + 1;
  - server_status = 1;
  - % server is busy
  - % Schedule a departure (service completion).
  - time_next_event(2) = sim_time + exprnd(mean_service_time);
- end

Figure 7.19 The Simulation's Arrive( ) Routine.

- num_in_s = num_in_s - 1;
- delay = sim_time - time_arrival_system(1);
  - % delay experienced by the departing message
  - % time_arrival_system(1) identifies the time of arrival of the head-of-the-line (HOL) message.
- total_of_delays = total_of_delays + delay;
  - % Increment the total by adding the delay incurred by the currently departing message
- num_delay_custs = num_delay_custs + 1;
- for k = 1 : num_in_s
  - time_arrival_system(k) = time_arrival_system(k+1);
  - % pushes up message arrival times in the array so that the current HOL message arrival time is at entry 1
- end
- if num_in_q == 0
  - % The queue becomes empty, the server is set to idle and we eliminate the next departure event from consideration.
  - server_status = 0;
  - time_next_event(2) = 1.0e+30;

- else
  - % The queue is nonempty, so decrement the number of customers in queue.
  - num_in_q = num_in_q - 1;
  - % Compute the wait of the customer that is starting service and update the total wait accumulator.
  - wait = sim_time - time_arrival_queue(1);
  - total_of_waits = total_of_waits + wait;
  - % Increment the number of waiting customers and schedule the next departure.
  - num_waiting_custs = num_waiting_custs + 1;
  - time_next_event(2) = sim_time + exprnd(mean_service_time);
  - % Move each customer in queue up one place upwards (FIFO service assumed):
  - for k = 1:num_in_q
    - time_arrival(k) = time_arrival(k+1);
  - end
- end

Figure 7.20 The Simulation's Departure( ) Routine.

The report( ) routine shown in Fig. 7.21 illustrates the calculation, recording and display of certain performance functions. They are calculated by using the performance data collected throughput the simulation process. While only average performance function levels are shown, one may similarly calculate and display other performance functions, such as those that preset probability distribution histograms of a multitude of performance functions.

- display(['E[D] = 'num2str(total_of_delays/num_custs_delay)']);
  - % Average delay in the system $E[D]$
- display(['E[X] = 'num2str(area_num_in_s /sim_time)']);
  - % Average number in the system $E[X]$
- display(['E[W] = 'num2str(total_of_waits/num_custs_delay)']);
  - % Average delay in the queue $E[W]$
- display(['E[Q] = 'num2str(area_num_in_q /sim_time)']);
  - % Average number in the queue $E[Q]$
- display(['Server utilization = 'num2str(area_server_status/sim_time)']);
  - % Server utilization = fraction of time that the service channel is busy;
  - % At steady state, for M/M/1 system, it will be equal to $\rho = \lambda/\mu$.

Figure 7.21 The Simulation's Report( ) Routine.

Problems

7.1 Describe the structure of a basic queueing system model. Identify the models and processes used to characterize the following system components and provide the definitions of the shown parameters:
a) The message arrival process; $A(t)$, λ, $E(T) = \alpha$.
b) Message service times; $B(t)$, $\beta = E(S)$, $\mu = 1/\beta$, $\rho = \frac{\lambda}{m\mu}$.
c) System capacity and admission threshold, N.
d) Service policy.

7.2 Define the following system parameters:
a) Message arrival rate λ.
b) Erlang loading f.
c) Channel (server) service rate μ.
d) Number of servers m.
e) Traffic intensity ρ.

7.3 Define the following system variables and processes:
a) The system size at time t, X_t, the system size process **X**; and the system size at steady-state X and its probability distribution $\{P(X = i)\}$; its average value $E(X)$.
b) The number of messages in service at time t, $N_S(t)$, and the number in service in steady-state N_S; its average value $E(N_S)$.
c) The unfinished load at time t, L_t, the unfinished load at stead-state L; its average value $E(L)$.
d) The queue size at time t, Q_t, the queue size process **Q**, and the queue size at steady-state Q; its average value $E(Q)$.
e) The waiting time of the n_{th} message W_n, the waiting time process **W**, the message waiting time at steady-state W and its probability distribution function $W(t) = \{P(W \leq t)\}$; its average value $E(W)$.
f) The delay time (or system time) of the n_{th} message D_n, the delay time process **D**, the message delay time at steady-state D and its probability distribution function $D(t) = \{P(D \leq t)\}$; its average value $E(D)$.

236 | *7 Queueing Systems*

    g) The virtual waiting time at time t, V_t, the virtual waiting time at stead-state V; its average value $E(V)$.

    h) Duration of the n_{th} busy period G_n; its distribution function $G(t)$; its average value $E(G)$.

    i) Duration of the n_{th} idle period I_n; its distribution function $I(t)$; its average value $E(I)$.

    j) Duration of the n_{th} cycle period C_n; its distribution function $C(t)$; its average value $E(C)$.

    k) The number of messages served during the n_{th} busy period, N_n; its average value $E(N)$.

7.4 Consider a single server queueing system model. Draw realizations of the system size and of the virtual waiting time processes, spanning 3 cycle periods. Identify in the figures the corresponding times of message arrivals, denoted as $\{A_0 = 0, A_1, A_2, A_3, \dots \}$, and message departure times $\{R_1, R_2, R_3, \dots \}$. Assume specific values for message inter-arrival times and service times, aiming to attain $N_1 = 3, N_2 = 4, N_3 = 5$. Indicate in the drawing the waiting time and delay time values incurred by served messages. Also identify the duration of incurred busy periods, idle periods, and busy cycles.

7.5 Consider a service system that is modeled as a basic queueing model. System measurements show the message throughput rate to be equal to $\lambda_D = 20$ [messages/s]. The average duration of a busy cycle is shown to be equal to 1.5 [s]. Calculate the average number of messages that are served during a busy period.

7.6 Consider a queueing system the is loaded by a Poisson message process. The message arrival rate is equal to $\lambda = 10$ [mess/s].

    a) Calculate the average duration of an idle period.

    b) Calculate the probability that an idle period is longer than 0.2 [s].

7.7 Consider a system the is modeled as an $M/G/1$ queueing system with message arrival rate $\lambda = 30$ [mess/s] and average message service time $\beta = 25$ [ms].

    a) Calculate the traffic intensity value.

    b) Calculate the average number of messages served during a busy period.

    c) Calculate the average lengths of a busy period, an idle period, and a busy cycle.

    d) Calculate the probability that at a given time (at steady state) the system will be observed to be idle (i.e., containing no messages).

7.8 Consider a multi-server queueing system that has reached steady-state conditions. The message throughput rate is measured to be equal to $\lambda_D = 25$ [mess/s]. The average message delay is recorded to be equal to $E(D) = 2$ [s].

    a) Use Little's formula to calculate the average number of messages held in the system (i.e., its average system size), $E(X)$.

    b) Assuming the average message service time to be equal to $E(S) = \beta = 0.32$ [s], calculate the average number of messages that are held in service.

7.9 Consider an outgoing stage of a packet-switching router that is loaded by packets that arrive in accordance with a Poisson process at a rate of λ [packets/s] and is served by a radio module that transmits packets on a FIFO basis across an outgoing communications channel. Each packet is of random length. It contains an average of 2500 [bits], including overhead of 400 [bits]. The buffer size is sufficiently large. The transmission data rate across the outgoing channel is equal to 10 [Mbps]. The packet transmission time is assumed to be exponentially distributed. The occurrence of packet errors while transmitted across the outgoing communications channel is neglected.

a) The packet arrival rate at the output stage is measured currently to be equal to $\lambda = 3400$ [packets/s]. Calculate the probability that the message delay time (its waiting time plus transmission time) in this output stage is longer than 3 ms.

b) The system engineer wishes to design a flow control mechanism that limits the rate of packets that arrive to this output stage so that 95% of the served packets experience a total delay time that is no higher than 3 [ms]. Calculate the maximum allowable packet arrival rate that meets the designer's objectives.

c) For the above-designed system, the designer evaluates the sufficiency of its buffer capacity by calculating the 99-percentile system size level N, which is exceeded by at most 1% of the packets that reside in the system (whether waiting or in service). Calculate the value of N.

7.10 Consider a protocol processor that is loaded by messages that arrive in accordance with a Poisson process at rate of 18 [packets/s]. The message service time follows an exponential distribution. The processor's service rate is equal to 20 [packets/s]. The system is limited to accepting only N messages (waiting and in service).

a) Calculate the lowest value that should be set for N so that the blocking probability is not higher than 10%.

b) Calculate the blocking probability when N is set to the above calculated value.

c) Packets that depart from this processor are fed to a subsequent processor that uses 3 service channels and that provides no queueing space. Packets that arrive to the subsequent processor when all of its 3 service channel are occupied are blocked. The performance of the subsequent processor is evaluated by assuming packets to arrive in accordance with a Poisson process and each packet service time to be exponentially distributed with average service time that is equal to 0.3 [s/packet]. Calculate the packet throughput rate at the output of the subsequent processor.

7.11 Consider a statistical multiplexing node that is served by two parallel communications channels. Each is operating at a data rate of 800 [Kbps]. The packet transmission time across a communications channel is exponentially distributed. Each packet consists of an average of 2000 [bits]. The node's storage capacity is equal to 3 packets, so that when saturated, two packets are in-service and a single packet waits in the buffer. Packets arrive at the node in accordance with a Poisson process at a rate of 760 [packets/s]. A flow admission control scheme is designed so that when the number of packets in the system is equal to 2, an arriving packet is admitted into the system with probability 0.5. Otherwise, it is blocked and lost. When the total number of packets in the system is equal to 3 (i.e., when there exists a waiting packet), no new packets are admitted.

a) Calculate the probability that (at steady state) there are no packets residing in the system (either in service or in the queue).

b) Calculate the average number of packets residing in the system.

c) Calculate the packet departing rate.

d) Calculate the packet blocking probability.

e) Use Little's formula to calculate the average packet time delay in the system.

238 | *7 Queueing Systems*

7.12 Consider a telephone exchange switching system that is served by N communications links (or trunks), where each link operates at a data rate of 1 [Mbps]. Calls arrive to the system in accordance with a Poisson process at a rate of 0.02 [calls/s]. A call holds its assigned communications link for a period of time that is exponentially distributed with an average value of 180 [s].

a) Determine the minimal number of communications links (N) that should be used to assure a call blocking probability that is not higher than 1%.

b) Assuming the number of links to be set to the value calculated above, calculate the average number of active calls held in the system.

7.13 Consider a multiplexing module that is modeled as an $M/M/1$ queueing system. Write a discrete-event simulation of an $M/M/1$ queueing system, following an approach similar to that presented in Section 7.9.2. Assume a message arrival rate that is equal to $\lambda = 4$ [mess/s], and a service rate that is equal to $\mu = 5$ [mess/s]. Run you simulation so that at least 100,000 messages depart the system.

a) Compare the simulation results obtained for $E(X), E(Q), E(D), E(W)$ with the corresponding ones that are derived by using mathematical formulas.

b) Assume that the service rate is held fixed at its value of $\mu = 5$ [mess/s], but that the message arrival rate λ is gradually increased in a manner that leads to a range of traffic intensity levels, $\rho = \{0.1i, i = 1, 2, \dots, 9\}$. Draw graphs that depict the variation of $E(X), E(W), E(D)$ as a function of the traffic intensity parameter ρ as obtained under the simulation runs and through analytical calculations.

7.14 Consider a switching module that employs two processors. Assume its buffer facility to be highly limited so that, when both processors are occupied, it is able to store only a single message that is waiting for service. The switch's service system is modeled as an $M/M/2/3$ queueing system. Assume the system to be loaded by packets that arrive at a rate of $\lambda = 9$ [packets/min]. The average service time of a message by each one of the processors is equal to 0.2 [minutes]. Write a discrete-event simulation of this queueing system by following an approach similar to that presented in Section 7.9.2. Run you simulation so that at least 200,000 messages depart the system.

a) Compare the simulation results obtained for $E(X), E(Q), E(D), E(W)$ with the corresponding ones that are derived by using mathematical formulas.

b) Assume now that the service rate is held fixed at its value of $\mu = 5$ [mess/s], but that the message arrival rate λ is gradually increased in a manner that leads to the following range of traffic intensity levels: $\rho = \{0.1i, i = 1, 2, \dots, 9\}$. Draw the performance graphs that depict the variation of $E(X), E(W), E(D)$ as a function of the traffic intensity parameter ρ, comparing the results obtained under the simulation run with those calculated by using the corresponding mathematical formulas.

7.15 Consider a radio module included in a user device that serves to transmit packets across a wireless channel to a base station node. The transmission data rate is equal to 5 [Mbps]. Packets served by this module are classified into two priorities. Priority-1 (i.e., high priority) packets arrive at the module in accordance with a Poisson process at an arrival rate of 2000 [packets/s], and the average packet length is equal to 1000 [bits/packet]. Priority-2 (i.e., lower priority) messages arrive at the module in accordance with a Poisson process at an arrival rate of 2500 [packets/s], and the average packet length is equal to 5000 [bits/packet]. The service system is modeled as a 2-priority $M/M/1$ queueing system.

a) Calculate the average packet waiting time level for class-1 and class-2 packets.
b) Calculate the factor by which the average class-2 packet waiting time in the queue is higher than incurred by a class-1 packet.
c) Check if it is feasible to double the value of the latter factor. If so, present an adjustment in the system parameters that would lead to such a result.

7.16 Consider a network that is modeled as a 4-node Jackson-type queueing network. The nodes are denoted as $\{N_1, N_2, N_3, N_4\}$. The network is structured to assume a ring topological layout. Thus, bi-directional links connect nodes N_1 and N_2, N_2 and N_3, N_3 and N_4, N_4 and N_1. Each link operates at a data rate of 100 [Mbps]. External messages arrive to each node (in accordance with a Poisson process) at a rate of 40 [Mbps]. A message received at a node is processed by the node and then departs the system with probability 0.5, while with probability 0.5, it is routed to a neighboring node located in a clockwise direction. Thus, a message that completes service at node N_1 will depart the network system with probability 0.5, while with probability 0.5, it is routed to node N_2. Each node is modeled as an $M/M/1$ queueing system. Each message is assumed to contain a total of 1000 [bits/mess].

a) Calculate the message flow rate across each network link.
b) Calculate the average delay incurred by a message at each node.
c) Calculate the average message system size at each node.
d) Calculate the network's message throughput rate.
e) Regarding the whole network as a service system, use Little's formula to determine the average message delay, averaging over all network nodes and over all messages served by the network system.

7.17 Consider a 3-node tandem queueing network such as that shown in Fig. 7.12. Assume the service rates of the corresponding nodes to be set as $\mu_1 = 10$ [mess/s], $\mu_2 = 5$ [mess/s], $\mu_3 = 8$ [mess/s]. Each message is assumed to consist of a single segment of fixed length 2000 [bits]. Messages arrive to the network in accordance with a Poisson process at rate 4 [mess/s]. Calculate the average message end-to-end aggregate waiting time and average aggregate delay time.

8

Multiple Access: Sharing from Afar

Objectives: *In a multiple access system, geographically distributed stations, or end users, share the resources of a service system such as a communications medium. A **multiple access algorithm**, forming the basis of a **Medium Access Control (MAC)** protocol, is used to manage, coordinate, and schedule the use of the shared medium resources. Examples of networks that use multiple access communications include cellular wireless networks (where mobiles share uplink communications channel resources), local area wireless systems such as Wi-Fi nets, satellite networks, and peer-to-peer wireless networks such as those used to connect autonomous vehicles.*

*Methods: Under a fixed allocation MAC protocol, an admitted station is allocated time, frequency, code, or space slots, or a hybrid thereof, on a dedicated long-term basis. Under a Demand-Assigned Multiple Access (DAMA) scheme, a station goes through a reservation process by sending a reservation message to a resource manager. Alternatively, a central **polling scheme**, or a distributed token-passing algorithm, may be employed. Under a **random access** scheme, active stations contend among themselves in attempting to capture shared medium resources. A Wi-Fi wireless local area network (WLAN) uses a version of such a protocol, while also enabling the use of an access mechanism that uses a hybrid of DAMA and spatial-reuse processes. For the latter, directional antenna beams are dynamically configured and employed for the transmission of data messages to targeted stations.*

8.1 Multiple Access: Sharing from Afar

Sophie and Tess are working at home in remotely located rooms. They are engaged in online interactions with their friends, streaming, downloading, and uploading video and data files, transported through Internet communications. To connect to the Internet, they use a Wi-Fi-based wireless local area network (WLAN) that covers their home's space. They connect to their shared Wi-Fi router, which has been placed at their home at a different floor. Often, they are simultaneously active in such communications, having then the need to share their sole Wi-Fi wireless communications medium. The data that they wish to communicate is stored separately in Sophie's and Tess's computers. How do these computers go about sharing the resources of the Wi-Fi communications medium *from afar*? Such a configuration is identified as a **multiple access communications** system.

A multiple access network configuration is illustrated in Fig. 8.1. Stations 1–3 are located at geographically distinct places. They share a wireless communications medium to communicate to a central node. The latter may be identified as a Hub or as an Access Point (AP). The central node may be remotely located and may communicate with stations across a satellite communications

Principles of Data Transfer Through Communications Networks, the Internet, and Autonomous Mobiles, First Edition. Izhak Rubin.
© 2025 The Institute of Electrical and Electronics Engineers, Inc. Published 2025 by John Wiley & Sons, Inc.

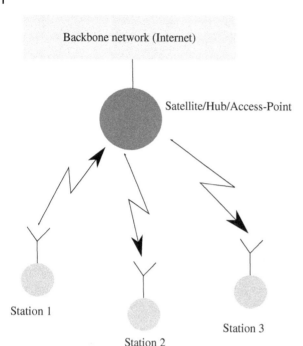

Figure 8.1 A Multiple Access Network.

network. The central node be attached to a base station (BS) node that connects it to a cellular wireless network. An AP may use a Wi-Fi WLAN to communicate with near-by user stations.

In the figure, station 1 is shown to be granted permission to transmit its messages to a central node, which then relays these messages to stations 2 and 3. The central node may use a wide beam antenna that covers all network stations, or in turn employ multiple narrow antenna beams that are directed to targeted regions or stations. The central node may be able to receive, across a prescribed uplink frequency band, only a single message transmission at a time, one that has been sent by one of the stations. The central node is also shown to be connected to a backbone network, such as the Internet. As such, it acts as an AP connecting the net's stations to the Internet. It disseminates the data flows that it receives across the multiple access channel from its client stations over the Internet aiming them to reach their identified destinations. Data that it receives from sources across the Internet is sent (in a multiplexing manner) across a shared downlink medium to targeted stations.

Definition 8.1 **Multiple Access** schemes are used to regulate, coordinate, schedule, and manage the sharing of a resource, such as a communications medium, among multiple geographically distributed stations, so that **each station separately employs its own transmission module and separately keeps in its own buffer the messages that it wishes to send across the shared medium**.

As another example, consider messages that are stored in the buffers of their own remotely located satellite earth stations, as they wait to be granted access to the satellite's uplink communications channel.

Also, consider mobile users that are managed by a BS node as they share the wireless radio channels that are used by them to communicate with their BS node. Each station keeps its messages in its own buffer, as it awaits for the scheduling of its transmission across the medium.

Examples of communication network systems that use multiple access communications channels as key elements include wireless cellular networks, satellite networks, Wi-Fi systems, mobile radio public safety networks, local area networks (LANs) such as Ethernet LANs, communications networks that are used to connect autonomous highway (self-driving) vehicles, Unmanned Aerial Vehicle (UAV)-based networks, and many other wireless and wireline network systems.

In a vehicle transportation system, a segment of a highway is shared among multiple vehicles. Vehicles that wait at distinct ramps for accessing the highway are regulated for highway admission by adhering to a multiple access protocol.

To illustrate the differences between multiplexing and multiple access schemes, consider a classroom setting, as shown in Fig. 13.3. The teacher and his students share the medium across which they communicate, visually and via audio. The teacher checks attendance by calling the names of students. In doing so, the medium is shared by the teacher on a multiplexing basis, noting that the teacher's prompting messages originate at a **single** source.

In contrast, response messages that are sent by students share the medium on a multiple-access basis. Messages that distinct students produce and wish to convey to the teacher are *separately generated and sent by sources that are situated at multiple distinct (i.e., geographically distributed) locations.* If currently several students have messages (such as questions) that they are interested to convey to their teacher, they will do so by sharing the medium through the use of a multiple access scheme. The visual dimension of the medium may be shared by having an interested student raise her hand and in this manner send a signaling (reservation) message to the teacher that indicates her interest to talk. Subsequently, through prompting by the teacher, the student will be permitted to deliver her message (question) across the multiple access shared medium.

Under a typical *multiplexing* scheme, a resource is shared by messages that reside in a single station in a **common** local buffer facility. A multiplexer's controller schedules the transmission of its local messages for across a shared communications medium. In doing so, it takes advantage of its direct local access to its local buffer, in determining the sharing needs of currently active flows.

In contrast, in a *multiple access network system*, each user's device, identified as a **station**, holds its active messages in its own buffer. Messages that are associated with different stations are held in distinct (geographically distributed) buffers. To coordinate the use of the medium among distinct stations, a busy station may wish to learn about the current transmission resource needs of other stations. Since stations are situated in distinct locations, a station would have to make use of cross-station coordination means, such as by using a control channel through which stations can inform each other about their current message transmission states and requirements.

In comparing with multiplexing-based sharing schemes, multiple access networking mechanisms are noted to generally be more complex, demand coordination resources, and induce longer message delays and lower resource sharing utilization levels. The efficient operation of many modern network systems critically depends on the quality of the processes that are employed to operate, manage, and regulate their multiple access schemes.

What are the different approaches that are used to share resources, such as communications media, among geographically distributed users? In this chapter, we describe the principles of key commonly employed multiple access protocols, schemes, and algorithms.

The access of distributed busy clients to the shared system is regulated either by using a central controller or by employing a distributed scheduling algorithm. When employing a **fixed multiple access** mechanism, a station, or a message flow, is allocated shared resources on a fixed, relatively long-term, basis. In turn, under a **demand-assigned multiple access (DAMA)** scheme, shared resources are dynamically allocated. Under a **random access** scheme, an active user that wishes to access a shared resource must determine on its own the proper time and resource segment that it should use without going through explicit coordination with other active users or without being regulated by a central manager.

8.2 Fixed Multiple Access Schemes

Under a fixed multiple access scheme, each station that shares a common service system, such as a communications medium, is allocated a fixed resource on a dedicated long-term basis. While this allocation is fixed over a period of time that allows the allocated resource to be used for the transport of many messages, a system manager is able to adjust the allocation through the use of a network management scheme. As it is not frequently adjusted, the station may at times be allocated a resource level that it may not use in an efficient manner. The station's allocated resource may then stay unused for relatively long periods of time. It would remain unused in periods during which the station is not active.

Channel resources can be shared across multiple dimensions that include time, frequency, code and space, and their hybrids. The corresponding fixed multiple access schemes are identified as Time Division Multiple Access (TDMA), Frequency Division Multiple Access (FDMA), Code Division Multiple Access (CDMA), Space Division Multiple Access (SDMA), and their hybrids. These schemes are discussed in the following.

8.2.1 Time Division Multiple Access (TDMA)

Under a **TDMA** scheme, each station is assigned, on a fixed basis, time periods during which it is permitted to occupy the shared medium.

Example 8.1 To illustrate, assume that a class consists of 10 students. At the end of the lecture, the teacher requests each student to provide a very brief summary of the main points presented in her lecture. For this purpose, the teacher addresses each one of the students, providing the student a time slot of 2 minutes to present a summary. The teacher prompts each student at the start of the student's allocated time slot to let the student know that her presentation can start at that time. It may also point out the termination time of the time slot. In this manner, the multiple access medium is time-shared by the students. If of interest, the teacher can repeat this cycle several times to allow students more time to express their views. The time at which a student may initiate her question is coordinated across the shared medium through the use of a prompting (control/signaling) operation.

To simplify the implementation of the time-sharing operation, the following TDMA system structure is generally adopted. Time slots and time frames are established, declared, and recognized by the sharing stations. Each **time slot** is announced to be of fixed duration, say τ [s]. A collection of M successive time slots is recognized as a **time frame**. The frame duration is thus equal to $T_F = M\tau$ [s]. Such a configuration is illustrated in Fig. 8.2. In this system, two active stations time share the medium. The time frame consists of two fixed size slots. Station 1 is allocated the first time slot in each time frame, while Station 2 is allocated the second time slot in each time frame. Stations are assigned a frequency band. Its allocated bandwidth is noted here to be equal to the aggregate bandwidth of the indicated four frequency bands ($W = F_1 + F_2 + F_3 + F_4$). Other frequency band allocations can also be made.

A station fragments its messages into segments, whereby the transmission time of a segment fits within the station's allocated time slot. For example, if a time slot's duration is equal to 1 ms, and if the transmission data rate across the shared channel is equal to 10 Mbps, then the maximum segment bit length is equal to 10 Mbps × 1 ms = 10 kbits. If the station wishes to transmit across this channel, a message whose length is equal to 25 kbits, then the message is fragmented into

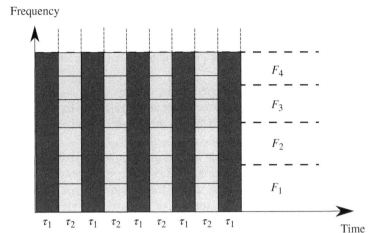

Figure 8.2 An Illustrative Time Division Multiple Access (TDMA) Network Whereby a Medium is Time-Shared by Two Stations.

three segments. For the illustrative system, it will take the station 3 time frames to complete the transmission of this message, as it is permitted in this setting to send a single segment in each time frame. Since each time frame contains two time slots, the time frame duration in this case is equal to $T_F = 2$ ms.

Each station transmits its segments during its allocated time slots. It is essential that transmissions made by different stations do not overlap in time as they are received at their intended receiver(s). Consequently, the system must implement a sufficiently precise system-wide **time synchronization scheme**. Each station adjusts its identification of the time instants at which each one of its time slots should start in a manner that induces each segment's transmission to reach its targeted station at the start of its allocated time slot as measured at the targeted station.

For this adjustment, noting that different stations may be located at different distances from their targeted receivers, each station determines the time delay incurred by its signal as it propagates from its location to the location of its targeted receiving station. For example, assume stations 1 and 2 to share a communications channel in aiming to reach a common central station. Station 1 is closer to the central station; its signal incurs a propagation delay of 0.2μ [s] in traveling to the central station. The propagation delay incurred by the signal issued by station 2 is equal to 0.4μ [s]. Consequently, if a segment's transmission should arrive at the central station at time t, station 1 will initiate the transmission at time $t - 0.2\mu$ [s], while station 2 will initiate the transmission at time $t - 0.4\mu$ [s]. A commonly used method to measure a sending station's signal propagation delay to a targeted receiving station is for a sending station to send a test signal which the receiving station reflects back to the sending station, enabling the measurement of the signal's round-trip time delay.

Often, a certain node is selected as the system's controller or manager. This is the role played by a Wi-Fi AP or by a BS that manages a wireless cell. The managing node is then used to disseminate, on a periodic basis, a control signal that serves as a **beacon**, conveying timing information to network stations. Stations use this information to synchronize their clocks. The local timing source (or clock) at each station tends to *drift* between updates, leading to timing errors. Timing updates must thus be provided at a sufficiently high rate to avoid timing drift deviations that exceed a specified level. The duration of a time slot is often extended to accommodate an inter-slot time drift margin, or guard time, serving to prevent signal reception overlaps. In systems that do not employ a central controller, one of the stations may be nominated to assume this role. It disseminates timing information that is used by stations to synchronize their time clocks.

Different stations can be allocated different numbers of time slots in a time frame. A station that is scheduled to transmit a larger number of data segments in a time frame would be assigned, if available, a larger number of time slots.

The **delay-throughput performance behavior** experienced by messages at a station that shares a communications medium on a TDMA basis is characterized as follows. The **throughput capacity** available for the transmission of messages by a TDMA station is determined by the number of slots allocated to the station in each time frame and by the data rate that it uses to transmit its segments during these time slots. For example, consider a time frame that consists of $M = 100$ time slots, whereby each time slot is of duration $\tau = 1$ ms. The corresponding time frame is of duration $T_F = M\tau = 100$ ms. Consider a station that is allocated $K = 10$ time slots in each frame. Assume that the transmission data rate across the shared medium (also identified as the channel transmission data rate) is equal to $R = 100$ Mbps. The station is then able to send at most $RK\tau = 1000$ kbits, or 1 Mbits, during each time frame. The time frame duration is $T_F = 100$ ms (or 0.1 second). Hence, this station is scheduled to transmit its data at a station data rate of 1000 kbits/0.1 s $= 10$ Mbps.

The delay time incurred by messages that arrive to a station and are transmitted by it across a medium that is shared on a TDMA basis consists of the following components. Assuming a First-In First-Out (FIFO) service policy, an arriving message experiences a queueing delay as it waits at the station for the transmission of messages that have arrived to this station earlier and are still waiting for their transmission across the channel. The message queueing delay at the station can be evaluated by using the approaches described for the analysis of multiplexing schemes and queueing systems. For example, an approximate analysis can be carried out by modeling the system as a service module, whereby the server operates at a rate that corresponds to the throughput capacity rate (or station data rate) allocated to the station. For example, if the station is allocated K time slots in each time frame of duration T_F [s], then the module's service rate is equal to $\mu = K/T_F$ [segments/s].

In addition to accounting for the incurred queueing delay, it is also necessary to account for a message delay component that is identified as its **frame latency**. It accounts for the time delay incurred by a message that is positioned at the head of the queue but must still wait for transmission at the occurrence time of the next time slot allocated to the station. For example, if a station is allocated $K = 10$ time slots in a time frame of duration $T_F = 100$ ms, the average frame latency is equal to $T_F/K = 10$ ms, as this station is allocated a time slot on an average of every 10 ms. At higher traffic intensity levels, messages tend to incur longer queueing delays so that the queueing component dominates. In turn, under a lower traffic intensity loading level, the frame latency component plays a more prominent role.

8.2.2 Frequency Division Multiple Access (FDMA)

Sophie and Tess are currently listening to different radio stations, station A and station B, respectively, which broadcast music that plays at their individual radio receivers. In broadcasting its music, each station's transmissions occupy a distinct frequency band. Sophie and Tess tune up their radio receivers to the corresponding frequency bands used by station transmitters A and B, respectively.

Under a **FDMA** scheme, each station sharing the medium is allocated a separate frequency band. To illustrate, assume that the bandwidth of a shared communications link is divided into M frequency bands. The bands can be of the same or of different widths. A station is allocated

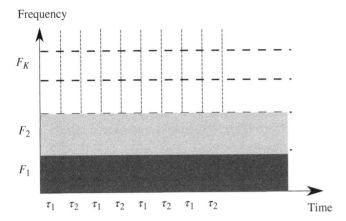

Figure 8.3 An Illustrative Frequency Division Multiple Access (FDMA) Network Whereby Each Station Is Dedicated a Frequency Band.

a frequency band for use in transmitting its messages or streams. In this manner, multiple stations are able to transmit messages at the same time, provided each shapes the spectrum of its transmitted signal so that it is restricted to occupying only (or mostly) its allocated frequency band.

The sharing of a medium on an FDMA basis by two stations is illustrated in Fig. 8.3. Station 1 is allocated frequency band F_1 while Station 2 is dedicated frequency band F_2.

Consider a communications channel that shares a bandwidth of W [Hz] among M stations. Assume first that an FDMA scheme is used to share the medium. Assume stations to be allocated equal frequency bands, so that each is assigned a band that is equal to W/M [Hz]. Assume that the Modulation/Coding Scheme (MCS) that is used by each transmitting station provides a spectral efficiency that is equal to η [bps/Hz]. Then the data rate realized by a station when it transmits its messages is equal to $R_F = \eta W/M$ [bps]. For example, assume $W = 100$ [MHz], $M = 10$, and $\eta = 2$ [bps/Hz]. Then, under the above described FDMA scheme, each station is able to transmit its messages at any time, by occupying its allocated frequency band, at a data rate $R_F = \eta W/M = 20$ Mbps.

In comparison, under a corresponding TDMA scheme whose time frame consists of M slots, assume that each station is allocated a single time slot in each time frame. The transmitting station's signal is set to occupy the full frequency band of W [Hz]. Assuming its MCS to achieve a spectral efficiency that is equal to η [bps/Hz], a station transmits its messages during its allocated slots at an instantaneous (channel) data rate that is equal to $R_T = \eta W = MR_F$ [bps]. Under the above-mentioned numerical values, the TDMA station transmits its messages at a data rate that is equal to $R_T = 200$ Mbps, but it does so during only 1/10th of the time. In this manner, it achieves a long-term average throughput rate that is equal to $R_T/10 = 20$ Mbps. A corresponding FDMA station transmits its messages continuously in time at a data rate of $R_F = 20$ Mbps, occupying 1/10th of the bandwidth. Under either FDMA or TDMA scheme, ideally the same longer term average throughput rate is realized.

It is essential to limit signal interference at intended receivers that is caused by transmitting stations that share the channel. Under a TDMA scheme, it is crucial to implement a precise time synchronization system. Time guard periods are included so that transmissions that occupy neighboring time slots do not cause excess signal interference that impacts the quality of segment reception. Such time gaps are typically included into the specification of established time slot duration, making it somewhat longer. These gaps are taken into consideration when calculating the net throughput efficiency level.

In turn, under FDMA, it is essential for a transmitting station to filter the spectral components of its signal transmissions so that they occupy, in great part, its allocated frequency band, and so that they do not cause excessive interference to signal receptions performed over other frequency bands.

The MCS is configured in accordance with the Signal-to-Interference plus Noise Ratio (SINR) measured at the targeted receiving node. The configured MCS then realizes a spectral efficiency level that is denoted as η [bps/Hz]. Accordingly, the frequency bandwidth allocated to a station under FDMA is based on the data rate R_F [bps] that the station wishes to realize, which is equal to its provided average throughput capacity rate, $TH_F = R_F$ [bps]. The bandwidth used is then equal to $W_F = R_F/\eta$ [Hz].

For a station to achieve a targeted throughput rate (TH_T [bps]) under a TDMA scheme, when assuming that the station is allocated an average of one slot every M slots, the instantaneous transmission data rate is set to $R_T = M \times TH_T$. The bandwidth that should be allocated to this station is then equal to $W_T = R_T/\eta$ [Hz].

Similarly, under FDMA, depending on the SINR incurred at the targeted receiving station and the desired throughput rate, different transmitting–receiving station pairs would require the allocation of different frequency bands.

There are at times reasons to reduce the data rate used by a station in communicating over a shared medium, while accordingly reducing the frequency bandwidth allocated to the station. For example, when transmitting a message at a high data rate, the message transmission time is very short. In many wireless communications channels, a transmitted message may propagate to the intended receiving node by a direct path as well as via other paths, such as those that result through reflections from objects located along the way. Such *multipath* reflections may interfere with each other (at a receiver) and can consequently prevent the correct reception of a message. The magnitude of such signal interference can be reduced by lowering the message transmission data rate.

There are other incentives to reduce the message transmission data rate, including such that relate to the physical implementation of the ensuing transmitter, receiver, and the channel signal and its propagation across the communications channel.

It is therefore of interest to often implement a **hybrid TDMA/FDMA** scheme. Under such a configuration, a station is allocated a specified number of time slots per frame whereby a transmission in an allocated time slot is set to occupy a prescribed frequency band. Such a scheme is employed by a multitude of communications network systems, including various mobile wireless networks.

Under a **multicasting** configuration, messages are destined for reception by a specified class of nodes that are recognized as members of an underlying *multicast group*. Under a **broadcasting** dissemination, all nodes that belong to an underlying local network, or that are located in a specified geographical neighborhood, are identified as targeted destinations. In these cases, it is necessary to set the FDMA or TDMA system parameters in a manner that assures an acceptable quality of transmission while communicating with all targeted nodes.

The design of a TDMA modem and transmitter is challenging in that it must operate in a **burst** mode, as a station turns-on its transmitter only during its assigned time slots. During these periods, a station sends data at typically a high data rate. Peak power requirements then impose design limitations. In turn, an FDMA module transmits its active messages at a reduced data rate in a time-continuous fashion. Such an operation induces a less restrictive implementation requirement on the transmission module, imposing average power (rather than peak power) limitations.

In sharing a medium by a high number of stations, an FDMA system would have to assign a large number of finely tuned (relatively narrower) frequency bands. The implementation of such a system can be challenging. In turn, a TDMA system is generally designed to assure resource sharing among many stations in a very fine manner, as it is capable of allocating access times to stations on a time slot by time slot basis.

Frequency division-based resource allocations are often used to regulate signal interference that is caused by stationary or mobile stations that are active in neighboring regions. For example, in cellular wireless networks or autonomous highway-networked transportation systems, transmissions carried out in neighboring cells are often assigned distinct frequency bands.

8.2.3 Space Division Multiple Access (SDMA)

I am attending a conference party with colleagues. As I often do, I am standing in a large conference room having a conversation with several colleagues that stand near me. No too far away, other groups of colleagues have their own group discussions. While these conversations are conducted at the same time so that conversation sounds may spill over, I am generally able to tune out the interfering sounds produce by conversations conducted in other groups. Yet, every so often, a nearby conversation is conducted at a relatively high sound intensity, preventing me from properly hearing the discussion that takes place within my group.

In an analogous manner, multiple groups are often able to share the spatial medium surrounding a conference room in simultaneously conducting their own message communications. To properly work, it is necessary to assure that the SINR detected by each listener is sufficiently high, so that the power of the signal received at an intended listener is not over-masked by the aggregate power of noise and interference signals.

This sharing scenario illustrates the notion of **SDMA**. Under this process, geographically distributed users, or user groups, use different **spatial** segments to provide for effective sharing of the medium.

To attain higher efficiency in the use of communications channel resources, an SDMA scheme is employed to allocate the same time slot and frequency band for use over different space segments. When a **frequency reuse** pattern is employed, the same frequency band is reused across distinct space segments. When a **time slot reuse** pattern is employed, the same time slot is reused across distinct space segments. The resource reuse configuration is planned in a manner that assures an acceptable induced signal interference level.

Such an assignment is highly valuable in wireless communications systems, for which frequency spectrum is a valuable limited resource. It can produce significant frequency and time reuse factors, leading to higher message throughput rates, or reduce the required frequency bandwidth levels.

To illustrate, consider a cellular wireless network that is employed to support communications in a specific area of operations. The area is divided into N cells. Each cell is managed by a single BS node. Assume each BS node to employ an omnidirectional (i.e., one that radiates energy equally in all directions) antenna for communications (uplink and downlink) with the mobiles located in its cell.

Assume an overall available frequency bandwidth of W [Hz]. Assume first that a single frequency band, spanning the bandwidth W, is repeatedly employed in each cell and that transmissions in each cell are carried out at the same time. Depending on the positions of the involved sending and receiving nodes, such simultaneous transmissions may cause non-negligible signal interference. Lower data rate values may therefore be employed, leading to reduced throughput rates. When no such signal interference effects are induced, spatial reuse of channel resources would lead to enhanced throughput performance. In the same manner, rather than making use of a frequency reuse scheme, a time slot reuse scheme may be configured so that the same time slot can be used for data transmissions at multiple cells.

Under a reuse scheme, the system designer would allocate distinct frequency bands (or time slots) for use across different spatial segments (or cells). For example, if $k \geq 1$ frequency bands (or time slots within a time frame) are used, each one may be set to occupy W/k [Hz]; a corresponding time interval is set to be T_F/k long. To reduce intercell interference, neighboring cells may be allocated distinct frequency bands (or time slots).

A reuse pattern can be represented as a **graph**, which represents a layout of nodes connected by lines. Each node (or vertex) of the graph represents a cell. Two nodes are connected by a line (or edge) if they represent neighboring cells. The process of **coloring** the graph, using a specified number of colors, entails the assignment of colors to the nodes such that neighboring nodes (i.e., nodes that are connected to each other by a line) are assigned distinct colors. Hence, the process of assigning frequency bands to different cells, or spatial segments, is identified as a *coloring* process. The use of k colors signifies the allocation of k distinct frequency bands to the area's cells or spatial segments. Similarly, when nodes are assigned time slots, the coloring process proceeds by assigned distinct colors (i.e., time slots) to neighboring nodes (such as cells).

As the number k of allocated colors increases, the average signal interference induced by transmissions that occur in cells that use the same color is reduced as such cells are located further away from each other. This leads to the selection of a MCS that achieves a higher spectral efficiency factor η [bps/Hz]. However, the per-spatial segment bandwidth level that is then available for use, which is equal to W/k, is reduced. Hence, depending on the underlying topology and network system conditions and parameters, an optimal coloring level can be determined. It usually aims to produce a high aggregate throughput rate (at designated quality level).

Under a frequency division method, the number of employed colors (k) represents the number of frequency bands into which the frequency band resource is segmented, whereby the same frequency band segment is reused in different cells (or spatial regions). For sharing via a time division scheme, the number of colors (k) represents the number of groups of time slots per frame that are reused at the same time for transmissions that are carried out in different cells (or spatial regions). At times, the number of colors (k) is identified as the **coloring index**.

The **spatial reuse advantage** is expressed by the ratio $1/k$, identifying the **spatial reuse factor**. The lower is the coloring index, the higher is the spatial reuse advantage. For example, if a single color is used, $k = 1$, a very high spatial reuse advantage is attained: the same (overall available) band of bandwidth W [Hz] (or a given group of time slots) is used in every cell, leading to a high-**frequency reuse factor**. In comparison, if three colors are used, $k = 3$, 1/3rd of the area cells employ the same frequency band, the bandwidth of which is equal to $W/3$ [Hz]. Similarly, under a time division technique, 1/3rd of the area cells simultaneously use the same time slot.

A 3-color reuse pattern is illustrated in Fig. 8.4. Central cell-0 employs the same frequency band (or time slot) as that used by cells 8, 10, 12, 14, 16, and 18. It is noted that neighboring cells are assigned distinct colors and thus use distinct frequency bands (or time slots). Message transmissions that are destined to users that dwell in cell-0 may experience signal interference at their respective receivers from signals produced by message transmissions that take place at the same time and that are issued by transmitters located in cells of the same color. The power levels of the interference signals induced by ensuing simultaneous transmissions are, on the average, lower than those that would be caused by transmissions that originate at neighboring cells.

Figure 8.4 A 3-Color Cellular Space Division Multiple Access Network.

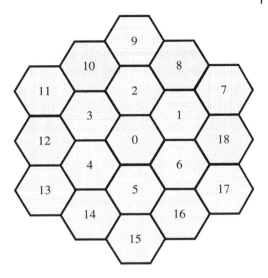

Higher spatial reuse gains are attainable when **directional antennas** are employed. In cellular wireless networks, BS nodes tend to employ directional antennas such as those that produce radiation beams that span four 90-degrees wide antenna sectors. Assuming transmissions in each sector is associated with a separate transceiver, the BS node is then capable of transmitting or receiving signals simultaneously in time over different quadrants of the BS's cell. In Fig. 8.5(b), we depict such simultaneously carried uplink communications from mobiles located in the four quadrants of a cell to their BS node. The system can be configured to provide for frequency (or time slot) reuse over the quadrants.

By using directional antennas at the transmitters and at the receivers, it is possible to aim the antenna beam of a transmitter to have its peak transmit radiation power pattern so that it covers effectively the antenna beam of the targeted receiver. The receiver's antenna beam may also be managed to be directed to a targeted transmitting station. Nearby stations may configure their directional beams to properly communicate with their targeted stations. Using directional antennas and antenna arrays, it is possible to realize a high spatial reuse factor, leading to a substantial increase in the system's throughput rate. Such simultaneously performed directional communications is shown in Fig. 8.5(a), whereby three sending nodes (T_1, T_2, T_3) transmit

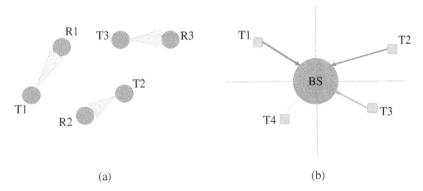

Figure 8.5 Directional Communications: (a) Peer-to-Peer Directional Communications; (b) Simultaneous Uplink Communications in Four Quadrants of a Cell.

messages to three, respectively, targeted receiving nodes (R_1, R_2, R_3). To provide for successful receptions of intended messages, the used MCS, employed data rates, allocated bandwidth, and transmitted power levels are properly configured, based on the induced SINR levels that are measured at the intended receivers. It is necessary to take into consideration the impact of interference signals, noting that the sending and receiving beams are generally not perfectly aligned, so that some cross-interference signals can be produced.

A further enhancement in the ensuing data rates is induced by the high (transmit and receive) antenna gains that are realized by using directional antennas. An antenna gain is measured as the ratio of the power intensity radiated by the antenna in the targeted direction (achieving highest power radiation) divided by the intensity radiated by a hypothetical isotropic antenna producing the same total power. An omnidirectional antenna produces electromagnetic radiations that assume equal power in all directions. As the antenna's main lobe of its radiation pattern becomes more narrow, higher (transmit and receive) antenna gains are achieved, leading to higher receive and transmit effective signal power levels.

While an antenna that forms a narrower beam yields a higher antenna gain factor and enables the system to achieve a higher spatial reuse measure, its operation is more demanding in the performance of the underlying aiming and alignment processes. For mobile entities, the aiming process must be repeated at a rate that increases as a narrower beam is configured and as the mobility speed of the node increases.

The size of a basic antenna element is of the order of 1/4th to 1/2th of the corresponding wavelength being used. The wavelength associated with an electromagnetic signal wave is equal to $\lambda = c/f$, where λ represents the wavelength, f is the carrier frequency, and c designates the speed of light. Accordingly, when the carrier frequency is set to 200 MHz, or 1 GHz, or 60 GHz, the corresponding wavelengths are equal to 1.5 m, or 0.3 m, or 5 mm, respectively. Thus, a significant reduction in antenna size is realized as higher-frequency levels are employed. Under mmWave communications, such as those performed at carrier frequencies that range around 60 GHz, antenna elements assume millimeter lengths. In using such elements to form highly directional communications beams, one attains high spatial reuse factors, leading to high-throughput rates. Applications include the use of mmWave communications networks for mobile cellular wireless networks and for communications with and among autonomous highway vehicles.

In many cellular wireless and other communications systems, multipath propagation effects are commonly incurred. Multiple reflections of a signal transmitted by a sending node may be received at a targeted receiving node. The instants of time at which such reflections are detected at the intended receiver, during an underlying window of time, form a characteristic signature of the underlying propagation channel. **Multiple-In Multiple-Out (MIMO)** systems employ multiple transmission and reception antenna beams to exploit multipath propagation, as illustrated in the following.

Consider a transmitting node that acquires the underlying pattern of time delays (or corresponding signal phase shifts) characterizing signal propagation between a transmitting–receiving pair of nodes that employ multiple antennas. It is identified as Channel State Information (CSI). For the purpose of sending a data stream across a multipath communications channel to a targeted receiver, the transmitting node uses its acquired CSI to construct distinct signal formations of the message stream that it feeds to its antennas. The ensuing propagated signals are combined constructively as they are received at the targeted receiving node. Such a process serves to increase the throughput capacity of the underlying communications channel.

A MIMO antenna configuration is also used to achieve **spatial multiplexing** features. For this purpose, a message stream is split into multiple lower rate streams, whereby each stream is transmitted (using the same frequency band) by a different antenna. A receiver acquires the CSI that characterizes the arrival time signatures of these streams and uses it to accordingly de-multiplex them. Such an operation, when carried out at a sufficiently high SINR level, leads to the realization of a higher channel capacity level. The number of employed spatial streams is determined as the minimum of the lower of the number of transmit and receive antennas. This mechanism is also employable as an SDMA scheme, known as multiuser MIMO. Each receiver provides to the transmitter its own distinctive CSI based on the detected spatial signature of each. The transmitting node uses the corresponding CSI signatures and employs them for configuring the formation of its transmitted message streams.

MIMO-based operations have been widely employed. Applications include cellular wireless systems (such as 4G LTE and later generation cellular wireless systems), Wi-Fi WLAN systems (such as IEEE 802.11ac), and power line communications.

8.2.4 Code Division Multiple Access (CDMA)

Under **CDMA**, users share a communications medium for the transmission of their messages by jointly occupying the same temporal, spectral (i.e., frequency band), and space segment resources. To this end, a sending station encodes its message by applying to it a signature code, known as a **CDMA code**. The identity of the code is commonly known to the sending and receiving stations. It enables the receiving entity to discriminate the reception of message signals that are intended for its reception from signals produced by other messages that it may detect at the same time and that occupy the same frequency, time, and spatial segments.

As an analogy, consider a party setting whereby several groups of conversing persons form and multiple persons simultaneously speak to their nearby group colleagues. A listening person is often able to discriminate between such multiple ongoing conversations, identifying the one that is directed to her or that she is interested in. For this purpose, the listener may focus on specific characteristics that identify the speech of the speaker of interest, whether involving the speaker's accent, tonal characteristics, or language. As far as this listener is concerned, the cumulative signals produced by the other conversations then become background noise. As long as the power of this noise (as detected by this listener) is not excessive, the listener is able to comprehend the conveyed information of interest.

How is such a method implemented in communications networking systems?

Commonly employed CDMA systems apply unique signature codes to transmitted data messages by modulating them with a **CDMA code sequence** along the temporal or spectral dimensions. An overview of such a coding operation, known as a **Direct Sequence (DS)** scheme, forming a **DS-CDMA** system, is outlined in the following.

The generated signature CDMA code consists of code symbols that are produced at a much higher rate than the data symbols. Consider a data symbol that is represented as a pulse of duration T_s [s]. The data message rate is then equal to $R_s = 1/T_s$ [symbols/s]. Under an M-ary MCS, each data symbol carries $\log_2 M$ bits of information, so that the message data bit rate is equal to $R_b = (\log_2 M)/T_s$ [bits/s]. In turn, the signature CDMA code consists of a sequence of symbols where each symbol is represented as a much narrower pulse. Each code symbol, identified as a **chip**, assumes a chip duration $T_c \ll T_s$ [s].

To illustrate, assume henceforth binary code symbols to be employed. The bandwidth occupied by a pulse of duration T is proportional to $1/T$. Hence, while the frequency bandwidth required to carry the message data symbol is of an order of $1/T_s$ [Hz], the frequency bandwidth required to carry a code symbol is of an order of $1/T_c$ [Hz]. The bandwidth required to accommodate the

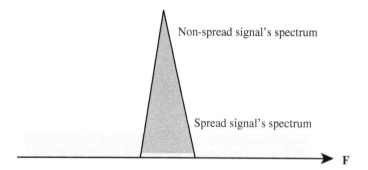

Figure 8.6 Illustrative Baseband (Non-spread) Spectrum and Spread Signal Spectrum in a CDMA System.

transmission of a CDMA-encoded signal is therefore higher than the bandwidth occupied by the original data, its *baseband*, by a high factor. The spectral band occupied by the CDMA-encoded symbol is thus observed to have been **spread**. The corresponding **spreading factor (J)** is equal to $J = R_c/R_s = T_s/T_c$.

The spectral spread incurred under a CDMA operation is graphically illustrated in Fig. 8.6. The spectral span of the baseband data message signal is narrow in comparison with the spectral span of the CDMA-encoded signal.

As an example, consider the following CDMA system parameters (as used in WCDMA 3G cellular wireless UMTS systems for the transport of voice messages). The chip rate is set equal to $R_c = 3.84$ [Mchips/s]. Hence, a chip duration is equal to $T_c = 1/R_c \approx 0.26$ μs. Consider a message data rate of $R_s = 15K$ [symbols/s]. The symbol duration is then equal to $T_s = 1/R_s \approx 66.67$ [μs]. The spreading factor is equal to $J = R_c/R_s = 256$. If a QPSK MCS is used, which is an M-ary MCS for which $M = 4$, so that $\log_2(M) = 2$ [bits/symbol], the data bit-rate is equal to $R_b = 2R_s = 30$ [kbits/s].

For the CDMA system to properly work, it is necessary to carefully select the CDMA codes that are assigned to the stations that share the medium. In particular, as will be noted from the following description, the codes are selected so that they mutually display no, or very low, **cross-correlation**. Such codes are said to be **orthogonal** (or nearly orthogonal). Then, a receiver which is interested in successfully decoding a message encoded by a particular CDMA code is not prevented from doing so in spite of its simultaneous reception of messages that are encoded by other CDMA codes. In addition, it is desirable to select codes that exhibit low **auto-correlation**. This aids in the prevention of signal interference induced by the reception of multiple time-shifted copies of the same chip signal, which is triggered by multipath reflections.

The following two types of chip sequences have been commonly employed: synchronous sequences and pseudo noise (PN) sequences, leading to **synchronous-CDMA (also identified as CDM)** and **asynchronous-CDMA** systems, respectively.

When the shared medium is used to transport multiple CDMA-encoded signals, the input signal received at a targeted destination station consists of the sum of CDMA-encoded signals that include the message that is targeted for reception by the station and possibly several other messages. The receiving node must be able to separate between a desired message signal and interfering message signals. For this purpose, the receiver employs a **correlator** circuit, which serves to extract the desired signal. It decorrelates the other signals and produces an output that consists solely of the baseband data of the desired message. The correlator's output signal spans the much narrower baseband bandwidth. Hence, it is said to **de-spread** the received signal. The circuit correlates the input with the CDMA code that is associated with the targeted message. In contrast, when processing messages (or multiple simultaneously received messages) that are not targeted for reception by

Figure 8.7 Illustrative Message and Chip Symbols in a CDMA System.

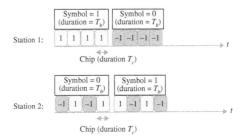

the receiving station, the station's correlator output indicates that the input does not convey to the underlying station any data message that is targeted for reception by this station.

In Fig. 8.7, we show illustrative messages that are transmitted by stations across a shared communications channel using a CDMA scheme. We assume that 4-dimensional Walsh codes are used as the signature CDMA codes, realizing a CDMA system with a spreading factor value that is equal to $J = 4$. These codes are assigned to four stations, one code per station. A 0 data bit is represented as a signal with amplitude -1 and is therefore identified as equal to -1. A 1 bit is represented as a signal with amplitude 1 and is identified as 1.

The corresponding four CDMA codes are represented as the following vectors: $c_1 = \{1, 1, 1, 1\}$, $c_2 = \{1, -1, 1, -1\}$, $c_3 = \{1, 1, -1, -1\}$, and $c_4 = \{1, -1, -1, 1\}$. Thus, code 1, c_1, consists of four chips, and each chip carries a data bit that is equal to 1, etc. We use $c_i(n)$ to denote the nth chip of the CDMA code of station i, or of an underlying message that is associated with station i. Then $c_1 = \{c_1(1), c_1(2), c_1(3), c_1(4)\} = \{1, 1, 1, 1\}$.

The baseband data bit that is produced by station i is denoted as m_i, where $i = 1, 2, 3, 4$. To transport m_i across the channel, station i uses it to modulate its signature code c_i, so that it actually transmits across the channel the sequence of chips x_i, where $x_i = \{x_i(n), n = 1, 2, 3, 4\}$, so that $x_i(n) = m_i c_i(n)$ denotes the nth chip that station i is transporting across the shared medium.

Assume that J stations transmit synchronously and simultaneously their message-based chip symbols across the shared channel. The stations are assumed to be time synchronized at the chip level. Consider a node that is the intended destination of messages sent by station 1 but that is also receiving signals corresponding to transmissions of encoded messages that are sent by other stations. Assume that these signals are received at this node at about the same power level. We assume that the reception at the intended node is dominated by signal interference rather than by background noise. The amplitude of the signal received in the nth chip period, denoted as $y(n)$, is equal to the sum of the amplitudes of all signals transmitted during this chip period, and is therefore expressed as:

$$y(n) = \sum_{i=1}^{J} x_i(n) = \sum_{i=1}^{J} m_i c_i(n).$$

Upon reception, the destination node correlates the received signal with the CDMA signature code of the intended message, which we assume here to be identified by, say, the CDMA code of Station 1, which is c_1. The **correlation** of two sequences (each one of which is mathematically represented as a vector) is performed by calculating the sum of the chip-by-chip products. The correlation of the received sequence $\vec{y} = \{y(n), n = 1, 2, 3, 4\}$ with the CDMA code $\vec{c}_1 = c_1 = \{c_1(1), c_1(2), c_1(3), c_1(4)\}$ is thus calculated as:

$$\vec{y} \cdot \vec{c}_1 = \sum_{n=1}^{4} y(n) c_1(n).$$

256 | *8 Multiple Access: Sharing from Afar*

If the outcome of this correlation is a positive number, it is interpreted as a 1 bit; if negative, it designates the receipt of a 0 bit; if it is equal to 0, no data bit is said to be received.

Example 8.2 For the CDMA system configuration described above, for which $J = 4$ and the 4-dimensional Walsh codes are used, assume that, as shown in Fig. 8.7, two stations are currently transmitting data bits across the shared communications channel. In the first data symbol period, Station 1 transmits a 1, $m_1 = 1$, while Station 2 transmits a 0, $m_2 = -1$. The corresponding transmitted chip sequences are then equal to $x_i = \{m_i c_i(n), n = 1, 2, 3, 4\}$, so that $x_1 = \{1, 1, 1, 1\}, x_2 = \{-1, 1, -1, 1\}$. The output of the channel is equal to the sum of these two transmissions, yielding

$$\vec{y} = x_1 + x_2 = \{0, 2, 0, 2\}.$$

When correlating this output with the signature CDMA code c_1 of Station 1, we obtain

$$\vec{y} \cdot \vec{c_1} = \sum_{n=1}^{4} y(n)c_1(n) = 0 + 2 + 0 + 2 = 4,$$

which is interpreted as a data bit that is equal to 1, $m_1 = 1$, as expected. Similarly, when correlating this output with the CDMA signature code c_2 of Station 2, we obtain

$$\vec{y} \cdot \vec{c_2} = \sum_{n=1}^{4} y(n)c_2(n) = 0 - 2 + 0 - 2 = -4,$$

which is interpreted as a 0 data bit, $m_2 = 0$, as expected.

For the above-described DS-CDMA system that employs synchronous CDMA codes to operate effectively, it is important that simultaneous chip transmissions are received in a synchronous manner, enabling the system's operation to decorrelate and eliminate interfering signals. Such a system operation has been realized in wireless systems, such as 3G cellular systems, when simultaneous downlink transmissions are executed by a BS node. Each such message transmission is destined to a targeted mobile station that resides in this cell. Each symbol transmission is encoded by the CDMA code that is associated with its targeted destination mobile station. As such messages, prior to their transmissions, reside in a common buffer managed by the BS's controller, this is a multiplexing system and is therefore identified as a Code Division Multiplexing (CDM) system.

In comparison, when considering uplink message transmissions from mobiles to a BS node, multipath reflections and other channel propagation effects make it more challenging to realize a DS-CDMA synchronous operation. An alternate CDMA-encoding system employs PN codes instead of synchronous codes. A **pseudo-noise (PN)** code is a sequence of J binary chips that to an outside observer seems like a random sequence of bit symbols. However, it is in effect (as far as its sending and receiving stations are considered) a deterministic (nonrandom) sequence that is readily generated at the transmitting and receiving nodes by applying a parameter key that is used to drive a PN production circuit. Although it seems to lack any definite pattern, the pseudo-random noise generating circuit produces a deterministic sequence of bits that repeats itself after a long (cyclic) period. The key parameter used to produce the PN code is exchanged (securely when desired) between the message sender and its intended receiver. Through the use of proper PN sequence generation algorithms, it is possible to produce PN sequences that exhibit desirable cross-correlation and auto-correlation properties, even when received in an asynchronous manner.

Employing PN codes as CDMA signature codes, an asynchronous-CDMA system is implemented, enabling many-to-one multiple access communications. Yet, since the employed PN codes are asynchronously received so that they are not precisely orthogonal, the reception of a targeted message is now subjected to signal interference that is caused by the cumulative signal power generated by interfering transmissions. The performance of the system is related to the corresponding SINR level detected at a receiving node.

Effective operation of a CDMA system requires transmit power regulation. As multiple signal transmissions take place at the same time, the extraction at a receiving node of a targeted message requires the different signals to be received at similar power levels. This is identified as the *near-far problem*, as transmitting nodes that are located closer to the receiving node may overwhelm the receiver with much stronger signals. Hence, an effective CDMA system performance operation involves the use of transmit power adjustment scheme.

A CDMA system that uses a spreading factor that is equal to J is effectively supporting the transmission of J message flows. Hence, its attained spectral efficiency (representing its realized throughput rate per unit bandwidth) is equal to about the same value as that achieved in a multiple access cell that employs corresponding FDMA or TDMA schemes. However, when considering a regional network that consists of multiple cells, CDMA network operations were noted to be beneficial in earlier cellular systems, such as second-generation (2G) and third- (3G) cellular systems. Benefits include the simplification of the coloring process used to regulate cross-cell signal interference that is achieved by reusing frequency bands across non-neighboring cells. Due to the graceful degradation in voice message reception quality incurred by random fluctuations in signal interference, higher voice throughput rates may be achieved when accounting for the statistical variation of signal interference processes that are induced by the random activity of voice calls.

Medium access control schemes used in key 2G cellular wireless systems include TDMA schemes, as used by GSM systems in Europe and other locations and (using the IS-136 standard) TDMA based systems in the United States, and (IS-95 standardized) CDMA systems in the United States and other countries. Third-generation (3G) medium sharing schemes include TDMA as well as CDMA, whereby the latter followed a set of standards that have been identified as CDMA 2000. Subsequent generation, 4G cellular wireless systems provide higher data rate and throughput operations in support of high speed Internet access for multimedia packet flows. Quality of Service (QoS) provisions are offered in transporting flows across wireless cells. Such systems are based on Long-Term Evolution (LTE) Standards developed by the 3rd-Generation Partnership Project (3GPP). Medium-sharing schemes are based on demand-assigned joint FDMA/TDMA resource allocations, and the use of Orthogonal FDM (OFDM) techniques at the physical layer. Advanced next-generation systems employ higher-frequency bands, including mmWave bands. They make extensive use of SDMA techniques, employing demand-assigned TDMA/FDMA/SDMA mechanisms, as they dynamically synthesize antenna beam configurations that produce multiple directional communications links, yielding high spatial reuse factors.

Spread spectrum schemes were employed over many years in a multitude of communication systems. They have been commonly used in military systems, for the purpose of protecting communications that are subjected to intentional jamming interference undertaken by an adversary. By spreading the frequency spectrum occupied by a message transmission, in using, for example, a *PN secret code*, one forces a jammer to spread its transmit power over a much wider spectrum span and consequently be less effective in interfering with the transmission of intended signals. The corresponding procedures that are employed to this aim are based on using PN codes across the time domain (represented as chip level symbols), identified as **Direct Sequence**

Spread Spectrum (DSSS) systems, or using PN sequences to operate the system in a **Frequency Hopping Spread Spectrum (FHSS)** mode.

Under FHSS, transmitting and receiving stations coordinate the mutual selection of a PN code. The available frequency band is divided into smaller sub-bands. Signals rapidly change their carrier frequencies, hopping across the center frequencies of a selection of these sub-bands, following an order described by a selected PN sequence. In this manner, data bit transmissions hop randomly among selective frequency sub-bands. Interference from other signals can impact reception within certain sub-bands during corresponding short intervals. Forward error correction (FEC) codes are employed to detect and correct reception errors that occur due to such interference events. Other applications have also been made. For example, such FHSS-based transmissions can share frequency bands that are used by other transmission systems, as they cause (and incur) limited interference to (from) narrowband communications systems. Civilian systems and consumer devices have also been using such a method for multiple access communications.

Cognitive radio systems have been designed to rapidly detect and temporarily capture and use spectral bands that are determined to be unused. Under such frequency-agile operations, certain bands can be affiliated with primary users. A cognitive radio device would vacate a band that it has occupied as a non-primary user, moving its communications operation to another band when the currently occupied band is needed for use by its primary user.

8.3 Demand-Assigned Multiple Access (DAMA) Schemes

8.3.1 Demand- Assigned Schemes

The sharing of a resource, such as a multiple access communications medium, by using a fixed multiple access scheme is efficient if the stations to which resources are assigned utilize them over a high fraction of the time. In contrast, if the stations sharing the multiple access medium are active in a **bursty** fashion, as data and other traffic sources often behave, a medium resource assigned to a station on a relatively fixed (or long term) basis will be highly underutilized. Bursty sources are active during randomly occurring times. Their activities often last for random periods of time. Such sources thus occupy medium resources on a relatively low duty-cycle basis.

Under such conditions, it is more effective to allocate shared multiple access medium resources to stations only when they are determined to actually require them, and then assign them only for a period duration that corresponds to their actual transport activity. We identify such procedures as **DAMA** schemes. Under a DAMA mechanism, resources are assigned to a station based on its need or demand.

In the following, we classify these schemes into two categories: reservation and polling. *Under a reservation scheme, it is the responsibility of the user to request resources, when needed. In turn, under a polling scheme, it is the responsibility of the network system to determine the resource requirements needed by its users.*

8.3.2 Demand-Assigned Reservation Schemes

Under a **reservation scheme**, a station that wishes to communicate across a multiple access medium sends first a reservation request to a network controller. In its request, the station expresses (explicitly or implicitly) the level of resources that it desires and may also include information about its activity profile and traffic statistics, as well as specify its quality-of-service (QoS) performance requirements. A station may request the system to provide it with a preferred data rate, specify its desired delay-throughput performance, and possibly also identify its flows to be governed by the characteristics of a specific traffic or application class. Resources may be requested to be allocated over a specific period of time, or for a period that continues until the station signals the system that the resources are not needed any longer.

While the following description assumes that the allocation of resources is performed by a system controller, other resource management mechanisms may be used. For example, under a distributed system architecture, station requests may be disseminated to all stations and then, using an agreed-upon resource allocation algorithm, each station could individually calculate the allocation to be made for accommodating (when feasible) requesting stations. Each station employs a copy of the same scheduling algorithm and holds a copy of the latest record of allocations.

Allocation of resources to requesting stations, when available, can be performed by using a TDMA scheme, so that a requesting station is allocated a number of specified time slots within identified time frames. Alternatively, FDMA-oriented allocation could be used, so that a requesting station is assigned a frequency band. Also, a station may be allocated (in an SDMA manner) spatial resources or CDMA codes. Such techniques are respectively identified as **Demand-Assigned TDMA (DA/TDMA)**, **Demand-Assigned FDMA (DA/FDMA)**, **Demand-Assigned SDMA (DA/SDMA)**, and **Demand-Assigned CDMA (DA/CDMA)**.

A hybrid scheme combines several of these methods. Under a **DA/FDMA/TDMA/SDMA** scheme, as employed by 4G LTE, other cellular wireless network systems, and advanced Wi-Fi networks, in sharing communications resources across a Radio Access Network (RAN), a station is assigned a frequency band for use within specified time slots and may also be allocated a spatial segment in utilizing specified antenna beams to cover targeted spatial segments.

When transporting a realtime flow across the shared communications medium, such as constant bit rate (CBR) voice or video packet flows, the station may request the allocation of a realtime circuit across the medium for the duration of a connection (under a circuit switching oriented operational mode), or for the duration of an activity burst (under a burst oriented packet switching mode). In support of variable bit rate (VBR) and other bursty data packet flows, a user may request and be allocated resources that vary in accordance with its QoS requirements and the stochastic fluctuations produced by its traffic flow activities.

Artificial intelligence (AI)-based algorithms can be used to assist a system controller in rapidly evaluating the application type and traffic profile of each flow, and in deducing a flow's QoS requirements and priority level. This evaluation is used to determine the best resource allocation to be made for the support of system flows.

In addition to accounting for data included in reservation messages, a DAMA system may continuously observe the statistical performance and traffic parameters of flow activities, characterizing the flow type produced by each active user. This data is used to guide its resource allocations.

The reservation mechanism that is implemented is a critical element in determining the performance efficiency of a reservation-based DAMA scheme. **Reservation and resource allocation communications channels**, also often identified as **signaling or control channels**, are assigned for this purpose. In reference to a system controller, such as a BS that manages a cell in a wireless cellular network, we identify **forward or downlink channels** as those that are configured for the transmission of data from the controller to the stations. In turn, **reverse or uplink channels** are configured for the transmission of data from stations to their controller. **Reverse or uplink signaling channels** are used by stations to transmit reservation, control, and other signaling messages to their controller. Such signaling channels are also identified as order-wire channels. **Forward or downlink signaling channels** are used by the controller to transmit resource allocation and other control messages to its managed stations.

We differentiate between in-band and out-of-band signaling channels. **Out-of-band signaling** channels are established in frequency/code and/or spatial bands that are distinct from those used for the transport of data messages, which are identified as **traffic or data channels**. In turn, **in-band signaling channels** are configured by sharing bandwidth resources with **traffic channels**, which carry data messages. For example, a frequency band can be employed to provide for the transport of signaling and data messages, whereby certain groups of time slots in each time frame are announced as allocated for the transmission of signaling and control packets while other time slots in each time frame are used as traffic channels, which accommodate the transmission of data messages.

Figure 8.8 depicts out-of-band uplink and downlink signaling and traffic channels.

Reservation requests arriving at the controller are buffered, arranged in a queue, and served by the controller's scheduler on either a blocking or queueing basis. Under a **blocking service**, the controller will allocate resources to a requesting station only if such resources are available at the time. Otherwise, the controller will block the request, providing no allocation. It could then send a blocking message (such as a busy tone) to the requesting station across a downlink signaling channel. The station may then cancel its reservation request or send it again at a later time.

Under a **queueing service**, if resources are not available at the time, the controller will store the request and place it in a queue of buffered requests. The latter are served in accordance with the **scheduling policy** employed by the controller. Such a policy may take into account attributes such as station and message priority identifiers, resource size requirements, service level agreements (SLAs) that a user may have negotiated a priori, fairness in the allocation of resources among users over identified periods of time, cost-revenue trade-offs, resource availability, as well as other considerations.

Communications channel resources are used for the transport of data as well as control messages. Resources used for the support of information-bearing data messages (whether representing data, voice, or video flows) are identified in the following as **service resources**. Resources used for the transport of reservation, control, and management messages are identified in the following as **control resources**. To operate the system in an efficient manner and at the same time provide user flows their requested QoS performance, it is necessary to efficiently apportion available bandwidth resources for allocation to these resource types. If control resources are not allocated a sufficient magnitude of communications capacity, users may not be able to gain access to the reservation subsystem, or not send their reservations in a timely manner, even when there are sufficient service resources. In contrast, if excessive communications resources are allocated

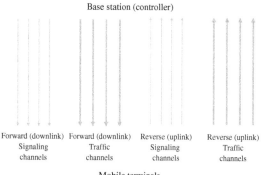

Figure 8.8 Uplink and Downlink Signaling and Traffic Channels in a DAMA System.

8.3 Demand-Assigned Multiple Access (DAMA) Schemes | 261

to the control segment, there may not be adequate service resources left for accommodating data flows at acceptable performance levels.

Thus, the system capacity levels allocated for the support of service and control communications should be adapted to underlying system and traffic conditions. This is illustrated by the following Example.

Example 8.3 Consider a shared radio access communications channel used in a wireless network. Consider the allocation of uplink communications resources. Assume that, on the average, every 10 seconds, old sessions are terminated and 1000 new sessions are established and carried by the medium. The produced initiation and termination actions load the uplink with the transmissions of call request and call termination signaling packets. Each session is assumed to require an average data rate of 20 kbps, yielding an overall required average service data rate of 20 Mbps. Also, assume that each session produces uplink transmissions of two 600 bits-long reservation/termination signaling/control packets, which require control communications capacity of the order of 1200 bps per each session, leading to a total control capacity requirement of 1.2 Mbps. For illustrative purposes, we assume that the data rate value of the control channel is set to a value that keeps the control message rate to a level that is no higher than 10% of this value. (See a forthcoming Section for a discussion of the throughput efficiency of a random access scheme, which is often used as the medium access control technique for regulating the access of signaling and control packets.) Hence, a total of about 5.7% of the uplink capacity (of 21.2 Mbps) must be allocated for the transport of signaling and control messages while the remainder of the capacity would be available for the service segment.

Assume that at a later period of time, the system is used to support shorter sessions. During every period of 2 seconds, old sessions are terminated and 1000 new sessions are established and carried across the medium. Each session produces uplink transmission of call request and call termination signaling packets. Each session is assumed to require an average data rate of 20 kbps, yielding an overall required average service data rate of 20 Mbps. Assuming parameter values for the control system that are similar to those noted above, we find that the required control communications capacity level is of the order of 6000 bps per session, leading to a total control capacity requirement of 6 Mbps. Consequently, a total of about 23% of the uplink capacity (of 26 Mbps) must now be allocated for the transport of signaling and control messages while the remainder of the capacity would be available for use by the service segment.

The additional control capacity required to support the system under the second scenario, which is induced by the support of rapidly changing flow conditions, may be obtained by using stand-by (reserved) resources, if any.

If available uplink capacity resources are exhausted, it will become necessary to reduce the loading of the communications uplink channel. A lower number of sessions may then be accommodated.

Signaling and control messages are typically generated by stations on a low duty-cycle basis at relatively lower message rates. A station would generally send resource allocation and termination requests to the controller at the start and end times of its session period. Unless the number of stations that can potentially share medium resources is quite low, it is not effective to dedicate channel resources to stations for use in transmitting their signaling messages. Hence, it would be inefficient for the system to employ fixed multiple access schemes (such as TDMA, FDMA, CDMA, and SDMA) for the transmission of uplink signaling messages. Also, using a reservation scheme would not be effective, as it would require the establishment of another reservation channel to be used for sending packets for the purpose of reserving resources for the transmission of primary reservation messages. Hence, **random access** multiple access techniques, as those described in a later section, are often used for this purpose.

However, random-access-based multiple access algorithms tend to inefficiently utilize signaling channel capacity resources. They effectively make use (when a short transport latency is required) of just 5–15% of the channel's communications capacity, while wasting the remainder capacity.

It is generally of interest to structure the reservation mechanism in a manner that promotes efficient utilization of communications capacity resources. To this end, we note the following techniques.

1) The multiple access system controller identifies the applications that the station is employing to produce its traffic flows across the shared medium. It accordingly determines the capacity level of signaling and control resources that should be allocated to the station for accommodating the transmission profile of ongoing reservation messages, in relation to available system's capacity resources and resource demands imposed by active stations.
2) The controller observes the activity process of a station, as it performs **station activity sensing**. Using such data, the controller synthesizes statistical models of activity, for each user, or per user-class, for each application type. This is coupled with performance metric values that are used by the controller for each application and traffic class. These measures are used by the controller to calculate the signaling/control and data/service channel resources to be allocated for the support of active stations.
3) These dynamically performed characterizations are also used for *predictive pre-allocation* of data resources to an active station. For example, under bursty traffic flow conditions, the controller could allocate resources to the station upon its detection of the start of burst activity and terminate this allocation when the corresponding burst activity is detected to end.
4) To conserve multiple access bandwidth resources, a station that has already been allocated channel data resources could use them for the joint transmission of data, signaling, and control messages. A station could share its allocated medium resources among its data, signaling, and control messages. A station could use a data slot to insert a signaling message that carries a request for the (same or modified) continued allocation of data resources, or to inform the controller that its activity (at the call, session, or burst level) has ended so that allocated resources can now be reused by the controller for the support of other clients.

Reservation schemes can provide highly effective performance behavior as multiple access communications schemes and are therefore frequently implemented. The allocation of resources to mobiles to be used for sharing a radio access network of a cellular wireless network is governed by the operation of reservation protocols. As noted above, to operate such systems in an efficient manner that also meets performance objectives, system communications capacity resources must be properly split between the service and control segments, preventing system operations from being limited by bottleneck overloading of one of these segment types.

8.3.3 Polling Schemes

8.3.3.1 Polling Methods and Procedures

Under a polling scheme, it is the responsibility of the system to determine user requirements and in response allocate system resources, when and if feasible. The system uses a control process to determine, over time, which station or stations are active and require the allocation of shared resources.

To describe the methods used for polling operations, we consider a system of user stations that share a resource such as a bandwidth of a multiple access medium. This resource is used by the stations for the purpose of communicating with each other, or with a controller or an AP. We assume that a station that is granted access to the medium occupies a prescribed shared bandwidth during a specific (preset or dynamically assigned) period of time.

The basic architecture for such a scheme is formed by implementing a *centralized polling* scheme, known as **hub polling**. A station, undertaking the role of the system's hub, is set to act as a *polling controller*, or manager. The controller sends a control message, identified as a poll, or a polling message, in a specific order, to network stations, checking with each station, in turn, whether it requires the allocation of shared resources. If the station is inactive or does not require at the moment any resources, it will send to the controller a negative reply message. In this case, the controller would proceed to send a poll message to the next station in its **polling list**.

In turn, if the polled station is interested at the time in using the shared resource, it will send the controller a positive reply message, indicating the level of resources that it desires to receive. It can also provide the controller with its traffic descriptor and performance (QoS) requirements, which will be used by the controller to calculate the resource levels that it would allocate to the station.

The controller would respond by sending the station an assignment message, which specifies its allocation, including the length of time (or frequency bandwidth) that the station is allowed to use. Alternately, to reduce the delay incurred in the transmission of messages, upon receipt of a poll, a ready station (i.e., a station that has messages ready for transmission across the shared medium) is given permission to access the shared medium (or an assigned resource such as a prescribed spectral band) for a specified or preset period of time. It would then proceed to accordingly send its messages across the medium.

Ring networks often employ a polling scheme in sharing the communications link resources of the network. In such a network, stations are actively inserted into the medium. Stations are directly connected to their neighboring stations by point-to-point links. Bidirectional links may be used, so that message transmissions can be carried out in both directions. Alternately, unidirectional links are used, as is the case when fiber-optic links are employed. A cyclic topological layout is formed. A ring configuration is illustrated in Fig. 8.9(a).

In a ring network that employs a hub-polling scheme, one of the stations is configured (or elected) to act as the polling controller. It issues polling messages to stations, permitting a polled station that has messages to send to go ahead and use the medium. When a station completes its transmissions, the controller proceeds to poll the next station. In the ring network shown in Fig. 8.9(a), station S1 serves as the polling controller. It polls the other stations in a cyclic order.

In Fig. 8.9(b), we show a *multi-dropped network layout* where stations S_2, S_3, S_4, and S_5 are attached to a common communications bus, sharing it as a medium for communicating with a central cluster controller station S_1. The latter serves as the polling controller. It polls cyclically its

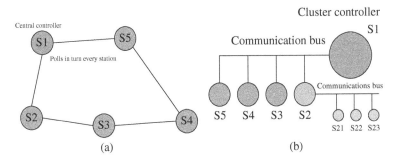

Figure 8.9 (a) Hub Polling in a Ring Network; (b) Hub Polling in a Multi-dropped Tree Network.

264 | *8 Multiple Access: Sharing from Afar*

client stations. When a station receives a poll message from the controller, it either proceeds to send its own messages across the bus or responds that it is currently idle, requiring no communications means.

As demonstrated in Fig. 8.9(b), a polling network can be physically (or logically) synthesized by using a *hierarchical polling architecture*. Station S_2 acts as a secondary polling controller, managing the access of stations S_{21}, S_{22}, and S_{23} into the shared bus. When polled by station S_1, station S_2 will proceed to poll the stations whose access it manages, allowing them (in turn) to send their messages, which it will relay across the primary bus to the central station. It is noted that station S_1 regulates its own access to the bus, as it multiplexes the messages that it directs downstream to other stations during the time period that it sets aside in the polling cycle for its own use.

A secondary polling station, such as station S_2, may instead be configured to act as a store-and-forward relay. In this case, station S_2 is set to regulate the access of messages from its client stations via a polling process that it operates simultaneously in time with the polling process managed by S_1. Messages sent to station S_2 from its clients are stored in its buffers and are forwarded to S_1 when the latter polls station S_2.

The delay-throughput efficiency of a polling operation is impacted by the *efficiency of the operation of the control mechanism* that is used to disseminate polling messages. The effective dissemination time of a polling message, also identified as the poll's **walk time**, is a key performance parameter.

The following approach is used to calculate the throughput efficiency of a polling system. Averaging over the activities of the stations that share the multiple access medium, we define the following parameters. Consider two stations, say Stations A and B that are consecutively polled. Assume Station A to be idle, so that at this time, it has no messages to transmit across the shared medium. Following the polling of Station A, identifying it to be idle, the polling of Station B is initiated. The time latency elapsed between the polling of an idle Station A and the presentation of a polling message to Station B is identified as the *per-station walk time*. It represents the time taken by a polling message in "walking" between (i.e., being transmitted to) consecutively polled stations. Its average duration is denoted as $E(W_s)$.

Each time that a given station receives a poll, it spends time, denoted as a poll visit service time, in reacting to it. If at this time the station is idle, the station takes time to respond to the poll by informing the system that it is idle so that the next station can be polled. If busy at this time, the station proceeds to transmit its allowed quota of messages and then it informs the system that its transmission process has been completed so that the next station can be polled.

Let $E(S_s)$ denote the average time that a station spends in transmitting its allowed quota of messages across the shared medium, per each time that it receives a poll, in averaging over all the system's stations. The service time is set to 0 if a polled station has no messages to transmit.

The throughput efficiency of the polling system, denoted as s, measures the average fraction of system capacity that is used for the transport of data messages. It also represents the average fraction of time that the channel is used for the transport of data messages. It is calculated as follows:

$$s = TH = \frac{E(S_s)}{E(S_s) + E(W_s)}. \tag{8.1}$$

Example 8.4   Reservation vs. Polling in the Classroom. As I am presenting a lecture to my class, I am allowing students to use a reservation process to ask questions. A student that has a question to ask would raise her hand. Noting this signal, which represents the student's reservation message, I will subsequently signal to the student that she can proceed to ask her question.

8.3 Demand-Assigned Multiple Access (DAMA) Schemes | 265

I have noted that students may be at times shy to ask questions. I would then poll students individually, asking one-by-one each student whether they have any question to ask or comments to make.

To demonstrate the relative efficiency of a polling scheme, consider the following case. Assume that there are 30 students in the class and that it takes 10 seconds to poll a student and receive a response as to whether the student is interested to ask a question or not. If the response is negative, I proceed to poll the next student. If the student is interested in asking a question, she will indicate it to me and will then (after receiving my go-ahead signal) proceed to have a conversation with me. To illustrate, assume that only 10% of the students (i.e., three students) are interested to ask a question during each cycle time. A cycle time is defined as the time period that it takes to poll and engage, as proper, all the students in the class, polling once each student. Assume identical polling and activity related parameters per student (i.e., per station). We note that during a single cycle time, the aggregate per cycle walk time is equal to $30 \times 10\,s = 300\,s$. Assume that during a cycle time, each one of the three active students spends an average of 1 minute with me in conducting our discussion. The average cycle duration is thus equal to $E(C) = 300 + 3 \times 60\,s = 480\,s$. The fraction of time that is used in conducting discussions (i.e., in executing data message transmissions) is then equal to $180/480 = 37.5\%$.

In comparison, when I let students engage with me by using a reservation scheme, I note the following. Assume that each reservation process takes an average period of 10 seconds (used by me to observe that the student has raised her hand and then to grant her the floor). Assume that I am engaged for an average of 1 minute in a discussion with each student that has reserved the floor. The relative efficiency of using my classroom time through this reservation process increases to $60/70 \approx 85.7\%$.

In turn, if 90% of the students in the class are interested in having a brief discussion with me (each lasting an average of 1 minute), and the associated per student polling walk time is only 2 seconds long, the efficiency of the ensuing polling scheme is enhanced, as it is now equal to $\frac{27 \times 60}{30 \times 2 + 27 \times 60} \approx 73\%$.

Thus, a polling scheme can be quite (bandwidth) efficient when the polling control overhead and the polling process induced control latency are relatively low, as can be the case when there is a high probability that a polled station will respond positively to its received poll.

As noted above, polling schemes are also useful when it is essential to reduce the operational complexity of user devices, as is the case when considering small devices with limited energy resources. Such illustrative operations include embedded sensor systems, such as Internet of Things (IoT) systems.

As noted from the latter calculation, a low system throughput rate is incurred when the polling time is not much lower than the average station transmission time triggered by the receipt of a poll. This would typically be the case when the shared medium provides for a high data transmission rate and the polling message dissemination speed is relatively low. For high-speed communications systems, this would be the case when the polling messages needs to traverse a long distance, noting that then the propagation delay component can be quite long, when compared with the time it takes to transmit a data message. Hence, polling schemes are more effective when employed to regulate the sharing of multiple access resources in a LAN system. The range span is then of the order of several meters to kilometers.

The system's throughput rate can be improved if the walk time is reduced, leading to shorter poll times. To this end, a *distributed-control polling mechanism* is employed. Under this scheme, the controller (following its own transmissions across the shared medium, if any) initiates a polling cycle by issuing a polling message that it sends to a specified station. Upon receipt of a polling message, a busy station is permitted to transmit its ready messages (if any) across the shared medium for a specified period of time or until it has sent a designated number of messages.

For a station to capture the medium upon the receipt of a poll, this station removes the polling message from the medium and proceed to send its messages. Following the transmission of its messages, this station creates a new polling message that it then sends to the next station to be polled, rather than sending a control message to the controller to signal the completion of its transmission period. In this manner, the walk time parameter is reduced. Rather than distributing polling messages back and forth between the control station and its client stations, polling messages are now sent directly between a station and its designated next station in the polling order. Often, the next station to pass the poll to is a close-by neighbor of the station that passes the poll.

Such a distributed polling scheme is known as a **token-passing** procedure. The polling message is regarded as a *token*. It is identified as such by displaying a token symbol in its header.

When receiving a token, a station that has messages to send removes the token from the medium and then proceeds to send its messages. When completing its transmissions, it regenerates a new token. The latter is then sent to the next station in the polling list. If upon receiving a token, a station is idle (having no messages to send), it immediately (following a very brief delay that allows the station to identify the received message to be a token message) passes the token to the next station on the polling list.

The throughput efficiency realized by a token-passing scheme is given by Eq. 8.1 whereby now the per-station walk time parameter $E(W_s)$ measures the average time it takes to pass the token to the next station in the polling list (rather than the time incurred in sending a polling message back and forth to the controller station).

Such an operation is often used to regulate access across a ring network. The latter system is identified as a **token-ring** network. Typically a single token is employed, and a **source removal** mechanism is used. A message transmitted by a station circulates around the ring, passing by the interface of each **Ring Interface Unit (RIU)**, and is removed from the ring's medium by the source station that is placed it on the medium. Such networks have been widely implemented as *Token Ring LANs*. Examples include the IEEE 802.5 token ring and the FDDI token ring standardized LANs. The IEEE 802.5 protocol standard for token-ring LAN systems was set to operate at transmission data rates of the order of 4 Mbps or 16 Mbps, and has also been upgraded to 100 Mbps. The ANSI standardized Fiber Data Distributed Interface (FDDI) token-ring network has served as a corporate backbone network. It operates at a nominal data rate of 100 Mbps. It can span a one way distance that is as long as 100 km, and it uses two counter rotating fiber rings (Fig. 8.10(b)).

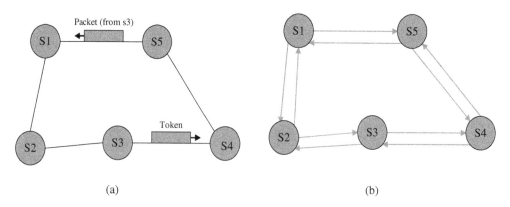

Figure 8.10 (a) Token-Passing Ring with Early Token Release; (b) Dual Counter Rotations Ring Network Layout (as for FDDI).

Under a *late token release* operation, the station that holds the token releases it after it has received its own frame. A positive acknowledgment (PACK) indicator is included (by the intended recipient) in a message trailer, following the circulation of the frame around the ring. In turn, under an *early token release* operation, as illustrated in Fig. 8.10(a), a station releases its token immediately after the transmission of its message. When synthesizing ultrahigh-speed networks, higher efficiency is attained when implementing a protocol that allows several data messages to simultaneously circulate around the ring medium.

Under the IEEE 802.5 token-ring protocol, provision is made for stations that have high-priority messages to signal their urgent need for the receipt of a token. Under this protocol, a station that captures a token for the purpose of transmitting its messages tags it as a busy (rather than idle) token and releases it for reception by other stations. Stations that receive a busy token can insert into it a signal that identifies their urgent need for the receipt of a high-priority token. If such a need has been advertised, then when a new token is generated, it is created at a priority level that permits only stations with high-priority messages to capture it.

As noted above, the polling operation is bandwidth efficient when the time it takes to poll a station is relatively low in comparison with the average time taken by a station to send its data messages when polled. Hence, when operating a network at a high data rate, the impact imposed by the poll's walk time becomes highly critical, as the time it takes to transmit messages is very short. The poll's propagation delay component can then become a critical factor. Its value is independent of the employed data rate. It is proportional to the distance traversed by the polling message. It is calculated as the product of the propagation speed (typically of the order of 75% of the speed of light, which is equal to about 5 [μs/km]) and the distance along which the signal propagates.

When transmitting a message that consists of 1000 bits at a data rate of 10 Mbps, 100 Mbps, or 1 Gbps, the corresponding message transmission times are equal to 100 [μs], 10 [μs], and 1 [μs], respectively. The propagation delay along a distance of 1 km is equal to 5 [μs], which can thus become a distinctive fraction of the time that it takes to transmit a message. Hence, token-ring and hub-polling systems are typically implemented as LANs, which span limited ranges.

Token-ring networks require the use of a somewhat involved RIU, as it needs to examine and buffer messages that circulate the ring and insert into the medium its own ready-to-send frames at the proper time. Furthermore, as the RIU is actively inserted into the medium, its failure can result in the disconnection of a network link. Consequently, mechanisms are used at the RIU to short-circuit it upon the detection of local failure. Such component complexity and failure sensitivity contributed to the preferential use of more robust layouts, including Ethernet hub-based networks.

To ease installation, control, and management operations, LAN systems have been configured to use a central control and management node. A token ring system then assumes a **physical star and logical ring** configuration. Every RIU is connected by a link to a central hub, often placed in a wiring closet.

Hub-polling mechanisms are advantageous in that the complexity of the access control scheduling is nested at the hub, which serves as the controller station. The latter tends to possess higher energy, storage, processing, computing, and stand-by resources that are activated upon the detection of failure events. The controller may dynamically collect and analyze status data concerning the activity and performance requirements of its client stations.

Such an operation is highly desirable when regulating the access of devices that run critical applications that must gain communications access to essential data bases and other entities in a timely fashion. For example, medical instruments that collect critical data of patient vital statistics and sensors embedded in autonomous vehicles and IoT devices must often be able to periodically communicate across the medium in a rapid and timely manner. Access permissions can then be provided by having the controller issue polls at proper time instants. Such mechanisms have been planned for use in advanced Wi-Fi wireless LAN (WLAN) systems, such as Wi-Fi-6 LANs, autonomous highway vehicles that are regulated by Road Side Units (RSUs) and next-generation cellular wireless network systems.

To assure stations with **fair** allocation of resources, it is of interest to limit the level of system resources that are allocated to a station. For example, consider N stations that share a prescribed frequency band on a *time division* basis. When a busy station with ready messages is polled, the time period that it is allowed to use the shared medium for the purpose of sending its own messages, identified as the station's per poll **dwell time**, can be programmed by the controller, or the system manager, to not be longer than a prescribed **station dwell time limit**. For example, under the IEEE 802.5 token-ring specification, a RIU is programmed by the manager to subscribe to a Token Holding Time (THT) limit.

The ANSI FDDI token-ring network uses fiber-optic links, nominally operating at a data rate of 100 Mbps. A fiber link connects a RIU with a neighboring one in a unidirectional manner. As shown in Fig. 8.10(b), the FDDI network uses two counter-directional fiber links between each neighboring pair of RIUs, forming thus two counter-directional ring sub-networks. One of the rings is generally used as the primary networking medium, while the other one is employed under failure conditions to re-establish connectivity.

The FDDI network was targeted for use as a corporate backbone network. It spans a diametrical range of about 100 km and is thus often categorized as a metropolitan area network (MAN). To regulate access fairness, the system's manager prescribes a value to a parameter identified as a Target Token Rotation Time (TTRT). When receiving a token, a station calculates its permitted dwell time by basing it on the difference between the value of the TTRT and the most recent value that it measured for its Token Rotation Time (TRT), which expresses the time elapsed since it has last received the token after releasing it. In this manner, a station's dwell time (the time that a station can take for transmitting its messages upon the receipt of a token) is increased when the network is not too congested (as a shorter TRT is then recorded). Such a dynamically adaptive dwell time algorithm has been shown to assure stations with an average message access latency of the order of TTRT and a maximum access latency that is of the order of 2TTRT.

For simplified implementation and analysis, it is often useful to regulate a station's dwell time through specifications that relate to the number of messages that a polled station is allowed to send. The following schemes have been used to approximate the operation of various dwell time regulation mechanisms:

1) **Exhaustive Polling**: A station is permitted to send all the ready to transmit messages that it possesses at the time that it is polled, plus all the ready messages that it generates or that arrive to the station while it still maintains access control (such as still holding the token under a token passing protocol).
2) **Gated Polling**: A station is permitted to send only the ready messages that it stores at the time that it is polled.
3) **Limited-to-k (Chained) Polling**: A station is permitted to send no more than a specified number, say **k**, of messages at the time that it is polled. As a special case, under a limited-to-1 scheme, a polled station is allowed to transmit no more than a single message, when polled.

A polling controller may use a network management scheme that employs a **Service Order Table (SOT)**. The latter specifies the frequency at which different stations are polled. Associated dwell time limits per-station may also be specified. Different stations can be polled at different rates.

Stations that support higher traffic loads are polled a higher number of times during each polling cycle or during another specified time window.

The polling order of a station and its allocated transmission rate can be adjusted in accordance with the *station priority* level. Priority levels can also be associated with different polling cycles. A high-priority polling cycle can be used to poll only high-priority stations and/or trigger service of only messages whose precedence level is equal to or higher than that identifying the underlying polling cycle. The following are illustrative multiple access schemes that employ **implicit polling** procedures:

1) A **Shuttle Networking** operation can be described as follows. Consider a shuttle bus that has capacity to carry N passengers. It follows a specific route and stops to pick-up and release passengers at identified stations along the route. Passengers wait at stations along the route. They access the shuttle bus (analogous to a token) when it arrives at their station. The number of passengers that are admitted to the bus depends on its current residual (i.e., leftover) capacity.

2) An **elevator system** operates in an equivalent fashion, whereby waiting persons access the elevator at the time that it visits their floor, when the residual capacity of the elevator allows.

3) In a **slotted bus** network system, a controller station generates idle time slots and disseminates them (like tokens) along the bus's medium. Each slot is identified in its header field as being either idle or busy. Stations are attached to the bus medium in a passive manner, so that they are able to write (i.e., transmit) a message onto the bus as well as read and copy messages that traverse the bus. In turn, a station cannot remove messages from the bus. A ready station (i.e., one that has a message to send) examines incoming slots and captures an idle slot. It then tags it as a busy slot, and inserts its message segment into the captured slot.

4) Under a **positional priority** polling algorithm, stations are granted priority for access to the medium in accordance with their explicit or implicit position in relation to a reference location, such as the position of a system controller. To illustrate, consider such a hub-polling scheme. A controller station polls station 1 and if busy (i.e., having ready messages to send across the medium), it permits it to send its messages. It then polls station 2 and if busy, it authorizes its transmissions. Next, it polls station 1 again, etc. Thus, following access by a station, the controller resumes its polling process by re-starting at the head of the polling list. Stations are ordered in the polling list by their relative position. When implemented across a token ring, the order of stations in the list is determined by their distance from the controller.

5) A mechanism that is similar to the slotted bus system, and also assumes positional priority oriented features, is employed by the MAN system that was specified by the IEEE 802.16 standard. This network uses a dual bus-shared medium. Slot generators are placed at the edge of each bus, producing and disseminating slots along each bus, in counter directions. A station is attached to both buses. To send a packet, a station needs to capture an idle slot. For this purpose, it selects a bus that can be used to reach its destination. The underlying MAC layer protocol has been identified as Dual Queue Dual Bus (DQDB). A station that is located closer to a slot generator has a better chance of capturing an idle slot. Algorithms have been introduced to induce a more fair distribution of resources among stations. For example, a station would limit the rate at which it captures idle slots in accordance with its position on the bus.

8.3.3.2 Performance Behavior of Polling Systems

How often is a station being polled when considering a polling system that is loaded at a prescribed traffic intensity level? What is the average cycle time, identifying the average time taken for a polling message between successive visits to a specific station? In the following, we provide answers to these questions by analyzing a common polling system layout.

Consider a polling system that employs a repetitive polling cycle during which each one of the N system stations is polled at least once in each cycle. The polling order is repeated in each cycle. Assume station i to be loaded by messages that arrive (or are generated) at a rate that is equal to λ_i [messages/s]. The average transmission (or service) time of each message served by station i is equal to β_i [s/message]. The Erlang loading of station i is thus equal to $\rho_i = \lambda_i \times \beta_i$ [Erlangs]. The overall traffic loading of the medium is equal to $\rho = \sum_{i=1}^{N} \rho_i$.

We assume the system to allow no more than a single message transmission to be initiated at a time. Hence, if a shared data rate that is equal to R [bits/s] is available, the throughput bit rate cannot exceed a value of R [bits/s]. Therefore, the system's throughput rate cannot exceed a value of 1 [s/s] or 1 [Erlang]. Consequently, to not overload the system, we limit the overall loading rate as follows:

$$\rho = \sum_{i=1}^{N} \rho_i = \sum_{i=1}^{N} \lambda_i \beta_i < 1. \tag{8.2}$$

When considering dwell time disciplines that do not explicitly limit the service times performed by stations, such as when using Exhaustive, Gated, and other (including those that employ prescribed polling lists, rates, and schedules) the above condition would guarantee the stability of the system's operation. Under this condition, the system's stochastic service process reaches equilibrium and messages experience finite average waiting times while residing in their station buffers.

Additional loading rate limits may have to be imposed when service restrictions are imposed. For example, under a limited-to-1 dwell time discipline, a polled station is not permitted to send more than a single message. Consequently, for a stable operation, the message arrival rate λ_s [mess/s] at a limited-to-1 station must be regulated so that the average number of messages arriving at this station in an average cycle duration $E(C)$ is lower than 1:

$$\lambda_s E(C) < 1. \tag{8.3}$$

In the following, we calculate the average cycle time for a polling system that permits the issue of only a single polling message at a time. We assume that traffic loading regulations are imposed so that the system is not overloaded, and equilibrium conditions are reached. We set $E(W_p)$ to represent the mean walk time of the polling message over a cycle. It measures the average aggregate time that is incurred in passing polling messages during a cycle, in accordance with the order at which stations are set to be polled (including the possibility of polling certain stations multiple times within each cycle). We assume, as noted above, station i to support at equilibrium a message arrival rate that is equal to λ_i [mess/s], providing each message an average service time (message transmission time) that is equal to β_i [s/message]. At equilibrium, at each station, the message transmission (or service) rate must be equal to the message arrival rate. Hence, the average number of messages transmitted by station i during an average cycle time is equal to $\lambda_i E(C)$ [messages]. The average dwell time by station i during an average cycle time is thus equal to $\lambda_i E(C)\beta_i$ [s]. The mean cycle time is calculated by adding the average dwell times of all stations and the mean walk time $E(W_p)$, yielding at equilibrium:

$$E(C) = E(W_p) + \sum_{i=1}^{N} \lambda_i E(C)\beta_i.$$

Solving this equation for the average cycle time, we obtain the following formula, when the polling system is not overloaded so that it reaches equilibrium:

$$E(C) = \frac{E(W_p)}{1 - \rho}, \qquad \rho = \sum_{i=1}^{N} \lambda_i \beta_i < 1. \tag{8.4}$$

To calculate the mean cycle time for a limited-to-1 polling system, it is not sufficient to impose the condition $\rho < 1$. It is necessary then to require the stricter condition stated in Eq. 8.3 to hold. To illustrate, consider a polling system in which all stations are symmetrically loaded. We assume the message arrival rate at each station to be equal to λ_s [messages/s], and the average transmission time of a message at each station to be equal to β [s/message]. Using the formula for the mean cycle time expressed by Eq. 8.4, we conclude that the limited-to-1 polling system will reach equilibrium if its loading level satisfies the following condition:

$$\lambda_s < \frac{1 - \rho}{E(W_p)}. \tag{8.5}$$

Substituting in the latter formula the traffic intensity expression $\rho = N\lambda_s\beta$, we conclude that the following condition must be satisfied by the message arrival rate at each station:

$$\lambda_s < \frac{1}{E(W_p) + N\beta}. \tag{8.6}$$

Equivalently, we write:

$$\rho < \frac{N\beta}{E(W_p) + N\beta}. \tag{8.7}$$

Under a **limited-to-k** polling scheme, the corresponding limit on the system's traffic intensity rate is concluded by replacing β with $k\beta$:

$$\rho < \frac{1}{1 + \frac{E(W_p)}{kN\beta}}. \tag{8.8}$$

As the *polling message walk time overhead ratio*, expressed as $\frac{E(W_p)}{kN\beta}$, assumes a lower value, a higher traffic intensity loading level can be accommodated. This is illustrated by the following example.

Example 8.5   Illustrative Limited-to-One Polling System. Consider a limited-to-one polling system that serves (at an equal rate) $N = 10$ stations, for which the polling walk time per station is equal to 1 ms, so that the system walk time is equal to $E(W_p) = 10$ ms. Assume the mean message transmission time to be equal to $\beta = 5$ ms. We then observe the polling overhead ratio to be equal to $\frac{E(W_p)}{N\beta} = 0.2 = 20\%$. Using Eq. 8.6, we determine that for this limited-to-1 polling system to not be overloaded, and thus reach steady-state behavior, we must require the traffic loading level ρ to be lower than $1/1.2 \approx 0.83$ [Erlangs]. Hence, the system must not utilize its data transport bandwidth for more than about 83% of the time. The residual capacity is utilized by the polling system's control mechanism. In comparison, under a limited-to-2 policy, the maximum attainable service utilization level increases to $1/1.1 \approx 90.9\%$.

Assume that we set the limited-to-1 system's overall loading level to be equal to 0.75 [Erlangs], noting that $\rho = 0.75 < 0.83$. The system would then use its shared data medium for an average of 75% of the time. The mean polling cycle time is calculated by using Eq. 8.4. We obtain it to be equal to $E(C) = \frac{E(W_p)}{1-\rho} = 40$ ms. Hence, a station served by this system receives a polling message at an average rate of once every 40 ms.

In comparison, during a period of lighter traffic activity, when the traffic loading is reduced to $\rho = 0.25$ [Erlangs], so that the data service medium is utilized for an average of only 25% of the time, the average cycle time is now reduced to $E(C) \approx 13\,\text{ms}$. It is noted that the average cycle duration is not affected by the dwell threshold parameter k, as long as the system is regulated to not operate in an overloading regime.

As observed above, the throughput rate produced by a polling network can be increased by employing **spatial reuse** methods. This was illustrated when discussing a network that employs a hierarchy of store-and-forward-based polling controllers. As another illustration, consider a ring network that consists of dual counter rotating fiber-optic links. Each station's interface (identified as a network node) is connected by two directional fibers to each one of ts neighboring interfaces. A slot generator produces time slots that it circulates in both directions around the dual ring network. A station that has a message segment that is ready for transport, waits to insert it within an idle slot that it receives. An addressing field identifies a segment's destination node. The segment is transmitted across the ring's directional route that offers the shortest path to the destination node. Upon receipt at a non-destination node, a busy slot is repeated and its segment is transmitted across the outgoing link that meets the direction of its selected path. When received at a destination node, the segment is removed from its slot.

Thus, in contrast with the source removal mechanism used by the previously described token ring system, whereby a message that circulates the ring is removed by its source node, a destination removal approach is employed in the underlying spatial reuse network. The vacated slot is then either directly used to carry a local segment or is repeated as an idle slot, available for capture by a busy downstream node.

For the latter ring system, it is noted that if source–destination nodal flow pairs are uniformly distributed across the ring system, the average path length is equal to one-fourth the length of the ring. Consequently, an average throughput utilization of 400% is attainable, in comparison with a maximum 100% throughput utilization level attainable by a system that uses a single polling token (permitting the initiation of a single segment transmission at a time). Such a spatial-reuse operation requires a more complex nodal implementation, while, to simplify, is made to capitalize on the regularity of the topological layout. Such methods have been used to implement ultrahigh-speed ring and meshed-ring electro-optical and all-optical communications networks.

8.4   Random Access: Try and Try Again

8.4.1   Uncoordinated Transmissions Using Random Access

Consider a scenario under which a medium is shared among a relatively large number of stations. Each station is usually infrequently active. When active, a station may have a limited number of data messages that it wishes to send. At that time, the station wishes to gain rapid access to the medium.

Illustrative cases abound. In a wireless cellular network system, a mobile user needs to send a call setup message to a BS controller to request the allocation of wireless resources. The allocation is employed by the user for the transmission of data messages or streams. The latter may occupy a high-speed communications resource for a noticeable period of time. In contrast, the transmission of a call request message requires the use of a communications channel, such as a signaling channel, for only a brief period of time. Call request messages are generated by users at the times that they become active and wish to communicate their data messages. The total number of system users that may potentially produce active signaling (such as call request) messages is often very large. However, only a small number of users would typically become active at about the same time, or within a brief period of time. The number of such active users tends to randomly vary. A wireless cell can be populated by thousands of users, but only few users would usually become active during a short period of time (such as a duration that is of the order of the length of a user activity burst).

Such a scenario is applicable to the operation of a typical *demand-assigned reservation* based network system. User reservation (such as call request) messages must be transmitted in a timely manner across a shared signaling channel. It is also applicable to the operation of a data network whose (wireless or wireline) communications medium is shared by a relatively large number of bursty user stations. Such a user station is active at a low duty-cycle. It becomes active at random times after staying inactive for relatively long period of time.

Illustrative wireless networks, identified also as wireless nets, which may at times contain such low duty-cycle user stations are shown in Fig. 8.11. In Fig. 8.11(a), we show a wireless net whose nodes employ a centralized connectivity pattern. User stations are managed by a net's central station, which is identified as its AP or BS. User stations associate with their AP and are regarded as its clients. Message transmissions across the shared wireless medium are directed either from a client station to the AP or from the AP to a client station.

Assuming the AP node and each user station to employ an omnidirectional antenna, and also assuming stations to be within effective communications range from each other, **a message transmission by one station will be received by all other stations**. The communications system is then said to be characterized as a **broadcast net**.

In Fig. 8.11(b), we show a wireless net which is used by stations to communicate directly with each other. Connectivity is on a *peer-to-peer* basis, directly from a station to another station (or stations) through the shared wireless medium. The stations are regarded as being members of a common peer layer, in terms of their connectivity association. The latter peer-to-peer connectivity pattern contrasts the star connectivity configuration illustrated in Fig. 8.11(a), with the AP serving as the star's central node.

By employing a physical layer configuration that is similar to that noted above, we obtain the peer-to-peer net to also possess the *broadcast property*. Thus, a message transmission by net station A that may be destined to net station B (or to several other net stations) would be detected (overheard) by all other net stations. A station that receives a message whose header does not identify it as an intended destination station will not proceed to deliver the message for higher layer processing and may drop it.

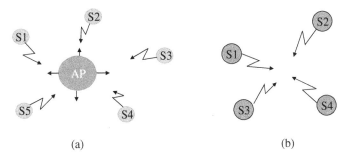

Figure 8.11 (a) Wireless Net Whose Subscriber Stations Communicate with Their Managing Access Point (AP) Station. (b) Wireless Net with Peer-to-Peer Communicating Stations.

To illustrate, consider colleagues that meet at a conference, gathering as a group during a session break and having a group discussion. At times, a single person jumps in and expresses his opinion. Once he is finished, another person may jump in and express her own opinion. At other times, following a break in the discussion, two persons may jump in at about the same time. Subsequently, these persons will realize that their messages did not go through. They may then wait for a random period of time and jump in again.

This communications protocol illustrates a mechanism for sharing a medium by using a protocol that is not based on employing a coordinated means for the allocation of medium resources. Users (or stations) are not assigned specific time slots to use to convey their messages. Users do not request a coordinating entity to manage (and schedule) their speaking times. Rather, a user that wishes to send a message, just jumps-in and sends it. If the message transmission is not completed successfully (as it may be determined by a receiving user to be garbled, in being subjected to overlapping transmissions), the sending users could try to send it again at a later time.

Such a multiple access procedure is identified as being in the class of **random access** schemes. Under the network conditions described above, the use of this scheme is advantageous in that it does not require stations that are ready to send messages to a priori be scheduled through the pre-allocation of access resources. A disadvantage of such an operation is that the message transmission process may not proceed in an orderly manner. If multiple users are interested to initiate message transmissions at about the same time, their transmission attempts may be garbled and deemed unsuccessful.

Under the stated network scenario, the use of a fixed assigned scheme (such as TDMA/FDMA/SDMA/CDMA) to share medium resources would generally not be effective. A station is active infrequently, so that allocating to it a resource on a dedicated basis will lead to inefficient utilization of shared medium resources.

To illustrate, consider a multiple access medium that is shared among a large number of bursty user stations (such as 1000 user stations). Only a small number of stations would become active or inactive within a brief period of time. Assume the medium to be shared by using a TDMA scheme. Clearly, it is not practically effective to dedicate a distinct time slot to each one of the system's 1000 user stations, as they will largely remain unused.

Also, the deployment of a demand assigned scheme is not practically efficient. To implement such a scheme, reservation messages must be transmitted to the controller across a shared signaling channel. However, for the reasons noted above, it is not effectively feasible to employ a fixed assigned scheme to serve as a reservation channel. Instead, if a polling scheme is employed, one needs to be able to poll periodically all user stations. Such a process will be highly inefficient as most stations would not have a new message to send (or a connection to initiate or terminate) when polled, leading to the conduct of a highly lengthy and inefficient polling process.

Under such a network composition and user activity scenario, a *random access* scheme is often employed. As noted above, applications include the transmission of reservation messages across signaling channels in demand-assigned network systems, or data transmissions over wireless networks for which it is not effective or desirable to pre-allocate to users shared medium resources.

In the following sections, we describe key classes of random access-based medium access control (MAC) schemes. Associated procedures include those employed by Ethernet LANs, Wi-Fi Wireless LANs (WLANs), packet radio nets, mobile ad hoc networks (MANETs), MANs, data communications among autonomous highway vehicles, including Vehicular Ad hoc Networks (VANETs), and signaling and control channels that are employed in wireless networks, including such that are used for the transmission of reservation messages.

The message entity that is transmitted across a shared communications channel is a MAC frame. It carries, as payload, higher layer messages, including application layer, transport layer, and packet layer data units. Depending on restrictions imposed on the maximum length of a message that is transmitted across the underlying shared communications channel, the sending station may need to fragment its messages to produce MAC frame segments that meet imposed frame size requirements. In a slotted operation, the segment length is set so that its transmission time fits within a specified slot duration. The corresponding MAC segment is often referred (including in the following) to also as a packet or a message.

8.4.2 Pure Random Access: The ALOHA Protocol

The Method: *Do not coordinate. Just jump in and talk. If your reception is interrupted, wait for a random period of time, and jump in again.*

The Outcome *You may be successful in getting your message to the destination entity provided not too many users are active during the same period of time.*

The basic and simplest random access scheme employs the **ALOHA protocol**. It is based on the following method. A station with a message to transmit does not request a medium resource to use for this purpose, nor does it coordinate its transmission with other stations. It just goes ahead and transmits its message. It then waits to find out whether its message was successfully received by the intended destination. If it receives no PACK of such a reception within a designated response time, it assumes that its message transmission may have overlapped the transmissions of other messages and may have thus **collided** with some or all of the latter messages and consequently (as multiple messages may then be simultaneously received at the destination station) was not successfully received. In this case, a colliding station will retransmit its message. However, if those stations whose messages have collided proceed to retransmit their messages after waiting for a fixed retransmission backoff time delay, they will collide again. Hence, a colliding station is programmed to select a *random retransmission backoff time* period and then retransmit its colliding message.

The ALOHA protocol was used in 1971 for sharing a wireless medium by member stations of the ALOHAnet, which was a computer communications network developed at the University of Hawaii. Communications between terminals situated at geographically distributed locations and a computer center were characterized by low-intensity message rates and therefore did not require a sophisticated multiple access scheme. For the same reason, such a scheme was also used by terminals for sharing a cable medium, as well as in sharing the uplink of a satellite communications channel by low duty-cycle stations.

Two versions of the ALOHA algorithm have been employed. One version is used in asynchronous nets, where stations do not need to maintain time synchronization, while the other one is used in synchronous nets. The asynchronous version employs an **Unslotted ALOHA** algorithm. It uses the following procedure.

Definition 8.2   Unslotted ALOHA Algorithm.

Under the Unslotted ALOHA protocol, a station that has a packet (which is formatted as a MAC frame, and at times also referred to as a frame or a message) ready for transmission across the shared medium executes the following steps:

Step 1: The ready station proceeds to transmit its packet across the channel at any time, while saving a copy of the packet in its buffer.

Step 2: The station waits to find out whether its transmitted packet was successfully received at the intended destination station.

Step 2.1: If it finds out that the transmitted packet was successfully received, it removes its copy from its buffer. It can then proceed, restarting at Step 1, to transmit another packet, if any.

Step 2.2: If the station does not receive (within a specified period of time) a conformation that its transmitted packet has been successfully received, it proceeds as follows:

Step 2.2.1: If the number of previous retransmissions of this packet has not exceeded a specified threshold, the station backs off (i.e., waits) for a random period of time and then it retransmits the packet, and proceeds to perform Step 2 from its starting point.

Step 2.2.2: Otherwise, if the packet has already been retransmitted a number of times that is equal to the specified threshold, the transmission process of the current packet is terminated. The station's MAC entity may then inform a higher layer protocol entity of its failure to transmit this packet and may then be instructed to terminate its transmission process for this packet or to restart it again.

The above operation does not require the net's stations to be time synchronized. In turn, under a **synchronous net operation**, net stations are synchronized, so that they maintain common timing. A commonly used method for maintaining network synchronization is to configure one node in the net to act as a *synchronization master*. The latter disseminates timing data through the periodic transmission of messages that contain timing information. Such messages are identified as *synchronization beacons*. Net stations use the timing data embedded in these beacons to synchronize their clocks to the master clock.

In a synchronous net, a **slotted channel** operation is maintained. A time slot of prescribed duration, which is set in the following to be equal to τ [seconds], is announced. Discrete time marks, and underlying slots, are then defined and recognized by the net's stations. In a slotted system, events are recorded by a receiving entity at such discrete time marks, denoted here as $\{t_n = n\tau, n = 0, 1, 2, \dots \}$. The n_{th} time mark is associated with the time instant $t_n = n\tau$.

A slotted version of the ALOHA algorithm is defined as follows. The duration of each time slot, τ, is set equal to the time that it takes to transmit a maximum length packet (or the corresponding MAC frame segment). The employed multiple access algorithm is similar to that described above for the Unslotted ALOHA scheme except that the transmission of a packet must start at a time mark, which is a starting time of a certain time slot, rather than at any time. Accordingly, the following algorithm describes the Slotted ALOHA protocol.

Definition 8.3   The Slotted ALOHA Algorithm.
Under the Slotted ALOHA scheme, a station that has a packet (formatted as an MAC frame) that it is ready to transmit across a shared medium performs the following steps:

Step 1: A ready station proceeds to transmit its packet across the shared medium at the start of a time slot, while saving a copy of the packet in its buffer.

Step 2: The station waits to find out whether its transmitted packet was successfully received at the intended destination station and then follows the same steps as those used by the Unslotted ALOHA scheme, except that if a packet needs to be retransmitted, the station would back-off for a random number of slots and select a time mark to use for the start of its retransmission.

How does a station determine whether its transmitted packet has been successfully received?

8.4 Random Access: Try and Try Again | **277**

If a full-duplex radio transceiver is used, the station is able to transmit and receive messages at the same time. In this case, the station can listen to the channel while transmitting its packet and determine whether it hears a transmission whose data is the same as the data that it has transmitted.

To simplify this determination process, certain implementations just compare the received signal power with a specified threshold level. If the received power is measured to higher than this level, it is conceivable that multiple signal transmissions may have been executed at about the same time (relative to their reception times at an intended receiver) so that the sent packet is then likely to have collided with other packets. In this case, the intended destination node may not be able to successfully receive this packet. Such a power-based comparison mechanism, often accompanied by other measurements, has been commonly used by Ethernet LANs that share a wireline medium as well as by other systems.

When a **half-duplex radio transceiver** is employed, the station is able to transmit and to receive signals and packets but not at the same time. This is commonly the case when using packet radio wireless radio modules, including Wi-Fi-employed modules. In this case, the transmitting station will wait to receive a positive *Acknowledgment (ACK)* packet (PACK) from the destination station. This approach is used by Wi-Fi WLAN and other systems.

A typical backoff process is illustrated by considering a slotted random access protocol. Upon the retransmission of a packet for the n_{th} time, the station selects at random a slot from within a contention window W_n that consists of the integers: $W_n = [0, 1, \ldots, CW_n]$. By selecting an integer j, the station is set to retransmit its packet after waiting for j backoff slots. Under an **exponential backoff scheme**, we set:

$$CW_{n+1} = \min\{PF \times CW_n + 1, CW_{max}\}, n = 1, 2, \ldots, K, \tag{8.9}$$

where *PF* is identified as the *persistence factor*. The largest window value is set as CW_{max}, while the initial value is set as $CW_{min} \leq CW_{max}$. The window's upper value is not expanded beyond its specified maximum value. The system may set a value $N_T(\max) = K$ for the maximum number of transmissions of a message that the MAC layer entity should execute. Under a **binary exponential backoff scheme**, we set $PF = 2$. An illustrative sequence of corresponding backoff period lengths, when we set $CW_{min} = 3$, $CW_{max} = 63$, and $K = 10$, would be: $\{3, 7, 15, 31, 63, 63, 63, 63, 63, 63\}$. In Fig. 8.12, the transmission dynamics across an ALOHA communications channel is illustrated. In Fig. 8.12(a), it is noted that under the Unslotted ALOHA scheme, a packet collision can be caused by partial overlap of the transmission times of two packets. Under a Slotted ALOHA operation, it is noted in Fig. 8.12(b) that a collision is induced by two packet transmissions initiated at the start time of a commonly selected slot. The shown packet transmission process dynamics, as illustrated over a short period of time, demonstrate a successful retransmission of one of the colliding packets.

How efficient is the ALOHA random access scheme?

When the system is lightly loaded, a station may have a higher probability of being the only one that actually transmits a message across the net's medium. This message is therefore expected to be received successfully at the intended receiver, assuming that no communications channel noise-induced message errors are detected by the receiving station. As the net's traffic intensity level increases, two or more packets may be received at a targeted station during overlapping time periods, which may prevent the intended receiver from correctly receiving the intended packet(s). In this case, packets are said to collide. Colliding packets will be retransmitted by their source stations after random backoff times. They may collide again and would then be retransmitted again.

During a certain slot or time period, a single packet is transmitted and received by the intended receiver. It may then be successfully received. During certain time periods, collisions may take

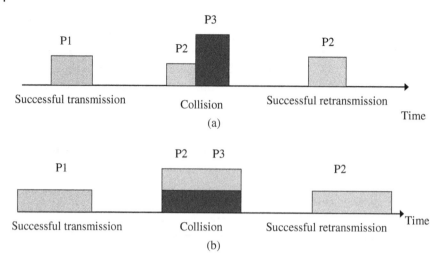

Figure 8.12 Illustrative Packet Transmission Dynamics Across (a) an Unslotted ALOHA Channel; (b) a Slotted ALOHA Channel.

place, while during other time periods, the medium is observed to be idle, so that no transmissions are executed.

To simplify the following performance analysis, we assume henceforth all packets to assume the same length, which is set as equal to the maximum allowable frame length, so that a packet's transmission time is equal to the slot's duration.

What is the maximum normalized throughput rate, denoted as S_{max}, that can be achieved under the ALOHA scheme? This value is identified as the **throughput capacity** of the communications medium. It represents the *maximum value that can be attained under this multiple access sharing protocol for the average number of successful packet transmissions per slot*.

As it takes a single time slot to transmit a single packet, we note that $0 \leq S_{max} \leq 1$. When $S_{max} = 1$, every slot carries a single packet transmission, so that 100% of the channel capacity is fully utilized for the support of message transmissions. A random access scheme does not employ a procedure that is used to perfectly allocate specific slots for specific packet transmissions. Hence, we expect the value of S_{max} to be lower than 1, as the randomness of the transmission process is likely to induce packet collisions, so that not all slots are expected to carry successful packet transmissions.

In the following, we first calculate the throughput capacity S_{max} of the **Slotted ALOHA** scheme, following an approach presented in [7]. We assume that the system has been running for awhile, and has entered steady-state mode.

Selecting a slot at random, we set N_S to denote the number of packets that are successfully transmitted (implying henceforth that they have been successfully received) across this slot and set N_G to denote the number of packets that are transmitted (whether successfully received or not) across the slot. The average values of these random variables are denoted as $S = E(N_S)$ and $G = E(N_G)$. Thus, S denotes the average number of successful packet transmissions per slot, representing the system's (normalized) throughput. In turn, G represents the channel's load, expressing the average number of total packet (successful or colliding) transmissions per slot.

To simplify the following analyses, we consider a broadcast net under which a packet transmission by a station can be successfully received by any other station only if its transmission does not overlap at its receiver with its reception of other transmissions. For a slotted system, it implies that a packet transmitted in a time slot will be successfully received only if no other packet transmission occurs within the same time slot. Within the boundary of the net, the packet's propagation delay is assumed to be much shorter than the packet transmission time.

Since a successful packet transmission will happen at a slot if and only if there is a *single packet transmission* executed at this slot, we write:

$$\{N_S = 1\} \quad \text{if and only if} \quad \{N_G = 1\}. \tag{8.10}$$

Consequently, we have:,

$$P(N_S = 1) = P(N_G = 1).$$

Using this relationship, the throughput value S is expressed as:

$$S = E(N_S) = 1 \times P\{N_S = 1\} + 0 \times P\{N_S = 0\} = P\{N_S = 1\} = P\{N_G = 1\}. \tag{8.11}$$

We model the statistical behavior of the variable N_G by assuming it to follow the distribution of a Poisson random variable with mean G, so that we have:

$$P\{N_G = k\} = e^{-G} \frac{G^k}{k!}, \qquad k = 0, 1, 2, \dots . \tag{8.12}$$

Noting that $P\{N_G = 1\} = Ge^{-G}$, following result is deduced:

S vs. G expression for the Slotted ALOHA protocol:

$$S = Ge^{-G}, \quad G \geq 0, \quad 0 \leq S \leq 1. \tag{8.13}$$

In analyzing an **Unslotted ALOHA** scheme, consider again packets whose transmission duration is equal to a slot's duration. Under the Unslotted scheme, a packet's transmission can start at any time (rather than only at a slot's start time). Hence, the condition for the occurrence of a successful packet transmission is stated as follows.

A given packet transmission is successful only if there are no other packets whose transmission time periods overlap its own transmission period. Assume that the transmission period of an underlying packet takes place in the interval $(t, t + \tau)$. For a successful reception, there should be no other station that initiates a packet transmission during the latter interval as well as in the interval $(t - \tau, t)$.

Hence, the probability that a slot will carry a successful packet transmission is expressed as the product of the probability that there will be no station generating any packet transmission in a preceding period of length τ, which is equal to e^{-G}, and the probability that a single station will be transmitting its packet during a subsequent period of length τ. The latter probability is equal to Ge^{-G}. This calculation yields the following result:

S vs. G expression for the Unslotted ALOHA protocol:

$$S = Ge^{-2G}, \quad G \geq 0, \quad 0 \leq S \leq 1. \tag{8.14}$$

The above-derived relationships between the system's throughput rate (S) and the channel traffic load (G) for the Unslotted and Slotted ALOHA schemes are plotted in Fig. 8.13. The performance graphs show that the throughput (S) under the Unslotted ALOHA scheme is upper bounded by the value $\frac{1}{2e} \approx 0.184$ [packets/slot]. The peak value is attained under channel traffic load $G^* = 0.5$ [packets/slot]. In turn, under the Slotted ALOHA scheme, the throughput (S) is upper bounded by the value $\frac{1}{e} \approx 0.368$ [packets/slot]. The peak value is attained under channel traffic load $G^* = 1$ [packet/slot]. These bounds can be effectively reached in system operations, leading to the following statement.

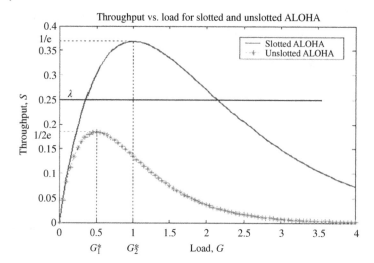

Figure 8.13 Throughput (S) vs. Channel Load (G) Performance Curves Under Slotted and Unslotted ALOHA Schemes.

Theorem 8.1 Throughput capacity attained under the Unslotted and Slotted ALOHA schemes. The throughput capacity (S_{max}) levels attained by the Unslotted and Slotted ALOHA schemes are equal to $\frac{1}{2e} \approx 0.184$ [packets/slot] and $\frac{1}{e} \approx 0.368$ [packets/slot], respectively.

Thus, under the Unslotted ALOHA scheme, only about 18.4% of the channel's resources are used to carry successful transmissions. Remaining resources are occupied by collisions or by idle periods. *Under the Slotted ALOHA scheme, the throughput capacity rate is doubled*, yielding a maximum throughput rate of about 36.8%.

Assume the system loading to be such that the system has reached a steady-state mode. The packet arrival rate, denoted as λ [packets/slot], is then equal to the packet departure rate, representing the rate at which packet transmissions are successfully completed. Thus, $\lambda = S$ [packets/slot]. To avoid overloading conditions, the packet arrival rate must be limited, satisfying: $\lambda < S_{max}$. Under this condition, the operating point of the dynamic scheme is calculated as the intersection of the packet arrival rate λ with the S vs. G performance curve, setting $S = \lambda$.

As shown in Fig. 8.14(a), under a packet loading rate $S = \lambda < S_{max}$, one notes the existence of two equilibrium (intersection) points of operation, corresponding to channel traffic loading levels G_1 and $G_2 > G_1$. Operating at the lower G_1 point leads to lower rates of packet collisions and retransmissions and thus to lower packet delays. In contrast, when the system is operating around the second (G_2) equilibrium point, the channel process incurs a higher rate of packet collisions and retransmissions, leading to higher packet delays.

Consider the behavior of the system when subjected to a traffic loading burst, while the long-term packet arrival rate is equal to λ. When operating at the second equilibrium point and its surrounding, the system behaves in a locally unstable manner. An incremental increase in the channel traffic rate G caused by a traffic burst would lead to reduction in the throughput rate (as it then yields a lower S level), which consequently induces an incremental decrease in the realized throughput rate. In turn, when operating at the first equilibrium point, the system behaves in a locally stable manner. A limited stochastic burst that causes an increase in the number of message arrivals would lead to an incremental increase in the channel traffic rate G, resulting in a higher throughput (S) level. Such a change would in turn lead to reduction in the G level, and the system would transition back to operate around its lower equilibrium point. When subjected to longer-term stochastic loading fluctuations, the system tends to oscillate between periods spent around each equilibrium point.

In observing the shape of the S vs. G performance curve for the Slotted ALOHA scheme (with similar observations applicable to the Unslotted ALOHA scheme), we note the following. As the channel traffic rate G increases, while being kept to the range $G < G_1^* = 1$, spanning the first (monotonically increasing) part of the performance curve, higher channel traffic rates (and correspondingly higher packet arrival rates) induce higher packet departure (and throughput S) rates. The peak throughput rate is reached at $G = G_1^*$.

When the channel traffic loading level G exceeds G_1^*, any further increase in the channel rate leads to diminishing throughput returns. On the one hand, the system is then loaded at a higher packet rate so that more packets arrive and more may have the chance to depart. On the other hand, the higher traffic rate leads to much higher packet access contentions and consequently to a higher rate of collisions, and a consequent reduction in the throughput rate. Furthermore, active stations continue then to contend for access with newly arriving packets. Consequently, stations then contend for access with a rapidly increasing number of packet transmissions, leading to further increase in collisions and retransmissions and to consequent reduction in the realized throughput rate.

As illustrated in Fig. 8.14(b), to avoid reaching the undesirable second equilibrium point, it is beneficial to incorporate into the system's operation a **flow admission control** mechanism. Such a procedure serves to regulate the admission of new packets into a congested system. For example, system performance observations can be used to estimate the channel loading rate G and determine when it has reached or exceeded a threshold level. Under the latter condition, new packet transmissions would be blocked, and certain ongoing transmissions may also be terminated. The blocking state is relieved when the channel traffic loading rate is reduced. Other stabilization mechanisms can also be invoked. For example, stations may be dynamically (when the system enters a congestion state) prevented from executing more than a specified number of retransmissions.

The performance analysis model presented above assumes that a large number of stations are involved in sharing the communications channel. The cumulative packet arrival and transmission processes have been modeled as Poisson processes. In comparison, consider the following slotted ALOHA system model under which a **finite, and lower, number of N stations** share

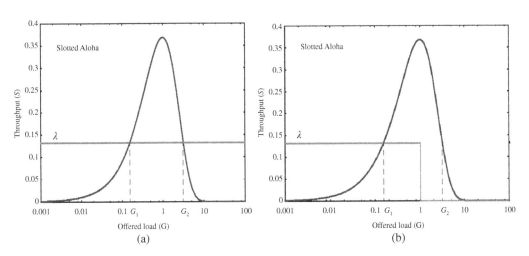

Figure 8.14 Throughput (S) Performance Dynamics Under the Slotted ALOHA Scheme: (a) Without the Use of Flow Admission Control; (b) When Flow Admission Control Is Applied.

the medium. Assume that each station, independently of other stations and of its past behavior, would transmit a packet in any slot with probability p, where $0 < p < 1$. Then, the probability that k packets are transmitted in a given slot is expressed by the Binomial distribution $P\{N_T = k\} = \binom{N}{k} p^k (1 - p)^{N-k}, k = 01, 2, \ldots, N$. The throughput rate is now calculated as $S = P\{N_T = 1\} = Np(1 - p)^{N-1}$.

An upper bound on the attainable throughput rate is derived by setting the transmission probability parameter to $p = 1/N$. Under such an operation, one assumes that a system controller (or each station individually) is able to attain or calculate (and disseminate to other stations) a good estimate of the number N of stations that are currently active. In this fashion, each station dynamically regulates its access rate into the medium by adjusting it to a value that is inversely linear relative to the estimated number of currently active stations. The ensuing throughput rate is then given by the following expression:

Throughput rate for a Slotted ALOHA channel loaded by a finite number of N stations:

$$S = \left(1 - \frac{1}{N}\right)^{N-1}, \quad N \geq 2. \tag{8.15}$$

Using this expression, we note that the throughput rate S varies from a value of $S = 0.5$ for $N = 2$ stations, to $S \approx 0.387$ for $N = 10$ stations to $S \approx 0.377$ for $N = 20$ stations. Thus, the realized throughput rate is monotonically decreasing as the number (N) of sharing stations increases, but the throughput decrease level is relatively modest as the number of contending stations grows beyond 10. As the number of these stations increases, the maximum achievable throughput rate converges to $1/e \approx 0.368$, which is equal to the throughput capacity rate calculated above for a system that is loaded by a large number of stations:

Throughput rate for a Slotted ALOHA channel loaded by a large number of N stations:

$$S = \lim_{N \to \infty} (1 - \frac{1}{N})^{N-1} = \frac{1}{e}. \tag{8.16}$$

It is noted that as the number of contending stations increases from 2 to 10, the throughput rate decreases by about 22.6%. In comparison, as the number of contending stations increases from 10 to a very large number, the throughput decreases by only about 4.9%. It is also observed that to incur the latter moderate throughput degradation, one needs to expand the backoff time period in a manner that increases linearly with the number of actively contending stations.

Assuming the system to operate at a prescribed equilibrium operating point characterized by the rate pair (S, G), we calculate the average number of transmissions incurred by a packet until it is successfully received as equal to $E(N_T) = G/S$. Hence, the number of retransmissions incurred by such a packet is equal to $E(N_R) = G/S - 1 = \frac{G}{Ge^{-G}} - 1 = e^G - 1$.

Assuming a large number of net stations, whereby each station holds in its buffer at most a single packet at a time, and assuming an average packet retransmission delay of d_R [slots], the average time delay $E(D)$ [slots] incurred by a packet until it is successfully transmitted is approximately expressed as $E(D) = (\frac{G}{S} - 1)d_R + 1$ [slots], noting that the transmission time of a successfully transmitted packet is equal to 1 slot. Hence, we have:

$$E(D) = (e^G - 1)d_R + 1. \tag{8.17}$$

Under a prescribed packet loading rate $\lambda < 1/e$, the value assume by G is calculated by setting $S = \lambda = Ge^{-G}$.

The variation of the average delay experienced by a packet as a function of the throughput parameter S is related to the variation in the average number of transmissions (to successful reception) incurred by a station in transmitting a packet. As we gradually increase the channel

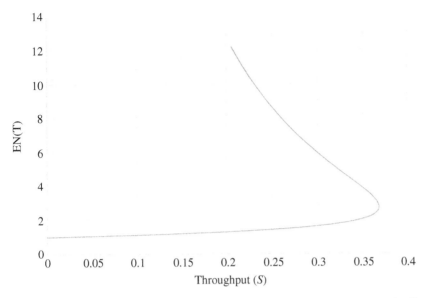

Figure 8.15 Average Number of Packet Transmissions vs. Throughput (S) Under the Slotted ALOHA Scheme.

loading rate G, the corresponding increase in the average number of packet transmissions (and thus in packet delay) is shown in Fig. 8.15. We observe that over the range $G \leq G^* = 1$, as the channel loading rate G increases, the packet delay monotonically increases while the realized throughput rate metric S increases as well. In turn, in the range $G > G^* = 1$, the packet delay increases rapidly with G while the network throughput rate S decreases. Thus, while for the other discussed multiple access schemes, including those that use fixed assigned and demand assigned algorithms, the delay–throughput performance behavior is represented by a monotonically increasing performance curve, under a random access scheme, the delay–throughput performance curve follows the form of the bending curved displayed in Fig. 8.15. This is explained by noting that at higher loading rate ($G > G^*$) as the packet arrival rate increases, the throughput rate decreases while the packet delay keeps increasing.

Example 8.6 Consider a highway segment that is managed by a Road Side Unit (RSU) that supports communications from $N = 50$ vehicles while they travel along this segment. A control channel (CCH) is provided, using a dedicated frequency band, for the purpose of allowing vehicles to transmit status and safety data packets, operating at an (uplink) data rate of $R = 10$ Mbps. Each status packet contains $L = 2000$ bits (including overhead). A Slotted ALOHA protocol is used by the vehicles to transmit their status packets. The slot duration is set equal to the packet transmission time. It is assumed that packet transmission error rate is negligible. It is desirable to set the system's offered loading rate to a value that will allow each one of the 50 stations to have an average of one successful transmission during each 50 ms long period. We employ a mathematical model that is applicable for a large number of stations sharing the medium and assume packet arrivals and transmissions to follow a Poisson process.

a) We calculate the overall loading level (G) that the channel has to carry, including successful and colliding packet transmissions.

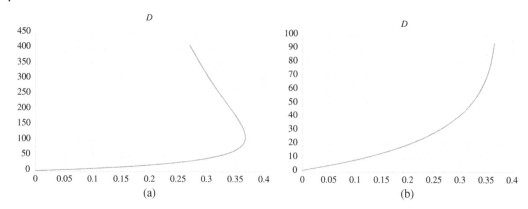

Figure 8.16 Average Packet Delay (D [slots]) vs. Throughput (S) Under the Slotted ALOHA Scheme: (a) Under Unrestricted Load; (b) Under Flow Control.

Solution: The desired packet throughput rate is equal to 50 packets every 50 ms, amounting to 1000 [packets/s]. The channel's transmission rate is equal to 10 Mbps, which is equal to 5000 [packets/s]. Hence, the desired normalized throughput rate is equal to $S = 1000/5000 = 0.2$. The corresponding channel loading rate G is calculated by solving the equation $S = 0.2 = Ge^{-G}$. The resulting solution is $G \approx 0.26$ [packets/slot].

b) Using the targeted channel throughput rate calculated in (a), we calculate the average number of times that a packet is transmitted until it is successfully received (including its retransmissions and final successful transmission).

Solution: $E(N_T) = G/S = G/0.2 = 5G \approx 1.30$. Hence, under the loading rate calculated in (a), a packet will be transmitted an average of 1.3 times, so that it will be retransmitted an average of 0.3 times.

c) Assuming an average retransmission delay of $d_R = 64$ slots, we calculate the average delay incurred by a packet.

Solution: The average packet delay is calculated by noting that the average number of packet retransmissions is equal to $E(N_T) - 1$, so that we write $E(D) = 1 + [E(N_T) - 1] \times 64[slots] = 20.2$ [slots]. The slot duration, denoted as T, is equal to the average packet transmission time, $T = L/R = 0.2$ ms. We consequently obtain the average packet delay to be equal to $E(D) \approx 4.04$ ms.

The corresponding variation of the packet delay (D [Slots]) vs. the system's throughput (S) is shown in Fig. 8.16. In Part (a), we display the delay–throughput curve incurred when the system is subjected to a wider channel loading (G) range. The performance behavior shown in Part (b) is attained by using a flow control scheme to restrict the incurred channel loading rate (G) levels. It is noted that when realizing a peak throughput rate of around $S = 0.368$, an average packet delay of the order of about 100 slots is incurred. As shown in Part (b), to restrict the average delay of admitted packets to about 12, 15 or 19 slots, the channel loading rate must be sufficiently restricted so that the realized throughput rates (S) are equal to only 0.15, 0.2 or 0.25, respectively. A loading level of $S = 0.15$ corresponds to a throughput rate that is equal to a ratio of $0.15/0.368 \approx 40.7\%$ of the peak throughput rate. The average fractional number of packet retransmissions would then be equal to about $0.16 = 16\%$.

8.4.3 Carrier Sense Multiple Access (CSMA): A Listen Before Talk Protocol

The Idea: *Listening before talking. If the medium is not occupied, jump in and talk. If you are interrupted, wait a random period of time, listen, and when idle, try again*

The Outcome: *This scheme performs effectively provided the time that it takes other users to determine that you are currently talking, so that they should not interrupt, is much shorter than the time that it takes you to send your message over the medium.*

Under a pure random access scheme (such as that implemented by the ALOHA algorithms), a ready station is not equipped to sense the communications channel to determine whether the medium is occupied with the transmission of a packet by another station. Stations that employ a **Carrier-Sensing Multiple Access (CSMA)** protocol are equipped with a carrier sensing (CS), also known as carrier sense, module that enables them to determine whether the channel is currently idle or busy.

In using Carrier Sensing (CS), also known as **Clear Channel Assessment (CCA)**, a station employs Physical Layer processes to decide whether the medium is currently occupied with the transmission of message signals. If so, **a busy channel** status is recognized. Otherwise, it is said to be **an idle channel**. One of the key determining factors is the power of the signal measured by the listening station, also identified as the **Received Signal Strength Indicator (RSSI)**, or the **Receive Channel Power Indicator (RCPI)**, which is based on measuring the signal's receive RF power. It is often used in conjunction with other measurements and calculations to determine whether a signal observed on the channel is induced by noise and/or interference signals rather than by a station's message transmission.

A setting performed by the network management entity configures a value for a **Carrier Sensing Threshold** parameter, TH_{CS}. The station declares the channel to be busy if the sensed signal is associated with a message and its measured received power P_r is higher than or equal to the configured carrier sensing threshold level, $P_r \geq TH_{CS}$.

The shared medium and the stations that use it to communicate form the **shared medium net**. For a carrier sensing-based multiple access protocol to function well, the net configuration must be such that a transmission by a net's member station, identified often in short as a station, can be received by any other net station, in that at least its signal presence is detected. Such a property is identified as a net's **broadcast property**. We generally assume that a network of stations that employs a CSMA scheme functions as a **broadcast net**, though at times message transmission across the medium may not be detected by all stations.

When using a CSMA protocol to share a multiple access communications medium, a station that is ready to transmit its packet must sense the channel prior to initiating its transmission. In doing so, the station aims to determine whether the channel is busy or idle. If it determines the channel to be busy, the station will not proceed at this time with the transmission of its packet across the medium.

Thus, if you hear another person talking, you are required to not interrupt. The station would then act in one of the following ways. Under a **Non-persistent CSMA** protocol, the station will then delay its access action by a deferral time (whose value may be configured through a station management setting) and then proceed to sense the channel again. When a ready station determines that the shared channel is idle, it will then proceed to transmit its packet.

Under a **1-Persistent CSMA** protocol, the station will keep sensing a busy channel until it determines the channel to have changed from busy to idle state. It would then proceed to transmit its packet.

Under either protocol, following the transmission of its packet, a station waits to find out whether its packet's transmission has been successful. If no such confirmation is obtained prior to the expiration of a timeout period, the station will backoff, waiting for a random period of time and then restart the transmission process by sensing the channel again. The waiting time period can be set to follow a binary exponential backoff algorithm.

Consider a station that senses an idle medium and proceeds to transmit its packet across the shared medium. After a period of time, the transmission issued by this station will reach all other net stations, so that the latter declare the medium to be busy. To determine how long to wait before making this determination, a channel slot time, often also identified as a channel minislot, is defined. Its length is set to span a sufficient period of time, aiming to allow the underlying station's transmission to reach all net stations.

The access algorithm followed by the CSMA multiple access scheme is then described as follows.

Definition 8.4   The Carrier Sense Multiple Access (CSMA) Algorithm.
A station that has a packet that it is ready to transmit across the shared medium follows the following steps:

Step 1: A carrier-sensing process is undertaken; the ready station senses the medium to determine whether it is busy or idle.

    a) If sensed idle: The station proceeds to transmit its packet across the medium. Under a variation, such as when using a p-persistent oriented CSMA operation, the station will wait for a random period of time to find out whether the medium continues to stay idle, and then, if still idle, transmit its packet.

    b) If sensed busy:

        i) Under a non-persistent CSMA operation, the station's MAC entity defers its action by using a random backoff time delay. It then restarts at Step 1 to schedule its packet's transmission.

        ii) Under a p-persistent oriented CSMA operation, $0 < p < 1$, the station will wait for a random period of time (which lasts for an average of $1/p$ minislots) to find out whether the medium has remained idle, and then, if still idle, transmit its packet. If in the meanwhile the medium becomes busy, it will resume sensing the medium until it becomes idle again. Alternatively, it may then defer its action by using a random backoff time delay.

        iii) As a special case ($p = 1$), under a 1-persistent version, the station keeps sensing the busy medium and when it determines it to become idle, it proceeds to immediately transmit its packet.

Step 2: Upon completing the transmission of its packet, the station waits to find out whether its packet was successfully received at the intended destination station.

    a) If it finds out that the transmitted packet was successfully received, it removes its copy from its buffer. It can then proceed, starting at Step 1, to transmit another packet, if any.

    b) If the station does not receive within a specified period of time a confirmation that its transmitted packet has been successfully received:

        i) If the number of previous retransmissions of this packet does not exceed a specified threshold, the station backs off for a random period of time and then restarts at Step 1 to schedule its retransmission.

        ii) If the packet was already retransmitted a number of times that is equal to the specified threshold, the current transmission process is terminated. The station's MAC entity may then inform a higher layer protocol entity of its failure to send the packet and may then be instructed by it to terminate its transmission attempt or to restart it.

A CSMA MAC protocol entity that is ready to transmit a frame (which carries a packet as its payload) waits to receive a carrier sensing reading, also known as clear channel assessment (CCA) notification, from its physical layer entity. If the latter informs it that the channel is idle, it would then proceed to initiate the transmission of its frame (also identified here as a packet). Its transmission is however not instantly sensed by other stations. It takes time for another net station to determine that a packet is currently being transmitted across the medium so that it should defer any plan for initiating its own packet transmission to a later time.

If another station (or several ones) is ready to transmit its packet at about the same time that the underlying station transmits its packet, and if it then proceeds to sense the channel, it may not be able to detect the ongoing transmission until the latter propagates across the channel and reaches its receiver at a power level that is sufficiently high to induce carrier sensing determination. The period of time that is required by a packet transmission to induce carrier-sensing determination at other net stations is identified as the **medium acquisition time**, and its duration is often represented as a **minislot** time. The duration of this acquisition time duration is denoted as t_a.

If the underlying station is the only one that is transmitting a packet across the medium over a minislot period that starts at the initiation time of its transmission, other non-transmitting net stations that sense this transmission will declare the channel to be in a busy state (following this minislot period) and consequently avoid the initiation of their own transmissions. In this case, the transmitting station *is said to have acquired the channel*, and its transmitted packet is then likely to not collide with packet transmissions carried out by other stations.

To assess the *performance effectiveness of the operation of the CSMA scheme*, one needs to compare the acquisition time t_a [s] and the packet transmission time $T = L/R$ [s], where L is equal to the average packet's length (expressed in [bits/packet]) and R represents the data transmission rate across the medium (expressed in [bits/s]). The corresponding ratio, calculated as $a = t_a/T$, is identified as the **acquisition factor**, or acquisition parameter.

The duration of the channel acquisition time t_a is calculated by accounting for several components. A key component is the time taken by the transmitted message signal to propagate across the medium so that it reaches other net stations that share the underlying medium. The propagation speed depends on the medium that is used and is often estimated (for wireline and over the air wireless media) to be of the order of 75% of the speed of light, which is equal to about 5 [μs/km]. Other components include the packet's signal build-up time at the transmitting station, and at the receiving station, the time taken to process the packet's preamble. The latter is a leading field that must be analyzed at the receiver to assert the signal as being part of a valid packet rather than just interference, as well as at times to derive targeted signal parameters and achieve signal related synchronization. In addition, time is required at the receiver to collect a sufficient number of signal symbols to allow the receiver to make a reliable carrier sensing decision, as it needs to determine whether the channel is busy with the transmission of valid messages or is just perturbed by background noise and/or signal interference, in which case the channel may be declared to be in idle state.

We note that the carrier sensing acquisition time value used by a Wi-Fi WLAN relates to minislot t_a lengths of the order of 5 [μs] to 20 [μs], for the CSMA/CA-based protocols employed by several versions of its medium access control (MAC) protocol. The corresponding values assumed by the acquisition factor depend on the packet's transmission time. To illustrate, consider a packet whose length (including overhead) is equal to $L = 1000$ bits. Consider several communications channel systems that provide for a transmission data rate across the medium that is equal to 10 Mbps, or 100 Mbps, or 1 Gbps. The packet transmission times ($T = L/R$) are then equal to 100 [μs], or 10 [μs], or 1 [μs], respectively. Assuming an acquisition time that is equal to $t_a = 20$ [μs], the corresponding acquisition factors (a) are equal to 0.2, 2, or 20, respectively. If the acquisition time is reduced to $t_a = 5$ [μs], the corresponding acquisition factors are reduced to 0.05, 0.5, or 5, respectively.

The delay–throughput performance of a CSMA network system is critically dependent on the underlying values assumed by the acquisition factor a. Under a setting of $a = 1$, the time it takes a station to determine that another station is transmitting, following the start of the other packet's transmission, would be equal to the average packet transmission time. In this case, the underlying CSMA operation would not enhance system performance beyond that exhibited under a Slotted ALOHA protocol. The latter scheme would then be preferred as it does not require a ready station to sense the medium at all, but simply jump-in at a start of a packet-slot and transmit its packet. For much lower a values, using the CSMA scheme can be advantageous. If the system is not too highly loaded, and t_a is sufficiently short, it is more probable that no other station will become active and initiate its packet transmission during the first t_a seconds that follow a packet's transmission time.

How low should the CSMA system's acquisition factor a be to yield an operation that exhibits a good throughput rate, one that is higher than that attained by using a much simpler protocol such as the ALOHA algorithm that does not require the station's radio to employ a carrier sensing (CCA) mechanism?

If you need to listen for a relatively long period of time to determine if somebody else is using the medium, will you be better off by not listening at all and just jumping-in and sending your message?

At the time that the other packet has reached you and caused your CCA mechanism to declare a busy channel, your transmission may have already been completed, so that you will not be able to defer your transmission!

To demonstrate the impact of the acquisition factor a on the system's throughput performance behavior, we consider a slotted non-persistent CSMA algorithm. The slot in this context is actually a minislot of duration that is equal to t_a. A minislotted operation implies that all stations are synchronized in identifying the start times of minislots. A station that senses the channel to be idle starts its packet transmission at the start of the next minislot. If two or more stations initiate transmissions at the same minislot, their packets are said to collide and each corresponding station will select a random backoff time after which it will sense the channel again. If a station senses a busy channel, it defers its packet transmission and will sense the channel again after a random back-off time.

The duration of each packet slot is equal to T. Following the modeling method described when analyzing the ALOHA protocol, we assume channel processes to be modeled as Poisson processes. The channel traffic loading process, including new and retransmitted packets, is assumed to be a Poisson process with intensity that is equal to g [packets/s]. The average number of packet transmissions per packet slot is denoted as $G = gT$.

Consider a single minislot and the duration of channel occupancy, identified as a cycle time C, induced by the events that occur during this minislot. With probability e^{-gt_a}, no stations initiate their packet transmissions in this minislot, so that this minislot remains idle.

In turn, with probability $1 - e^{-gt_a}$, at least one packet transmission is initiated at the start of this minislot, leading to occupancy of the medium for a period that is equal to $T + t_a$. Hence, the average cycle time period is equal to $E(C) = T(1 - e^{-gt_a}) + t_a$.

The normalized channel throughput rate is denoted as S. It expresses the fraction of time that a successful packet transmission takes place. In considering a single cycle time, we note that this period includes a successful packet transmission if and only if in the leading minislot, a single packet initiates its transmission. The probability of this event is equal to $gt_a e^{-gt_a}$. When such an event occurs, the cycle consists of a transmission of a single packet which occupies a period of duration T for the transmission of its payload bits. The normalized throughput rate S is then expressed as the ratio of $Tgt_a e^{-gt_a}$ and $E(C)$, yielding the following result [35]:

Property 8.1 Throughput Capacity of a Minislotted Non Persistent CSMA Scheme:

$$S = \frac{aGe^{-aG}}{1 + a - e^{-aG}}. \tag{8.18}$$

The performance behavior of the above discussed minislotted non-persistent CSMA scheme is illustrated in Fig. 8.17. In Fig. 8.17(a), we show the throughput (s) performance behavior vs. channel load (G) under acquisition factor $a = 0.2$. The bi-stable character of the operation, similar to that exhibited for the ALOHA scheme, is noted. The peak throughput, identified as the throughput capacity value (S_{max}), is noted to be equal to about 0.51 and is achieved at $G = 2.4$. Under a lower acquisition factor value, such as $a=0.05$, the throughput capacity value is equal to about 0.72 and is achieved at $G = 5.6$. For a higher acquisition factor values, such as $a=0.5$, $a = 1$, and $a=2$, we obtain the corresponding throughput capacity values to be equal to 0.347, 0.23, and 0.14, respectively, and the channel loading levels at which they are attained are equal to 1.3, 0.8, and 0.4, respectively.

The variation of the throughput capacity level vs. the channel's acquisition factor value is shown in Fig. 8.17(b), under the *series 1* curve. We note its decrease as the acquisition factor value increases. For acquisition factors that are higher than about $a=0.4$, the Slotted ALOHA algorithm, which does not require the station to employ a carrier-sensing mechanism, is noted to exhibit throughput performance behavior that is superior to that attained by the underlying CSMA protocol.

Furthermore, we note that at the loading levels that achieve the throughput capacity values, the average number of packet transmissions that are carried, $E(N_T) = G/S$, can be quite high, inducing high packet delay levels. For example, for acquisition factor values that are equal to 0.05, 0.2, 0.5, and 1, the corresponding average number of times that a packet is transmitted until successfully received is noted to be equal to about 7.78, 4.69, 3.74, and 3.45, respectively. Hence, to achieve good delay–throughput performance behavior, the channel loading level must be well reduced to a level that is lower than the throughput capacity rate.

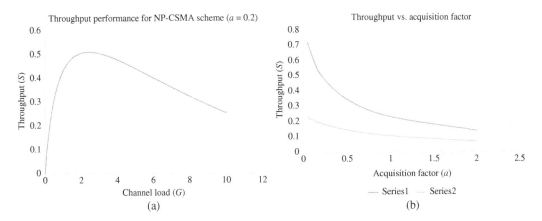

Figure 8.17 Performance Behavior of a CSMA Scheme: (a) Throughput Performance Under Acquisition Factor $a = 0.2$ (b) Throughput as Function of the Acquisition Factor (a): Throughput Capacity (Series 1) and Throughput Under $E(N_T) = 1.3$ (Series 2).

To illustrate, we consider, as a reference, a Slotted ALOHA scheme that produces a normalized throughput rate of 20%, $S = 0.2$. We then obtain that $E(N_T) \approx 1.3$. To attain the same packet delay performance, we calculate the maximum throughput rate that will be achieved under a slotted non-persistent CSMA scheme with acquisition factor a. The results are displayed by the *series 2* curve shown in Fig. 8.17(b). We find that under an acquisition factor value of $a = 0.2$, the maximum throughput rate produced by the CSMA scheme is equal to about 0.19, whose value is close to the corresponding throughput level permitted (in aiming to achieve the prescribed $E(N_T) = 1.3$ value) under the Slotted ALOHA scheme. We thus conclude that, under the specified packet delay constraint that limits the average number of packet transmissions to about 1.3, the incorporation of a carrier sensing mechanism in using a Non-Persistent-CSMA scheme provides no performance benefits when compared with a Slotted ALOHA protocol, when the acquisition factor value is equal to or higher than 0.2.

As noted above, when the system is characterized by a much lower acquisition factor a value, the CSMA scheme yields a much higher maximum throughput rate. However, to achieve this rate, a much higher channel loading rate G must be incurred, leading to higher packet delay levels.

To support a relatively high-throughput rate and low packet delay performance behavior when using a CSMA scheme, it is essential to synthesize the scheme to assume a sufficiently low acquisition factor value, $a < 0.1$; preferably setting it to not be higher than about 1–5%, when feasible. A key component of the acquisition factor is the signal propagation time across the medium, t_p. The value assumed by the acquisition factor is proportional to the ratio $t_p/T = Rt_p/L$, where L is the average packet length, R is the transmission data rate across the medium and $T = L/R$ is the packet transmission time. The propagation delay is calculated as the product of the propagation speed v_p and the distance span d, $t_p = v_p d$. The propagation speed is estimated to be equal to about 75% of the speed of light, so that we use $v_p = 5$ [μs/km]. For illustration purposes, we assume in the following that $L = 2000$ [bits/packet], $a = \approx 3t_p$, noting that t_p is a key component, yet not the only one, embedded in the acquisition factor a.

Clearly, to reduce the a value, it is essential to keep the medium range span (d) low. Consequently, CSMA schemes are effective only when used to regulate access to the medium over a relatively shorter range. Hence, *CSMA schemes can be effective only when employed as Medium Access Control (MAC) protocols that schedule access to a shared medium of a LAN, which spans a properly short distance range.*

To demonstrate, consider first the medium range to be equal to $d = 2$ km, while the data rate is set to be equal to 10 Mbps (which are similar to the settings used by the original wireline Ethernet LAN). Under these settings, we have $a \approx 3dv_p R/L = 0.015$, so that an advantageous delay-throughput performance is attained. As the data rate for this system is increased to 100 Mbps, the acquisition factor increases by a factor of 10, yielding $a = 0.15$. The corresponding performance behavior of the CSMA scheme is consequently degraded. We can upgrade the system's performance by, for example, reducing the network's shared medium span by a factor of about 10. Hence, by reducing the span to $d = 200$ m, we restore the acquisition factor's value to about $a = 0.015$.

If we further increase the transmission data rate across the medium, setting it to 1 Gbps, the above calculation yields an acquisition factor value that is increased by a factor of 10, yielding $a = 0.15$ over a span to $d = 200$ m. We can reduce the value of the acquisition factor under this setting by again reducing the span of the medium by a factor of 10, setting it equal to about $d = 20$ m. Thus, CSMA systems that use a very high transmission data rate are set to span a much shorter shared medium range.

Consequently, to accommodate the operation of CSMA schemes, such as those used by Wi-Fi and Ethernet systems, under a much higher transmission data rate, as used by many current systems, we tend to modify the CSMA scheme by using one of the following adjustments. When acceptable, we reduce the distance span of the shared medium network, so that communicating stations are quite close to each other. Alternatively, or when the later adjustment is not sufficient, we change the CSMA protocol to employ a MAC scheme that does not require the use of a carrier sensing operation. Such alternate schemes can be selected from classes of fixed assigned and demand assigned FDMA/TDMA/SDMA schemes. If a distributed (noncoordinated) access control scheme is desired, a non-carrier-sensing scheme, such as an ALOHA protocol, is often employed. Another approach is to limit the broadcast net dissemination requirement and integrate into the operation spatial division components, as realized through the use of directional (or sectorized) antenna beams.

The basic operation of CSMA-oriented scheme makes use of a shared medium that induces a broadcast net configuration, so that a station's transmission can be received at all other net stations. However, at times, certain stations may not be able to receive transmissions issued by several (or all) other stations. Such stations are identified as **hidden stations or hidden terminals**. Under such a condition, the CSMA protocol operation may not function effectively.

For example, consider the following scenario. Station A senses an idle medium and starts its packet transmission. Station B is not able to receive the transmissions issued by station A. Hence, even after station A's packet has been in transmission for a period of time that is longer than t_a, station B is still not able to detect it. Assume that both stations can directly communicate with a common destination station. Then, it is possible for station B to initiate the transmission of its own packet while station's A packet's transmission is ongoing (and has been initiated at a time that is longer than t_a time units earlier) and therefore station B's transmission may be interfering with the successful reception of station's A packet at the common destination station. Such an issue may also arise when stations aim their packets for reception by distinct stations.

When detected to be disruptive, methods are used to resolve issues caused by hidden stations. One such approach makes use of a central station that has wide access to all, or most, net stations. To visualize such a scenario, consider a net that consists of stations that reside on different sides of a hill, whereby those on one side cannot hear transmissions issued by stations locating on the other side. A central gateway station is place on the top of the hill. The central station is able to directly communicate to and hear all net stations. The central station can then transmit a control signal to inform other stations that the channel is currently busy (and may also specify for how long it will continue to stay busy) so that no other station should attempt access during this time. Also, a reservation process can be enacted. A ready station uses the central station to inform other net stations that it hereby reserves the medium for a specified period of time. We will illustrate the operation of such a scheme when discussing a Wi-Fi multiple access protocol.

Under an alternative setting of a broadcast net, the gateway station is used as a **physical layer relay**. It relays any packet received from any station. After possibly performing signal restoration and amplification, it retransmits each received packet, aiming it for reception by all net stations.

8.4.4 Carrier Sense Multiple Access with Collision Detection (CSMA/CD) and Ethernet Local Area Network (LAN)

The Idea: *Listen before talking and if the medium is not occupied, jump in and talk. While talking, you should continue listening to evaluate transmission activity across the medium to determine if some*

292 | *8 Multiple Access: Sharing from Afar*

other station's transmission overlaps your transmission. If so, stop your data transmission and try again after a random backoff time.

Performance Behavior: *This access scheme can be effective provided that the time needed by others to hear your transmission and the time that you need to determine whether your transmission is experiencing collision are each much shorter than your packet transmission time.*

Under a Carrier Sense Multiple Access (CSMA) protocol, a station whose packet transmission is colliding with other packet transmissions is not immediately aware of this situation and therefore will continue with the complete transmission of its packet. Each involved station then waits to receive a PACK signal. If not received by a specified deadline, the station proceeds to reschedule the retransmission of its packet. Under a **CSMA/CD** protocol, a station uses a carrier-sensing mechanism, as it does under CSMA. However, the station is now assumed to employ a **full-duplex radio module**, so that it is capable of receiving signals and messages that use the medium at the same time that it is engaged in the transmission of its own message frame. Making use of this capability, a station is assumed to be able to carry out **carrier detection (CD)** processing. Such processing enables it to rapidly determine whether its transmission is colliding with other transmissions, hopefully making this determination while its transmission is still ongoing.

In the following, we consider a *shared-medium broadcast net*. The broadcast property implies that a transmission issued by a station propagates across the medium so that it is detected by other net stations. Each station employs a full-duplex radio module. The stations gain access to the medium by using the CSMA/CD multiple access protocol described in the following. Such an operation has been used by the Ethernet LAN. Wireline media are often used by such a system, including coaxial cable, twisted pair copper wires such as Unshielded Twisted Pair (UTP) wires, and fiber-optic lines.

Such a broadcast net, in which only a single message frame transmission can be carried out successfully at a time, is often identified as a **LAN segment**, or briefly as a segment. Time simultaneous transmissions by multiple stations would generally result in unsuccessful frame receptions. Hence, such a network span is referred to as a *collision domain*.

The time that it takes a station to make a determination that its frame transmission is incurring collision, measured from the time that it has started the transmission of the frame, is identified as the *Collision Detection Time*, and is denoted as t_{CD}. The time that it takes the station to transmit its frame to completion (when not interrupted) is denoted as T. If $t_{CD} < T$, the station would generally be able to determine that its transmission is colliding with other transmissions while it is still in the process of transmitting its frame. In this case, upon CD, the station could immediately *abort its transmission*, even if its transmission has not yet been completed. The CSMA/CD protocol is similar to that used by the CSMA algorithm, supplemented by the inclusion of a CD process and its transmission abortion initiative.

Accordingly, a commonly used version of the **CSMA/CD multiple access protocol** employs the following algorithm.

Definition 8.5   The CSMA/CD multiple access Algorithm.

A station that has a packet that it is ready to transmit across the shared medium uses the following steps:

Step 1: The ready station performs carrier sensing by listening to the medium to determine whether it is busy or idle. The station then follows the same process as that described under Step 1 of the CSMA algorithm.

Step 2: The station performs CD as it monitors the medium to determine whether its ongoing packet transmission is being interfered, such as by simultaneously encountering transmissions by other stations and thus experiencing collision. If such an interference (or collision) determination is made while the packet is being transmitted, the station proceeds to immediately abort its ongoing packet transmission. It then reschedules the retransmission of its packet following a random backoff time, restarting the process at Step 1.

Step 3: Following the completed transmission of its packet, the station waits to find out whether its packet was successfully received at the intended destination station. The station then follows the same process as that described under Step 2 of the CSMA algorithm.

As noted above, to take advantage of CD, it is essential that $t_{CD} < T$. Hence, this operation is performed by a **full-duplex radio module**. Such a module is able to, simultaneously in time, transmit as well as receive messages. While the station is in the process of transmitting a frame, it is able to also monitor the medium for message signal receptions.

Several techniques can be employed to perform CD. A CD determination can be performed by a station by measuring the cumulative power of the received signal and comparing it with a threshold level. The latter corresponds to the receive signal power level that would be measured if a single message transmission were occupying the medium. If more than a single message occupies the medium, the monitoring station will detect a higher receive power level and will then declare the medium to currently carry multiple message signals, indicating a collision state.

Traditional Ethernet LAN system implementations tend to employ the latter power measurement-based technique. For this process to properly function, it is necessary to limit the shared medium's distance span. If stations A and B are at a relatively long range from each other and are both currently transmitting their frames across the medium, the power of the signal transmitted by station B may be highly attenuated when measured at station A. Station A may then not be able to detect an ongoing collision condition. For this reason, when a CD mechanism is employed, the maximum distance span between stations sharing the medium is restricted.

Initial implementation versions of the Ethernet LAN assumed a distance span that is lower than about 0.5 km across each segment (which forms a broadcast domain). This range can be extended through the use of physical layer repeaters that interconnect multiple segments and provide signal power amplification. As discussed in the following, it is noted that while this configuration helps in upgrading the signal power level required for the implementation of the CD function, the overall length of the network, when aiming to provide network-wide physical layer-based broadcast feature, is still limited by the necessity to reduce the network's medium acquisition factor and CD times.

An alternative CD mechanism would be to have a transmitting station compare the data contents of a packet received from the medium with that embedded in the packet that it has been transmitting. A collision state is declared if these two data streams are found to not be identical. Such an operation is however more demanding in terms of implementation complexity.

Clearly, the CD component of the algorithm is effective only if a transmitting station is able to determine a collision state before it has completed the transmission of its packet. Under which conditions would it be feasible by a station to make such a rapid determination?

We first estimate the time that it takes a station to make a CD determination, denoted as t_{CD}, measured from the start time of its packet's transmission. Assume that the distance span of the medium, identifying the longest distance between any two net stations, is equal to d_{span}. The corresponding one-way propagation delay incurred by a message signal in traversing the shared medium is equal to $t_p = v_p d_{span}$, where v_p is the signal propagation speed. This speed is typically assumed to be equal to about 75% of the speed of light, so that it is equal to about 5 [μs/km].

8 *Multiple Access: Sharing from Afar*

Consider the following scenario. Station A senses an idle medium and initiates the transmission of its packet at time $t = 0$. This packet's transmission starts to reach station B at time $t = t_p$. A short time thereafter, station B will be able to determine that the medium is occupied by a packet's transmission issued by another station and would then defer is access action. However, assume that a brief time prior to this CS determination, station B becomes ready to transmit its own packet. It then senses the medium and finds it to be idle. It consequently proceeds with the transmission of its own packet. The latter transmission will take t_p [s] to propagate across the medium and then be detected by station A. Thus, after a period of time that is somewhat longer than $2t_p$ (as one needs to include subsequent time needed by the station to confirm the detected signal to in fact be induced by multiple frame transmissions), station A will find out that the medium is now occupied by multiple packet transmissions, confirming a CD state. It would then proceed to immediately abort its packet transmission, provided the latter is still ongoing. The CD time is thus estimated to be of the order of $t_{CD} = 2t_p = 2v_p d_{span}$.

The performance efficiency of a CSMA/CD scheme is thus observed to depend on the following parameters:

1) To produce an effective CSMA operation, the system's **acquisition factor** $a = t_a / T$ should be quite low, $a << 1$.
2) To produce an effective CD operation, the CD time should be much shorter than the packet transmission time, $t_{CD} << T$, so that we should have $2t_p << T = L/R$.
3) For a station to be able to detect collision of its packet transmission with a packet transmission issued by another station located in the same collision domain, the signal transmitted by the other station must reach the station at a sufficiently high-power level. Accordingly, the distance span of a network segment must be appropriately limited.

As discussed when evaluating the performance of a CSMA scheme, to assure a low acquisition factor a, the distance span of the broadcast net must be limited and the transmission data rate could not be set to a corresponding value that is too high. The net propagation delay t_p plays a key role in impacting the value assumed by the acquisition factor as well as the CD time.

To illustrate, consider the following cases. Assume a packet length of $L = 1500$ bits. When using a transmission data rate of $R = 10$ Mbps, the packet transmission time is equal to $T = 150$ [μs]. For a broadcast net span of $d = 2.5$ km, the propagation delay is equal to $t_p = 12.5$ [μs]. The CD time is equal to about $t_{CD} = 2t_p = 25$ [μs], which is lower than the packet transmission time, yielding an effective CD operation. Such a setting is of the order of the setting that was undertaken by initial implementations of the Ethernet LAN.

In turn, when the transmission data rate is increased to 100 Mbps, as commonly used by later implementations, the packet transmission time is reduced to $T = 15$ [μs], while the CD time is not changed (as it includes the net propagation time as a key component, and the value of the latter does not depend on the data rate's level), so that it is still estimated to be equal to $t_{CD} = 25$ [μs]. Hence, $t_{CD} > T$, so that the CD operation is not effective (noting that by the time that the station has determined that its packet is experiencing a collision, the packet's transmission has already been completed). To restore the effectiveness of the CD process, one may reduce the span of the net. For instance, it may be shortened to a range that is of the order of 100–200 m.

As the transmission data rate is further increased, say to 1 Gbps, we note that $T = 1.5$ [μs]. To reduce the CD time to a value that is much lower than this time, the corresponding span needs to be shorter than 100 m. Assuming an acquisition time that is of the order of $3t_p$, we obtain $a \approx 3t_p / T = 1$. As noted when discussing the CSMA scheme, such an a value is too high, as it leads to inefficient CSMA throughput performance. To enhance performance, the broadcast net's

Figure 8.18 Ethernet Local Area Network (LAN) Single Broadcast Domain Segment Configurations: (a) Broadcast Bus Segment; (b) Repeater Hub Base Segment.

span would have to be shortened to a much shorter range, possibly spanning a range that does not exceed several meters.

The CSMA/CD medium access control (MAC) protocol has been employed by the Xerox Corporation in its development of the **Ethernet LAN**. It was commercially introduced in 1980 and standardized in 1983 as an IEEE 802.3 Standard recommendation. The Standard specifies the physical and medium access control (MAC) protocol layers. It has become over the years the dominant system in use as a wireline LAN. It has gained favor in comparison with other LAN implementations that include Token Ring and FDDI. It is a shared medium network that uses wireline media, including coaxial cable and twisted-pair wires. In addition, fiber-optic wires can be employed in conjunction with Ethernet switches. Transmission data rates have been increasing from around 1–10 Mbps to multiple 100 Gbps rate levels.

An Ethernet implementation over a (bi-directional) bus medium is illustrated in Fig. 8.18(a). Each station is attached passively to the shared medium bus through its medium access unit (MAU). A message frame transmitted by a station propagates across the bus segment in both directions so that it is broadcast to all attached stations. The bus is terminated at its edges so that signal reflections are avoided. The bus segment forms a broadcast collision domain so that time-simultaneous frame transmissions would incur collisions. Each frame includes destination and source MAC addresses.

A station that detects the transmission of a frame across the medium would copy it and perform a filtering action by searching a match between the frame's destination MAC address and its own address. When such a match is detected, the station proceeds with copying and processing of the frame.

In Fig. 8.18(a), station S1 is shown to transmit a frame across the bus. The frame propagates across the full span of the bus medium and is received by all attached stations, S2–S5, which are members of a single broadcast domain. This **broadcast property**, induced by the physical layer character of the shared medium, is a key feature of such LAN segments. Thus, a frame placed on the bus by a station is automatically received by all other net stations attached to the bus medium.

A wide range of applications that make use of this broadcast property have been developed for use by such LANs. If the source station wishes all segment stations to pass its frame to a higher protocol layer, such as an application layer entity, a broadcast address is included as the destination MAC address (typically by setting an all 1's address). The broadcasting process is rather straightforward here as a single frame transmission induces reception of the frame by all attached stations.

For ease of implementation of media cabling in office, industrial, and other facilities, the shared wiring layout is typically implemented in the form of a **physical star that induces the features of a logical bus**. Such a configuration is demonstrated in Fig. 8.18(b). Every station is connected to a central module that acts as a **repeater hub**. A frame transmitted by a station across the wire that connects it with the repeater is reflected by the hub module across all other interface wires.

In Fig. 8.18(b), the frame transmitted by station S1 is reflected by the repeater module across all other wires so that it is received by all other stations. Simultaneous transmissions by several stations would collide at the repeater module, so that this configuration forms a single collision domain. The reflection process performed by the repeater is a physical layer function, so that no packet-oriented store-and-forward operation is enacted. The repeater just extends the broadcasting span. It does not serve to isolate transmissions executed across the hub's branches. Such a star configuration simplifies the wiring plant and the implementation of management, expansion, and maintenance functions.

Ethernet implementations that use higher data rates tend to employ configurations that do not induce packet access contentions and collisions. The CD mechanism is then not exercised (as no collisions take place) and the effectiveness of the carrier sensing function, and hence the impact made by the value assumed by the acquisition factor, is not at issue. The Ethernet protocol is then often used to implement a transmission scheme across a link that provides for the interface between two stations, say station A and station B. To avoid collisions, the interface between the two stations is often implemented by using two half-duplex links, one from A to B and the other one from B to A. In this case, each link is dedicated for use by just a single active station so that access contentions and collisions are avoided. Under such a *full-duplex Ethernet* connection, the link's span and throughput rate can be upgraded substantially.

As discussed above, when implementing a LAN that operates at a much higher data rate or that requires a wider message dissemination span, the underlying carrier sense-induced acquisition and CD factors would assume higher values, leading to significant performance degradation. It is then beneficial to subdivide the LAN's layout by forming subnetworks.

Multiple LAN segments are interconnected by a switching node. Messages produced by a station are transmitted across its shared medium LAN segment and are then forwarded to a switching node. At the switch, messages are processed in a store-and-forward manner. Thus, an arriving message is stored in a buffer, joining a queue of messages as it awaits the switching processor to examine each frame's destination address and switch it for transmission across the proper outgoing segment. The MAC protocol required for accessing the outgoing medium is then employed for managing the access of the message across the medium of the outgoing segment.

This is illustrated in Fig. 8.19. A bus-based LAN that provides shared medium access to stations S3–S6 is also used to transport messages to/from a switch node (SN). This switch node is shown to attach to the latter LAN at port P4. Frames that are broadcast across this LAN are also received by the switch node, and may subsequently be switched by it for distribution across one or several outgoing segments. For example, a message transmitted by station S5 whose destination is station S1 will be broadcast across its LAN segment, captured by the switch node, and then (in a store

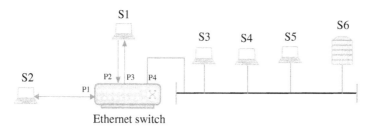

Figure 8.19 Switched Ethernet Local Area Network (LAN).

and forward manner) be switched to port P3 for direct transmission to S1 across a non-shared link. A response message issued by station S1 would be transmitted to the switch node across a direct non-shared link that is attached to the switch at port P2. The switch node would then switch it to port P4. It would then be handled by the MAC module of the switch node, which manages transmissions across the shared medium LAN segment, using the LAN's access protocol (such as an IEEE 802.3 Ethernet protocol). It will be broadcast across the LAN's bus and be captured and processed by the intended destination station.

It is noted that the underlying LAN segment operates over a reduced distance span, leading to improved performance behavior of the underlying multiple access net.

We show stations S1 and S2 to be connected to the switch node (SN) through half-duplex and full-duplex link attachments, respectively. The full-duplex LAN segment that connects S1 and the Switch Node consists of two half-duplex links: one that is employed for directional transmissions from S1 to SN, connecting to SN at port P2, and the other one that is used by directional transmissions from SN to S1, connecting to SN at port P3. As the resources of each link are exclusively used by a single station, no frame collisions take place. Therefore no performance limitations are induced by the involved acquisition factor and CD time parameters, and consequently no induced span and data rate limitations are imposed.

Hence, the access segment involving SN and station S1 can be implemented to operate at a significantly higher data rate and span longer distance ranges. Though the related interface is often identified as one that follows the Ethernet specification, no limitations that are related to the performance efficiency of the CSMA/CD MAC protocol are involved. High-speed Ethernet segments are thus often configured in this manner, whereby each segment accommodates the access of only a single station. Using fiber-optic links, such full-duplex Ethernet access links are set to operate at higher rates and accommodate longer spans.

8.4.5 The Carrier Sense Multiple Access with Collision Avoidance (CSMA/CA) Protocol and the Wi-Fi Wireless Local Area Network (WLAN)

The Idea: *Listen before talking and if the medium is not occupied, wait for a designated period of time and then when your turn is reached jump in and talk. Wait to receive an immediate ACK to confirm that your message was received. If not confirmed, try again after a random backoff time.*

Performance Features: *As a CSMA-based protocol, a lower acquisition factor leads to a higher-throughput rate. Higher-priority messages can be granted faster access. Dense wireless LANs that produce high-throughput rates are synthesized by using a hybrid DA/FDMA/TDMA/SDMA multiple access scheme.*

8.4.5.1 WLAN Layout and Shared Wireless Medium Resources

IEEE 802.11 Standards have been widely used since around 1997 for the implementation of a wireless WLAN. Wireless frequency bands have been allocated for this purpose. Allocations include the 2.4 GHz ISM frequency band which is segmented into 5 MHz wide channels, and the 5 GHz band that consists of 20 MHz wide channels. Multiple channels can be combined to allocate higher bandwidth for message transmissions and achieve higher throughput. In the 2.4 GHz range, the 802.11a or 802.11g Wi-Fi versions can bond four channels to offer stations 20 MHz wide bandwidth, while 802.11n also allows the bonding of eight channels to provide a 40 MHz band. In the 5 GHz frequency band, typical channel bandwidth allocations include 20, 40, 80, and 160 MHz. Various versions of IEEE 802.11 protocol specifications at the physical and link layers have been used to form wireless LANs that have been promoted by the nonprofit Wi-Fi Alliance as Wi-Fi networks, using their trademarked Wi-Fi identifier.

A Wi-Fi WLAN system provides for wireless communications of station members. Computers, smart phones, embedded sensor modules such as those used by IOT systems, and a wide range of other stationary or mobile devices can become members of Wi-Fi (and other) WLAN nets. Each member contains a Network Interface Card (NIC) or an embedded built-in on-chip module that implements the physical and link layer entities specified by the corresponding IEEE 802.11 protocols, often identified as the device's station (STA). The latter implements the relevant modulation/coding, multiple-access, and link layer protocols that are used to communicate messages (formatted as frames) across the WLAN.

As a LAN, the distance span between member stations is limited, so that stations that are members of a single WLAN reside in close proximity. A typical transmit power used by a Wi-Fi radio module is of the order of 100 mW (or 20 dBm). The distance span per WLAN net is longer when using the lower frequency band. Coverage is typically limited to 100's of meters. In comparison, shorter ranges are involved when considering low power wireless personal area network (PAN) systems, such as Bluetooth.

The expectation is for the stations that are members of a single net to form a single **broadcast domain**, so that a transmission by a station would be generally detected at all other stations that are members of the net (and possibly even such that are members of a nearby net). At times, certain stations may be *hidden* from other stations in that they are not able to receive message transmissions performed by certain other stations.

A Wi-Fi WLAN can operate in one of the two configuration modes. When employed in an **infrastructure configuration** mode, its operation is managed by a central station, known as the net's **Access Point (AP)**. The AP effectively acts as a BS in the sense that every member of a net must then be able to directly communicate with the AP. Net stations do not communicate directly with each other. Rather, they communicate directly only with their net's AP station. The AP is often connected to a Router that provides it with the capability to access a backbone network, identified as the underlying Infrastructure, such as the Internet. A Wireless Router embeds the functionality of an Internet Router and of a Wi-Fi access module. The latter implements the corresponding protocols that allow it to function as a Wi-Fi AP. As such, it is able to communicate with station members of its WLAN.

Wi-Fi WLANs can also operate in an **Independent or Ad Hoc mode**. Under such a setting, net station members can communicate directly with each other without having to go through an AP. The corresponding ad hoc networking protocol is however much more complex and is not commonly employed. Our discussion here focuses on a WLAN net that is configured to operate in an Infrastructure mode, whereby each WLAN net is managed by an AP station.

As is the case for wireline LANs such as Ethernet LANs, communications between stations that are members of a single LAN are performed by using MAC layer addresses, which are included within each MAC frame. MAC addresses are assigned by device manufacturers. They are also referred to as hardware or physical addresses. MAC addresses are formed according to the principles of two numbering spaces based on Extended Unique Identifiers (EUI) managed by the Institute of Electrical and Electronics Engineers (IEEE), EUI-48 and EUI-64. The commonly employed EUI-48 is a 48 bits long MAC address, typically represented in six groups, whereby each group contains two hexadecimal digits. Groups are separated by hyphens or colons. For example, the MAC address FF:FF:FF:FF:FF:FF represents in hexadecimal notation the all 1's broadcast address (noting that in hexadecimal notation, F represents the value 15, which in binary notation is represented as a series of four 1s).

The MAC address of an AP station identifies the WLAN net that it manages. A net managed by an AP station is defined as a **Service Set**. Its name is identified by a **Service Set Identifier (SSID)**. A frame transmitted by an AP across the WLAN shared wireless medium that is destined

Figure 8.20 A Station Hears Two AP's and Associates with a Selected One.

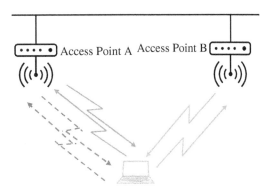

to a member station includes the net's AP's MAC address as the transmitter's address and the destination station's MAC address as the receiver's address. A frame transmitted by a station across the WLAN's wireless medium that is destined to the AP includes the source station's MAC address as the transmitter's address and the AP's MAC address as the receiver's address. Stations use the receiver's MAC address that is included in a header field of a frame to filter frames transmitted across the net's wireless medium, as they match this address with their own MAC address. When matched, a station would process and pass the received frame to higher layer protocol entities.

As shown in Fig. 8.20, a station may be able to hear announcements made by two (or more) APs. The station would send a Probe message to each AP. Upon receiving probe responses, the station would select an AP to associate with, sending it an Association Request and subsequently receiving an association response. At a later time, the station may move and re-associate with another AP station.

8.4.5.2 Frame Types

The following frame types are transmitted across the shared medium of a Wi-Fi network: Data frames, Control frames, and Management frames.

A *data frame* carries in its payload field the data packet that the transmitting station transports across the WLAN. As a MAC layer PDU, the data frame's payload is typically a Logical Link Control (LLC) (generally formatted in accordance with IEEE 802.2) link layer PDU.

The LLC frame header field includes a Destination Service Access Point (DSAP) identifier, which is typically followed by a Subnetwork Access Protocol (SNAP) header. These identifiers are used by the receiving entity to pass a received frame to a device driver that can intelligently handle the designated protocol. The LLC payload contains the data message, such as an IP packet.

Control frames are used to facilitate the transport of data frames between stations. Included are ACK, RTS, and CTS frames (which are defined in the following).

Management frames are employed to maintain communications and to provide for the formation, authentication and disconnection of connections. Commonly included are authentication and association (of a station with an AP) frames, as well as a multitude of action frames.

A key management frame is the **Beacon frame**. It is transmitted periodically by the AP, every Beacon Interval. It contains essential information about the WLAN net that its station members must have. The time at which an AP sends a beacon is known as a Target Beacon Transmission Time (TBTT). It is used to announce the presence of this WLAN (with its identified SSID), and it contains a preamble that is used by stations to synchronize their clocks to that of the AP, as well as other management data.

When using a Power-Save mode of operation, the beacon identifies, in a Traffic Indication Map (TIM) data set, the stations that should awaken themselves and proceed to fetch data that is stored for them at the AP. A beacon also contains information about the capability attributes of the network and its devices, encryption configuration, its possible support for polling mode operations, and a multitude of parameter sets such as those used for Contention-Free (CF) operations.

8.4.5.3 Distributed Coordination Function (DCF): The Basic CSMA/CA Medium Access Control Scheme

The fundamental multiple access scheme defined by the 802.11 standard and used by Wi-Fi nets is identified as the **DCF**. It employs a **Carrier Sense Multiple Access with Collision Avoidance (CSMA/CA)** algorithm, which is based on the following principle of operation.

Definition 8.6   The Distributed Control Foundation (DCF) – a CSMA/CA multiple access algorithm.

A station that is a member of an infrastructure-based WLAN and has a packet (formatted as a MAC frame) that it is ready to transmit across the shared wireless medium carries out the following operations [23]:

1) The station performs Clear Channel Assessment (CCA) by sensing the medium to determine whether it currently carries a frame transmission (and is then declared to be busy) or not (and is then declared to be idle).
   a) As noted when discussing the functioning of a CSMA scheme, a frame transmitted by a net station across the shared medium will be sensed by other net stations within a period of time of duration t_a, which we have identified as the medium acquisition time. Under DCF, the corresponding time is defined as the **Slot Time**, also identified at times as a Slot or a Minislot. The slot's duration t_S is a parameter, denoted in the Standard as aSlotTime. Its value depends on the protocol's version. For example, for version 802.11b, $t_S = 20$ μs while for versions 802.11a and 802.11ac, $t_S = 9$ μs.
2) If the medium is sensed to be busy, the station defers its transmission, scheduling its subsequent sensing of the medium to take place after a period of time that is identified as a **Network Allocation Vector (NAV)**. The NAV is set equal to the time period specified in the **duration field** of the sensed frame. In this field, the transmitting station specifies the period of time that it reserves the medium for its own use. This period accounts for the time that is needed to transmit its frame, and possibly several consecutive frames, when allowed, supplemented by the time occupied by the transmission of an Acknowledgment (ACK) frame by a destination station.
3) If the medium is sensed to be idle, the station waits for a period of time that is identified as **Inter Frame Space (IFS)** before taking any further action. The IFS duration that a terminal uses depends on the type of frame that it wishes to transmit:
   a) To send a highest priority frame, such as an Acknowledgment frame (ACK), the station uses a **short IFS (SIFS)**.
   b) To send a medium priority frame, such as a time critical frame that is used when employing the protocol's Point Coordination Function (PCF) mode of operation, the station uses a **PCF IFS (PIFS)**.
   c) To send a regular frame, such as an asynchronous data frame that is transmitted in accordance with the DCF scheme, the station uses a **DCF IFS (DIFS)**.
   d) Typical settings of IFS parameters are noted as:

$$PIFS = SIFS + SlotTime,$$

$$DIFS = SIFS + 2 \times SlotTime.$$

8.4 Random Access: Try and Try Again | **301**

For example, for version 802.11b, $SIFS = 10\,\mu s$, $DIFS = 50\,\mu s$ while for versions 802.11a and 802.11ac, $SIFS = 16\,\mu s$, $DIFS = 34\,\mu s$.

4) If the medium remains idle for the corresponding IFS period, and if the station is currently not in Backoff Phase, then the station selects a random backoff slot from the interval $[0, CW]$, where CW is the duration of the current Contention Window (CW). A **Backoff Timer** is configured, where its value is set to equal the residual time (within a CW period) that the station must wait until its selected slot is reached.

The contention window parameter (CW) is set equal to $CW(n)$ for calculating the backoff for a frame that is transmitted for the nth time, $n \geq 1$. The initial value is set equal to the parameter CW_{min}. Its maximum value is specified by the parameter CW_{max}. The selected CW assumes an integer value and the actual CW time is calculated as CW $\times$ SlotTime. When transmitted for the $(n + 1)$st time, the contention window is calculated by using a binary exponential backoff formula ($n \geq 1$):

$$CW(n + 1) = \min\{2 \times CW(n) + 1, CW_{max}\}.$$

5) The backoff timer is updated by an active station by reducing its value by 1 for each idle slot that is sensed during the backoff phase. If the medium turns busy during the backoff phase, the timer is stopped. Based on a duration field specified within the sensed frame, a new NAV is configured. The station must refrain from scheduling a transmission during this NAV period. Subsequently, the backoff process resumes, setting its timer's initial backoff value to be equal to the timer's value that was recorded at the recent time that the timer's countdown was stopped.

6) When the backoff timer value reaches the value 0, the station proceeds to transmit its frame.

7) Following its frame transmission, the station waits for a SIFS time as it expects to receive by then an ACK frame from the net's destination station. Since an ACK frame is recognized as the highest priority frame, the destination station is able to access the medium after a SIFS period and transmit then its ACK frame. This creates a contentionless access to the medium, noting that the underlying data frame is assumed to be a unicast frame, so that only a single destination station is targeted with its reception. A broadcast transmission of a multicast frame is not acknowledged.

8) If an ACK is received (within the specified SIFS period), the station removes a replica of its transmitted data frame from its buffer. It may be allowed to subsequently transmit another frame. Otherwise, it restarts the above process in determining the access time of its next frame, if any.

9) If no ACK is received, the station schedules the retransmission of its frame. It follows then the above stated process, whereby the selected contention window parameter is now set to CW(n) if this is the nth transmission of the underlying frame. The system prescribes a parameter that specifies the maximum number of frame transmission trials that a station can undertake. The previous state of the countdown backoff timer is reset, assuming now a new value that is randomly selected from the new contention window.

An illustrative frame transmission process that follows the contention-based access mode of the CSMA/CA DCF and EDCA protocols is depicted in Fig. 8.21. The shown $AIFS(i)$ value is an IFS parameter that is used by a frame that assumes a priority-i level under the Enhanced Distributed Channel Access (EDCA) scheme that is described in the following. A busy station defers access to the medium in accordance with its associated IFS period. It then configures a contention window and selects at random a backoff slot. Starting with this value, it decrements its backoff timer by 1 each time that it senses a CW idle slot. The scenario illustrated in Fig. 8.21 shows that following deferral by a DIFS period, frame $(n + 1)$ is transmitted at the start of a selected slot located within

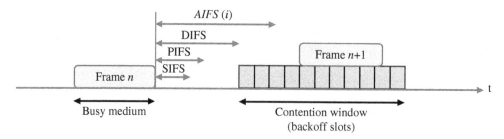

Figure 8.21 Illustrative Frame Transmission Process Under the Contention-Based Access Mode of the CSMA/CA DCF and EDCA Protocols.

its contention window as it has sensed no other frame transmissions to occupy any of the window's preceding time slots. In this case, no other frame transmissions have occurred during its selected time slot, so that frame $(n + 1)$ is noted to incur no collisions.

8.4.5.4 Point Coordination Function (PCF): A Polling-Based Contention-less Access Scheme

The IEEE 802.11 standard also defines an optional access method that is identified as a **Point Coordination Function (PCF)**. Under PCF, the AP station acts as a polling manager. The AP uses its priority access IFS parameter (PIFS) to access an idle medium when it desires to do so in a contention-less manner. It sends a control message as a **beacon** that notifies the net stations that it is initiating at this time a **contention-free access period (CFP)**.

It then sends a *polling frame* that permits a designated station to gain contention-free access to the medium for a specified period of time. In this manner, it is able to provide timely access to the medium to stations that desire to access the medium at particular times (or within specified periods). The AP notifies net stations when it wishes to terminate a currently ongoing CFP operation. This notification cancels a previously issued NAV announcement that prevented net stations from accessing the medium by using the contention based access scheme. As noted in Fig. 8.22, this termination time is selected in a manner that does not interrupt an ongoing transmission. The CFP is followed by a contention access period during which busy stations (including the AP) determine their access times by following the CSMA/CA DCF protocol.

8.4.5.5 Alleviating the Hidden Terminal Problem: An Optional RTS/CTS Scheme

To alleviate (but not fully resolve) the hidden terminal problem, DCF specifies an optional virtual carrier sense mechanism under which source and destination stations exchange short Request-To-Send (RTS) and Clear-To-Send (CTS) frames prior to the start of frame transmission.

To illustrate, assume that WLAN net stations can hear and have direct access to the AP station but that certain stations cannot hear each other. Before scheduling the transmission of its data frame, a ready Station A schedules a transmission to the AP of a short Request-To-Send (RTS) frame. If the AP is currently sensing no other transmission activity, it will respond with the transmission of a

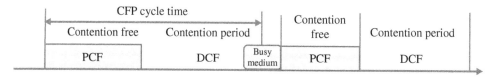

Figure 8.22 Alternating Point Coordination Function (PCF) Contention Free (CFP) and DCF Contention-Based Periods.

Clear-To-Send (CTS) frame. This CTS frame transmission is received by all net stations, informing them that Station A has reserved the medium for a specified NAV time. This prevents other stations from initiating data transmissions that can interfere with the reception of the frame sent by Station A. Assume that Station B cannot hear the data transmissions issued by Station A. Without the use of the RTS/CTS process, Station B may initiate its own data transmission in a period during which the AP is occupied in the reception of a Station A's data transmission.

An illustrative hidden terminal scenario and the associated conduct of a RTS/CTS dialog are shown in Fig. 8.23. Stations A and B cannot hear each other's transmissions but each can hear AP's transmissions. Under the illustrative scenario, station B hears the CTS message sent by the AP in response to the RTS message received from station A and consequently station B avoids accessing the medium at that time.

It is noted that even when the RTS/CTS dialog is used, it is possible for a frame transmission by Station B (including the transmission of its own RTS frame) to collide with Station A's RTS frame. However, the time window that allows such a collision is much narrower.

The RTS/CTS process may reduce the throughput performance efficiency exhibited by the WLAN net, as bandwidth is used in executing the RTS/CTS reservation dialog. Hence, Wi-Fi WLAN systems tend to typically not employ this process. Potential benefits may be realized in situations under which hidden station conditions are frequently incurred, and when relatively longer data frames are involved. Therefore, Wi-Fi systems provide for a network management parameter that specifies the minimum data frame length under which the RTS/CTS process is activated. When this parameter is set equal to the maximum allowable length of a data frame, the use of the RTS/CTS dialog is eliminated.

An illustrative dialog that consists of RTS/CTS transmissions followed by Data/ACK transmissions is shown in Fig. 8.24. Assuming the AP to be available to receive a new frame, the RTS frame transmission is followed, within a SIFS period, by a CTS frame transmission. It is then followed, within a SIFS period, by a data frame transmission. Assuming the latter to be received successfully, an ACK frame transmission would then be received by the sending station within a subsequent SIFS period. It is noted that by using the shortest IFS (SIFS) values, a station that transmits a frame during this dialog process is granted preferred and uncontested access to the medium.

Also shown in Fig. 8.24 are the Network Allocation Vector (NAV) values included in the RTS and CTS frames. These values serve as **virtual carrier sensing** indicators, informing other net stations that the transmitting station is reserving the medium for the specified time duration. The lengths set for these periods account for the duration of time required by the transmitting station to complete

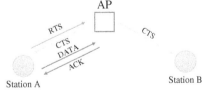

Figure 8.23 Illustrative Hidden Terminal Scenario and the RTS/CTS Dialog: Stations A and B Do Not Hear Each Other but Each Can Hear the AP.

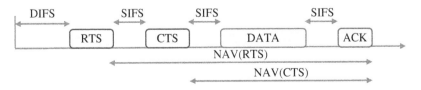

Figure 8.24 Illustrative RTS/CTS and Data/ACK Transmission Dialog in a Wi-Fi WLAN.

its dialog. If a station does not engage in the RTS/CTS component of the corresponding dialog, as is often the case, the employed dialog process consists of just data and ACK frame transmissions. The NAV value is then used to reserve access for the transmissions of the data frame and the succeeding ACK frame.

8.4.5.6 Hybrid Coordination Function (HCF): Providing QoS to Designated Traffic Categories (TC)

A medium access scheme called Hybrid Coordination Function (HCF) was defined by an IEEE 802.11e amendment to the standard. It specifies two methods for medium access: Enhanced Distributed Channel Access (EDCA) and HCF-Controlled Channel Access (HCCA). Aiming to provide QoS-based medium access service, Traffic Categories (TC) are defined. Access priority levels are associated with Access Categories (ACs). Higher-priority messages are provided faster access and higher throughput. For instance, voice and video frames can be configured to be given higher priority in gaining access to the medium.

The EDCA scheme is an extension of the basic DCF CSMA/CA scheme. The following modifications are included:

1) A shorter arbitration inter-frame space (denoted as AIFS) is used for access by higher-priority frames.
2) Contention window (CW) slot assignments are set in accordance with the traffic rate levels that are expected in each access category. The set values for aCWmin and aCWmax depend on the underlying priority level. They are configured in a manner that provides an earlier access for the transmission of a higher priority frame. By setting a contention window interval [aCWmin, aCWmax] that is allocated to higher-priority frames to span lower boundary values, one provides faster access to such frames.
3) EDCA provides contention-free access to the channel during a Transmit Opportunity (TXOP) period. A station can send multiple frames during a time period that does not exceed the TXOP time interval. When TXOP is set to 0, a station is allowed to send only a single MAC frame at a time.

In Fig. 8.25, backoff window default settings are shown per Access Category (AC), based on IEEE Standard 802.11 document specifications [23], with TXOP values presented under Clauses 15 and 18. The values shown for the CW_{min} and CW_{max} parameters are configured by a system management entity. Illustrative access parameters that are configured in a specific system are shown to assume the values $CW_{min} = 15$ and $CW_{max} = 1023$, as used by 802.11a and 802.11n.

For the illustrative system, four Access Classes (AC) are configured. Preferred access is granted in the following (decreasing priority) order, per traffic type: Voice(AC_VO), Video (AC_VI), Best-Effort (AC_BE), and Background (AC_BK). The underlying parameters configure the location of the contention window (whereby an earlier window location is set for higher priority

AC	CW_{min}	CW_{max}	AIFSN	TXOP limit
AC_BK	aCWmin	aCWmax	7	0
AC_BE	aCWmin	aCWmax	3	0
AC_VI	(aCWmin+1) / 2 − 1	aCWmin	2	6.016 ms
AC_VO	(aCWmin+1) / 4 − 1	(aCWmin+1) / 2 − 1	2	3.264 ms

Figure 8.25 Default Access Parameters Under 802.11e per Access Class(AC): Backoff Window Ranges and Illustrative Access Parameter Settings. Source: Adapted from [23].

traffic classes), the IFS length (recalling that a shorter IFS provides higher access priority) and the Max Transmit Opportunity (TXOP) duration, which limits the total time that a station can continuously use, once it has gained medium access, to transmit its frames.

The hybrid coordination function (HCF)-controlled channel access (HCCA) is a polling oriented access scheme that is similar to PCF. Its implementation is not mandatory and it is not used as often as EDCA or DCF. In contrast to PCF, where the interval between two beacon frames is divided into two periods, contention free period (CFP) and contention period (CP), the HCCA allows for CFPs to be initiated at nearly anytime during a CP. Such a CFP is identified as a Controlled Access Phase (CAP). A CAP is initiated by the AP whenever it wishes to send a frame to a station or to receive a frame from a station in a contention-free manner. During a CAP period, the AP, identified as the Hybrid Coordinator (HC), controls access to the medium. During the CP, all stations function in an EDCA operational mode.

A PCF scheme can determine access of a frame based on the frame's Traffic Class (TC). Traffic Streams (TS) are defined and supported. The HC can provide a flow-based access service. Stations can provide information to the HC about the lengths of their queues for each Traffic Class (TC). The HC uses this data to plan the order and rate at which it sends polls (identified as CF-Poll frames) to busy stations. It schedules access times, allocates medium resources, and grants TXOP duration levels to busy stations in regulating the access of their frames and streams.

Under IEEE 802.11e, the **Power-Save** Polling mechanism that was defined under previous 802.11 versions is extended. It is identified as an **Automatic Power Save Delivery (APSD)** scheme. Under previous versions, stations in power saving mode are required to wake up to listen to beacon messages. A station then finds out whether the AP holds messages that need to be delivered to the station. If so, the station will send a message to the AP requesting it to send it these messages.

Such an operation may induce collisions among request messages, leading to reduced throughput rates. Under 802.11e, two methods are defined for initiation of the delivery of frames by the AP to a station that operates in a Power Saving mode: scheduled APSD (S-APSD) and unscheduled APSD (U-APSD).

A station that is configured to operate in a power saving mode enters a *doze state*. During a doze period, the station is not available to receive frames but is still able to receive certain control and management signals. To send a frame, a station would wake up to initiate the transmission of the frame. In this manner, the bulk of its energy resources are saved.

Under S-APSD, which can be used by both EDCA and HCCA, a predetermined time schedule is set for the initiation of service periods, at which times a station will automatically wake up. During a service period, the AP can transmit all the frames that it has buffered for the underlying station. Under U-APSD, which is available under EDCA, the sending of a frame by a station to the AP triggers the start time of a service period.

Due to the current widespread embedding of Wi-Fi chips in many types of devices, there exists an increasing demand for supporting much higher throughput, requiring the implementation of higher bandwidth efficient methods as well as the allocation of higher bandwidth resources. Many public and private facilities make use of multiple WLAN nets, each managed by its AP station.

As demonstrated in Fig. 8.26, it is often the case that multiple WLAN nets are located in close proximity to each other. Recent extensions of Wi-Fi network systems physical layer and MAC layer mechanisms accommodate such dense layout configurations. Bandwidth efficient cross-layer methods are employed. Included are high-performance physical layer methods and access schemes that dynamically configure multiple beam antenna arrays. Also included are advanced MAC layer scheduling schemes that make use of joint FDMA/TDMA/SDMA-oriented demand-assigned algorithms. Several such developments are discussed in Chapters 17–21.

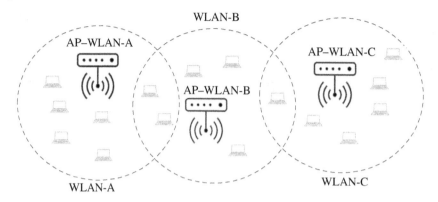

Figure 8.26 LANs in Close Proximity.

Problems

8.1 Describe the operation of each one of the following fixed multiple access schemes. Also, identify the key parameters and structural elements for each scheme.
  a) Time Division Multiple Access (TDMA).
  b) Frequency Division Multiple Access (FDMA).
  c) Hybrid FDMA/TDMA.
  d) Code Division Multiple Access (CDMA).
  e) Space Division Multiple Access (SDMA).
  f) Hybrid FDMA/TDMA/SDMA.

8.2 Describe and discuss the differences in the operation and performance behavior of fixed multiple-access schemes vs. demand-assigned multiple-access procedures.

8.3 Describe the concept of operation of reservation multiple access schemes. Identify the roles of the networking protocols used to schedule transmissions within the data plane and the reservation control plane.

8.4 Describe the operation of a hybrid DA/TDMA/FDMA/SDMA protocol.

8.5 Describe the operation of hub polling and token-passing polling schemes and compare their performance features. Illustrate their operation in acting to regulate access to the shared medium of a ring network.

8.6 Identify the operational and performance features of the following dwell-time methods that are used by polling schemes when regulating the access of stations to a shared medium:
  a) Exhaustive polling.
  b) Gated polling.
  c) Limited-to-k (chained) polling.

8.7 Provide examples of systems that employ polling and implicit polling methods.

8.8 Describe the principle of operation of the following shared medium networks:
a) Slotted bus.
b) Positional priority ring.

8.9 Define the **walk time** parameter of a polling network system. Describe how its value impacts the throughput efficiency attained by the network system.

8.10 Consider a token-passing ring network that connects 10 stations. The stations are loaded by statistically independent message arrivals. Stations are loaded at equal message arrival rates. Assume an exhaustive dwell time method to be employed. The data rate of a communications link that connects two neighboring stations is equal to 10 [Mbps]. The token walk-time between two neighboring stations is equal to 0.01 [ms]. The average packet length is equal to 1000 [bits/packet].
a) Calculate the maximum attainable level for the message throughput rate.
b) Calculate the average cycle duration by assuming that the system is (symmetrically) loaded at 80% of its total throughput capacity.
c) Assume that the dwell time method is changed so that each station now uses a limited-to-1 protocol. Calculate the ring system's total throughput capacity. Determine the average cycle time when the system is loaded at 80% of its total throughput capacity rate.

8.11 Define the following multiple access algorithms and identify the performance behavior features of each:
a) Unslotted ALOHA.
b) Slotted ALOHA.
c) CSMA.
d) CSMA/CD.

8.12 Consider a multiple access wireless medium that is shared among a relatively large number of stations by using a Slotted ALOHA algorithm. The slot duration is equal to a packet transmission time (assuming fixed size packets). The normalized message loading rate is measured to be equal to $S = 0.3$ [packets/slot].
a) Calculate the medium's carried load G [packets/slot].
b) Calculate the average number of times that a packet is transmitted until its transmission is successfully received.
c) Estimate the average delay time (expressed in slots) of a packet, which is measured from the instant of the start of its first transmission to the time that its successful transmission ends, assuming an average retransmission backoff time that is equal to 12 slots.

8.13 Describe the operation of a CSMA/CA protocol such as that used by the Wi-Fi DCF protocol.

8.14 Describe the operation of the Point Coordination Function (PCF) protocol used by a Wi-Fi system. Also describe the method used to switch from a CSMA/CA contention based mode to a polling mode, and vice versa.

8.15 Describe the operation of the RTS/CTS scheme that can be used by a Wi-Fi net and discuss its potential for alleviating the hidden terminal problem. Provide an example of a network configuration that illustrates the functioning of this scheme. Discuss conditions under which the RTS/CTS process would be configured to not be performed.

8.16 Describe the operation of the Hybrid Coordination Function (HCF) protocol of a Wi-Fi net. Provide examples of settings of AIFSN, TXOP limit, and (CW_min, CW_max) parameters that enable the system to achieve a priority-based medium access operation.

9

Switching, Relaying, and Local Networking

9.1 Switching

The Function: *A network supports the transport of many different flows. Different message flows may traverse different routes to get to their destinations, while sharing network links. To enable such an operation, switching nodes are used. A switch moves a message that arrives at an incoming port via an incoming communications link to a proper outgoing port and forwards it to the next node by transmitting it across an outgoing communications link.*

The Process *To execute its function, a switch must determine for each incoming message the outgoing port to which it should be directed. For this purpose, the switch maintains a switching table that contains forwarding entries. Under a circuit switching method, a signaling system is used to select ahead of time, prior to the start of data transport, the route to be used for a given call, as well as to allocate to it communications resources, forming a circuit. If the call is active in producing messages on a steady basis, the bandwidth resources of the configured circuit will be efficiently utilized. In turn, under a packet switching method, a packet switch that acts as a router determines the best route to use for routing a packet to a specific destination by making use of network performance data that it continuously collects. This information is used by the router to determine and update the routing entries that it keeps in its routing table. An entry identifies the output port that should be selected for reaching a specified destination entity. By examining the destination address specified in the header of an incoming packet, the proper routing entry is selected and the incoming packet is accordingly switched to an output port, where it joins a transmit queue. A packet flow occupies communications networking resources only when it is active in producing packets for transport across the network. In a local network that consists of interconnected local area networks, a layer-2 switching device, known as a bridge, is often used. It determines the output LAN to which an arriving MAC frame should be switched. For this purpose, it examines the source and destination MAC addresses included in the incoming frame and determines the output port by making use of observations of traffic flows that traverse attached links.*

Network systems make use of switching nodes for the transfer of message flows between source and destination nodes as they are transported across selected routes.

Why do we use switching nodes in a network system?

Consider a transportation system. I am scheduled to make a presentation at a conference session. I plan to drive from my office in Los Angeles, California, to a conference that is held in San Francisco, California. I will have to use multiple local roads and several high-speed highways (identified in California as Freeways) to reach my destination. On my way, I will encounter multiple road intersections. At such a junction, I will **switch** from an incoming road to an outgoing road, aiming to drive along the roads that make up my selected route. As illustrated in

---

Principles of Data Transfer Through Communications Networks, the Internet, and Autonomous Mobiles, First Edition. Izhak Rubin.
© 2025 The Institute of Electrical and Electronics Engineers, Inc. Published 2025 by John Wiley & Sons, Inc.

Figure 9.1 Switching Highways at an Intersection.

Fig. 9.1, when I cross an intersection of highways, I continue on the currently used road or switch to another road in following my preferred route.

How does the topological layout of a network look like if no switches are used?

Consider a system that consists of six edge nodes, to each of which an end-user station is attached. In Fig. 9.2, we show a network whose topological layout follows that of a *fully connected graph*. Each node is connected by a direct communications link to every other node. Assuming bidirectional communication links, a fully connected network of n nodes requires the use of $n(n-1)/2$ full-duplex links. Hence, the illustrated 6-node network employs 15 bidirectional communications links. If the number of edge nodes is equal to 100, the number of bidirectional communications links required by a fully connected network layout is equal to $9900/2 = 4950$.

Clearly, this is not a realistic implementation. It is generally too costly. Furthermore, an edge node does not usually require to communicate with every other edge node. In fact, most stations tend to communicate a high fraction of the time with a limited number of other stations. Furthermore, stations do not generally need to communicate simultaneously with more than a limited number

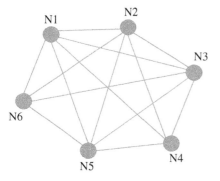

Figure 9.2 A Fully Connected Network Requiring the Use of No Switches.

of other stations. Consequently, if we were to use a fully connected network layout, many of the communications links will remain unused during a high fraction of the time.

The practical solution is to make connections between edge nodes **on demand**, as they would be used only when needed to accommodate current flows. To construct an affordable topological layout, a limited number of communications links would generally be employed. The resources offered by a communications link would then be **shared** by multiple message flows that traverse it.

A communications link is often referred to as a communications line. We use here these terms interchangeably.

The route used by a flow that originates at a source node and leads to a destination node usually consists of multiple links. A flow's message is transmitted across a link that connects two neighboring nodes and then, as it arrives to the subsequent node, it is switched to the proper outgoing link. In this manner, flow messages are navigated across their desired route by being switched at each intermediate node to the proper outgoing link. At each node, this redirection function is performed by a **switching module**, identified simply as a switch.

The process taking place at a vehicle intersection is similar to the one that is performed at a **switch**. Messages (or message flows) and communications links play the role of vehicles and roads, respectively. Messages arrive to the switch across a single or multiple *incoming links*. Each incoming link is attached to the switch at a *switch port*. A single or multiple *outgoing links* are attached to the output of the switch. Each outgoing link is attached to the switch at a switch port (identified in the following as a port).

An incoming message aims to depart the switch across a specific outgoing link. When messages are multicasted or broadcast to multiple destinations, a message may wish to depart the switch across multiple outgoing links.

A message that arrives to a switch across a specific incoming line may have to (temporarily) wait in a message queue for its turn to be selected and processed by the switch's engine (i.e., the switching processor). The latter determines the targeted destination of the message and accordingly proceeds to **switch** the message to the appropriate output port(s) that connects the switch to the targeted outgoing link(s). A switched message may have to wait in an outgoing queue for processing and service by a transmission module prior to its transport across an outgoing link.

At a vehicle intersection, the switching process may be controlled by a management module that uses traffic lights to regulate the transit of vehicles. Vehicles may have to wait at an incoming road until permitted to enter the intersection and be switched to their designated outgoing road. Such a switching process may also be performed without the explicit use of traffic lights by programming vehicles to use sensor data to autonomously *self-navigate* themselves across an intersection in reaching their targeted exit roads.

Thus, the main function of a switching module is to *switch* messages that arrive across incoming communications lines to specified (or selected) outgoing communications lines.

An illustrative diagram of a switch module is shown in Fig. 9.3. Communications lines L1–L6 are shown to be attached to ports P1–P6. The **Switching Matrix** identifies the inter-port connections

Figure 9.3 An Illustrative Switch Module.

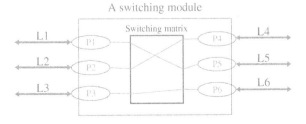

made by the switch. Bidirectional lines and connections are illustrated. Line L1, which is attached to the switch at Port P1, is shown to be configured (at this instant of time) to connect to Port P5, to which line L5 is attached. Under this setting, messages or flows that arrive across line L1 are switched to line L5. As will be discussed in subsequent sections, the setting of the Switching Matrix can be fixed or programmable, so that it can be modified at a relatively low rate, as typically performed by a network system manager or by a network's Signaling System. In contrast, switch connections can be dynamically configured so that switching configuration adjustments are performed at a high rate.

A communications network that employs switching modules to connect communications links and networks is identified as a **switching network**. An illustrative switching network is shown in Fig. 9.4. It displays switched pathways that are used by end user stations to communicate with other stations.

Terminal stations S11–S15 are connected to a shared medium Local Area Network (LAN1). Assume LAN1 to operate as an Ethernet local area network (LAN), employing a wireline bus as a shared medium. Switch node SW1 is attached to the Ethernet bus via port P3. It makes use of an Ethernet station module S_{SW1}. This module enables SW1 to access LAN1, sharing its medium resources. In this manner, SW1 receives Ethernet frames that are broadcast across the shared medium bus of LAN1. This module also enables SW1 to transmit designated frames across LAN1 by using the underlying Ethernet protocol.

The transport of frames between stations that are directly attached to LAN1 is carried out by using the broadcasting feature of the Ethernet LAN. As illustrated in the figure, we note that a MAC frame transmitted by station S11 across the medium is broadcast across the bus so that it is received by all the LAN's attached stations. A MAC frame contains a header that identifies its destination and source MAC addresses. Each station receiving a broadcast frame performs destination address filtering, checking for a match between its own MAC address and the destination MAC address included in the broadcast frame. When a match is detected, the station copies the frame and processes it. If the destination LAN station is associated with the targeted end-user terminal, the frame's data payload will be extracted. Alternately, the frame could be destined to a station that is not attached to the LAN. In this case, the frame is handled by switch module $SW1$, which will capture the frame and then switch it for transport across a path that leads to its ultimate destination.

A MAC address is typically formatted as 48-bit IEEE type address. It is also known as the underlying module's physical address. Each station is identified by a unique per-port MAC address and a station ID. A destination MAC address of a frame can also be set to identify a multicast or a broadcast address. The latter is typically configured to assume an all 1's format (denoted in hexadecimal

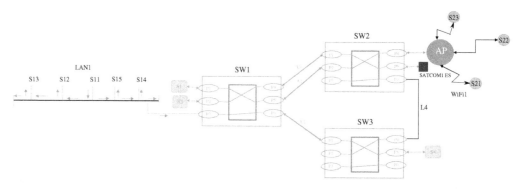

Figure 9.4 An Illustrative Switching Network.

notation as a succession of 12 hexadecimal 'F' symbols, as each such symbol represents a sequence of four 1-bits). Certain addresses serve as multicast addresses. Each multicast address identifies a specific multicast group, consisting of stations that have joined the group. All stations will copy and process a broadcast message sent across the medium, while only stations that belong to the designated multicast group will proceed to copy and process multicast frames that are associated with this multicast group.

While LAN stations are identified by MAC addresses, they are often also identified by IP addresses. The latter addresses identify each port of a network element and of an end-user station, in connection with the routing of IP packets that are transported across a packet switching network that employs the IP protocol, such as the Internet. An application is available to facilitate the ability of a station to attain a mapping between the IP address of a station's port and its corresponding MAC address. Such a service is provided by the Address Resolution Protocol (ARP) and by the inverse-ARP protocol. These schemes enable a station to find the MAC address of a destination station that is attached to the same LAN, given that it knows the IP address of that station, and vice versa. Similar services are available when networking protocols other than IP are employed.

Frames sent across LAN1 by station S11, which are intended for reception by a station that is not directly attached to LAN1, such as frames that are destined to station S4, are captured by a switching module, which typically acts a layer-2 bridge or as a layer-3 switch or router. Accordingly, switch SW1 is set to capture LAN1 frames that are not destined to a station that is directly attached to LAN1. As such, SW1 is assigned to act as the LAN's designated bridge or router, noting that there may be multiple switch modules that are attached to the LAN.

The frames that arrive at port P3 of SW1 are switched to port P6. Depending on the format of an incoming message processed by SW1, and on the configured format of its output message, incoming frames may keep their layer-2 format or may be mapped to a different format. In the illustrated scenario, outgoing messages are transmitted across line L2 to SW3. Depending on the types of messages that are handled by SW3, another reformatting transaction may be performed. Port P6 of SW1 is shown to connect across link L2 to port P1 of SW3. The latter is shown to be configured to switch the underlying messages to outgoing Port P5. The destination end-user terminal S4 is attached to the latter port and would consequently receive the destined messages issued by S11.

As an illustration of another end-to-end pathway, consider a message flow that originates at station S1 and is destined to wireless station S21. Under the shown configuration, S21 may represent a device that is a member of a WLAN denoted as Wi-Fi1. The Access Point (AP) of this Wi-Fi LAN is attached to port P4 of switch SW2. Messages produced by S1 are shown to be switched by SW1 from incoming port P1 to outgoing port P5. These messages are then transmitted across link L1 to port P2 of SW2. The latter switches these messages to outgoing port P4 of SW2. They are then passed to the AP module of Wi-Fi1, which will deliver them to the destination wireless station S21 via their transmission across its shared radio medium, in following the MAC protocol of the Wi-Fi system.

If, in turn, rather than a Wi-Fi system, the underlying wireless medium represents the Radio Access Net of a cellular Wireless Network, the pathway will be similar, except that the AP module will now be replaced by a Base Station (BS). A wireless medium sharing protocol will be used by the BS to schedule downlink message transmissions to each member station, including to station S21. A reverse pathway would be used for transporting messages originating at wireless device S21 to a destination station such as S1.

A pathway used by a message flow originating at station S2 is shown to be switched by SW1 and then by SW2, reaching outgoing port P5 at SW2. The latter is attached to a multiple access module (such as one that employs a DA/TDMA MAC scheme), which provides access to a satellite communications system, across which it is set to reach the destination Earth Station (ES) that is displayed as SATCOM1 ES.

How does a switch determine to which one of its outgoing lines should it switch an incoming data unit?

The answer depends on the character and scope of the information that an incoming data unit is providing to the switch upon its arrival as it relates to its targeted destination. It also depends on the process used by the network system to configure the forwarding entries that are embedded in the switching table. An arriving data unit is stored in an input buffer, joining a queue of arriving messages. When it reaches the head of the queue, the switch processor examines it and/or inspects the entities that it has used to reach the switch (such as the identity of its incoming line or frequency band). It uses this data, and at times also other information, to determine the outgoing port to which the incoming data unit should be switched.

To illustrate the key switching concepts discussed below, consider the routing and switching processes involved when navigating a vehicle that is moving along an autonomous highway. The vehicle enters the highway at an access ramp and it moves along the road to its targeted destination location, at which point it may depart across an exit ramp. A vehicle's preferred route to a destination may traverse several highways. At an intermediate intersection between highways, the vehicle is switched from an incoming highway to an outgoing highway.

1) Under a **connection oriented transportation process**, a vehicle's driver informs the system as to its travel objectives. In case that a flow of vehicles is involved, the flow's manager would inform the system as to its transit objectives. The system uses a planning application to select an end-to-end route that meets the user's objectives. The route consists of a tandem (consecutive) sequence of interconnected highways. At intersections, switching controllers are informed as to the layout of the selected route so that a vehicle belonging to a user's flow is autonomously switched upon its arrival at an intersection to the next highway on its route. Once the route planning phase has been finalized and the involved switching controllers are notified, the mobility phase starts. Two different approaches can then be invoked:

    a) Under one approach, identified below as a **Circuit Switching (CS)** method, the system employs a resource allocation planning tool to reserve resources for the user's exclusive use. Capacity resources are set aside along the route's highways and intersections for accommodating the movement of vehicles that are members of this flow. In this manner, spatial (such as lane oriented spaces) and/or temporal, such as time slot based, resources would be set aside for use by this flow's vehicles.

    b) Under a second approach, identified below as a **Virtual Circuit Switching (VCS) or Connection-Oriented Packet Switching** method, a route planning and flow management system is involved as well. It determines a priori (before vehicles start their movement across the highway) a preferred route. It provides this data to the intersection controllers that are located across its selected route, which would then configure accordingly their switching tables. However, no resource capacity reservations are made. Rather, the user flow's vehicles occupy highway and intersection resources across the pre-planned route only when they actually need to do so, recognizing that vehicles may incur random delays across their paths and be impacted at times by unexpected road conditions. Highway (link) and intersection (nodal switching) resources are dynamically and statistically shared among vehicle flows. Such a sharing technique is identified below as a *Packet Switching* method.

2) Under a third approach, the planning-ahead process is NOT performed, so that no route, link or nodal resources are set prior to the start of a flow. No routing set-up process (noted latter to be carried out by a signaling system) takes place, so that there is no need to inform the switching nodes in advance about the outcome of an underlying pre-planning process. A vehicle is guided at its access ramp and at each intersection that it visits, at the time that its visit takes place, to a subsequent highway to follow. Such a process is identified below as a **Connection-less or Datagram Packet-Switching** method.

As illustrated above, we note the following widely employed switching disciplines.

1) Under a **CS** method, a **connection setup phase** is followed by a **data transport phase**:
    a) A **connection setup phase**, also identified as a **signaling phase**, is invoked prior to the start of the data transport phase.

    During the signaling phase, the end-user signals a communications sub-system known as the **Signaling System (SS)** its transport and resource requirements for an impending call. The latter can be identified as a flow that consists of a prospective stream of messages or data units (occurring in a stochastic or regular manner). Accordingly, the signaling system selects an end-to-end network route across which the flow's data units will be transported. In addition, communications capacity resources are selected across the route's links and nodes and assigned to this flow on a dedicated basis for the complete duration of the call.

    In this manner, communications links are shared among multiple flows so that the data units of each flow are dedicated link resources on a static (fixed) multiplexing basis. Such a transfer method is identified as a *Synchronous Transfer Mode (STM)*. The corresponding entity that consists of the resources that have been assigned across the configured route is called a **Circuit**.

    If the system is unable to configure at that time such a circuit, the call request is rejected. If a circuit is identified and the call is admitted, a **connection** is set up in the network for accommodating the flow of data units that are produced and transported as part of this call. Forwarding entries are accordingly created in the switching table of each switch that is used across the selected route. Each call and its preset circuit is identified in the switching table by a **Circuit ID**, or an equivalent designation. Data units that belong to a flow produced by a specific call for which a circuit has been configured, are identified as members of the configured circuit. For example, a Circuit ID, or an equivalent attribute, is included in such message headers.

    b) During the subsequent **data transport phase**, the flow's data units are switched and transmitted by the circuit switches that are located across the configured circuit until they reach their destination node. As a data unit enters a CS node, its Circuit ID (CID, or equivalent designation) label is used by the switching engine to extract from the Switching Table the corresponding forwarding entry. This entry identifies to the switch fabric the output port and outgoing line that the data unit should be switched to. The switched data unit is then transmitted across the outgoing line to the next circuit switch located along the route. When reaching the circuit's terminating switch, a call's data unit is switched to a line that connects it to destination node.

2) Under a **Packet Switching** method, the flow's data units are not dedicated resources across their route's data links. Data units of flows that share a communications link occupy link resources on a **statistical multiplexing** basis. Such a transfer process is also identified as an *Asynchronous Transfer Mode (ATM)*. No link resources are reserved for accommodating the data units of a specific flow, as performed under a CS method. Rather, when a flow

becomes active, its data units would then occupy communications link resources, as well as the processing and queueing resources of the packet switches that they visit while flowing along their route.

The configuration of a switching table entries and its switching operation are performed by using either connection-oriented or connection-less processes, as delineated in the following and discussed in further detail in subsequent sections and chapters.

a) Under a **Connection-Oriented Packet Switching** method, also known as **Virtual Circuit Switching (VCS)**, a **connection setup phase** is invoked prior to the start of the data transport phase, as performed in a CS network system. For this purpose, a *Signaling System (SS)* is used. In processing a new flow or call, the signaling management system selects an end-to-end route. It is used for setting up a forwarding data entry at each switching table of each switch that is located along the route. These forwarding entries are used by the switch system to direct (inside the switching node) a call's data unit from its input port to an output port. As a packet switching technique, communications links are shared by the traversing data units on a statistical multiplexing basis.

The connection entity configured across the network for supporting the transport of a call connection's flow is now identified as a **Virtual Circuit (VC)**. It consists of the connection's spatial (i.e., link) elements but does not prescribe specific dedicated communications resources in each link that is included in the flow's route.

A specific Virtual Circuit is identified by a Virtual Circuit Identifier (VCI). The signaling system informs the VCS switches along the route as to the association between a connection, its assigned VCI and the incoming and outgoing links that are used by this VC. This information is used to configure the forwarding entries of the switching table. Protocol Data Units that are created to carry a call's payload include header fields that identify the underlying VCI. As it enters a VCS switch across its route, a data unit is identified by its VCI label (aided often also by the identity of its incoming line), leading to the selection of a forwarding entry that specifies the output switch port that connects to the outgoing line to which the data unit is directed.

b) Under a **Connection-less Packet-Switching** method, also known as **Datagram Switching**, no connection entities are created and no a priori connection setup process is carried out. Hence, services provided by a Signaling System are not required. As a packet switching method, communications links are shared by data units on a statistical multiplexing basis.

To determine routes, datagram packet switches (identified henceforth as packet switches) exchange control data units that convey network system state status data. Using this data, packet switches calculate effective routes. A **routing algorithm** is used to configure and dynamically update the forwarding entries included in the switch's **Routing Table**. As it calculates routes, such a packet switching node is identified as a **Router**.

A forwarding entry is selected by a router following its examination of the destination address included in the header of an incoming data unit. The selected forwarding entry indicates the output port to which the incoming data unit should be switched, and consequently also the corresponding outgoing line that should be used to forward it to the next packet switch along its route.

Under a Layer-3 datagram packet-switching operation, the involved data units are Layer-3 Protocol Data Units, identified simply as packets. This is the method used for the execution of switching and routing processes across networks that use the Internet Protocol (IP).

9.2 Extending the Coverage Span: Repeaters and Relays

The Issue: *A station may not be able to receive messages sent to it by a transmitting station over a communications link. This can happen when the communications path is blocked or when the transmitted signal travels over a relatively long distance and is received at a too low power level. A repeater or a relay station is used to extend the communications reach of a data message.*

The Solution: *A repeater uses physical layer means to enhance communications. Incoming signals are reflected and re-sent. Enhanced relay operations also provide filtering and amplification services. Digital relay operations include demodulation/decoding, filtering, and re-encoding/modulation. More intelligent relaying can also provide switching services by using higher layer protocols.*

A relay device is used to enhance the ability of a transmitting station to reach a targeted receiving station, or a group of stations, often without using intelligent processing and switching operations.

The power of a transmitted signal attenuates rapidly as it propagates away from the source station, so that it may become too weak when it reaches a destination station that is located farther away. A relay device enhances reception by amplifying the signal. A relay is also employed when a direct path to the receiving node is blocked. A receiving station is then not able to detect messages that are transmitted to it across a direct, such as Line-of-sight (LOS), path. Such a station is said to be **hidden** from the transmitting station.

In Fig. 9.5, we show a radio relay node that is placed at the top of a hill, serving to enable communications between station A and station B which are hidden from each other due to terrain blocking.

A **Repeater or Relay** station is used to resolve such issues. A repeater station takes the signal that it receives from transmitting station A and retransmits it to receiving station B. The basic relaying operation is performed at the Physical Layer, so that no higher layer switching and routing intelligence is required. Physical layer PDUs that are received by a basic repeating relay node are retransmitted across all outgoing communications links connected to the output of the relay node.

When higher layer operations are involved, a relay can provide more intricate relaying and switching services, such as those provided by a layer-2 frame relay device. A frame relay node examines incoming frames, which are layer-2 PDUs, and based on control or addressing data included in their headers, it performs a filtering service, as it determines whether an incoming frame should be retransmitted across an outgoing link, or not. Furthermore, when multiple outgoing links are considered, it may also provide a layer-2 switching service, in determining across which outgoing link a frame should be transmitted. Such functions are further examined when reviewing the operation of bridges, as discussed in a later section. Layer-3 relays make use of control information included in incoming packets, which are layer-3 PDUs, to relay and switch incoming packets.

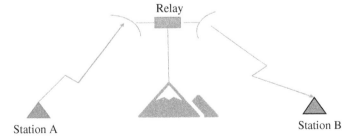

Figure 9.5 A Radio Relay Node Placed on a Hill to Enable Communications.

The following commonly employed relaying services are noted:

1) A **Reflecting Repeater** transfers incoming physical layer PDU's signals across a configured spatial segment of an outgoing communications medium. The repeating device can employ a static or a configurable antenna array.

2) A **Non-decoding Amplifying Repeater** performs signal processing and amplification operations on an incoming physical layer PDU prior to its retransmission. It does not however engage in demodulation/decoding and subsequent encoding/modulation operations. Radio signals are handled through the use of Radio Frequency (RF)-based processing and amplification operations. Optical signals may be processed in the optical domain.

3) A **Baseband Processing Repeater** performs demodulation/decoding operations on incoming physical layer PDUs. The produced baseband signal is processed to improve its quality, removing noise and interference components. The ensuing PDU is undergoing encoding and modulation operations that serve to improve the reception quality of the retransmitted PDU.

   When a non-demodulation-decoding repeater is used, both information bearing signal and noise components are amplified. Yet, provided the incoming signal-to-noise ratio is sufficiently high, the amplification operation is useful in improving reception at a destination entity when the latter is impacted by interference signals or background noise.

   In turn, when baseband processing is performed, significant enhancement can be made in the quality of the retransmitted signal, separating the processing functions applied on incoming signals and on outgoing ones. For the latter, the best modulation/coding schemes and power levels are selected in a manner that corresponds to the communications parameters of the outgoing communications channel and of the targeted receiving entity.

4) A **Reflecting Repeater Hub**: Consider a multi-port reflecting repeater that is connected at each port to a wireline channel. When functioning as a reflecting hub, a PDU received on an incoming line is retransmitted across all other lines. This is the service rendered by an *Ethernet Hub*. In this manner, a message transmitted by a station attached to the reflecting hub is broadcast to *all other* stations that are attached to the hub.

5) When considering a relay that is attached to incoming and outgoing multiple access wireless channels, **frequency translation** operations may be performed. Multiple frequency bands may be used across incoming and outgoing channels. A message received by a relay module across frequency band F1 is retransmitted across frequency band F2. Under a full-duplex relaying operation, simultaneous bidirectional relaying and translations are conducted.

To illustrate the multitude of ways that are implemented by relaying systems, consider the following modes of operation of a **satellite communications system**. As discussed in Section 1.7.8 and illustrated in Fig. 1.18, a satellite node is used to provide communications between Earth-Stations (ESs). When using a satellite node that is assigned an orbit at a high altitude, communications between stations that use the relaying services of the satellite are able to disseminate data over wide geographical spans, avoiding potential terrestrial blockages. Assume that the satellite employs a broad-beam antenna that covers communications to/from all earth stations that are members of a specific network system. An **uplink** frequency band is used for communications between an earth station and the satellite, while a **downlink** frequency band is used for communications from the satellite to earth stations. The following modes of operation are noted:

1) A **Reflecting and RF Processing** satellite operates as a "mirror in the sky." A message is transmitted across the uplink by using the uplink frequency band. To reflect it for transmission

across the downlink, its uplink frequency span is translated at the satellite to the downlink frequency span. As the transmission is broadcast across the downlink, it can be received by all network's stations, but it is processed by only those stations that are specifically identified as destination stations. Onboard processing and amplification are performed at the RF level, as no demodulation/decoding operations are undertaken.

2) An **Onboard Baseband Processing Satellite** performs demodulation/decoding operations on incoming message signals that arrive across the uplink. The ensuing baseband messages are processed and amplified so that the impact of noise and interference signals is resolved or reduced. The produced signals are then transmitted downlink, using the downlink frequency band. No onboard message queueing delays are induced as only a single stream of message signals is handled at a time.

Many satellite systems employ antenna arrays that form multiple uplink and downlink narrow beams. A multiple beam satellite system may employ n beams, where the footprint of a beam covers a specific terrestrial region. Uplink and downlink frequency bands are allocated for beam oriented communications. Frequency resources used for uplink (downlink) communications that are used by a beam may be reused for communications across another beam, provided limited cross-beam interference is maintained.

An earth station located in region-i, which is covered by beam-i, may wish to communicate a message to a station that is located in region-j, where $i, j = 1, 2, \ldots, n$. To accommodate such a flow, a control and management station may be used to assign beam capacity resources. A message that is directed to a destination station that is covered by a specific beam makes use of assigned (time/frequency) downlink beam resources.

An **onboard switching** table is configured and used to switch messages that arrive across an uplink beam to a designated downlink beam.

To provide wide terrestrial coverage, particularly when satellites are placed in a low earth orbit, inter-satellite communications links are established and used to route message flows across a mesh network of satellite nodes. The onboard switching matrix at each satellite is then extended so that outgoing links include uplink and downlink beams as well as inter-satellite links.

Onboard multi-beam satellite switching systems operate in the following modes:

1) When using an onboard **RF Switching** operation, a **CS**-oriented procedure is undertaken. A signaling and control system is used to determine the routing requirement of each flow and then configure the switching matrix so that incoming data units of a flow are switched in Radio Frequency (RF) domain in real time, so that no queueing delays are incurred. The synthesized switching schedule must assure a non-blocking realtime operation. Accordingly, messages that arrive at the same time across distinct uplink beams can be simultaneously switched, effectively in real time, across the switching module to their targeted downlink beams.

2) An **Onboard Store-and-Forward Packet Switching** operation may be employed. When using a datagram packet-switching operation, an **Internet in the Sky** system is realized. Uplink signals are demodulated and decoded. Received packets are examined onboard to determine their destination and consequently deduce the outgoing beam to which they are switched.

Multiple messages that need to be switched to the same downlink beam may be arriving at about the same time. Arriving messages are queued in input buffers as they await their turn to be switched to an output buffer. They are then queued in the corresponding output buffer

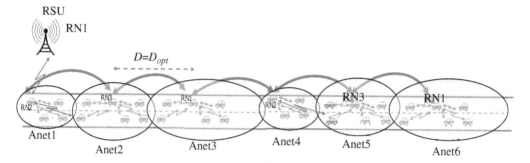

Figure 9.6 Dissemination of Data Messages by Relay Nodes Across a Vehicle Highway.

as they wait for transmission across their targeted downlink beam or inter-satellite link. When reaching the head of their output queue, packets are encoded and modulated and transmitted across their output beams or links.

Such a baseband-oriented switching operation allows a more versatile end-to-end operation, accommodating a wider range of traffic flow rates and patterns. A VCS-based system can be configured as well, implementing a connection-oriented packet-switching operation.

A **network of interconnected mobile vehicles** that move along a two lane highway is illustrated in Fig. 9.6. It serves to demonstrate the functioning of a link-layer-oriented relay network. The illustrated network system is based on a method that we have developed for disseminating data message flows among highway vehicles [12, 44, 45]. Certain vehicles (shown in the figure to be a distance of about $D = D_{opt}$ from each other) are elected to act as **relay** nodes.

In the illustrated scenario, a Road Side Unit (RSU) generates a flow of messages that it aims to disseminate to vehicles moving across a segment of the highway. Networking is based on a geo-routing protocol that we have identified as a **Vehicle Backbone Network (VBN)** scheme.

An elected relay vehicle stores each message that it receives for a brief period of time, sending such a message when its retransmission is required (which can be determined based on vehicle positions along the highway) at a designated time slot.

To illustrate, a reuse-3 spatial TDMA multiple access scheme is employed. Each relay vehicle is assigned, every three slots, a single time slot that it can use for the transmission of a packet that it has received. These transmissions are locally broadcast so that they are received by vehicles located nearby as well as by a downstream relay node. The spatial reuse mechanism enhances the system's data throughput performance, noting that relay nodes that are at a distance of about $3D$ from each other are able to execute packet transmissions in a time simultaneous manner.

Relay nodes engage in the reception and filtering of link-layer frames (which carry data packets as payloads) and then, on a store-and-forward basis, retransmit selected frames at designated time slots.

The RSU may pace the transmission of data frames, setting the time intervals between its transmissions of subsequent frames to be sufficiently wide so that queueing delays of frames at intermediate relay nodes that they traverse along their tandem route are reduced or essentially eliminated.

We have employed such protocols in studying methods for data message dissemination across communicating autonomous highway vehicles. Further details of VBN systems and of methods for geo-routing operations in communicating between autonomous highway vehicles are provided in Chapter 21.

9.3 Local Networking Across a Switching Fabric: Bridging of MAC Frames

9.3.1 Local Internetting Using Bridges and Layer 2 Switches

The Concept: *In a user or organizational facility, user and system devices are attached to LANs. They often inter-communicate through a fabric of interconnected LANs by disseminating messages that are formatted as layer-2 MAC frames. Such facilities include data centers, and networks used by medical, industrial, manufacturing, and educational institutions. Network nodes that function as bridges and layer-2 switches are employed. A multi-port bridge is attached to two or more LANs. It forwards messages across LANs by examining the MAC addresses embedded in observed frames. A network of interconnected layer-2 switches is used to transport MAC frames to their destination client and server stations.*

The Method: To perform its forwarding and switching actions, a bridge or a layer-2 switch configures a forwarding table. The table contains forwarding entries, which are used to determine the output port to which an arriving frame is switched, based on its destination MAC address or on a label that it carries. The operation must avoid forwarding frames in a manner that leads them to flow in cycles, forming transit loops across the network. A commonly employed method is based on using a spanning tree protocol (STP). It produces an embedded tree subnetwork across which all frame transmissions are sent out. To enhance the utilization of communication capacity resources, multiple spanning trees may be synthesized. Each spanning tree instance can be used to support a distinct group of Virtual LAN (VLAN) flows. For larger facilities, it is beneficial to further enhance system performance, increase utilization of communications capacity resources and attain post-failure resiliency. For this purpose, network layer oriented control intelligence is integrated into the link layer forwarding structure. A control plane underlay is employed, providing for dissemination of network-wide link and nodal states that are used to enhance the operational efficiency of the data plane overlay. Link state dissemination schemes are used to select loop-free high-quality routes and configure switching tables. Such a method, when further extended to accommodate a flow dissemination scheme that is responsive to Virtual LAN (VLAN) and Virtual Service network (VSN) layouts, as implemented by a Shortest Path Bridging (SPB) protocol, provides rapid, high performance, and robust transport of unicast and multicast data flows.

A **LAN** serves to connect stations that are typically used in a single organization and are located in close proximity to each other. This is the case when considering the basic use of an Ethernet LAN, as discussed in Section 8.4.4. Ethernet bus and Ethernet repeater hub configurations are illustrated in Fig. 8.18. An Ethernet LAN segment forms a broadcast domain, whereby a MAC frame transmitted by a station across the shared medium is received by all stations that are attached to the same segment, though it is only picked-up and processed by those stations whose destination address is cited.

As noted in Fig. 9.7(a), following a preamble field, the header of an Ethernet frame contains the MAC address of the station to which the frame is destined. It also contains the MAC address of the station that issued this frame.

In many facilities, users are distributed over diverse positions. Their stations may be attached to geographically distributed distinct LAN segments. These LAN segments are often located in a single building or facility, or within a campus that includes multiple buildings. Users are often divided into diverse groups based on a multitude of criteria, including their function in the organization. They may accordingly be granted specific group-oriented permits to access to certain categories of data types, clients, or servers.

9 Switching, Relaying, and Local Networking

Preamble 8 B	Destination MAC 6 B	Source MAC 6 B	EtherType /Size (2 B)	Payload 42-1500 B	FCS (CRC) 4 B	Inter frame gap 12 B

(a)

Preamble 8 B	Destination MAC 6 B	Source MAC 6 B	802.1Q header 4 B	EtherType /Size (2 B)	Payload 42-1500 B	FCS (CRC) 4 B	Inter frame gap 12 B

(b)

Figure 9.7 Format of Ethernet II Frame (a) Without a VLAN Tag and (b) With a VLAN Tag.

Users that are members of specific groups or departments tend to be interested in receiving messages that pertain to their function. They are also often interested in keeping their communications separate from users that are members of other groups. Certain classes of group messages are set for dissemination (or *multicasting*) to all group members. Stations that are associated with a specific group of end users may be attached to distinct LAN segments or connected to distinct switching nodes.

The concept of a **Virtual Local Area Network (VLAN)** is used to facilitate the transport of messages among members of a group that are attached to distinct LANs. A **VLAN-ID (VID) tag** identifies members of a specific group. A station may become a member of one or several VLANs. A message sent by a station that is a member of a given VLAN will often be targeted for reception by stations that are members of the same VLAN, whether attached to the same LAN or attached to other LANs or switching nodes.

In Fig. 9.7(b), we show a version of the format of an Ethernet frame that includes a VID tag. The shown tag is formed by using the format recommended by the IEEE 802.1Q standard. The VID is represented as a 12-bit long number, allowing up to 4094 distinct VLAN groups. The added tag also includes a 3-bit *Priority Code Point (PCP)* field which refers to a class of service defined by the IEEE 802.1p specification. It is used to identify a frame's priority level.

A shared medium Ethernet provides a switching service to its attached stations. A source station that wishes to communicate with another station that is attached to the same LAN segment broadcasts its MAC frame across the medium. The intended destination station detects the frame's destination MAC address to be the same as its own address and subsequently copies the frame and passes it to a higher layer entity for processing.

If the source station has not yet acquired the MAC address of a targeted destination station that is attached to the same LAN, but it possesses its IP address, a service provided by employing the *ARP* is used to secure the needed MAC address. Using ARP, the source station broadcasts across the LAN an ARP frame that cites the IP address of the targeted station. The latter then responds by broadcasting across the LAN the association between its IP address and MAC address.

While located in the same facility, a bridge device may at times be situated at a relatively longer distance from a LAN that it serves. In this case, a wide area network (WAN)-based link protocol is used. This can be accommodated by using devices that are known as "half bridges" that are located at the two edges of the latter link. Each half-bridge device provides, at one edge, access to/from a LAN and at the other edge access to/from the WAN link.

As noted in a previous section, when discussing the configuration of Ethernet systems, it is frequently more efficient to use **Ethernet Switches** to replace an operation that employs Ethernet buses or repeater hubs. An Ethernet switch consists of multiple ports, whereby each of its edge ports is connected, typically in a full-duplex manner, to an end-user station. Each attached link uses an Ethernet access protocol.

9.3 *Local Networking Across a Switching Fabric: Bridging of MAC Frames* | **323**

Such access links are often operated in a full-duplex mode. In this case, the attachment employs separate parallel links, where each termination is assigned its own link, as assumed henceforth. Each such link is shared only by a single station in the user-to-switch direction and by a single station when used in the switch-to-user direction. Hence, no frame collisions take place when communicating across the access links. When ready, frames are sent between an end-user station and the switch port that terminates the employed access link. The switch operates in a **store-and-forward** manner so that transmissions in either direction that involve distinct switch ports can be conducted at the same time in a collision free manner. At each switch, multiple incoming and outgoing frame transmissions can take place. Such a switch handles (receiving, processing, and forwarding) layer-2 MAC frames and is identified as a **Layer-2 Switch**.

Which methods are used to inter-network among LANs such as those that involve Ethernet LANs and/or layer-2 switching nodes?

A station that is attached to a shared medium LAN and wishes to communicate with a station that is attached to another LAN would start by broadcasting its data frames across its attached LAN. Noticing the destination of these frames to be an entity that is associated with another LAN, the frames will be captured by a *designated gateway* node. The latter, functioning as a store-and-forward bridge or layer-2 (L2) switch, uses the information embedded in its forwarding table to forward a captured frame for transmission across another LAN (or L2 switch) to which it is connected. In this way, a frame would be transported from one LAN (or L2 switch) to the next one until it reaches its destination LAN (or L2 switch) and station.

The corresponding store-and-forward switching modules that are used to provide local internetting functions isolate aspects of the operations conducted across individual LANs. They therefore accommodate the use of distinct data rates and the occurrence of variable traffic loading conditions across attached LAN segments or switch access links.

A store-and-forward switching node reviews a message that it observes to be transmitted across a LAN that one of its ports is attached to, examining its destination address, or associated VLAN tag, and accordingly switches it to a selected output port(s).

Layer-3 (L3) routers are often employed to provide internetting services, enabling message communications among local stations and networks as well as among networks that are distributed over wide geographical regions. A router operates as store-and-forward packet switch. As it operates at the (L3) network layer, it is able to learn the layout of the underlying network. It can dynamically select its routing paths in adapting to variations in the network's topological layout and traffic loading conditions.

A layer-2 bridge has been traditionally used to selectively forward frames that it observes to be transmitted across the LAN links that are attached to its ports. It is not adapting as effectively to network status variations. It uses less sophisticated and lower complexity processes in determining which incoming frames it should forward and to which ports it should switch them. Hence, its synthesized internetting configuration is not as throughput-delay performance efficient as that provided by a packet router. However, its configuration and processing simplicity can induce ultrafast-switching operations. As further noted in the following, when integrating a layer-3-oriented mechanism into the setting of forwarding tables that are used in the operation of a layer-2 switch, one can enhance the performance of the system in a significant manner.

In comparing the L2 operations of bridges vs. those of L2 switches, we note that bridges tend to accommodate a small number of ports. In turn, L2 switches are driven by VLSI-based hardware engines, enabling them to support a larger number of ports in performing their filtering, switching and forwarding operations at high speed. They can induce very low frame latency and offer high-throughput rate performance.

In the following, we first describe a variant of a method that is identified as a **STP** and describe its use for internetting among shared-medium LANs. Subsequently, we outline the use of essentially the same protocol for the internetting of frames among L2 switches. We then briefly describe the use of efficient internetting approaches, including a method that employs an L3-oriented control plane to upgrade the operation of an L2-based frame forwarding data plane.

9.3.2 Building a Frame Forwarding Table via a Flooding Protocol

Consider a network that consists of three LANs that are interconnected by bridges, as shown in Fig. 9.8. Bridge B1 is attached to LAN1 via port P11 and to LAN2 via port P12. A frame that is transmitted across LAN1 (LAN2) is broadcast to all attached stations and is thus also received by bridge B1 at port P11 (P12). Bridge B2 is attached to LAN2 via port P21 and to LAN3 via port P22. A frame transmission across LAN2 (LAN3) is broadcast to all stations attached to the LAN and is thus also received by bridge B2 at port P21 (P22).

How does a bridge determine which ones of the frames that it detects on a LAN to copy and queue and then forward to another LAN?

A network management application that resolves this issue would allow a bridge to compose a forwarding table that identifies, for each frame that arrives at a given input port, using its destination MAC address, the outgoing port(s) to which it should be forwarded.

Certain frames should not be handled by a bridge, including those that are destined to a station located on the same LAN. Also, certain ports can be specified to not process certain unqualified frames or be **blocked**. In addition, another bridge may be assigned to be responsible for handling frames sent across a LAN.

We describe in the following a process under which a bridge observes the frames detected at its ports and uses their embedded control data to compose its forwarding table. Such observed control data include each frame's source MAC address and may also include group membership tags and other attributes.

Certain ports can be specified to engage in the reception/transmission of only frames that are destined to an address included in a configured address list or that are tagged as associated with a certain collection of VLANs and thus identified by a specified VID group attribute, and/or a specific Service-ID (SID) attribute.

A manual composition of a forwarding table can be highly demanding if the number of stations is relatively high and if station migrations, additions and deletions occur often. A more manageable

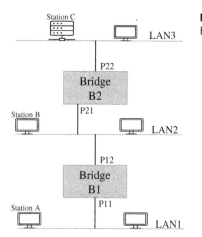

Figure 9.8 Bridges B1 and B2 Are Used for Internetting Frames Between LAN1, LAN2, and LAN3.

9.3 *Local Networking Across a Switching Fabric: Bridging of MAC Frames* | **325**

solution would be to configure the bridge to learn on its own the relative location of involved stations (or targeted LANs or user groups) to the extent needed by it to configure the underlying forwarding entries and automatically adjust forwarding entries when changes occur. Such a bridge is said to operate as a **learning bridge**. How does a bridge learn which frames it should forward and, if so, across which port(s) should they be transmitted?

A commonly employed method is illustrated by considering the network layout shown in Fig. 9.8. The process consists of two main phases: (1) the learning phase, during which the bridge creates its current version of the forwarding table; (2) the frame forwarding phase, during which the bridge forwards data frames.

The learning process is carried out by having the bridge record the source addresses of frames that it monitors at each one of its ports. For example, consider station A that is attached to LAN1 and that has just initiated its activity by transmitting a frame across LAN1. Assume the frame to be destined to station C that is attached to LAN3. Bridge B1 receives this message, noting its source address and thus recording in the forwarding table that it maintains that this station is reachable via its port P11.

Assuming the address of destination station C to not yet include a corresponding forwarding entry in its forwarding table, bridge B1 proceeds to flood the received frame message across the network. The *flooding protocol* requires a participating bridge to transmit such a received frame across each one of its ports except the one across which it has received this frame. Accordingly, this frame will be transmitted via port P12 across LAN2. Following a similar process, bridge B2 records in its Forwarding Table that it can reach station A via port P21. It will then proceed to flood the frame by broadcasting it across LAN3 via port P22. Destination station C will then identify itself as the destination entity and will copy the frame.

Consider next the following scenario. Following the receipt of the latter message, station C is assumed to send a response frame to station A, which is still attached to LAN1. In following a similar procedure, bridge B2 will create an entry in its forwarding table to indicate that station C is reachable via port P22. It has already learned that destination station A is reachable via port P21, so that it will forward the response frame by transmitting it via P21 across LAN2. Subsequently, bridge B1, whose forwarding table already contains an entry that links station A with port P11, will proceed to store and forward the frame by broadcasting it across LAN1 for reception by destination station A. At the same time, bridge B1 will create a forwarding entry that identifies the use of port P12 when aiming to forward a frame to destination station C.

In this manner, as frame transactions take place, the bridges automatically compile forwarding tables. The flooding process need not be repeated once a forwarding entry for a destination has already been learned. Yet, as changes in the layout may occur, each forwarding entry is given a **time aging** limit, after whose expiration, the learning process is repeated. Following the learning phase, each bridge is ready to use its forwarding table to execute the forwarding of data frames.

Will this process always work? We will note in the following that the answer is negative. We will then describe a method that has been frequently used for implementing a workable solution. It involves the synthesis of a spanning tree layout and is known as a *spanning tree protocol (STP)* scheme.

9.3.3   Spanning Tree Protocol (STP) Methods for Constructing a Forwarding Table

The procedure illustrated above for building a forwarding table may not work properly if the network's connectivity graph contains cycles (also known as "loops"). In this case, there may be two or

9 Switching, Relaying, and Local Networking

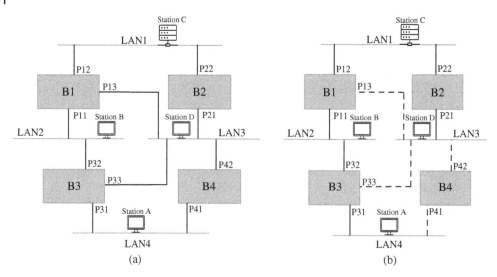

Figure 9.9 (a) Topological Layout of a Network of LANs Interconnected by Bridges; (b) an Embedded Spanning Tree Layout.

more distinct routes that messages can take to reach certain destination nodes from certain source nodes. Then, during the learning phase, flooded frames issued by a station may take multiple routes across the network and would then be received at multiple ports of certain bridges. Such a bridge may therefore not be able to select a unique port to use for forwarding future data frames to certain destinations.

Such a scenario is illustrated in Fig. 9.9(a). The underlying network layout includes cycles. Frames issued by station A that is attached to LAN4 and that are (during the learning phase) flooded in the network will be received at bridge B1 at each one of its three ports! One frame will use a route that passes through bridge B3, reaching bridge B1 at port P11; another copy of the same frame will reach bridge B1 through bridge B4, arriving across port P13. A different copy of the frame will be routed through bridges B4 and B2, reaching bridge B1 at port P12. As a result, bridge B1 is unable to determine which forwarding entry to use for forwarding data frames that aim to reach Station A.

This matter is resolved by using a protocol that synthesizes a subnetwork that includes all the network's LANs but that contains no cycles, known as a *spanning tree* layout or graph. The flooding-based learning phase would then be conducted across this embedded tree network. It is further explained as follows.

A network topological layout is represented as a graph, which consists of nodes and lines. Nodes are connected to neighboring nodes by lines. The bridged network's layout is modeled as a graph by representing each LAN as a node (or vertex) and each bridge as a line (or edge). A **tree graph is a connected graph that contains no cycles**. A **spanning tree graph is a connected subgraph that contains no cycles and that includes all of the network's nodes and a subset of its lines**.

A spanning tree of the network shown in Fig. 9.9(a) is illustrated in Fig. 9.9(b). The latter displays a subnetwork topology for which ports P41, P42, P33, and P13 are blocked so that the corresponding lines (or links) are not used. The resulting layout forms a spanning tree subgraph as it includes all network nodes (i.e., LANs), a subset of its lines (or links), and it contains no cycles. Consequently, each pair of nodes is connected by a single (unique) path.

During the learning phase, the flooding of message frames is limited to using the spanning tree layout. A message issued by station A that is attached to LAN4 is flooded in the network during the learning phase across the spanning tree, so that it will now be received at bridge B1 only at port P11 by following the path {LAN4, B3, LAN2, B1}. Hence, bridge B1 will be recording a forwarding entry in its forwarding table that instructs it to forward to port P11 those received frames whose destination address is A.

A multitude of algorithms may be employed for the autonomous synthesis of a spanning tree graph. IEEE standard 802.1d was created in 1990 and updated in 1998. Its recommended protocol is often identified as a **STP**. Such a scheme is also known as **Transparent Spanning Tree Protocol**, as it is set to autonomously reconstitute the construction of the tree graph, when feasible, upon the occurrence of LAN or bridge failures that cause the current spanning tree to become disconnected. Other standards include IEEE 802.1q and IEEE 802.1w. The latter is known as Rapid Spanning Tree Protocol, enabling faster post failure re-constitution of the spanning tree. It has been incorporated into the 802.1d-2004 protocol, replacing STP. Other versions that are known as *Multiple Instance Spanning Tree Protocol (MISTP)* and *Multiple Spanning Tree Protocol (MSTP)* were defined to handle the synthesis of multiple spanning trees in a bridged network and for facilitating the operation VLAN switches. In the following, we address certain key ideas involving the operation of such schemes.

The basic STP process is based on the following approach. Each Bridge is assigned an identifier that consists of a priority level and its MAC address as components. Each bridge port is assigned a port ID. To synthesize the spanning tree, network bridges exchange control packets, known as bridge protocol data units (BPDUs). Through this exchange, the bridges are able to determine which one possesses the lowest ID value. The latter bridge is selected as the **root bridge**. Each port is assigned a cost metric, which identifies the desirability of including this port in a synthesized tree.

A commonly used port cost metric is set so that it is inversely proportional to the speed of the attached line (in case that a layer-2 switch is employed as the bridge node) or of the shared medium (in case that a shared medium LAN is used). By exchanging control messages, each bridge determines the least cost path to the root bridge, noting that the cost of a path is calculated as the sum of the costs of the ports used in the path. At each bridge, the corresponding bridge port that is used to access the least cost path that leads to the root bridge is identified as the *root port*. When there are multiple bridges attached to a LAN, the bridge with the lowest cost path to the root bridge is selected as the one to use for forwarding flows. This bridge is identified as the *designated bridge* and the port of this bridge that attaches to the LAN is identified as the *designated port*.

Following this process, the least cost path from each node (i.e., a LAN) through the designated port of its designated bridge to the root node is identified. Assuming the network to be connected (so that each node is able to communicate with any other node), the superposition of these least cost paths yields the synthesized spanning tree.

The first phase thus consists of synthesizing the spanning tree layout. The learning process takes place during the second phase, whereby dissemination of frames along the spanning tree allows each employed bridge to construct its forwarding table. In the third phase, data frames are forwarded by the bridges across the network. Following the expiration of an aging parameter, the process is refreshed.

Control messages are periodically disseminated across the network, allowing the system to detect failure conditions and to trigger a fast recovery process that results in the synthesis of a new tree. Priority port weights are often employed to prioritize the use of certain ports in constructing a modified tree layout.

In the following summary of a version of the spanning tree algorithm, L2 switches play the same role undertaken by bridge nodes, observing that Ethernet shared-medium LAN entities are often implemented by employing L2 Ethernet switches.

Outline of a Version of the Spanning Tree Protocol Process:

1) At initiation, each switch sets all of its ports to reside at a **Blocking State**. It listens to and processes only BPDUs and drops all other frames. As noted below, it configures the states of its ports by selecting a root port, designated ports and blocked ports. Ports may stay at this state, often also identified as a *Listening State*, for a specified period of time before a refresh process is initiated. The process results in the selection of a spanning tree topology.

   a) The switches elect a switch to act as the root bridge. All ports of the root bridge are configured to operate in the forwarding mode (and are thus set as Designated ports). A port that resides in a forwarding state receives and sends BPDUs, gets MAC addresses, and forwards user traffic.

   b) Each non-root switch selects a **root port**, which is the port that is part of the least cost path to the root switch. The root port is used by a non-root switch to forward data frames upstream to the root bridge. The root bridge does not have a root port.

   c) In considering the ports of a switch, whereby a port is used to connect it with a network device, or to attach it to a LAN, select the port(s) to be used to forward data downstream and receive data flowing upstream (while the root port is used to forward data upstream to the root bridge), if any. Such a port is identified as a **designated port**. Specifically, if the least cost path of a neighboring downstream switch traverses the underlying switch (as it passes through the root port of the downstream switch), this path will access the underlying switch at a designated port of this switch.

   d) The switch's remaining ports, which are not configured to perform as designated ports or as the root port are set to operate at a **blocking or discarding state**. At this state, a port is set to receive and send BPDUs, but it does not process MAC addresses or forward user traffic.

2) For each switch, the root port and its designated ports are set to enter into the **Learning State** from their previous Listening State. Using these ports, switches listen and process BPDUs and user frames. Each switch examines the source MAC address of user frames that it receives and uses them to update its forwarding table. It does not yet forward any user frame. The system requires ports to stay in this state for a prescribed period of time before transitioning to the next state.

3) Subsequently, the switch enters a **Forwarding State**. The switch processes and forwards BPDUs and user data frames. It uses BPDUs to monitor the network topology. It updates its forwarding table by reading the source MAC address of user frames.

An illustrative network of interconnected layer-2 switches is shown in Fig. 9.10(a). End-user stations, networking devices, or servers may be attached (not shown) to several of the switch nodes. For example, switch node 5 has five ports that connect across five communications lines to five neighboring switch nodes. The cost weight of each port can be set as a value that is inversely related to the speed of the associated line. Under the IEEE 802.1D-1998 standard, the SPT cost weights of ports that are attached to (Ethernet) links that operate at rates 100 Mbps, 1 Gbps, and 10 Gbps, were set equal to 19, 4, 3, and 2, respectively. Under the IEEE 802.1D-2004 (RSPT) standard, the cost weights of ports that are attached to (Ethernet) links that operate at rate R were modified to be calculated in proportion to 1 Tbps/R, so that port costs for 100 Mbps, 1 Gbps, 10 Gbps, 100 Gbps, and 1 Tbps are set equal to 200,000, 20,000, 2000, 200, and 20, respectively.

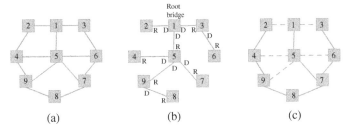

Figure 9.10 (a) A Network of Interconnected Layer-2 Switches; (b) A Spanning Tree for the Network in (a); (c) A Shortest Path Tree (SPT) Rooted at Switch 8 Which May Be Employed by a Shortest Path Bridging (SPB) Protocol such as IEEE 802.1aq.

A layout of a spanning tree created for the network shown in Fig. 9.10(a) is exhibited in Fig. 9.10(b). Switch 1 was elected to act as the root bridge. The corresponding Root ports (denoted R) and Designated ports (denoted D) for each switch are also identified. It is noted that all ports of the root bridge are designated ports. Note the alternate use of root and designated ports along paths in the tree that lead to the root bridge. Each such path was selected for inclusion in the tree since it was calculated to be a corresponding least cost path. For example, the tree path leading from switch 8 to the root bridge traverses intermediate switches 9 and 5. The port of switch 9 that connects it with the link that connects to switch 5 is noted to be a root (R) port as it leads to the root bridge by taking a least cost path. The port of switch 5 that attaches it to the link that connects with switch 9 is noted to be configured as a Designated (D) port, as the latter link is on the least cost path from switch 1 to Switch 9. Equivalently, the least cost path from switch 9 to switch 1 traverses the link that connects to switch 5 at the corresponding Designated port.

As noted above, the autonomous construction of a transparent spanning tree results in the formation of unique forwarding entries at the bridges. Yet, using a tree topology implies that nodal and link bandwidth resources may not be well utilized. Communications bandwidth resource utilization is enhanced when *multiple spanning trees (MSTs)* are constructed. For this purpose, a multiple spanning tree protocol (MSTP) has been introduced. For load balancing of traffic flows among multiple trees, flows associated with different VLAN IDs can be assigned to different trees.

A system may contain a large number of VLANs while synthesizing a much smaller number of trees (also identified as multiple spanning tree instances). Hence, groups of VLANs can be associated with each instance of a spanning tree.

The MSTP protocol also provides for the construction of *common trees* that are used for the dissemination of control messages and for accommodating VLANs that are not assigned to specific MST instances. The protocol also provides for the setting of multiple areas, so that distinct MSTs are employed in distinct areas. Provisions are also provided for inter-area communications and the use of a backbone area.

While located in the same facility, a bridge device may be at times situated at a relatively longer distance from a LAN that it serves. In this case, a WAN link protocol is used for the long distance transport of messages. This is accommodated by using devices that are known as "half bridges" at the two edges of such a WAN link. Each half-bridge device provides, at one edge, access to/from a LAN and at the other edge access to/from the WAN link. The operation of the STP is transparent to the use of such WAN links. In turn, switches connect through their ports with an external communications link such as an Ethernet full-duplex link.

Often, interface devices, such as *Application Delivery Controllers (ADC)*, are used at a network facility to regulate outgoing and incoming flows associated with end-user devices (such as computer hosts). Such a device, located at the interface of a facility and a backbone network, provides additional services, such as cybersecurity protection, fail-safe operation, local and global load balancing, and network address translation (NAT). Under the latter, internal company addresses are translated to/from public network addresses (such as Internet's public IP addresses), facilitating the ability to keep private a company's internal addressing configuration and networking layout.

9.3.4 Multipath Networking Across Local Switch Fabrics: Shortest Path Bridging (SPB)

9.3.4.1 Shortest Path Bridging (SPB)

As noted in Fig. 9.10(b), a spanning tree layout is utilizing only a limited number of network links, so that a significant portion of the system's communications bandwidth remains unused. As routes are restricted to using the tree topology, certain end-to-end paths become quite long, inducing excess loading on the core's links, leading also to longer end-to-end frame transit delays. In a large facility, such as a data center or a cloud computing facility, the network system often assumes the form of a very large **switching fabric**.

Such a fabric may consist of hundreds or thousands of switch nodes that are interconnected as a **mesh**. A switch node may then be directly connected via communications links to multiple other nodes.

To illustrate, consider a mesh network that contains n nodes. For a network graph to be connected, we need to use a total number of lines, denoted as m, that is not lower than $n - 1$, so that we require $m \geq n - 1$. A spanning tree graph contains exactly $m = n - 1$ lines. In a tree graph, every two nodes are connected by a single path.

Assume the average degree of a node, measuring the average number of neighboring nodes that a node is connected to, to be equal to k. In a mesh network, whereby two nodes may be connected by more than a single path, nodes are connected on the average to several other neighboring nodes, so that $k \geq 2$.

For example, for a mesh network that assumes the topology of a cycle, or a ring, using bi-directional (full-duplex) links (also identified as lines), we have $k = 2$. It consists of $m = n$ lines.

A mesh network of n nodes whose average nodal degree is equal to k, contains a total number of lines that is equal to $m = kn/2$ (a result that is known in Graph Theory as Euler's Theorem). To illustrate, consider a mesh network that supports $n = 400$ nodes, for which the average nodal degree is equal to $k = 4$. This network contains a total number of lines that is equal to $m = nk/2 = 800$ lines. Yet, a tree subnetwork that spans all nodes of this network consists of only $m = n - 1 = 399$ lines. Thus, by using a spanning tree layout to forward data frames, the system does not make use of a large number of communications links.

As noted, the use of a spanning tree layout to disseminate data frames may cause messages to traverse excessively long paths. This results in increased message delays and higher internal loading of the network's communications links. Note that a flow whose data rate is equal to 1 Mbps applies a total load of 2 Mbps on the network's links if it traverses a path that consists of two links. In turn, if the path consists of six links, the total internal link load imposed by the flow would increase to 6 Mbps.

Therefore, when considering large facility networks, it is highly advantageous to route messages across short paths, and preferably across *shortest paths*. To synthesize a stable and robust networking scheme, it is also essential to avoid the creation of networking loops and to use frame forwarding schemes that are highly resilient. As such, forwarding operations would rapidly adapt to network element additions, deletions, and failures, and enable predictable

9.3 Local Networking Across a Switching Fabric: Bridging of MAC Frames | **331**

message delay-throughput performance behavior. Such layer-2 networking approaches have been developed, serving to replace STP oriented methods.

Standardized such methods include the IEEE 802.1aq specification, known as **Shortest Path Bridging (SPB)**. It enables multipath shortest path routing across a fabric of layer-2 switches. A conceptually similar approach is used by an IETF-defined protocol known as *Transparent Interconnection of Lots of Links (TRILL)*. The **Internet Engineering Task Force (IETF)** develops voluntary Internet standards. Included are the standards that comprise the Internet TCP/IP protocol suite. Modified versions and proprietary protocols that employ related methods have also been under development by several organizations. Our description here will focus on the SPB method.

As further explained and illustrated in the following, a key feature of SPB is that **network services are virtualized and decoupled from the physical infrastructure**.

The approach enables the use of layer-2 (L2) and layer-3 (L3) unicast and multicast networking operations. A **control plane** is used to calculate shortest paths through the fabric of interconnected core switches; consequently delivering sub-second failover and recovery operations. Edge switches are used to interface customer terminal and server stations, so that edge-only provisioning and reconfigurations can be performed in realtime.

Dynamic autonomous configurations are performed via attachments to authenticated edge nodes, serving to connect seamlessly across the fabric to specified groups of service stations, forming **Virtual Service Networks (VSNs)**. Networking connectivity functions such as those used for realizing Virtual LANs (VLANs) and **virtual routing and forwarding (VRF)** are treated by the SPB backbone network as **network services**, forming VSNs.

VLAN extensions across a network backbone infrastructure are identified as L2 VSNs, and VRFs extended across the network are identified as L3 VSNs. Both **unicast** flows (directed to a specific station) and **multicast** flows (aiming for reception by stations that have joined the underlying multicast group) are accommodated.

An **L2 Virtual service Network (VSN)** is configured so that any station, through its attached bridge that requires a particular instance of service (such as access to a particular subset of servers) can obtain this service by directing its data flows to this VSN. Backbone Edge Bridges (BEBs) are locally provisioned so that an end-user VLAN is mapped into a corresponding Backbone Service Instance Identifier (I-SID). BEBs that configure the same I-SID are served by the same L2 VSN.

As explained below, an IP unicast flow transport through an SPB network is executed by encapsulating the IP data packet payload within an Ethernet frame and then the Ethernet frame is further encapsulated with a header that contains a destination B-MAC address. The encapsulated frame is then switched in the layer-2 backbone network by the intermediate backbone switches, identified as Backbone Core Bridges (BCBs), which reside along the backbone path. As noted in the following, the switching and forwarding action undertaken at each intermediate core switch is determined by the forwarding entries embedded in the forwarding table of each switch, which are calculated by using a *link state routing protocol*.

As a frame is received across an incoming line, a BCB reads the B-MAC label that is included in the header of the received frame. It then rapidly (at hardware based speed) searches for a matching address in its forwarding table, which points out the outgoing port(s) to which the frame is switched, avoiding the need to perform hop-by-hop routing calculations.

For the transport of layer-2 or IP multicast flows, an (L2 or L3) VSN is configured and identified by an Instance of a Service ID (I-SID). The employed link state routing protocol advertises this I-SID to backbone network nodes. It is processed only by nodes that have joined as members of the underlying multicast service group.

9 Switching, Relaying, and Local Networking

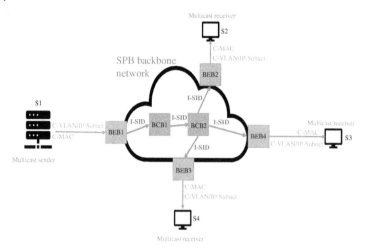

Figure 9.11 A Multicast Flow in a SPB Network Using VSN I-SID Involving Group Members S1, S2, S3, S4.

To join a multicast group, when considering IP flows, users can employ the Internet Group Membership Protocol (IGMP). Multicast subnetworks (such as multicast trees) are formed in the SPB backbone and used for the efficient dissemination of multicast messages along shortest paths. Such an illustrative multicast tree configured in an SPB network is shown in Fig. 9.11.

For the shown customer-VPN (C-VPN), server station S1 multicasts messages to customer stations S2–S4. This multicast VPN arrangement is served by a VSN that is identified by a service ID instance I-SID. At BEB1, based on VSN advertisements disseminated across the backbone in the control plane by the employed link state protocol, such as an IS–IS) protocol, this I-SID is associated with a backbone MAC address (B-MAC).

The data payload message (being a L2 frame or a L3 IP packet) is encapsulated, so that an external frame header that identifies the B-MAC address is included. A multicast backbone tree network is shown to employ backbone switches BCB1 and BCB2. At BCB1, incoming frames are forwarded to switch BCB2. Frames that enter BCB2 from BCB1 are replicated and switched to three outgoing ports, and forwarded to edge switches BEB2, BEB3, and BEB4. At each destination edge switch, the payload message is extracted (and the frame decapsulated so that those extended frame headers are removed) and delivered to the destination station. At each backbone switch, switching is controlled by the forwarding tables in accordance with a joint I-SID/B-MAC labels included in arriving frames.

In the following, we further overview and illustrate key concepts related to the operation of a version of a SPB networking method, while not discussing the detailed operation of a particular implementation of the protocol.

The **network topological layout** consists of an end-user access level and a core network level. End-user stations, such as computers, terminals, smart devices, and servers, are attached to **Edge Switches** at an interface that is known as the User–Network Interface (UNI). **Backbone switches** are used to form the backbone (or core) network. Direct connections between network switches, including edge and backbone switches, use the Network-to-Network Interface (NNI). Five end-user stations are depicted in the illustrative SPB network shown in Fig. 9.12. S1–S3 are data stations while S4 and S5 are server stations. These stations connect directly to edge switches, which are identified as **Backbone Edge Bridges (BEBs)**. Switches 1–9 serve as core switches, identified as **Backbone Core Bridges (BCBs)**.

9.3 Local Networking Across a Switching Fabric: Bridging of MAC Frames

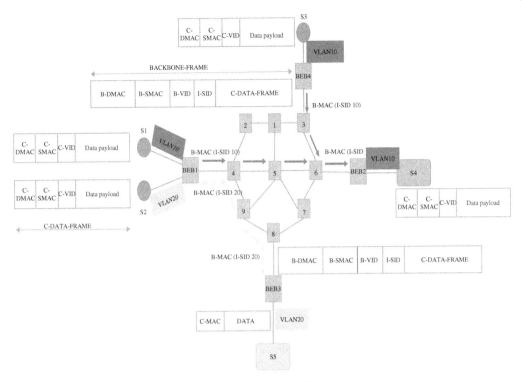

Figure 9.12 Illustrative Layout of a Shortest Path Bridging (SPB) Network.

The SPB network can use a SPBV operational mode or a Shortest Spanning Bridging MAC (SPBM) mode. Under the SPB-VID (SPBV) mode, tagging is employed to perform shortest path forwarding. Message flows produced by end-users that are members of a Virtual LAN (VLAN) are identified by a VLAN ID (VID) tag, such as that defined by IEEE 802.1Q. The use of VLAN tagging facilitates the realization of a scheme to virtualize layer-2 broadcast domains, leading to **VLAN Virtualization**. The 12 bits that are available in the 802.1Q specification enable the configuration of 4096 individual virtual LANs. To improve scalability, the frame header has been extended (following the IEEE 802.1ad specification). In addition to the end-user VID tag included in the message payload, an additional header is defined and used to include a carrier tag. The carrier tag is the one that the network switches use in their forwarding tables to forward the data frames that they receive.

Another operational SPB mode, identified as noted below as SPBM, provides enhanced scalability and isolation of end-user MAC addresses. We will henceforth focus our description on this mode.

Under the SPBM mode, the end-user's frame, which contains the **customer's MAC address (C-MAC)** is encapsulated by an header that contains an additional **backbone MAC header (B-MAC)**. It follows the IEEE 802.1ah Standard and is identified as **MAC-in-MAC encapsulation**. In this manner, customer MAC addresses are hidden from backbone switches whose forwarding decisions are based on B-MAC addresses and service identifiers (I-SID), leading to enhanced network scalability and operational efficiency. The use by SPB of a layer-3-oriented routing method prevents network looping issues, so that the limiting use of a STP to synthesize a spanning tree which is used to regulate the flooding of user frames and the consequent construction of forwarding tables is avoided.

9.3.4.2 Illustrative SPB Network

An SPB network and associated frame formats are illustrated in Fig. 9.12. End-user station frames are noted to contain (among other fields that are not explicitly shown) data payload fields and customer MAC (C-MAC) address fields that include end-user destination and source MAC addresses. The encapsulated backbone frame is illustrated as well. It contains the client's frame (including its payload data and C-MAC fields), encapsulated by a VSN service ID (I-SID) field and encapsulated again with a backbone MAC (B-MAC) address field, containing backbone destination and backbone source addresses, B-DMAC and B-SMAC, respectively.

Also illustrated in this figure is the mapping of VLAN IDs at the BEBs to VSN IDs (I-SIDs) and then to B-MAC addresses. These addresses are used by network switches to route user data frames to their destination edge bridges. End-user provisioning is performed only at the BEBs. Performing such provisioning at core SPB nodes is not required. Such an architecture provides a robust and flexible mechanism in supporting the addition or modification of services.

Backbone switches learn and use B-MAC addresses to make forwarding decisions while an edge BEB switch learns both B-MAC and C-MAC addresses for each VSN. A Backbone Service Instance Identifier (I-SID) is assigned at the BEB to each VLAN. All VLANs in the network that share the same I-SID participate in using the same VSN.

The use of two VSNs to guide the distribution of message frames is illustrated in Fig. 9.12. Stations S1 and S3 communicate with server S4, as part of VLAN 10, across a VSN which is identified as I-SID 10. Assuming each link to have a cost metric that is equal to 1, shortest paths between S3 and S4 and between S1 and S4 are noted. Communications from S4 to S1 and S3 would proceed across the same paths (in the reverse direction). This communication can be arranged to consist of two unicast paths (one from S4 to S3 through their edge bridges, BEB2 to BEB4, and the other from S4 to S1, as BEB2 to BEB1). Alternatively, a **multicast tree** can be used. It is set as a **shortest path tree (SPT) rooted at BEB2**, trimmed to include only involved backbone nodes. In this case, frames that originate at BEB2 and are multicasted to VSN members are split at node 6. A copy of each frame is sent from node 6 to BEB4 across the illustrated shortest path and another copy of each frame is sent from node 6 to BEB1 across the illustrated shortest path.

Under VLAN20, a different VSN is illustrated. It is identified by I-SID 20, and it involves communications between station S2 and server S5. The shortest path that is used by a corresponding unicast flow between edge bridges BEB1 and BEB3 and the corresponding encapsulated frames are also noted.

It is possible to produce, in a similar manner, an SPB L3 VSN topology. A Backbone Service Instance Identifier (I-SID) is then assigned at a Virtual Router (VRF) level instead of at a VLAN level. All VRFs in the network that share the same I-SID can share the same VSN. An IP-based link state routing scheme could then be employed.

9.3.4.3 The Control Plane: Link State Dissemination and SPT Constructions

At the **control plane**, a layer-3 (L3) routing protocol is employed to calculate the shortest paths from any network node to any other network node. Various L3 routing protocols can be used. Such protocols and the routing algorithms that they use are discussed in the Routing chapter. It is helpful to the reader to review first this chapter. Included are **Link State Routing** protocols such as the IS–IS protocol that is used by SPB. Other such employed routing schemes include Open Shortest Path First (OSPF), a link state routing scheme, and distance-vector-oriented routing schemes, including Border Gateway Protocol (BGP), which is employed in IP-based Internet systems.

The backbone network can be shared among several virtual networking tiers, each of which is identified by a backbone-VLAN ID (B-VID). For each node to acquire the state of the network's backbone layout, the B-MAC addresses of each node are advertised by the IS-IS scheme in the context of one or more B-VIDs. Each SPBM network instance is associated with at least one B-VLAN in the core network. This B-VLAN can be used by both control plane and data plane traffic flows.

Using IS–IS hello messages, each node learns the topology of the network as well as the affiliation of nodes with VSNs, B-VLANs, and the corresponding end-user stations that are attached to the BEBs. BEBs learn from other BEBs in their B-VLAN as to the stations attached to them so that it can direct frames to a specific targeted station.

Using an employed link state routing scheme, each backbone node calculates the shortest paths from itself to every other backbone node, for each VSN.

Under the use of a link state routing protocol, each switch node disseminates to all the network nodes, periodically and in response to state changes, the status of its attached links. Link states identify for each node the current status of nodal neighbors and of each one of the associated links connecting the underlying node to each one of its neighbors. State information may include each link's current capacity, traffic loading level, delay, throughput, and reliability performance metrics. Cost metrics that reflect a weighted composite of several performance parameters may also be used. A path cost (also identified as a path "length", metric, distance, or weight) is calculated as the sum of the costs of its links.

To demonstrate the types of state data disseminated across the control plane of a SPB network by a link state routing algorithm such as IS–IS, we note this protocol to define, among other, the following state parameters:

1) The IS–IS Hello (IIH) Protocol is used by a node to detect its SPB-capable neighbor nodes. It is also used to exchange state digest data. The data is used to validate the commonality of Link State Databases (LSDB) in the same SPB domain. Digest data includes:
   a) Bridge identifiers, concatenating bridge priority and bridge sysID.
   b) Link metrics involving neighboring nodes that share the same link.
   c) Associations between Virtual LAN IDs (VID) and equal-cost tie-breaking algorithms.
2) SPB instance that includes the following nodal information:
   a) SPSourceID: 20-bit value of nodal nickname. It is used to form multicast destination address (DA).
   b) Bridge Priority: 16-bit value combined with 4 bytes System ID, forming the Bridge Identifier.
   c) Number of Trees: Set to the number of VLAN-ID tuples.
   d) VLAN-ID tuples: Consists of 4 bytes ECT-Algorithm, 12-bit Base VID, 12-bit SPVID, and U, M, A flag bits.
3) Adjacency information extensions. Including:
   a) SPB-Link Metric: 24-bit unsigned number. Administrative weight of a link. Smaller number indicates lower weight. When two neighbor nodes advertise different SPB-Link Metric value on the same link, the maximum value is considered as the right value of the given link.
   b) Port Identifier: 16-bit.
4) Service information extensions that are carried in SPBM Service Identifier and Unicast Address (SPBM- SI). Including:
   a) B-MAC address: A unicast address of the node. Each SPB node has a unique nodal Backbone MAC address.
   b) Base VID: It is also included in VLAN-ID tuples in SPB-Inst. B-MAC address, Base VID, and ECT-Algorithm are thus associated.
   c) I-SIDs: 24-bit group identifier. This I-SID set is assigned the Base VID in the same SPBM-SI. If two different nodes advertise the same I-SID, intermediate nodes between the two nodes

will configure themselves to carry traffic generated from the two different nodes. The intermediate nodes will create a Filtering Database (FDB) for underlying unicast and multicast addresses.

IS–IS packets disseminated across the control plane include the following types:

1) Hello packets are exchanged periodically between two switches to establish adjacency. One of these switches is selected as the DIS (Designated IS). When IS–IS is activated on a routing device's interface, the device first sends some IS–IS hello packets (IIHs) to its neighbors to ensure that the circuit is capable of transporting packets in both directions. In the IIHs, the router embeds information about the designated router.
2) An Link State Packet (LSP) carries routing-oriented data, including a node's list of neighbors and the metrics of the node's attached links. One of the roles undertaken by a designated router placed on an IS–IS broadcast circuit is to synchronize the link-state databases used by associated LANs. The designated router sends periodically a directory of its link -state database, which is received by all routing devices associated with the underlying LAN.
3) A complete sequence number PDU (CSNP) packet is sent periodically only by the DIS. It contains the list of LSP ID's along with sequence number and checksum.
4) A Partial Sequence Number Packet (PSNP) is sent by a switch when upon receiving a CSNP packet it discovers discrepancy in its own database. It uses this packet to request the DIS to resend a specific LSP.

A shortest path between a pair of nodes is a path whose length (or cost) is equal to the lowest cost of any path that connects these nodes. A **SPT** rooted at node $B1$, denoted also as $B1 - SPT$, is a tree graph whose nodes represent the backbone network's nodes and whose lines are selected so that the cost of the unique path connecting root node $B1$ with any other network node in the network domain, say $B2$, is equal to the lowest feasible cost, so that no other path that connects these two nodes displays a lower cost. Under a layer-3 IP routing scheme, nodes are identified by their IP addresses. In an L2-oriented system, nodes are identified by L2 MAC addresses. By selecting paths in this manner, no network looping issues need to be resolved. It is thus not necessary to execute flooding and source learning processes, as used in the operation of a STP.

Under the link state scheme, whereby the link state of each node is broadcast across the control plane of the backbone network , each backbone node acquires the complete topology of the backbone network as well as the status of each one of its links. Using this information, each backbone node employs a routing algorithm (such as Dykstra's algorithm, as described in the Routing chapter) to calculate, on its own, the SPT and consequently determine a shortest path (SP) between itself and any other node in the network domain.

Forwarding computations are performed in a distributed fashion. Each node computes its forwarding behavior independently, based on a synchronized common view of the network and of its customer and service attachment points. **Forwarding tables**, also called **Forwarding Data Bases (FDBs),** are populated locally by each network node. They are used to implement the node's forwarding behavior. As discussed below, these tables are composed in a manner that is consistent with shortest path calculations performed by other nodes. A **Content-Addressable Memory (CAM)** structure is often used in performing the forwarding operation. This hardware-driven module is designed to perform at ultrahigh speed the operation of matching the B-MAC address (and other labels, when applicable) included in the header of an incoming frame with a corresponding content word nested in the forwarding table, consequently producing (at very low latency) the identity of the forwarding port.

To illustrate, we show in Fig. 9.10(c) an SPT that is rooted at node 8. Using this SPT, the shortest path from node 8 to node 6 consists of the two links that connect these two nodes through node 7. In contrast, when using a spanning tree graph derived by a STP, as shown in Fig. 9.10(b), the path employed to connect node 8 with node 6 traverses five links, transiting intermediate nodes 9, 5, 1, and 3. This realization demonstrates the significant enhancement that can be attained in the network's delay-throughput performance behavior by using SPTs rooted at each source node to designate the routes to be used by MAC frames, rather than using a single SPT graph. The frame forwarding tables at the switches are accordingly synthesized.

9.3.4.4 Multitier Overlay: Data Transport Across Multiple Equal-Cost Paths

Performance efficiency can be further enhanced by allowing a source node to split different flows across multiple paths that connect to a destination node(s). This involves the construction by a node of **fat multipath trees**. This process accommodates the use of **multiple equal-cost paths** employing multiple equal-cost SPTs. Such an approach is the basis of *Equal-cost multipath (ECMP) routing*, under which message forwarding to a destination can occur over multiple best paths.

An SPB network supports up to 16 B-VLANs and each node can build a distinct (equal-cost) SPT for each B-VLAN. The latter is identified by a B-VID. Load balancing is accomplished by mapping different services (I-SIDs) to different B-VLANs. In the following, we refer to each B-VLAN level as a **tier**. Different equal-cost unicast paths or multicast trees are linked to different tiers.

Across each tier, a unicast message flow between a pair of backbone nodes uses a single tandem path. The forwarding entries at the intermediate nodes along this path are set in such a manner that the identity of this path, as determined by its source backbone node, is preserved along the path. For example, if source node B1 determines the tier's path to destination node B4 to traverse nodes B1–B2–B3–B4, then the forwarding entries in core node B2 are set to forward frames to the outgoing port that connects it to node B3 and the entries in core node B3 are set to forward frames to the outgoing port that connects it to node B4. Thus, within a tier, the source edge bridge effectively selects the path's intermediate core nodes. An intermediate node does not alternate the route. In this way, we observe that *network paths are preconfigured and frames are delivered in the order that they were sent.*

Such a path preservation feature is generally identified as *path congruency*. It applies to the configuration of SPTs at distinct backbone nodes within each tier. A backbone node may configure multiple **equal-cost trees (ECTs)**, each linked to a distinct tier, identified by its corresponding B-VLAN ID (B-VID).

As noted in IETF RFC 6329, an important property of SPB networks is **path symmetry**. In a tier, between any source–destination pair of nodes, routes are *forward and reverse path symmetric*, so that the path from node B1 to node B2 is identical to the path from node B2 to node B1.

Also, such routes, when used for either unicast or multicast flows, are characterized as being **congruent** in the following sense. The SPT to node B is congruent with the **multicast distribution tree (MDT)** from node B. *The MDT for a given B-VLAN is a pruned subset of the complete MDT for a given node that is identical to its SPT.* Such *Symmetry and congruency* properties serve to preserve packet ordering and proper functioning of regular Operations, Administration, and Maintenance (OAM) schemes. This is not the case for certain IP-based technologies, such as Multiple Protocol Label Switching (MPLS) that is discussed in a later section, in which the reverse path may differ from the forward path.

Enforcing congruency and symmetry restricts the topological layouts of acceptable trees that should be constructed to guide the dissemination of a group of flows; hence, limiting link utilization levels. However, utilization is improved through the employment of (up to 16) tiers and the distribution of traffic among tiers.

Each backbone node constructs its forwarding table based on topological connectivity, link costs, and VSN data broadcast across the control plane. To assure these constructions preserve congruency, an **equal-cost tie-breaking algorithm, named ECT-ALGORITHM** is defined and used to determine the selection by each backbone node of equal-cost paths and trees.

A path between two nodes is represented as a sequence of 8 byte Bridge Identifiers, whereby 2 bytes of each Bridge Identifier is Bridge Priority. Among equal-cost paths, a path with lower hop counts has the higher priority. If two paths have the same cost and hop count, a value is calculated for a function of two parameters: a per-tier ECT-ALGORITHM-based value and the corresponding path Bridge ID value. A path that is associated with a lower is deemed to be preferred. This calculation is performed at each involved backbone node in each tier, leading to the construction of congruent unicast and multicast forwarding entries at each node, in each tier.

As an example, illustrating the configuration of consistent (congruent and symmetric) paths and multicast SPTs, we consider the network shown in Fig. 9.13(a). We assume the cost metric of each line to be equal to 1. In this case, the path cost is equal to the number of hops (or links) traversed along the path, whereby a hop represents a link that connects two neighboring switching nodes. Assume the underlying VSN to involve stations {S1, S2, S3}, which are connected at corresponding edge bridges {1,6,5}. The system defines two tiers, associated with equal-cost trees for this VSN and its I-SID = 0001, identified by B-VIDs =0100 and 0300.

In Fig. 9.13(b), we show a SPT rooted at node 1. This SPT, which we identify as T_1^{0300}, and which is configured by node 1, is associated with tier 1 that is identified by B-VID=0300. This tree represents

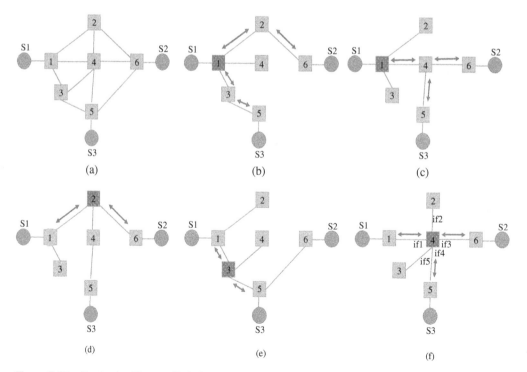

Figure 9.13 Illustrative Shortest Path Trees in a SPB Network: (a) Network Layout, (b) SPT1/B-VID=0300 Rooted at Node 1, (c) SPT2/B-VID=0100 Rooted at Node 1, (d) SPT1/B-VID-0300 Rooted at Node 2, (e) SPT1/B-VID-0300 Rooted at Node 3, and (f) SPT2/B-VID-0100 Rooted at Node 4.

9.3 Local Networking Across a Switching Fabric: Bridging of MAC Frames | **339**

the tier 1 shortest paths from node 1 to other nodes in the backbone network. Another SPT rooted at node 1 is presented in Fig. 9.13(c). Assume that this SPT, identified as T_1^{0100}, is associated with tier 2 that is identified by B-VID=0100. This tree represents shortest paths that are used in tier 2 from node 1 to the other nodes in the backbone network.

In Fig. 9.13(d), we show a SPT rooted at node 2, associated with the tier identified by B-VID=0300. It is denoted as T_2^{0300}. It is noted that the corresponding lowest cost paths induced by this tree are congruent and symmetric with those of tree T_1^{0300}. For example, under T_1^{0300}, the shortest path from node 1 to node 6 consists of nodes {1,2,6}. The shortest path between node 1 and node 6 under tree T_2^{0300} remains the same. Consequently, the forwarding table configured at node 2 in accordance with T_2^{0300} switches frames of a flow with node 6 as its destination, which arrive to node 2 from node 1, to depart at the port that connects it to node 6. Thus, the configuration of the forwarding table at node 2 is such that the path from S1 to S2 configured at node 1 is conserved at node 2.

In Fig. 9.13(e), we show a SPT rooted at node 3, associated with the tier identified by B-VID=0300. It is denoted as T_3^{0300}. It is noted that this SPT conserves the path between S1 and S3.

In Fig. 9.13(f), we show a SPT rooted at node 4, associated with the tier identified by B-VID=0100. It is denoted as T_2^{0100}. It is noted that the corresponding shortest cost paths induced by this tree are congruent and symmetric with those of tree T_1^{0100}. Under these trees, the forwarding table entries set at nodes 1 and 4 preserve the corresponding paths used to connect associated pair of stations. We observe the congruency of the paths used for frame distributions along the multicast trees rooted at edge nodes 1 and 4.

According to Fig. 9.13(c), multicast frames produced at S1 and targeted to reach stations S2 and S3 use the SPT (and corresponding forwarding entries) generated at edge node 1. These frames are forwarded by node 1 to node 4. There, the multicast flow is split. One copy of each frame is forwarded to node 6 while a second copy is forwarded to node 5. This is exactly the forwarding action undertaken by node 4. Its forwarding entries are determined by the topology of the tree rooted at node 4 (calculated by node 4 in a manner that preserves the paths planned at source node 1), as confirmed by the layout of the SPT rooted at node 4 that is shown in Fig. 9.13(f).

9.3.4.5 The Forwarding Data Base (FDB)

For this network system, an illustrative format of the forwarding table, also known as a Forwarding Data Base (FDB), configured at node 4 for the underlying I-SID linked to the tier identified by B-VID=0100, in accordance with the SPT depicted in Fig. 9.13(f), is shown in Fig. 9.14. The first five rows of the table contain entries that are used to forward frames that belong to a unicast flow that arrives at one of the input interfaces (IN_IF) of node 4. For example, the first row indicates that frames that arrive at any input interface to node 4 whose frame header identifies their destination, as specified by a backbone destination MAC address (B-DMAC), to correspond to edge bridge named as node 1, under B-VID = 0100, should be forwarded to the output identified by interface OUT_IF = if1.

The last three rows of the forwarding table at node 4 show the entries that are used for forwarding multicast frames. Source specific multicast trees are used. Such a tree is often denoted as (S,G), where S designates the source node and G denotes the multicast group, which consists of the nodes that have joined as members of the multicast group. This information is encoded in the destination B-MAC address, which is configured as the concatenation of a 20-bit SPB-wide unique nickname of the source root node (referred to as the SPSourceID) and the 24-bit I-SID together with the B-VID that relates to the defined ECT-ALGORITHM.

For example, multicast frames that arrive at interface 1 (if1), IN_IF = if1, are encapsulated by a header that identifies their destination address by specifying the combined SPSourceID/BVID

U/M	IN_IF	Destination address (DA)	BVID (I-SID)	Out_IF(s)
U	ifxx	B-DMAC (xx-001)	0100	if1
U	ifxx	B-DMAC (xx-002)	0100	if2
U	ifxx	B-DMAC (xx-003)	0100	if5
U	ifxx	B-DMAC (xx-005)	0100	if4
U	ifxx	B-DMAC (xx-006)	0100	if3
M	if1	SPSource(N1)-BVID(0001)	0100	If3, if4
M	if3	SPSource(N6)-BVID(0001)	0100	if1
M	if4	SPSource(N5)-BVID(0001)	0100	if1

Figure 9.14 Forwarding Table/Forwarding Data Base (FDB) at Node 4 Based on Multicast Shortest Path Tree (MSPT) Rooted at Node 4 at Tier B-VID=0100.

(I-SID), whereby here the SPSource corresponds to the nodal name of node 1, and BVID(0001) = 0100 that corresponds to the B-VID associated with I-SID=0001, noting that, for brevity, reduced identifiers are used. The outgoing interfaces, OUT_IF(s), to which such arriving frames are forwarded by node 4 are if3 and if4, accounting for the frame replication that is executed at node 4 as frames arriving from node 1 are split, so that copies are sent to nodes 5 and 6. The last two rows in the forwarding table indicate that frames that arrive from nodes 6 and 5 are forwarded to outgoing interface if1, and are thus sent to node 1. It is noted that frames that are sent from node 6 to node 5, or vice versa, use a shortest path that consists of the direct line that connects nodes 6 and 5. Hence, such frames do not traverse node 4, and consequently the illustrated table does not contain forwarding entries for such flows. Hence, frames that arrive on input interface if3 (if4) are forwarded just to interface if1.

Following the construction of the network's layout enabled by the link -state data dissemination process, source switches proceed to construct (unicast and multicast) shortest paths and trees and assign traffic flows to these paths. In summarizing the operations, the methods that are used for the forwarding of data flows, and also illustrated by the table shown in Fig. 9.14, are outlined as follows:

1) **Data Paths for Unicast Flows:** A customer's data frame, such as an Ethernet frame, is encapsulated. The added header includes the following fields: Backbone Source Address (B-SA) that consists of the B-MAC address of the source edge backbone BEB; Backbone Destination Address (B-DA), which identifies that B-MAC of the destination edge node BEB; Backbone VLAN ID (B-VID); VSN ID, I-SID. At a backbone node, in guiding tandem NNI to NNI unicast flows, the forwarding table (that has been pre-populated through dissemination and computations carried out in the control plane) is used to forward incoming frames by mapping {B-DA, B-VID} to an outgoing port. An ingress check is used to mitigate looping. It confirms, by using the entries included in the forwarding table, that the reverse path with destination parameters set to {B-SA,

B-VID} points to an outgoing port that is the same as the input port at which the frame has arrived.

2) **Data Paths for Multicast Flows Performed by Head-End Replication:** Under this scheme, the frame is replicated at the source edge node, once for each receiving edge node that is registered as a member of the same I-SID. Each copy is then encapsulated and sent as a unicast frame. Such a multicasting operation is effective when the multicast membership is relatively low or when the underlying multicast traffic flow rate is low.

3) **Data Paths for Multicast Flows Performed by Tandem Replication:** A SPT rooted at the source node is constructed. Frames are replicated (i.e., copied and transmitted) at *tree forks* that require splits. For such a dissemination, customer frames are encapsulated at the source BEB node with a Backbone Destination Address (B-DA) that is configured as a multicast address. The B-DA concatenates the source node ID, denoted SPSourceID (20-bit long) and the associated I-SID (24-bit long), which identifies the backbone nodal group (G) of this multicast group. This multicast tree is denoted as (S,G) = (SPSourceID, I-SID). This DA label is contained in the forwarding tables of all backbone nodes that reside on this multicast tree. Ingress check by using B-SA/B-VID entries may also be performed here, as done for unicast flows.

4) The multicast forwarding tables are created based on computations that account for every node that is on the shortest path between a pair of nodes which are members of the same service group. Only nodes located on a shortest path create forwarding database (FDB) entries, so that the multicast flow dissemination process makes efficient use of network resources.

5) Among equal-cost paths, a path with the lowest number of hops is preferred. If still the same, an ECT-ALGORITHM is used to break ties. ECT-ALGORITHMs are mapped to B-VIDs. Services that are assigned to these B-VIDs. Base VLAN Identifiers (SPB B-VID) provide information as to which SPT sets are used by which VIDs. The ensuing architecture is used to provide a service assignment to each VID that is network-wide end-to-end consistent. It can be used to regulate traffic loading rates to manage targeted performance metrics.

Problems

9.1 Consider the switch module shown in Fig. 9.3. Write the switching matrix that corresponds to the displayed switch connections.

9.2 Consider the illustrative switching network shown in Fig. 9.4. Assuming the switches to be configured to have the input–output connections shown in the figure, indicate the shortest paths, considering those that traverse a minimum number of inks, that would be selected for routing messages that are transported between the following source and destination stations:
   a) Source station = S13; destination station = S14.
   b) Source station = S12; destination station = S4.
   c) Source station = S1; destination station = S4.
   d) Source station = S2; destination station = S21.
   e) Source station = S21; destination station = S23.
   f) Source station = S21; destination station = S24.

9.3 Discuss and illustrate the operation of a radio relay that is placed at a top of a tall tower in serving to relay messages sent between two stations.

342 | *9 Switching, Relaying, and Local Networking*

9.4 Describe and discuss the different modes of operation of repeater modules.

9.5 Describe and discuss the operation and compare the performance features of the different modes of operation of the following satellite network systems:
a) Onboard processing satellite using RF switching.
b) Onboard processing satellite using store-and-forward packet switching.

9.6 Consider the configuration illustrated in Fig. 9.6, depicting a realization of a Vehicle Backbone Network (VBN) of interconnected vehicles that move along a two-lane segment of a highway.
a) Identify the path taken by packets transmitted by source node RN_1 as they are disseminated across the shown segment of the highway.
b) Identify the time slots used by each relay node to transmit these packets, in using a time frame that consists of three slots, and assuming the RSU to employ the described flow control process.
c) Assume now that the employed time frame consists of four slots. The RSU employs a corresponding flow control scheme that aims to reduce message queueing delays. Each relay node transmits a packet in its allocated time slot. Describe the operation of this spatial reuse-4 scheme and discuss its advantages and disadvantages in comparison with the presented spatial reuse-3 scheme.

9.7 Describe the operation of a layer-2 bridge node.

9.8 Consider a layer-2 network, such as one that involves LANs that are interconnected by bridges. Describe the operational principles and performance features of the following concepts:
a) Virtual Private Network (VPN).
b) Virtual Service Network (VSN).
c) Backbone Edge Bridges (BEBs) vs. Backbone Core Bridges (BCBs).
d) User-to-Network Interface (UNI) vs. Network-to-Network Interface (NNI).
e) Shortest Path Bridging (SPB).
f) Forwarding Data Base (FDB).
g) Multicast Distribution Tree (MDT).
h) Equal-Cost Tree (ECT).

9.9 Describe and illustrate the operation of a multitier L2 network overlay that employs multiple equal-cost paths.

9.10 Consider a network that consists of three LANs that are interconnected by bridges, as shown in Fig. 9.8. Describe a method that can be employed to determine a path that is used by frames issued by Station A for routing to Station C.

9.11 Consider the network shown in Fig. 9.10(a) and specified therein. Derive and present the topological layout of a SPT that is rooted at node 6.

9.12 Consider the network shown in Fig. 9.13(a).

a) Explain the process used to configure consistent (congruent and symmetric) paths and multicast SPTs as shown in Fig. 9.13(b)-(f).

b) Explain the formation and contents of the forwarding table configured at node 4 that is shown in Fig. 9.14, noting that it involves the underlying I-SID linked to the tier identified by B-VID=0100, in accordance with the SPT depicted in Fig. 9.13(f).

10

Circuit Switching

10.1   Circuit Switching: The Method

The Objective: *Certain calls require their message flows across a network route to be transported at specified quality of service (QoS) levels. This is often the case when streaming in real time at high-quality voice and video calls, as well as when supporting the transport at high data rate of high-throughput data flows issued by continuously active sources. For this purpose, it is essential that the network accommodates such a call by assigning to it sufficient communications resources on a dedicated basis. A circuit switching (CS) network is designed to provide such a service. In a CS communications network, users that initiate a call, request the network to configure a connection that is assigned sufficient resources so that it can be used for the effective realtime accommodation of the call's flows. The network uses a signaling system (SS) for facilitating the connection set up process. Network resources are allocated to the call on an exclusive basis, and the connection is established before the user is signaled to start the transport of its call's messages.*

The Method: *Per request by an end user, the SS sets up a connection and assigns it a **circuit** across the communications network. The circuit is used to accommodate the transfer of the connection's messages. A circuit spans the configured end-to-end network route. The circuit is assigned dedicated communications bandwidth resources across each one of its route's links. These resources remain allocated to the circuit even when the call is inactive and produces no message flows. An advantage of this operation is the high QoS that the connected end user is provided. This is realized by the exclusive allocation of link resources and often also associated nodal and processing resources, to the user's circuit. If such resources cannot be made available, the end-user's request for accommodation is blocked. The forwarding entries in a CS table that correspond to a configured circuit are set by the SS prior to the start of communications. The ensuing switching operation can be conducted by using hardware-based operations, enabling the support of ultrahigh-speed switching rates. A key disadvantage of such an operation materializes when the user's connection is active in a statistically bursty manner, as is the case for many data flows. In this case, allocated resources remain underutilized, leading to low utilization of communications capacity resources. In the latter case, to increase the utilization level, it would be preferred to employ a packet switching method.*

Under a *circuit switching (CS) method*, as employed by a traditional Public-Switched Telephone Network (PSTN) and by other systems, a **signaling system (SS)** is used to configure a **connection** across the network, which is used to accommodate a user's **call**. An end-user transfers its call's messages across a configured network's route to a destination end-user station. A circuit is established to support the transfer of the call's messages.

The establishment of a circuit entails the joint setting of two elements: across the **network spatial dimension**, an end-to-end route is configured; and across the **communications resource**

Principles of Data Transfer Through Communications Networks, the Internet, and Autonomous Mobiles, First Edition. Izhak Rubin.
© 2025 The Institute of Electrical and Electronics Engineers, Inc. Published 2025 by John Wiley & Sons, Inc.

dimension, communications resources are allocated to each link of the connection's route. These communications resources are assigned at each one of a circuit's links on a dedicated basis and are reserved exclusively for the transport of messages that are disseminated by the connection's user.

The transport of the call's data messages is initiated only after its circuit has been successfully established. If no sufficient spatial and capacity resources are available for configuring such a circuit, at a quality level that accommodates the user's request, the call is blocked.

If a call is admitted, the call's user makes use of its allocated circuit's resources. These resources are typically assigned for the duration of a user's call. When a call ends or is terminated, the call's connection is terminated as well. The associated circuit is dis-established and its allocated communications resources are released. They then become available for use in configuring new circuits that are used to accommodate new call connections.

In a highway transportation system, an operation that is analogous to a CS discipline entails the following. A driver plans his trip ahead of time, selecting its network route, identifying the roads and intersection turns that will be taken. The driver would then make *reservations* of highway resources for her exclusive use along the planned travel route. Such resources can include time slots and space segments associated with the use of roads and intersections that the driver will be traversing along her selected path. In an air transportation system, a passenger would reserve seats on consecutive flights that lead to her destination.

10.2   The Circuit Switching Network System Architecture

A **SS** is used to facilitate the establishment and termination of a connection and the setting up of a network circuit that is used to carry the messages sent by a user (also identified as an end user, customer, client, or subscriber). In a traditional wireline-based **PSTN** system, also known as a *Plain Old Telephone System (POTS)*, the calling user lifts up the handset of the telephone device and receives in response a dial tone from an edge switch, known also as a **local exchange** or an **end office**.

The end user is connected to an edge switch through a **client access network**, known as the **local loop**. A wireline telephone device is connected via a fixed line to the local loop system. This is done via an analog copper access line, or via digital access channels that may employ copper wires, coaxial cables, fiber-optic lines, or a hybrid of such and other media. Multiple voice/data/video streams can be multiplexed across a digital access line. Such subscribers connect to a network that is designed to support multimedia flows by following standards based protocols, as those defined by Integrated Services Digital Network (ISDN) standards.

The user's device that enables its communications across the network, also known as **customer premises equipment (CPE)**, may be a plain old telephone set or an intelligent computerized terminal or handset. In Fig. 10.1, we show subscribers S1–S3 to connect to the end office via a local loop access network. The communications capacity resources of the local loop system are shared among local users. A user's local access connectivity to an edge switch can be realized by various communications means, including copper wires, copper links that are conditioned via signal processing methods to accommodate higher data rates and forming the Data Subscriber Line (DSL) system, CATV access system that uses a cable plant to access a central hub known as an Headend, fiber-optic, and wireless access channels. High-capacity local loop systems are often synthesized through the use of ring networks that employ fiber-optic links and their interconnects.

In a POTS network, when using a legacy telephone device to initiate a call, the source user dials the telephone number of the destination party. The address of the called party is included in a

10.2 The Circuit Switching Network System Architecture

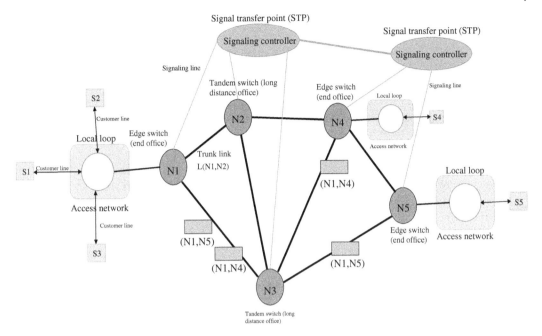

Figure 10.1 Architectural Depiction of a Circuit-Switched Network System.

signaling message, forming a *call request message*. In advanced implementations, it is possible to include in signaling messages additional information, such as data relating to the quality of service (QoS) requested and information that identifies the use's preference for specific call features.

A computerized end-user device can be employed and used by a customer to specify the destination address and identify requested features for the ensuing connection. A client uses a *Send* button to send her call request message across a customer access line or a shared medium access network. The message is received at a local Edge Switch (End Office). In Fig. 10.1, the call request message initiated by a local user, such as S1, would be transported across the local loop to edge switch N1.

Upon receiving a call request message, the edge switch initiates a signaling process. The SS triggers the setting of a circuit that will be used to support the call's connection, serving to carry the data messages produced by a caller. In Fig. 10.1, the circuit-switched network includes circuit switches N1–N5. Trunk links are used to connect neighboring circuit switches. A trunk link can consist of multiple lower speed lines or employ higher speed communications lines that are shared among multiple connections on a fixed multiplexing (such as TDM) basis. A call may be assigned a fixed number of time slots during each time frame, across each network trunk that it traverses, for as long as the call has not been terminated by the user.

The establishment of a circuit, as performed by a SS, entails the following key processes:

1) **Routing:** The SS determines which end-to-end path to use in routing the call's messages. The route may be used by messages produced by full-duplex end-to-end communications, so that it is employed for the simultaneous exchange of messages in both directions. It is also possible to configure unidirectional paths.

   To illustrate, we show in Fig. 10.1 the configuration of two routes:
   (a) For a call from user S1 to user S4, the corresponding path that is configured between end-office N1 and end-office N4 traverses tandem switch N3 and the corresponding trunks

(N1,N3) and (N3,N4). It is noted that a telecommunications line or link attached to a switching office is often identified as a trunk.

(b) For a call from user S1 to user S5, the corresponding path that is configured between end-office N1 and end-office N5 traverses tandem switch N3 and the corresponding trunks (N1,N3) and (N3,N5). End users are connected to their edge switches through direct communications means across their corresponding local loop systems.

The SS and the associated call setup management system are kept aware of the availability of capacity resources, including residual capacity levels at system trunks, as well as the congestion status and other QoS metric values associated with the network's nodes and links. Accordingly, routes can be selected in a manner that meets call QoS requirements and network load balancing objectives. While it is generally advantageous to configure short routes (i.e., routes that employ a lower number of links), it is essential to not overload nodes and links.

2) **Communications Capacity/Bandwidth Assignment:** Under a CS scheme, a circuit assigned must provide the communications capacity required by the connection. For example, a circuit that carries a voice call connection at toll quality level, may accommodate a voice stream that is produced by using Pulse Code Modulation (PCM), and would thus provide the flow a data rate of 64 kbps. Such a data rate level must be provided across each trunk that makes up the corresponding circuit's path.

For example, consider the circuit established in Fig. 10.1 from N1 to N5, using tandem switch N3. Assume that network trunks are shared on a Time Division Multiplexing (TDM) basis, using a time frame that contains four time slots. As illustrated in Fig. 10.2(a), the circuit that connects edge switches N1 and N5 through tandem switch N3 is allocated slot 1 in each time frame (for the duration of the call) across link (N1,N3) and slot 1 in each time frame across link (N3, N5). In turn, as illustrated in Fig. 10.2(b), the circuit that connects edge switches N1 and N4, through tandem switch N3, is allocated slot 2 in each time frame (for the duration of the call) across link (N1,N3) and slot 4 in each frame across link (N3, N4).

Consider the TDM multiplexing arrangement that is made across each trunk. For illustration purposes, assume all trunks to have the same TDM configuration and data rate parameters. Let the peak data rate used across a trunk be equal to R_p [bps] and the duration of each time slot be denoted as τ [s]. Then, each slot carries a segment that is b [bits] long, where $b = R_p \tau$. Assume the TDM time frame to consist of N [slots], and a circuit to be allocated k slots in each time frame. The circuit's data rate, denoted as R_c, is then equal to $R_c = kR_p/N = k\frac{b}{N\tau} = k\frac{b}{T_F}$ [bps], where $T_F = N\tau$ [s] is the duration of a time frame.

For the illustrative circuits of Fig. 10.1, we have set $k = 1$ [slot/frame] and $N = 4$ [slots/frame]. Hence, a circuit is allocated a data rate that is equal to $R_c = R_p/4$ [bps] across each trunk that is included in the circuit's path. Consider a circuit to be used for the transport of a PCM voice stream

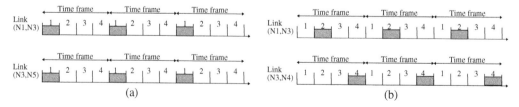

Figure 10.2 Illustrative Circuit Capacity Allocation in a Circuit Switched Network: (a) For a Circuit Connecting Edge Switches N1 and N5 Through Tandem Switch N3; (b) For a Circuit Connecting Edge Switches N1 and N4 Through Tandem Switch N3.

10.2 The Circuit Switching Network System Architecture 349

for which we set $b = 8$ [bits/slot] and $T_F = 125$ μs. $R_c = 64$ [kbps/circuit] is the data rate required to support a PCM voice stream, which is equal to its peak rate, assuming a voice activity factor of 100%. Since a CS system involves the employment of a **fixed multiplexing** scheme, the voice circuit is assigned a data rate that is equal to its peak rate. In comparison, in a packet switching system, we will note the system to utilize the capacity of communications links on a statistical multiplexing basis, so that a link's bandwidth is consumed by a flow only when it is active.

To accommodate a new circuit's fixed TDM multiplexing arrangement, each link included in the circuit's route must have a sufficient number of available (i.e., unused) slots per frame. The SS takes this information into account in selecting a route to use for supporting a new call connection.

Once the SS has selected a route to be used for establishing a circuit for the support of a new call, it proceeds to inform the signaling entities embedded in the network switches that are included in the route. For example, in setting the circuit in Fig. 10.1 from N1 to N5, using tandem switch N3, the SS informs switches N1, N3, and N5 about the setting of this circuit as relevant to each switch. These switches then insert this information into their CS tables. At each switch, this information identifies the incoming and outgoing links that are employed by messages that are transported as part of this connection. The switch then acts accordingly to switch messages that belong to the underlying connection.

For example, for the call connection that originates at switch N1 and terminates at switch N5, we note the following switching operations that are undertaken by involved switches. At edge switch N1, messages that are identified as members of this call are switched to an output port that connects with link (N1,N3) and are then transmitted across the outgoing trunk to node N3. At tandem node N3, messages that are members of this call arrive to the switch at a port that interfaces link (N1,N3). These messages are switched by N3 to an output port and are transmitted across trunk (N3,N5) to edge switch N5. At N5, this call's messages arrive at a port that interfaces trunk (N3,N5). They are then switched to an output port that connects them across the local loop to the local destination user.

The specification of the time slots to be used at a circuit switch for multiplexing a circuit's messages across a designated output trunk is performed locally at each switch. Following the designation by a switch of the output slot that it will use for an underlying circuit, the switch sends this information to the signaling subsystem entity embedded in the neighboring downstream switch.

For example, consider the time slot settings made by the switches located across the route selected for supporting the N1 to N5 circuit. As shown in Fig. 10.2(a), at edge node N1, slot 1 in each time frame is used to multiplex the data segments that it transmits across outgoing trunk (N1,N3). At tandem node N3, data segments are scheduled to arrive in slot 1 of each time frame, and are scheduled to be transmitted in slot 1 of each subsequent time frame (and thus experiencing frame latency delay of about 1 frame duration) across outgoing trunk (N3,N5). At destination edge switch N5, the circuit's data segments are scheduled to arrive in slot 1 of each frame; they are then transported across the local loop to the destination user.

The complexity of the circuit management configuration scheme is much reduced by setting the selection of the identity of the time slots to be performed locally at each switch node rather than carried out by a single controller. Each switch node keeps track of its own slot assignments across each trunk, saving locally the identity of its outgoing slots that are currently not assigned and that are available for allocation to new connections.

The format of a switching table employed by a circuit switch involves the elements illustrated in Fig. 10.3. For each connection, the switching table identifies the incoming line (and corresponding switch's interface port) and time slot(s) used for receiving data segments and the outgoing line (and corresponding switch's interface port) and time slot(s) used for the transmission of the circuit's data

Connection ID	Line-In (in_if)	Time Slot-In	Line-Out (out_if)	Time Slot-Out
Set by signaling system	Set by signaling system	Set by signaling data received from preceding switch	Set by signaling system	Set by the switch in selecting available slots
C(N1,N5)	L(N1,N3)	TS#1	L(N3,N5)	TS#1
C(N1,N4)	L(N1,N3)	TS#2	L(N3,N4)	TS#4

Figure 10.3 Illustrative Circuit-Switching Table.

segments. Two rows of entries in the table used at N3 show the corresponding data that identify, in accordance with Fig. 10.2, the setting of circuit C(N1,N5) that connects edge switch N1 with edge switch N5 and the setting for C(N1,N4).

When establishing a conference connection that involves several end-user stations, data segments may arrive across multiple lines and time slots and may be switched for transmission across multiple outgoing links and time slots. For audio conference connections, *conference bridges* are used to combine the voice segments produced by multiple conference participants and then disseminate the combined segments to conference parties.

In the early days of telephone telecommunications, a telephone switchboard operator would manually make connections between a subscriber line and a telephone trunk, in response to the originating user's spoken request of a connection with a named destination party. When the exchange operator determined (or be informed) that the call has ended, the physical wires that connected the incoming and outgoing lines would be extracted, tearing down the connection.

Over time, the manually driven switching operation was replaced by an automated electrical switching mechanism, which was used to configure a circuit that connects the corresponding incoming and outgoing lines. Initially, analog modulation schemes were used to transport signals across communications links. Connections were made through the use of a direct dialing process. The dialed consecutive characters that form the telephone number of the destination party were mapped into data symbols that were represented as consecutive signals, causing electromechanical switching systems to automatically form the intended connection's circuit across a selected path.

The performance capability and capacity of a modern telecommunications switching network system have been significantly upgraded as advances were being made, leading to the *digitization of the telecommunications system in its two key domains: the communications links and the switching systems*. Voice, data, and video message signals were produced in digital format, or converted to digital format, and transported across digital communications systems. Signaling messages that are used to identify the called and calling parties, and additional connection features, were digitally included. Upon receiving a signaling message, the SS would employ microprocessor based computerized systems that rapidly determine the best network path that should be taken to reach the targeted destination node, confirm that the requested communications capacity levels can be met, and then manage the insertion of corresponding entries in the CS tables of involved switches.

An advanced circuit switch module is designed to operate as a digital system. For example, the large telephone network's *Electronic Switching System (ESS)*, such as the AT&T 5ESS-2000 system, introduced in the 1990s, is able to interface and switch a large number of data-multiplexed communications links, including optical links. It consists of multiple modules, including an Administrative Module (AM) whose computers control the operation, a Communications Module (CM), functioning as a time-division switch and a Switching Module (SM) that performs switching, multiplexing and processing and provides interface attachments. Advancement in the design of ultrahigh-speed packet switching modules have led to the integrated use of such modules into CS systems.

10.3 The Switching Fabric

To demonstrate the principle of operation of a CS module, we illustrate in Fig. 10.4 the architecture of a **Time–Space–Time (TST)** switching fabric, which employs at times a *Module Controller/Time Slot Interchange (MCTSI) circuit*. We show the switching operation that takes place for a circuit that carries a flow whose data segments access the switch at input line 2 and are switched to output line 4. The illustrated circuit is assigned time slot 2 in each time frame that is configured across input line 2. Outgoing segments are inserted in slot 1 of each time frame configured across output line 4.

The TST switching scheme operates as follows. An input TDM demultiplexing stage examines, for each input line, the segments that arrive in each time slot of each time frame. In the illustrative case, each segment that arrives in slot 2 of a time frame is scheduled for transfer from its input line 2 to a buffer module that is used by an output transmitter to feed output line 4. Such a transfer is realized by a **space switching** module.

For low frame latency, the rate at which the space switching process is performed is targeted to be higher than the aggregate sum of the speeds of the input lines. When the incoming lines are not highly occupied, so that the utilization factor of the switching fabric over time is statistically lower than 100%, the space switching fabric may be configured to operate at a lower rate. The system may, for example, strive to achieve a segment latency across the switch that is lower than a time delay of about one to several time frame duration(s).

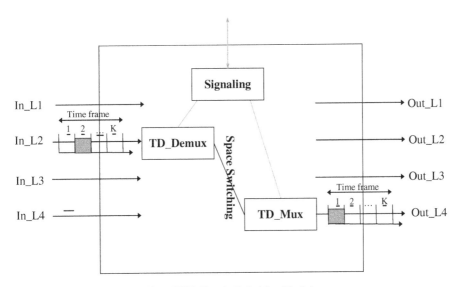

Figure 10.4 Time–Space–Time (TST) Circuit-Switching Module.

10 Circuit Switching

Once a segment has been transferred by the switch fabric to an output buffer that feeds its designated outgoing port, the segment joins an output queue. The segment is then multiplexed across its designated outgoing line and transmitted to the next switch that is located across its path, or to an attached station if the segment has reached its destination node.

As shown in Fig. 10.4, the switched segments of the illustrated circuit are multiplexed into slot 1 of each time frame for transmission across an output line. The time frame duration is typically set to a low value (such as 125 μs when handling the 8-bit segments of a PCM voice stream), so that the ensuing frame latency is low. When transporting realtime audio signals, a very low segment latency is imperceptible to the user's ears.

For an established circuit, the space switching fabric moves a segment that arrives (in each time frame) across input line i, for each input line and input time slot, denoted as $In\_Line(i)$, at time slot k_i, denoted as $In\_Slot(k_i)$. This segment is switched to a buffer that feeds segments for transmission across designated output line j, denoted as $Out\_Line(j)$, whereby each such segment is multiplexed into slot k_j, denoted as $Out\_Slot(k_j)$. In the following, we note several designs that have been employed for the implementation of a space switching stage.

1) **Space Division Switching Fabric**: The switching fabric is synthesized by interconnecting a *matrix of elementary switching modules*. These modules are interconnected in a specific layout across a *grid of rows and columns*. For example, a basic $\{2 \times 2\}$ single module can be configured to connect its inputs to specified outputs. These connections can be changed at different time slots. Using signaling messages, a switch controller can be used to set the state of each switch module so that it switches a specific input to a specific output during each time slot. In this manner, a circuit is established by setting an input line to connect to an output line during a specified time slot, whereby the connectivity pattern can be repeated in each time frame for the duration of a connection. To achieve a high switching rate, a multistage matrix fabric is employed, enabling the switch to sustain, simultaneously in time, multiple circuit connections. It is desirable for such an implementation to avoid using segment buffering at the switch's modules, in aiming to provide **non-blocking** transit, so that simultaneously active distinct circuit paths are accommodated, while avoiding internal blocking (incurred when different circuits are set to use the same elementary switching module at the same time).

   Matrix-type configured switches have been commonly employed by legacy telephone systems. Originally, such systems used horizontally and vertically positioned metal bars that were moved by a switch control system to form targeted intersection joints, forming a desired path through the switch fabric. Such systems have been identified as **cross bar** switching systems. Original implementations that employed electromechanical elements were replaced over time with solid-state-based elementary switches. Higher dimensional switching modules were synthesized by implementing a fabric that consists of a multi stage matrix that interconnects smaller size per-stage switching systems.

   To simplify the underlying control mechanism, certain designs have implemented *self-routing switching fabrics*. A message segment that arrives to the switch in slot k_i across a specific incoming line is automatically recognized by the switch, as delineated by the information included in the CS table (configured by the SS), as belonging to a specific connection and thus associated with a specific circuit. As such, the switch control system would configure a corresponding path within the switch fabric; for example, by setting the proper states of involved elementary switch modules.

   Alternatively, a control system may attach to each data segment, which is recognized as belonging to a given circuit, an internal source routing header. The data included in this header enables the switch modules to route the segment across the switch to its designated output line. For example, assume the internal routing header to be designated as {1,0,0,1}. In this case, when this segment is received by an elementary switch located at a first column of the matrix structure, this module would switch the segment to its output port identified as level 1 output. Elementary switches located in subsequent columns 2, 3, and 4, would in this manner switch this segment to outputs 0, 0, and 1, respectively. This way, each segment, will *autonomously navigate* itself through the switch fabric to the proper output.

As noted above, it is desirable to manage this scheme so that it leads to a non-blocking switching process. Under certain designs, higher switching rates are attained by allowing a limited measure of internal segment queueing (when multiple segments arrive at the same time at an input of an elementary internal switch), yielding however a more complex system design. Higher internal basic operational rates are used at times to enhance the system's switching rate performance.

In striving to implement switching fabrics that provide low transit latency, high switching speeds and energy efficient, matrix-based structures have been commonly replaced by shared medium and shared memory-based architectures, which function as outlined below. The space switching fabric subsystem is effectively designed to operate as a packet switching system. Yet, a circuit-switched operation is maintained as each connection is dedicated capacity resources (such as designated time slots in each time frame) across each one of the communications lines that its messages traverse over their path.

2) **Shared Medium Switching Fabric**: A shared medium multiple access scheme is employed to provide a switching service. This operation is similar to that described when discussing the functioning of shared medium local area networks (LANs). Circuit-switched or packet switched-oriented connectivity can be employed in performing the module's switching action. The shared medium resource can consist of a system of single or several parallel buses. A bus module consists of a conductor to which several ports are attached. To illustrate, consider the following configurations.

Multiple ports can be attached to a bus and share its capacity on a TDM basis. As delineated above, a local signaling controller is used to identify a data segment that arrives in a given time slot at a given input line as a member of a preset connection. According to the information placed during the signaling phase in the CS table, the segment is scheduled to be switched to a specific output line. Accordingly, the TDM bus controller schedules a specific set of time slots in each time frame to be dedicated to this circuit. An arriving segment is multiplexed across the bus, so that it is broadcast in one of its designated time slots. A TDM de-multiplexer is programmed to copy the segments sent in these time slots and deliver them to the output module(s) that feeds the output line(s) designated for this circuit.

Multiple bus modules, operational in parallel, are used to enhance the overall capacity of the switching system. Such systems are frequently employed in computer and multiprocessor driven systems, serving to internally switch data flows among system processing, memory and Input/Output (I/O) modules.

Rather than allocate internal capacity resources on a circuit switched fixed multiplexing basis, it is also possible to share resources on a statistical multiplexing basis, as performed by a packet switching system. This becomes more performance effective when the activity duty cycle of a connection is low. If a circuit activity factor is equal to 10%, the time slots allocated to this circuit would be occupied during an average of only 10% of the time. If, in turn, time slots are allocated for the transfer of segments from an input line to an output line only when the corresponding input connection is determined to be active, the shared capacity of the bus system can be used to accommodate a much higher number of data flows.

To perform the de-multiplexing process, the switch control system can attach a header to each data segment to identify the intended output line module. Incoming segments are broadcast across the bus system so that each output module is able to detect them. The intended output module will then copy the broadcast segment.

Under such a statistical multiplexing operation, it is possible for the bus system to become overloaded over a period of time during which the overall segment arrival rate is high. In this case, a segment may have to be delayed prior to being switched. If the incurred delay level is deemed excessive, the segment may be discarded.

As an alternative, a fast circuit setup operation can be synthesized. As soon as a connection is detected to be active, the circuit is established through the setting of TDM slots across a designated bus to be used in connecting the targeted input and output lines. Segment delays or discards can occur under such an operation as well.

3) **Shared Memory Switching Fabric**: Incoming segments that belong to established connections are stored in a shared memory system. The memory space is organized in a manner that enables the switch control system to keep record of the locations in memory that are used to store the segments that belong to distinct connections. Using the data embedded in the CS table, the switch controller accesses the memory spaces used to store the segments of each connection at its designated output rate. If at this access time, the corresponding memory space of a connection contains a segment, it will be directly fed to its designated output module. At a fast memory access rate, such an operation is highly effective.

As noted for the shared medium system, shared resources, such as memory storage assets, can be allocated to connections on a fixed multiplexing basis to avoid data segment discards. Alternatively, when memory resources are shared on a statistical multiplexing basis, a much higher data throughput rate can be sustained, at the risk of having to discard or delay data segments.

10.4 The Signaling System

A CS network system employs a signaling subsystem and a data transport subsystem. The primary function of the signaling subsystem is to set up and tear down connections. It also enables the user to send to the system its request for desired service features and performance objectives.

In older telephone systems, signaling messages were transmitted within the same circuit as that established for transporting the audio messages. Such an operation is called **channel-associated signaling (CAS)**. Under **in-band signaling**, signaling messages are transported across the same channels as those used to carry voice calls; the latter are known as *bearer channels*. In the USA, Signaling System No. 5 (SS5) and earlier systems use in-band signaling. Later systems transport signaling messages by using **out-of-band signaling** channels.

In telephony, the statistical characteristics of signaling messages are distinctly different than those exhibited by audio messages. For each call, few signaling messages are issued. A call request message is a signaling message that is used to request the system to configure a circuit to carry a call to a specified destination party. When the call is completed, a call termination signaling message is issued. Several other signaling messages may be associated with a call when needed to specify or modify various call parameters.

Thus, for a call, typically a much higher number of audio messages are produced than signaling messages. It is therefore not capacity efficient to set aside circuit resources for the sole use of transporting signaling messages. It is more performance efficient to design a separate network for the transfer of signaling messages and employ a packet switching scheme to share this network's nodal and link resources. In this way, a signaling process associated with a call will have its messages occupying communications link resources only when signaling messages are produced and transported.

Accordingly, an advanced SS, such as the US **Signaling System No. 7 (SS7)**, uses a *separate network* to carry signaling messages. This system is known in the USA as a **Common Channel Interoffice Signaling (CCIS) system**. Associated telephony signaling protocols were developed in 1975 and have been widely used, making it feasible for diverse telephone systems to exchange signaling messages, enabling them to set up and tear down telephone calls that are set across the worldwide PSTN. This signaling protocol also performs multiple other services, such as call forwarding, voice mail, call waiting, conference calling, calling name and number display, call screening, and other. The SS7 protocol has been defined by Q.700-series recommendations of 1988 by the International Telecommunications Union (ITU-T). National variants of the SS7 protocols have been defined, including by the American National Standards Institute (ANSI) and the European Telecommunications Standards Institute (ETSI).

Also of interest are non-facility-associated signaling (NFAS). It is not directly associated with the call's path and involves information located at a centralized database, accommodating subscription and modification of various features and service functions.

Signaling messages are transported between SS nodes and the network's switches, also identified as offices. Signaling messages that are transported between the network and a user's device are carried across the local loop system and involve the associated edge switching offices. The protocols and schemes used by such systems are referred to as **common-channel signaling (CCS)**. Using SS7, the connection is not established between the end points until all nodes located across the route confirm their availability. If the destination party is busy, the caller receives a busy signal, and no bearer network resources are occupied.

The SS7 signaling architecture can be operated in an associated or quasi-associated modes. Under the associated mode, the signaling process progresses from switch to switch through the telecommunications network, using the same path as that used by the associated facilities that carry the telephone call. Under the quasi-associated mode, which is the one that dominates US implementations, the signaling process progresses from the originating switch to the terminating switch, following a path that is established across a separate CCIS signaling network.

The CCIS signaling network is implemented as a packet switching network. Its signaling packet switches are known as **signal transfer points (STPs)**. A user wishing to establish a call connection, sends a request message across the local loop to its associated end-office or edge switch. The edge switch forms a call request packet whose format follows SS7 protocol specifications. As shown in Fig. 10.1, the switch is connected via CCIS signaling lines to an STP packet switch via a signaling end-point module (not shown). The signaling packet switching network enables wide transport of signaling packets, covering the scope of the underlying network system.

Edge switches in the circuit switched network connect via signaling lines to SS7 end-nodes that are identified as signaling end-points (SEPs), which include service switching points (SSPs) and service control points (SCPs). These nodes are interconnected via signaling links, using the switching services provided by the STPs. The latter switch a received signaling packet in accordance with the address data included in its header.

A signaling protocol standard, such as SS7 (whose specifications are developed by the International Telecommunications Union (ITU), a UN-based organization, defines the formats of signaling messages and the procedures and processes that configure, control and manage connections across a telecommunications network system. By adhering to such a specification, the system is able to proceed with the establishment and management of a connection that supports a call that traverses multiple switching systems, including such that are operated by distinct telecommunications service provides around the world.

Computer servers that are attached to a signaling packet switching network provide signaling services that follow the protocols defined by a signaling standard. For example, using a received call request packet, a signaling server will proceed to set up a connection and configure the associated circuit. The signaling packet is transported across the CCIS signaling network from its originating node to the terminating node. A routing algorithm that is employed by the SS is used to select an end-to-end route across the bearer telecommunications network. The CCIS SS nodes interact with SS entities embedded in the network's circuit switches, as they keep track of bandwidth availability and effectively configure the switching table entries of the circuit switches located across a selected path.

The SS7 protocol architecture uses protocol entities at multiple OSI layers. The Network Service Part (NSP) consists of the Message Transfer Part (MTP) and the Signaling Connection Control Part (SCCP). It provides services at OSI layers 1–3, identified as MTP levels 1–3, respectively. Signaling message formats and processes that are used to manage circuit connections are specified by application layer entities, which are identified as the User Part (UP). The employed CS oriented protocol is used to establish, maintain, and terminate call connections. The Transaction Capabilities Application Part (TCAP) uses the SCCP entity as it employs connectionless networking operations. It is used to form database queries and implement advanced intelligent networking functions. A connection-oriented SCCP operation is used to accommodate the transport of signaling messages over an air interface.

The Internet Engineering Task Force (IETF) has defined the SIGTRAN protocol suite, which implements protocols compatible with SS7. Also called Pseudo SS7, it is layered on the Stream Control Transmission Protocol (SCTP) transport mechanism for use in Internet Protocol (IP) networks, such as the Internet.

10.5 Performance Characteristics of a Circuit Switching Network

In the following, we outline the key performance characteristics of a CS network system.

1) **Call Connection and Circuit Configuration Using a SS**: A CS system requires the use of a SS. This system is used to set up a connection for an impending call. A successful establishment of a circuit for this connection is essential. Only after the circuit is successfully configured, the associated end-users are signaled to start the information transport phase.

2) **Provision of a SS**: The provision, operation, design, and maintenance of a SS is a critical component of a circuit-switched system. Its access to client databases allows the system to provide a multitude of intelligent service features. Yet, it is also a major system cost driver and can be subjected to security issues in guarding subscriber data and records.

3) **Quality of Service (QoS)**: Using the connection data collected by the SS enables the network control system to employ dynamically adaptive connection oriented network management services. Included are capacity, connectivity, failure recovery, security, and accounting performance management services. For telecommunications networks, Telecommunication Management Network (TMN) system modules are employed.

10.5 Performance Characteristics of a Circuit Switching Network | 357

4) **Call Admission Control (CAC)**: To guarantee to an admitted connection its desired QoS performance, the SS performs **CAC**. During the signaling phase, based on call performance requirements requested by the user and the call's specified or estimated traffic and data rate statistical features, and incorporating the current availability of link and nodal capacity resources across the selected call's path, a management and control entity determines whether the system has sufficient unused resources that it can allocate to satisfy the requirements imposed by a new call connection request. If so, the corresponding new call is admitted. If no sufficient resources are available, including when considering alternate routes, the **new call is blocked** and not admitted into the system.

In this manner, communications capacity resources allocated to an admitted call are preserved for exclusive use by such a call. As the network management system allocates a circuit's capacity resources based on assuring the call's performance even when active at its peak data rate, the call's data performance behavior is preserved, independently of activity fluctuations exhibited by the call's data source and by other calls.

5) **Network System Performance Metrics**: Key performance metrics of a CS network system include:

   a) **Circuit Setup Time**: It measures the time that it takes for the SS to set up the circuit in response to a call request message. The involved delay relates to the delay-throughput performance efficiency of the underlying packet-switched signaling network (such as CCIS) and the signaling delay incurred across the local loop access network that connects subscribers with their edge switches.

   b) **Data Message Delay Time Across the Telecommunications Network**: Since each circuit is assigned dedicated communications resources across its network path's links, disruptive message queueing delays are usually avoided. Considering a TDM multiplexing approach in sharing network links, a key delay component incurred by a message segment as it waits to be transmitted across the output link is attributed to its **frame latency**. Considering a message segment that arrives at an idle switch and thus does not incur queueing delay. Yet, it experiences a delay component identified as its frame latency, which accounts for the time delay elapsed between its arrival time position of its assigned output slot(s) within the TDM frame. Such a delay level is of the order of a single time frame duration. Consequent message delay and delay variation values are generally very low.

   The end-to-end message transmission time is calculated as the aggregate sum of the aggregate message transmission times across each traversed link. It is nominally of the order of 1 time slot per link. Hence, the overall end-to-end delay experienced by a message across its circuit is generally quite low. It is of the order of a time frame duration multiplied by the number of links spanning the circuit's path. It thus also exhibits low delay variability. Processing delays at the switching nodes are nominally low, as the operation of the circuit switch is generally hardware based.

   c) **Call Blocking Probability (CBP) as the system's Grade of Service (GoS)**: The CBP (or ratio), denoted as P_B, is defined as the probability that an arriving new call that requests admission into the system is blocked by the system due to lack of available communications capacity resources that can be made available for the support of the call. It represents the long-term fraction of arriving calls that are blocked, and are thus not admitted into the system. The call admission probability is calculated as $P_A = 1 - P_B$. For example, under a CBP of $P_B = 1\%$, the system blocks 1% of arriving calls, while admitting $P_A = 1 - P_B = 99\%$ of arriving calls. The call blocking ratio that a CS system induces is a key performance metric and is identified as the system's **GoS**.

358 | *10 Circuit Switching*

d) **Utilization of the System's Communications Resources**: As each call is provided a circuit, for which communication link capacity resources are dedicated, the utilization of these communications resources, which measures the average fraction of time during which these resources are used for the transmission of call messages, is directly related to the call's activity factor. For example, a voice call may keep unidirectional circuit resources active for an average of 30% of the time. In this case, the circuit's level of utilization, denoted as ρ, is equal to only $\rho = 0.3$. Therefore, the communications resources of this circuit remain unused for an average of 70% of the time.

A circuit-switched system exhibits a higher communications capacity utilization level when used to transport flows that exhibit relatively high and steady message activity levels, as observed to be the case for typical voice and uncompressed video calls. In turn, dedicated circuit communications capacity resources are not well utilized when used to carry flows produced by highly bursty data sources. Such sources could produce data at high rates during certain short periods of time, while at other times the sources are idle or require much lower data transfer rate resources.

e) **Preservation of message ordering**: Messages that belong to a call's connection are generally transported in a sequential fashion across the circuit's route. Hence, when correctly received, data units are delivered to the destination node in the same order that they were transmitted by their source node. In contrast, under a connectionless (datagram) packet switching transport process, messages may be delivered to a destination node in a misordered (or randomly ordered) manner. Different data units may be traversing different network paths.

f) **Transport reliability**: A reliability metric relating to the transfer of data units characterizes that rate of errors that are detected at receiving nodes. Under a connection oriented transport service, error control measures may be invoked. Upon reception at a node, encoded messages are decoded to determine whether they contain errors. Messages that contain errors may be discarded or retransmitted. The end-to-end message error rate incurred across a circuit may be configured to meet a specified target level. Induced by message discards, gaps that occur in the reception of a flow's data units are detected and may be compensated by interpolation processes that are performed at the receiving entity when certain realtime streams are invoked, or otherwise by triggering a retransmission process.

6) **The Nodal Switching Platform**: Intelligent software driven processes are performed by the common channel SS during the signaling phase. During this phase, call management functions such as routing and call set-up processing operations are performed. The CS table entries are calculated and placed at involved switching nodes prior to the start of the data transport phase. These operations can limit the speed of the switching process if not carried out at sufficiently high speed. The switching operation carried out during the data transport phase involves fast (such as TST) transactions that are typically implemented by using ultrahigh-speed oriented hardware based mechanisms.

The following model is a common element of the performance analysis and design of a CS system. It is used to size the required capacity of the system in terms of determining the minimal number of circuits that should be assigned to service a given call traffic level and accordingly evaluate the number of needed trunks. The CBP, $P_B = GOS$, is used as the key performance metric. A basic calculation proceeds as follows. It is assumed that each call requests the assignment of a single circuit for the duration of the call. A circuit offers a prescribed data rate for the support of the call's message peak flow rate. For example, if the peak traffic data rate produced by a call is equal to

1 Mbps, it is assumed that under a CS service, an admitted call will be dedicated a circuit that offers the call a data rate of 1 Mbps. The call's data activity factor, denoted as ρ_{AF}, represents the fraction of time that the call produces data. Data units will occupy the capacity resource of the allocated call's circuit as they are transmitted at the provisioned peak rate. A voice call may exhibit an activity factor of the order of $\rho_{AF} = 0.35 = 35\%$, while many data calls tend to exhibit activity factors that are much lower, often lower than 1% or even 0.1%.

An M/M/m/m (or M/G/m/m) queueing system model is commonly used to calculate the blocking rate. Such a model is exhibited in Fig. 7.9. This system employs a total of m circuits. Calls arrive to the system in accordance with a Poisson process at a rate of λ [calls/s]. Each admitted call occupies its allocated circuit for a random period of time. The average holding time of a circuit by a call is equal to $h = \frac{1}{\mu}$ [s/call]. The system's loading rate is equal to $f = \lambda h = \lambda/\mu$ [Erlangs]. Under these parameters, the CBP is calculated by using **Erlang's Loss Formula**, as expressed by Property 7.13.

Example 10.1 Illustrative performance values are given in Example 7.9. In the following we consider a different scenario to present another illustrative system design. The above noted formula is applied for the calculation of the CBP P_B. Assume the system to be loaded by call traffic load of $f = 10$ [Erlangs]. This implies that there is an average of 10 simultaneously active calls supported by the system. The actual number of active calls is randomly varying and can be at times much higher than 10. A loading at this level can, for example, be produced by a system supporting an average of $\lambda = 1$ [call/s], whereby the average circuit holding time by a call is equal to $h = \frac{1}{\mu} = 10$ [s/call].

For this case, assume that the system's manger wishes to assure clients that the call blocking probability (GOS) is not higher than 1%, so that the system provides a call admission rate of 99%. Using the blocking formula, we find out that the system should then deploy a total of $m = 21$ circuits, while aiming to observe on the average an offered demand for only $f = 10$ circuits. Since only a fraction $1 - P_B$ of the calls will be admitted, the system will effectively observe an average circuit occupancy level of $f(1 - P_B)$. Hence, the average fraction of actively used circuits is equal to $f(1 - P_B)/m$, which for this case is equal to $10 \times 0.99/21 \approx 47.1\%$. However, noting that a circuit that is allocated to an admitted call is occupied for only a fraction of time that is equal to ρ_{AF}, we conclude that the fraction of the system's allocated circuit capacity that is occupied by user data is equal to $\rho_{AF}f(1 - P_B)/m$. Using the above-noted system parameters and this formula, we calculate the system bandwidth utilization ratio, when assuming the call's activity ratio to be equal to 1%, 10% or 35%, to be equal to 0.47%, 4.71%, or 16.48%, respectively. Thus, medium to good level of bandwidth utilization is attained when using the circuit allocation process to support a call, such as an audio or video session, that exhibits a steady flow of data at a relatively high activity factor. In contrast, the dedication of circuit capacity to a call that produces data at a low duty cycle, as is the case for many data calls, results in very low utilization of the system's bandwidth resources. It is preferable then to use a packet switching networking method.

10.6 Cross-Connect Switching and Wavelength Switched Optical Networks

Communications trunks that are used to connect switches in a circuit switched telecommunications network, as in a public telephone network, tend to carry a large number of circuits. A circuit associated with a user's call may carry a single digital voice stream. Such a PCM-coded stream consists of a single 8-bit sample that is sent (during voice activity) once in every time frame, which

360 | *10 Circuit Switching*

is 125 μs long, at a consequent rate of 8000 [frames/s] or 64 kbps. At times, 1 bit data that is normally used to represent the quantized value assumed by the sample, is used for synchronization and management purposes. If a trunk is used to carry 24 voice circuits across a link that connects two switches, each requiring support at a data rate of 64 kbps, the trunk would have to operate at a data rate that is sufficiently high to allow 24 voice samples to be transmitted during each time frame. This can be achieved by using a synchronous TDM scheme.

In telephony, a single PCM voice stream is formatted and represented as Digital Signal 0, denoted as **DS0**. It is assigned a circuit operating at a data rate of 64 kbps. When using a communications line to synchronously multiplex (using, say, TDM) 24 DS0 voice signals, so that each time frame carries 24 8-bit samples plus an extra bit that is used for frame synchronization and signaling maintenance purposes, we observe the line to carry 193 bits per frame. It is thus operating at a data rate that is equal to $193/125\mu = 1.544$ Mbps. Such a TDM signal arrangement is defined in telephony as one that follows a **DS1** protocol. The line that carries such a digital signal is identified as a *T-1 line or T1 carrier*.

As we delve deeper into the core of a circuit switched network, we observe trunks that carry an even larger number of circuits. Using TDM, 28 DS1/T1 signals are multiplexed into a **DS3 signal**. A single DS3 signal carries 44.736 Mbit/s of data, as supported by a **T3 trunk**. It is used to carry $28 \times 24 = 672$ DS0 signals (or PCM voice streams). The formats and protocols used in this fashion to operate electrical trunks that multiplex several TDM telecommunications channels over a single communications circuit were standardized around 1957 by AT&T Corp. and are recognized in the USA as the **T-carrier** system. It uses copper wires, employing one pair for transmitting, and another pair for receiving. Signal repeaters are often used to extend coverage. In other countries, including Europe, a similar (but not the same and incompatible) E-carrier system was standardized.

Synchronous optical networking (SONET) and **synchronous digital hierarchy (SDH)** are standardized protocols that have been defined to accommodate the multiplexing of several digital streams across optical fiber trunks. These two standards are essentially similar, with SONET more prevalent in the USA. They define containers (or frames) that are used to multiplex data units that can be produced by distinct networking protocols, such as IP packets, ATM cells, and Ethernet frames, as well as circuit-based audio or multimedia digital streams.

The basic unit of transmission under SONET is defined as STS-1 (Synchronous Transport Signal 1) or OC-1 (Optical Carrier 1). It operates at a data rate of 51.84 Mbps. At this rate, an OC-1 circuit can carry the data units of a DS-3 channel. In SONET, the STS-3c (Synchronous Transport Signal 3, concatenated) signal is composed of three multiplexed STS-1 signals; it may be carried via an OC-3 signal. The basic unit of framing in SDH is a STM-1 (Synchronous Transport Module level 1), which operates at 155.520 Mbps, which is the unit referred by SONET as STS-3c.

The Standard defines each frame to consist of header bits, which aid in performing control and management functions, as well as payload bits. Header bits are interleaved among payload bits and placed at standardized frame locations. The following channel rates are noted: Under SONET, optical carrier 1 (OC-1) assumes a frame format that is defined as STS-1. It provides a payload rate of 50,112 kbps and requires a line rate of 51,840 kbps. It is identified as STM-1 in SDH. The corresponding parameters for OC-3/STS-3 under SONET (and STM-1 under SDH) are a payload rate of 150,336 kbps and line rate of 155,520 kbps. The rates for the corresponding parameters for OC-N have been conveniently defined to be equal to N times the corresponding OC-1 rates. For example, for OC-48/STS-48 (STM-16) the payload rate is 2,405,376 kbps and the line rate is 2,488,310 kbps, which are equal to the respective OC-1 rates multiplied by a factor of 48.

A Virtual Concatenation (VCAT) scheme can be employed to provide for better bandwidth utilization of trunk resources. A quite general assembly of lower-order multiplexing containers can be employed, resulting in larger containers of desired sizes. For this purpose, Generic Framing Procedure (GFP) protocols are used to map payloads of any bandwidth level into a targeted virtually concatenated container.

Consider a switch node in a circuit switched network that acts to switch incoming T-carrier (or E-carrier) bit streams and/or SONET/SDH bit streams to outgoing trunks and multiplex them into their circuit designated T-carrier or SONET stream formed across the outgoing trunks. For example, consider such a switch node that is fed by input signals on incoming optical lines LI-1, LI-2 and LI-3, whereby across each line, streams are arranged in accordance with SONET optical carrier OC-1/STS-1. It is required to multiplex these incoming streams into a single outgoing line LO-1 and format the latter in accordance with SONET OC-3/STS-3. A device that performs such a multiplexing operation is called an **Add-Drop Multiplexer (ADM)**. A switch node that provides for such a transcode multiplexing operation (also called transmuxing) is identified as a **cross-connect device**. It is also identified as digital cross connect (DXC or DCS) when it performs on electrical signals, and as optical cross connect (OXC) when acting on optical signals.

DCS devices can be used for *grooming* telecommunications traffic, in managing the grouping and de-grouping of traffic streams for transmission within the framing structures of trunk systems. In this manner, the number of circuits employed within the core network can be reduced and failure recovery mechanisms can be effectively employed.

In fiber-optic communications, an optical fiber is used to conduct the light signal emitted by an optical transmitter, such as a modulated laser or an LED device. Optical fiber communications signals are carried at high data rates across long distance optical fiber lines. An optical cross-connect (OXC) device is used to cross-connect optical signals, providing trans-multiplexing and switching from input to output optical fiber lines in an optical mesh network. Several methods are employed by OXC devices:

1) **Opaque OXCs** perform switching in the electronic domain. Incoming optical signals are de-multiplexed and converted into electrical signals (denoting this translation as an O/E process). The signals are then switched electronically and are subsequently converted (via an E/O process) into optical signals. The signals are then optically multiplexed onto outgoing optical links. This is known as an Optical-Electrical-Optical (OEO) process. The speed of this operation is constrained by the limited operational rate of the electronic switching process. The latter rate is much lower than the transmission rate used across optical fibers. This creates a *mismatch between the high transmission speed executed across the optical fiber and the lower processing and switching rate employed when using an electrical processor*. The use of Wavelength Division Multiplexing (WDM), under which streams that operate at different wavelengths and are then multiplexed across an optical fiber, is helpful in reducing the impact of this mismatch as the message stream carried by a single wavelength channel occupies lower bandwidth, which is of the order of the electrical switching and processing operational bandwidth. A dense WDM system may multiplex hundreds of streams, each operating at a data rate of 1 Gbps, across a single fiber-optic channel. While the multiplexed stream flows at an ultrahigh data rate, each individual wavelength based stream operates at a data rate that is of the order of the processing rate performed by an underlying electrical processor or switch.

An all-optical cross-connect switch module can achieve a switching rate that is of the order of the much higher optical line rate. The conversion process makes the operation to not be transparent to the employed networking protocols. However, using electronic switching is advantageous in that *a more intelligent and dynamically adaptive* signal regeneration and amplification operations can be undertaken. Once converted to electrical domain, processing is used to remove

362 | *10 Circuit Switching*

interference signals. Also, message headers can be examined and processed electrically to implement dynamically adaptive switching functions. Consequently, control plane and associated signaling messages are processed and switched electronically, while message transmissions are often handled and transported in an optical domain.

2) **Transparent OXC** is an all-optical device. Switching is performed directly on optical signals. Such a switch is often called transparent OXC or photonic cross-connect (PXC). Incoming optical signals are de-multiplexed, switched and multiplexed onto outgoing optical lines, all in the optical domain. The operational switching rate can be very high, but no intelligent processing of the data is performed, so that processing functions involving monitoring and regeneration of optical signals are not carried out.

3) A **translucent OXC** includes a stage that uses an optical switch as well as a stage that uses an electronic switch. While in certain scenarios optical switching may be used more frequently, the electronic switching stage may be employed when signal regeneration is required and/or when an optical switch is not available.

A special OXC module is an **optical add-drop multiplexer (OADM)**. This device is used in **wavelength-division multiplexing (WDM)** systems for multiplexing and routing different channels of light into or out of a single-mode fiber (SMF). A WDM fiber link multiplexes multiple channels, each operating at a distinct wavelength and frequency band, as performed in a Frequency Division Multiplexing (FDM) system. Each wavelength channel is modulated, enabling the transport of a data stream (or multiplexed combination of data streams) at a very high data rate (of the order of 1 Gbps or higher). As noted above, a dense WDM (DWDM) link multiplexes a high number of wavelength channels. Using an OADM device, wavelength channels can be added to or removed from an incoming WDM signal.

The OADM device de-multiplexes the wavelengths channels included in an input fiber, attaching them to internal ports. A fiber patch panel or an optical switch connects the wavelengths to a multiplexer or to drop ports. Together with wavelengths that are added, the remaining wavelengths are multiplexed onto the output fiber. A remotely reconfigurable optical switch module can be included as an intermediate stage, yielding a device that is called a **reconfigurable OADM (ROADM)**. While the term OADM applies to both types, it is often used interchangeably with ROADM. It is noted that ADMs are typically employed in conjunction with SONET/SDH networks, while OADMs are employed in conjunction with WDM links.

A network system that employs cross-connect modules is illustrated in Fig. 10.5. Add-drop multiplexer ADM1 is driven by three fibers at its input, each carrying data streams that are arranged in accordance with the protocol defined by SONET's Optical Carrier 1 (OC-1). One of the OC-1 streams is noted to carry a data flow produced by source station S1. The three streams are multiplexed by ADM1 and transmitted across an output fiber in accordance with an OC-3 SONET format. This fiber is connected to cross-connect module OXC1. At this module, the OC-1 stream produced by source station S1 is de-multiplexed and switched to drive a fiber that follows an OC-1 format. The latter fiber is directed to a facility that operates at the electrical (rather than optical) domain. The OC-1 optical signal is shown to be converted (via an O/E module) to a corresponding electrical signal. It is then shown to be multiplexed electronically in a T-3 line, which is connected as an input to the digital cross connect module DXC1. At the latter module, the incoming signal is de-multiplexed into three streams, which are transmitted across T-1 lines to their respective destination stations D11, D12, and D13.

Also shown is a signal produced by source station S2 that enters cross-connect module OXC1 via transmission across an OC-3 optical fiber line. At OXC1, the latter OC-3 optical signal, plus

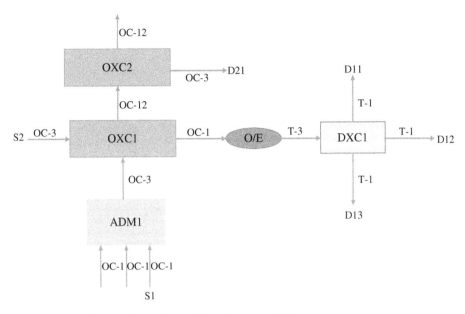

Figure 10.5 A Network System Using Cross-Connect Modules.

the two OC-1 optical streams arriving from ADM1 (excluding the OC-1 stream that was dropped and switched for transport to DXC1) are switched for transmission across an OC-12 output fiber to OXC2. There, the OC-3 stream produced by S2 is shown to be switched to an OC-3 fiber that transports it to its destination station D21. The remaining streams are transmitted across an outgoing OC-12 fiber.

All-optical networks that use WDM lines and optical cross-connect nodes are shown in Fig. 10.6. As a circuit switched network, a SS is used to configure a circuit for each end-to-end flow. The circuit enables a source station, which is attached to an edge OXC node, to route its stream in the optical domain to its designated destination edge OXC node which extracts the designated wavelength circuit for connection to the destination station. Each circuit carries its data stream along a route, called a **lightpath**, while staying unaltered end-to-end in the optical domain. A light stream is distinguished by the identity of its wavelength channel, whose wavelength is denoted as λ.

As a circuit switched network that is often called a *wavelength switched optical network (WSON)*, it makes use of a control plane and a data plane. The data plane uses WDM fiber links that connect optical cross-connect (OXC) nodes. A wavelength channel often operates at a rate of 10–40 Gbps, or higher. In the control plane, a SS is implemented as a separate electrically based network.

In Fig. 10.6(a), the shown network is topologically configured as a ring. It is also shown to use two counter-directional fibers, used to connect neighboring nodes. This provides the system a choice of two counter-directional paths between any pair of edge nodes. Due to the regularity of the topological layout, each optical cross-connect node needs to operate just as an OADM. An incoming flow is either carried through the output port as it is wavelength division multiplexed across the output fiber together with other transit flows, or it is dropped for delivery to an attached client station.

In Fig. 10.6(b), an optical mesh network is illustrated. A topological mesh layout is shown, requiring the use of optical cross-connect (OXC) nodes that are capable of performing cross-connect optical switching. At such a switching node, a flow that arrives across a specific line may have to be switched to one of several output lines, depending on the identity of its wavelength. For illustration purposes, certain links have been assumed to use just a single fiber, providing certain neighboring

364 10 Circuit Switching

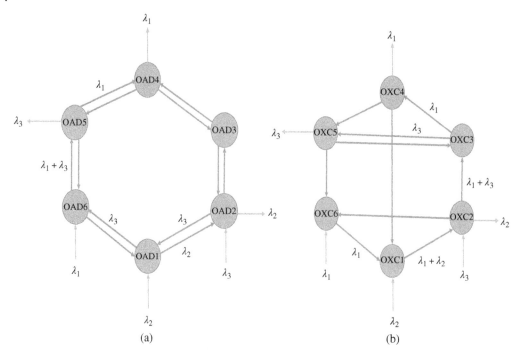

Figure 10.6 All-Optical Cross-Connect Networks Using WDM Lines: (a) Ring Network Using Optical ADM (OADM) Nodes; (b) Optical Mesh Network Using Optical Cross-Connect (OXC) Nodes.

nodes with unidirectional rather than bidirectional connectivity. Generally, it is advantageous to implement bidirectional links (using counter-directional fibers to connect neighboring nodes), as the average path length between end nodes is then shorter, leading to enhanced bandwidth utilization and upgraded throughput performance. For example, assuming a uniform traffic matrix, for a ring network that connects N nodes, the average path length is equal to $N/2$ if directional links are used. In turn, if a bidirectional ring system is used, and the shorter path is selected to route internodal traffic (with random selection of paths when considering equal length paths), the average path length is equal to $N/4$, leading to an increase in the system's index of utilization and throughput rate by a factor of 100%.

To demonstrate the functioning of an all-optical cross connect network, consider the meshed network system shown in Fig. 10.6(b). The stream that is assigned to use the lightpath that uses wavelength λ_1 enters the optical mesh network at node OXC6. It aims to reach a destination station that is attached to node OXC4. Its lightpath traverses the fibers that sequentially connect the following nodes: OXC6, OXC1, OXC2, OXC3, and OXC4. A second data stream is shown to use a lightpath that is arranged to route the flow that uses wavelength channel λ_2. It enters the optical mesh network at node OXC1. It aims to reach a destination station that is attached to node OXC2. Its lightpath traverses the fiber that connects OXC1 with OXC2. A third data stream is shown to use a lightpath that is arranged to route the flow that uses wavelength channel λ_3. It enters the optical mesh network at node OXC2. It aims to reach a destination station that is attached to node OXC5. Its lightpath traverses the fibers that connect OXC2 with OXC3 and OXC3 with OXC5. The figure shows the wavelength channels that are multiplexed across each utilized fiber link. For example, the fiber that connects (in a directional way) OXC1 with OXC2 multiplexes the streams whose channels are designated as λ_1 and λ_2. At OXC2, the λ_2 channel is dropped, and the λ_3 channel is added to the traversing λ_1 channel to produce a WDM signal across the output fiber that multiplexes channels λ_1 and λ_3.

Each optical cross-connect (OXC) module in the *WDM-based optical mesh network* performs as a circuit switch in an all-optical manner. We recall that in a CS network, a SS is used to set up a connection, which entails the selection of the proper end-to-end route and the dedication of communications capacity to the circuit at each one of the route's links. A corresponding forwarding entry is set in each circuit node's CS table. For example, for the circuit-switched WDM network shown in Fig. 10.6(b), the connection used by the flow that originates at a station that is attached to node OXC6 and terminates at a station that connects to node OXC4 consists of the path that involves nodes OXC6, OXC1, OXC2, OXC3, and OXC4. Bandwidth is dedicated to this connection across each link that carries wavelength λ_1.

Each switching entry included in a CS table of each node that is located along a circuit's route is identified by the circuit's input line and input wavelength, forming an *input label* that consists of the pair $(\lambda\_in, line\_in)$ for the underlying lightpath. It points to the circuit's output pair that is defined as $(\lambda\_out, line\_out)$. In the illustrative networks of Fig. 10.6, we set $\lambda\_out = \lambda\_in$, so that no input-to-output wavelength translation is performed. To illustrate, consider the CS entries embedded in the switching table of node OXC1, when considering the optical meshed network shown in Fig. 10.6(b). One switch entry points to the stream that enters the node at the input line attached to the source station at input wavelength channel λ_2. This entry directs the corresponding optical stream to be switched to the output link that connects to neighboring node OXC2. The corresponding output data stream uses the same wavelength channel while being multiplexed across the output link together with wavelength channel λ_1.

We recall that for a T-S-T-based circuit switch, the switching entry (as set by the SS) is identified by an input pair $label\_in = (line\_in, slot\_in)$ and the corresponding output pair $label\_out = (line\_out, slot\_out)$. For a WDM-based optical mesh network, the switching operation at the circuit switch, whose role is now taken by the OXC module, is determined by the input pair $label\_in = (line\_in, wavelength\_in)$ and the corresponding output pair $label\_out = (line\_out, wavelength\_out)$.

In both cases, the underlying operation can be regarded as a *label switching* operation. Under a CS process, the label that identifies a call's circuit connection, which is assigned to a data stream or a lightpath, is characterized by a vector whose entries are *physical entities* such as those that identify a communications line (or its attachment port) and a time slot or a wavelength. A SS is used to configure the circuit and to guarantee the feasibility of allocating the bandwidth resources required by the configured circuit.

Example 10.2 To illustrate the concept of the operation of a lightpath over a WDM-based optical cross-connect network, consider the transportation of two distinct streams (or platoons) of (possibly autonomous) vehicles moving from location A (identified as source node A) to locations B and D through several roads and connecting intersections. It is illustrated in Fig. 10.7. A signaling phase precedes the transportation phase. The signaling manager sets up an end-to-end transport circuit by selecting an end-to-end route and by reserving spatial resources along the route's roads and intersections. Assume each road to consist of multiple lanes, and regard each lane resource to represent a distinct wavelength channel. A reserved end-to-end circuit consists of an assigned route, consisting of a tandem sequence of roads coupled with an assigned dedicated lane across each road. This circuit is used to navigate the underlying flow of vehicles.

For example, assume the routes used to connect source node A with destination nodes B and D to traverse through intersection node C. Also assume that the first flow, from A to B, is allocated lane-1 along road (A,C). Its vehicles are switched at intersection C to reserved lane 2 of road (C,B). At intersection C, the corresponding node operates as an optical cross-connect switch in that it automatically switches vehicles that arrive in lane-1 across road (A,C) to depart along lane-2 of road

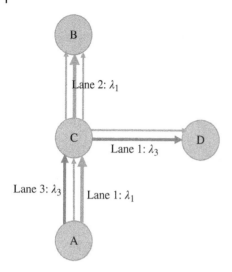

Figure 10.7 A Highway Transportation Path Analogous to an All-Optical Cross-Connect WDM-Based Lightpath.

(C,B). The road and lane identifiers of an arriving flow are used as a label that identifies this flow. As such, it also identities its outgoing road and lane forwarding entries. Similarly, cross-connect node C automatically switches vehicles that arrive in lane-3 across road (A,C) to depart along lane-1 of road (C,D). In this manner, each stream of vehicles is transported across its pre-configured end-to-end circuit. By using the SS to provide traffic engineering (TE)-based resource allocations, one assures the availability of sufficient capacity resources across the routes. Vehicle flows can then be transported across their paths in a timeline that induces very low queueing delays at the intersections.

In comparison, in a connection-oriented packet switching network, also identified as a virtual circuit switching (VCS) network, a SS is used to select an end-to-end route for a call's connection, but no bandwidth resources are dedicated. Communications links are shared on a statistical multiplexing basis. The SS sets the switching entries at the involved virtual packet switches. An incoming packet is identified as a member of a specific flow by its virtual circuit identifier that is included in each packet's header field, serving as the corresponding switching **label**. The packet switch reads the label of an incoming packet and then searches for a match of this label with a corresponding label included in its switching table, also known as a **label switching table**. A label field is shorter than the packet header field used by a datagram, leading to higher bandwidth efficiency and higher switching rates. A similar method is employed for the operation of a connection oriented routing mechanism known as **multi-protocol label switching (MPLS)**. The message payload is carried within an envelope (or container) which is identified by a label. Such an operation, whereby the message, including its header and payload fields, is attached an external header, is regarded as a **tunneling** process. As the message is switched from a source node to a destination node, each switch across the route uses the label (rather than a packet's destination address) to rapidly identify a match within its label switching database and accordingly switch the packet to the identified output port. Subsequently, the incoming packet's label is swapped with an outgoing packet's label, which is used by the downstream neighboring switch to carry out its switching and routing operations. The message payload carried across the tunnel can be formatted in accordance with one of multiple networking protocols.

MPLS is thus protocol-independent as packet-forwarding is based on the external label alone. In this manner, one can configure end-to-end circuits that traverse different media and accommodate multiple layer 2/3 protocol technologies. It can provide integrated support for flows that are configured through the use of CS (such as SONET or WDM-based optical networks), connection-oriented packet switching networks (such as frame-relay or ATM networks) as well as connectionless (datagram) packet switching (such as IP) protocols.

A datagram packet switching network, such as an IP network, does not employ a separate control plane that implements a SS. It does not configure physical or virtual circuits that are used to route flows or calls. Routing protocols are dynamically used to determine paths for individual messages. To adapt the selection of routes to network conditions, switching nodes interact with each other to learn about network connectivity, loading levels and performance conditions. A switching node that acts as a router calculates routes based on status information that it collects.

Under MPLS, a label switching router (LSR) switches and forwards an incoming packet by examining its external label. Such a packet is represented as a frame that carries routing information. It may be regarded as a layer 2.5 PDU. Labels are distributed among MPLS routers by using a Label Distribution Protocol (LDP), a Resource Reservation Protocol (RSVP), or by using the RSVP-TE protocol noted below. The data transport plane of an underlying datagram packet switching network is used (in-band or out-of-band) by LSRs to exchange labels and reachability information with each other. In this way, it serves as a signaling oriented system that is employed to configure virtual circuit connections. The edge LSR, known as a Label Edge Router (LER), is used to attach labels to incoming packets. When exiting an MPLS domain, the exit LER strips the label and forwards the payload message to its destination client.

Multiple stacks (or tiers) of MPLS labels may be used when transiting multiple MPLS domains. This is analogous to stuffing a letter (the payload message) in an envelope that is identified by Label A and then stuffing the later envelop in a larger envelope that is identified by Label B. The stuffed external envelope is transported across its tier in following the path linked to Label B until it reaches the exit node of this tier. At this point, the external envelope is discarded and the inner envelope is used to transport the data message through its next MPLS domain, following the routing scheme linked to label A. The MPLS method has been extended under the **Generalized Multi-Protocol Label Switching (GMPLS)** protocol to include the use of **Generalized Labels**. Such a label can represent entities such as: a fiber in a bundle of fibers; a wavelength or a collection of wavelengths within a fiber; a group of time slots embedded within a wavelength channel. The GMPLS protocol can be used to manage a multitude of switching schemes, such as TDM, layer-2 switching, wavelength switching and fiber switching. Several GMPLS signaling protocols have been developed and used. They provide for the allocation of paths and associated path resources in a manner that takes into account the current availability of spatial and bandwidth level resources. Circuits and connections can then be set up in a manner that meets the capacity and performance requirements requested for a flow.

Such a *QoS*-based resource aware setting of switching entries, including the allocation of lightpaths in a WDM-based circuit-switched optical mesh network, is used to perform **Traffic Engineering (TE)** oriented resource allocations. GMPLS employs signaling protocols that perform Traffic Engineering-oriented allocations, such as the Resource Reservation Protocol with Traffic Engineering extensions (RSVP-TE), and the routing protocol that is known as Open Shortest Path First with Traffic Engineering extensions (OSPF-TE). A *separate out-of-band control-plane network* is often employed to distribute signaling messages, generally through the use of electrically processed and switched operations. These processes are used to configure lightpaths and the corresponding cross-connect entries used by OXC nodes. A signaling protocol such as GMPLS RSVP-TE is often used to configure the underlying circuits.

368 | *10 Circuit Switching*

Problems

10.1 Describe the principle of operation of a circuit switching network system.

10.2 Describe the role of a signaling system in the operation of a circuit switching network system.

10.3 Describe the operation and performance features of the following signaling systems and signaling entities:
   a) In-band signaling.
   b) Out-of-band signaling.
   c) Common Channel Inter-Office Signaling (CCIS).
   d) Signaling System No. 7 (SS7).
   e) Signaling Transfer Points (STPs).

10.4 Consider the circuit switching network system shown in Fig. 10.1.
   a) Describe the operation of a circuit switching arrangement that serves to connect circuits across the route N1–N2–N4–N5.
   b) Describe the operation of a circuit switching arrangement that serves to connect end-user S1 with end-user S5.

10.5 Consider a communications link in a circuit switching network that is shared by multiple circuits on a Time Division Multiplexing (TDM) basis. Assume that a time frame consists of 10 slots and that each slot provides for the transmission of a single message segment that contains 1000 [bits]. The transmission data rate across the communications link is equal to 100 [Mbps]. Assume that 10 different circuits are sharing the link and that each circuit is assigned 1 slot per frame.
   a) Calculate the time duration of each time slot.
   b) Calculate the duration of a time frame.
   c) Calculate the data rate that is assigned to each circuit.
   d) Consider a station that is allocated such a single circuit. Assume that this station sends data over the link as part of a flow activity. The flow initiates its request for transmission across the link in the middle of a time frame and is subsequently assigned slot-1 in each time frame for the transmission of its segments. The flow's activity lasts for a period of time that requires segment transmissions over 100 time frames. Calculate the total elapsed time measured from the instant of initiation of its request and ending at the instant that the flow's activity terminated.

10.6 Explain the structure of the Circuit Switching table shown in Fig. 10.3.

10.7 Describe the switching operation that is performed by using a Time–Space–Time (TST) switching module, following the illustrative configuration shown in Fig. 10.4.

10.8 Describe the operation of the following spatial switching fabrics that are employed by a circuit switch:
   a) Space Division Switching.
   b) Shared Medium Switching.
   c) Shared Memory Switching.

10.9 Define and discuss the following performance measures associated with the operation of a circuit switching network system:

a) Circuit set-up time.

b) End-to-end network message delay.

c) Grade of Service (GOS), also identified as Call Blocking Probability (CBP).

d) Resource utilization ratio.

10.10 Consider a circuit switching node that is served by m communications circuits in providing pure-blocking service. No queueing facility is available. Calls that arrive when all m circuits are busy are blocked. Each admitted call occupies a single circuit. Calls arrive to the system in accordance with a Poisson process at a rate of λ [calls/s]. The average time that a call holds an allocated circuit is equal to h [s]. The holding time is exponentially distributed. Assume that $h = 4.5$ [s]. The operation of the switch's system is modeled as an M/M/m/m queueing system.

a) Configuration 1: Assume that $m = 5$, and $\lambda = 1$ [call/s]. Calculate the probability that an arriving call is blocked. Then compute the switch's throughput rate.

b) Configuration 2: Considering the same values for h and λ, calculate the number of circuits (m) that the system should use if the system's manager wishes to attain a call blocking probability that is lower by 50% than that calculated above under Configuration 1.

c) Calculate the average number of active calls that are simultaneously supported by the system for each one of the configuration cases.

d) Assume that $m = 5$, and $h = 4.5$ [s]. Calculate the maximum value for the call arrival rate that the system should support in order for the system to provide a call blocking probability that is not higher than the level specified for Configuration 2.

10.11 Describe the concept of operation of a cross-connect switching system.

10.12 Identify several different circuit configurations that are included in the SONET/SDH system.

10.13 Consider cross-connect network systems.

a) Describe the operations of *opaque* and of *transparent* Optical Cross Connect (OXC) systems.

b) Describe the cross-connect networking operations shown in Fig. 10.5.

10.14 Consider all-optical cross-connect networks that employ WDM links.

a) Describe the operations of the ring network that is shown in Fig. 10.6(a) that employs OADMs.

b) Describe the operations of the ring network shown in Fig. 10.6(b) that employs optical cross-connect (OXC) nodes.

10.15 Describe the concept of operation of a Multi-protocol Label Switching (MPLS) system.

10.16 Describe the concept of operation of a Generalized Multi-protocol Label Switching (GMPLS) system.

10.17 Describe the concept of operation of a RSVP-TE system.

11

Connection-Oriented Packet Switching

The Objective: *Consider a user that wishes a network system to support its message flow by pre-assigning it an end-to-end route prior to the start of its activity. Furthermore, it may also require that this route enables its messages to be transported to their destination at a specified quality-of-service (QoS) performance level. These requirements are specified by the user by interacting with the network management system via a connection process. As typical for many data flows, the user's flow's data message traffic is produced in a bursty manner, alternating between periods of high activity and relatively long periods of low or no activity. Hence, it is not efficient for the network to allocate communications capacity resources across the selected path for dedicated use by this flow's messages. Rather, a **connection-oriented packet switching**-based operation is required, so that the flow's messages share the communications resources of the links that they traverse on a **statistical multiplexing** basis with messages that belong to other flows.*

   The Method: *During an initial call connection process, a signaling protocol is employed, enabling a user to specify the QoS parameters that are requested for supporting the transport of its flow's messages. If sufficient resources exist, the network would then configure an end-to-end network transport entity that will be used for supporting the flow. This entity is identified as a **Virtual Circuit (VC)**. Using a packet-switching method, messages that are transported along a call's VC **share** on a statistical multiplexing basis the communications resources of the links that they traverse with other flows. Each flow's message contains a header field, also known as a label, that identifies the VC-based route that has been configured. A flow's messages are routed across a network path in a store-and-forward basis aided by intermediate nodes that act as VC switches. A VC switch uses the label of a message that arrives across a specific link to determine the outgoing link to which this message should be switched, as it follows its predetermined path to its destination. A key advantage of this operation, when compared with a circuit-switched transport method, is that when a flow is active in a statistically bursty fashion, no bandwidth resources are wasted during inactivity periods. Hence, higher communications link bandwidth utilization levels are attained. In turn, messages may incur variable queueing delays at network nodes, particularly when the communications links that they traverse experience high traffic loading conditions. To accommodate diverse flow types, whereby different flows require different QoS-oriented networking treatments, statistical multiplexing schemes that share communications and nodal resources on a class-of-service basis, such as multitier-weighted round robin and priority-oriented scheduling algorithms, are employed.*

Principles of Data Transfer Through Communications Networks, the Internet, and Autonomous Mobiles, First Edition. Izhak Rubin.
© 2025 The Institute of Electrical and Electronics Engineers, Inc. Published 2025 by John Wiley & Sons, Inc.

11.1 Connection-Oriented Packet Switching: The Method

Under a packet switching networking method, communications links are statistically shared among packets that belong to diverse flows. An active user station feeds the network data packets, generated by applications such as audio, data, video, or multimedia that are associated with a flow that she aims to transport across the network to a destination station. A bursty flow would carry messages only during certain periods of activity, while no data may be produced during relatively long periods of inactivity. The activity duty cycle, expressing the fraction of time that a flow generates data traffic, may thus be quite low. It is clearly not bandwidth efficient to dedicate communication capacity resources for the transport of messages produced by a *bursty* flow. Dedication of link capacity resources to a flow, as performed by using a circuit switching method, can be efficient when a flow presents messages to transport during high fraction of time. Under a packet switching method, a communications link is assigned for use by a flow's packets only during an activity period.

To illustrate, consider a flow that is active for an average of 1 ms during each period of 100 ms. The activity duty cycle of this bursty flow is thus equal to $a_{dc} = 1\%$. Assume that the flow requires transmission by a communications channel to occur at a peak rate $R_{peak} = 100$ Mbps. The average traffic rate produced by the source of this flow is thus equal to $R_{average} = 1$ Mbps. Under a circuit switching method, assuming availability, the flow will be assigned a circuit that provides it with transmission data rate that is equal to its peak rate of 100 Mbps. In turn, under a packet-switching method, a communications link that operates at a peak rate of 100 Mbps will be occupied by the transmission of the flow's packets only during 1% of the time. Therefore, this flow will utilize link capacity resources at an average rate of just 1 Mbps. The communications links that the flow traverses can therefore be shared among many flows.

Under a packet switching operation, a flow's packet that arrives to a link that is situated across its path may find a random number of packets that belong to the same or to other flows already waiting for transmission across the same link. Hence, packets may experience queueing delays as they wait for their turn for transmission across each one of the links that they traverse.

To assure a flow's packets with limited queueing delays, the system manager configures the average fractional loading of a traversed link, denoted as ρ, to be sufficiently lower than 1. By setting this loading level to a properly selected low value, and limiting accordingly the number of flows that share a link, the system's management scheme can regulate the end-to-end delays incurred by flow packets across their routes. Using queueing models and employing a flow admission control scheme, packets produced by an admitted flow can be guaranteed to incur end-to-end delays that are no higher than a specified level.

A packet-switching network that sets up an end-to-end connection for each flow (also identified as a call or session) is known as a **connection-oriented packet switching network**. A signaling system is used to configure a connection across a selected end-to-end route. A **virtual circuit (VC)**, also known as a *logical circuit*, is established and used to support the transport of a flow's messages. The route associated with a VC is configured prior to the start of message transport. No link capacity resources are dedicated to the VC over the path. Flow messages share the route's links on a statistical multiplexing basis with other flows. Such an operation is identified as **virtual circuit switching (VCS)**.

A connection-oriented packet-switching network offers the feasibility of granting admitted connections their desired performance levels. This is accomplished during the VC setup phase, making use of the signaling system. It is coupled with the employment of an intelligent call admission control algorithm. Ongoing loading states of network nodes and links are monitored. A route and associated capacity resources are selected for a new connection so that the new connection's requested performance levels are met, while preserving the performance levels reserved for ongoing admitted flows. If such an assignment cannot be accomplished for the support of a new call, the call is not admitted, and is regarded as blocked.

When a call ends, the call's connection is terminated. The VC is then dis-established by the signaling system. The ensuing impact of this release on network traffic and on the utilization of nodal and link resources is noted by the call admission management system. The consequent availability of additional networking resources can be used for the admission of new calls.

A key advantageous feature of a connectionless (datagram) packet-switching operation is that it provides for dynamic fast adaptivity to failures of communications link and nodal elements. Routers in a datagram packet-switching network tend to employ packet-oriented dynamic routing algorithms. Forwarding entries at network routers are dynamically adjusted to avoid transporting packets across congested or failed network segments. In comparison, under a connection-oriented packet switching operation, failure events would require the resetting of impacted end-to-end VC connections (VCC).

A datagram packet router is engaged by its processing of packet headers and the ongoing adaptations of its routing data base, which are software-driven operations. In turn, a VC switch can be effectively implemented by using high speed hardware-oriented mechanisms, as it employs efficient label switching-oriented procedures, though high-speed datagram switching mechanisms have also been implemented.

In a highway transportation system, an operation that is analogous to a VC switching method entails the following. A manager that wishes to transport a platoon of vehicles plans the trip ahead of time, prior to its start, selecting its network route. The manager does not reserve explicit highway resources (over specific time periods and spatial segments). To provide performance guarantees, an admission control scheme may be used. A new flow, if admitted, is guided across a route that meets its latency and throughput requirements.

11.2 The Virtual Circuit Switching and Networking Processes

Messages that are transported along a VC share on a statistical multiplexing basis the communications resources of the links that they traverse. Each flow's message contains a header field, also known as a **label**, that identifies the VC based route that has been configured. A flow's messages are routed across a network path in a store-and-forward basis by intermediate nodes that act as **VC switches**. Such a switch examines the label of a message that arrives across a specific link to determine the outgoing link to which this message should be switched, as it follows its predetermined path to its destination. It swaps the incoming label with a preset output label.

Certain VCS technologies, such as Asynchronous Transfer Mode (ATM) networks, identify a VC as a Virtual Channel (also denoted as VC). An ATM network can configure a **Virtual Path (VP)**. The associated **Virtual Path Connection (VPC)** is a logical connection that carries a bundle of VCs that are switched in the same manner along a network path as that used by the VPC. This is particularly useful in switching flows inside the core of a network when many flows (each associated with its own VC) use the same highly traveled path, for at least a portion of their end-to-end routes. The switching actions that are performed as their data units travel within a VP are enabled by using a VP label that is known as a **Virtual Path Identifier (VPI)**. A corresponding switch is also identified as a Virtual Path Switch. A packet arriving at the switch

11 Connection-Oriented Packet Switching

has its incoming VPI label swapped with an outgoing VPI label. The swapping configuration is set during the signaling phase, in a manner that is similar to the swapping of VCI labels.

The terminating switch of a VP examines the embedded VCI of each packet that traverses the VP path and performs individual routing decisions for each data unit based on the VC setting specified during the signaling phase. Multiple VP tiers can be configured.

To illustrate, consider a highway transportation system in which a long-distance high-speed route that consists of a serial connection of several highways is used to connect City A with City B. Different vehicles may use different local roads to access the highway at City A, and may also use different local roads to drive to their distinct local destinations within City B. The traversal of vehicles through a commonly used long distance high speed highway illustrates the operation of a Virtual Path (VP). When using a VCS oriented scheme, a vehicle's VPI label serves to trigger the proper turns at intersections, directing it to properly traverse the roads that make its desired long distance route. A VCI-type label may be used by a vehicle to navigate itself along a local path.

As a message may be routed along a path that involves both a VPC and a VCC, the routing label included in the header of a data unit would then include both VP and VC identifiers. A VPI/VCI-combined label is used in the header of an ATM data unit.

VP and VC connections may be permanent or dynamically switched. A **permanent VP or VC**, denoted PVP and PVC, respectively, can be configured by using a *network management application*. A connection that is held static for a relatively long period of time is identified as permanent. As conditions change, it can be reconfigured by using a network management application. Therefore, it is also often regarded as a *semipermanent* connection. Virtual paths tend to be set to govern transit along semipermanent configurations.

A **Switched Virtual Circuit (SVC)** is dynamically configured to follow the logical connection that is applicable to a specific VC. Such a connection is denoted as SVCC. A network's signaling system is employed for configuring an SVCC.

An illustration of message flows in a connection-oriented packet-switching network is shown in Fig. 11.1. A VCC, denoted A(N1,N5), is configured to connect a flow from station S1 to station S5 across a route that serially traverses nodes {N1, N3, N5}. The packets that are transported as part of this connection carry a VCI label that is denoted as VCI-A1, which is swapped at switch node N3 to VCI-A2. Station S1 requests the edge switch N1 to also configure a VCC for a second flow whose packets are transported to destination station S4. The VCC that is configured by the network for accommodation of the latter flow is denoted as B(N1,N4); it is set across a route that serially traverses nodes {N1, N3, N4}. A third flow is shown to originate at source station S1 and to be destined to station S8. A fourth flow originates at source station S2 and is destined to station S7. For each flow, a source station interacts with its edge switch across its User-to-Network Interface (UNI) to specify its transport performance requirements. Based on this information, the edge node uses the services of the signaling subsystem (and at times a management system) to configure the associated VPC/VCC and to produce the VP/VC labels.

The switching operation that takes place at switch node N3 is deduced from Fig. 11.1. It follows the switching entries included in the VC Switching Table shown in Fig. 11.2.

Also shown in Fig. 11.1 is the use of a VPC to aid in the networking of flows. The third flow, routed from S1 to S8, and the fourth flow, from S2 to S7, are bundled in sharing a virtual path that traverses the path {N1, N2, N6}. At N6, the individual flows, whose associated virtual channels are denoted as VC's C and D, are de-bundled at the termination of the VPC and are then independently switched and transmitted across the corresponding links.

11.2 The Virtual Circuit Switching and Networking Processes

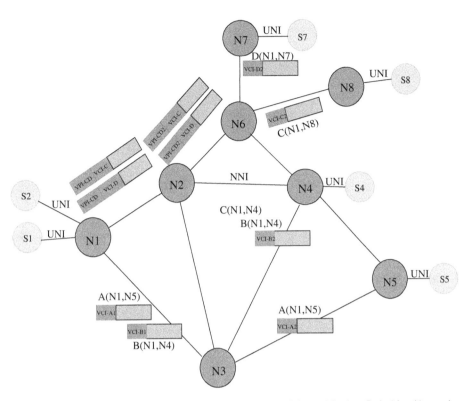

Figure 11.1 Message Flows in an Illustrative Connection-Oriented Packet-Switching Network.

Connection ID	Line_In	VCI_In	Line_Out	VCI_Out
Set by signaling system	Set by signaling system	Set by signaling data received from preceding hop switch	Set by signaling system	Set by the switch; use signaling data to inform next hop switch
A(N1,N5)	L(N1,N3)	VCI-A1	L(N3,N5)	VCI-A2
B(N1,N4)	L(N1,N3)	VCI-B1	L(N3,N4)	VCI-B2

Figure 11.2 An Illustrative VCS Switching Table at Node N3.

11.3 Technologies That Use a Connection-Oriented Packet-Switching Method

A multitude of wireline and wireless networking technologies have employed connection oriented packet switching techniques. Wireline networks that employ this technique through the use of layer-2 to layer 3 associated protocols, including X.25, Frame Relay, ATM, and MPLS/GMPLS-based networks. In the following, we review certain aspects of such implementations.

Connection-oriented procedures have also being implemented at higher layers. In implementing layer 4 services, Transmission Control Protocol (TCP) is used to provide a connection-oriented transport service that involves interactions between transport layer entities embedded at communicating end-nodes. These entities use a signaling protocol to set up transport layer connections. Message units that are exchanged between these end-user entities are identified are associated with a configured connection. They are transported across the network by using the services provided by protocol layers 1–3. The operation of TCP is discussed in Chapter 16, where we also discuss the User Datagram Protocol (UDP) as a connectionless transport layer protocol. The **ATM** network technology was developed in the 1980–1990 time period. We note the following features of an ATM network technology in aiming to illustrate the operational and performance characteristics of a connection oriented packet switching network:

1) Messages are divided into segments that are identified as ATM cells. Each cell carries a 48 Byte (with 8 bits per byte) payload and a 5 Byte header.
2) A connection-oriented packet-switching networking method is used. To transport a flow, virtual path and/or virtual channel connections are configured between the flow's end nodes. Cells are exchanged over the connection in a full duplex manner. VPI/VCI label identifiers are included in the cell's header.
3) Under ATM, both semipermanent and switched VCCs are used. VCCs are configured and used for data flows and for signaling and control message exchanges across the **User-to-Network interface (UNI)**. VCCs are also configured for the transport of messages between nodes, and between networks in accordance with network–network interfaces (NNIs), involving interchanges of data, network management, control, and routing message units.
4) To support of Quality of Service (QoS) for a flow, a call admission control process is enacted across the UNI. The user specifies its request for QoS metric values, including performance measures such as cell loss ratio and cell delay variation (CDV). The user specifies a **Traffic Descriptor** that characterizes the statistical parameters of the traffic flow processes that it expects to produce across the connection, including parameters such as average traffic rate, peak traffic rate during burst activity, average burst duration. The network traffic management scheme admits the flow if it can satisfy the user's QoS requirements. The network runs a traffic policing algorithm across the UNI to confirm that the traffic flows generated by the user across the network interface conform to the agreed upon traffic contract. Nonconforming cells may be discarded by the system.
5) Signaling messages are used to establish SVCs. A virtual channel is established for the transport of signaling messages. A meta-signaling channel is made available and used to set up a signaling VC.
   a) The signaling process is performed via hop-by-hop communications, whereby a VP is counted as a single hop.
   b) The signaling protocol is run at the application layer over a reliable transport layer.

c) VC setup and release processes are employed. Signaling processes and messages are similar to those used to configure and release circuits in a telephone system.
d) Information Elements (IEs) carry the data used in a signaling message exchange process. Included are the following entities. A call request message is denoted as a SETUP message. It is sent fy the source switch to the next hop switch, and similarly along the path from switch to switch. Once a connection is established, a CONNECT message is sent. The caller sends a CONNECT ACKNOWLEDGE message upon receiving a CONNECT message. To close a connection, A RELEASE message is issued; it is followed by a next hop RELEASE COMPLETE message.

The formats assumed by an ATM cell across the UNI and across the NNI are shown in Fig. 11.3(a) and (b), respectively. The 5 Bytes header includes VPI/VCI fields, noting that a longer space is allotted for the VPI at the NNI as it is used more often within the network's core. The payload-type (PT) field is used to identify the cell as being either a network management or a data cell. In this manner, in-band control messages can be accommodated across a VC. The header error control (HEC) field enables the receiver to determine if a cell contains errors that have occurred during transmission; it can be used at times to correct certain error patterns. A cell loss priority (CLP) bit is used as a measure of priority accommodation of a cell, whereby low priority cells are subjected to discard if they cannot be accommodated.

The ATM network was aiming to support multi media (voice, data, and video) communications. The very small data units (i.e., its cells) were deemed important for keeping the end-to-end latency low in supporting realtime voice communications, and in facilitating statistical multiplexing of multi media data units. The use of a fixed size cell, as well as the adopted connection-oriented packet switching method, were regarded of importance in aiming to implement hardware-based high-speed switching engines. In time, highly effective high-speed connection-oriented as well as connectionless packet-switching engines were developed, eliminating the incentives to use such small size cells. Such use contributes to excessive loads on the system and on the underlying QoS oriented processing, traffic management, and policing environment. ATM operations over virtual paths in the core of high-speed networks, including such backbone networks that carry IP-based aggregate Internet flows, can be performed in a more effective manner. ATM virtual path tandem switches have been employed to support core flows. The underlying ATM principles of operations are valuable in that they also point to useful methods for various networking designs.

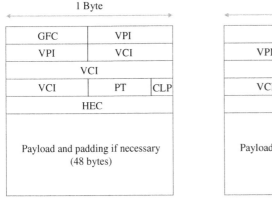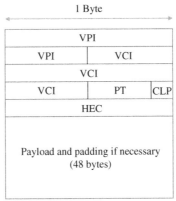

Figure 11.3 Formats of Asynchronous Transfer Mode (ATM) Cells: (a) Across the UNI; (b) Across the NNI.

11 Connection-Oriented Packet Switching

In reviewing other connection-oriented packet-switching technologies that have been employed in the past, we note the following. Under X.25, data packets were transported across a wide area network (WAN) by using a connection-oriented packet-switching method. The network provided for reliable packet transport in that received packets were checked for errors and retransmitted if needed.

A Frame Relay network is a WAN that transports layer-2 frames as its data units. Fiber-optic links are used, so that the link error rate is low, and error control fields in the frame can therefore be eliminated. Instead, an expanded link layer header is used to carry the label of the configured logical connection, identified as a Data Link Control Identifier (DLCI). It is capable of performing flow and congestion control regulations, and of offering dynamic bandwidth allocations in providing QoS-oriented support to user flows. It largely has been replaced by broadband architectures and higher capacity IP-based Internet and wireless flows.

As presented when discussing cross-connect networking, **Multi-Protocol Label Switching (MPLS)** and Generalized MPLS (GMPLS) protocols perform as a connection-oriented packet-switching systems. Message units that are formed by using various protocols, including IP packets, are switched and routed end-to-end across an MPLS domain by using labels that are attached to messages during a signaling phase. Such systems have been widely used by carrier and service providers. During the signaling phase, which precedes the data transport phase, when configuring a VCC, an end user interacts with an MPLS capable edge switch, a **Label Edge Router/Switch (LER)**. It informs the LER about the flow's destination node's address.

Under Traffic Engineering (TE) operations, it also specifies its performance requirements for the configured VC, as carried out when using GMPLS-TE. As a router, the edge node uses a routing protocol to determine an end-to-end route that will be used to transport the flow's packets (or other data units) across the MPLS domain. An interior gateway routing protocol (IGP), such as OSPF or IS–IS, is often used by MPLS networks that span a single administrative domain (as will be discussed when reviewing IP-based routing algorithms).

In extending coverage and traffic engineering to inter-area spans, the employment of an inter-domain routing scheme, such as a Border Gateway Protocol (BGP), is applicable. Under traffic engineering-based performance requirements, constrained routing algorithms are used to discover a route that satisfies the flow's requested performance levels, rather than just using a basic routing algorithm that determines an unconstrained "shortest" route. The latter route may not provide sufficient capacity, reliability, or acceptable message delay performance to satisfy flow requirements. Traffic engineering extensions are used to distribute QoS and Shared Risk Link Group (SRLG) information on each link in the network. This enables the routing algorithm to select routes that offer guaranteed QoS parameters, and synthesize backup label-switched paths (LSPs) that traverse links and/or network elements that are distinct from those used across a primary path, enhancing the robustness of the transport process.

Once a route is determined, a label distribution protocol (LDP) is employed across the control or data planes to preconfigure a **Label Switched Path (LSP)** across the MPLS network domain. This signaling/control protocol configures the label switching entries of each switch or **label switched router (LSR)** located across the route. Commonly employed signaling protocols include the LDP, when traffic engineering is not required, while RSVP-TE is used for MPLS transport when traffic engineering is required, and for GMPLS transport. Also, BGP is used as a signaling protocol for certain MPLS services, such as for BGP/MPLS Layer 3 VPNs.

Resource Reservation Protocol (RSVP) and RSVP-TE are transport layer control plane protocols that are used to reserve resources across a network. They operate over IP. Resource reservations are initiated by receivers. Requests produced by hosts and routers specify QoS metric levels for associated application flows. Schemes are employed for placing and terminating resource reservations. Requested resources are reserved at nodes located across a path, which is selected by a separate routing protocol that is not part of the RSVP scheme. Other corresponding signaling protocols have also been investigated (including the Next Step In Signaling (NSIS) scheme).

As MPLS and GMPLS transport services are often provided by a telecommunications service provider or carrier, a packet arriving from a customer edge (CE) router at a provider edge (PE) router, has labels applied to it by the latter router. The packet is routed across its LSP to its designated egress PE router, at which point it exits the provider's MPLS service domain. It is then routed across a destination end-user domain by using the latter's domain routing protocol.

As noted when discussing cross-connect networks, MPLS GMPLS technologies accommodate data units that can be created by multiplicity of protocols, including IP, ATM, and Frame Relay. GMPLS extends the MPLS method by providing also for the use of implicit labels, including wavelength identifiers that are used in a dense WDM (DWDM) optical transport systems, or time-slot-based channel identifiers that are used in SONET/SDH cross-connect systems. In these cases, a physical layer observable is used, replacing the need of a switch to read a label. A traffic engineering capable signaling protocol, such as RSVP-TE, is often used to configure the label-switched path.

As described earlier, GMPLS can also be used to establish LSPs for connection oriented circuit switched traffic. TDM or WDM identifiers can be used to switch LSP data stream across a pre-configured end-to-end path, rather than switching a flow in a packet by packet manner, by having each LSR that is located across the route use its label switching database.

Also noted is the use of **Software-Defined Wide Area Network (SD-WAN)** technology. It is employed to form a virtual WAN architecture that is able to connect users to applications by combining several transport technologies, such as MPLS, cellular wireless systems and broadband internet services. Employed software-driven control schemes are kept separate from hardware-based switching and networking entities. For this purpose, cloud-based implementations are often employed. Internet-based virtual network layouts are dynamically configured by using *centralized intelligent software*-driven algorithms, traffic and system monitors, and control mechanisms. Network overlays are often configured over multiple wireline and wireless networking systems, as they are topologically synthesized and sized to meet each client's QoS performance objectives in a robust, reliable, and secure manner.

11.4 Performance Characteristics of a Virtual Circuit Switching Network

The performance characteristics of a connection-oriented packet-switching network stem from its packet-switching operation and its use of a signaling system to configure connections. The signaling system is used to set up a connection and configure an associated VC. A routing algorithm is used for the selection of an end-to-end route. No capacity resources are associated with this VC. The flow's data units are identified by a VCI, or a combined VPI/VCI, labels. These labels, and associated links, are used by each VC switching table for switching data units along the configured route.

A key performance feature of connection-oriented networks is the ability to provide **QoS** guarantees to admitted flows. Intelligent signaling and control algorithms can be employed for enacting a **call admission control (CAC)** scheme that provides for *traffic engineering (TE)*.

Using services provided by the network signaling and management systems, an end-user that wishes the network to configure a connection for supporting a flow would provide the system, by interacting with an edge VCS switch, its *performance requirements* and its *traffic descriptor specification*. The network management system proceeds to determine the best path to select for accommodating a satisfactory VCC. If no such VCC can be configured, the call is blocked. If admitted, QoS performance guarantees are preserved **on a statistical basis** for the duration of the connection.

Employed performance metrics typically include end-to-end packet latency, packet loss ratio (PLR) and packet delay variation (PDV), also known as packet delay jitter. The underlying data unit that is transported across the VCC can be a layer-3 packet, a layer-2 frame, or another message type. A 90-percentile performance objective specifies the value of a metric that is provided for 90% of the time, or for 90% of the messages. For example, when a VC is set to guarantee its flow's packets 90-percentile performance values of PLR of 1%, a packet latency of 100 ms and a PDV of 20 ms, these performance values are guaranteed to hold for 90% of the packets, so that they may be exceeded by at most only 10% of the packets.

During the signaling phase, the following processes are performed. First, a constrained routing process is undertaken, resulting in the selection, when feasible, of a QoS-based VCC. Second, a distribution protocol is used to configure the entries of the VC switching tables of the VCS switches that are used across the selected route.

The packet transport phase follows the signaling phase. Flow packets are switched across the configured route by encountered VCS switches in accordance with the entries placed in the VC table of each switch, as illustrated in Fig. 11.2 for switch N3. A packet that belongs to a specific connection that arrives to the switch at a specific input line and that is identified by a specific VCI (or VPI/VCI) is switched to the designated output line and is attached an output VCI (or VPI/VCI) label. Assuming that the underlying node is not the packet's destination node, the switched packet will be stored at an output buffer and join an output queue. When reaching the head of the queue (or when selected based on using another queueing discipline), the packet will be transmitted across an outgoing communications line and transported to the next VCS located along its route.

Since in a packet switching network, communications links are shared via statistical multiplexing, flow's packets are not dedicated output link communications resources. Rather, packets utilize the capacity resources of a link only when they show up and require transmission across the link. A packet arriving at a switch's output buffer will often find a random number of other packets waiting in the same buffer for their turn to be selected for transmission across the same outgoing link. Therefore, an arriving packet may incur a random queueing delay while waiting in the output buffer of a switch. This is the case whether flow packets are routed in a connection-oriented manner, as discussed herein, or in a connection-less manner, as will be discussed in Chapter 12.

Key performance features of data flows that are transported across a connection-oriented packet switching network are thus related to the statistical multiplexing processes that are performed in sharing the resources of switch processors and of communications links, inducing random packet queueing delays. When higher traffic loads are applied, longer queueing delays are incurred, leading to higher packet delay levels and often also to PDVs. The system may impose limits on the number of packets that can be queued in its buffers, so that un-accommodated packets are discarded.

By using call admission control and other traffic management schemes, the network system strives to provide performance guarantees to admitted flows. Flows may be classified into different **Class of Service (CoS)** categories, so that an admitted flow that belongs to a certain COS is provided class distinctive values for its required QoS metrics. Flows that require no QoS provisions are supported on a *best effort* basis. Depending on current network loading conditions, such flows may be supported or blocked. Flows that request performance guarantees may not be admitted if the system currently lacks the resources that are required for their support.

11.4 Performance Characteristics of a Virtual Circuit Switching Network

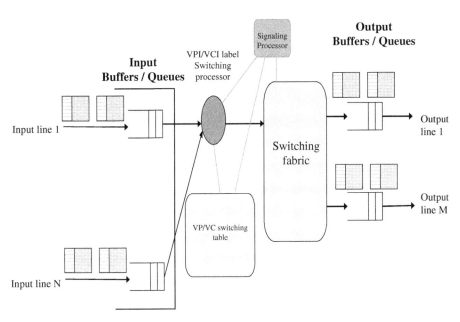

Figure 11.4 Functional Modules in a VCS-Switching Node.

We examine in the following the performance behavior of packets as they traverse a VCS switching node. The diagram shown in Fig. 11.4 depicts the key functional elements of such a node.

1) **Signaling and control subsystem:** The signaling processor represents the control module that interacts with end users, with other network nodes and with signaling subsystems.
   a) When functioning as an edge node, the node interacts across the User–Network Interface (UNI) with signaling and control entities embedded in attached user stations. Control interactions across the UNI are used to import end-user specifications of flow features, such as a flow's traffic descriptor, CoS, application type, and connection performance requirements. The node transports control and signaling messages to its end-users, including connection parameters and system regulation, management, control, and alert messaging.
   b) Control and signaling messages are communicated with corresponding entities embedded in other nodes. In-band or out-of-band signaling channels are formed and used to transport these messages across the packet-switched network.
   c) Interactions with other network nodes and with a signaling subsystem are used to obtain inputs for routing calculations and for the selection of a QoS based route for a new flow. A VP/VC connection is established across the selected route. The ensuing (VPI/VCI) labels are communicated to associated neighboring nodes. They are used to form the switching entries included in the nodal VP/VC switching table.
   d) A management interface is provided to connect to a network management station that enables a network manager to configure desired VCs (or LSPs) for designated connections.
   e) **Performance Elements:** Time latency and throughput performance metric levels are also characterized for the operation of the signaling and management systems and processes. A key performance measure is expressed by the connection set-up time. It represents the time that it takes the signaling system to configure a VC, including the setting of associated switching entries.

382 | 11 Connection-Oriented Packet Switching

2) **Input Queueing and Processing** modules consist of input buffering and queueing schedulers, the switching processor, and the data base that contains the switching table.

a) Messages such as labeled packets or frames are stored in an input buffer as they await their turn for processing by the switching processor. Waiting messages can be stored in separate buffers, in accordance with their identified CoS, flow type, requested performance metric values, or other attributes. Alternatively, a single shared buffering module is used to provide for temporary storage of messages. Identifiers, such as software based pointers, are used to keep track of the location of messages in memory.

b) Queueing service policies are employed to determine the order used in scheduling messages for service by the switching processor. A multitude of queueing disciplines may be used, as described in Chapters 6 and 7. Scheduling schemes can make use of priority designations and/or incorporate weighted fair queueing (or round-robin) methods.

c) The switching processor examines the label included in the header of a message and then reviews the switching table to find a switching entry whose incoming label attribute matches the one included in the processed message. The matching switching entry designates the output port to which this message should be switched.

d) **Performance Elements:** In connection with the operation of the input queueing and switching system, it is of interest to assess the time latency and throughput rate capacity levels of this stage. The switching operation tends to be carried out at a very high rate as it involves a label matching process that may be executed by using a VLSI-based design. Hence, the switching rate is often higher than the maximum rate of aggregate message arrival rate at all input links (which is limited to the aggregate sum of the data rates offered by the input links). For example, if the switching node is loaded by 10 input lines, each of which operates at a data rate of 100 Mbps, then the aggregate input rate is equal to 1 Gbps. If each message is on the average 1000 bits long, then the input rate is equal to 1 M [messages/s]. In this case, if the label-switching module operates at a rate that is higher than the latter rate, messages that are stored in the input buffer would experience very low queueing delays. Due to variations in the message arrival rate, while expected to be low, message delays incurred at the input stage may still be observed.

3) **The transfer of the message across the switching fabric** of a switching node is handled by an internal mechanism. An input message whose label has been matched by a switching entry is transferred to the designated output module. The latter provides a buffer for storing this message as it awaits transmission service by an output transmission module. As discussed when describing methods for transferring messages across a circuit switching node's switching fabric, a multitude of approaches can be employed. Such mechanisms can make use of space based, shared medium, shared memory, and methods that combine such means.

4) **The output queueing and transmission** modules are used to serve switched messages in transmitting them across output communications links or media to either a local destination end user or a downstream switching node. Typically, a dedicated output buffer is used for storing the messages that are awaiting service by a transmission module such as a radio transceiver. Messages stored in an output buffer are organized to form one or several queues, which identify the order used to submit stored messages for service by the output transmitting module.

Various scheduling schemes can be employed in forming the queues and in employing a service mechanism. Such schemes have been described in the Multiplexing Section. Scheduling algorithms can make be priority oriented, CoS based, and/or employ weighted fair queueing methods. Once a packet has reached the head of the output queue, it is delivered to the output module for transmission across the corresponding outgoing link.

11.4 Performance Characteristics of a Virtual Circuit Switching Network | **383**

The transmission module follows the access protocol imposed by the corresponding output communications medium. For example, an output transmission module may be sharing the resources of a multiple access channel, such as is the case when using a Wi-Fo-based wireless local area network (WLAN), or it may share the resources of a Radio Access Network (RAN) operated by a cellular wireless network system.

5) **Performance Elements:** The queueing delays incurred by messages while waiting for service at an output buffer represent a critical performance component. The number of active flows whose VCs share a given output module can vary in a significant manner, noting that VCs may carry highly bursty data flows. Hence, during certain periods of time, an output buffer may be loaded by a message traffic rate that is much higher than its service rate (i.e., the output transmission rate). During such a period of time, a message arriving at an output buffer may find a large and random number of messages that are waiting in the buffer, incurring high queueing delays.

Example 11.1 The following example illustrates the calculation of the random queueing delays incurred by messages that wait for service at an output buffer. Consider a (10×10) switch with 10 input links and 10 output links. Assume that the data rate of each input and output communications link is equal to 100 Mbps. The total communications throughput capacity of the system is thus equal to 10×100 Mbps = 1000 Mbps = 1 Gbps. Assume that the average traffic loading of the system is equal to 80% of its capacity. Also assume that over the long run, input traffic is switched in a uniformly distributed manner from input to output links, and that all flows under consideration are granted the same CoS and that their messages are served by the transmission module on a First-In First-Out (FIFO) basis.

In this case, the average data rate loading of a single output buffer and its transmission module, modeled as the output queueing system, is equal to 80 Mbps. The average normalized loading of an output queueing system is thus equal to $\rho = 0.8 = 80\%$.

Assume that the traffic process loading an output queue and the corresponding transmission module, which is the server of the output queue, is randomly fluctuating and is governed by the statistics of a Poisson process. During certain brief periods of time, the loading of the transmission server would be higher than its service rate of 100 Mbps. Over such a period, the message arrival rate is higher than its service rate. During such a time, the number of messages queued in the buffer will continuously grow, so that arriving messages will experience longer and longer waiting time delays.

The traffic loading rate is expected to decrease after a period of high activity, so that messages would then start to experience shorter waiting times. Thus, messages would experience output queueing delays that vary randomly and widely. High delay levels tend to degrade the performance of realtime voice and video streams as well as of data flows that require fast and consistent message latency levels.

Messages can assume stochastically variable lengths. Message transmission times would then vary in a random manner. To illustrate, assume a message length to follow an Exponential distribution. Assume that the average number of bits in a message be equal to 2000 [bits]. Hence, the average transmission time (or service time) of a message is equal to $\beta = 1/\mu = 2000$ [bits]/100 Mbps = 20 [μs]. The average message service (transmission) rate is then equal to $\mu = 50$ K [messages/s]. The message arrival rate at the server (i.e., transmitter) is equal to $\lambda = 80$ [Mbps]/2000 [bits/message] = 40 K [messages/s]. It is noted that the fractional loading of the system, identified as the system's traffic intensity, is equal to $\rho = \lambda/\mu = 0.8 = 80\%$, as expected.

Using the formulas presented for such a service system, which is modeled as an M/M/1 queueing system, as described in Section 7.5.1, we obtain the following results. The average message delay,

11 Connection-Oriented Packet Switching

which represents the message waiting time in the queue plus its transmission time (i.e., its service time), is obtained by using Eq. 7.23 to be equal to 100 µs. The probability that the message delay is longer than 100 [µs] is calculated by using Eq. 7.26 to be equal to $0.018 = 1.8$ %. The probability that the message delay is longer than 120 [µs] is obtained to be equal to about 0.823%. Hence, about $1 - 0.823\% = 99.177\%$ of the messages will incur a message delay time that is lower than 120 [µs], while $1 - 1.8\% = 98.2\%$ of the messages will incur a message delay time that is lower than 100 [µs]. The probability that a message that arrives at the output buffer will find upon its arrival no other message in the queue is equal to $1 - \rho = 0.2 = 20\%$. In this case, the message will incur no queueing (i.e., waiting time) latency.

These calculations illustrate the statistical fluctuations that may occur in the queueing delays incurred by messages served by an output service module. As noted in Chapter 7, a message queueing delay time increases in a significant manner as the relative loading of the service module increases.

Such queueing calculations are used by the network management system to determine the maximal allowable value of the server's traffic loading rate. When evaluating a request for configuring a connection for a new flow, the system's traffic management scheme calculates the ensuing system performance. If the admittance of a new flow would cause the queueing delays experienced by currently served flows to become higher than their assured levels, the system's call admission control manager would block the admission of the new flow.

As illustrated above, to guarantee strict performance requirements for a flow whose messages are generated in a statistically bursty fashion, it is necessary to make sure that sufficient communications capacity is available to serve active flows even during periods of time when a relatively large number of flows are active, though such periods may occur at a low rate. When impending flows are categorized as requiring the same CoS, the longer-term average utilization of the capacity of an output communications link may become quite low.

However, in many cases, different flows, and their corresponding configured VCC, are classified as requiring different service parameters and performance metric values. It is highly effective to use service policies that employ class-oriented service protocols, such as multitier weighted round robin (or weighted fair queueing) and/or priority-based scheduling algorithms. Such algorithms are described in Section 6.4. Such a scheduling policy may reserve bandwidth resources to different groups of flow classes in a multitier hierarchical manner. Capacity resources that are currently left unused by one flow class are then allocated for use by messages that belong to another class.

Problems

11.1 Describe the method of operation of a Virtual Circuit Switching (VCS) network system.

11.2 Define the concepts of Virtual Circuit (VC) and Virtual Circuit Connection (VCC). Also provide examples of a virtual circuit connection in a network system.

11.3 What is the role served by a signaling system in the operation of a Virtual Circuit Switching (VCS) network system.

11.4 Describe the following entities and identifiers that are used in a Virtual Circuit Switching network system:

a) Virtual Circuit Identifier (VCI).
b) Permanent Virtual Circuit (PVC); Switched Virtual Circuit (SVC).
c) Virtual Path (VP); Virtual Path Connection (VPC); Permanent Virtual Path (PVP). Under what conditions would a configuration of a Virtual Path (VP) be preferred to that of a Virtual Circuit (VC)?
d) Virtual Path Identifier (VPI).
e) VPI/VCI.

11.5 Using the network configuration shown in Fig. 11.1, configure and illustrate the setting of VPI/VCI connections that are used to accommodate flows between S1 and S5 and between S2 and S4.

11.6 Compose a VCS switching table that assumes a format similar to that presented in Fig. 11.2 and that accounts for the connections described in the previous problem.

11.7 Describe the functions of the fields included in UNI ATM cells and in NNI ATM cells.

11.8 Describe the operation of an MPLS network system in reference to the architecture of a Virtual Circuit-based network system.

11.9 For a VCS network system:
a) Discuss the performance features of the system.
b) Explain the use of Call Admission Control (CAC) as means for providing connections with requested quality-of-service (QoS) performance guarantees.

11.10 Consider a VCS switching node, as described in Fig. 11.4.
a) Describe the roles of the different modules in supporting the operation of a VCS node.
b) Considering the operation of the input queueing module and output queueing modules, describe the service time parameters associated with each one of these service systems.

11.11 Consider an output queueing module in a VCS switching node that serves to transmit packets arriving at output port 1. Its transmissions are performed across its assigned outgoing communications link. Assume the service function of this module to be modeled as an M/M/1 queueing system. The transmission data rate across its communications link is equal to 10 [Mbps]. A packet contains an average of 6000 [bits]. The queueing system that feeds packets for transmission at output port 1 is loaded at a packet arrival rate of 1480 [packets/s]. In selecting queued packets for service, the output module is assumed to employ a FIFO service policy.
a) Calculate the average waiting time of a packet in the output queue.
b) Calculate the 90-percentile level of the delay time (i.e., its waiting plus service time) of a packet, so that 90% of the packets experience delay times that are no longer than this level.
c) Calculate the average number of packets residing in this output system (in-queue and in-service).
d) Calculate the probability that, at any selected instant of time at steady state, the output link would be observed to be idle, so that it is not busy with packet transmissions.

12

Datagram Networking: Connectionless Packet Switching

The Concept: *Data flows tend to present high traffic message activities during relatively short burst periods. Between burst periods, during relatively long pause periods, many data flows tend to produce no messages, or feed the network message traffic at a low rate level. It is not efficient to dedicate communications capacity resources across network links for the support of such a flow as these resources will remain under-utilized for much of the time. Hence, network links are shared among flows on a statistical multiplexing basis. It is more efficient to set flow messages at the network layer, identified as packets, to occupy communications resources across their route's links only when they need to use these links. Furthermore, it is often essential to guarantee a flow a highly robust and rapidly configurable network transport process. In response to fluctuations in the loading levels of communications links and of switching nodes, it is desirable for the networking system to rapidly direct impacted packets across alternate links and routes. To this end, a connectionless operation is preferred, avoiding the pre-assignment of a network route for use by an end-to-end flow, as employed when a connection-oriented circuit or PS mode of operation is used. Hence, a connectionless PS-networking operation, also known as datagram PS, is employed for network transport of bursty data. When properly conditioned, it can also be used for the transfer of voice and video multimedia streams. This networking technology is used for message transport over a multitude of networks, including the Internet, Wi-Fi, and cellular wireless networks.*

    The Method: *Under a datagram PS operation, the system avoids the need of implementing a sophisticated, and generally costly, signaling system. Packets that belong to a specific flow are not committed to travel the network along a preconfigured route and to be assigned pre-provisioned communications bandwidth resources. As changes in the operational integrity and performance characteristics of network nodes and links occur, packets that are sent to a given destination node may be directed to travel across different routes. Switching and routing operations are performed at the network layer, making it feasible for the operation to promptly react to changing network conditions. As each communications link is shared on a statistical multiplexing manner, adaptive scheduling mechanisms are used, serving to dynamically and efficiently utilize available communications capacity resources. The operation at the network layer is connectionless. The system does not pre-allocate resources to a flow prior to the start of its network transport activity. Network flows are not provided performance guarantees through the pre-allocation of transport resources. During their transport across the network, packets may be discarded or lost, and their contents may be perturbed by noise. Datagram PS-schemes often do not provide end-to-end error control, flow control, multiplexing, reliability, or congestion control services. To provide such services, higher layer transport layer protocols are employed.* **Transmission Control Protocol (TCP)** *is a connection oriented transport layer protocol that provides end-user to end-user error and flow control services. The ordering of packets transported*

Principles of Data Transfer Through Communications Networks, the Internet, and Autonomous Mobiles, First Edition. Izhak Rubin.
© 2025 The Institute of Electrical and Electronics Engineers, Inc. Published 2025 by John Wiley & Sons, Inc.

across a TCP connection is preserved. TCP also provides for multiplexing of distinct application flows across the network through the use of service access points (SAPs), serving to identify the interface between the underlying transport and application layers. A connectionless transport layer protocol, such as **User Datagram Protocol (UDP)** *is employed to provide end-to-end application oriented multiplexing. Under either transport layer protocol, the ensuing packets can be carried across the network by using a connectionless PS method. Such a service is implemented by the Internet, which uses the Internet Protocol (IP) to identify source and destination packet addresses. Combining such transport layer and network layer protocols, the Internet employs TCP/IP and UDP/IP operations. Instead of using TCP, client server flows across the Internet may use the QUIC protocol, which is a connection oriented protocol that runs on top of UDP. The latter runs on top of IP at the network layer.*

12.1 Connectionless Packet Switching: The Method

Under a *packet switching (PS)* networking method, as employed by the Internet and other communications systems, no signaling system is used to configure network layer end-to-end connections within the network prior to the start of data communications. Under a **connectionless** or a **datagram** PS method, messages are segmented at an end-user device into network layer protocol data units (PDUs) called packets. Each packet carries information bits produced by an application, producing voice, video or data units, which we refer to in the following as the packet's payload data bits. Since no physical or virtual circuit connection is preconfigured to serve for the transport of flow messages, it is necessary to include in the packet's header addressing information, enabling its routing across the network. A packet's header includes the packet's *destination address (DA)* and *source address (SA)*. Header data that signals to network nodes the quality of transport that the packet prefers is often also included. The Internet's network layer packet addressing formats are based on the **Internet Protocol (IP)**.

A message produced by an application layer entity at an end-user device is processed by a transport layer protocol entity, forming a message that is identified as a transport layer PDU (TPDU). It is then submitted to a network layer protocol entity. The latter is responsible for enabling the switching and routing of a message across a datagram PS network. For this purpose, the message is segmented into network layer fragments, known as packets.

Each packet carries a header field, which includes networking related data, and a payload field. A packet is disseminated across the network in a **store and forward (S&F)** manner, as it is transported from the source node to its destination node (to which the destination user station is attached), traversing the route's intermediate packet routers. Each network router (or switching node) is typically directly involved in determining the best route to use in sending an incoming packet to a specified destination node, through interactions with other network nodes. Each node is actively involved in configuring the entries included in its routing table, and in dynamically updating it. In this manner, each PS node acts as a router.

Each packet arriving at a router is processed by the router and, based on the control data that it conveys in its header, switched and forwarded to the next router or node that is selected by the router. The capacity of each communication link is shared by packets on a **statistical multiplexing** basis. When reaching its destination router, the latter delivers the packet to an attached destination station. At a destination end-user station, the payload data units of received packets are submitted, as transport layer messages, for processing by a transport layer protocol entity. The latter submits the application payload data units of successfully processed transport layer messages to the properly identified application layer entity.

In a highway transportation system, an operation that is analogous to a datagram PS discipline entails the following process. No end-to-end route is preconfigured and no highway space assignments are made. Rather, as the driver reaches the next intersection (analogous to a packet router), it is briefly engaged (analogous to a packet being delayed or stored), in using the most recent highway status data, to make a decision as to which is the next road that is best for it to take (analogous to selecting the route's next hop and then queueing and forwarding the packet to the next router along its dynamically selected path). In this manner, if a specific outgoing road (or ensuing route) is noted at the time to be congested or under repair, an alternate road (or route) is selected.

Packet flows in a datagram PS network are illustrated in Fig. 12.1. For a flow that originates at station S1, whose messages are destined to station S5, messages are segmented into packets whose headers include the indicated SA and DA, identified in the figure as (S1, S5). As different packets belonging to the same flow may traverse different routes to the destination, it is noted that certain packets follow the path {N1, N3, N5}, while other packets are directed across the route {N1, N2, N4, N5}.

For a flow that originates at station S2 whose messages are destined to station S4, messages are segmented into packets whose headers include the indicated SA and DA that are identified in the figure as (S2, S4). Different packets are shown to traverse different routes to the destination, as certain packets follow the path {N1, N3, N4}, other packets are directed across the route {N1, N3, N2, N4}, while other packets are routed across the path {N1, N3, N5, N4}.

Packets are assembled into their original messages at their destination nodes. Sequence numbers that are inserted into packet headers aid in this assembly process. Packets received by a node are checked for errors. When errors in a data packet are detected and cannot be corrected by the

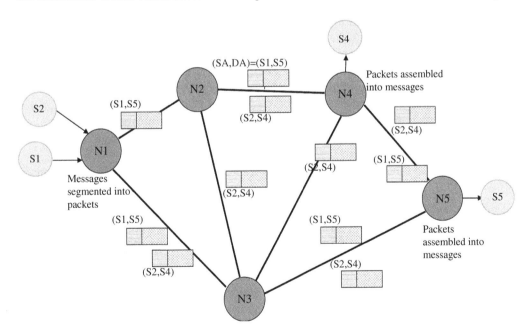

Figure 12.1 Packet Flows in a Datagram Packet-Switching Network.

12 Datagram Networking: Connectionless Packet Switching

receiving module, the packet is discarded. If upon message reassembly at the destination node, certain packets are determined to be missing (e.g., as they were discarded due to transmission errors or lost along the way), the complete message is declared to be in error and is often discarded.

12.2 Packet Flows and the Packet Router

As described in Section 1.5, a layered networking reference model is used to construct PDUs that are communicated between end users across a communications network. The corresponding TCP/IP reference model employed by the Internet is shown in Fig. 1.6. It illustrates the following message format and flow processes. An end-user message is produced at the Application Layer. The application message is carried as a payload by a TPDU, which provides transport services between end-user entities, such as end-to-end multiplexing, error, and flow control. The transport of a TPDU message across a network is executed by using the services of protocol entities at layers 1, 2, and 3. A layer 3 PDU, identified as a **packet**, carries a TPDU as its payload. Layer 3 PDUs are switched and routed by network routers and accordingly transported to their intended destinations. A data link PDU, known as a frame, is constructed for the purpose of transporting a packet, as its payload, across a network link. A frame's information units are processed by a physical layer entity, forming physical layer PDUs. A modulation/coding scheme (MCS) is used to produce signals that are transmitted across a communications link.

At a router, as noted in Fig. 1.6, physical layer (L1) PDUs and the corresponding signals are processed (such as demodulated and decoded) by the physical layer (L1) entity. Received information bits are then grouped into data link (L2) frames. The payload unit of a data frame is an L3 packet, which is processed by the router's layer-3 entity. Upon examining the DA included in the packet's header, the router switches an incoming packet to a properly selected output link. The L3 packet is then included as a payload in an L2 frame, whose information units are processed by the physical layer entity that produces L1 PDUs and signals that are transmitted across the selected output link to the next network node. As shown in Fig. 1.6, within a router, an arriving data unit is processed sequentially by L1–L2–L3 entities, and then once switched at L3, the processing follows the L3–L2–L1 sequential layering process as packets are forwarded to the next router across their route. In this manner, as a packet travels across its network route, it is processed at each intermediate router as it traverses protocol layer entities in the order L1–L2–L3–L2–L1. When the packet reaches its destination router, the latter connects it to the attached destination end-user station, where it is processed by traversing L1 to L7 protocol layer entities, ending at the targeted application layer protocol entity.

A functional diagram of a packet router is illustrated in Fig. 12.2. The following functional modules are noted:

1) **Communication interface modules** are used to perform communications system and protocol processing functions in interfacing with input and output communications lines:
   a) In interfacing an input communications line, layers L1–L2 processing operations are performed. Physical layer processing operations include detection, demodulation and decoding of the signals received across an input communications link, which are performed by the receiver section of the transceiver module. If completed successfully, the L2 entity examines the ensuing L2 frame, providing link layer services and, when relevant, medium access control (MAC) sublayer functions are carried out. If successful, the packet payload is extracted and submitted to the L3 protocol entity. This entity performs routing and switching functions.

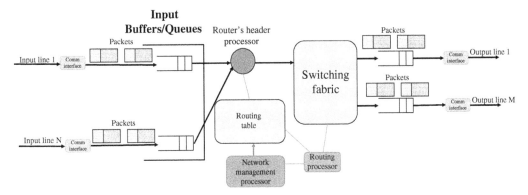

Figure 12.2 Functional Modules of a Packet Router.

b) When a switched packet reaches the head of its output queue, the system controller moves it for processing by an output communications module. When required, control data embedded in the L3 packet may be adjusted. An L2 data link control frame is constructed, carrying the outgoing packet as its payload. When sharing an outgoing multiple access channel (such as one used across a multiple access wireless medium), a MAC layer frame is synthesized. It carries the L2 data frame as its payload. Subsequently, the physical layer entity performs its functions in constructing a communications link PDU and signal, including relevant coding/modulation operations. The output transceiver is directed to transmit the packet across the output communication link at the proper time, by using the proper allocated link resources.

2) A **routing engine** implements the underlying routing algorithm. For this purpose, the router proactively and dynamically interacts with other routers and possibly other entities to calculate the best path that an arriving packet should take to reach its destination node based on the DA, and possibly other parameters, specified by this packet. The corresponding routing (forwarding) entries are stored in a Routing Table. Identifying the link connecting two neighboring routers as a **hop**, the selected output link that connects this router to the next router is accordingly identified as the next hop. Routing algorithms are discussed in Chapters 15–16.

3) A **network management** processor is used to interact with the system's management system. A network management station may be used. Such interactions are used to monitor, test, and update the employed routing algorithms and their parameters. For example, a network management entity can be employed to specify preferred routing entries and/or processing for incoming packets that are identified by specific source and/or DA sets.

4) A **Routing Table** is used to store the configured routing (or forwarding) entries. These entities are derived by the routing processor (that implements the employed routing algorithms) or are configured by a network management entity.

5) The **input buffering and queueing** stage is used to store the packets that arrive at the input ports of the router. Stored packets are arranged in queueing formations. They are served by the underlying header processor at an order that is determined by an employed input-queueing scheduling policy, such as a FIFO or a priority-based scheme.

6) The **header processor** reads the header of an incoming packet that is positioned at the head of the input queue, and then searches for a match of the packet's DA (and possibly other parameters) with a forwarding entry stored in the Routing Table. It uses this entry to identify the output port to which this packet is switched.

7) The **switching fabric** enables the transport of a packet that has been processed by the header processor to an output buffering and queueing module. As outlined when discussing connection-oriented switching nodes, fabric operations may employ a spatial layout of elementary switches that support input–output auto-navigation of packets. Other fabric structures include shared medium (bus oriented) modules, shared memory switching mechanisms, and hybrids of such structures.

8) The **output buffering and queueing module** associated with an output link holds the packets that have been switched by the header processor for transmission across this output link. These packets are served by a transmission module (such as a radio transceiver). For such an operation, a frame that carries the packet as its payload is constructed. It follows the format that is associated with the protocol used for link layer transport. As noted above, the output communications interface module performs the involved L3 to L2 to L1 PDU format transformations. When sharing an outgoing multiple-access channel, the transmission module schedules the transmission of a MAC frame in accordance with the employed multiple access protocol. A link signal is synthesized in accordance with the physical layer protocol that is used by the output communications link.

12.3   Performance Characteristics

In the following, key performance characteristics of a datagram PS operation are noted:

1) As observed for connection oriented PS network systems, the sharing of communications lines in a **statistical multiplexing** manner implies that a flow is not dedicated capacity resources across the system's communications links. Consequently, communications bandwidth resources are utilized in a more effective manner when shared among bursty packet flows. In turn, packets may incur random delays in waiting for their turn for transmission across the network's communications links.

2) A network layer connectionless operation is used so that no signaling system is employed. This implies much simplification in the implementation and operation of the networking system. In turn, as connection setup and resource allocations for it are not performed prior to the start of the data transport phase, it is more demanding to assure flow's packets with desired **quality of service (QoS)** performance for their network transport experience.

3) A pure network layer datagram operation does not provide **error control** and is thus often identified as an **unreliable datagram** service. Packets that are detected by network routers to contain errors that they cannot correct are discarded. No retransmission operation is enacted when using an unreliable datagram service. Error control retransmission schemes can still be employed at the link layer and at the transport layer, as will be noted in a later section.

4) A basic network layer datagram operation does not provide a **flow control** service. Consequently, the rate at which an end-user source feeds packets into the network is not regulated in accordance with the congestion status of the network system. A flow control scheme can be implemented at the User-to-Network Interface (UNI). In this case, the rate of packets that are admitted into the network is regulated. Flow and congestion control schemes are discussed in Chapter 14.

5) A basic network layer datagram service does not guarantee an **orderly delivery** of packets. The order at which packets are received at a destination network node may be different from their order when fed into the network by a source entity. Following their transport across the network,

packets that are associated with a single flow may reach their destination in an out-of-order manner. A process is used at the destination node to assemble arriving packets that belong to a single message or stream in correct order. This way, a destination node is able to properly reconstruct a message prior to its delivery to a higher layer entity at the destination end user. Such an operation is identified as a **reassembly** process. It is aided by the inclusion of a sequence number at the header of each packet that identifies the orderly position of the packet within its message. In contrast, it is noted that under a virtual circuit switching (VCS) service, packets that belong to a single flow connection are directed to follow the same end-to-end route, enabling the orderly delivery of packets.

6) By segmenting a message into multiple packets and transporting each packet independently across the network, the store-and-forward transport of packets can lead to reduced packet delay levels, prompted by the underlying **pipelining** process. This is illustrated in Fig. 12.3, and is explained as follows. In Fig. 12.3(a), a store-and-forward message switching flow is shown. An end-to-end tandem route is used, so that messages traverse sequentially switching nodes {A, B. C, D}. The nodes are assumed to be connected by point-to-point links. The end-user message (M) is not segmented. At each intermediate network router (or switch), the forwarding of a message is delayed until the whole message is received, stored, and reaches the head of the line in its forwarding queue.

In comparison, a store-and-forward PS network layer flow is demonstrated in Fig. 12.3(b). The message is now segmented into packets (denoted as packets P1, P2, and P3) and each packet is individually and independently routed across the network (noting that in this illustrative scenario, each packet follows the same end-to-end route) and forwarded by each intermediate router across the output link when reaching the head of its queue. Routers do not wait for the arrival of all packets that belong to a single message before processing and forwarding them. In this manner, end-to-end message delay levels are reduced. The packet-based transport process effectively reuses the network's spatial–temporal dimensional resources. For example, while

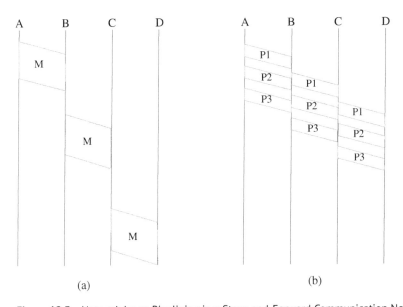

Figure 12.3 Network Layer Pipelining in a Store-and-Forward Communication Network: (a) Message Switching Transport; (b) Packet Switching Transport Illustrating the Impact of Packet Pipelining Across the Network Layer.

packet P1 is transmitted across link (C,D), packet P2 is transported across link (B,C) and packet P3 is sent across link (A,B). It is however noted that a reassembly delay component is incurred as packets are reassembled into the corresponding original message at the destination node.

The employment of a packet pipelining process that provides spatial reuse gains is also illustrated in Fig. 9.6, where a network of interconnected mobile vehicles is illustrated. This networking process is based on our **Vehicle Backbone Network (VBN)** networking protocol. A Road Side Unit (RSU) generates a flow of messages that it disseminates to vehicles moving across the highway. A reuse-3 spatial TDMA multiple access scheme is employed. The RSU segments messages into packets. It implements a **source pacing** mechanism, adjusting the rate used to feed packets into the network. It transmits, in a designated time slot, a single packet every time-frame period (which consists of three time slots). Mobiles that are elected to act as relay nodes (RNs) transmit received packets in their designated time slots, at a rate of a single packet per frame. By using such a networking arrangement, a pipelining process is realized. Packets are observed to be transmitted in a time simultaneous manner across distinct network links.

7) **Performance at the input stage:** The input stage includes modules that provide for input buffering, queue formation, service scheduling, and header processing. It connects with a data base that contains a routing table. Input packets are stored in an input buffer, as they await processing by the header processor. The latter examines the header of each incoming packet, observing its DA and possibly other parameters that may provide transport directives. It then searches a list of DA included in a routing table (and possibly certain directive oriented parameters), encountering a match with its own specified DA, and consequently identifying a corresponding forwarding entry stored in the routing table. The latter specifies the output port (and corresponding link) to which the packet should be transferred across the router's switching fabric.

The highest rate at which this processing is carried out represents the maximum **number of packets per second (PPS)** that the header processor can transfer from the input buffer to the proper output buffer. For example, a router that is capable of performing this switching process at an average latency of 1 μs per packet, operates at a switching rate of 1 Million packets/s. if the header processor's switching rate is much higher than the packet arrival rate, only limited queueing delays are incurred by packets that wait in the input buffer for their header to be processed.

8) **Performance at the output stage:** As described for a connection-oriented packet switching (VCS) module, queueing delays incurred by packets while waiting for service at an output buffer constitute a key factor in impacting the performance behavior of transported packets. The traffic loading rate of an output buffer tends to statistically fluctuate. Consequently, packets may arrive at an output module, at least temporarily, at a traffic rate that is higher than the transmission rate of the output module (such as a radio transmitter). During such a high traffic activity period, packets may incur high queueing delays in waiting for service.

The calculations described in Example 11.1 illustrate the statistical fluctuations in queueing delays incurred by packets while waiting for transmission at an output buffer. As noted in Chapter 7, message queueing delays increase in a significant manner as the relative traffic loading of a congested service module increases. This is well illustrated by the performance graph shown in Fig. 7.5.

Packets that belong to different flow classes would typically require the network to provide them different transport performance levels. To this end, policies that employ class oriented service protocols are employed. Multitier weighted round-robin (or weighted fair queueing) and/or priority-based scheduling algorithms are often used. Such algorithms are described in Section 6.4. An employed scheme may dynamically, possibly using AI-based classification and adaptation modules, observe and categorize active flow types, evaluate their activity parameters and performance objectives, and accordingly allocate bandwidth and processing resources for their accommodation. Resources that are left unused by flows that belong to a specific class can be allocated for use by packets that belong to flows of another class. Input rate flow controls and admission schemes are employed by management algorithms to regulate network loading conditions in a manner that offers desirable delay-throughput performance behavior to designated classes of packet flows.

Under proper circumstances, in aiming to reduce the time delay incurred by a packet at a switching node, a process known as **cut-through switching** may be used. Consider a router that is in the process of processing the header of an incoming packet. Assume that it rapidly performs its header processing, determining the output link to which this packet should be switched, and completing it even before the full packet has arrived and properly received at the router's input port. Consider the case when the input speed of a packet arrival process is higher than the speed of the output line. The node may then proceed to start transmitting the packet across its outgoing line (following the completion of the header processing function) even before the full packet has been received. The residual data embedded in a packet manages to arrive and handled by the output module promptly, avoiding the occurrence of transmission gaps, a condition that is known as **underrun**. It is noted that such an operation may not be acceptable in many commonly used situations. Such an operation is applicable for a switching node that does not perform error control processing, which requires the packet to be received in full and checked for errors prior to its output transmission. Typically, PS nodes check the validity of a received packet before forwarding it. A received data packet that is determined to contain errors is often discarded. A packet switching module may also make modifications to the control contents of a received packet, which may require reception of the full packet prior to its transmission processing by the output module.

Problems

12.1 Describe the principle of operation of a datagram (connectionless) packet switching network. Demonstrate the operation by describing the store-and-forward process used for the transport of packets from source station S2 to destination station S4 across the network shown in Fig. 12.1.

12.2 Consider the model for a datagram packet switch that is shown in Fig. 12.2.
   a) Describe the function and method of operation of each one of its components.
   b) Define the parameters of a queueing system model that represents the operation of the input stage. Involved packets are stored in an input buffer and served by the header processor.
   c) Define the parameters of a queueing system model that represents the operation of the an output stage that consists of an output buffer and is served by an output transmitter.

12.3 Consider the input stage of a packet switching module modeled as an M/M/1 queueing system. Assume that the header processor is loaded at a packet rate of $\lambda = 50,000$ [packets/s]. Each packet contains an average of 2000 [bits]. The header processor serves incoming packets at a rate of $\mu = 75,000$ [packets/s].

a) Calculate the fraction of time (ρ) that the input processor will be busy in processing incoming packets.

b) Calculate the average time delay spent by a packet at the input stage in waiting and in being served.

c) Calculate the probability that a packet will spend in the input stage a time delay that is longer than 50 [μs].

12.4 Consider an output stage of a packet switching module modeled as an M/M/1 queueing system. Assume that the transmission module at this stage is loaded by packets at a packet rate of $\lambda = 4000$ [packets/s]. Each packet contains an average of 2000 [bits]. The transmitting module transmits outgoing packets across a communications link at a data rate of 10 [Mbps].

a) Calculate the fraction of time (ρ) that the transmission module will be busy in transmitting outgoing packets.

b) Calculate the average time delay spent by a packet at this output stage in waiting and in being transmitted.

c) Calculate the probability that a packet will spend a time delay at this output stage that is longer than 1 [ms].

12.5 Describe the performance characteristics of a packet switching network system.

12.6 Describe the principle of operation of cut-through switching and draw a space–time diagram that illustrates the flow of a message across a path that consists of three cut-through packet switches. For this illustration, assume that each message is a data file that consists of 20,000 [bits], and that it is processed and switched at each node following the reception of segments that contain 4000 [bits].

13

Error Control: Please Send It Again

The Concept: *A message transported across a communications network may be detected to contain errors, upon its reception at a destination device. Error detection and error correction techniques are used to regulate the occurrence of message errors. The selection of an error control method depends on the application type and the tolerance of users to the occurrence of errors. Users that receive realtime voice and video streams may tolerate a limited error rate level but are sensitive to the occurrence of random variations in the times at which messages are received; hence, they strive to avoid message retransmissions. In turn, for many data flows, a receiving entity is required to accept a message only if it is verified to not contain uncorrectable errors, while certain levels of message transport latency and delay variations could be tolerated. Such transactions could include message retransmissions. Under a basic Automatic Repeat Request (ARQ) scheme, an error detection code (EDC) is used to enable a receiving entity determine whether a message contains errors. If so, the receiver requests the sender to retransmit the message. Under a Forward Error Correction (FEC) scheme, the message is encoded by the sender by using an Error Correction Code (ECC). The receiving entity uses a decoding mechanism that enables it to detect whether the received message contains errors, as well as, when the number of errors is below a threshold level, to correct these errors without having to request for the retransmission of the message. Powerful FEC schemes are advantageous in that they reduce the occurrence rate of message retransmissions and are therefore employed for transporting time-sensitive message flows and realtime streams. However, the inclusion of FEC code bits reduces a flow's throughput efficiency.*

   The Methods: *Error control mechanisms may be invoked at one or at several communications layers. A physical layer entity typically implements a modulation/coding scheme (MCS) that employs a FEC code. It is configured so that signal transmissions across a communications link are received at an acceptable error ratio. An error control scheme that operates at the transport layer, such as one that uses TCP, assures the reliability of message transactions between end users. A receiving end user may induce the sending end user to resend a TCP message that it determines to contain errors. For systems that use communications media that experience relatively high error rates, as is often the case for transmissions over wireless media, error control schemes are jointly employed at the transport, link and Medium Access Control (MAC) protocol layers. Schemes that dynamically combine FEC and ARQ procedures, known as Hybrid ARQ (HARQ), dynamically adapt to ongoing conditions by adjusting their error control processes and parameters to achieve enhanced delay-throughput performance behavior.*

13.1   Error Control Methods

Messages sent across a communications network are liable to get corrupted by noise or interfering signals, or by discards of message segments. Consequently, they may be determined upon reception

Principles of Data Transfer Through Communications Networks, the Internet, and Autonomous Mobiles, First Edition. Izhak Rubin.
© 2025 The Institute of Electrical and Electronics Engineers, Inc. Published 2025 by John Wiley & Sons, Inc.

to contain errors or to be missing sub-message fragments. Error control schemes regulate the rate at which messages are received in error and are used to resolve error occurrences. Upon receiving a message, a determination is made as to whether it contains errors (or is incomplete). In response, a correction action may be triggered. Data messages associated with many applications demand reception by an application entity at a high level of **accuracy**, as is the case for many financial transactions, e-commerce interactions, critical file transfers, medical scans, and emergency notifications. Certain voice, image and video streams require high-quality replays, demanding receptions at their destination application layer protocol entities at sufficiently low error ratio levels.

To illustrate, consider the following scenario. Person-A communicates a message to person-B by shouting it over the air. Background noise and sound signals produced by nearby conversations that reach person-B may interfere with the reception of the intended message. To improve reception by person-B, person-A may encode the message. Then, if interfering signals impact the reception process by person-B at a sufficiently low power level, using a decoding process, person-B may be able to extract the correct message without having to request it to be resent. Otherwise, person-B may request person-A to re-send the message.

As another illustration, consider the following scenario. A company is using a transportation network to ship to Chloe a computer system that consists of N interconnected modules. Due to mobility issues, the shipped system may incur damages that cause M modules, $M \leq N$, to malfunction upon arrival at Chloe's location. To handle such a situation, the company may ship a sufficient number of redundant modules. Such modules, when determined to be properly functioning, are used by Chloe to replace damaged modules. To reduce costs, the company may ship only a limited number of redundant modules. It may not therefore be possible for Chloe to fix the damaged system on her own. In this case, Chloe may request the company to re-ship the damaged modules that it cannot fix or replace.

This illustrative scenario points out two key error-control approaches that are used in communications network systems. Under a *Forward Error Correction (FEC)* method, the message is encoded by an **Error Correction Code (ECC)**, enabling the receiving entity to detect and correct errors, provided the number of such errors is not higher than a threshold level. Under an **Automatic Repeat Request (ARQ)** method, a message is protected by an **Error Detection Code (EDC)**, which enables the receiver to determine whether the message contains errors but not necessarily to correct them. In the latter case, the sender is requested to retransmit the message. Systems often use a method that combines the use of FEC and ARQ methods by using a **Hybrid ARQ (HARQ)** scheme.

Error control schemes can be used simultaneously at different layers. FEC methods are employed at the physical layer by implementing a modulation/coding scheme (MCS), which is used in the transmission and reception of message signals across a communications channel. ARQ schemes are often used across the link layer, and its embedded MAC and LLC sublayers, in particular when using communications media that induce relatively high message error rates, as is the case for many wireless links. A transport layer error control scheme such as that employed by Transmission Control Protocol (TCP) makes use of ARQ algorithms to provide reliable end-user to end-user transactions of transport layer messages.

For data transmission systems that employ communications media that exhibit relatively high error rates, as is often the case for wireless links, error control mechanisms may be conducted jointly at the link and transport layers. By using an ARQ scheme at the MAC/link layer, a frame (i.e., a link-layer PDU) transmitted across a link that is determined by a link-layer entity at a receiving device to contain (uncorrectable) errors, may be resent by the link-layer entity at the sending device. Under an ARQ scheme that is performed across the transport layer, a message that is formatted as a transport layer PDU by an end-user device and is sent across a network to destination end-user device, may be resent by the TCP entity at the sending device if its correct reception is not confirmed in time.

A transport layer error control scheme is often also carried out when it is essential to confirm the correct reception of messages on an end-to-end basis. When the end-to-end route used for the transport of TCP messages between source and destination end-users includes high error ratio links, it is beneficial to employ link/MAC layer error control schemes across these links, serving to avoid the propagation of messages that contain errors across subsequent network links. In turn, if an end-to-end network path used for communications between end users consists of links that induce very low error ratio levels, such as fiber-optic links, it could be sufficient for many flow types to just perform transport layer error control and not employ link-layer error control schemes.

The following key parameters and metrics are used in the performance analysis and design of an error control scheme:

1) The **PDU Error Ratio** measures the average fraction of protocol data units (PDUs) that are determined, over a specified period of time, to contain errors upon processing by the corresponding protocol layer entity. When calculated before applying the error control scheme employed at this layer, it is often known as the **Raw PDU Error Ratio**, while when measured following the application of this layer's error control scheme, it is known as the **Uncorrectable PDU Error Ratio**, or simply as the PDU Error Ratio. In the latter case, only errors that cannot be corrected through the use of the error control scheme employed by this protocol layer entity are accounted for.

   Unless stated otherwise, we henceforth assume that the determination of whether a received message contains errors is made following the application of the error control scheme used at the underlying layer.

2) The **Bit Error Ratio (BER)** expresses the average fraction of information bits, out of the total number of transported bits, which are received in error over a specified period of time, as determined by using the underlying error control scheme. It is at times also identified as the **uncorrectable BER**. The **raw BER** measures the error ratio impacting the receiving entity before applying the underlying error control scheme. The complement of the error ratio represents the **reliability** of the message transport system. A system that induces a lower error ratio value is said to be more reliable. The **Bit Error Rate**, denoted as BER as well, expresses the average number of information bits received in error per unit time.

   The **bit error probability** denotes the average value of the BER, expressing the long-term average level of the BER.

a) At the physical layer, the data unit transmitted across a communication link is an information symbol. The corresponding **symbol error ratio** expresses the ratio of received symbols that contain errors. When translated to bit units, it yields the **BER**.

   Consider a communication channel that exhibits a relatively high **uncoded (or raw) BER** that is equal to 10^{-3}. Thus, an average of 1 out of 1000 transmitted bits are received in error. Consequently, a message that contains 1000 or more bits is highly likely to contain bit errors when examined at the receiver. To reduce the error rate, the underlying MCS is enhanced, including also a FEC code.

   Consider the use of an enhanced MCS that reduces the BER level to 10^{-5}. In this case, using the capability of the encoding mechanism to correct errors under certain conditions, an average of only 1 out of 100,000 transmitted bits remain uncorrected following the application of the decoding scheme. An average of about only 1% of transmitted (1000 bits long) messages would then contain uncorrected errors.

b) At the MAC and LLC link sublayers, the associated PDUs are MAC and LLC frames, respectively. The Frame Error Ratio expresses the fraction of frames that are in error as determined by the receiving link sublayer protocol entity.

c) At the network layer of a packet switching network, the associated PDUs are packets. The Packet Error Ratio (PER) expresses the fraction of packets that are determined to contain errors. This is often measured for a packet flow between a network source (gateway or edge) node and a network destination (gateway or edge) node.

d) At the transport layer, the associated PDUs are transport layer messages (TPDUs). The Transport Message Error Ratio expresses the fraction of transport layer messages that contain errors. TPDUs are included in an end-to-end flow between a transport layer entity resident at a source end-user station and a transport layer entity embedded at a destination end-user station.

e) At an application layer, the associated PDUs are application layer messages. The Application Message Error Ratio expresses the fraction of application layer messages that contain errors, as measured on an end-to-end basis between interacting application layer entities.

3) The **Throughput Rate** measures the average number of PDUs that are successfully transported between the corresponding protocol layer entities, per unit time, during a specified time window. Net throughput rates are calculated by excluding overhead bits, while gross throughput rates include them. Typical (gross or net) throughput units that are used in measurements performed at different layers include the following ones: at the physical layer—symbols per second, or bits per second; at the link layer (MAC and LLC sub-layers)—frames per second; at the network layer—packets per second (PPS); at the transport layer—transport messages (TPDUs) per second; at the application layer—application messages per second. The throughput rate can be defined to include or exclude messages that contain uncorrectable errors as measured at the targeted destination protocol layer entity.

4) The **error control scheme's complexity** relates to the computational, processing, cost, and energy resources and ensuing latency levels involved in implementing and executing the underlying error control procedures. Sophisticated FEC schemes may require the implementation of complex, costly, and often software-based processes. Simpler higher speed encoding/decoding schemes use hardware-dominated implementations.

5) An error control scheme may be applied to a data unit that consists of one or several PDUs. Such a data unit is identified as an **error control block**. For example, a MAC layer block may consist of several MAC frames. It is sent as a single *error control data unit* to a receiving MAC entity. The later checks the block for errors. If the received block is determined to contain uncorrectable errors, the system may request the retransmission of the entire block. As an alternative, sub-block segments may be encoded for error processing at the receiver, so that only sub-block segments that are determined to contain errors are retransmitted.

In the following, we review the operation of several commonly used error control methods.

13.2 Error Control Using Forward Error Correction (FEC)

Under a *Forward Error Correction (FEC)* scheme, a message is encoded by an *Error Correction Code (ECC)* prior to its transport across the system. This method enables the receiving entity to perform *error detection*, determining whether the received message contains errors, as well as to perform *error correction*, allowing it, under proper conditions, to correct erroneously received bits.

The extent to which corrections can be made depends on the code rate (as noted below) and on the number of bits (or symbols) in error that the message contains (and at times on other factors, including those noted below). To enable correction, the latter number should be lower than a threshold value that depends on the correction power of the code. Also, to enable correction, it is often essential that bit errors are randomly spread within the received message so that not too many errors are bunched next to each other, forming a long error burst. Under proper conditions, a received message that is detected to contain bit errors can be corrected by a receiving device (the "forward" entity), avoiding the need to resend it. Hence, this method is identified as a **FEC** method.

Consider an error control operation that is used in communicating a segment of bits that is identified as an **error control block**. Prior to its error control-oriented encoding, this segment is identified as an **uncoded block**. Encoding and decoding operations are applied to this block. The sending entity employs a specific code to encode the block, as it transforms the block to another block that is identified as the **encoded block**. To provide error correction and/or detection services, the encoding process uses the code's algebraic structure, or other means, to embed information *redundancy* into the pattern of bits included in the encoded block. Hence, the encoded block is longer, containing a larger number of bits than the number included in the uncoded block.

Each sequence of bits of specified length that is included in an uncoded block is transformed by the encoder to a coded sequence of bits, identified as a **codeword**. Each encoded word represents an error free mapping of a corresponding uncoded word. The mapping yields a *dictionary of codewords*.

A coding scheme whose structure uses an algebraic encoding process and which introduces a higher measure of redundancy by forming a longer encoded block, enables the receiver to correct a number of block errors. This is explained by noting the following. If no channel noise is present, a codeword is received error free at the intended receiver, so that the decoding process at the receiver is able to readily produce the correct uncoded word issued by the sender. In turn, if the channel is noisy or reception is perturbed by signal interference, several of the encoded block's bits may be received in error. Yet, under proper circumstances, when codewords are structured to be sufficiently different from each other, a mildly corrupted received codeword may be determined by the receiver's decoder to be closer to the correct uncoded word than to any other codeword, resulting in FEC. In turn, when many block bits are received in error, the decoder may determine the received word to be closer to an incorrect codeword. It then fails to accomplish its FEC task.

The transformation processes and segments involved in communicating source words (as sequences of information bits) in error control blocks through encoding and decoding operations are illustrated in Fig. 13.1. The following notations are used. A word in an uncoded block is expressed as consisting of k source information bits, denoted by the vector $x = \{x_1, x_2, \ldots, x_k\}$, where x_i represents the ith source bit (being equal to 0 or 1). Applying an encoding algorithm, we obtain an encoded block. A codeword in this encoded block consists of n code bits, $n > k$, denoted by the vector $y = \{y_1, y_2, \ldots, y_n\}$, where y_i represents the ith codeword bit (being equal to 0 or 1) formed at the sending device. Following the transport of the encoded block across the communications channel, the version of the codeword detected by the receiver is denoted as the

Figure 13.1 Encoding and Decoding Error Control Blocks Transported Across a Communications Channel.

n-dimensional vector $z = \{z_1, z_2, \ldots, z_n\}$, where z_i represents the ith codeword bit (being equal to 0 or 1) recorded at the receiver. If the ith bit of a source codeword is received in error, then $z_i \neq y_i$. Using the channel output sequence z, the decoder produces an estimate of the source sequence x which is denoted as

$$\hat{x} = \{\hat{x}_1, \hat{x}_2, \ldots, \hat{x}_k\},$$

where $\hat{x}_i$ represents the decoder's estimate of the ith source bit (being equal to 0 or 1). Clearly, if $\hat{x}_i \neq x_i$, then the decoder has made an error in recovering the ith bit of the sent codeword.

The distance between distinct codewords is defined as equal to the number of bits in which they differ. Higher error correction capability can be attained when the code is configured to have a codeword table whereby the minimum distance between any two codewords assumes a higher value. This is illustrated as follows.

Consider a simple encoding scheme known as a *repetition code*. Consider an uncoded block that consists of a single information bit. The underlying code maps the uncoded bit by repeating it five times. In this case, the uncoded block that consist of the 1 bit, {1}, is mapped to the encoded block $CW_1 = \{1, 1, 1, 1, 1\}$. The uncoded block that consist of the 0 bit, {0}, is mapped to the encoded block $CW_0 = \{0, 0, 0, 0, 0\}$. The corresponding codeword dictionary then consists of the two codewords CW_0 and CW_1. Assume that the following decoding mechanism is employed: The decoder counts the number of received 1 bits and compares it with a threshold that is set to be equal to 2. If the receive count is lower or equal to 2, it declares the received codeword to represent the uncoded 0 bit; otherwise, it decodes it to represent a 1 bit. This coding scheme works well, providing FEC that eliminates all received error bits if the received block contains less than or equal to 2 bits in error; so that only 0, 1 or 2 bits in the received block, out of its 5 bits, have been corrupted by the transport process. When this is not the case, the decoder's outcome may turn out to be incorrect.

The encoding and decoding mechanisms used above to implement an illustrative version of a repetition code are simple to implement. However, the code is inefficient, attaining a low throughput rate. The illustrative code transports only 1 information bit in each block of 5 bits. It thus yields a throughput efficiency rate of only 20%. Hence, 80% of the capacity of the underlying transport medium is utilized for coding purposes alone, a high penalty in the utilization of the channel's bandwidth resource.

Error correction power and throughput efficiency metrics associated with a coding scheme are related to the actual structure of the code itself and to its code rate. The **code rate**, C_R, is defined as the ratio between the length (i.e., number of bits) in the uncoded block divided by the length of the encoded block. Thus, it expresses the ratio of the number of information bearing bits included in the error control block to the total number of bits (information and code bits) carried by the block. For the illustrative repetition code presented above, the scheme's code rate is equal to 0.2. For the model shown in Fig. 13.1, we have $C_R = k/n$, as each codeword is n bits long while containing k information bits.

Consider the *systematic* FEC block illustrated in Fig. 13.2(a). The block is depicted to consist of an information field and a code field. Under a systematic encoding configuration, the uncoded block

Figure 13.2 Information and Code Fields in a Systematic Error Control Block for: (a) Forward Error Control (FEC) Block; (b) ARQ Error Control Block.

is included in an unmodified manner in the information field of the encoded block. Codes that do not follow the latter configuration are identified as non-systematic. The encoded block shown in Fig. 13.2(b) illustrates the structure of an encoded block that is used by an ARQ scheme that is discussed in Section 13.3.

A wide range of error correction coding schemes have been devised and used. Included are algebraic or probabilistic type codes, turbo codes, low-density parity-check (LDPC) codes, and other. We also differentiate between fixed (length) block codes, and convolutional codes, whereby for the latter the encoding/decoding process is applied on a sliding window basis, while the former is based on using a fixed block by fixed block encoding/decoding operation. It is noted that for a given family of error correction coding schemes, the higher is the redundancy built into the code, so that effectively a higher minimum distance is induced, the lower is the ensuing code rate (C_R), implying a lower throughput rate and a more reliable message reception process, and an ensuing lower *BER*.

As it relates to encoding information for transmission across a communications channel, Shannon's Channel Capacity Theorem is discussed in Section 3.7. The formula presented by Eq. 3.11 provides an upper bound on the spectral efficiency attainable in the transmission of information across a communications channel, where the transmitted signal is perturbed by noise and signal interference processes that are modeled as additive white Gaussian stochastic processes. The spectral efficiency metric represents the realizable value of the data rate per unit channel bandwidth, expressible as $\eta = C/B$ [bps/Hz], where C denotes that attainable data rate capacity level and B represents the channel's bandwidth. This result incorporates the requirement that the error ratio achieved under a coding scheme that operates at the calculated rate is very low, asymptotically converging to 0 as the length of the encoded block is increased. While this theory points out the asymptotic power of a random coding method, it does not provide prescriptions for constructing practical codes that yield low BER while also offering the highest feasible code rate levels, it can be used by a code designer to measure the rate of an implementable coding scheme against the ideal Shannon upper bound, under a prescribed spectral efficiency level. It also points out the advantage gained in combining data segments for transmitting longer encoding blocks of data. This is induced by taking advantage of statistical averaging effects, as the actual number of errors that are detected at the receiver when a long encoded block is used, becomes asymptotically quite close to the corresponding average value.

Examining Shannon's capacity formula shown in Eq. 3.11, we note the **logarithmic** dependence of the spectral efficiency on the Signal-to-Interference-plus-Noise Ratio (SINR) measured at the receiver. The SINR value required to sustain a targeted spectral efficiency level, increases in an exponential fashion as a higher spectral efficiency level is prescribed. This has been observed to be the case when specific coding schemes are used under various channel conditions. Assuming a communications channel that spans a prescribed bandwidth level, the attainable value of the data rate decreases fast as the SINR value decreases (so that the receiver is encumbered with higher noise and/or signal interference levels). Consequently, many systems employ an **Adaptive Modulation/Coding Scheme (AMCS)**, under which the configured MCS is determined based on the SINR level measured at the receiving entity. A table that represents the various MCS configurations that are employed by an AMCS used by a Wi-Fi wireless LAN system is displayed in Fig. 3.6. Under high noise and interference conditions, which lead to lower SINR values, lower efficiency combined modulation and coding schemes need to be employed, yielding lower rate (and lower spectral efficiency) encoding operations. In turn, under lower noise and interference conditions, the SINR level recorded at a receiver is higher, so that an employed modulation and coding scheme can be configured to exhibit a higher data rate and achieve higher spectral efficiency. For example, under MCS

404 | *13 Error Control: Please Send It Again*

Index 16, which is employed under relatively high SINR conditions, 64-QAM modulation and ECC code rate 5/6 are used, achieving a high spectral efficiency level of about $\eta = 130/20 = 6.5$ [bps/Hz]. In contrast, under high noise/interference conditions, when MCS Index 1 is used, QPSK modulation and ECC code rate 1/2 are used, leading to a much lower spectral efficiency level of about $\eta = 13/20 = 0.65$ [bps/Hz]. Similar behavior is exhibited in Fig. 3.7, which relates to an AMCS configuration that is used by a radio access network (RAN) segment employed by a version of the LTE cellular wireless network.

Under excessively high noise and/or signal interference conditions, the underlying SINR levels detected at the intended receiver may be too low, not permitting recovery of the information bits under any employed encoding mechanism. When such conditions occur, the channel is said to experience an **outage**. No successful reception is achieved. If desired, messages transmitted during an outage period will be retransmitted at a later time, when favorable communications conditions are observed.

FEC schemes are of particular importance in situations that require the avoidance of message retransmissions, as is the case for deep space communications, where data transmissions incur ultralong propagation delays, making it impractical to retransmit message segments. This is also often the case for broadcasting or multicasting message flows for which no message acknowledgments and selective re-transmissions are carried out. FEC schemes that use strong ECCs are also employed in systems that require relatively low time delay performance in the transport of high fraction of user messages. This is the case for high-speed systems for which messages experience relatively long propagation time delays across the transport medium, including satellite communications systems, and various streaming flows associated with applications of critical importance. FEC mechanisms are also used to provide error correction for data stored in the memory of many systems.

13.3 Automatic Repeat Request (ARQ)

13.3.1 Error Detection Coding

Under an *Automatic Repeat request (ARQ)* error control scheme, a message, formatted as an error control block, is encoded by an **EDC** prior to its transport across the system. Using an EDC enables the targeted receiver to determine whether the received message contains errors, a process that is known as **error detection**, but generally not to correct them. An ARQ algorithm is jointly used by the receiving and sending entities. Generally, if correctly received, the receiving entity confirms to the sending entity its correct reception of the error control block. Otherwise, the block would be retransmitted.

To illustrate, consider a classroom setting, as illustrated in Fig. 13.3. An instructor conducts a lesson, as he teaches his students. Relating to the concept of an ARQ scheme, consider the following. A student that wishes to clarify a point made by the teacher may raise her hand and when prompted to speak, she asks her question. If the teacher understands the question, he responds to the student by providing clarification. He may then also encourage the student to raise a follow up question. In turn, if the teacher does not completely hear or understand the student's question, he tells the student to repeat the question and possibly also to expand on it.

When using an ECC, the receiver is able to perform error correction, in addition to error detection. It is more efficient to employ an EDC for the sole purpose of detecting errors. EDCs are much simpler to implement. As shown in Fig. 13.2(b), an encoded ARQ block contains a code field that carries the bits calculated by applying the EDC-based encoding process to the data included

Figure 13.3 Communications in a Classroom.

in the information field. As its aim is limited to just detecting the existence of errors in the received block, the EDC code field contains a relatively low number of bits. This number remains low independently of the length of the information field. At the receiver, a decoding process is used to determine whether the received block contains errors or is error-less.

A **parity check (PC) code** is a simple and weak EDC. A single code bit, identified as the parity check bit, is calculated by the encoder. Under an even parity scheme, the code bit is set to 1 if the number of 1 bits in the error control block is even; it is set equal to 0, otherwise. At the receiver, the decoding algorithm consists of examining the received block and comparing the parity of the received information field with the received parity bit. The two variables will not match if the value of incurred bit errors is an odd number. Thus, this error detection scheme works properly when the number of block errors is expected to be low, essentially equal to either 0 or 1. Otherwise, a *mis-detection* event may take place, whereby the decoder may declare a received block that contains more than a single bit in error to contain no errors.

A more powerful EDC is employed for handling transmissions across noisy channels. A **Cyclic Redundancy Check (CRC)** is a strong EDC that is commonly employed in data communications networks. A systematic cyclic code is used to encode (map) the bits included in the uncoded error control block to produce a fixed length code field whose contents represent what is known as a check or a redundancy value. The encoder's outcome is based on calculating a long division of a polynomial, whose coefficients are based on the uncoded data bits, by a generator polynomial that identifies the code's structure. The remainder of this polynomial division yields the check bits included in the EDC field.

CRC encoding and decoding mechanisms are simple to implement by using a process that operates at high speed, entailing low computational delay and complexity. Employing hardware based processing, such commonly employed encoding and decoding chips are widely used in data communications and networking systems.

A CRC that produces an m-bit long check value is identified as an m-bit CRC. It uses a generator polynomial whose highest degree is equal to m. Using specific polynomials, the most commonly employed CRC codes are CRC-8, CRC-16, CRC-32, and CRC-64. For example, a CRC-32 code that is defined by IEEE as well as by other Standards organizations uses the following polynomial:
$x^{32} + x^{26} + x^{23} + x^{22} + x^{16} + x^{12} + x^{11} + x^{10} + x^8 + x^7 + x^5 + x^4 + x^2 + x + 1$.

13 Error Control: Please Send It Again

A CRC-encoding system provides excellent error detection performance. Consider a CRC whose information field is k bits long, the encoded block is n bits long, where $n > k$, and its check field is m bits long, so that $m = n - k$. When applied to a data block of arbitrary length, the CRC code will typically enable the detection of any single error burst that is not longer than m bits, as well as a fraction, which is equal to $1 - 2^{-n}$, of all longer error bursts. It thus generally exhibits an extremely low probability of misdetection.

13.3.2 The ARQ Process

The basic ARQ process employs the principle of *positive acknowledgment (PACK) plus timeout*. It is described as follows. Upon completing the transmission of an EDC block, a sending entity initiates an *ACK Timer* whose timeout duration is set to a prescribed time level, denoted as $T(TO)$ [s]. Upon receipt of this EDC block, the receiving entity determines whether the block contains uncorrectable errors or is error-less. In the former case, the received block is discarded. Under such a scheme, no feedback is then sent to the sending station. In turn, if the received block (following its processing by the underlying error control scheme, including the use of error correction when employed) is determined to be error-less, a PACK message is sent (across the network, unless other means are available) to the sending entity.

If the sending entity receives a PACK message in response to its sent block within a period of time (following the end of transmission of its block) that is not longer than $T(TO)$, so that its ACK timer has not yet timed out, it resets the timer to 0. In this case, the sender has successfully been notified and consequently confirmed that the underlying block has been correctly received at the destination entity. It can then proceed to delete a copy of the transmitted block that it has been holding in its memory.

In turn, if the sending entity does not receive the anticipated PACK prior to the expiration of its ACK Timer, it checks whether it is still interested (or allowed) to send this block, and if so, it proceeds to schedule its retransmission.

While ARQ schemes often follow the above-described process, various variations of this scheme are frequently incorporated. We will note several variants when discussing in the following various ARQ schemes.

Certain schemes make also use of *negative acknowledgment (NACK)* messages. In situations under which the receiving station is aware of the identity of the sending station, the receiving station may form a NACK message to promptly inform the sender about a reception issue, so that the sender would not have to wait until the expiration of the ACK timer prior to initiating the retransmission of a block. For many applications, including those that require the transport of messages of high value, the reception by the sender of positive confirmation of reception is essential.

An ARQ process that is based on the sole use of NACKs can be problematic. Noted are situations under which the receiving entity, when receiving a block that contains errors, may not be able to determine the identity of the sender, and would therefore be unable to send a NACK notification to the source.

In evaluating the performance features of an ARQ scheme in comparison with a FEC scheme, we note the following.

A FEC scheme is advantageous in that, when the induced error rate is within a targeted level, blocks would be retransmitted at a relatively low rate. Consequently, the message transport latency (to correct reception) is low. This is crucial when considering message transactions that require strict end-to-end time latency deadlines, and when considering communications network paths

13.3 Automatic Repeat Request (ARQ) | **407**

that include links that impose long propagation delays (as is the case over satellite links). However, to provide a strong error correction capability, a FEC scheme requires the message to include a relatively high number of code symbols. This reduces the throughput efficiency of the block transport process, lowering the utilization rate of the communications media resources.

For example, for FEC purposes, a system may append a FEC code trailer field of length of 800 bits to a message that carries 1200 data payload bits. The message is assumed to serve as the error control block. This message then contains a total of $1200 + 800 = 2000$ bits. It yields an effective throughput rate of $1200/2000 = 0.6$, so that only 60% of the capacity of the employed communications resource is used for data transport. The error-control overhead rate is equal to $800/2000 = 0.4$ or 40%. The employed code assumes a code rate of 0.6. As the noise and interference power levels impacting the receiving entity increase, a lower code rate (which uses a higher number of code symbols per block) is required, lowering the data throughput rate.

Under an ARQ scheme, higher-capacity utilization levels are achieved at the expense of possible degradation in the time latency incurred in delivering messages to their targeted destinations. EDCs use short code fields, noting that their function is to be used to just determine whether the message contains errors and not to perform error correction. For example, for ARQ purposes, using a CRC-32 EDC, a code field of length of 32 bits is appended to an information field that carries, say, 1200 data bits. This message contains a total of $1200 + 32 = 1232$ bits and yields an effective throughput ratio of $1200/1232 = 0.974$, so that 97.4% of the capacity of the allocated communications resource is used by this message for its data transport. The error-control overhead rate is then equal to just $32/1232 = 0.026$ or 2.6%.

In the following, we review three key ARQ algorithms that form the basis of many employed ARQ schemes: Stop-and-Wait, Go-Back-N (GBN) or Sliding Window, and Selective-Repeat. As noted above, the determination of whether a received block contains errors or is error-less is made following the application of the employed error control scheme. A correctly received block does not contain uncorrectable errors.

Simplified throughput performance analyses of these three schemes follow the models presented in [47].

13.3.3 Stop-and-Wait ARQ

Under a **Stop-and-Wait ARQ**, the sending entity can have at most a single outstanding un-acknowledged error control sent block at a time. Only when a sent block has been confirmed to be correctly received by the destination entity (or its re-transmission attempt is canceled), the sending entity will proceed to schedule the sending of another block that is waiting in its send queue, if any. In the following, we describe the key elements of this algorithm.

Definition 13.1   The Stop-and-Wait (SW) ARQ Algorithm.
Consider an error control block that has reached the head of the sending queue of a source entity. Under a Stop-and-Wait (SW) ARQ protocol, the sending entity carries out the following operations:

1) Once the sending entity has completed the transmission of the block, it turns-on an ACK timer whose timeout period is set to a prescribed value, denoted as $T(TO)$.
2) Following this transmission, the sending entity **stops**, as it does not initiate the transmission of other blocks that may be waiting in the send queue. The sending entity is allowed to have at a time at most a single outstanding sent block that has not yet positively acknowledged.
3) If the sending entity does not receive a PACK message at the expiration of its ACK timer, it schedules the re-retransmission of the outstanding block. A parameter value is set to specify the

maximum number of times that a block should be retransmitted. Once reached, the resending of the outstanding block is terminated.

4) If the sending entity receives a PACK message prior to the expiration of its ACK timer, it discards a copy of the block that it holds in its send buffer, resets the timer and proceeds to schedule the sending of the next block, if any.

Using a Stop-and-Wait protocol, an illustrative exchange of error-control blocks and ACK messages between a sending entity at Station A and a receiving entity at Station B is shown in Fig. 13.4. The ith sent error control block is denoted as $F(i), i = 1, 2, \ldots$. Such a block is formatted as a frame when sent between link layer or MAC layer neighboring protocol layer entities. It is a transport layer PDU message when sent between transport layer entities of interacting end users. Upon the correct reception of the ith block, the receiving entity sends a positive acknowledgment that is denoted as $A(i)$. Under the flow exchange shown in Fig. 13.4(a), the $(i-1)$st block is correctly received, so that PACK message $A(i-1)$ is accordingly returned. In contrast, the received version of block $F(i)$ is assumed to contain uncorrectable errors so that the receiving entity does not return a PACK message. The sending entity is shown to then wait for a time duration that is equal to $T(TO)$ for the ACK timer to expire, during which it sends no other blocks. Thereafter, it resends block $F(i)$, which is shown to then be correctly received, triggering the sending of PACK message $A(i)$ by the receiver. Upon receiving the latter PACK message, the sending entity will proceed to schedule the sending of the next block residing in its transmit buffer, if any.

One might think that exchanged blocks and PACKs do not need to be uniquely identified, as there can exist at most a single outstanding unacknowledged sent block at a time. However, as illustrated by observing the scenario depicted in Fig. 13.4(b), this is not the case. Block $F(i)$ is received correctly at the receiving entity but its PACK message $A(i)$ is assumed to have been lost (or corrupted by errors) during its transport across the communications system so that it is never received at the sending entity. Consequently, after a timeout delay period $T(TO)$, the sending entity re-sends block $F(i)$, which is assumed here to be correctly received and acknowledged. This scenario points out the following issue. If in a flow of blocks, individual blocks are not differentiated from each other,

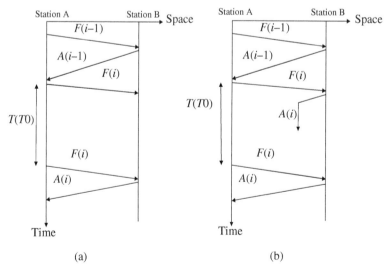

Figure 13.4 Illustrative Exchange of Error-Control Blocks and ACK Messages Between Two Stations Under Stop-and-Wait ARQ Scheme.

for example, by attaching to them distinct identifiers, such as consecutive **sequence numbers**, the receiving entity may consider consecutively received blocks to be distinct blocks, rather than resent copies of the same block! Such a confusion may lead to **duplicate reception** errors. This can have serious consequences. For example, an unidentified message informing a receiving entity of a deposit (or withdrawal) of funds from a specified account that has been resent three times, as the first three PACK transmissions have been lost, may be interpreted by the receiving entity to represent four distinct messages, and thus be interpreted as four separate (identical) deposits (or withdrawals).

To eliminate duplicate reception errors, consecutively sent distinct blocks are identified by attaching to each block a distinct sequence number. A resent block is attached a sequence number that is identical to that used to identify its previously sent version. In this manner, duplicate reception confusions are eliminated.

What should be the size of the sequence number field included in an error control block?

To reduce overhead, it is desirable to select a short sequence number field. Under the SW ARQ operation, since there can be at most a single outstanding unacknowledged sent block at a time, it will be sufficient to have a sequence number field that is 1 bit long. All is needed is to differentiate between blocks that are placed in a flow at even numbered positions in the sending order list and those sent at odd numbered positions. Consider a block that is identified by sequence number 0. If resent, its sequence number will stay equal to 0. When it is positively acknowledged, the next sent block will be identified by sequence number 1. In this manner, a receiving entity that detects the correct reception of several consecutively correctly received blocks identified by the same sequence number (whether equal to 0 or 1), would determine them to be replicas of the same block.

Key measures characterizing the *performance of block transport* between sending and receiving entities consist of throughput and packet delay metrics. A **block throughput** measure, denoted as TH_B, expresses the long-term average number of distinct blocks, per unit time, that are correctly received and are thus determined by the receiving entity to contain no errors. A **block delay** measure, denoted as D_B, expresses the time elapsed from the instant that the block is sent by its sending entity to the instant that this block is positively acknowledged. The latter event may relate to the time that the corresponding PACK is produced at the receiver or to the time that the PACK is received at the sending entity.

To calculate these measures, it is of interest to examine a typical realization that exhibits a characteristic flow of blocks (also identified as messages) and their corresponding ACKs (also identified as PACKs), as shown in Fig. 13.5. We assume here the underlying source and destination entities to operate in a half-duplex manner (as is the case for many wireless communications devices such as those used by Wi-Fi systems). Such devices are able to operate in receiving and transmission modes, but not at the same time.

The **Round Trip Time (RTT)** parameter represents the time that elapses between the instant that a message is sent by the sending entity to the time that a response (such as a PACK) is received at the sender, assuming that message and PACK transmissions are received correctly. Its duration is denoted as t_{RTT}.

Messages are assumed to each contain a total of n information and code bits. The average data rate at which messages are sent between the corresponding entities is assumed to be equal to R [bits/s]. Hence, the message transmission time (across the entity to entity connection) is equal to n/R [s]. An ACK message is assumed to contain $n(a)$ bits, and its transmission time is assumed to be equal to $n(a)/R$ [s]. The average message propagation time, one way in either direction, between the sending and receiving entities, is denoted as $t(p)$ [s].

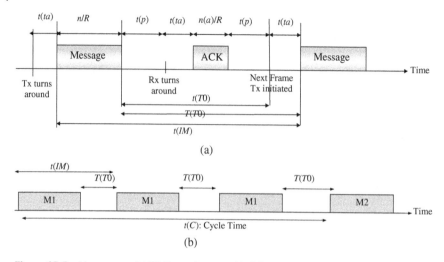

Figure 13.5 Message and ACK Flows Across a Half-Duplex Channel Connection Under a Stop-and-Wait ARQ Scheme.

We assume here the underlying source and destination entities to operate in a half-duplex manner (as is the case for many wireless communications devices such as those used by Wi-Fi systems). Such devices are able to operate in receiving and transmission modes, but not at the same time. The time that it takes the half-duplex device to switch between these modes is identified as its *turn around (ta)* time, and its value is denoted as $t(ta)$ [s].

Observing the flow depicted in Fig. 13.5(a), we note the duration of an inter-message period, denoted as $t(IM)$. It expresses the nominal time elapsed between the time that the sending entity transmits a message to the time that it is able to transmit its next message, whether it is a retransmission of a previously sent message or a transmission of a new message, assuming its send queue to be continuously occupied. As a conservative estimate, the next message transmission (or retransmission) is assumed to occur just before the ACK timer $T(TO)$ period expires. As discussed below, the setting of the timeout time is determined by the RTT t_{RTT}. The inter-message time period is given by the following expression:

$$t(IM) = (n + n(a))/R + 2t(ta) + t(RTT). \tag{13.1}$$

The probability that a block is received in error is denoted as P_{BE}. For example, if $P_{BE} = 0.1$, 10% of the sent blocks are observed (over the long term) to be received in error. Consider a flow that incurs a BER that is equal to p_b. For example, if $p_b = 10^{-5}$, an average of 1 bit will be received in error per every 100,000 transmitted bits. Assuming bit errors to be modeled to occur in a statistically independent fashion. Assume that a block, consisting of n bits, is declared to be in error if at least one of its bits is received in error. The blocks error probability is then calculated to be given as:

$$P_{BE} = 1 - (1 - p_b)^n. \tag{13.2}$$

It is noted that under low message error rate conditions, so that $np_b \ll 1$, we can estimate the message error probability by using the following approximation:

$$P_{BE} \approx np_b, \text{ for } np_b \ll 1. \tag{13.3}$$

To illustrate the corresponding error rate computation, consider a block that contains $n = 1000$ bits. Assume a transport connection (such as that provided by a communication channel across a physical layer, a MAC layer, a link layer, or a transport layer) to exhibit a BER that is equal to $p_b = 10^{-4}$. Using Eq. 13.2, we compute the block error probability to be equal to $P_{BE} \approx 0.0951671$. The approximate calculation, carried out by using Eq. 13.3, yields an error probability value of 0.1. The approximation is tighter when the product np_b is much lower than 1, as is often the case.

The probability that a block will be received in error increases as the product np_b of the message length and the connection's bit error rate increases. Hence, to reduce the rate at which a block is retransmitted, which depends on the probability that a message is received in error, P_{BE}, it is necessary to design the system so that the np_b product is kept low. The number of block retransmissions is denoted as N_R, while the number of block transmissions to correct reception is denoted as N_T. These are random variables that are related to each other, noting that $N_T = N_R + 1$. Assuming block error events to be statistically independent, and a block error rate $P_{BE} < 1$, we obtain the following expressions for the average values of these variables:

$$E(N_R) = \frac{P_{BE}}{1 - P_{BE}}. \tag{13.4}$$

$$E(N_T) = 1 + E(N_R) = \frac{1}{1 - P_{BE}}. \tag{13.5}$$

To illustrate, consider a communications connection across which the block error ratio is equal to $P_{BE} = 0.1$. Using the latter formulas, we obtain $E(N_R) \approx 0.111$, and $E(N_T) \approx 1.111$, so that on the average about 11.1% of the blocks are retransmitted, and the average number of block transmissions to successful reception is equal to about 1.111. In comparison, consider a connection for which the block error ratio is higher, as may be caused by higher channel raw BER level and/or longer block lengths. For example, consider the excessively highly noisy connection scenario under which $P_{BE} = 0.5$, so that 50% of the blocks are received in error. We then conclude that $E(N_R) = 1$, so that on the average each block is retransmitted once, and the average number of times that a block is transmitted until it is successfully received is equal to $E(N_T) = 2$.

As shown in Fig. 13.5(b), a block (denoted as M1) may have to be transmitted several times before it is successfully received. Under the above-discussed model, the figure shows block transmissions to be initiated every $t(IM)$ units of time. Accordingly, we conclude that the connection yields a throughput rate that amounts to successfully transporting one message every $E(N_T)$ inter-message periods. Hence, the block (or message) throughput rate attained by the connection, denoted as TH_B, representing the average number of blocks that are successfully carried by the connection per unit time, is expressed as follows:

$$TH_B = \frac{1}{E(N_T)t(IM)}. \tag{13.6}$$

To illustrate, consider a low-speed connection with the following parameters: $n(a) = 48$ bits; $t(p) = 10$ ms, $t(ta) = 100$ ms, $R = 4800$ bps. Assume that each block contains $n = 4800$ bits, out of which 48 bits are protocol overhead bits. The BER is equal to BER $= p_b = 0.0001$. We then obtain the net (i.e., excluding overhead bits) throughput rate, denoted as TH_b [bps], to be equal to about $TH_b = 2390$ bps. The corresponding throughput efficiency (η), also identified often as the Normalized Throughput (NTH), expressed as $\eta = TH_b/R$, is equal to about 0.498. Thus, only about 49.8% of the channel's available data rate (and corresponding bandwidth resource) is utilized for the transport of user data.

To improve performance, the MCS (and possibly other parameters, such as the transmit power level) is upgraded to yield BER $= p_b = 0.00001$. The net information throughput rate carried by this channel connection is then calculated to rise to about $TH_b = 3682$ bps, so that throughput efficiency is improved to about 76.7%.

412 | *13 Error Control: Please Send It Again*

It is thus noted that the key factors that determine the delay-throughput performance behavior of the SW ARQ scheme are: a) the ratio of the Round Trip Time (RTT, also denoted as t_{RTT}) and the block transmission time; b) the block's error ratio. In the following, we discuss the impact of each.

The first key parameter is the ratio of the **RTT** (t_{RTT}) and the block transmission time ($T_B = n/R$). It is denoted as $\gamma_{RTT} = \frac{t_{RTT}}{T_B}$. It expresses the time it takes for the sender to receive an ACK to its sent block, relative to the block's transmission time, when this block is received with no errors.

Which value should be configured for the ACK timer's timeout value $T(TO)$?

If we set $T(TO)$ to a value that is shorter than the RTT, the ensuing retransmission of the outstanding block would be performed too early. The sender would then not be waiting sufficient time to give itself a chance to receive the corresponding PACK message. Such a setting leads to throughput degradation as it may cause un-necessary retransmissions. In turn, if we set $T(TO)$ to a value that is much longer than the RTT, the sender may have to wait for an excessively long period of time before retransmitting its message in case no PACK has been produced, or the PACK has been lost or incurred unusual network delay. Under realistic network scenarios, the *round trip time* that represents the actual realized time that it takes a block to be transported across a communications network from its sender to its receiver plus the time that it takes the PACK to transit the network and be received at the sender, is a *random variable*.

To prevent high-throughput degradation while securing a relatively low response time, it is useful to *set the T(TO) period to a value that is equal to the RTT level that is not frequently exceeded*. For example, it can be set to a value that is not exceeded by over 95% of the round trip scenarios occurring over a specified period of time. For this purpose, certain retransmission schemes, such as those used by a version of the TCP transport layer protocol that is employed across the Internet, continuously record incurred RTT values and use this data to calculate the mean and standard deviation values realized by the RTT variable over a recent period of time. The $T(TO)$ is then set equal to a level that is equal to the measured RTT's average value plus a multiple factor (such as two to three times) of its measured standard deviation value.

Noting such settings, assume that we configure $T(TO) = t_{RTT}$, where t_{RTT} is set equal to a desired RTT value. Consider first the system's throughput performance when no channel errors occur. In this case, within a period of time that is equal to the RTT that follows each block's transmission, the sender would receive its expected PACK. It then proceeds to send the next waiting block (assuming the sender's send queue to be continuously busy). The maximum achievable NTH rate, also identified as the NHT capacity, denoted as η_C (no errors), is then calculated as follows: η_C (no errors) $= \frac{T_B}{T_B + t_{RTT}} = \frac{1}{1 + \frac{t_{RTT}}{T_B}} = \frac{1}{1 + \gamma_{RTT}}$. This formula indicates that when no channel errors are incurred, or effectively when the block error ratio is very low, the parameter γ_{RTT} is the key factor dictating the connection's throughput rate.

For connections that provide fast ACK reaction times relative to the block transmission time, so that $\gamma_{RTT} < 1$, high-throughput efficiency levels are attainable. In turn, for situations under which a relatively long time is incurred to receive an ACK, so that $\gamma_{RTT} > 1$, low-throughput performance behavior would be observed. The following scenarios illustrate such varied behavior.

Example 13.1 **Stop-and-Wait ARQ over a satellite link** Consider a communication link across a geosynchronous satellite, for which the round trip signal propagation delay is equal to about 540 ms. The corresponding value assume by the RTT is dominated by the signal's propagation delay, so that we approximately have $t_{RTT} = 540$ ms. Assume a block to contain an average of 5000 bits and the data rate to be equal to $R = 500$ kbps, so that the block's transmission time is equal to 10 ms. The corresponding value assumed by γ_{RTT} is equal to 54. Hence, following each block's transmission time, the sender needs to wait for a period of time that is equal to 54 block transmission times for the receipt of a PACK. Assuming a negligible occurrence rate of block errors, the corresponding throughput efficiency factor is then equal to η_C (no errors) $= 1/55 \approx 1.8\%$. The relative long pause imposed on the sender by the system's long propagation delay induces the SW ARQ scheme to exhibit very low-throughput behavior, even when the occurrence rate of channel errors is very low.

When subjected to channel error events, the number of block retransmissions should be limited to a very low value to avoid even higher degradation of the throughput rate as well as to limit the incurred end-to-end block transport delay.

Example 13.2   Stop-and-Wait ARQ over a terrestrial link Consider a terrestrial communications link across which fast ACK response times are realized. Assume that $t_{RTT} = 1\,\text{ms}$. This is, for example, the case when the RTT is dominated by the signal's propagation delay and when the communications link spans a distance of 100 km, under a signal propagation speed of 5 [µs/km]. Following Example 13.1, assume that a block contains an average of 5000 bits and that the data rate is equal to $R = 5000\,\text{kbps}$. The block transmission time is thus equal to 10 ms. In this case, we have $\gamma_{RTT} = 0.1$, so that η_C (no errors) $= 10/11 \approx 90.9\%$.

Under the occurrence of channel errors, the number of block retransmissions should be limited to a low to moderate value, depending upon the desired delay-throughput performance behavior.

Example 13.3   Stop-and-Wait ARQ over a high-speed terrestrial link Consider a high-speed terrestrial link connection. Assume it to operate at a data rate of 1 Gbps, and to span a distance of 1000 m. For a block that is 5000 bits long, the message transmission time is equal to 5 µs. Assume the RTT to be dominated by the signal's propagation delay. Under a propagation speed of 5 [µs/km], the round trip propagation delay is equal to 10 µs. In this case, we have $\gamma_{RTT} = 2$, so that we obtain the throughput efficiency factor to be equal to η_C (no errors) $= 1/3 \approx 33.3\%$.

Assume next that the data transmission rate is increased by a factor of 10–10 Gbps. Then, we obtain that $\gamma_{RTT} = 20$, and the throughput efficiency is reduced in a significant manner to η_C (no errors) $= 1/21 \approx 4.76\%$.

Thus, when ultrahigh data rate links are employed, the block transmission time is very short. However, comparatively, the RTT value remains high even when communicating over shorter link distances, as the signal propagation speed is not affected by the data rate level, as the RTT parameter is dominated by signal propagation delay. Therefore, high-speed link configurations lead to higher γ_{RTT} values so that, under a SW ARQ scheme, a degraded throughput performance is experienced. To avoid further degradation induced by channel error events, it is of interest to synthesize MCS/FEC schemes that realize lower block error rates.

A second key parameter is the block's error ratio, denoted above as P_{BE}. It is calculated by using Eq. 13.2, or approximately by using Eq. 13.3. Its value determines the average number of block retransmissions, denoted as $E(N_R)$ and given by Eq. 13.4, and the average total number of block transmissions denoted as $E(N_T)$ that is calculated by using Eq. 13.5. As indicated by examining the formulas, the average number of block retransmissions increases as the block error ratio increases. The rate of increase accelerates for higher values of the block error ratio. As noted in Eq. 13.6, the block throughput rate (TH_B) decreases as the average number of block transmissions increases. Hence, to achieve a high-throughput rate, it is necessary to keep a sufficiently low $E(N_R)$ level, which in turn requires the system designer to configure a sufficiently low block error probability (P_{BE}).

13 Error Control: Please Send It Again

Keeping a low block error probability is also needed for achieving a low transport latency. To illustrate its calculation, consider a link-layer entity engaged in transmitting message frames across an outgoing link to a neighboring node, expecting to receive PACK messages from the link layer entity of the neighboring node. The HOL block delay, denoted as D_{HB}, is defined as the time taken by a block that is positioned at the head of the line (HOL) of the sender's transmit queue to successfully complete its transfer across the output link. This delay time is measured from the instant that the block has been transmitted for the first time across the output link to the time that a PACK message has been received at the sender, assuming the latter event to eventually take place. Not included is the queueing time spent by a block in waiting in the send queue to reach the HOL position.

An approximate expression for the average value of this delay is given as follows:

$$E(D_{HB}) = E(N_T)t_{IM} = \frac{t_{IM}}{1 - P_{BE}}. \tag{13.7}$$

Which steps can be taken to reduce the block error ratio (P_{BE}) so that the system exhibits good throughput and delay performance behavior?

By examining Eq. 13.2, we note that the block error ratio depends on the following two parameters: the block length ($L_B = n$) and the link's bit error probability (p_b). Under a given BER level, the block's error ratio will be decreased as the block's length is reduced. It is not however efficient to make the block too short for the following reasons:

1) **Protocol Overhead**: The number of control bits, including protocol and EDC bits, included in a block is basically fixed. It is not dependent on the number of the information bits carried in a block. Hence, as the number of information bits is reduced, the overhead ratio becomes a more prominent factor, leading to reduced throughput efficiency. This is of particular concern when using short blocks, such as those used for VOIP-based voice packets and query messages.

2) **Segmentation and Reassembly**: When using a short block, a user message would be segmented into a higher number of blocks. This leads to increased overhead inefficiency as each block contains its own overhead bits. Also, it increases the message processing and transport delay level as a larger number of blocks is handled. A larger number of blocks must then be re-assembled to produce the original message. Since blocks may be received at a destination entity at random times, the reassembly process, aside from imposing a measure of storage and processing complexity, may also cause time delays at a destination reassembly module. The latter processor must wait for the arrival of all message blocks prior to completing the re-assembly process.

Taking these factors into consideration, the system designer calculates the best value to select for the length of the error control block. Yet, at times, under a calculated or prescribed block length, one may find the following. The raw BER of the channel may be observed to be quite high, inducing a high raw block error ratio. For example, consider a channel whose raw BER is equal to $BER = 10^{-3}$, so that an average of about 1 bit out of 1000 bits is received in error. Assume that the designer configures the block length to be of the order of 1000 bits. Consequently, a very high fraction of the received blocks will contain errors and (if uncorrectable) will have to be retransmitted a large number of times, yielding a highly inefficient delay-throughput performance behavior. In this case, a preferred solution would be, if feasible, to upgrade the underlying MCS and/or the employed FEC scheme. Following the error control operation at the receiving entity, block errors will be corrected at a high rate, yielding a post-decoding improved BER level. Such an upgrade may require the setting of a higher transmit signal power and a more capable encoding scheme, which may entail higher encoding/decoding complexity. To limit complexity and resource expenditures, it may not be necessary to implement a scheme that attains a too low BER. For example, for certain terrestrial communications applications, it may be sufficient to synthesize an MCS/FEC scheme that achieves a BER level that is reduced from 10^{-3} to 10^{-5}. In this case, an average of only about 1% of the blocks will have to be retransmitted. Such a retransmission rate may yield acceptable delay-throughput performance for many application cases.

13.3.4 Go-Back-N ARQ: A Sliding Window Protocol

In this section, we consider send and receive entities that communicate over a **full-duplex** communications link or connection, so that they are able to simultaneously send and receive messages between each other. When a high data transmission rate is used, as is often the case, the RTT taken to receive an ACK is long relative to the block transmission time. In this case, the connection's throughput efficiency can be improved by authorizing the sender to continuously transmit multiple blocks.

To illustrate, consider a scenario where an end-user wishes to send consecutively a bunch of items. Rather than sending one item, receiving a PACK for it and then sending the next item, receiving a PACK for it, and so on, the user sends consecutively multiple items (one directly following the other). During the sending period, PACK messages are expected to be returned by the receiver and received at the sender. In this manner, as long as items are received correctly and the corresponding PACK messages are received by the sender in a timely manner, the sender may continue to continuously send more items, without having to pause the transmission process.

In a similar manner, under a **GBN ARQ** scheme, the sending entity is authorized to continuously send blocks. A **window threshold** (W) of N sent outstanding unacknowledged blocks is prescribed. When the number of such sent blocks reaches this threshold, without any of these blocks yet positively acknowledged, the sender stops temporarily the transmission of any further blocks. Transmission resumes when a PACK message is received, confirming prior sent blocks. The sending process is then continued, while still subjected to this window limit. It is noted that the Stop-and-Wait (SW) protocol is a Go-Back-1 version of this scheme, for which $N = 1$.

This scheme is generally implemented as a **Sliding Window ARQ** protocol. In this context, the transport of a block over a connection can be performed through the regulated transmission of sub-blocks identified here as **segments**. Such an operation enables the system to apply a finer grade regulation of data flow across the underlying link or connection. An error control block may consist of multiple segments whose transmission process is regulated in accordance with a window sliding mechanism. The receiving entity needs to receive and assemble all of a block's segments to be able to determine if this block contains errors. If correctly received, the block is positively acknowledged by the receiving entity. To simplify the description of the underlying ARQ algorithm, we assume henceforth that each block consists of a single segment, so that the same data units are used as error control blocks and as flow control segments.

The sending entity negotiates or is assigned a window span parameter, denoted as W [segments]. It value serves to limit the maximum number of outstanding unacknowledged segments that the sender can have at a time. When this limit is reached, the sender temporarily stops sending segments, resuming transport once the number of outstanding unacknowledged segments is reduced below the currently authorized window level. Such reductions take place when the sender receives PACK message confirmations.

In the following, we illustrate the working of a sliding window ARQ algorithm. We set the window threshold value to be equal to $W = N = 7$ segments. Assume that at the present time, the sender has already transmitted seven blocks, identified by their sequentially assigned sequence

416 | *13 Error Control: Please Send It Again*

numbers $\{0, 1, 2, 3, 4, 5, 6\}$. Assume that the first subsequently received PACK confirms the correct reception of three blocks, identified by sequence numbers $\{0, 1, 2\}$. At this point, assuming the sender has many more blocks to send, it is permitted to additionally sent three more blocks, as it is allowed to have at most seven outstanding unacknowledged blocks. The sequence numbers of blocks whose PACKs are permitted to be outstanding are accordingly sliding to include sequence numbers $\{3, 4, 5, 6, 7, 0, 1\}$, consisting of a window of seven blocks. Note that since the blocks identified by sequence numbers 0 and 1 have already been PACKed, these sequence numbers can now be reused to identify new subsequently sent blocks. It is noted that as explained for the SW ARQ scheme, to avoid duplicate reception conditions, the length of the sequence number field is set to $W + 1$, or $N + 1$, yielding here a sequence number field of length 8, consisting of sequence numbers $\{0, 1, 2, 3, 4, 5, 6, 7\}$.

A **batch acknowledgment** scheme is generally employed by the receiving entity. It accumulates in its buffer for a specified period of time all received segments, and then sends a batch PACK that confirms via a single message those blocks that have been correctly received. Communications protocols often use the following method. This method is coupled with a setting that induces the receiving entity to not provide PACK confirmation to blocks that are received in an out of order fashion. For example, assume that the receiving entity stores as correctly received blocks whose sequence numbers are $\{3, 4, 6, 7\}$. In this case, the produced PACK positively confirms the receipt of the blocks that are identified by sequence numbers 3 and 4. Since block #5 has not been received, the receipt of blocks #6 and #7 is not as yet confirmed. If block #5 is not received prior to a specified time period, blocks #6 and #7 would be dropped by the receiver, as it is expecting the sender to retransmit at a later time the missing blocks in an orderly manner. Under such an operation, the receiver is generally set to provide the sender with a batch PACK that informs the sender about the *sequence number of the next block that the receiver is expecting to receive*. The batch PACK message will then specify #5 as the sequence number of the next block that the receiver is expecting. In this way, coupled with a process that accepts only blocks that are correctly received in a sequential fashion, such an ACK confirms the correct receipt of all blocks whose sequence numbers are lower than that of block #5.

As noted, a GBN scheme is often implemented over a **full-duplex connection**, as assumed henceforth in this section. Each block contains a control field that is used to carry the batch PACK sequence number. The PACK indication is said to be **piggybacked** within a reverse flowing message, if such exists. Such a commonly used implementation tends to increase a flow's throughput efficiency as it prevents, when the connection is bidirectionally active, the need to form multiple separate PACK messages. The combination of the latter provisions is the basis for the implementation of a **piggyback batch ACK** operation.

As noted above, the size of the sequence number field for a GBN ARQ scheme is set equal to $N + 1$, where the use of an additional sequence number aids in resolving the duplicate reception issue. Hence, the bit size of the sequence number field when using a window parameter N is equal to the integer value that is larger or equal to $\log_2(N + 1)$. For $N = 7$, the sequence number field consists of eight numbers so that it is 3 bits long.

The key elements of the indicated version of the GBN algorithm are summarized as follows:

Definition 13.2   The Go-Back-N (GBN) ARQ Algorithm.

1) The sending entity sends continuously error control blocks that are identified by consecutive sequence numbers (modulo $N + 1$). The sender is authorized to send a maximum of N outstanding unacknowledged blocks, identified as its allocated window level W.

2) Once the sending entity has completed the sending of a block, it turns-on an ACK timer whose timeout duration is set to a timeout parameter value, denoted as T(TO).

13.3 Automatic Repeat Request (ARQ) | **417**

3) If the sending entity does not receive a PACK by the expiration of the corresponding ACK timer, it schedules the re-transmission of its outstanding unacknowledged transmitted block(s). Under an imposed provision that the receiver does not accept blocks out of order, when a sender retransmits a block, it also retransmits all subsequently sent (unacknowledged) blocks. A threshold parameter may be set to specify the maximum number of retransmissions of a block that the entity is authorized to perform.

4) If the sending entity receives a PACK message prior to the expiration of its ACK timer (for a transmitted block or for a group of transmitted blocks), it discards a copy of this block (or blocks) and refreshes accordingly its authorized window quota. It then proceeds to schedule the sending of the next block(s) in its send queue, if any.

Simplified implementations are often employed. For example, the sending entity uses just a single ACK timer, proceeding as illustrated in the following. Consider a block identified by sequence number i that when sent, the sender does not have any outstanding unacknowledged blocks. Following its transmission, the sender initiates its ACK timer. It continues sending subsequent blocks (limited by window size N). Assume that blocks $i, i + 1, i + 2$ were consecutively sent and that they are the only blocks currently stored in the send buffer (where $N > 2$). Assume them to be received correctly. For the illustrative case, the receiving entity waits (for a prescribed period of time) to receive and process these blocks and then sends a batch-ACK indicating that the next block to be received is a block whose sequence number is equal to $i + 3$. If this batch-ACK is received at the sender prior to the expiration of the ACK timer, the sender interprets it as a positive ACK of these three blocks. It then resets the ACK timer. Its residual quota (and window's span) slides three positions to account for the confirmed reception of the latter three blocks. The ACK timer is reset and triggered again when required to account for the transmission of subsequent blocks.

As an alternate scenario, consider the case that block i is received correctly but block $i + 1$ contains errors. Then, the receiver accepts block i but discards blocks $i + 1, i + 2$. The receiver then sends a batch ACK to the sender indicating that the next block that it is expecting to receive is block $i + 1$. Provided this ACK is received prior to the expiration of the ACK timer, the sender will interpret this ACK as positively acknowledging the reception of block i. Its window span is extended to account for the reception of this single block. The sender then resets and reactivates the ACK timer and retransmits blocks $i + 1, i + 2$, and possibly other blocks (while subjected to its window limit) that may have arrived to its send buffer in the meanwhile.

An illustrative flow of blocks that are sent from Station A to Station B, under a GBN ARQ scheme, and PACK messages that are sent from Station B to station A (in a full-duplex manner so that distinct channels are employed) is shown in Fig. 13.6(a). Block F1 is correctly received and acknowledged. In turn, block F2, which has been transmitted even before the ACK for block F1 has been received, is detected to contain errors. Consequently, station B issues no PACK message for this block, so that the associated ACK timer for this block expires after a time period of duration $T(TO)$. At that time, previously transmitted blocks that have not yet been acknowledged, identified here as blocks F2 and F3, are retransmitted.

The delay-throughput performance behavior of a Go-Back-N ARQ scheme is further discussed by examining the transmission process illustrated in Fig. 13.6(b). We focus on the transport of block 1. The sending entity is assumed to be constantly loaded with block messages that it aims to send to the receiving entity.

Which value should be used for setting the timeout parameter $T(TO)$?

The RTT taken to receive an ACK, denoted as t_{RTT}, is assessed by using system parameters, and/or estimated by dynamically collecting ongoing round trip measurement data. As noted for the SW

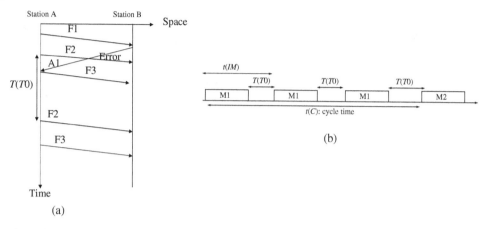

Figure 13.6 Block and ACK Flows Under a Go-Back-N ARQ Scheme: (a) Timeout Triggered Retransmission of Block F2; (b) Time Delay Incurred in the Transport of Block 1 That Is Retransmitted Three Times.

ARQ scheme, since the RTT involves transport across a communications network whose loading level may be stochastically fluctuating, RTT is modeled as a random variable. Under a conservative design, the RTT parameter is set to its 95th percentile value, denoted herein as $t_{RTT_{0.95}}$. The reception of 95% of PACKs, when issued by the receiver, takes place in a time that is not longer than this 95-percentile level. Other percentile values may also be specified.

The ACK time threshold can be set to a sufficiently long value, to make it unlikely that a block will be retransmitted prematurely. This happens when the sender does not wait for a sufficiently long time to give a chance to the corresponding PACK, when issued, to transit the network and be received at the sender. For this purpose, one may set $T(TO) = t_{RTT_{0.95}}$.

Which value should be used for setting the window parameter N?

The time it takes the sender to transmit N blocks is identified as the sender's window time and is hereby denoted as W_t. Under the illustrative configuration, it is calculated as $W_t = Nn/R$, where each block is n bits long and the transmission data rate across the sender-to-receiver link or connection is equal to R [bps]. If a relatively low value for N is selected, we may have $W_t \ll t_{RTT}$, so that the connection is kept occupied for a low fraction of time yielding a low-throughput rate, approaching the performance of a SW ARQ scheme (for which $N = 1$). The NHT across the connection is of the order of $TH_{CR} = W_t/t_{RTT}$, under a condition of Correct Reception (CR) of the underlying blocks.

To achieve higher-throughput performance, it is of interest to select an acceptably high N value. As illustrated in Fig. 13.6(b), following the transmission of block 1, the sender could send another $N - 1$ blocks before its window quota is exhausted, so that by the time that the transmission of these blocks is completed, which takes $(N - 1)n/R$ [s], the sender receives a PACK confirming the correct receipt of block 1, assuming that the latter is correctly received. Accordingly, the ACK timer's threshold parameter $T(TO)$ is set equal to the RTT time. In this fashion, as long as no block errors occur, the sender proceeds to continuously send its blocks. When feasible, the corresponding setting of N is then determined by the relationship $(N - 1)n/R = T(TO) = t_{RTT_{0.95}}$.

The window span N value may however have to be limited by flow control based considerations. At any point in time, the receiving entity may limit the maximum number of outstanding unacknowledged blocks N that it is willing to hold in its buffer on behalf of a sending entity. Induced by congestion and flow control regulation, the receiving entity may have to limit the number of blocks that it is able to store and process for a specific sender during a given time period and thus induce the sender to use a modified window size. For example, under the TCP transport layer protocol, which is commonly used for Internet based data transmissions, the receiver dynamically authorizes the sender to use a window size (N) whose value depends upon the receiver's current available resources.

As noted for the SW ARQ scheme, while RTT is a key parameter affecting the scheme's performance, another key parameter involves the probability that a received block will contain errors, denoted P_{BE}. It is calculated by using Eqs. 13.2 and 13.3. The connection's throughput and delay performance behavior is impacted by the values assumed by the average number of block retransmissions and transmissions, given respectively by Eqs. 13.4 and 13.5. As deduced from Fig. 13.6(b), the throughput rate attained under GBN is approximated by a formula that is derived in a manner that is similar to that used to obtain Eq. 13.6 for SW ARQ, by assuming that the parameter N is set in the fashion that is noted above. The figure illustrates a scenario under which block 1 is retransmitted three times, so that $N_R = 3$. The third retransmission of the block is correctly received. Consequently, the total time elapsed until block 1 is correctly received is calculated as $3(Nn/R) + (n/R)$, whereby the last component accounts for the time taken for the successful transmission of block 1. Accordingly, the attained throughput rate is expressed as follows:

$$TH_{B_{GBN}} = \frac{1}{E(N_R)(Nn/R) + (n/R)}. \tag{13.8}$$

During a period in which all blocks are received correctly, so that $N_R = 0$, the corresponding throughput rate is equal to R/n [messages/s], or R [bps], achieving throughput efficiency of 100%. In turn, when channel noise or interference effects cause received blocks to contain uncorrectable errors, inducing retransmissions, the throughput rate is degraded. The average delay incurred by a block across a connection is calculated by following the approach used to obtain the expression given by Eq. 13.7, yielding:

$$E(D_{HB_{GBN}}) = \frac{Nn/R}{1 - P_{BE}}. \tag{13.9}$$

In comparing the system's performance under a GBN ARQ scheme, where $N > 1$, with that attained under a Stop-and-Wait scheme, which is a Go-Back-1 protocol, we note the following. Under the condition that RTT, the time to receive PACK, is longer than the time to transmit a single block, it is efficient to set a higher window value, setting $N > 1$. In this case, the sender does not have to pause (for long time, or not at all) the transmission of waiting blocks. It can send consecutively multiple blocks even when not all outstanding blocks have yet been PACKed.

Assuming that the employed MCS/FEC schemes yield an operation under which the probability that a block contains uncorrectable errors is sufficiently low, the employment of an ARQ scheme that uses higher N levels leads to much higher throughput rates across high speed communications links and connections, under which it is feasible to send many frames at high rate within a single RTT period. This is also the case for connections that span long distances, such as satellite links, which induce long propagation delays and consequently long RTTs. This is illustrated by the following examples.

Example 13.4 GBN ARQ over a high-speed link. Consider a high-speed terrestrial link connection. Assume it to operate at a data rate of 1 Gbps and to span a distance of 1000 m. Assume the RTT to be dominated by the signal's propagation delay. Under a propagation speed of 5 [μs/km], the round trip propagation delay is equal to 10 μs. Assume thus that the RTT to receive a PACK is equal to 15 μs. A block is assumed to contain an average of 1000 bits, so that the block transmission time is equal to 1 μs. Hence, by setting $N = 15$, the sender would be able to fully utilize the RTT period by continuously transmitting 15 blocks.

420 | 13 Error Control: Please Send It Again

If no channel errors take place, the sender would be sending successfully 15 blocks every RTT period, achieving a throughput of 15 [blocks] or 15,000 bits every 15 μs, attaining 100% utilization at a throughput rate of 1 Gbps. In turn, if the probability that a block contains uncorrectable errors is equal to 10%, we conclude by using Eq. 13.4 that the average number of block retransmissions is equal to approximately 0.111. Hence, using Eq. 13.8, we obtain the throughput rate to be reduced to about 0.375 Gbps, a throughput reduction by about 62.5%. If the probability that a block contains uncorrectable errors is equal to 1%, attained by using a channel that is perturbed by lower noise power or by employing a more powerful MCS/FEC scheme, we find that the throughput rate is now upgraded in a significant way to about 0.87 Gbps.

Using Eq. 13.9, and assuming the block error ratio to be equal to 0%, or 1%, or 10%, the estimated average block delay level is equal to approximately 15 μs, or 15.15 μs, or 16.67 μs, respectively. It is thus noted that the block's delay level is dominated by the RTT value of 15 μs.

Example 13.5 GBN ARQ over a satellite link. Consider a geosynchronous satellite communication link for which the round trip propagation delay of a signal is equal to $t_{RTT} = 500$ ms. Assume a block to contain an average of 5000 bits and the data rate to be equal to $R = 1$ Mbps, so that the block's transmission time is equal to 5 ms. Hence, by setting $N = 100$, the sender would be able to fully utilize the RTT period by consecutively transmitting 100 blocks.

If no channel errors take place, the sender would be sending successfully 100 blocks every RTT period, achieving a throughput of 100 [blocks] or 500 kbits every 500 ms, achieving 100% utilization and a throughput rate of 1 Mbps. In turn, if the probability that a block contains uncorrectable errors is equal to 10%, we conclude by using Eq. 13.4 that the average number of block retransmissions is equal to approximately 0.111. Hence, using Eq. 13.8, we obtain the throughput rate to be reduced to about 82.64 kbps, yielding a throughput efficiency of only about 8.264%. If the probability that a block contains uncorrectable errors is improved to 1%, the throughput rate is equal to about 500 kbps, a significantly improved throughput efficiency of 50%. Using Eq. 13.9, we readily note that a reduction in the block's error ratio also yields significant improvement in the block's delay performance.

13.3.5 Selective-Repeat ARQ: Resend Only Uncorrectable Received Blocks

A **Selective Repeat (SR)** ARQ scheme, also known as a **Select-and-Repeat (SAR)** ARQ protocol, operates as a continuous ARQ scheme. In comparison with the functioning of the above described GBN (or sliding window) ARQ scheme, the following rule is used. Only those error control blocks that have not been positively acknowledged prior to the expiration of the ACK timeout timer are resent. Considering a full-duplex connection, the SR-ARQ scheme uses thus the following steps.

Definition 13.3 Select-and-Repeat (SAR) ARQ Algorithm.

1) The source entity sends continuously error control blocks, each including a sequence number that identifies its position within the sent flow of blocks. A maximum number N of sent outstanding unacknowledged blocks are authorized to be sent at a time.
2) Once the sending entity has completed the sending of an error control block, it turns-on an ACK timer that is set to have a timeout period of duration $T(TO)$. Under a batch timer setting scheme, a single timer can be configured to check for the timely reception of a batch of blocks.
3) The receiving entity positively acknowledges those blocks that have been received correctly, even if they are received out of order, identifying them to the sending entity by their sequence numbers.

4) The receiving entity keeps track of the sequence numbers of received blocks, identifying reception gaps induced by missing blocks.
5) If the sending entity does not receive a PACK message for a certain block (or a batch of blocks if a batch timer activation process is used) by the expiration of the corresponding ACK timer, it schedules the retransmission of selectively only those outstanding blocks that have not been positively acknowledged. A threshold parameter may be set to specify the maximum number of block retransmissions
6) If the sending entity receives a PACK message prior to the expiration of its ACK timer for a specific block or for a batch of blocks, it discards the copies of these blocks that it holds in a buffer. It then refreshes accordingly its authorized window quota and proceeds to schedule the sending of waiting blocks, if any.

A flow of error-control blocks and ACK messages under a SAR ARQ scheme is illustrated in Fig. 13.7(a). Block F2 is shown to be received in error. Following the expiration of an ACK timer, block F2 is retransmitted. At that time, block F3 is noted to have an outstanding PACK, as its PACK has not yet been received. However, in contrast with the operation used by a GBN ARQ scheme, block F3 is not retransmitted at the time that block F2 is retransmitted. The PACK message for block F3 is noted to have been received by the sender at a later time.

As noted, the receiving entity may have to assemble several blocks to reconstruct a message that it must deliver to its client. A relatively long network layer packet may be segmented into multiple link-layer error-control blocks, whereby each block is a link-layer frame. Such a segmentation procedure can reduce the link-layer block's error ratio, realizing a more efficient transmission of data units over a noisy wireless link. The link layer entity of the receiving node may be configured to deliver the received data units to a packet switching router, which operates at the network layer. For this purpose, it will wait to receive all the link layer blocks that are associated with a packet, assemble them in the proper order (using their identified sequence numbers) to reproduce the targeted packet, which it will then deliver to the network layer entity. If at any time, the link-layer-receiving entity finds that certain link-layer blocks that are require for its assembly

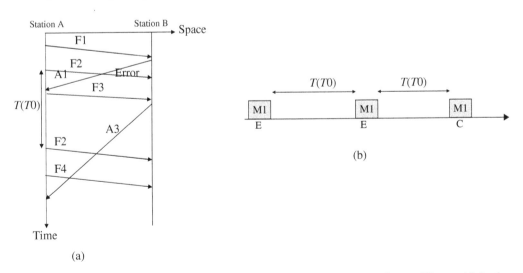

Figure 13.7 Block and ACK Flows Under a Selective-Repeat ARQ Scheme: (a) Timeout Triggered Selective Retransmission of Block F2; (b) channel Occupancy and Time Delay in the Transport of Block M1 That Is Retransmitted Two Times.

422 | *13 Error Control: Please Send It Again*

process are missing, or received in error, it must wait for the retransmission and subsequent correct reception of such blocks.

This situation points out the underlying complexity associated with the operation of the SAR-ARQ scheme. Due to the possible occurrence of gaps in the orderly reception of blocks, the receiving ARQ processor must engage in a more sophisticated storage, processing and assembly operations. In turn, a selective-repeat ARQ operation can offer improved throughput and delay performance behavior. The reception of a block that contains errors would trigger the selective retransmission of only erroneously received blocks.

The throughput and delay performance behavior of the SAR ARQ scheme can be expressed by observing the flow process illustrated in Fig. 13.7(b). Block M1 is shown to be received twice in error and is thus retransmitted twice, $N_R = 2$, and transmitted a total of three times, $N_T = 3$. Assume that the ACK timer's timeout threshold $T(TO)$ is set by using the same approach that was applied for the GBN scheme. Thus, $T(TO)$ is configured to be of the order of the 95-percentile value of the RTT. Block M1 is shown in 13.7(b) to be retransmitted after a period that is equal to $T(TO)$.

To calculate the throughput rate attained across an underlying link, we note that under the SAR-ARQ scheme, a block occupies communications resources only during the time periods that it is transmitted or retransmitted. Hence, the block throughput rate (expressed in units of [blocks/s]) is calculated as follows:

$$TH_B = \frac{1}{E(N_T)(n/R)} = \frac{1 - P_{BE}}{n/R} = \frac{R(1 - P_{BE})}{n}. \tag{13.10}$$

The throughput efficiency (also known as the NHT), denoted as η, expresses the fraction of the data rate that is utilized for producing data throughput across an underlying link or connection. Noting that the block throughput rate (expressed in units of [bits/s]) is calculated as $TH_B(n/R)$ and that the link's (or connection's) data rate is given as R [bps], we conclude that the throughput efficiency factor is calculated as $\eta = (n/R)TH_B/R$. Hence, the throughput efficiency attained by the SAR ARQ scheme is given by:

$$\eta_{SAR} = 1 - P_{BE}. \tag{13.11}$$

The HOL delay experienced by a block, accounting for the total time elapsed from the start of its first transmission to the reception of a PACK, given that a PACK will eventually be issued for this block, is noted to span a period of $(n/R) + T(TO)$ for each incurred transmission, so that it is therefore approximated as:

$$E(D_{HB_{SAR}}) = [(n/R) + T(TO)][E(N_T)] = \frac{(n/R) + T(TO)}{1 - P_{BE}}. \tag{13.12}$$

Example 13.6 SAR ARQ over a high-speed link. Consider a high-speed terrestrial link connection. Assume it to operate at a data rate of a 1 Gbps, and to span a distance of 1000 m. Assume the RTT to be dominated by the signal's propagation delay. Under a propagation speed of 5 [μs/km], the round trip propagation delay is equal to 10 μs. We assume the RTT to receive a PACK to be equal to 15 μs. A block is assume to contain an average of 1000 bits, so that the block transmission time is equal to 1 μs.

Assume the probability that a block contains uncorrectable errors to be equal to 10%, so that $P_{BE} = 0.1$. Using Eq. 13.4, we obtain that the average number of block retransmissions is equal to approximately 0.111. Using Eq. 13.10, we obtain the throughput rate to be equal to 0.9 Gbps, a throughput efficiency of 90%, as confirmed by using Eq. 13.11. The attained throughput efficiency under the GBN ARQ scheme represented in Example 13.4 is much lower, as it was shown there to be equal to only about 37.5%.

Under an upgraded block error ratio of 1%, the throughput efficiency under SAR GBN, as noted by using Eq. 13.11, is equal to 99%. The attained throughput efficiency under the GBN ARQ scheme represented in Example 13.4 is much lower, as it was shown to be equal to about 87%.

By examining Eq. 13.12, we note that the average time delay incurred in successfully transporting a block across the link (or connection), as noted in Example 13.4, is dominated by the involved RTT latency, provided that a low block error ratio level is maintained. The incurred 99-percentile delay values would generally be highly improved through the use of a SAR ARQ scheme, as retransmission events would not impact long batches of transmitted blocks.

Example 13.7   Transfer of large data records over a high-speed link. A scenario that exemplifies a case under which using an SAR scheme is of high utility involves the transport of very large data records across a high-speed link, which is characterized by having a relatively high RTT to block transmission time ratio. For example, consider the transmissions of sale records by an organization's branch to its headquarter office, carried out at specified times during the day. Each record is segmented into a large number of blocks. Even when a low block error ratio is realized, so that only a small fraction of sent blocks are received in error, the number of retransmitted blocks would still be high. For example, under a block error ratio of 1%, when a record that consists of 10,000 blocks is sent, the number of blocks that would be received in error will be equal to 100. Under a GBN ARQ scheme, each retransmission of a block will trigger the retransmission of many other blocks (that may have been correctly received, containing no errors), including blocks that have been sent within an RTT period, leading to an inefficient transport process. In contrast, for such an application, under an SAR ARQ scheme, only those blocks that are received in error would have to be retransmitted, yielding a much improved delay-throughput performance behavior.

13.4   Hybrid ARQ (HARQ) Error Control

Our discussion in the previous section well illustrates the need for the error control system's designer to jointly configure the parameters of the combined FEC and ARQ schemes. **Hybrid ARQ (HARQ)** schemes are used to configure and adjust the parameters of a combined FEC and ARQ error control process to meet targeted *delay-throughput performance objectives*.

A selected MCS combines modulation and error control schemes in providing the receiving entity with the capability to detect and at times also to correct errors. Using error detection and correction codes and ARQ schemes, blocks that contain uncorrectable errors are retransmitted. As demonstrated above, to attain acceptable delay-throughput performance, it is generally essential to configure the MCS scheme, including the settings of involved EDC and FEC, as well as the size of the error control block, to yield a sufficiently low probability (P_{BE}) that a received block contains uncorrectable errors.

The characteristics of the communication channel, including its associated noise and interference processes, are assessed to determine the MCS scheme that should be configured in aiming to achieve an acceptable block error ratio. As noted in Fig. 3.6 in Section 3.8.4 for the 802.11n based Wi-Fi system, and in Fig. 3.7 for the RAN of the LTE cellular wireless network, specific technologies include several pre-set candidate MCS settings in their system implementations. When the receiving entity is impacted by higher noise and/or interference power, leading to a degraded SINR, a more powerful MCS configuration is automatically selected. A lower rate code may then be employed, as it is used to provide enhanced block error ratio levels and upgraded message throughput and delay experience. The FEC capability that is used may be upgraded to reduce the block retransmission rate. As such a modification may induce a lower spectral efficiency level, the allocated channel's bandwidth and/or the employed data rate and transmit power levels may also have to be adjusted.

For a version of the LTE cellular wireless system, under a prescribed targeted block error ratio (BER), it is noted from Fig. 3.7 that as the SINR, and corresponding Channel Quality Indicator (CQI) value, is improved, a more powerful MCS configuration is employed. For example, under high SINR conditions, for CQI=12, a bandwidth efficient modulation scheme and a higher rate code are employed. The operation then achieves the relatively high spectral efficiency (i.e., data rate per unit bandwidth) level of $\eta = 3.9023$ [bps/Hz]. Under a prescribed channel bandwidth of 1 MHz, the realized data rate is thus equal to about 3.9 Mbps. In contrast, when the receiver is perturbed by higher noise and/or signal interference power levels, and the SINR value at the receiver then corresponds to a CQI=6 setting, we observe the system to select a modulation scheme whose bandwidth efficiency is lower; it may then also employ a lower rate forward error correcting code. The operation then achieves a spectral efficiency level of only $\eta = 1.1758$ [bps/Hz]. Under a prescribed channel bandwidth of 1 MHz, the realized data rate is then reduced in a significant manner to about 1.17 Mbps.

Under prescribed channel bandwidth and transmit power level conditions, the above-noted behavior indicates the following trade-off in configuring a HARQ scheme. Under low noise and interference channel conditions, higher rate modulation and FEC schemes are used, yielding higher spectral efficiency levels. As the reception quality of the intended signal deteriorates, lower rate modulation and FEC schemes are employed, leading to lower spectral efficiency. FEC code bits now constitute a higher fraction of a block's bits. These adaptations aim to achieve a targeted value for the block error ratio (P_{BE}). The parameters of the involved ARQ process are mutually adjusted accordingly. For example, if a more relaxed (or more strict) block delay is acceptable, a higher (or lower) average (and percentile) level for the number of retransmissions can be set, so that a corresponding higher (or lower) value or the block error ratio is specified.

In many systems, the noise and interference characteristics of the communications channel are fluctuating over time. When considering mobile wireless communication systems, signal interference and noise levels are impacted by user mobility. Consider a cellular wireless communications network. A mobile user that is located closer to its base station (BS) tends to detect at its receiver higher SINR levels in its reception of intended signals that it receives (downlink) from its BS. It also induces higher SINR values at the BS's receiver when transmitting data uplink to its BS. In turn, as the mobile is moving away (toward the boundary of a cell), the reception of a message sent to it by its BS is typically subjected to higher interference signals that are produced by mobiles that reside in neighboring cells. As it is now located further away from its BS, its own messages are received at lower power at its BS.

Under **non adaptive HARQ** schemes, retransmitted blocks use the same information payload and modulation/coding configurations and assigned resource parameters (such as bandwidth, power, data rate, and block composition) as those used for the first transmission. The MCS used in the first transmission may however be configured in accordance with the channel's signal propagation and interference conditions that are currently observed. They take into account the availability level of underlying system resources (such as bandwidth and transmit power levels). In a cellular wireless network, the BS often monitors resource availability and communications channel conditions and uses this information to instruct transmitting entities about the MCS and system parameters that they should use. Similarly, in advanced Wi-Fi systems, a resource management module managed by the Access Point (AP) is employed and used to allocate system resources for communications across the shared multiple access wireless medium.

Under a **soft combining** mechanism, a received block that contains uncorrectable errors is stored at the receiver rather than discarded. Retransmitted versions of the same block, each containing the same information and code bits, are combined with stored versions. Under a **Chase Combining (CC)** method, symbol-by-symbol combinations are performed. As distinct transmissions of blocks may experience difference noise and interference conditions, such combinations can enhance the quality of the combined signal. Various methods have been employed in setting the weight parameters used in calculating a weighted combination of multiple received blocks. Weights are often set in proportionality with the SINR value associated with each block's reception.

Other combining techniques have also been used. For example, retransmitted blocks may include different information fields and different codes and associated code fields. They can then be also appropriately combined at the receiving entity.

Under **adaptive HARQ** schemes, retransmission methods are employed to dynamically modify the parameters of retransmitted blocks. Such adaptations can be performed in a distributed manner, independently by each entity, or managed by a central controller. In the following, we illustrate such operations when undertaken in wireless network systems.

Under a **distributed adaptive HARQ** operation, dynamic adjustment of HARQ parameters and the underlying operation are performed individually and independently by each involved entity, based on its ongoing observations of the performance of its own HARQ processes. For example, consider a *Wi-Fi Wireless Local area Network (WLAN)* that employs a distributed medium access control (MAC) scheme, such as the widely employed Wi-Fi CSMA/CA access protocol. An adaptive HARQ scheme used by some implementations functions as follows. At its first transmission, a block's MCS is configured to use a primary setting. If the sender does not receive a PACK by the expiration of the ACK timer, it may use the same MCS to configure the first retransmission. If subsequently a PACK is still not received, the sender shifts its MCS setting to a lower (code and data) rate. Such an operation leads to a lower block error ratio as well as to lower data rate and a spectral efficiency. If such an adaptation still does not result in correct reception, the sender may transition to use even lower code and data rates. After operating successfully for a specified period of time at a lower rate, the sender will gradually transition to a higher rate transmission phase. The operation continues in this manner in autonomously transitioning to higher or lower rate modes, signaling to the receiver, in each mode, the employed MCS or Redundancy Version (RV) being used.

Under Wi-Fi, the basic ARQ scheme that is used by a station in transmitting its blocks as MAC frames across the MAC layer, employs a **Stop-and Wait (SW) ARQ** protocol. The sending station triggers an ACK timer upon the transmission of its frame, setting its timeout level to a value that accounts for the time taken by the receiving station to rapidly error-check the received frame and then, if error-less, transmit its PACK across the shared wireless medium. In accordance with the CSMA/CA access protocol, the receiving station is provided an exclusive access to the medium for the transmission of its PACK frame. If no PACK is received by the expiration of the ACK timer, the frame is retransmitted. Under an adaptive HARQ operation, the retransmitted frame may employ a different MCS. It is noted that multicast or broadcast frames are not acknowledged.

Using HARQ, a preliminary version can be selected in several ways. If good signal reception conditions are expected, and a high spectral efficiency and data rate values are of interest, the first transmission of a block may employ just an EDC, such as CRC. A low number of code bits is then included in each frame. A high fraction of transmitted blocks will be correctly received upon their first transmission, yielding high-throughput and high spectral efficiency rates and low block transport delays. *Incrementally higher rate* FEC schemes are employed for encoding block retransmissions. For applications that require lower block transport time delays, the first transmission is configured to use a higher rate FEC scheme. In this case, lower-throughput rates would be realized.

Advanced Wi-Fi versions (such as 802.11ac, 802.11ax) employ a **Selective Repeat (SR) ARQ** scheme, aiming to achieve higher throughput and lower message delay performance. As a continuous ARQ operation, a sliding window scheme is employed. A window of outstanding unacknowledged frames is maintained and updated as PACK messages are received. In addition to using the above-mentioned process in which ACK messages selectively PACK specific frames, a batch ACK message, known as a **Block ACK (BACK)** is used. This forms a more efficient way for the receiver to acknowledge a group of frames, indicating the ones that have been received correctly. The sender retransmits only those blocks that have not been positively acknowledged. Wi-Fi systems specify the maximum number of frame transmission retries that the MAC protocol entity should perform. Short and long retry parameters can be set. These parameters can be dependent upon the flow's application type, its access category and traffic class.

Under a **centrally managed adaptive HARQ** operation, a managing station, also identified as a **station controller**, is responsible for allocating transport resources and for specifying HARQ parameters. The controller monitors ongoing network and channel system conditions. It keeps aware of current resource states, including the availability of spectral resources and the progress of transmission and retransmission processes. Accordingly, it calculates the desired parameters of the HARQ processes that should be used by client entities, based on their targeted Quality-of-Service (QoS) levels and underlying application types. In the following, we illustrate such a HARQ process for certain implemented versions of wireless network systems.

Consider the RAN of a cellular wireless network system. For downlink (DL) communications from the BS to mobile users, downlink control channels (DLCCHs) and downlink service channels (DLSCHs) are established. A DLSCH is used to carry messages sent by the BS for reception by mobile users. A DLSCH is used to carry control messages produced by the BS and directed to users. The latter include control messages that are used for service channel assignment and for the specification of resources that are allocated to service channels. Included are specifications of MCS, antenna configurations and various transport parameters. For uplink transmissions by a mobile user, a control message specifies the allocated frequency span and time slots that the transmitting mobile user should use across the allocated uplink (UL) service channel, as well as the associated MCS and HARQ process parameters to be employed.

Consider downlink transmissions of error control blocks that belong to a specific information flow that originates at the BS and that is destined to a mobile user. A centrally managed non-adaptive or adaptive HARQ process is used for each flow. Multiple ongoing such HARQ processes can be supported. Each HARQ process is used for a flow whose messages consist of error control blocks. In the following, we outline certain key elements of such a HARQ process, as typically performed by certain versions of the LTE cellular wireless system. Using a DLCCH, the BS informs the flow destination user as to the parameters of the DLSCH that are used for the transmission of each message, including the employed frequency band and time slots and the Redundancy Version (RV) that characterizes the used FEC code. Across the MAC layer, a message is embedded in a frame that is identified in LTE as a *transport block*. The allocated parameters of the DLSCH can be dynamically adjusted by the BS to accommodate varying system and resource availability conditions.

Block transmissions across the downlink are executed in an *asynchronous and adaptive way*. *Soft combining* is performed at the receiver's physical layer, as the HARQ process pertains to the MAC layer. The receiving entity at the mobile user makes use of its uplink control channel, or (when available) an allocated uplink service channel, to inform the BS about the outcome of its reception of the message sent to it by the BS. Taking advantage of the availability of an uplink channel that was assigned to it, the mobile makes use of a provisioned 1-bit ACK, identified as an **ACK/NACK** bit. The mobile recipient sets this bit to either 1 or 0 to inform the sender whether the block is received correctly or incorrectly. The ACK signal serves as a PACK for a correctly received message. If the message contains uncorrectable errors, the user's signal to the BS is identified as a **Negative ACK (NACK)**. The BS is able to correlate these ACK/NACK signals with the corresponding blocks that it sent to the mobile by observing the underlying channels and their associated time slot and frequency band assets.

Upon reception of HARQ feedback signals, the BS may proceed to transmit a new block or retransmit a previously sent block. In the following, consider an implementation version under which distinct frequency bands are used for uplink and downlink communications between the BS and a mobile user. Such an operation is identified as Frequency Division Duplex (FDD). Under an alternate configuration, which is identified as Time Division Duplex (TDD), uplink and downlink communications channels are differentiated by using distinct time slots, while sharing the same frequency band. Accounting for processing times at the BS and at the user, and for a cell radius that is no longer than 100 km (limiting thus the round trip propagation delay), the HARQ round-trip time (for an underlying version of the system) is determined to be such that an ACK/NACK message is received within 4 ms and it takes another 4 ms to transport the subsequent block (as a MAC frame). In LTE, each pair of time slots (identified hereby as LTE slots) is combined to form a subframe (noted below as an ARQ-slot), and each subframe is 1 ms long. Thus, the next block is transmitted at a time that is 8 ms (or 8 subframes) later than the time at which the previous block was transmitted. Hence, a (downlink or uplink) block that is transmitted at LTE-subframe i and is negatively acknowledged can be retransmitted at slot $i + 8$. If a positive ACK is received, a new block can be transmitted at the latter slot.

To save mobile energy resources, ACK signals for multiple messages can be combined through the use of a logical AND function, by using a process that is known as *ACK/NACK bundling*. In this case, a positive ACK is sent only if it is valid for all involved blocks, so that when a bundled NACK is sent, all involved blocks need to be retransmitted by the BS.

In this manner, the BS and its mobiles are able to engage in **eight parallel HARQ processes**. The downlink transmission of blocks by the BS is performed in an *asynchronous* manner. Eight HARQ processes can be conducted in parallel and assigned asynchronously. Modified channel resource allocations that are made by the BS can be used for subsequent retransmissions of new message transmissions. A mobile user can also engage in eight parallel HARQ processes for its uplink transmissions, except that it is conducted in a synchronous manner.

For **uplink communications** from a mobile user to its BS, uplink control channels (ULCCHs) and uplink service channels (ULSCHs) are established. An uplink service channel is used to carry messages sent by a mobile user for reception by the BS. An ULCCH is used to carry control messages produced by a mobile user and directed to its BS, including channel assignment request messages. An error control block, formatted as a MAC frame, is transmitted uplink by a mobile user across an uplink service channel that is assigned by the BS. The BS identifies the latter channel by specifying its allocated frequency band and time slot(s). The transmission of an uplink frame is acknowledged by the BS in the following manner. Since the BS is aware of the user's allocated uplink service resources, it uses this information to identify the source and its ongoing HARQ processes. It is thus able to send to each user an acknowledgment message across a corresponding downlink control channel (DLCCH). It uses a 1-bit ACK, identified as an **ACK/NACK** bit. The BS sets this bit to a 1 or 0 value, which informs the mobile sender whether the block is received correctly or not. In LTE, this ACK control channel is identified as a Physical HARQ Indicator Channel (PHICH).

As noted above for downlink communications, multiple HARQ processes can be carried out also across the uplink, except that they are conducted synchronously in parallel. An HARQ process is assigned an LTE-subframe every eight LTE-subframes. Each LTE-subframe is 1 ms long and consists of two LTE-slots. The user can thus spatially multiplex two transport blocks in a subframe that it will concurrently transmit across the uplink, sharing the same HARQ number. Yet, each is assigned its own MCS parameters. These two transport blocks are individually acknowledged. The BS is able to readily identify which HARQ process is assigned to which subframe as it is cognizant the corresponding assignments.

Under an **adaptive HARQ** operation, the BS can specify distinct MCS parameters for different uplink transmissions. Under a **nonadaptive HARQ** operation, each retransmission can use a different RV value by following a prescribed sequence of RVs. A synchronous flow of the above described uplink HARQ processes is illustrated in Fig. 13.8. A specific HARQ process is shown to synchronously use a subframe slot (represented here as a time slot) every eight time slots. It is shown to receive an ACK/NACK message after four time slots and to send a subsequent message (whether a new message or a retransmission) after another four time slots. In this manner, as different HARQ processes that originate at a mobile are assigned by the mobile to distinct time slots, the uplink ARQ scheme at a mobile's sending MAC layer entity consists of eight *parallel Stop-and-Wait HARQ processes*. Since each block is selectively acknowledged, it is possible for blocks to be received across the MAC layer in an order that is different from their transmission order. Reordering and reassembly processes are performed at a higher layer, such as the Radio Link Control (RLC) layer that is used for protocol processing across the RAN of a cellular wireless network system.

For a version of the **5G cellular wireless network system**, whereby the 5G New Radio (NR) module is employed, the employed HARQ mechanism is an enhanced version of the above-described LTE method. Operations across both the uplink and downlink are adaptive and asynchronous. A mobile user can support a higher number of parallel HARQ processes. While under LTE, the time window used to receive an ACK/NACK message is equal to 4 ms, a more flexible time spacing configuration can be enabled in 5G systems. Faster ACK feedback can be configured for response time sensitive flows. A batch ACK message structure provides for the multiplexing of ACKs associated with different parts of a transport block, whereby each part is encoded with its own CRC code, and can thus be selectively retransmitted.

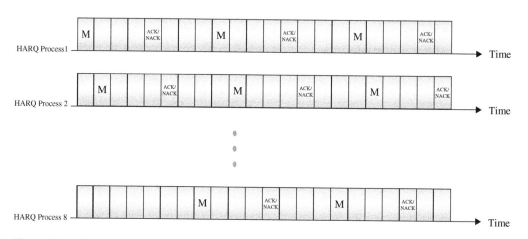

Figure 13.8 ARQ Configuration Consisting of Eight Parallel Stop-and-Wait ARQ HARQ Processes.

Problems

13.1 Describe the principle of operation and outcomes of an error control scheme that uses Forward Error Correction (FEC).

13.2 Discuss the outcome differences between the use of a Forward Error Correction (FEC) code and an Error Detection Code (EDC).

13.3 Explain the meaning of Bit Error Ratio (BER). What does $BER = 10^{-5}$ mean? What does a Packet Error Ratio (PER) that is equal to 10^{-3} mean? What is the connection between the latter two metrics?

13.4 Consider an error control system that makes use of Forward Error Correction (FEC) codes. Discuss how changing the code rate would impact the system's message transfer error performance.

13.5 Describe the operation of a system that employs Adaptive Modulation/Coding Scheme (MCS). Describe and illustrate the way that it should be used to adapt the employed MCS to the SINR value measured at the targeted receiver.

13.6 Describe the principle of operation of an Automatic Repeat Request (ARQ) scheme that is based on *PACK plus Timeout*.

13.7 Consider a Stop-and-Wait (SAW) ARQ algorithm.
a) Describe the operation of this scheme.
b) Describe how the timeout threshold level should be configured.

13.8 Consider a half-duplex communications channel that is used by station A to transmit messages to station B at a data rate of 1 [Mbps]. The channel's bit error ratio (BER) is equal to 10^{-4}. An error control block contains a total of 1500 [bits], including 160 overhead bits. The turn-around time at the employed transceivers is equal to 0.2 [ms] and the one-way propagation delay is equal to 0.4 ms. The retransmission timeout threshold is set equal to the RTT, measuring the average time that it takes for a PACK to be received by a transmitting entity following the transmission of a block, given that the block and its PACK are correctly received.
a) Calculate the block's error ratio.
b) Calculate the average number of times that a block is transmitted before it is correctly received.
c) Calculate the throughput rate across the channel.
d) Calculate the average head-of-the-line (HOL) time delay incurred by a block, measured from the instant of the start of its first transmission to the time that it is successfully received at its targeted node across the channel.
e) Determine the change induced in the attained throughput rate and in the corresponding normalized throughput ratio (i.e., the ratio of the throughput to the channel's data rate) if the channel's data rate is increased by a factor of 2 and the channel's BER also increases by a factor of 2, while all other parameters remain the same.

13.9 Describe the operation of a Go-Back-N (GBN) error control algorithm, also identified as a sliding-window algorithm. How is the timeout threshold determined?

13.10 A station uses a stop-and-wait ARQ protocol for the transmission of its frames across a communications channel. Each frame contains 8000 [bits]. The transmission data rate across the outgoing (half-duplex) channel is equal to 80 [kbps]. An acknowledgment (ACK) message contains 800 [bits] and is transmitted to the sender at a data rate of 8 [kbps]. The one-way propagation delay is equal to 100 [ms]. Assuming that no frame is lost (or received in error), calculate the throughput rate achieved by the sending station.

13.11 Compute the throughput rate attained across a communications link when using a stop-and-wait ARQ scheme, when assuming the following parameters: a 128 [kbps] link operating in a half-duplex mode; each frame contains 1700 [bits], including 100 [bits] header; an ACK message contains 84 [bits]. The one-way propagation delay is equal to 16 [ms], and the turn-around time is equal to 22 [ms]. The channel's BER is equal to 10^{-4}.

13.12 Describe the operation of a Select and Repeat (SAR) ARQ scheme and explain the derivation of a mathematical expression for the calculation of the ensuing throughput rate.

13.13 A communications link that operates at a data rate of 1 [Mbps] is used to send images from space ship to earth. Each image is of size $10,000 \times 10,000$ pixels. Each pixel is represented by 48 bits. The distance between earth and the space ship is equal to $384,000$ (km). The bit error ratio is equal to 10^{-5}.

a) Assume that we transmit each image message as a single error-control block and that we employ a Go-Back-N ARQ scheme. Determine a preferred setting for N and the ensuing attainable normalized throughput rate. The header of each error control block is equal to 128 [bits]. Assume that we neglect the impact of the ACK packet size, and of the receiver/transmitter reaction times.

b) As an alternative, consider breaking up an image into smaller blocks. Each block contains a 64-bit header and n data bits. A selective repeat ARQ scheme is used. Find a preferred value for n when aiming to maximize the normalized throughput rate.

13.14 Describe the operation of a Hybrid ARQ scheme.

13.15 Describe the ARQ error control operation that is used by a Wi-Fi system.

13.16 Consider the operation of a centrally controlled ARQ scheme that is used across a Radio Access Network (RAN) of a cellular wireless medium, managed by a base station node. Describe the operation of such schemes, including the use of protocols that employ ACK/NACK feedback signals and those that use parallel HARQ processes.

14

Flow and Congestion Control: Avoiding Overuse of User and Network Resources

The Objectives: Flow and congestion control schemes are employed to protect end users, network entities, and system resources from being overloaded by message flows. They also aim to protect admitted flows with targeted performance levels. **Flow control** *(FC) schemes regulate traffic flows between a sending entity and a receiving entity, whether across a user-network interface or between network or end-user entities.* **Congestion controls** *are employed to react to, or prevent, the occurrence of congestion at end users or inside a network.*

The Methods: Under a **closed-loop FC** *scheme, a receiving entity regulates the traffic rate of a flow produced by a sending entity. The receiving entity keeps a flow's sending entity informed as to a maximum quota of data segments that it is currently authorized to send to it. At an appropriate future time, the receiving entity may change this quota level in accordance with the level of resources that it is then able to allocate to the sending entity. Under an* **open-loop FC** *scheme, the sending entity adjusts the message traffic flow pattern that it uses to send to the network so that it conforms with an imposed or negotiated traffic boundary pattern. The sender may shape the profile of its traffic process to avoid violating a prescribed pattern.* **Congestion control** *schemes are used to resolve or prevent congestion occurrences inside the network and at user entities. Dynamically adaptive schemes make use of ongoing congestion monitoring techniques to adjust FC regulations at end users and at internal network entities.*

14.1 Flow and Congestion Controls: Objectives and Configurations

Flow and congestion control schemes are employed to protect end-user and network entities from being overloaded by message flows. **FC** schemes regulate traffic flows between a sending entity and a receiving entity. Included are message traffic processes that load user access nets and network links. Such traffic processes flow between network-edge nodes or between end-user entities. Excessively high message rates that load a receiving entity may cause its buffering resources to overflow. It may also overload its service resources and processing engines. This is the case when a source end user directs a data flow across the network to a destination end user that is currently (or predicted to soon be) congested. The destination entity may not have sufficient buffering capacity or service resources to accommodate the support requirements that are made by an ongoing or a newly admitted flow. Similarly, a communication link may be presently highly loaded and consequently not able to accommodate additional flows.

Flow admission control schemes also aim to maintain the quality of service (QoS) levels that have been granted to admitted flows. When the system cannot allocate sufficient resources to guarantee QoS target levels requested by a new flow, while preserving the performance levels

Principles of Data Transfer Through Communications Networks, the Internet, and Autonomous Mobiles, First Edition. Izhak Rubin.
© 2025 The Institute of Electrical and Electronics Engineers, Inc. Published 2025 by John Wiley & Sons, Inc.

that it has guaranteed to currently supported flows, the network traffic management system should block the admission of the new flow, unless it is regarded as a higher priority flow and it is acceptable for the system to block (or interrupt) lower priority admitted flows.

Congestion control schemes are used to react to, or prevent, the occurrence of congestion inside a communications network and at user entities.

Flow and congestion control mechanisms, algorithms and protocols operate in a reactive or proactive manner. Under a **reactive** flow or congestion control mechanism, congestion notifications are sent to involved traffic sources to induce them to adjust their traffic loading processes. Under a **proactive** flow or congestion control scheme, **congestion avoidance** means are employed, adjusting traffic loading processes to prevent impacted entities from reaching congestion conditions.

Illustrative spans of reactive flow and congestion control schemes are shown in Fig. 14.1(a). An end-user to end-user FC operation may involve the regulation of a traffic flow across a connection such as that established by using TCP. Under a **closed-loop** configuration, a destination entity dynamically paces the flow rate produced by a traffic source by either authorizing or restricting the transport of additional data units. At certain times, a destination entity may instruct a traffic source to throttle its traffic flow rate to prevent overflow or overloading, while at other times, it would authorize it to increase its flow rate. Such a **feedback**-based reactive FC operation may also take place across other network spans such as over access or network links. Feedback signals may also be produced on a *proactive* basis, as they are used preemptively to avoid the occurrence of congestion conditions.

As illustrated in Fig. 14.1(a), network nodes that experience congestion conditions may send *congestion notification* (CC) messages to source nodes whose traffic flows contribute to the occurrence of such conditions. These sources would react to these notifications by adjusting their traffic flow rates. They could act to lower these rates for as long as congestion condition indications are received from impacted network nodes. Congestion indications can also be produced proactively, as they are used to prevent the impending or forecasted occurrence of congestion conditions. The figure also shows the exchange of FC messages between neighboring network nodes.

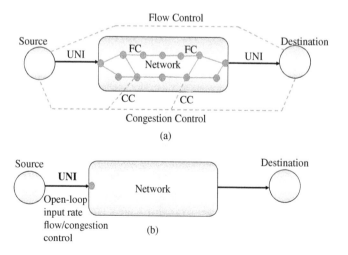

Figure 14.1 Flow and Congestion Control Methods: (a) Closed-Loop Feedback Based; (b) Open-Loop Input Rate Control.

14.1 Flow and Congestion Controls: Objectives and Configurations | 433

In reacting to notifications of congested network nodes, impacted flows may be directed to alternate non-congested paths.

An illustrative configuration of an **open-loop** flow and congestion control scheme, known as an **input-rate** regulation scheme, is shown in Fig. 14.1(b). A source node is connected to a network node over an access link spanning a **User-to-Network-Interface (UNI)**. A **traffic contract** is prescribed or negotiated between the source and the network. A source node that is governed by the underlying traffic regulation contract shapes the traffic process that it feeds into the network to conform with imposed regulations. The network system uses such open-loop input rate regulation of traffic flow processes to prevent the occurrence of unacceptable congestion states and to regulate the usage of its resources.

As an illustration, consider an organization that moves trucks over a highway the connects the Los Angeles harbor to a warehouse facility that is located in the outskirt of the city. Time delays are incurred in loading and unloading the trucks and in their round-trip travels. An harbor management node plans and coordinates these flows as it interacts with a management node located at the truck depot location. Accounting for space limitations and timing constraints, the depot manager implements a FC process, acting to regulate the times and rates at which trucks are authorized to depart the port. The depot truck processing (or service) rate would depend on its ongoing occupancy. It impacts the rate at which the unloading of trucks can be completed. Truck parking/wait space availability may fluctuate, depending on its occupancy in accommodating ongoing flows.

The following transport FC processes may be employed. Using a closed-loop reactive FC operation, the depot manager will send signals to the port manager informing it about its current availability and about the proper time for initiating the transport of additional trucks. In turn, to implement an operation under which trucks load, travel, unload, and return in a more predictable fashion, the managers could negotiate a traffic agreement that specifies the parameters of an open-loop input-rate FC operation. Under this agreement, assuming the transport process to be limited by the operation at the edge node rather than by highway resources (that the managers may not be able to control), the port manager will be able to move trucks at scheduled times.

To illustrate the operation of a congestion control scheme, assume that one of the above described FC schemes has been implemented. To provide a smooth and timely travel of the underlying trucks and of the other vehicles that share the same roads, a congestion control mechanism is employed. Congested intersection controllers that manage flows at highway interconnects would send congestion notification messages to source nodes that regulate the access of traffic into the highway. Sources that receive these notifications will react by regulating accordingly their traffic flow processes.

In Fig. 14.2, we show a commonly employed congestion control scheme. The access of vehicles into a road is regulated by using a traffic light. Alternately, the spacing between vehicles that merge into a road may be regulated. Such operations regulate the flow input rate to a road. Under a reactive operation, feedback signals obtained from highway traffic congestion monitors are used to adjust the length and frequency of occurrence of green light periods, adjusting the interval lengths between vehicle admission times.

The following are key objectives of the processes used by flow and congestion control schemes:

1) **Protection of Resources of Network Entities, Including Network Nodes and Links**: Induced by current limitations in the availability of resources, including communications, processing, and storage assets. Resource utilization levels are set at values that meet targeted performance objectives.

2) **Protection of End-User Resources**: Induced by current limitations in the availability of end-user resources. For example, a destination printing system may be currently overloaded as

Figure 14.2 Regulated Access.

it is tasked with carrying out a very high number of printing jobs. The system would then act to throttled down the rate of incoming printing tasks.

3) **Fair Allocation of Network System Resources**: Assuring end users and network flows with a "fair" allocation of network system resources. For example, highly active end-user sources or flows may be prevented from dominating scarce system resources that must be equitably shared among multiple sources or flows of the same service class and priority level.

4) **Rapid dynamic adjustments** in reacting to variations in traffic flow mixtures, in aiming to accommodate the support of critical flows at requested QoS. Upon the occurrence of loading bursts, lower-priority flows are throttled down so that higher-priority flows can be accommodated.

5) **Protection of the Performance Incurred by Admitted Flows**: Preventing new flows from degrading the performance of currently admitted flows to which performance guarantees have been made. Flow admission control schemes are employed to regulate the admission of new flows in a manner that provides such performance protection.

In the following, we present key FC regulation methods, algorithms and protocols.

14.2 Feedback-Based Closed-Loop Flow Control

Under a closed-loop FC scheme, the sending entity regulates the rate at which it sends messages to a destination entity based on feedback signals that it receives from the destination entity.

Consider a FC operation that is carried out across a communications link that connects two nodes. This link may provide for data communications between an end-user node and a network edge node, or connect two neighboring network nodes. Assume that the receiving node is able

14.2 Feedback-Based Closed-Loop Flow Control | **435**

to react rapidly to congestion occurrence and send promptly a congestion signal to the sending node. A FC scheme that is based on the use of **GO/NO-GO** control messages is often invoked. As it becomes congested, or when sensing that its buffer occupancy is ramping up and that it is going soon to experience a state of congestion, the receiving node sends a NO-GO control message to the sending node, instructing it to stop its transmission process. At a later time, when the receiving node observes its congestion state to have been relieved, it will send a GO control message, permitting the sending node to resume its transmission process.

A multitude of variations of the GO/NO-GO FC method have been implemented. When delayed beyond a specific time period, the sender may be allowed to probe the receiver to determine if its operation can be resumed. When considering a receiving node that processes multiple ongoing flows, whereby different flows may be attached different precedence, priority, or timeliness weights, a congested receiver may apportion its resources in a weighted manner among its currently active flows.

We note the similarity in operation and performance of FC schemes and of *ARQ error control* procedures. As discussed in Section 13.3, an *ARQ error control scheme* that is governed by a *Stop-and-Wait (SW) ARQ* protocol allows only a single outstanding unacknowledged sent message at a time. The next message in the send queue is sent only when the previous one has been positively acknowledged. Similarly, under a corresponding *Stop-and-Wait (SW) flow control (SW-FC)* scheme, following the successful processing of a received message, or when it becomes subsequently available, the receiver sends an **authorization** control message (or signal) to the sender, which allows it to then resume operation and send the next message that is waiting in the send queue.

As discussed in Section 13.3, such an operation is inefficient when the round trip time (RTT) is relatively high. This is the situation when considering the propagation delay incurred in sending signals to be relatively high and/or when a flow's end nodes communicate over a high-speed communications path. In comparison with the operation of an ARQ error control scheme, a corresponding FC process would involve the sending of an authorization message by the receiving entity when it has successfully processed a message and is able to accommodate a new message, rather than sending a positive acknowledgment (PACK) message when it has successfully received a message that contains no uncorrectable errors.

As discussed for error control schemes, a more complex control mechanism is required when the sender incurs a high RTT delay in waiting to receive a feedback message from the receiving entity, which induces a high RTT to message transmission time ratio. The error control operation is then enhanced by using a continuous ARQ scheme such as a *sliding window* based *Go-Back-N* ARQ.

A corresponding feedback-based **sliding window FC** scheme is described in the following. It is a version of the FC scheme that is employed by **Transmission Control Protocol (TCP)**. TCP is a connection-oriented transport layer protocol that has been commonly used at end-user devices as they communicate over the Internet. The underlying FC operation involves data message flows between two transport layer protocol entities. Generally, full-duplex communications takes place between the underlying transport layer entities, as assumed henceforth.

Consider a data flow for which entity A is operated by the source end user while entity B is used by destination end user. During the setup of the TCP connection, each source entity may negotiate (or set by default) parameter values to be used by the FC process. For example, data unit (segment) length constraints may be set, and a destination entity may specify window size limits as it aims to prevent its buffer from overflowing.

A sliding-window FC mechanism operates in a manner that is similar to that used by sliding window (or Go Back N) error control scheme. The transport layer entity at a source end-user A modifies the rate at which it feeds messages across a transport layer connection in accordance with feedback signals received from the transport layer entity at destination end-user B. The latter entity sends to the source, at certain times, an **authorization** message (signal) that directs the source to reconfigure its send **window quota**, also identified as the **Receive Window Size (RWND)**. For example, if the destination entity is limiting at this time the maximum number of Bytes that it is willing to receive and store for the underlying TCP connection to 100 KBytes, it sets RWND to 100 KBytes. In this manner, it is authorizing the sender to have the aggregate of its forthcoming transmissions across this connection to carry no more than 100 KBytes of data. The destination may increase or decrease this quota value depending on the buffering and processing resources that it is willing or able to allocate for this flow. When the destination end-user entity experiences an excessively high level of congestion, as can be reflected by the presence of high message occupancy level at its buffer, or excessively long waiting times, or excessive energy usage level, or a combination thereof, it reduces the quota that it authorizes the source to use. At a later time, when its congestion level is reduced, the destination entity may raise the authorized quota level by expanding its advertised sliding window. The use of the TCP sliding window protocol as a congestion control scheme is further discussed in Sections 14.4.1, 16.5.1.

14.3   Open-Loop Input-Rate Flow and Congestion Controls

In many network systems, feedback messages that are directed to a source node and serve to regulate its traffic flows may experience relatively long and randomly fluctuating time delays. Feedback message delays are at least as long as the time that it takes the signal to propagate between the destination and the source. When considering high speed networks, this propagation time delay can be longer than the message transmission time. Furthermore, the time delay incurred by a feedback message can be much longer than the time that it takes the source to transport multiple messages to the underlying destination. When communicating across a high speed path, by the time that the source receives an intended feedback alert message that informs it about a congestion event, requesting it to reduce its message flow rate (or halt the flow), the source would have already sent a large number of messages into the system. When receiving the alert message, many source messages may already be in transit inside the network, as they make their way to a destination node. Upon arrival at the destination, they could further aggravate its congestion state. Consequently, a large number of these messages may be discarded. Retransmissions of such discarded messages could further aggravate the network system's congestion state.

In the latter situations, it would be useful to employ a traffic flow regulation scheme that does not require a source node to react to feedback messages but rather operate in an open-loop manner. Under an **open-loop** flow regulation scheme, the statistical profile of a traffic flow process produced by a source that is admitted into a network is regulated so that it conforms with an agreed upon (specified or negotiated) pattern. The latter serves as a *boundary profile of a traffic process*, limiting traffic intensity and specifying other traffic parameters should be observed by a flow that is admissible into the network. While this profile boundary can be modified at times, such changes tend to be made infrequently, in contrast with the need for urgent fast flow throttling adjustment actions that must often be carried out when prompted by feedback-based congestion alert messages.

Open-loop FC is thus based on the specification of a targeted *traffic profile boundary* per flow. The boundary specification defines an upper bound on specified metric values of an admissible traffic flow. Typically, such metrics account for the highest average and variation levels of the rate of an admitted traffic process. They pertain to a specified range of the corresponding flow parameters that must be specified for the flow to be admissible into the system.

For example, such controls may be executed across an access network (or link, including a multiple access wireless communications channel) that serves to connect an end-user traffic source to a network edge node. The corresponding interface is identified in certain systems, such as Asynchronous Transfer Mode (ATM) connection-oriented packet switching networks, as the **User-to-Network Interface (UNI)**. Fig. 14.1(b) shows a UNI link that serves to connect an end user with a network edge node. The network edge node may require the end user to shape its traffic flow so that it does not violate the prescribed traffic profile boundary, as specified in a negotiated **traffic contract**. Similarly, when considering a traffic flow across a UNI that is directed from a network-edge node to an end-user node. The latter may have highly limited resources that are apportioned among multiple network nodes to which it may be connected. It may therefore impose traffic rate bounds on edge nodes to which it is connected across the UNI. Open loop flow regulation controls across a UNI are also available for use in other connection-oriented packet switching networks, such as Frame Relay and X.25 networks. In the latter, corresponding end-user and network edge entities are identified as *Data Termination terminal Equipment (DTE)* and *Data Communications terminal Equipment (DCE)*, respectively.

Open loop regulation can also be instituted between two network nodes, such as between neighboring network nodes, non-neighboring network nodes, or between network-edge nodes. Such nodes may agree to subscribe to specified traffic profile parameters relating to flows in which they are involved as source–destination pairs.

Similarly, **open-loop congestion controls** can employ traffic profile boundary specifications to limit traffic loads that would cause unacceptable congestion conditions at network nodes.

Open loop traffic profile boundary specifications are synthesized, disseminated, and enforced by using a network's **traffic management system**. It may involve a *central traffic management entity* or a *distributed management structure*. Interconnected and interacting local or regional management entities, operating autonomously in a distributed manner or through close interactions, are often employed. The management system is aware of the ongoing states of system resources, loading rates, traffic profiles, and of any QoS requirements by active flows. It determines the parameters of the specified traffic boundary profiles for those flows that are assigned open-loop traffic profile constraints.

A network management system may offer *QoS* performance guarantees for the support of certain flows. For this purpose, it must be able to configure the network operation in a manner that properly allocates resources to admitted flows. The network management system implements a *flow admission control* scheme that admits a new flow only when it is able to offer to it, and to existing flows, sufficient resources, as it aims to meet their guaranteed performance objectives. The management system protects the network from overloading by unauthorized or over-active traffic sources. This is achieved by admitting into the system, for each traffic source, traffic processes that satisfy, or are shaped to meet, their specified traffic boundary profiles.

For example, when regulating traffic flows that are produced by mobile users, such as those that carry messages that are transmitted uplink to a base station (BS), the sum total of the uplink traffic data rate must not exceed the level of the currently available uplink data rate. When considering traffic flows that are statistically active, higher throughput rates are attained by sharing uplink wireless resources on a statistical (demand assigned) multiple access basis. Typically, the BS acts as the resource manager for the radio access network (RAN). During each time/frequency slot, it could allocate access resources to a subset of the sources, delaying the transmission of messages queued by other sources. By specifying limits on the total number of supported traffic sources and by imposing a traffic profile boundary on each flow, the system manager is able to guarantee each admitted flow the proper resource levels that it requires to meet its desired QoS objectives.

In an Asynchronous Traffic Mode (ATM) network, **open-loop input-rate** flow and congestion control schemes are enabled across the UNI. A message source negotiates a **traffic contract** which it is committed to follow. In return, the network system is committed to keeping sufficient communications, storage, and processing resources for the networking support of a user's flow. To maintain this support, the traffic fed by the user into the network across the UNI must **conform** to the specified **traffic profile boundary** as defined by the traffic contract. Conforming messages may be offered QoS performance guarantees. For example, for a conforming flow, the network may assure admission to at least 95% of its conforming messages as well as guarantee that at least 90% of the flow's admitted messages would experience end-to-end delay levels that are not higher than a specified value.

Systems may implement a **traffic policing** scheme to determine whether a flow's message that arrives across the UNI at a network's edge node is conforming with the traffic boundary scheme. Nonconforming messages may be discarded. In certain systems, such as ATM networks, under certain network loading states, nonconforming messages may be admitted to the system but are tagged as lower priority messages. They may then be discarded along their route or at their destination node when they incur congested entities.

In the following, we overview versions of several algorithms that have been employed for defining traffic boundary profiles. The **Token Bucket** and **Leaky Bucket** algorithms are similar schemes that are commonly employed to provide *traffic metering*. As described below, they are also used in the operation of traffic policing and traffic shaping schemes.

The data unit that is employed in the traffic regulation process is identified henceforth as a data segment. This segment can correspond to 1 bit, 1 byte (such as an Octet of bits), 1 frame, 1 packet, a PDU of any layer entity, or any other data block. The boundary profile characterization of a conforming traffic flow includes the specification of the flow's average rate, denoted as R_a [segments/s], and its traffic peak rate, denoted as R_p [segments/s]. The corresponding boundary-based targeted average inter-arrival time between segments is set as $T_a = 1/R_a$ [s], while the peak inter-arrival time between segments is set to $T_p = 1/R_p$ [s].

To alleviate congestion occurrences across a flow's access link or a network path, it is desirable to induce a flow's source to drive its segments into the network in a close to a deterministic fashion. This is accomplished when the source's traffic process that is prepared for network transport is shaped to produce segments in a deterministic manner. This way, an active source would be feeding into the network a single segment every exactly $T_a = 1/R_a$ [s]. However, under realistic conditions, traffic flows are governed by stochastic variations, induced by their own production mechanism and by network system communications transport and processing effects. Consequently, segment arrival processes at a network edges or at internal network nodes are governed by random fluctuations. In particular, they tend to be impacted by activity bursts. During a burst period, times between segment arrivals are much shorter than the corresponding specified average times. Such bursty traffic processes can cause high congestion states at internal network nodes and links, and at destination nodes.

To protect network system resources, it is therefore of interest to also impose traffic profile boundary limits on the burst profile of the traffic process that characterizes a flow that is admitted into the network. To illustrate such a specification, assume henceforth that the network system prescribes the peak segment arrival rate R_p [segments/s] for an admitted flow.

Consider a traffic flow that is fed into the network at the UNI across an access link that operates at a link rate of R_l [segments/s]. If this link is dedicated for the exclusive use of the flow's traffic source, it will serve to constrain the latter's peak rate, setting it to be equal to $R_p = R_l$. Under realistic conditions, the monitored peak data rate may deviate from this rate due to variations in the transmission process. The traffic boundary specification accounts for such variations, identified as jitter. To simplify our presentation, we assume in the following that, during an activity burst, admitted segments arrive at the receiving entity at a rate that is equal to R_p [segments/s].

The following algorithm, identified as a version of a **Token Bucket Algorithm**, serves to illustrate the use of traffic boundary parameters to classify a flow segment wishing to access a network as conforming or as nonconforming. The boundary specification entails the use of a bucket into which tokens are regularly and periodically added as long as the bucket is not full to capacity. When a new message arrives at a regulated interface, the traffic manager examines the bucket to determine whether it contains a sufficient number of tokens to be used for accommodating the transport needs of the message. If so, the message is declared to be conforming with the traffic boundary specification (also identified as a traffic contract). Otherwise, it is said to be non-conforming. Used in this manner, the algorithm operates as a *metering* scheme, which can be implemented at a receiving and/or a sending entity. It is not employed here as a procedure for queueing and transmitting segments.

Nonconforming messages may be discarded at the sending or receiving entity, or tagged as lower-priority messages that are admitted but may be discarded when they transit congested links or nodes. Alternatively, the traffic source may avoid having its messages declared nonconforming (and possibly discarded) by delaying the sending of a message that its metering algorithm identifies as nonconforming. The source would keep such a message in a local queue until such a time at which the token bucket will have a sufficient number of tokens to accommodate the sending of this message. Such a message delay oriented operation modifies the statistical feature of the traffic process that the source feeds into the network. Such a scheme is said to perform **traffic shaping**.

Definition 14.1   A Token Bucket algorithm

1) A token is added to the bucket every T_a [s], so that tokens are added at rate of $R_a = 1/T_a$ [tokens/s].
2) The capacity of the bucket is set equal to $C(\max) = C$ [tokens]. When a token arrives to a full bucket, the token is not added to the bucket and is discarded.
3) At message arrival time, the current number of tokens residing in the bucket is observed. The number of tokens required for admitting a message, denoted as m, is determined. For example, assume each token to permit the entry of a single segment. When variable length messages are considered, messages that carry m segments would require to withdraw from the bucket m tokens to become eligible for admission.
4) If the bucket's occupancy at this time is such that it contains the required number m of tokens, the message is declared to be **conforming**. The message is then admitted into the system and m tokens are removed from the bucket's contents.
5) In turn, if the bucket's occupancy at this time does not contain a sufficient number of tokens, the message is declared to be **nonconforming**, and no tokens are removed from the bucket's contents.
6) The system may proceed in several different ways in handling nonconforming messages. They may be discarded, or admitted and then tagged as discard eligible and treated by the network as lower-priority messages.

The process of traffic shaping of a flow by using a Token Bucket Algorithm is illustrated in Fig. 14.3. The shown operation takes place within the traffic source. A similar process is used for

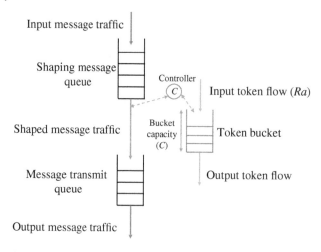

Figure 14.3 Traffic Shaping Using a Token Bucket Algorithm.

each flow. Messages at the traffic source that need to be transported across the network are stored in a Shaping Message Queue. The traffic management controller continuously monitors the status of the token bucket. The operation of the token bucket follows the dynamics of the Token Bucket Algorithm described by Definition 14.1. As long as the shaping message buffer contains incoming messages, the traffic controller monitors the occupancy of the token bucket. At the time that it determines that the bucket contains a sufficient number of tokens to meet the requirement imposed by the message positioned at the head-of-the-line of the shaping message queue, the controller acts to transfer this message to the message transmit queue. The program that runs the Token Bucket Algorithm, adjusts accordingly the state of the bucket.

The next message located in the shaping message queue, if any, is subsequently processed in a similar manner. Messages residing in the message transmit queue are transmitted across the outgoing link at a prescribed speed. For example, if the output link is dedicate for exclusive use by this station across a UNI, the station would be able to transmit its messages across the link at the speed of the link, R_l, offloading the message transmit queue by transmitting its messages at the peak rate $R_p = R_l$.

In the following, we illustrate the impact of a Token Bucket Shaper in regulating the stochastic behavior of a flow's traffic process. To simplify our discussion, assume a segment to represent a single protocol data unit (PDU), such as a single frame, packet, or transport layer message, depending on the layer of operation. For illustration purposes, we will identify each message as a fixed length packet, and regard it to consist of a single segment. Such a message requires the use of a single token to be declared conforming and to be then transferred to the transmit queue.

The impact of a traffic shaping operation conducted in accordance with the Token Bucket Algorithm is illustrated in Fig. 14.4. In Fig. 14.4(a), we show a profile of an admitted non-bursty steady traffic flow. It is generated as such by the source, or shaped to assume this profile. Packets are fed into the regulated path in a periodic fixed manner, so that following each period of duration $T_a = 1/R_a$ [s] a packet is sent across the flow's interface. The shaped traffic flow operates at a fixed (and average) rate that is equal to $R_a = 1/T_a$ [packets/s], identified as the specified **Sustained Information Rate (SIR)**. Each outgoing packet fully conforms with the regulation imposed by the Token Bucket Algorithm and is therefore admitted for transport across the flow's network path.

14.3 Open-Loop Input-Rate Flow and Congestion Controls | 441

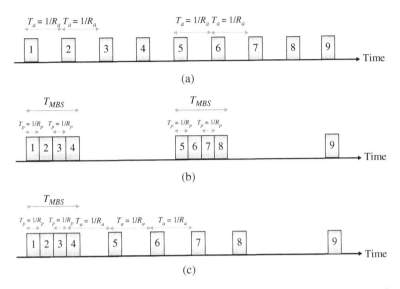

Figure 14.4 Impact of Traffic Shaping by Using a Token Bucket Algorithm. (a) Steady Flow, (b) Bursty Flow, and (c) Quasi-bursty Flow.

Under such a traffic profile, where a flow's stream is characterized by such a deterministic process, its packets are not delayed at the shaping queue.

In contrast, in Fig. 14.4(b), we show a profile of an admitted bursty traffic flow. The admitted traffic process consists of packet burst periods that are separated by pause (inactivity) periods. Averaging over a long period of time, and over each *cycle time* (designating the instants of time elapsed between consecutive burst start times), a conforming flow would send data at a rate that does not surpass the specified **SIR** of R_a [packets/s]. To protect network resources from being overloaded by a too high a number of consecutive packets driven into the system at a peak rate of R_p [packets/s], a **Maximum Burst Size (MBS)** duration metric, denoted as T_{MBS} [s], is imposed. The token bucket capacity parameter, denoted as C [tokens], in conjunction with its average and peak rate parameters, serves to shape the traffic so that this restriction is met.

Consider high burst activity condition. Assume that a pause in the traffic stream's activity that precedes the start of a subsequent burst period is sufficiently long to cause the token bucket to then fill up to capacity. Consider a burst that lasts for an MBS duration (identified as an MBS burst) of T_{MBS} [s]. To simplify the calculation of this duration, noting that typically $R_a \ll R_p$, the following approximate calculation is performed, not accounting for the fact that the number of packets served during an MBS, denoted as N_{MBS}, should be represented as an integer. A total of $R_p T_{MBS}$ packets are admitted by the system during an MBS period. Since each admitted packet makes use of a single token, a total of $R_p T_{MBS}$ tokens depart the bucket during an MBS period. During this period, the token bucket gains $R_a T_{MBS}$ new tokens (noting that the actual number gained is an integer portion of this value). Hence, starting from a full state of C tokens in the bucket, we conclude that the total number of tokens that were used by conforming packets during the underlying burst period, before the token bucket has exhausted all of its available tokens, is equal to about $C + R_a T_{MBS}$. This value also represents the number of packets (whereby each packet is represented as a segment) that have been sent during the burst period, which is equal to $R_p T_{MBS}$. By equating these two expressions, we conclude the following:

442 | *14 Flow and Congestion Control: Avoiding Overuse of User and Network Resources*

Property 14.1 **Under the Token Bucket Algorithm, an approximate expression for the time duration of a MBS period is given as follows for Rp > Ra:**

$$T_{MBS} = \frac{C}{R_p - R_a}. \qquad (14.1)$$

The number of conforming packets (denoted as N_{MBS}) admitted into the system during an MBS period is approximated as:

$$N_{MBS} = \lfloor T_{MBS} R_p \rfloor = \left\lfloor \frac{C R_p}{R_p - R_a} \right\rfloor. \qquad (14.2)$$

A flow's **duty cycle**, denoted as η, represents the fraction of time that the flow is active. The duty-cycle level of the regulated flow is equal to $\eta = \frac{T_{MBS}}{N_{MBS} T_a} = \frac{N_{MBS} T_p}{N_{MBS} T_a} = \frac{T_p}{T_a} = \frac{R_a}{R_p}$. Thus, the duty-cycle level is equal to the ratio of the average to peak rates, independently of the burst's duration. However, admitted flows that produce longer burst periods would generally impose higher resource requirements from the network system and can lead to higher congestion conditions. It is noted that the network system may not have sufficient resources, or will induce excessive packet delays, when it is loaded by several bursty flows that are simultaneously active in producing long bursts. It is therefore essential for the network's traffic manager to use FC schemes that regulate the time duration of incoming burst periods while also limiting the duty-cycle factor of each flow.

In Fig. 14.4(c), we show a profile of an admitted quasi-bursty traffic flow. The scenario leading to such a traffic flow profile corresponds to the following. After a relatively long time of inactivity, a regulated flow's traffic source produces traffic that generates packets at high data rate for a relatively long period of time, longer than the MBS period duration imposed by the regulating Token Bucket Algorithm. Consequently, as shown in the figure, the shaped traffic produces first a number of packets that is equal to the maximum number of packets that can be sustained during an MBS, as calculated by using Eq. 14.1. Following this period, the token bucket gains a single token every T_a [s]. Consequently, the remainder of the source packet stream is shaped into a deterministic flow by delaying the involved packets in the shaping queue.

Example 14.1 Consider a source whose packet stream is flow controlled as it is sent across a communications link that provides for a UNI. Packets are assumed to each contain 1000 bits. The access link operates at a peak data rate of 10 Mbps, so that the peak packet access rate across the link is equal to $R_p = 10$ K [packets/s]. The source is assumed to produce data traffic at an average rate of 1 Mbps, so that $R_a = 1$ K [packets/s]. The traffic managers that operate at the end-user station and at the network's edge node each employ a token bucket scheme to regulate the packet flow across the UNI. Assume that the network's traffic manager aims to restrict the duration of an admitted burst to 10 ms, setting $T_{MBS} = 10$ ms. Using Eq. 14.1, the value that should be set for the capacity of the token bucket is determined to be equal to $C = 90$ packets. Using this parameter and applying Eq. (14.2), we calculate the number of packets that are admitted into the system during an MBS period to be equal to $N_{MBS} = 100$ packets. The system capacity and performance management schemes would thus allocate sufficient resources to sustain a loading of 100 packets that arrive at a peak rate within a single burst period. The flow's duty cycle and the corresponding average to peak rate ratio are equal to $\eta = \frac{R_a}{R_p} = 0.1 = 10\%$.

Assume that at a later time the network management system further restricts the MBS duration, setting it to be equal to only $T_{MBS} = 1$ ms. Using Eq. 14.1, we find that the corresponding capacity of the token bucket must now be reduced by a factor of 10 to only $C = 9$ packets. The number of packets that are now admitted into the system during an MBS period is reduced by a factor of 10 to become equal to $N_{MBS} = 10$ packets. While imposing a much shorter burst period, no changes have been made to the corresponding values of the average and peak rates, so that the duty cycle of the admitted flow is still equal to 10%.

The **Leaky Bucket Algorithm** is another metering scheme that has been used to control the input rate of a flow. It is used to determine whether an arriving packet is conforming or non-conforming, as done by using a Token Bucket Algorithm. Based on this determination, a network receiving entity, such as a network edge node attached to an end-user station across an access link at the UNI, proceeds to admit a conforming packet, and delete or admit-and-tag a nonconforming packet. Also, by using this scheme, the source can shape its traffic stream so that its transmitted segments conform to the profile boundary. Packets would be then delayed at the source so that a conforming input process is formed.

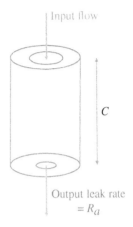

Figure 14.5 Illustration of a Leaky Bucket Algorithm.

The operation of the **Leaky Bucket Algorithm** is illustrated in Fig. 14.5. A flow of data segments that arrive to, or generated by, a traffic source is visualized as a water flow (fragmented to represent discrete data segments) that is fed into a bucket. Assume segments to be identified in the following as packets, though they can represent other data units, including bits, Bytes, frames, or message blocks. As a water stream is fed into a leaking bucket, as long as the bucket contains water, it will leak out at a rate of R_a [packets/s]. Once it becomes empty, it stops leaking. The bucket's capacity is set equal to C [packets], so that it can store a maximum of C packets at a time. An arriving packet aims to add an amount of water to the bucket that is related to its size. If the current water level of the bucket is too high to accommodate the addition of a new packet, the packet is declared to be nonconforming. As treated by a token bucket scheme, a nonconforming packet may be discarded or it may be stored by the source in a traffic shaping queue, delaying its admission until the bucket can accommodate it. A nonconforming packet does not modify the water contents of the bucket. The leaky bucket scheme induces a FC operation that is analogous to that provided by the token bucket scheme.

A **dual leaky bucket** scheme is used by an **Asynchronous Traffic Mode (ATM)** network for the purpose of regulating the input traffic into a network across a dedicated UNI link. The regulated data units are ATM packets, which are known as ATM cells. Each cell is 53 Bytes (Octets) long, including a 5 Byte header that contains the Virtual Circuit Identifier (VCI) used by the underlying flow across the ATM's connection oriented packet switching network. Two leaky bucket regulation schemes are used. The first one serves to regulate the average data rate (R_a), allowing it to vary within a specified limit level (that relates to the specified duration of a maximum bust size, T_{MBS}). The second leaky bucket scheme is used to check conformance of the flow's peak data rate (R_p), allowing a prescribed level of variation in the values of realized packet inter-arrival times that are incurred during a burst period.

The above-described open loop FC schemes can be employed in conjunction with the categorization of the traffic into multiple classes. The allocation of tokens to queued packets may be performed, for example, on a priority basis, or on a weighted round robin basis. Under another arrangement, a hierarchical arrangement of multiple token buffers is configured. Tokens that are currently not utilized by packets that are set to be served at a certain hierarchical level, are made available for use at a lower hierarchical level.

14 Flow and Congestion Control: Avoiding Overuse of User and Network Resources

To illustrate, consider the following priority-based input rate regulation scheme. Assume vehicles that are waiting on ramp to access a highway to queue in multiple lanes. Regulation of access to the highway is lane dependent. Access priority may be lane dependent. When no such vehicles wait at higher-priority lanes, vehicles that wait at lower priority lanes may be admitted.

Combined open loop and closed loop FC schemes are also employed. As resource availability and congestion conditions at a destination entity change, the parameters that are configured for use by an employed open-loop FC mechanism are changed. Such modifications may be invoked at a lower rate.

As an illustration, consider a truck depot that serves to unload cargo brought-in by trucks that travel to the depot from several ports. The depot traffic manager specifies regular traffic flow parameters to be used by each flow under normal operating conditions. Under exceptional conditions, a flow may be delayed at its source due to abnormal operating issues that may arise, requiring the destination controller to reallocate resources. The destination manager would then send control messages to a selected set of ports instructing them to modify certain FC parameters over a specified time period. Open-loop flow regulation would be performed during the latter period.

14.4 Congestion Control: Relieving Bottlenecks

A congestion control scheme is used to regulate, prevent, or react to the occurrence of congestion states inside a communications network. Network nodes may experience at times a high congestion state induced by high traffic loads or by failure events that cause processing or storage resources to be degraded. What should be done when the network system incurs such a condition? In the following, we describe several reactive and proactive congestion control methods.

14.4.1 Reactive Congestion Control

Under a **reactive congestion control** scheme, the network system reacts to the occurrence of congestion at the time that it occurs. This can be accomplished in many ways.

Tess is driving from work to her home during an evening rush-hour period. She usually follows a direct route, driving on Ventura Freeway. She often checks online to determine whether explicit congestion notifications (ECNs) that identify exception conditions along her route have been advertised. Such alarms provide congestion status updates for those traveling along a highway as well as along nearby alternate routes. Using these notifications before she starts her journey, she selects her route accordingly. One day, as she started to check such ECNs, she found out that they were slow to arrive. Her vehicle's navigator program was unable to receive timely traffic status data. She decided to examine implicit signs of congestion. For this purpose, she checked messages disseminated by other drivers regarding the travel times that they have been experiencing while using area roads. She used this data to estimate her prospective travel times across several candidate routes. She selected a route and proceeded to follow it. However, in the midst of her travel, she found out that the highway that she is taking is becoming highly congested. She proceeded then to evaluate highway travel time estimates disseminated by drivers. Using this information, she selected an alternate route. On another day, she found out that all candidate roads are expected to stay highly congested for a specific period of time. She postponed her departure time to a later time.

This example illustrates the use of several approaches when reacting to congestion. Under a first approach, Tess is searching for alternate, less congested, routes. This may be a workable solution when the system is not overly loaded, so that she is able to find a route that not too many other

14.4 Congestion Control: Relieving Bottlenecks | 445

travelers are interested in taking. However, if the regional transportation system is moderately to highly loaded, many other travelers may be looking to use the same alternate routes. In this case, upon driving along an alternate route, she may find out that many other drivers have made the same choice, causing this other route to become congested. In turn, decisions by her and others to delay their travel to less busy times could well contribute to reducing the traffic load, enabling smooth traveling experience for other drivers.

How does a traffic source determine that its messages are experiencing network congestion? Which methods are used to implement a reactive congestion control scheme? In the following, we discuss such explicit and implicit means.

Under an ECN scheme, a source receives congestion notifications that inform it about congestion states incurred by its flows. When such notifications are not available, the traffic source may monitor and evaluate selective network states and disseminated performance measurements and use them as **implicit indicators of congestion conditions**.

A congested router may send *ECN* messages or signals to neighboring routers or, when known, also to other routers that are traversed by flows that are involved in contributing to its congestion state. Such congestion notifications may be issued repeatedly at a certain rate for as long as congestion status continues. Informed network nodes may react by attempting to search for *alternate routes*.

In turn, rather than notifying other network nodes, a congested router may generate congestion notifications and send them to **end users** whose flows it serves. Such congestion notifications may be issued repeatedly at a certain rate for as long as the congestion status persists. A notified end-user entity would react to such notifications by reducing the rate at which it drives the network with messages that belong to a flow that has been notified to traverse congested nodes.

For fast resolution, for as long as congestion state persists, the input rate may be *reduced in an exponentially fast manner*: for example, by reducing the rate at a factor of 2 following every specified period of time. When the source is notified that the congestion state has been resolved, it would ramp up the input rate of its messages in a conservative manner, such as by raising it over time in a linear manner. Such an **exponential decrease and linear increase** rate adaptation function is commonly employed as a reactive measure for handling congestion conditions.

To illustrate such an operation, consider a TCP/IP network in which certain routers and end-users are configured to generate and react to *ECN* control messages. A packet that traverses a congested router is flagged (at the IP layer) to indicate *Congestion Encountered (CE)* state. As the packet reaches the TCP protocol entity at its destination end-user, the latter would proceed to notify the source TCP protocol entity of the flagged congestion state. The source would then react by reducing its message rate for as long as CE signals are being received.

A node may set the condition for declaring its system to be congested relative to a specific flow when the occupancy or loading level of resources that it has assigned for use by this flow, or by groups of flows of the same QoS category and priority class as this flow, are higher than a specified threshold level. For example, a router my set maximum targeted levels for packet storage occupancy, processor loading, and communications resource utilization, per traffic/service/application class and priority category.

A congestion control method that is based on **implicit congestion indicators** has been widely employed. It makes use of reactive acknowledgment, resource allocation and retransmission processes performed by TCP layer entities. Under TCP, the sending and receiving TCP protocol layer entities establish a connection for each flow, in each direction (when full duplex communications activity takes place). Once established, these entities exchange transport layer messages, identified also as message segments, which carry network layer packets as their payloads.

14 Flow and Congestion Control: Avoiding Overuse of User and Network Resources

When a sending end-user entity determines that its flow's data messages could be incurring congestion within the network, it acts, when configured to do so, to rapidly to reduce its flow's input message rate. As noted above, when a drastic reaction is appropriate, the source implements a **Geometric/exponential rate decrease** function. At the end of each specified period of time, such as a RTT period during which no positive confirmations are received, the source reduces its message rate by a prescribed factor. For example, the source may then reduce, during each RTT period, its network traffic loading rate by a factor of 2. It will keep doing it for as long as it determines the congestion state to persist. Once the source determines that its flow is not anymore incurring congestion, it will act to increase its input rate by following a prescribed rate increase function. When aiming to behave conservatively, the source would use a **linear rate increase** function to ramp up its input traffic rate. Ramping up is performed carefully, in trying to avoid contributing to network congestion, noting that an unknown number of sources may increase their input traffic rates at about the same time.

The operation of a TCP congestion control scheme, as used over the Internet and other networks, is based on an algorithm that transitions between several phases. A version of the involved phase transition process and associated parameters in shown in Fig. 14.6 and discussed in the following. It functions as a sliding-window congestion control scheme, similar to the operation of a

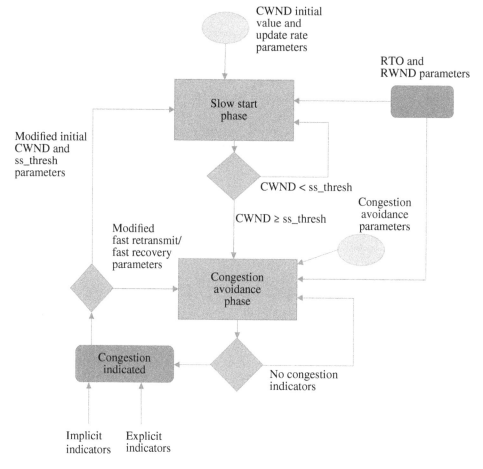

Figure 14.6 TCP Congestion Control.

sliding-window ARQ error control mechanism. A Selective-Repeat Sliding-Window type operation is employed. The number of outstanding unacknowledged sent packets that are in transit across an end-to-end TCP connection is limited by the size of a window that is identified as the **congestion window (CWND)**. Following the transmission of a segment, the sender initiates a **retransmission timeout timer (RTO)** whose duration is of the order of a **RTT**, which is set equal to the time that the sender must wait to receive an ACK for a segment, prior to triggering its retransmission.

Reviewing its error control operation, we note the following. The receiving entity of a connection reacts to the correct receipt of a segment by returning a positive ACK. It identifies the sequence number of the next segment that it expects to receive. In this manner, it acknowledges the in-sequence receipt of segments that have been identified by preceding sequence numbers. Segments that are received out of order are buffered and not acknowledged.

For example, if it has received segments identified by sequence numbers 1, 3, and 4, it reacts as follows. Upon receiving the first segment, its ACK cites the sequence number of the next expected segment to be #2. Upon receiving the next segment, which is identified by sequence number #3, its ACK cites the sequence number of the next expected segment to again be equal to #2. Upon receiving the next segment, which is identified by sequence number #4, its ACK cites the sequence number of the next expected segment to still be equal to #2. Such repeating ACKs are identified as **duplicate ACKs**. If segment #2 arrives promptly thereafter, the receiving TCP entity will then issue an ACK that identifies the sequence number of the next expected segment to be #5, provided it has buffered the other received segments.

The TCP source entity makes use of this acknowledgment process to implicitly determine the occurrence of network congestion. It reacts to the receipt of duplicate ACKs by proceeding to reduce its input flow rate. As TCP actions have an end-user to end-user scope, a TCP entity is not directly involved with network layer status monitoring. It assumes that the ensuing error control based ACK notifications are also indicative of congestion occurrences.

A version of the underlying congestion control algorithm function is described in the following. The maximum length of each segment is set by the specification of a **Maximum Segment Size (MSS)** parameter. Assume in the following that segments are interchangeably also referred to as packets.

Upon the initiation of traffic flow across a TCP connection, or after occurrence of a certain congestion induced state, a **slow start** phase is triggered. The initial CWND size is of the order of several MSS units. For every packet that is acknowledged, the CWND is increased by 1 MSS. This way, the CWND size increases exponentially fast over the duration of a slow-start phase. For example, starting with 1 MSS, the window expands to two segments following the receipt of an ACK. Subsequently, following the receipt of an ACK for the next two segments, the window will increase by two more segments to a window size of four segments, etc. In this manner, the window size doubles every RTT.

When the CWND exceeds a threshold value, known as the **slow-start threshold (ssthresh)**, the scheme enters a **congestion avoidance** phase. As long as non-duplicate ACKs are received, the CWND keeps increasing in an **additive** order, such as by 1 MSS every RTT.

This growth continues until the source determines that its flow has encountered congestion. Congestion is recognized when the TCP sender stops to regularly receive its PACKs, or upon its reception of duplicate ACKs. The congestion control scheme then transitions to a different operational phase. Under one version, the operation returns to a *slow start phase*, using a modified set of parameters. The CWND is reduced, for example to 1 MSS. The ssthresh is also reduced. For example, it is restricted to a value that is equal to half of the window size that was used in the most recent phase.

To protect its own resources, the receiving TCP entity manages a **RWND**. It limits the total number of segments that it is willing to store at one time on behalf of an underlying source and the associated TCP flow. The TCP source entity is kept informed as to the size of the receive window, which the destination TCP entity may change from time to time as its resource availability status changes. The source entity limits the data quantity that it is permitted to send at any time, during any phase, by taking into consideration the currently configured sizes of both the CWND and the receive window.

The described algorithm regulates the transmission rate across a TCP connection in a closed-loop manner by using an **Additive Increase/Multiplicative Decrease (AIMD)** method, as the CWND is expanded linearly during congestion avoidance periods and is reduced exponentially fast upon congestion recognition. Different versions of the algorithm employ different rules relating to the selection of the underlying parameters and the method that is being used for retransmission.

Under a **fast retransmit** procedure, packet retransmission is executed faster when the sender believes that the currently observed loss of its packets is caused by noise or signal interference processes incurred during network transport and not by network congestion effects. When a segment loss is recognized based on receiving several consecutive duplicate ACKs, such as the reception of at least three repetitive ACKs, the source may assume that the network path used by its packets may not be too congested, as packets keep reaching the destination entity. In this case, rather than waiting for the RTO to expire, the sender would proceed promptly to retransmit the unacknowledged packet. This operation is often coupled with a **fast recovery** procedure. Rather than transitioning to a slow-start phase, the sender transitions directly to a congestion avoidance phase. At this time, instead of reducing its transmission rate exponentially fast, the sender decreases its CWND to half its size while increasing its input flow rate in an additive manner.

14.4.2 Proactive Congestion Control

Under a **proactive congestion control** scheme, the system and its users regulate traffic flows so that imminent, pending and future occurrences of network congestion are avoided. We note several such approaches.

Under a **congestion forecasting** method, the source regulates the rate it uses to drive its traffic flow into the network to avoid the occurrence of network congestion in the near future in response to pending congestion forecasts.

To predict the occurrence times and duration of congestion periods, endusers may make use of historical congestion data and traffic forecasts. Also, observed network traffic loading trends are used as explicit and implicit inputs for the determination of occurrence times of network congestion hot spots. An end user would then act to stagger its network flows over time so that periods of high congestion are avoided.

Flow performance monitoring can be used to track the build-up of congestion. By having access to network management entities, an end user is able to monitor associated network performance behavior. Using such observations, a user analyzes and tracks ongoing trends of congestion processes. The user then proceeds to accordingly schedule and regulate its traffic loading processes.

To evaluate the congestion level incurred by its flow messages, an end user may continuously measure the delay levels experienced by messages that it drives into the network. It may record the period of time elapsed from the time that a message is fed into the network to the time that it receives a response or acknowledgment message from the destination, identified as a *RTT*.

Another proactive congestion control scheme is based on the concept of **Random Early Detection (RED)**. A network router that observes a congestion increase trend (e.g., by noting its buffer occupancy to continuously increase and surpass a specified threshold level) would act pro-actively by informing end users whose flows it serves to reduce their traffic loading rates. The router may then select at random a fraction of these flows, and associated messages, to discard. A traffic source whose message has been discarded will not receive its expected PACK and may consequently suspect that its messages have incurred network congestion. It would then proceed to throttle the rate at which it feeds messages into the network. By having the router discard messages by choosing them at random, the ensuing rate reduction actions may be more fairly distributed among flows, rather than impacting only a limited number of specific flows and sources.

Under a **Weighted Random Early Detection (WRED)** scheme, queue-size thresholds that are configured for triggering the dropping of packets are set to values that depend upon a packet's priority level. Lower-priority packets are dropped when lower queue levels are observed, while the dropping of higher-priority packets will take place when the queue size level surpasses a higher threshold value. In this manner, the dropping of lower-priority packets will be initiated as the nodal congestion level just starts its build-up, leading to earlier throttling of the associated traffic flow at its source.

Under an **active queue management (AQM)** scheme, a system management process continuously tracks the states of transmit data queues, acting accordingly to resolve congestion build-ups. Packet dropping, such as RED and congestion notification schemes, such as ECN, are used proactively in aiming to prevent buffer overflows.

A message transported across the network is at times discarded by a network node or by its destination end-user node not because of nodal congestion but due to corruption of its data contents that is caused by communications channel noise or signal interference. In this case, it would be more appropriate to react by using an error control process rather than a FC procedure. For such situations, as is often the case when noisy channels such as those used in wireless communication links are involved, a link layer error control scheme is used so that a hybrid protocol that employs forward error correction and ARQ is employed.

Network congestion can be proactively avoided when it is feasible to effectively use, in a timely manner, a management and control node to pre-allocate sufficient network resources for the support of targeted flows. Such a method is performed effectively by using **flow admission control** and by allocating network resources to satisfy a flow's request for **QoS performance guarantees**, when such a request is produced. A network system's management node decides whether to admit a flow based on a **service request** made by the end user that produces the flow. The request cites the desired class of service, provides its traffic descriptor, and states its desired QoS values. If QoS performance guarantees are requested, the network checks to determine whether it has sufficient (communications, switching, processing, and memory storage) resources to meet the requested performance metric levels, and at the same time not violate the performance objective levels guaranteed to currently supported admitted flows. Upon admitting such a flow, the network system adjusts its network control and management schemes to provide the flow with its desired QoS performance levels.

In packet-switching networks, such reservations are performed on a statistical basis, assuring a user whose flow has been admitted that during activity periods, the network will have, within a specified time latency, at a high probability of assurance, sufficient resources to support its transport performance objectives. To protect the performance of active flows that have already been admitted into the network and are supported under QoS guarantees, a new flow to which the network is not able provide its required transport resources is denied admission and is said to be *blocked*.

End-user data sources tend to often be active on a statistical basis. For example, consider a source that is active for only 1% of the time. During an *activity burst* period, the source may produce messages only during, for example, 5% of the time. When such data sources are admitted into the system, the system needs to provide their messages with QoS performance guarantees for only a fraction of the time. During other times, system resources can be used to provide QoS performance guarantees to other flows. Residual resources that are not used for the support of flows that are provided performance guarantees can be used for the support of flows that are provided no QoS guarantees. The use of such statistical means for the sharing of resources (as reviewed when discussing statistical multiplexing and multiple access schemes) are of prime importance as they contribute in a major way to the efficient utilization of communications media and network resources.

Cellular wireless networks provide at times (statistically based) guaranteed performance to specified call or flow types. Performance guarantees can include statistical specifications of maximum message delay, message throughput and message discard (loss) rate levels. For instance, messages that belong to an admitted call or flow may be guaranteed delay levels that do not exceed 100 ms and a message discard ratio whose level is lower than 1% for 95% of the call's messages. The other 5% messages may experience higher delay and discard ratio levels.

Methods for QoS performance guarantees in IP networks include the following approaches. Under an **Integrated Services Architecture (IntServ)**, reservation and flow admission control processes are used. A process is employed to enable a user to send the management system its service reservation request. For an admitted flow, the network system maintains proper service states in the routers that are located across the selected flow's route. This state specifies, at each router, the resources and service processes that should be used in support of the admitted flow. Such an operation is recognized as using a **Stateful model**.

In turn, under a **Differentiated Services (DiffServ) Architecture**, a **Stateless model** is used. Packets are marked with a *DiffServ Code* that identifies the level of service that a packet desires. The *Differentiated Service Code Point (DSCP)* label serves such a purpose. As a packet is routed across the network, each traversed network node strives to satisfy its desired performance levels. IP packets can also include **performance labels in their headers**.

As another approach, the edge router can identify the *Per Hop Behavior (PHB)* that a marked packet should receive across each subsequent link that it traverses across its network route. Routers that cannot provide a packet its requested PHB-based service may discard the packet.

Under a WRED scheme that accommodates packet DiffServ labels, so that packets are colored in accordance with their DSCP values, the setting of packet drop probabilities can depend upon the packet's DiffServ label. A lower throttling measure is executed for packet flows that require preferred delay-throughput service.

The uplink and downlink wireless communications resources of a *RAN* of a cell in a wireless cellular network are controlled and managed by the BS. The BS has access to a data base that lists all currently occupied and available such resources. Accordingly, a BS is able to reserve available resources for the support of admitted flows that require QoS assurance. When an end user becomes active and is interested to transmit packets to the BS over the shared uplink medium, it sends a signaling packet to the BS to express its request. If and when available, the BS node will issue a channel assignment packet that specifies the allocated resources.

Specialized mechanisms are used to regulate congestion in new ultrahigh-speed wireless networks. Key applications that use these networks require the network medium to support the transport of critical flows in providing them very low delay and high-throughput rates. This is often the case when supporting the dissemination of control and sensing packets among autonomous vehicles, mobile robotic platforms, management and control stations, and road side units (RSUs). It is then essential to implement proactive congestion control schemes that forecast congestion trends and act promptly to eliminate congestion bottlenecks.

A multitude of different approaches have been developed and tested in striving to further improve the delay-throughput performance effectiveness of congestion control algorithms. Certain methods are based on the dynamic synthesis at a source of a queueing model that represents the delay-throughput performance behavior experienced by the its packets. Ongoing collection of performance data, such as sliding window-oriented measurements of the delay times incurred by packets, are used to tune up the synthesis parameters of such a network performance model. Using this model while also monitoring the dynamics of the packet queues formed at a source, rather than just employing the packet loss dynamics observed when using a TCP-based congestion control scheme, enable the synthesis of an effective congestion control process. The Bottleneck Bandwidth and Round-trip propagation time (BBR) technique illustrates the operation of such a model based regulation scheme.

Problems

14.1 Describe the objectives and methods of operation of flow control and congestion control mechanisms.

14.2 Identify the processes used in the execution of closed-loop flow control and in carrying out open-loop flow control. Discuss the performance features of these schemes.

14.3 Describe the operation and parameters used by a Token Bucket Algorithm. Also address the following:
a) Configuration of the scheme's parameters.
b) How the setting of the Bucket Capacity (C) parameter and the specification of the average token supply rate impact the average and variation levels of the outgoing regulated flow process.

14.4 Draw figures that illustrate the flow processes produced at the output of a token-bucket regulator in response to loading by three different types of input flows: a regular flow, a bursty flow, and a quasi-bursty flow.

14.5 Consider a token bucket regulator.
a) How is the outgoing Maximum Burst Size (MBS) regulated?
b) Estimate the length of the outgoing maximum burst size when the flow's peak bit rate is equal to 10 [Mbps], the average bit rate is equal to 2 [Mbps] and the bucket capacity is equal to 1 [Mbits].

14.6 Consider a leaky-bucket flow regulation and shaping algorithm.
a) Describe the operation of the algorithm.
b) Explain the impact of system parameters, including the output leak rate and the bucket's capacity (C) level, in regulating incoming flows and in shaping outgoing flows.

14.7 Describe the objectives and principle of operation of *reactive congestion control* schemes. Also, explain and illustrate the following features:
a) Explicit Congestion Notification (ECN).
b) Implicit indicators of congestion.
c) Exponential decrease and linear increase functionality.

14.8 Explain the method used by TCP in performing congestion control.

14.9 Discuss the following parameters associated with a TCP type congestion control scheme:
a) Congestion Window (CWND).
b) Implicit indicators of congestion notification.
c) Retransmission Timeout Timer (RTO) and its relation to a measured sliding-window estimate of the Round Trip Time (RTT).
d) Maximum segment size.

14.10 Describe schemes that can be used as means for congestion avoidance. Include description of the following elements:
a) Slow-start phase.
b) Proactive congestion control.
c) Random Early Detection (RED).
d) Weighted Random Early Detection (WRED).
e) Active Queue Management (AQM).

14.11 Describe the following:
a) Differentiated Services Architecture (DiffServ).
b) Integrated Services Architecture (IntServ).

15

Routing: Quo Vadis?

15.1 Routing: Selecting a Preferred Path

The Objective: *A node that wishes to send a message to a destination node across a communications network uses a routing algorithm to determine a route that its messages should take. It then accordingly selects a router to which to forward its message. A node that is capable of determining network routes for its messages through the execution of a routing algorithm, aided by interactions with other nodes, is identified as a router. A network system may offer several routes for a message to take to reach its destination node. Different routes may provide messages different levels of performance. A route's performance metric may account for its throughput capacity, reliability, monetary cost rate, and time delay. The performance level targeted by a message as it is transferred across a route depends on the application class of its carried data. Interactive applications require short response times while large file downloads demand high throughput rates.*

   The Methods: *Several routing protocols are available for determining the best route to be used for navigating a message to its destination. For scalability reasons, the network system is divided into routing domains, whereby a domain is often identified as a routing area that is managed as an Autonomous System (AS). Routing inside a domain is determined by internal (or interior) routers. Intra-domain routing algorithms typically employ distance-vector or link-state protocols. Under a link-state routing protocol, such as Open Shortest Path First (OSPF), each router advertises periodically to each other router in its domain the state of its attached links. Using this information, each router constructs a weighted graph of its domain's network layout and uses it to calculate, for each specified performance metric, the best route from itself to any other router in the domain. Under a distance-vector routing algorithm, a router advertises periodically to only its neighboring routers the result of its current calculation of the performance metric levels that it can attain when using its best known routes. Using this information, other routers update their calculations of the metric values of the best routes that they can employ, compiled as distance vectors. Routers continue in this manner to disseminate their updated distance vectors to their neighboring routers. Inter-domain routing protocols employ border (boundary or gateway) routers. Such a router is able to connect interior and exterior routers and thus to interface multiple domain areas. When routing a message to a destination node that belongs to a different domain, an intra-domain routing algorithm is used to navigate the message inside the source domain to its preferred gateway. An inter-domain routing algorithm, such as the external Border Gateway Protocol (eBGP), is employed by a border gateway to route the message to its destination domain area. For this purpose, a path-vector type routing algorithm is employed. Border routers exchange routing data and use this information to determine the best inter-domain path. Once a message reaches a border router that interfaces its destination domain, an intra-domain routing algorithm is used to navigate the message to its targeted destination node.*

Principles of Data Transfer Through Communications Networks, the Internet, and Autonomous Mobiles, First Edition. Izhak Rubin.
© 2025 The Institute of Electrical and Electronics Engineers, Inc. Published 2025 by John Wiley & Sons, Inc.

A routing protocol is used to select a network route across which messages are navigated as they are transferred from a source node to a destination node. Often, there may be several routes that messages can take as they make their way to reach their destination. The network routing protocol employs a routing algorithm to determine a preferred network route.

A route often consists of a network path whose links (also referred to as lines) connect neighboring network nodes that act as routers or switches. It is also possible for a route to consist of multiple paths that lead to a destination node. To simplify, we assume in the following the route to use a single path, unless stated otherwise.

To illustrate, consider the network layout shown in Fig. 15.1. The topological layout of the network is described by a graph model. It consists of nodes (also known as vertices) and links (also known as lines or edges). Consider a source end-user $S1$ that is attached to router $R1$ which sends messages to destination end-user $D1$ which is attached to router $R3$. Assume that the system selects a route based on a minimum-hop optimization criterion. Each hop represents a link connection between two neighboring routers. A "min-hop" route traverses the least number of hops. No other route offers a path that traverses a lower number of hops, though other routes may use the same number of hops. Path $P1$ that traverses routers ($R1$, $R2$, $R3$) is such a route. Using this route, a message issued by $S1$ is switched by $R1$ for transmission across communications link $L1$, which connects to router $R2$. The latter switches the message for transmission through link $L2$ to reach destination router $R3$ to which destination end-user $D1$ is attached.

An alternate route that leads from $S1$ to $D1$, which consists of 4 hops, rather than the above noted 2 hop route, directs messages to traverse routers ($R1$, $R4$, $R5$, $R6$, $R3$), along a path that we identify as $P2$. An alternate 4-hop route traverses routers ($R1$, $R4$, $R5$, $R2$, $R3$) along path $P3$.

If the system is not highly loaded, it usually prefers to select a minimum hop path such as $P1$. This is advantageous as messages directed across a shorter path would use a lower number of links and therefore occupy less communications capacity resources, leaving more capacity available for the support of other network flows. For example, if the flow's traffic loading level is equal to 1 Mbps, it will be occupying by using a 2-hop route, an overall link capacity of 2 Mbps in aggregate. In turn, when routed along path $P2$ or path $P3$, a total link capacity of 4 Mbps will be occupied. Assuming candidate routes to be similarly loaded, it would thus be advantageous for the system to select routes in accordance with the principle of *shortest path routing*.

In turn, under different loading conditions, it can be advantageous to route a new flow along a less congested yet longer route. For example, assume the current conditions along route $P1$ are such that the residual (i.e., leftover) capacity levels of several of the communications links included in $P1$ are rather low, inducing higher congestion levels. It is then possible that an alternate route, such as one that uses path $P2$ or path $P3$, would offer the new flow a higher residual capacity level, so that it can efficiently carry its full message load. Such an alternate route may offer the new flow

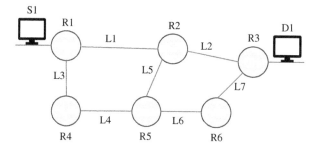

Figure 15.1 Network Layout and Routes.

a lower end-to-end message delay value, even though messages now traverse a path that includes 4 links rather than 2 links.

As another illustration, consider the following. Under the first scenario, a route needs to be selected for the transport of a caravan of 10 trucks from the Los Angeles port to a warehouse depot. It is essential for the selected route to accommodate the full load imposed by the flow. The system manager selects a longer route, such as $P2$, as its current capacity supports the required load. In turn, if at a certain time it is determined that the trucks are fuel limited, the use of a shorter path may be preferred.

15.2   Route Metrics

Message flows that belong to different application classes may use different route performance metrics to assess the suitability of a route for their transport. A **route's metric** expresses the underlying performance measure that is used to select a route. This metric is identified at times as a **route's length**, a **route's distance**, a **route's cost**, or as a **route's weight**. *We will use them interchangeably.* A wide range of performance measures can be used to specify route metrics. Using a given *route metric*, a route that provides a better desired *metric value* is preferred to one whose metric value is not as good. In the following, we present several commonly used metrics. They relate to the performance values offered to messages that are transported across a route.

As noted, we assume in the following, unless stated otherwise, that a route is represented as a path that consists of serially interconnected links. Routes and paths are used interchangeably. A route's metric value is typically set to be equal to the sum of the metric values assumed by its links.

1) **Delay Metric**: Characterizing the delay incurred by a message as it traverses a route. Delay components include queueing time (or waiting time) delays experienced by a message as it traverses the route's path and transits times that are incurred in its transmission across each one of the route's links. For example, for a vehicle that follows a route that consists of several roads, the route's delay metric includes the time incurred in transit across each road as well as the time delay experienced in waiting at traffic lights as it crosses junction points that connect the roads.
2) **Throughput Capacity Metric**: Characterizing the maximum flow rate that a route can handle. It identifies the maximum number of packets (or bits) that can be accommodated across a route per unit time (as expressed, e.g., in units of [packets/s], or [bits/s]). For a vehicle transportation system, the corresponding metric measures the vehicle flow rate capacity, expressing the maximal average number of vehicles that can be accommodated across a highway route per unit time (as expressed, e.g., in units of [vehicles/hour]).

    In reviewing the use of such a metric for a highway system, we note the following relationship to hold. The flow rate of vehicles across a lane of a highway, denoted as f [vehicles/s], can be expressed as the product of the average density of vehicles moving along the highway's lane, denoted as d [vehicles/m], and the average speed of a vehicle, denoted as v [m/s]. The average distance between two neighboring vehicles is noted to be equal to $1/d$ [m]. The following transportation equation holds:

$$f = vd. \tag{15.1}$$

3) **Reliability Metric**: Characterizing the reliability of the route, expressing the probability that it will stay operational during a specified period of time.
4) **Monetary Cost Metric**: Characterizing the total **monetary** cost incurred in transporting a unit of data along the route. Note that if the monetary descriptor is not attached, a link's or a route's

cost value parameter would hereby refer to a value that is calculated be using a nonmonetary metric, such as one that involves delay or throughput measures.

5) **Hybrid Metric**: The route is selected based on a mix of performance metrics. For example, an end-user may require its flow's packets to be routed across a path that provides a prescribed upper bound on the incurred end-to-end packet delay as well as a lower bound on the granted throughput rate. A sufficiently high reliability level may also be prescribed.

6) **QoS Routing**: The route metric value requested by an admitted flow is preserved (for a high fraction of time) during the flow's activity period. In aiming to preserve sufficient network resources for the support of admitted flows, a flow admission control scheme is employed, serving to regulate the admission (or blocking) of new flows and induce the selection of routes that meet requested QoS levels.

Computationally, it is generally assumed that a *preferable route is associated with a lower route metric value*. Generally, the **metric value of a route is calculated as the linear sum of the metric values of its links**. Thus, a path metric value, also identified as its cost, or length, or weight, is calculated by adding the metric values, or costs, or lengths, or weights, respectively, of its links.

A route's basic path consists of communications links that are serially connected. To illustrate, consider route $P2$ of the network layout shown in Fig. 15.1. The path traverses router nodes $\{R1, R4, R5, R6, R3\}$ and intermediate links $\{L3, L4, L6, L7\}$. The performance metric value for a message transported across link L when considering metric m is identified as the link's cost $C(L, m)$, or length/distance $l(L, m)$, or weight $w(L, m)$. For example, $m(L, Delay) = C(L, Delay) = l(L, Delay) = w(L, Delay)$ corresponds to the average delay incurred by a message transported across link L. The delay metric value assumed by path $P2$, $m(P2, delay) = C(P2, delay) = l(P2, delay) = w(P2, delay)$ is calculated as the sum of the corresponding link metric values. Hence:

$$m(P2, delay) = m(L3, delay) + m(L4, delay) + m(L6, delay) + m(L7, delay). \tag{15.2}$$

Under a prescribed metric, a routing algorithm calculates the metric values assumed by candidate routes in consideration for the accommodation of a given flow and selects the route that provides the best route metric value. Metric values can thus be set to represent *relative rather than absolute* performance-related values. For example, a route whose delay metric is equal to 5 provides for message transport at a delay level than is lower than that incurred when using a route whose delay metric is equal to 7. The corresponding metric values do not have to directly account for actual delay levels.

Similarly, under a monetary cost metric, the path cost is often calculated as the sum of the corresponding link costs.

A sum operation can also be used for the calculation of a path reliability metric by proceeding as follows. Let the probability of failure of link L be denoted as $p(L)$. Assume that the link's reliability index is set to $1 - p(L)$, expressing the probability that the link will stay intact (i.e., not fail). Assuming that failure events of the path's links are statistically independent, we evaluate the reliability index for the path as the product of the reliability indices of its links. The path is assumed to stay operational if and only if all its links are operational. To replace the product function by a linear sum, we define the reliability metric of link L as $m(L, reliability) = -Log(1 - p(L))$. Then, the reliability cost metric of path $P2$ is calculated as a sum of its link metrics:

$$m(P2, reliability) = m(L3, reliability) + m(L4, reliability) + m(L6, reliability) + m(L7, reliability).$$
$$\tag{15.3}$$

The throughput metric value of a path is related to the throughput metric of each one of its links. Under a throughput objective, a path that provides higher-throughput capacity is more desirable and would therefore be associated with a lower cost (or length, or weight) value. Consequently, a link's cost value may be represented to be inversely proportional to its throughput capacity level. For computational convenience, the associated path cost metric value is often still calculated as the sum of its associated link cost values.

Such an additive calculation often provides just an approximate characterization of a route's metric value, as illustrated by the following. Consider an operation under which the throughput capacity incurred by a flow along a path is determined by the capacity of the path's bottleneck link. The latter makes available to the flow the lowest-throughput value of any other link belonging to the path. In this case, we may perform the following calculation:

$$m(P2, throughput) = minimum\{m(L3, throughput), m(L4, throughput), \\ m(L6, throughput), m(L7, throughput)\}. \tag{15.4}$$

In addition to accounting for link metrics, we can also take into consideration *nodal states as performance metrics*. For example, a nodal metric can be used to account for the capacity of a node in performing processing, queueing, and storage operations for the transiting messages that it serves, as well as accounting for the availability of energy resources. Nodal metric values can be integrated into the calculation of a route's metric value by, for example, representing the node as an intermediate weighted link whose weight is equal to its metric value.

As a special case, if we set the delay metric value of each link of a path as equal to 1, the path's route metric value then represents the number of links included in the path. In a packet switching network, a link that connects two neighboring packet-switching routers is identified as a *hop*. To simplify, assuming henceforth a hop to correspond to a single link, the delay metric of a route is then equal to the number of its hops. Consequently, the selection of a corresponding minimum delay path is equivalent to finding a path whose number of hops is lower than or equal to the number of hops included in any other alternative path.

The determination of a minimum hop route from a source node to a destination node has been used as a popular objective of many routing algorithms. It is often identified as a **min-hop routing** scheme. For the network shown in Fig. 15.1, when considering routes the lead from $S1$ to $D1$, path $P1$ is 2-hops long while paths $P2$ and $P3$ are each 4-hops long, so that path $P1$ is a corresponding min-hop route. It is noted that it is possible at times to find several paths that assume the same minimal number of hops (or any other preferred metric value). The system would then employ a *tie-breaking* rule to select one of the candidate paths.

Communication links are represented as directional (also identified as unidirectional) or as bidirectional (or undirectional). A directional link (N_1, N_2) provides communications from a transmitter at node N_1 to a receiver at node N_2. Its link's metric (or cost, or weight) is represented as $m(N_1, N_2)$. In turn, a bidirectional link (N_1, N_2) is used for communications from N_1 to N_2 at a link cost $m(N_1, N_2)$ and for communications from N_2 to N_1 at a link cost $m(N_2, N_1)$. Unless stated otherwise, the latter link costs are assumed to be equal to each other. An undirected link can also be represented as a parallel composition of two directed links, one in each direction. The metric values assumed by the latter may be correlated or uncorrelated. In either direction, the link accounts for a single hop.

15.3 Routing Domains and Autonomous Systems

Many networks, such as the Internet, support a high number of end users and make use of a large number of network nodes and links. The determination of a preferred route may require the examination of the states of many nodes and links. The suitability of many different routes may have to be assessed. Thus, finding high-quality routes in a large network is a computationally complex and nonscalable task.

Consequently, the determination of routes and their management in a large network is performed by using a segmentation approach. The network system is divided into distinct routing areas. Route segments are configured separately in different areas. Different routing protocols may be employed in different areas, as well as for routing between and across areas.

The network system is thus divided into **routing domains**. For Internet systems, **a single routing domain consists of networked systems that employ common routing protocols that are managed by a single administration**. Link and nodal metric functionalities and their dissemination processes are performed in a well-defined manner within each domain. A routing algorithm that is used for routing inside a specific routing domain is identified as an **intra-domain or interior routing algorithm**. A routing algorithm that is used to route messages between and across domains is identified as an **inter-domain or exterior routing algorithm**.

A multi-domain architecture provides a more scalable routing system. It also accommodates the ability of different domain administrations to select their own preferred routing algorithms, protocols and operations.

Over the Internet, an **Autonomous System (AS)** serves as such a routing domain. Often, an AS is controlled by a single entity such as an **Internet Service Provider (ISP)** or by a large organization that administers multiple inter-connected network systems. Generally, multiple network operators may provide management and control services for an AS while serving as a **single administrative entity**. Over the Internet, intra-domain routing protocols are identified as **Interior Gateway Protocols (IGPs)** while inter-domain routing protocols are identified as **Exterior Gateway Protocols (EGPs)**. Within each domain, messages are routed by domain routers. In turn, **boundary routers**, also identified as *Autonomous System Boundary Routers (ASBR)*, are used as **border routers**, which serve to navigate messages across domains.

Consider driving long distance from Los Angeles, California, to Las Vegas, Nevada. As noted in Fig. 15.2, two possible routes of interest are identified. The first route is shorter and will take less time to drive. The alternate route is longer but it transits through geographical regions that

Figure 15.2 Driving Map to Las Vegas.

may be of particular interest to the driver. Different highway administrators are used to control the flow of traffic across different intermediate regions. Consider each region as analogous to a routing domain, or an AS. In a specific region, such as the Los Angeles metropolitan domain, the driver could be tuning to receive traffic reports disseminated by a local administrator, using them to select a preferred intra-domain path. No matter which one of the two candidate routes is taken, it is crucial for the driver to use a high-performance route to traverse the Los Angeles domain as this region tends to be highly congested and the time that it will take to transit it will become a significant component of the realized end-to-end delay. The driver may dynamically adjust its intra-domain and inter-domain routes by making use of local, regional, and inter-regional traffic report broadcasts.

Road conditions may change during the travel period. A driver may find out that it is not feasible at this time to follow a preselected route due to temporary closures. The driver would then employ a *dynamic routing* approach to select an alternate route.

Often, a network domain configuration is such that a transmission by an end user across a shared medium communications channel is directly received by all other end-users that are members of this network. Such a connectivity feature is identified as a **broadcast property** and the corresponding channel is said to be a *broadcast channel*. End-user nodes are then identified as members of a broadcast domain.

In a broadcast net (typically a local network system), the routing of messages among network members is readily accomplished. A transmission of a message by a net member node is received by all other net nodes. This system is thus identified as a *logical link*. Each message contains the address of its destination net node. By examining the intended net recipient address included in messages transmitted across the broadcast net, each node is able to determine whether each message is destined to itself. If this is the case, it will copy it and use it; otherwise, it will disregard it.

Accordingly, a message issued by a source node that is a member of a broadcast **local area network (LAN)** (such as an Ethernet LAN or a Wireless LAN) is broadcast across the shared medium so that it is received by other nodal members of the LAN. A local network may consist of multiple LANs that are interconnected by layer-2 bridges. In this case, routing of messages within such a local network domain does not require the use of routers that possess network wide intelligence. A **local network** is typically used for communicating messages within a local facility. In general, routers, switches, or bridges may be used to manage the routing of messages within an enterprise.

If a destination end-user is not located within the boundaries of a local network, a *designated router* is used as a gateway node. It is configured to interface both the local network and a backbone network or an employed wide area network (WAN). It is programmed to capture a wide area message issued by a local user and route it across a backbone network or an external **WAN** such as the Internet. The message will be routed across the backbone network to its destination by possibly transiting multiple domains.

A large organization may have multiple office branches that are nationally and often also internationally geographically distributed. Each branch may have its own internal private local communications network that serves to connect its local network nodes and end users. In communicating messages between branches, the organization may employ a private backbone core network, also known as an enterprise network, or it may use a public network such as the Internet. Corporate IT administrators configure and manage the routing protocols that are employed within their private networks. They manage the allocation of internal and interface addresses. At the borderline between a private branch network and a public network, interface devices are used to accommodate the required addressing and networking translations that need to be performed. Such devices provide for Network Access Translation (NAT), cybersecurity and firewall protections,

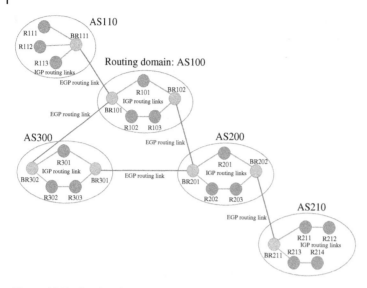

Figure 15.3 Routing Across Multiple Domains: Intra-domain Routing Schemes Employ Protocols such as RIP, OSPF, EIGRP. Inter-domain Routing schemes Employ Exterior Gateway Protocols (EGPs) such as BGP. Boundary Routers (BRs) Are Used to Interconnect Domains.

congestion and flow control regulations, and failure recovery adaptations. *Application delivery controller (ADC)* modules are employed for this purpose, often serving data centers and server farms.

Protocol processes that are used for routing messages across multiple routing domains, as used across the Internet, are illustrated in Fig. 15.3. Each routing domain is identified by an Autonomous System Number (ASN). Routing of packets inside a routing domain is performed by using an intra-domain routing protocol, identified as an IGP. Commonly employed intra-domain protocols (IGPs) used over the Internet and other packet switching networks include RIP (Routing IP), OSPF (Open Shortest Path First), and EIGRP (a CISCO Corp.-based routing protocol). In the figure, an intra-domain routing protocol is used within routing domain AS100 to select an interior route between interior router R102 and interior router R103 or boundary router BR102.

In turn, if router R102 wishes to send packets to Router R214, the selected route would traverse multiple routing domains. An IGP will be used to select a route for R102 packets to reach boundary router BR102. A boundary router is capable of selecting a preferred route that traverses routing domains. It selects the subsequent domain that the packet will traverse to reach its destination routing domain. Boundary router BR102 may select an inter-domain path that consists of domains AS100–AS200–AS210. Using the selection made by an inter-domain routing algorithm, following an EGP, messages may be guided by BR102 across an EGP managed link to reach boundary router BR201 in entering routing domain AS200. Using an intra-domain routing protocol for the selection of a route inside AS200, messages use the latter route to reach boundary router BR202. The latter guides the underlying messages to BR211 as they enter AS210. Within the latter domain, an intra-domain routing protocol is used to route messages to their destination router R214.

Similarly, as vehicles are navigated across multiple regions (such as those associated with distinct neighborhoods, cities, counties, or states), intra-regional and inter-regional paths are selected based on the specific routing schemes employed in each region and across regions. Highway congestion conditions in each region and across regions are monitored and used by impacted routing algorithms for the selection of preferred routes.

15.4 Route Selection Methods

In the following, we describe key approaches and methods that are invoked in selecting and controlling the routing of messages and message flows.

1) **Centralized Routing**: A network management system, or a network administrator, is responsible for selecting routes that are used for communications between nodes located inside the routing domain that it manages. Through communications with administrators of other routing domains, a system manager may also be used to specify its requirements for the selection of routes over a multi-domain system, as well as to select such routes. End-users and network routers interact with domain and/or system managers in setting preferences for the selection of routes, and in their routing implementations.

2) **Distributed Routing**: Under a distributed routing method, route selections are made by subsystem, nodal or user entities rather than by a central administrator. End users and routers may make their own selection of a route to use for sending a message or a flow of messages toward their destination. Such a determination is generally based on information received from other entities, including other routers. Information received from management stations or proxies may also take into consideration.

   An effective routing operation aims to configure a **robust** set of routes. In doing so, it tends to avoid the use of a layout that possesses a single point of failure, reducing the chance of occurrence of drastic performance degradation that may be triggered by failures of network elements. A distributed routing approach leads to a more failure safe operation as route selection does not depend on receiving state information and directives from a central management station.

   A central controller can continuously collect status data from network system elements and fuse this data to intelligently calculate routes that exhibit good performance behavior under currently observed conditions. However, a centralized route management system imposes higher communications and processing capacity requirements, as status and control data need to be continuously transferred between a central administrative authority and widely distributed system entities.

   To enhance scalability, adaptivity, performance effectiveness, and robustness to failure, a hybrid hierarchical architecture that combines the use of central and distributed methods is often implemented.

3) **Static and Dynamic Routing**: Under a **dynamic routing** method, the route selection process makes ongoing use of monitored system state observables. It accordingly adjusts its operation in adapting to changes in network system conditions as they are observed to occur or as they are determined as pending. Route adaptations may be executed on a per-message of per-flow basis, while an impacted message or flow is in transit. Under **static routing**, the route's specification is fixed for a specified period of time, or for the duration of a flow, or for the time that it takes a message to reach its destination.

   It is advantageous to employ a dynamic routing discipline when it is essential to rapidly adapt to variations in nodal and link conditions. Routing a message across a path that includes failed, degraded, or congested nodes and links may then be prevented. Yet, if the likelihood of such occurrences during an underlying message or flow transit period is low, the use of a static route can be appropriate. In this situation, a static setting can yield a higher level of efficiency due to the ensuing reduction of involved computational complexity. Control data overhead rates produced by monitoring processes would be reduced, and higher speeds in performing computational, processing, and switching operations could be realized. Under a hybrid implementation, route specification is dynamically adjusted only upon the occurrence of critical events; it is held static otherwise.

4) **Connectionless Datagram Routing**: When employing a connectionless (or datagram) oriented routing mechanism, forwarding decisions at routers are determined individually for each message, or *datagram*. For a TCP/Internet Protocol (IP) network system such as the Internet, a datagram routing approach is commonly employed. When a packet is received by a router, the router checks its destination address, represented as an IP address. It then examines a *Routing Table* to find the best outgoing hop to use for forwarding the packet to the next router along the selected route. Forwarding entries are kept by the router in a routing table. The entries in the routing table may be updated on a quasi-stationary basis, or dynamically based on status data received from other routers and possibly other entities, including a management entity. Route computation is performed by using a routing algorithm, using methods such as those described in Sections 15.2–15.7.

A **next hop routing** method is frequently employed in routing packets across the Internet. A router uses a routing algorithm to configure its routing table. It uses the table to select a forwarding entry in accordance with the addressing data included in the packet's header and possibly other attributes or metrics. The router then forwards the packet over the route's next hop to a neighboring router. Though the calculation of the route may identify a complete multi-hop path to be traversed to reach the destination node, under this method, the router does not force the transport of the packet to follow its calculated end-to-end path in completion. Rather, it leaves it to the next router to calculate the subsequent hop. This next-hop store-and-forward-based forwarding operation continues until the packet reaches its destination node.

5) **Connection-Oriented Routing**: Under a connection-oriented routing method, a *signaling subsystem* is used to set up a connection in the network, identifying a route that starts at a source node and terminates at a destination node. This configured route is used for routing messages that belong to an underlying flow or call. A routing algorithm is used to select a desired end-to-end route. Following this selection, a signaling subsystem configures the routing tables that reside in each switching node that is located along the selected path. Circuit Switching and Virtual Circuit Switching networks employ such a connection oriented routing method. Only after the route has been selected and configured, and the corresponding forwarding entries placed in the involved routing tables, the source node is notified, and the data transfer process is initiated.

Connection-oriented routing is advantageous in that during the route setup phase, sufficient resources can be reserved (statically or statistically) along the route, in aiming to provide admitted flows their desired quality of service (QoS) performance objectives.

When compared with connectionless routing operations, we note its following key disadvantages. A connection-oriented operation requires the establishment, operation, and maintenance of a signaling subsystem. It is not as rapidly responsive to link or nodal failure events. Following a failure event, the signaling system would generally have to select a new end-to-end path. In contrast, under a datagram operation, when a router detects the failure of an outgoing link, it may be able to act rapidly to select an alternate outgoing link.

6) **Source Routing**: Under a **source routing** approach, the source node calculates a desired end-to-end route for its prospective flow and then includes the complete end-to-end composition of the path in the header of each one of its flow's messages. A switch (or router)

15.4 Route Selection Methods | 463

located along the path examines the path information included in the header of each incoming message and uses it to switch the message. Under a **partial source routing** operation, the source node specifies only a portion of the end-to-end route, citing specific switching nodes that the route should include in any selected path.

A key advantage of a source routing operation is that the switching operation at each router or switch located along the route can be performed at a much higher speed, often using a hardware-based switching mechanism, as there is no need to search for forwarding entries embedded in a routing table and to engage in routing updates and computations. Such an operation is often used for routing packets along a path that requires low latency and high nodal switching speeds.

7) **Label Switching**: Routing via a **label switching** operation can combine the elements of connection-oriented routing and source routing. Employing a routing algorithm, a signaling or control subsystem is used to select a route prior to the start of data transport. It then accordingly communicates the proper forwarding data for insertion in the switching table of each switch that is located across the route. Each switch thus maintains a *Label Switching Table*. A packet received at such a switch from an incoming link contains in its header a *label*, serving as a routing identifier. The label switching table specifies for an incoming packet that arrives across a specific line and that carries a specific label, the outgoing line to which it should be switched and forwarded and the label that should be then inserted into the outgoing packet. An incoming label is thus swapped with an outgoing label.

The ensuing switching operation can be implemented in hardware and performed at ultra high speed. Such a switching process is often employed in optical networks. An optical fiber shares its resources by a large number of flows, where each may be transported across a light path, under a **Wavelength Division Multiplexing (WDM)** operation. Packets that belong to different flows are transmitted across each link by using a distinct wavelength. The wavelength identity used by a packet that is received at a switch is instantly sensed and treated as a label. Using such a wavelength-oriented label switching process, an incoming packet is instantly switched to a prespecified outgoing link and its wavelength is either maintained or translated to a different one.

8) **On-Demand Routing: Find a Route When Needed**: Under a **proactive routing** approach, a router calculates routes to multiple destination nodes, and often to all nodes in a routing domain, and keeps the results in a routing table. This calculation is performed, and repeated periodically or dynamically.

In contrast, under a **reactive routing** method, also known as **on demand routing**, a flow's source router proceeds to determine a route to the destination of a flow's messages at the time that the flow is initiated. The source router does not determine routes to destinations that its current messages have no interest to reach. An on-demand routing approach is useful when routing packets across a **Mobile Ad Hoc Wireless Network (MANET)**. Such a network consists of mobile nodes that communicate across wireless communications links. Ad hoc networks do not use a fixed infrastructure based backbone as that employed by cellular networks. Cellular networks use fixed base station (BS) nodes that communicate with each other across a fixed backbone network. Nodes in an ad hoc network system may move frequently, inducing a continuously changing network layout. An established route may therefore stay valid for only a limited period of time. Messages that arrive at a mobile node are often interested to reach only a limited group of destination nodes. Under such an operational scenario, it can be more effective to discover routes and configure routing entries for them, by using an on-demand process. Such routing algorithms are presented and discussed in Chapter 9.

15.5 Shortest Path Tree (SPT): Mapping the Best Path to Each Node

Consider a node, representing a switch or a router, hereby often referred to as a router, that wishes to find a least cost path from itself to every other node in its routing domain. The metric of a path that connects nodes X and Y is also identified as its length, weight, or cost and is denoted respectively as $m_X(Y) = l_X(Y) = w_X(Y) = C_X(Y)$. These metrics are used henceforth interchangeably. As noted above, the metric value of a path is calculated as the sum of the metric values of its links.

Considering all the paths that connect nodes X and Y in a network domain, the cost of certain paths is lower than the costs of other paths. A lowest cost path is also identified as a **shortest path**. The cost (or length) of a **shortest path** between nodes X and Y is defined as the **distance between nodes X and Y** and is denoted as $d_X(Y)$. Consider a specific path between nodes X and Y, denoted as $P_X(Y)$, also identified as an X–Y path, that is determined to be a shortest path between these nodes. Its metric value is equal to the distance between X and Y, so that $m(P) = C(P) = l(P) = d_X(Y)$.

Consider a network domain whose topological layout is represented by the weighted graph shown in Fig. 15.4(a). The graph's nodes identify routers while its lines represent (bidirectional, also known as nondirectional) communications links. The shown link weight values represent the link metric or cost values (applicable in either one of the link's directions). A path's cost value is calculated as the sum of its link costs. Several alternate paths are available for router A, to communicate to router E. Consider path A–C–E, an A–E path, $P_A(E)$ that passes through intermediate router C. Its cost is equal to $4 + 3 = 7$. In comparison, each one of paths A–B–C–E and A–B–E assumes a path cost value of 6, while the cost value of path A–B–D–E is equal to 5. We conclude that the latter path is the shortest (i.e., least cost) path from A to E, so that the A-to-E distance, which identifies the length of a shortest A-to-E path, is equal to $d_A(E) = C(A–B–D–E) = 5$.

One can proceed to determine shortest paths that serve to connect router A to each one of the other routers in its domain and consequently compile a distance vector D_A that represents the

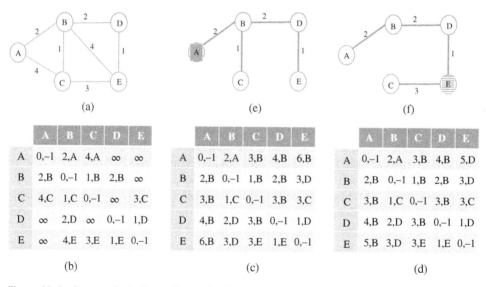

Figure 15.4 Shortest Paths for an Illustrative Network Using Distance Vector Routing Algorithm: (a) The Weighted Network Graph Showing the Link Cost Values; (b) Routing Table Showing Shortest Paths That Are No Longer Than 1 Hop per Path; (c) Routing Table Showing Shortest Paths That Are No Longer Than 2 Hops per Path; (d) Routing Table Showing Shortest Paths That Are No Longer Than 3 Hops per Path; (e) Shortest Path Tree (SPT) Rooted at Router A; (e) SPT Rooted at Router E.

distance values of the shortest paths from A to each other router in the domain. This distance vector consists of the entries $\mathbf{D_A} = \{d_A(Y), Y = A, B, \dots, E\}$, where $d_A(Y)$ denotes the distance from router A to router Y.

The following Property is highly useful.

Property 15.1 A SPT rooted at a source node. A collection of shortest paths that lead from a source node in a connected network domain to each one of the other nodes in the domain can be represented by a **spanning tree graph rooted at the source node**. This graph is called a **SPT** rooted at the underlying source node.

A connected graph provides at least a single path between any pair of its nodes. A spanning tree graph is a subgraph that spans (i.e., includes) all the graph's nodes and is a tree graph. A tree graph is a connected graph that contains no cycles. If a graph contains a cycle, there would be two distinct paths connecting a specific pair of nodes. If one of these paths is longer, it can be removed from consideration as a shortest path. If both paths offer the same length, then one of them can be selected without affecting the distance vector.

For the graph shown in Fig. 15.4(a), we show in Fig. 15.4(e) a SPT rooted at node A. As a tree graph, it identifies a **unique path** between any pair of nodes. For example, examining the SPT rooted at node A that is shown in Fig. 15.4(e), we note the shortest A-to-E path to traverse the nodes A–B–D–E. As noted above, its path distance is equal to 5. For the same graph, a SPT rooted at node E is shown in Fig. 15.4(f).

In Sections 15.6–15.7, 16.6.2, we describe methods that are used for calculating intra-domain routes by using distance-vector and link-state routing algorithms. For inter-domain routing purposes, the Border Gateway Protocol (BGP) is widely employed. It is used across the Internet to set inter-domain routes, in exchanging reachability information among AS. This information, coupled with network policies and rules set by network administrators, is employed in selecting preferred paths. It is also identified as a path-vector routing protocol. It will be further described in Section 16.6.3, when discussing Internet networking methods.

15.6 Distance Vector Routing: Consult Your Neighbors

In this section, we review a method for the selection of intra-domain routes. It is used by routers that are located within a routing domain to communicate with each other. Under a **distance vector routing algorithm**, each router announces to its neighboring routers, periodically over time, its current knowledge of the best metric values that can be achieved by using routes from itself to other routers (or nodes) in the domain. Over time, as each router keeps sending such route metrics to its neighbors and keeps receiving route metric values from its neighbors, routers learn the best metric values (i.e., distances) that they can achieve for reaching other nodes in the domain. Each router compiles a list of the latter distance values, known as a **distance vector**. A router sends periodically over time its current distance vector to its neighboring routers. It updates its distance vector by using the latest distance vector updates that it receives from its neighbors.

To determine its neighborhood connectivity graph, each router sends periodically *hello* messages to each one of its neighbors across each one of its attached links. Similarly, it receives hello messages from its neighboring nodes. Using such hello message exchanges, a node is able to continuously confirm the operational status of its attached links. Also, a router observes (and possibly measures)

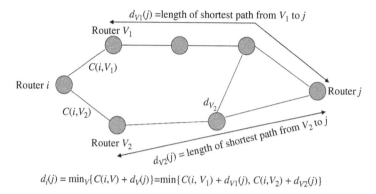

$$d_i(j) = \min_V\{C(i,V) + d_V(j)\} = \min\{C(i, V_1) + d_{V_1}(j), C(i,V_2) + d_{V_2}(j)\}$$

Figure 15.5 Distance Vector Routing.

cost metric values of each one of its attached links. The node forms a list of these link costs values for each attached link. It is known as its **link state** vector.

The principle of operation of a distance vector routing algorithm is based on the Bellman–Ford algorithm. It yields the distance values of paths that connect a router and other routers in the routing domain. The concept behind this scheme is illustrated in Fig. 15.5. Source *Router − i* wishes to determine the best route to use for forwarding a packet to destination *Router − j*. *Router − i* has two neighboring routers: *Router − V_1* and *Router − V_2*. *Router − V_1* announces to *Router − i* that its currently known best route to *Router − j* assumes the cost value $d_{V_1}(j)$. *Router − V_2* announces to *Router − i* that its currently known best route to *Router − j* assumes cost value $d_{V_2}(j)$. *Router − i* then calculates its current estimate of the best cost value of a route to *Router − j* as follows. If it selects a path that traverses router V_1, the ensuing cost of such a path would be equal to $C(i, V_1) + d_{V_1}(j)$, whereby $C(i, V_1)$ represents the cost value of link (i, V_1). Similarly, *Router − i* calculates its current estimate of the cost value of a path to *Router − j* that traverses router V_2. The ensuing cost value of the latter is equal to $C(i, V_2) + d_{V_2}(j)$. *Router − i* then determines the lowest of the latter two cost estimates and accordingly selects the preferred neighboring router to use based on information received at this time, and calculates the corresponding best-known distance value.

In Fig. 15.5, if we assume that the link cost for each link is set equal to 1, then the minimum distance route is a path from *Router − i* to *Router − j* that has the least number of hops. In this case, we note that the route that traverses *Router − V_1* spans 4 hops while the route that traverses *Router − V_2* spans 3 hops. Hence, the latter route is selected.

The principle used by a source router, identified as router V_S, to select the best route to a destination router, identified as router V_D, is thus described as follows. The source router considers in turn each one of its neighboring routers. The set of all of its neighboring nodes is denoted as A_{V_S}. It calculates the distance value to the destination node by considering all its neighboring routers. It uses the distance advertisements received from its neighbors and the cost values of its attached links. It selects the path that offers the lowest distance to the destination node. This calculation is repeated as nodes received updated distance vectors from their neighbors. If link weights do not change too rapidly, this operation will converge, so that each router then determines its best distance vector. The latter provides for the its setting of the forwarding entry members in its switching table. The ensuing calculation at each update stage is thus expressed as follows:

$$d_{V_S}(V_D) = \min_{v \in A_{V_S}}\{C(V_S, v) + d_v(V_D)\}. \tag{15.5}$$

Processes and data sets kept by router v as it executes a distance-vector routing algorithm include the following:

1) The router's **link state vector**, denoted as $\mathbf{C_v}$, consists of link costs (per each link metric) of each link attached to node v. Thus, $\mathbf{C_v} = \{C(v, u)$, for each neighbor $u \in A_v\}$, whereby $C(v, u)$ represents the cost value of link (v, u).

2) The distance vector $\mathbf{D_v}$ displays the current value derived by node v of the distance (per each metric) to each node in the underlying domain N. It contains the following entries (per each metric): $\mathbf{D_v} = \{d_v(z)$, for each $z \in N\}$, where $d_v(z)$ is the distance to destination node z (i.e., the metric value of the shortest path from v to z).

3) The distance vector $\mathbf{D_u}$ (per metric) announced by each node u that is a neighbor of node v, $u \in A_v$.

4) The metric value entries in distance vector $\mathbf{D_v}$ are calculated at each node in the domain by using the distance vector algorithm, using updated distance vectors received from its neighbors and its latest link state values, through application of the calculation delineated by Eq. 15.5.

5) This process and ensuing disseminations and calculations are repeated at each node on a periodic basis and as it receives updated distance vectors from its neighbors. When the performed distance vector exchanges and calculations converge, the targeted best distance vector is obtained, leading to an updated setting of the corresponding forwarding entries in the routing table.

6) The routing table at node v, denoted as $\mathfrak{R_v}$, contains entries that represent the outgoing port to be used for reaching destination node z for each such node in the domain. It often also includes the corresponding distance level $d_v(z)$, a perishability value, and other route attributes. Using the routing table, a node is able to determine the outgoing port (and thus also the outgoing link and the next router to be visited, if any) to which an incoming packet should be forwarded. If the destination node is outside the domain, a boundary node will be selected as an intra-domain destination node.

To further illustrate the functioning of a distance vector routing algorithm, consider again the network shown in Fig. 15.4(a). Consider an operation that initiates communications across a network domain, assuming all nodes to have just started compiling entries for their distance vectors and routing tables. Initially, each node is aware of the distance that it would traverse to communicate with each one of its neighboring nodes, using a path that is 1-hop long. For such a link, say (v, u), the single hop path consists of this link alone so that its distance value is equal to $d_v(u, 1) = C(v, u)$. Set $d_v(v) = 0$ as the distance between a node and itself. Define $d_v(u, h)$ as the h-distance from node v to node u, setting it equal to the length (or cost/metric value) of the shortest path from node v to node u that is no longer than h hops. If there is no such path from v to u, we identify the underlying distance by using the ∞ symbol.

The result of the calculation by node v of its h-distance values forms the **h-distance vector** $D_v(h)$, whose entries are equal to $d_v(u, h)$ for each node u in the domain. Through iteration over the hop parameter h, the $(h+1)$st distance values are calculated by node v by using the h-distance values that it receives from its neighbors:

$$d_v(z, h + 1) = \min_{u \in A_v} \{C(v, u) + d_u(z, h)\}, \quad \text{for each node } z \text{ in the domain.} \tag{15.6}$$

The progress of this calculation by iterating on the number of hops parameter (h) for the illustrative network of Fig. 15.4(a) is shown in Fig. 15.4(b)–(d). It is noted that each table shows the results of the calculations performed by each domain node. The table entries included in a row corresponds to those calculated by a specific node. For example, the entries in the first row represent the distance values calculated by node A.

The table in Fig. 15.4(b) contains the 1-hop ($h = 1$) distance entries, which represent the shortest paths between each node and its 1-hop domain neighbors. As noted above, $d_v(u, 1) = C(v, u)$. For this calculation, a node simply uses its link state values. It does not need to rely on advertisements received from its neighbors. Each entry in the shown distance tables which corresponds to a valid path consists of a pair of entries: the first entry states the distance (or cost metric) value of the associated h-path, while the second entry identifies the preceding node, which is the node that precedes the underlying destination node along the associate best path. It is set to -1 if no such node is relevant, such as when considering a degenerate path that represents communications between a node and itself. For example, the 1-hop path between node A and node C consists of the link (A,C), whose distance value is equal to 4 (its cost value), and for which the preceding node is node A. Similarly for the other 1-hop entries.

In Step 2 of the algorithm, each node receives from each one of its neighbors its 1-hop distance vector, as well as sends to each one of its neighbors its own 1-hop distance vector. Using this information, each node calculates its 2-hop distance values by using Eq. 15.6. The resulting values are included in the 2-hop table that is shown in Fig. 15.4(c). For example, Node A receives an announcement from its neighboring node B that the latter is 1 hop away from node C and that its distance to C is equal 1. Consequently, node A determines that a 2-hop path to node C that goes through node B has an A-to-C distance that is equal to 3. This distance is lower than the distance value assumed by the 1-hop A-to-C path, whose distance is equal to 4. Hence, the (A,C) distance entry that was set as (4,A) in the 1-hop table is replaced by the entry (3,B) in the 2-hop distance table, indicating that a better path has been found by using a path that assumes a distance value of 3 and that reaches node C via preceding node B.

In Step 3 of the algorithm, each node receives from each one of its neighbors its 2-hop distance vector as well as sends to each one of its neighbors its own 2-hop distance vector. Using this information, each node calculates its 3-hop distance entries using Eq. 15.6. The resulting 3-hop table is shown in Fig. 15.4(d). For example, node A receives an announcement from its neighboring node B that the latter has calculated a 2-hops least cost path to node E whose distance is equal 3. Consequently, node A determines that a 3-hop path to node E that goes through node B would assume an A-to-E cost value that is equal to 5. This cost is lower than the one that leads to node E through neighboring node C whose distance is equal to 4+3 = 7. Hence, the (A,E) distance entry in the 2-hop distance table that was set in the previous step to (6,B) is replaced in the 3-hop distance table by the entry (5,D). It indicates that a better path to use to reach node E would go through preceding node D, and that it assumes distance value 5.

The iterative process terminates after the execution of the above noted 3 steps since the diameter of this graph (representing the hop length of the longest shortest path) is equal to 3. Hence, after 3 steps, the hop-oriented iteration has exhausted all possible paths. The corresponding SPT graph rooted at source node A (as derived by router A following interactions with its neighboring routers after the 3 described iterative steps) is shown in Fig. 15.4(e), while the SPT rooted at router E is shown in Fig. 15.4(f).

Thus, if a packet arrives at router A indicating that its destination address is that of router E, router A examines its SPT to find out the shortest path to E to be A–B–D–E. It sets the forwarding entry of this packet in its nodal routing table to point to the output interface port that connects router A to router B. This packet will be switched to the latter output port. As this packet reaches router B, the latter will examine its own SPT (and corresponding entry in the routing table) and proceed to switch the packet to an interface that leads to router D. Similarly, at the latter router, the packet will be switched to an output port that connects to router E.

A key advantage of a distance vector routing algorithm relates to the requirement that each router sends its distance vector to only its neighboring routers rather than to all of the domain's routers. The involved computational process is quite simple. This method was used for routing in the early Internet, noting that the system was then link communication capacity and nodal processing limited. In turn, the computational complexity of the algorithm, expressing the worse-case number of computations required by a router to build up its distance vector and routing table through interactions with its neighbors, can be observed to be of the order of n^3, where n denotes the number of domain nodes or routers. While this is a relatively high computational effort, it is noted that often the underlying domain's graph topology is sparsely connected so that the actual computational complexity level is lower.

Issues relating to the functioning of a distance vector algorithm include the possible occurrence of routing instabilities, route cycles, and time delays incurred in percolating around the network link and nodal status changes and occurrence of failure events. The distance vector routing method requires a router to process distance vectors that it receives from its neighbors. Vectors are dynamically updated as nodes process modified distance vectors that they receive from their neighbors. As updates may be induced by widely distributed change events, the time that its takes to converge to a final stable distance vector and routing table can be relatively long. A change in the status of a link connecting a pair of routers would take time to propagate across the network. It will thus induce delays in updating the distance vectors of routers that are located multiple hops away. While nearby routers may promptly detect a change event and updated distance values, identifying that a link attached to a neighboring node has become highly congested, or has failed, it will take longer time for this change to propagate to further away nodes, impacting the accuracy of their calculations and the contents of their routing tables during certain periods of time. The times elapsed between state variations can be short while longer time delays may be incurred in propagating changes to farther away routers. Such temporal variations cause state changes to be taken into consideration at different times by different routers, resulting in possible inconsistencies between distance vector calculations and updates that are performed at different routers, possibly leading to mis-routing actions and to routing cycles.

The following simple scenario illustrates the occurrence of a routing loop in response to a failure event, leading to what is known as *counting to infinity* calculation of a route's cost. Consider, as shown in Fig. 15.6, a path A–B–C of 3 routers connected in tandem (i.e., serially). Assume that the link cost for (A,B) is equal to 1 and that the cost of link (B,C) is also equal to 1. Each node proceeds to determine its distance vector. Node A will use the advertisement sent by node B indicating that it is at a distance 1 from C. Node A is at distance of 1 from B. Therefore, it determines its distance to node C to be equal to 2, using node B as the next hop node. Assume now that the link connecting nodes B and C has failed. Consider the path to be selected by node B in aiming to reach node C. The distance of this path used to be equal to 1, using the direct link. Assume that node A is not aware of the failure of link (B,C). Hence, it keeps advertising a distance of 2 to node C, going through node B. Following the failure of the latter link, node B uses this advertisement to determine that it is at a distance of 3 from node C, as it selects a path that goes through node A. Subsequently, receiving the latest distance vector sent by node B, node A adjusts its distance to node C (through node B) to increase to 4, and so on. Proceeding this way, the underlying distance keeps increasing, and the routing algorithm fails to converge and to yield the desired SPT.

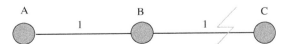

Figure 15.6 A Tandem Path That Connects Routers A, B, and C.

470 | *15 Routing: Quo Vadis?*

The cause for such instability is rooted in a number of factors, including the lack of detailed path information associated with the distance advertised by a neighboring node to a destination node, and the ability of the mechanism to allow the selection of a path from node B that goes through node B itself, creating a routing loop.

Several approaches have been developed over time to mitigate the occurrence of routing loops. Approaches include: Holding down the dissemination of updates as the distance values keep increasing, which delays the speed of convergence of the algorithm; use of a *split horizon* scheme under which a router does not advertise a destination through its next hop (so that A would not advertise to node B its route to destination C); use a *poison reverse* approach, so that node A advertises to B its distance to C to be very high.

In a large domain, the implementation of a distance vector-based routing scheme can induce high communications, processing, and storage requirements. Each router would have to store large distance vectors rooted at itself and at each one of its neighbors. The router needs to periodically engage in distance computations (from itself to every node in the domain), storing its results and disseminating them periodically to its neighbors.

Intra-domain routing algorithms that are based on a distance vector method include Routing Information Protocol (RIP) and a protocol developed by CISCO Corp. known as EIGRP. The latter employs a scheme for the prevention of routing loops, reduces communication rates by only disseminating status changes, and allows the use of multiple distance metrics as well as the use of combined (hybrid) metrics.

Frequently employed capable intra-domain routing algorithms tend to be based on the concept of link-state routing, which is described in Section 15.7. A distance vector routing-type approach is used in the implementation of inter-domain routing schemes, such as the BGP algorithm. The latter forms the basis for inter-domain routing across Autonomous System (AS) domains in the Internet, and is discussed in Section 16.6.3.

15.7 Link-State Routing: Obtain the Full Domain Graph

Under a **link-state routing algorithm,** a router obtains the complete topological map of the routing domain's graph and uses it to determine the best routes from itself to each other node in the domain. To obtain the full map of the domain's graph, each router **broadcasts its link-state vector to each other router located in its domain**.

A router's link-state vector contains the cost (metric) values associated with each one of the links attached to the router. For example, for the network domain that is represented by the weighted graph shown in Fig. 15.7(a), two links are attached to router node A: link (A,B) displays a cost value that is equal to 2 and link (A,C) shows a cost value that is equal to 4. The link-state vector disseminated by node A to all the other domain nodes would then identify these two links (i.e., identifying B and C as nodal neighbors of node A) and specify the cost associated with each link. It is possible for the cost metric to be represented as an array (vector) of different metric values; for example, specifying values for link metrics that can represent throughput capacity, delay and reliability. As performed by a router when using a distance-vector routing algorithm, each node continually identifies its neighbors and records the status and performance features of each link that connects itself to each one of its neighbors and accordingly record the corresponding metric values for each one of its attached links.

A message that carries a router's link-state vector and is disseminated to all domain router nodes is known as a **Link-State Advertisement (LSA)**. A broadcasting protocol is employed to disseminate an LSA message produced by a router to all other domain routers.

15.7 Link-State Routing: Obtain the Full Domain Graph

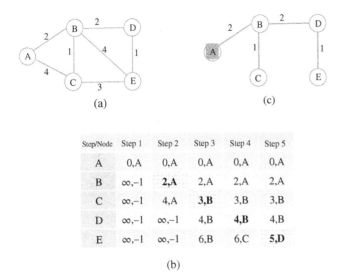

Figure 15.7 Calculation of the Shortest Path Tree (SPT) by Using a Link State Routing Algorithm: (a) The Weighted Graph Representing the Network's Domain Layout; (b) Step-by-Step Calculation Performed at Source Node A by Using Dijkstra's Algorithm; (c) The Resulting SPT Rooted at Router A.

A **flooding** scheme is a simple routing method that can be used for this purpose. Under this method, a router that receives an LSA message across an attached link proceeds to transmit this message across each one of its output links to all its neighboring routers. When a router detects that it has already received the same message, such a duplicate incoming message is discarded.

LSA messages are broadcast by their source routers periodically and when specified status variations take place. LSA messages issued by the same source router at different times are differentiated by marking them to having distinct source-based sequence numbers. A message that carries a source-based higher sequence number has been issued more recently than one that carries a lower source-based sequence number. A router replaces a *stale link-state information* with a source-based LSA message that is identified by a higher sequence number. Since LSA messages may experience random time delays as they are disseminated across a network, it is possible for a router to receiver an old LSA message after it has received a newer one. The older message will be deleted.

Other broadcasting methods may also be employed in aiming to reduce the data loading level imposed on the network by the use of a flooding dissemination process. For example, a network system may configure and announce a subnetwork (such as a spanning tree graph) that is employed for the purpose of broadcasting control data.

In a wireless data network, where nodes use omnidirectional antennas (so that a single message transmission by a node can reach multiple neighboring nodes), a selective set of neighboring nodes can be elected to relay an LSA message issued by a router. Under one such selection algorithm, such a selected node is identified as a *multi port relay (MPR)*. Another approach is to synthesize a subnetwork whose nodes form a *connected dominating set* and to use nodes in the latter set to disseminate LSA messages. Nodes that are members of this set are connected to each other while each node that is not a member is a neighbor of a member. Such algorithms will be discussed in Chapter 19 in which methods for networking in mobile Ad Hoc wireless networks are presented.

A router that receives LSA messages from all other routers in its domain is able to construct a mathematical model of the network's weighted graph, such as the one displayed in Fig. 15.7(a). The router then uses a routing algorithm to calculate a SPT rooted at itself, which identifies the shortest (i.e., least cost) path from itself to each other node in the domain.

472 | *15 Routing: Quo Vadis?*

Distinct SPT graphs are derived for each link cost metric. Using its derived SPT, the router composes the forwarding entries of its routing table, enabling it to switch an incoming packet to an outgoing port that connects it to the next router that is located across a lowest cost path that leads to the packet's destination node. Under a **next hop routing** process, the next router would perform a similar calculation in determining the subsequent next hop. Under a source routing method, the source router includes, in the header of each packet, the identity of the routers that are located across its selected end-to-end path. Under a **partial source routing** method, the source router would specify a subset of the routers that a packet's end-to-end path should include.

A commonly employed algorithm that is used by a router to calculate the SPT for its acquired domain weighted graph has been developed by Dijkstra, a Dutch computer scientist, and is known as **Dijkstra's Algorithm**. This algorithm performs as shown in Fig. 15.8, whereby the SPT rooted at source node $s = A$ is calculated.

The algorithmic process is explained as follows. At each step of the algorithm, each node is identified by a label that consists of two components: the first component at a node, say node v, is denoted as $d(v)$ and it expresses the length (or any selected metric value) of a path from source node s to this node v, as learned from the most recent calculations. Before the termination of the algorithm, the latter value expresses the length of currently selected path from s to v. At the termination of the algorithm, this label component's value would identify the length of the shortest path (i.e., lowest metric value) from node s to node v, identified as the distance value $d(v) = d_s(v)$. The second component of a temporary or final label at a node v is denoted as $pred(v)$. It represents the node that is (at the current step of the algorithm) the predecessor of node v across the underlying (temporary or final) path from node s to node v. It is the node that precedes destination node v across the underlying path from node s to node v. The label at each node v thus consists of the pair $(d(v), pred(v))$.

As shown in Fig. 15.8, the algorithm starts by going through an initiation setting, under which the label of source node s is set to $(0, 0)$, indicating that node s is at distance 0 from itself and that the predecessor node is the node itself. The label of each one of the other nodes is set to $(\infty, -1)$, indicating that at the start of the algorithm, the distance to any node that is not the source node s and its predecessor node are not known. A set of nodes G is defined. It consists of all nodes that, at the start of each step of the algorithm, have not yet been selected by the algorithm as pivot (or visited) nodes, so that their final labels have not yet been calculated. Initially, all nodes are included in G, so that we set $G = V$, where V denotes the set of all network nodes (modeled as vertices in the domain's graph).

Initiation:
G = V
d(s) = 0; pred(s) = 0.
For all other nodes u,set d(u) = ∞ and pred(u) = –1.

Main Routine:
As long as set G is not empty, proceed as follows:
      Select node u in G such that its current temporary label d(u) assumes a *minimal* distance value;
      This node is identified as the *currently visited* node.
      Update G by removing the selected currently visited node u from the set G.
Update the temporary labels of all nodes v in G that are neighbors of the currently visited node u as follows:
      If d(v) > d(u) + C(u,v) then
         replace d(v) with the value d(u) + C(u,v) and set pred(u) =v. (*)
  If set G is not yet empty, select the next node in G to be visited, extract it from G,
  and update the temporary labels of its neighboring nodes, if any, following (*).
The algorithm terminates when set G becomes empty.

Figure 15.8 Dijkstra's Algorithm for Calculating the Shortest Path Tree (SPT).

At the start of each step of the algorithm, set G is examined. A node whose current label's distance component is minimum, in comparison with the corresponding distance components of the current labels of all other nodes that are currently included G, is selected as the next pivot node. The selected node, say node u, is then removed from set G. All the nodes that are neighbors of this node u, and are still members of G are then examined. Let v be such a neighboring node. The label of neighboring node v is examined. If its current length component $d(v)$ can be reduced by selecting a path that traverses through visited node u, so that $d(u) + C(u, v) < d(v)$, then the current value of $d(v)$ is replaced by the lower value $d(u) + C(u, v)$, and the identity of the predecessor on the improved path is identified as $pred(v) = u$.

As the algorithm progresses, and all nodes have been selected as pivot nodes, set G becomes empty and the algorithm terminates. The final labels attached to each node identify the distance from s to this node and the predecessor node along a minimum length path to this node. The resulting final labels yield the SPT.

This process, including all the steps followed in executing Dijkstra's Algorithm, is illustrated in Fig. 15.7(b) for the weighted graph shown in Fig. 15.7(a). The source node is identified here as node A. By examining the final step (Step 5 in this case), one identifies the structure of the SPT, which is shown in Fig. 15.7(c). Observing the SPT, it is noted that the shortest path from source node A to node E assumes the label $(5, D)$, so that its length is equal to $d_A(E) = d(E) = 5$, and the predecessor node along the shortest path (which consist of the tandem set of nodes A–B–D–E) is node D.

The computational complexity that characterizes this algorithm, relating to the number of key computations that are performed by a router whose domain graph consists of n nodes, is of the order of n^2. By using a more efficient algorithmic structure, the computational complexity can be reduced, becoming of the order of $m \log n$, where m denotes the number of lines in the network's graph. Often, the domain's layout is represented by a sparse graph so that its number of lines may be of the order of n, rather than of the order of n^2.

QoS routes can be determined by using a modified link state routing method. For example, using the acquired up-to-date-weighted graph of the routing domain, a router can prune the graph by removing communications links that do not satisfy prescribed performance objectives, and then calculate shortest length paths by using the pruned graph and applying the performance metric of interest. For example, to route packets of a new flow that require a route that offers an acceptable data throughput rate while also providing a low packet delay level, the routing algorithm could prune all links in the domain graph that do not currently offer a residual communication capacity level that satisfies the new flow's requirement and then employ Dijkstra's algorithm to calculate the SPT for the pruned graph, with link weights that correspond to values assumed under a delay metric.

Dijkstra's algorithm has been widely employed by link state routing protocols. For IP-based packet-switching networks, OSPF is a commonly employed link-state routing protocol. IS-IS is another link state routing protocol. It has been commonly employed for configuring routes in also layer 2 networks.

Problems

15.1   Define and discuss each one of the following route metrics. Also, provide examples of flow types that would benefit by being routed by using these metrics.

a) Delay.

b) Throughput capacity.

474 | 15 Routing: Quo Vadis?

c) Reliability.
d) Monetary cost.
e) Weighted hybrid metric.

15.2 Explain the concept of routing by segmentation into multiple routing domains. Why does it result in a more scalable routing operation? What are the ensuing advantages and disadvantages when using such an approach?

15.3 Define the following:
a) Autonomous System (AS).
b) Interior Gateway Protocol (IGP).
c) Exterior Gateway Protocol (EGP).
d) Boundary router.

15.4 Describe the principle of operation of the following routing schemes:
a) Next Hop Routing.
b) Source routing.
c) Partial source routing.
d) Routing via the setting of Label Switching Tables.
e) Routing across a light path in a Wavelength Division Multiplexing (WDM) network.
f) Reactive on-demand routing vs. proactive routing.

15.5 Define the concept of a Shortest Path Tree (SPT). How do you compute an SPT for a given network system?

15.6 Explain why a collection of shortest paths from a source node to other nodes in a routing domain can be represented as a Tree Graph. Can you prove it?

15.7 For the graph shown in Fig. 15.4(a), confirm that a Shortest Path Tree rooted at node E can be represented as the graph shown in Fig. 15.4(f).

15.8 Explain the principle of operation of Distance Vector Routing.

15.9 Use a Distance Vector Routing method to derive the routing table presented in Fig. 15.4(d).

15.10 Explain the principle of operation of Link-State Routing.

15.11 Consider an intra-domain routing scheme that employs a link state routing method.
a) Define the contents and purpose of *Link-State Advertisements (LSAs)*.
b) Discuss methods that can be used for LSA dissemination over a network domain.
c) Describe how the use of a Multi-Port Relay (MPR) can be used to reduce the loading of a network by LSA message flows.

15.12 Use Dijkstra's algorithm to derive the Shortest Path Tree (SPT) rooted at Node C for the network graph shown in Fig. 15.7(a).

16

The Internet

The Concept: *The Internet is a worldwide public communications network that operates as a connectionless packet-switching network. Messages are segmented into layer-3 packets that are transported across the Internet in a store-and-forward (S&F) manner. Communications links are shared on a statistical multiplexing basis, so that link capacity is used by a flow only when it is active. A router examines the address field of a received packet and switches it to an outgoing network. A flow starts at a source end user that connects to a designated router. Flow packets traverse intermediate routers and networks until they reach their destination router, which transfers the packets to the destination end user.*

    The Method: *The Internet's protocol (IP) architecture is based on the Transmission Control Protocol (TCP)/IP-layered protocol model. An end-user device wishing to communicate over the Internet with another end-user device accesses a designated router across a communications link, a LAN, a layer-2 switch or a local switching network. While communications across the latter is carried out by transporting link-layer frames by using MAC addresses, message transport across the Internet is carried out by layer-3 messages known as packets. The format of packet data units and the methods employed for their transport across the Internet are governed by the addressing and routing rules imposed by the IP. IP Version 4 (IPv4) packets use IP source and destination addresses that are 32 bits long, while IPv6 addresses are 128 bits long. For scalability purpose, routing domains, known as Autonomous Systems (AS) are defined, where each employs a well defined routing policy. Intra-domain routing protocols (such as Open Shortest Path First (OSPF) that employs a link-state routing algorithm) and inter-domain routing schemes (such as Border Gateway Protocol (BGP) that employs a path-vector-based routing algorithm) are used by routers to dynamically route packets. The employed networking scheme does not provide a reliable packet transport service. Corrupted or lost packets are not automatically re-sent at the network layer. When such end-to-end reliability is desired, a connection-oriented transport layer protocol (TLP), such as TCP, is employed. Otherwise, a connectionless TLP, such as User Datagram Protocol (UDP), is employed. These TLPs include application port numbers, which identify the destination and source application layer entities that respectively consume and produce the application data that is carried across the network between a flow's source and destination end-user entities. The QUIC protocol provides a connection-oriented transport layer operation over UDP. It can be used by application layer entities that reside at interacting client and server end users, such as clients that employ a secure HTTP protocol to communicate over the Internet with a web server. QUIC transport connections include data encryption and offer lower transport latency levels. Multiple application streams can be multiplexed across a single QUIC connection, while separately managed. QUIC packets are transported across the network within UDP datagrams.*

Principles of Data Transfer Through Communications Networks, the Internet, and Autonomous Mobiles, First Edition. Izhak Rubin.
© 2025 The Institute of Electrical and Electronics Engineers, Inc. Published 2025 by John Wiley & Sons, Inc.

16.1 The Internet Networking Architecture

The Internet provides worldwide connectivity among networks, computers, and end-user devices. Its layered protocol architecture is based on the TCP/IP reference model, which is illustrated in Fig. 1.6. The concept and functioning of a layered communications network architecture, including the OSI and TCP/IP protocol reference models, is discussed in Section 1.5.

A network that employs the TCP/IP protocol architecture, such as the Internet, provides a **connection-less network layer service using a datagram packet-switching method**. The network layer addressing and routing scheme, and the format of the data units that are routed across the network, identified as **packets**, are based on the **Internet Protocol (IP)**.

End-user devices access the Internet by sending their packets to an Internet Router across an access network to which they are attached. As shown in Fig. 1.6, Physical Layer and Link/MAC Layer transmission and networking services are provided by systems that employ protocols that reside at the lowest two layers of the TCP/IP model. Layer-2 networks are often used to network within a user's facility as well as provide access to an Internet router, which in turn provides access to the Internet.

A layer-2 network may consist of just a single point-to-point communications line, as is often the case. Layer-2-based mesh network configurations are also commonly used, including layouts of interconnected layer-2 switches or bridges. Wireline and wireless Local Area Networks (LANs) are frequently employed, including Ethernet (often synthesized by using an Ethernet switch or hub) and Wi-Fi wireless LANs (WLANs). Access nets that consist of multiple access media are also commonly used. Included are wireless radio access networks (RANs) such as those used by satellite networks, cellular wireless networks, and autonomous vehicle networks.

A flow of messages between end users across a communications network that is governed by the TCP/IP protocol architecture is illustrated in Fig. 1.6. The application layer entity at a source end-user formats its message in accordance with an underlying application layer protocol. Several application layer protocols used by end users that communicate across the Internet are shown in Fig. 16.1. Highlighted are long standing applications, including such that have been in use across

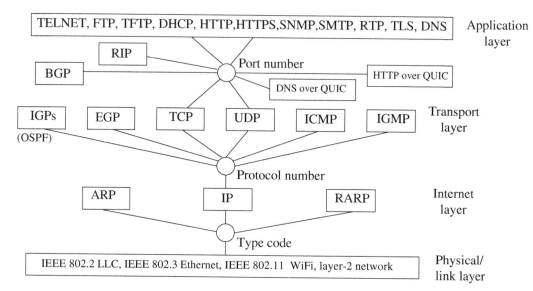

Figure 16.1 The Internet's Protocol Architecture and Commonly Employed Protocols.

the ARPANET and subsequently employed over the Internet. A wide range of many application layer protocols have been developed and are being constantly introduced, encompassing all scopes of individual, social, commercial, educational, medical, governmental, scientific, technical, and other scopes of everyday life.

Application layer messages are exchanged between end users. Telnet is an application that supports terminal to computer interactions across a network. File Transfer Protocol (FTP) is used for the transfer of files across a network, and Simple Mail Transfer Protocol (SMTP) is used for email messaging. The Dynamic Host Configuration Protocol (DHCP) is employed by an end-user device to attain an IP address, which is needed to provide for communications with other end-user devices across the Internet. Hypertext Transfer Protocol (HTTP) enables browsing of resources that are available Internet wide, known as the Web; HTTPS is its secure version. Simple Network Management Protocol (SNMP) is used to manage network entities, including the monitoring of a Management Information Base (MIB) of a network device (such as that associated with a switch or a router) and the setting or modification of its prescribed operational parameters. Real-Time Protocol (RTP) is used to add control data (such as sequence numbers and time stamps) to packets that are included in realtime streams such as voice and video. Transport Layer Protocol (TLP) provides for secure interactions between a client and a server over the Internet. The Domain Name System (DNS) provides for naming of Internet computers and other resources and services. It associates domain names with numerical IP addresses.

As shown in Fig. 1.6, the product of an application layer entity at a source end-user device is an application layer message, which is formatted in accordance with the rules of the underlying protocol. This message is passed down the protocol stack to the transport layer entity. The latter uses a TLP to provide transport layer services. Commonly employed such protocols include the connectionless User Datagram Protocol (UDP) and the connection-oriented Transmission Control Protocol (TCP). Also noted is QUIC, which is a TLP that is run over UDP; it embeds a security protocol such as TLS. The figure notes the use of HTTP and DNS over QUIC, observing that a specific Protocol Number identifies UDP while separate Port Numbers are associated with HTTP over QUIC and with DNS over QUIC.

Also passed to the transport layer entity, are tags that identify the source and destination entities residing at the application layer that are involved with the respective generation and consumption of the underlying application layer message. Tags that are used to mark the interface between layers are identified by the Open Systems Interconnect (OSI) Reference model as Service Access Points (SAPs). Certain such numbers, called well known SAPs, have been reserved for use in identifying interfaces that involve commonly used protocols. Under the Internet TCP/IP protocol model, SAP tags that identify interfaces between the application and transport layers are identified as **Port Numbers**. Well-known port numbers are employed when relating to commonly used application layer protocols.

UDP is a connectionless TLP. It provides for end-user to end-user multiplexing of application layer message flows by including at the header of each UDP message the relevant source and destination port numbers, which identify the corresponding application layer entities involved in the production and consumption of the message.

TCP is an end-user to end-user connection-oriented transport layer protocol. As provided by UDP, also TCP messages carry in their headers source and destination port numbers. Using these numbers, TCP is able to provide for multiplexing of application layer message flows. In addition, TCP provides end-user to end-user error and flow control services. It is therefore able to add reliability to a datagram-based network layer operation that is used in routing packets across the

Internet. Further details relating to the operation of these protocols and to the formats of UDP and TCP messages are included in Section 16.5.

The transport layer message produced at a source end-user device is submitted down the protocol stack to a network layer entity. The latter provides for switching and routing of transport layer payloads across a network. This layer is identified as the **Internet Layer**. It serves to switch and route packets among networks. End-user devices access the Internet at an associated internet router, such as that provided by an Internet Service Provider (ISP), by being directly connected to it or by reaching it across a private access network. The IP provides network layer services across the Internet. A SAP characterizing the interface between transport and network protocol entities is commonly identified by the Internet architecture as a **Protocol Number**. The message produced by a source network layer entity is identified as a **packet**. The packet's header includes information that is employed by Internet routers for routing the packet across the Internet to its destination node.

As noted in Fig. 16.1, the protocol number also identifies other processes/utilities whose entities reside at the transport layer, such as routing algorithms (e.g., Interior Gateway Protocol (IGP) and Exterior Gateway Protocol (EGP)), providing for interactions between routing protocols employed by routers distributed across the Internet (e.g., exchanging link-state or path-state data units). Also noted is the Internet Control Message Protocol (ICMP), which provides for the creation and dissemination of error messages and for performing network testing and loop-back processes.

As noted above, IP provides a connectionless (datagram) based packet-switching service in routing and switching packets across the network. An IP packet, also known as IP datagram, carries as its payload the transport layer message that it receives from the associated transport layer protocol entity at the source end user. Its header includes source and destination IP addresses that identify the packet's source and destination nodes (and, more specifically, access ports at these nodes). Also included are the underlying source and destination protocol numbers that identify the involved source and destination transport layer entities, such as UDP or TCP entities. The format of an IP packet is further discussed and illustrated in Section 16.4.

An IP packet is submitted at the source network/inter-network layer end user to the link-layer entity. A link-layer protocol entity constructs a link-layer frame. It provides services concerning the transport of a frame across a link's medium. Depending on the nature of the data that it carries and the reliability of the underlying communications channel, link-layer protocols can provide services such as multiplexing, error control, and flow control. When a multiple access link is involved, a Medium Access Control (MAC) layer protocol is employed as a lower sub-layer of the link-layer protocol, forming MAC frames. The payload of a MAC frame consists of a link-layer frame, at times formed by using the **Logical Link Control (LLC)** protocol, which carries an IP packet as its payload. The header of a MAC frame includes source and destination **MAC addresses**, also known as **physical addresses**. These addresses are employed for communications networking across a link or across a network that is configured as a logical link, as is the case for LANs whose frames are broadcast across their shared media. This is the case for versions of Ethernet and Wi-Fi LANs, a Cable-TV (CATV) network, a RAN of a cellular wireless network, or a communications uplink of a satellite communication system.

As noted in Section 9.3, MAC addresses are assigned by device manufacturers. They are formed as Extended Unique Identifiers (EUI), such as EUI-48 and EUI-64, and are managed by the Institute of Electrical and Electronics Engineers (IEEE). They are commonly represented in six groups of two hexadecimal digits separated by hyphens or colons.

In hexadecimal notation, a symbol represents a value that ranges from 0 to 15. Symbols with values that range from 10 to 15 are represented by the (capital or lower case) letters A through F.

For example, the MAC address FF:FF:FF:FF:FF:FF represents in hexadecimal notation the all-1's broadcast address, noting that F identifies the value 15, which in binary notations is represented as a series of four 1s. Since each hexadecimal digit represents 4 bits, each group of two hexadecimal digits represents a 8 bits value, also identified as an Octet or Byte. The EUI-48 MAC address consists of 6 such groups, assuming a 48 bits value.

The format of an Ethernet II frame is shown in Fig. 9.7. As noted in Fig. 9.7(a), 48 bits long destination and source MAC addresses are contained in the Ethernet frame following a preamble field.

Across a shared medium LAN such as Ethernet or a Wi-Fi wireless LAN (WLAN), a station that wishes to communicate with another station broadcasts its MAC frame across the medium. The intended destination station detects the frame's destination MAC address to be the same as its own address and subsequently copies the frame and processes it. If a source station has not yet acquired the MAC address of a destination station, but it possesses its IP address, the *Address Resolution Protocol (ARP)* is used to secure the needed MAC address. The Reverse Address Resolution Protocol (RARP) can be used to map a station's MAC address into its IP address.

Devices, computers, and servers in a private user or a company facility are often connected through a layer-2 network. Layer-2 switches and bridges then serve as the network's switching nodes. Frame forwarding tables are used to forward an incoming frame based on its MAC addresses to an identified outgoing link. As discussed in Section 9.3, a Shortest Path Bridging (SPB) algorithm may be used to route frames to their local destination devices.

Computers that are located within a facility, use a wide area network to communicate with computers located in remotely located facilities. This is the role played by a packet-switching core network such as the Internet. Routing of packets across the Internet is based on using source and destination IP addresses. Hence, to communicate across the public Internet, an end-user device must first acquire the IP address of its destination end-user device. It then directs the message to an IP router that can access the public Internet.

An end user connects to its designated router via a network that may consist of just a single point-to-point link, or via a LAN such as an Ethernet LAN (or switch), or via layer-2 switches or bridges across a layer-2 network.

The structure and operational principles and performance features of a connectionless packet-switching network such as the Internet are discussed in Chapter 12. As depicted in Fig. 16.2, routers are used to connect networks. Computers C1, C2, and C3 communicate with

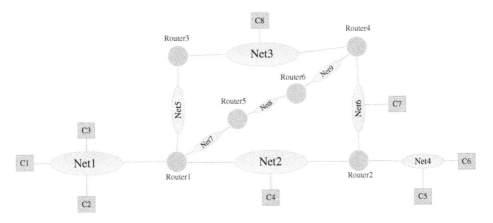

Figure 16.2 Illustrative Configuration of an IP Internetwork System.

each other across a local network such as Net1. To communicate with a destination device that is located outside Net1, an end-user device sends its packets to a designated router such as Router1. This router is shown to interface three other networks: Net2, Net5, and Net7. Consider a packet that is issued by computer C1 and is targeted to reach computer C7. The packet is sent across Net1 to a designated router, Router1. Assume C7 to have Router4 as its designated Internet access router, which it accesses by transmitting its packets across local network Net6. Router1 determines C7 to be attached to network Net6, and the latter to use Router4 as its designated router. Router1 then selects a preferred route across the IP network to Router4. For example, one possible route traverses Router5 and Router6, which are connected across the corresponding attached networks (or point-to-point links) that connect the involved routers. Another possible route traverses Router3 and associated networks Net5 and Net3.

Internet services are provided by ISPs. These organizations provide access to the public Internet as well as implement and manage the many networks that are responsible for packet transport across the Internet. They also provide other services that are essential to the functioning of the Internet, including domain name registration, web hosting, storage, as well as certain virtual server and cloud services. Their networks make up the **Internet Backbone**, which is responsible for the transport of packet flows across the Internet. The underlying core routers and data routes are hosted by a wide range of networking organizations, including commercial, government, and academic organizations.

The Internet backbone consists of a large number of networks owned by many companies of diverse size, bandwidth capacity, and coverage span. As illustrated in Fig. 16.3, ISP networks are hierarchically organized. The largest providers, known as *Tier 1 Service Provides*, or Network Service Providers (NSPs), operate *Tier 1 Networks*, in offering wide scope coverage.

Peering arrangements allow packet flows to traverse several NSPs. The latter are interconnected at Internet Exchange Points (IXs), also identified as Network Access Points (NAPs), and Metropolitan Area Exchanges (MAEs). Each NSP is generally connected to multiple NAPs. Network service providers include telecommunications companies, data carriers, wireless communications service providers, ISPs, CATV and fiberoptic based network operators, and a multitude of governmental and commercial companies.

NSPs sell bandwidth to smaller ISPs, which offer limited bandwidth and reduced coverage. Tier 3 ISPs provide local Internet access to end-user computers and devices and to end-user local facilities. Locally based end-user facilities may be able to interconnect across a network operated by a tier 3 ISP. In turn, if the destination address of an end user cited by a packet is not handled by a local

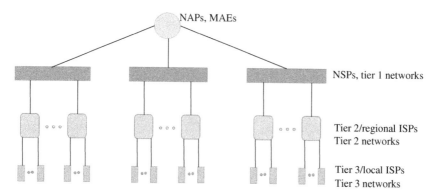

Figure 16.3 Hierarchical Organization of Internet Service and Backbone Networking Providers.

ISP, the packet is routed to a regional tier 2 ISP. The latter offers wider scope coverage and may be able to provide a network connection through its tier 2 network, which leads to a tier 3 network that serves a tier 3 ISP that is able to connect with the destination end user. If the regional ISP of interest is not able to reach the destination of interest, the regional ISP would route the flow to a tier 1 network, which can provide flows with high capacity global Internet coverage, often through peering with other tier 1 networks. Under such an hierarchical arrangement, tier 3 ISPs purchase bandwidth and regional coverage from tier 2 ISPs, while the latter purchase bandwidth and global Internet coverage from tier 1 ISPs.

In 2023, a total of more than about a couple of thousands ISP businesses were noted in the United States, an annualized increase of about 4% over a period of 5 years. Worldwide, a large number of ISP businesses have been recorded in 2023, an increase of about 5% from 2022. There are only about a dozen of global tier 1 ISPs.

To illustrate, the following businesses, among others, operate large tier 1 networks: AT&T (Autonomous System (AS)# 7018) employs fiber routes across a distance span of 660,000 km; Verizon Enterprise Solutions (formerly UUNET, Autonomous System (AS)# 6461) covers a fiber based span of about 805,000 km; Lumen Technologies (Autonomous System (AS)# 3356) provides fiber routes across 885,139 km.

The ISP connectivity architecture is in reality more complex than that illustrated in Fig. 16.3. An ISP may have multiple access nodes, each known as a *Point of Presence (PoP)*. An ISP may have multiple connections to an upstream ISP at multiple PoPs. An ISP's PoP may be connected to multiple PoP nodes of multiple upstream ISPs. Multiple redundant connections are used to provide higher operational reliability and data integrity. This structure also enables the use of adaptive congestion control and dynamic routing algorithms.

End users, including private users through their home based computers or mobile wireless devices, as well as business users that are attached to an organizational network, access the Internet backbone through a myriad of access links and networks. This is illustrated in Fig. 16.4. Users located in a company facility may use a facility router to connect to a router operated by a tier 3 ISP network via a point-to-point communications fiber or Ethernet link, when such a connection is available. Otherwise, access networks are used to facilitate the access of end users and organizational facilities to the Internet. Such networks may use communications media that include fiberoptic links, wireless channels, CATV plants, high-speed communications lines leased from a telecommunications carrier, and satellite links. It is noted that connections between a user

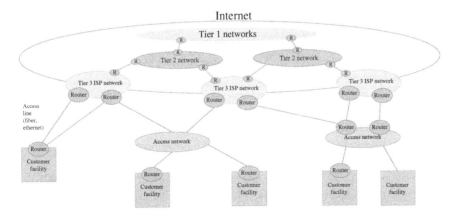

Figure 16.4 Illustrative Internet Access and Backbone Networks.

482 | *16 The Internet*

facility and its access network, as well as between the latter and its ISP may be layer-2 based, using for example CATV links, wireless system links, telephone links, and others. They may also be layer-3 based, employing routers.

A router attached to an ISP network is capable of routing across the backbone of the (public) Internet packets that it receives from end-user devices through access networks or links. It also serves as the destination of packets routed across the Internet that it will deliver to a router or layer-2 networking device located in an access network to which the destination end user is connected. Routers that are located at access networks are used to connect with ISP routers. Routers that are located at an access network and at a customer network are used for communications between an end user and an access network. Routers may also be used at user facilities to route packets across a customer's private layer-3 packet-switching network. In Sections 16.2–16.6, we provide further details of the principles of communications networking across the Internet.

16.2   HTTP: Facilitating Client–Server Interaction Over the Internet

As a client of a website server, you use a web browser to request data from the server. In response, the server would send you a message that contains the requested data. Such client–server request-response interactions are facilitated by using an application layer program known as HTTP. Its secure version is identified as HTTPS. This protocol defines the formats and associated semantics of such request and response messages.

An HTTP message created by a client or by a server is transferred over the Internet by using the services of a transport layer protocol, such as TCP or QUIC (that runs over UDP), which forms a transport layer connection across which transport layer messages are sent between communicating end users. Transport layer messages are routed across the Internet by using the IP network layer protocol.

HTTP was initially developed in 1989 by Tim Berners-Lee at CERN. HTTP protocol specifications have been subsequently defined by the Internet Engineering Task Force (IETF) though the publication of Request for Comments (RFC) documents.

HTTP/1 was presented in 1996 as version 1.0, and then as version 1.1. Advanced versions, including such that make use of transport layer security (TLS) programs, have then been developed. HTTP/2, which was published in 2015, was developed to produce faster interactions, and it has been widely employed. HTTP/3 has been developed in 2022 to provide even faster, secure, more robust interactions and to support mobility induced dynamically changing connections.

A user wishing to interact with a server initiates a session via a login process. During a session, a user may produce multiple requests followed by the reception of corresponding responses. The operation is *stateless*, so that the server does not keep a state for an ongoing dialog. To maintain data that may be needed throughout the session, HTTP cookies or other variables may be kept. Interactive authentication is undertaken via the login process.

A typical **HTTP request** message consists of the following fields: version type, a URL, an HTTP method, headers (including identification of the client's browser and the data requested), and the HTTP body, which a client uses to include the data passed to the server, including security parameters such as user name and password.

A Uniform Resource Identifier (URI) identifies a resource on the Internet. Its name and/or location can be used. A Uniform Resource Locator (URL) is a URI which enables a client to reach a resource on the Internet, as it specifies the protocol to be used and its name/number. For example, UCLA.edu is a URI that identifies the name of a resource. In turn, http://ucla.edu is a URI that is also a URL. It points to a network host that can be reached by using HTTP. For example, the URL https://www.ee.ucla.edu/izhak-rubin indicates that the file izhak-rubin that is located at a network host whose domain name is www.ee.ucla.edu is reachable by using the HTTPS protocol. The URL components are defined in an IETF RFC 2396 document (http://www.ietf.org/rfc/rfc2396.txt).

A typical format of a URL for HTTP or HTTPS is structured as follows:

$$scheme : //host : port/path?query.$$

It thus includes the following key parts:

1) A **Scheme**, which identifies the protocol that is used across the Internet for accessing the target resource, such as HTTP or HTTPS.
2) A **Host**, whose name identifies the entity (computer) that holds the target resource. It can be followed by a port number, which identifies access to the application entity residing at the destination end point. For access to HTTP and HTTPS, well-known port numbers are typically used so that they do not need to be specified explicitly as part of the URL.
3) A **Path**, which identifies a specific resource in the host that is being requested.
4) A **Query string** may be used to specify search parameters for the cited resource or to specify data to be processed by the targeted resource.

An **HTTP response** includes an HTTP status code (such as indicating whether the request was performed successfully, is in error, or has been redirected), HTTP response headers, and optional HTTP body. An HTTP response header includes identification of the format of the data carried by the response. The body of an HTTP response to a request for a webpage (known as a GET request) would include HTML data that is used by the web browser of the requesting client to produce and display the received webpage.

Thus, a client that wishes to access a resource at a server creates a request message that it sends to the server. A resource may consist of a file that the client wishes to download. Under HTTP, actions requested by a client to be performed by the server are identified as **methods**. The following methods have been specified in IETF RFC 9110, except that the PATCH method has been specified in IETF RFC 5789:

1) **GET**: Retrieve resource data.
2) **HEAD**: Retrieve just the representation of the data and not the data itself.
3) **POST**: Post data specified in the request message.
4) **PUT**: Update or create a state at a location specified in the request based on the data included in the request.
5) **DELETE**: Request to delete a state.
6) **CONNECT**: Request to establish a connection via an intermediary, such as through an HTTP proxy.
7) **OPTIONS**: Request for the transfer of supported HTTP methods.
8) **TRACE**: Request for the destination resource to include the request inside the response.
9) **PATCH**: Request for the destination resource to make partial changes in the state in accordance with the data included in the request.

HTTP/2 has introduced several modifications to HTTP/1.1, leading to enhanced operation and faster interactions, including the following:

1) **Stream Multiplexing**: Upon receiving a request from a client, a server would often proceed to send several streams of data to the client. The streams are multiplexed over a transport layer

connection, which in the case of HTTP/2 would be a TCP connection. Each stream is linked to its associated application. Different streams can be associated with different priority weights, which are used to indicate to the client the order at which data streams should be rendered.
2) **Header Compression**: To reduce the size of HTTP messages, message headers are compressed. Under HTTP/2, a more efficient compression scheme, known as HPACK, is used.
3) **Server Push**: A client's request would often require a server to download multiple resources. To accelerate the download process, the server can invoke a "push" action, proceeding to send associated resource data to the client before the client requests for it.

HTTP/3 was presented in IETF RFC 9114 as a proposed Standard in June 2022. It uses semantics that are similar to those employed by HTTP/1.1 and HTTP/2. Its message formats consist of the same fields and methods. In turn, it uses **QUIC as the transport layer protocol**. As shown in Fig. 16.5, QUIC runs over UDP rather than over TCP. Under QUIC, interactions of a client with a server are executed much faster. It has been implemented by many leading websites.

The use of QUIC improves the performance of HTTP/3 in several ways, as outlined when describing the QUIC protocol. The following features are noted:

1) **Faster Handshaking Process**: For a client to download desired data from a website's server, the client configures a connection with the server. The corresponding handshaking process is used by the communicating end-user entities to exchange transport layer parameters.
2) **Faster and More Efficient Security Process**: Data should often be transported in a secure manner. The handshaking connection that is used by transport layer end points can also be employed for the exchange of security data, such as passwords, cryptographic keys, authentication codes, in following a transport security process such as that offered by the TLS protocol. The encryption process is embedded within the operation of QUIC and is thus carried out more efficiently. Automatic encryption of data and meta-data (including control data embedded in packet headers) enhances the security and robustness of end-user to end-user transactions.
3) **Flexible and Efficient Multiplexing of Streams Across a Transport Layer Connection**: QUIC provides for a more flexible transport layer multiplexing of multiple end-to-end streams, enabling the implementation of more efficient error control and flow control schemes. Loss of a stream's packet does not block the reception of a subsequently sent packet that belongs to another stream, resolving Head of the Line (HOL) blocking issues.
4) **Connection Transportability**: Enabling an efficient process for transferring a transport layer connection from one entity to another one.

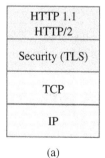
(a)

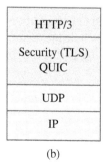
(b)

Figure 16.5 Application, Transport and Network Layer Protocol Services for HTTP under (a) HTTP 1.1 and HTTP/2; (b) HTTP/3.

16.3 Internet Protocol (IP) Addresses

16.3.1 Internet Protocol Version 4 (IPv4) Addresses

Internet protocol addresses, called **IP addresses**, are used to uniquely identify end user devices, also known as **hosts** or computer hosts that communicate across the public Internet, and hence are also identified as *public IP addresses*. Public IP addresses are assigned by the Internet Assigned Numbers Authority (IANA), which allocates address blocks to Regional Internet Registries (RIRs), including ISPs.

Private IP addresses are used for internal IP networking across an in-house (corporate or home) network. Private IP addresses of hosts within a private network are assigned by a network administrator or a network administration program. Private IP addresses have a structure that is similar to that of public IP addresses, though they assume different values and cannot be used for networking across the public Internet. In the following, we generally discuss communications across the public Internet, so that when mentioning IP addresses we refer to public IP addresses.

A host (i.e., a computer or another device that is capable of Internet communications) that wishes to communicate across the Internet needs to possess or acquire a public IP address. A host may be assigned a *static IP address* **for each one of its interfaces**. In this case, the assignment is fixed so that the host uses it for all of its flows. Most end users are however assigned a temporary IP address, known as a **dynamic IP address**. A dynamic IP address is assigned to an end user device as needed and is reassigned over time.

The assignment of a dynamic IP address is typically performed by using the DHCP. The assignment is granted as a lease with an expiration period. The host can renew its lease, if needed. Since a host possesses a unique MAC address, a network administrator can configure DHCP to allocate IP addresses based on MAC addresses.

In addition to assigning IP addresses to each interface of a host that communicates over the Internet, IP addresses are also assigned to each interface of an Internet router. This assignment is typically made on a fixed basis.

Under **Internet Protocol version 4 (IPv4)**, the IP address is a 32-bit long number. This structure has been employed since around 1983. Induced by the growth of the Internet, a newer version of IP, known as **Internet Protocol version 6 (IPv6)**, standardized in 1998, sets the IP address to be 128 bits long.

Under IPv4, the binary 32-bits long IP address is displayed in a human-readable format as 4 groups of binary-coded decimal (BCD) numbers. The number displayed in a group is a decimal representation of an 8 bits sequence. This is known as a dot-decimal notation, as group numbers are separated by dots. The number in each group ranges from 0 (corresponding to a sequence of 8 bits (an Octet) where each bit is equal to 0) to 255 (corresponding to a sequence of 8 bits where each bit is equal to 1). A group that consists of a sequence of eight 1s, 11111111, is represented by the decimal number 255, as $1 \times 2^7 + 1 \times 2^6 + 1 \times 2^5 + 1 \times 2^4 + 1 \times 2^3 + 1 \times 2^2 + 1 \times 2^1 + 1 \times 2^0 = 128 + 64 + 32 + 16 + 8 + 4 + 2 + 1 = 255$. The leftmost bit of the sequence represents the most significant bit so that the value of a binary-1 at this location is computed as the decimal number $2^7 = 128$. In Fig. 16.6, a dot-decimal representation of a binary number is illustrated, showing the representation of a binary IP address as the dot-decimal IP address 126.20.10.2.

In addition to **identifying interfaces of hosts and routers**, IP addresses are structured in a manner that serves to **identify the logical location of the addressed entity**. For this purpose, the IP address, which is represented as a numerical label such as 126.20.10.2, is divided into two parts. The **prefix** part of the address identifies the network or subnetwork in which the addressed

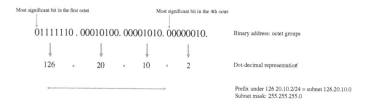

Figure 16.6 Illustrative Internet Protocol Version 4 (IPv4) Address Representation in Binary and Dot-Decimal Formats.

host resides. It is also known as the **routing prefix**. The remainder of the address is identified as the **host identifier** or as the **interface identifier** in IPv6.

How is the borderline between the two parts of an IP address identified?

Originally, the first 8 bits of the IP address were used as a network identifier. This proved to be insufficient as it allows to identify only $2^8 = 256$ networks. Then, in 1981, a *classful addressing* structure was introduced. A, B, and C addressing classes were defined for use in unicast networking. The first bit of the first Octet was used to identify an IP address as a class-A address and the remaining 7 bits were used to identify the network. This provided for the definition of only $2^7 = 128$ large networks, while the remaining 24 bits were used to identify the host members of the network. Hence, each network could contain as many as $2^{24} = 16,777,216$ hosts.

The first two octets were used to identify the network part of a class-B address, whereby the first 2 bits identified the IP address as a class-B member. This provided for the definition of $2^{14} = 16,384$ networks, while the remaining 16 bits were used to identify the host members of a network. Hence, each class-B network could contain as many as $2^{16} = 65,536$ hosts.

The first three octets were used to identify the network part of a class-C address, whereby the first 3 bits identified the IP address as a class-C member. This provided for the definition of $2^{21} = 2,097,152$ networks, while the remaining 8 bits were used to identify the host members of a network. Hence, each class-C network could contain as many as $2^8 = 256$ hosts.

Recognizing such a classification to be nonscalable, a classless addressing system was introduced in 1993, identified as a *Classless Inter-Domain Routing (CIDR) addressing system*. It introduced the concept of *variable-length subnet masking (VLSM)*, providing for flexible specification of the prefix length. The length of the prefix part is identified by the notation /x where x specifies the number of bits that constitute the packet's prefix.

Under IPv4 addressing, it is also possible to specify the prefix part by a *subnet mask*. The latter is a binary sequence that displays consecutive 1 digits to identify the span of the prefix field and displays 0 digits otherwise. For example, for the IP address shown in Fig. 16.6, which is represented in dot-decimal notation as 126.20.10.2/24, the first 24 bits of the address (i.e., the first 3 octets) represent the prefix, so that the address 126.20.10.0 serves to identify the network in which the host resides. In this case, the number 2 in the fourth octet is used as the host interface identifier. Another end-user device connected to the same network may be assigned the IP address 126.20.10.6. Fig. 16.6 also shows the corresponding subnet mask to be represented in dot-decimal notation as 255.255.255.0, which corresponds to the 32-bit binary sequence {1,1,1,1,1,1,1,1 1,1,1,1,1,1,1,1 1,1,1,1,1,1,1,1 0,0,0,0,0,0,0,0}, noting that the decimal value 255 corresponds to a binary sequence of 8 consecutive 1 digits. Either one of the prefix boundary specifications is referred to as a **subnet mask**.

Internet routers examine the destination address that is specified in the header of an incoming IP packet in determining the route to use for disseminating the packet. *A router uses the prefix part of the packet's destination IP address to find a matching forwarding entry in its routing table*. Once a match is found, the matching entry points out the output port to which the incoming packet should be switched. Hence, the prefix part of the destination address is known as the *routing prefix*.

A subnet mask is not included in the address field of an incoming (or outgoing) IP packet. How would an IP router then know which part to use as the routing prefix?

The forwarding entries in a router's routing table are linked to a destination IP address and a corresponding subnet mask. For example, a routing table may contain a forwarding entry that points to the address group 126.20.0.0/16. Consider an incoming packet whose destination IP address is 126.20.10.2. This address matches the latter entry in the table as the first two octets of both addresses are identical. As a second example, assume that the routing table contains entries for IP address groups 126.20.0.0/16 and 126.20.10.0/24. In this case, two matches are detected (with the destination address specified by the latter packet) as the first three octets of the IP address included in the address field of the incoming packet match the second entry. If only a single match is detected, the output interface number specified for this match is used for determining the outgoing link to which this packet will be switched. In turn, if, as is the case under the second example, when several matches are detected, then the longer match is preferred. For the second example, the output port corresponding to the entry 126.20.10.0/24 would thus be used.

The above noted example illustrates the way that the addressing structure is used to enable a scalable hierarchically based routing architecture. A router located further away from the destination entity may use a routing entry 126.20.0.0/16 to direct to the same output port packet flows that aim to reach hosts in networks identified as 126.20.x.0/24, where x may range between certain values (such as, e.g., 1–100).

In turn, a router that is located closer to the destination network may have configured a route that is more specific for use by flows that aim to reach the destination. The corresponding illustrative forwarding entry has been therefore set to match the address 126.20.10.0/24, pointing to a preferred route to use when aiming to reach hosts located in network 126.20.10.0. The latter route may be different than the one that will be used by packets that aim to reach destination hosts located in the network 126.20.20.0.

For example, a router located in New York City may forward packet flows that wish to reach networks located in a Los Angeles to a specific output port. Once such a packet has reached a router that is located within the Los Angeles region, the latter may use a selectively suited local route to reach a specific destination host computer that is located in a UCLA laboratory, while using another route to reach a computer that is located in a CalTech laboratory.

Several ranges of public and private IP addresses are shown in Fig. 16.7. Public IP addresses are used by routers to disseminate packets across the public Internet. In turn, the use of private

Public 1Pv4 address ranges:

- 1.0.0.0 – 9.255.255.255
- 11.0.0.0 – 126.255.255.255
- 129.0.0.0 – 169.253.255.255
- 169.255.0.0 – 172.15.255.255
- 172.32.0.0 – 191.0.1.255
- 192.0.3.0 – 192.88.98.255
- 192.88.100.0 – 192.167.255.255
- 192.169.0.0 – 198.17.255.255
- 198.20.0.0 – 223.255.255.255

Private 1Pv4 address ranges:

- 10.0.0.0 – 10.255.255.255 (10.0.0.0 /8)
- 172.16.0.0 – 172.31 .255.255 (172.16.0.0/12)
- 192.168.0.0 – 192.168.255.255 (192.168.0.0./16)

Figure 16.7 IPv4 Public and Private Address Ranges.

IP addresses is limited to networking packets within the boundary of a private network. Private IP addresses can be reused within distinct private networks and are not used for public Internet routing. A packet that is networked across a private IP network and needs to depart the private network and access the public Internet for transport across the Internet to a specified destination entity must employ source and destination public IP addresses.

To extend an operation that involves packet flows originated by hosts located in a private network that use the Internet to reach a remote data server, a **Network Address Translation (NAT)** process is often employed. To illustrate a version of such a translation operation under IPv4, consider the scenario depicted in Fig. 16.8. The address of the private network is 192.168.50.0. Three host computers and a router are shown to be members of and located within this private network. Assume that Host 1 is the source of a packet flow that it wishes to transport across the Internet to a remote server whose interface IP address is 200.100.10.4, and that Host 1 directs the server to provide the transported packets to a web browser application entity identified by application layer port 80. The application layer source port identified by Host 1 is selected as 2710. Hence, the packet created by Host 1 contains the source address and port values (S-ADD, S-Port) = (192.168.50.6, 2710) and the destination address and port values (D-ADD, D-Port) = (200.100.10.4, 80).

An outgoing packet that assumes the latter source and destination identifiers is received at the input to a NAT router that serves to perform NAT and to connect the private network to the Internet. Under the illustrative source-NAT (SNAT) process, the router modifies the address field of the packet (and may then also make other changes) as it translates the source address and source port number of the packet to (122.10.20.8, 4820), while not making any changes to the destination address and port number. At the same time, it keeps a corresponding entry in a NAT Table that it maintains, recording the details of this translation. The outgoing packet is then sent by the NAT router across the Internet for transfer to the targeted server.

A response packet produced by the server would then be identified with the following source and destination address and ports: (S-ADD, S-Port) = (200.100.10.4, 80), (D-ADD, D-Port) = (122.10.20.8, 4820). As this inbound packet reaches the NAT Router, the latter will use the corresponding entry recorded in its NAT table to translate the packet's addresses and ports to yield: (S-ADD, S-Port) = (200.100.10.4, 80), (D-ADD, D-Port) = (192.168.50.6, 2710). This translation ensures the proper routing of the packet to Host 1.

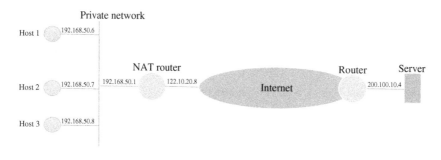

S-ADD, D-ADD for outbound packets at input to NAT router: 192.168.50.6, 2710, 200.100.10.4, 80
S-ADD, D-ADD for outbound packets at output of NAT router: 122.10.20.8, 4820, 200.100.10.4, 80
S-ADD, D-ADD for inbound packets at input to NAT router: 200.100.10.4, 80, 122.10.20.8, 4820
S-ADD, D-ADD for inbound packets at output of NAT router: 200.100.10.4, 80, 192.168.50.6, 2710

Figure 16.8 Illustrative Source Network Address Translation (S-NAT) Under IPv4.

Such a source NAT process also serves to hide from the outside the internal structure of the private network. It also helps in conserving the utilization of IPv4 public address resources. Such operations are often executed by a device that is identified as an Application Delivery Controller (ADC). The latter is generally used to also perform other functions, such as failure recovery (and the ensuing use of multi homing arrangements) and security protection of the interface of the private network with the public Internet world.

Several disadvantages have been observed. Included are the needs to maintain NAT tables and to account for the complexity of the operation. Address and port number modifications may also require the system to carry out check sequence recalculations and other adjustments of packet fields. It also violates a key fundamental element of the Internet networking operation under which much of a flow's networking states would be kept at the end-user entities rather than requiring active states to be maintained at internal network elements. Consequently, as IPv6 provides a large addressing space, associated systems often avoid the employment of NAT processes that involve the use of IPv6 addresses.

Several ranges of IPv4 addresses have been reserved for use for special networking purposes. For example, certain addresses are reserved for use for multicasting dissemination. The address block 169.254.0.0/16 has been reserved for **link-local addressing.** The latter addresses cannot be used for routing across the Internet. They are used for communications across a link, such as a point-to-point link or a logical link the represents a broadcast LAN segment.

An IP network configuration and associated illustrative IPv4 addresses are shown in Fig. 16.9. The specification of the network part of the IP address of each interface is demonstrated to properly indicate its network association. Router R1 is shown to have one interface that is attached to a local network whose IP address is defined as 50.4.1.0/24. Link local communications across this LAN is performed by the three noted attachments consisting of computer C11 at an interface whose IP address is 50.4.1.6, computer C12 at interface 50.4.1.5, and router R1 at 50.4.1.1. The other interface of R1 is attached to a local network characterized by IP address 60.4.3.0/24, whose attached entities are shown. Router R2 ports are attached to 3 networks: network 64.4.3.0/24, network 70.8.0.0/16, and the subnet 12.7.2.0/24 that represents a link segment. Router R4 interfaces local network 70.8.0.0/16 at interface 70.8.4.1. It also interfaces at interface 80.8.9.1 the switch module of the switch-hub network 80.8.0.0/16. The switch module also connects to 3 end-user computers. Router R3 is shown to be attached to three networks with interface IP addresses accordingly assigned. R3

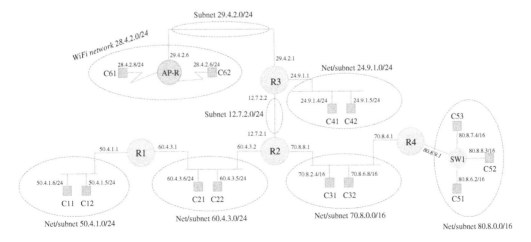

Figure 16.9 IP Network Configuration and Illustrative IPv4 Addresses.

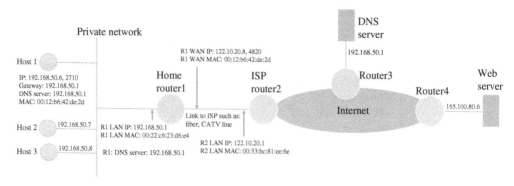

Figure 16.10 Home Network Configuration Showing DNS and Web Servers.

connects across a link segment that is identified as subnet 29.4.2.0/24 to a Wi-Fi router. The latter is integrated with a Wi-Fi AP module that is denoted as AP-R and is resident in Wi-Fi network 28.4.2.0/24.

In Fig. 16.10, an illustrative home network configuration is shown. An end-user HOST1 is establishing an IP flow, which it aims to reach a web server identified by the name Sophie_Stitchery.yahoo.com. It needs to first find the destination's public IP address. For this purpose, HOST1 sends a DNS query packet to its associated Home Router1, which is the designated router of LAN 192.168.50.0/24. Router1 is configured with the IP address of the local DNS server, represented here as 192.168.50.1. Home Router1 then sends a DNS query packet across the access line (implemented, e.g., as a fiberoptic or CATV link) to ISP Router2. IP packets that are transported across the latter link are included as payloads within MAC frames and transmitted across the link.

ISP Router2 transports the DNS query packet across the Internet to Router 3 to which the DNS server is attached. The latter performs the requested *name resolution* as it determines the IP address of the web server of interest to be 165.100.80.6. This information is included in a response packet that the DNS server sends back to HOST1 via a path that consists of an Internet route to ISP Router2, followed by transport to Home Router 1 and then to Host1.

Using this information, Host1 composes the packets that it aims to transport to the underlying web server by setting their source address to 192.168.50.6 (and port 2710) and their destination address as 165.100.80.6 (and port 80). Home Router 1 performs the proper NAT address transformation, serving to mediate the packet flow between Host 1 and ISP Router2, converting the corresponding private address to a public address. For example, it converts the private address (192.168.50.6, 2710) of HOST1, to a public address (122.10.20.8, 4820), and keep this transformation in its NAT database (which would be used for performing the reverse-NAT transformation applied to a response packet). Hence, an outbound packet that is received at the input of ISP Router2, destined for delivery to the Web Server, will then contain the addresses (SADD; DADD) = (122.10.20.8, 4820; 165.100.80.6, 80).

A response packet from the web server that is received at Router 2 is identified by (SADD; DADD) = (165.100.80.6, 80; 122.10.20.8, 4820). This inbound packet is delivered across the link to Router1. The latter performs a reverse-NAT transformation, which maps the destination public address (122.10.20.8, 4820) to the private address (192.168.50.6, 2710) of HOST1. The produced packet that is then sent to HOS1 contains the addresses of the web server as its source and of HOST1 as its destination: (SADD; DADD) = (165.100.80.6, 80; 192.168.50.6, 2710).

16.3.2 Internet Protocol Version 6 (IPv6) Addresses

An IP version 6 (IPv6) address is 128 bits long, which is 4 times longer than an IPv4 address. Using the significantly extended address space enables the synthesis, management, and operation of a wide range of hierarchically architectured networks and subnetworks, leading to higher scalability and improved routing efficiency. The addressing design can be much simplified to realize a more end-user-oriented global flow routing characterization. Yet, the use of IPv4 addressing has remained highly prevalent as IPv4 network implementations have been optimized. While continuing to make progress, as of 2020, it is estimated that about 35% of US Internet flows, and about 22% of out of the US flows, employ IPv6 addressing.

An IPv6 address is represented as eight groups of four hexadecimal digits. Each group thus represents 16 bits. The groups are separated by colons (:). For example, the following is such an address: 1924:2ce5:ac31:bfd2:0000:0000:0000:0000. This address is also represented in brief as follows by setting the notation :: to represent a sequence of trailing 0s: 1924:2ce5:ac31:bfd2::. It is noted that in hexadecimal notation the following letters (written in either lower case or capital case) are used to represent the decimal values ranging from 10 to 15: a = 10, b = 11, c = 12, d = 13, e = 14, f = 15. Hence, the ac31 group represents the following sequence of bits: 1010110000110001, as a = 1010, c = 1100, 3 = 0011, 1 = 0001. A group that consists of ffff, thus represents a sequence of bits that consist of 16 consecutive 1's: 1111111111111111.

The IPv6 address allocation process has been carried out by the IANA. Large address blocks are assigned for distribution to RIRs. The IANA has maintained the official list of allocations of the IPv6 address space since December 1995. Certain blocks of IPv6 addresses are assigned to service providers and specific ones are reserved.

IP version 6 (IPv6) addresses are categorized as unicast, multicast, or anycast addresses. When a **unicast address** is specified as the destination address of a packet, the packet is destined to reach a single specific interface. An **anycast address** identifies a group of interfaces. Any unicast address can be used for this purpose, with the difference being that this address appears in multiple interface points. A packet that is sent to an anycast address is delivered to just one of the group interfaces. Typically, the delivery is made to the nearest host. A **multicast address** is used for the dissemination of packets to a group of users that have joined as members of an underlying multicast group. Broadcasting in IPv6 can be implemented by using a multicast address that involves an all-nodes link-local multicast group ff02::1. Typically, IPv6 systems use a dedicated link-local multicast group for this purpose.

As done for IPv4 addresses, IPv6 unicast and anycast addresses are also divided into two logical parts, but the division arrangement is different, aiming to simplify the conduct of routing and switching processes. As shown in Fig. 16.11, the most significant 64 bits are used to form the **network prefix**. The network prefix can be subdivided, so that the network administrator can use the least significant bits of this part to define a **subnet ID**, and in this way subdivide the network into multiple subnets. The host identifier part of the address is called an **interface identifier** and is set to be 64 bits long. The host identifier can be assigned manually by a network administrator, configured randomly, obtained from a DHCPv6 server, or generated from the MAC address of the interface.

The CIDR notation is used to define the boundary between the network address and the subnet address parts. For example, a network identified by the address 1924:2ce5:ac31:bfd2::/48 starts at address 1924:2ce5:ac31:bfd2:0000:0000:0000:0000:0000 and ends at 1924:2ce5:ac31:bfd2:ffff: ffff:ffff:ffff:ffff. Such a network thus consists of $2^{128-48} = 2^{80}$ addressable entities.

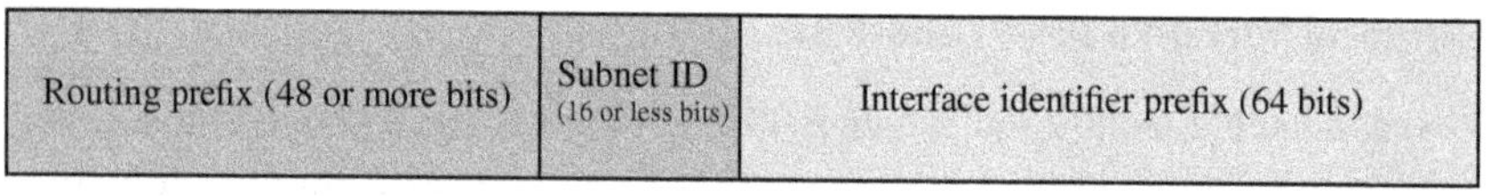

Routing prefix (48 or more bits)	Subnet ID (16 or less bits)	Interface identifier prefix (64 bits)

Figure 16.11 IP Version 6 (IPv6) Address Fields.

Unique local IPv6 addresses (ULAs) are configured for use as private addresses. They are identified by using a specific prefix. The routing prefix fc00::/7 has been reserved for this purpose. The addresses include a 40-bit pseudo-random number, aiming to minimize the risk of address collisions. Using unique local addresses in combination with network prefix translation can be employed to attain NAT like operations. Yet, due to its large addressing space, end-user devices can generally be assigned unique globally routable addresses so that such NAT-oriented operations are often not required.

Link-local addresses are defined in a manner that is based on the interface identifier. They consist of the prefix fe80::/64 and are non-routable. They are used for communications between IPv6 hosts attached to a local link. In this way, they are also used by the Neighbor Discovery Protocol.

A zone identifier (such as zone 5 denoted as %5) is often added to a non-global address to restrict the location where it is used, as it can be re-used. A special format is also defined for multicast addresses, including the use of a specific bit sequence (consisting of 8 consecutive 1 bits) in the prefix field (ff00::/8). The sending source sends a single packet from its unicast address to the multicast group address. Intermediary routers make copies of the packet, sending them to multicast group members. Under IPv6, broadcast addressing is carried out by multicasting to a multicast group that consists of all nodes. An IPv4 address that is transformed to assume the format of an IPv6 address is identified by the prefix ::ffff:0:0/96.

16.4 Internet Protocol (IP) Packets

16.4.1 Internet Protocol Version 4 (IPv4) Packets

As shown in Fig. 16.12, the IPv4 packet consists of a packet's header part and a data payload's part. The data field carries the payload, which consists of a transport layer message, typically being a UDP or TCP message. The data field has been set to not exceed 65,536 bytes (whereby each byte is an 8-bit Octet). The header is shown to consist of five required words, whereby each word is 32-bits or 4 bytes long. Hence, the basic header is 20 bytes long. The use of the options field is optional.

The first word includes the following five fields. The Version field (4 bytes) identifies IP addressing version 4. The Internet Header Length (IHL) field (4 bytes) specifies the length of the header (in words), whose size may vary due to the possible use of the Options field. The header's length can vary from a minimum of 20 bytes (when no Options field is used) to a maximum of 60 bytes. The (6 bytes long) Differentiated Services Code Point (DSCP) field was originally identified as Type of Service (ToS). It is used to specify special or preferred treatment that is desired for the underlying packet. It has then been defined as a Differentiated Services (DiffServ) identifier. For example, it may be employed when realtime streaming treatment is requested, as is the case for realtime Voice over IP (VoIP). The (2 bytes) Explicit Congestion Notification (ECN) field provides for the possible implementation of end-to-end proactive congestion notification. The (16 bits) total length field specifies the length of the entire packet in bytes, including its header and data parts. It varies from a minimum of 20 bytes (when no data payload is included) to a maximum length of $2^{16} = 65,535$ bytes, also represented as 64 kbytes, as 1 kilobyte (kbyte) is equal to 1024 bytes.

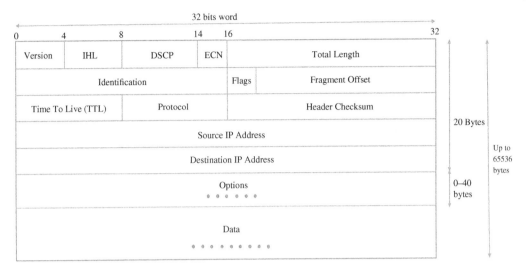

Figure 16.12 IP Version 4 (IPv4) Address Fields.

The packet sent by a source end-user host may need to be segmented into multiple fragments since the attached link may restrict the maximum length of a datagram that it accepts for transport. For example, an Ethernet frame does not carry a payload packet that is longer than 1500 bytes. Length restriction can also be imposed on a packet while it is transported across the network, as it may then need to traverse links that require it to be fragmented. In this case, fragmentation will be carried out by the corresponding intermediate router. The receiving end-user host assembles the fragments of a packet to form the original packet created at the source end-user host. The assembled packet would then be sent upward the protocol stack to the transport layer protocol entity (typically, a TCP or UDP entity) and subsequently (if properly received and processed by the latter) to the designated application layer protocol entity.

To assure the integrity of this assembly process, the fragments of a specific packet are identified as segments of this packet. In addition, their fragmentation order is specified, so that when re-assembled, they can be pieced together in the proper order. Missing fragments are then identified. Such fragmentation information is provided in the packet's second 32-bit word. Fragments that display the same number in their Identification field are recognized as fragments of the same packet. The fragment offset field identifies the placement order of a fragment within its packet, as it specifies its offset from the start of the original IP packet in units of eight-byte blocks. The first fragment has an offset of zero. The 3-bits Flags field includes a Do-Not-Fragment (DF) bit and a More-Fragments (MF) bit. The latter indicate whether this packet is the final fragment of the packet.

The 8-bit **Time To Live (TTL)** field is used to restrict the time that a packet spends in transiting the network. While the specification has originally intended it to specify the packet's residence time limit in seconds (with a period lower than 1 second recorded as equal to 1 second), it is practically used to set a limit on a packet's incurred hop count as it travels the network. It is named as **hop limit** in IPv6. As a packet transported across a network arrives at a router, the router decrements its TTL value by 1. When the TTL value reaches 0, the router discards the packet. The ICMP is then used to send an error message to the sender, informing the latter about the underlying time-exceeded event.

494 | 16 *The Internet*

The program Traceroute sends messages with adjusted TTL values, making use of the induced ICMP time-exceeded messages to identify the routers traversed by a packet as it flows across its Internet route.

For example, under a setting of TTL = 4, the packet would be discarded following the traversal of 4 hops, upon reception by the fourth router along its route. A TTL value of 1 is used to limit the dissemination scope of a packet, forcing it to stay within an organizational boundary. The TTL value was originally used as a measure that prevents packets from cycling around the network for an excessive period of time. As the TTL value is specified by an 8-bit number, its maximum value is equal to 255. Often an initial TTL value is set to 64. Such a value is often also used to limit the dissemination scope of multicast packets to avoid overloading of network regions.

The *Protocol field* specifies the protocol that is used to express the data payload that follows the header of the IP packet. For network transport purposes, it commonly specifies the transport layer protocol that is used to produce the data payload carried by the packet. Frequently employed such transport layer protocols are TCP and UDP. Protocol numbers are maintained and published by the IANA. TCP is assigned the Protocol Number 6 (denoted as 0x06 in Hex) and UDP is assigned Protocol Number 17 (denoted as 0x11 in Hex). Protocol numbers are also used to identify other protocol types. Protocol Number 8 identifies the EGP and Protocol Number 9 is used for IGP, which are routing protocols. Protocol Number 1 is assigned for ICMP, which is used to produce testing and error induced type messages. Protocol Number 2 is assigned for Internet Group Message Protocol (IGMP), which is employed for managing and configuring multicasting and multicast groups.

The 16-bit IPv4 header checksum field is used for error-checking of the header. When a packet arrives at a router, the router calculates the checksum of the header and compares it to the checksum field. If the values do not match, the router discards the packet. Errors in the data field must be handled by the encapsulated protocol. Both UDP and TCP have separate checksums that are applied to their data. Since a router decreases the TTL field in the header of an arriving packet, it needs to calculate a new header checksum.

The subsequent two 16-bit fields display the IPv4 source address of the sending entity and the IPv4 destination address of the receiving entity. The options field is often not used. Option examples include Time Stamp, identified by Option Type Code 68 (in Decimal) or 0x44 (in Hexadecimal) and Traceroute identified by Option Type Code 82 (in Decimal) or 0x52 (in Hexadecimal).

16.4.2   Internet Protocol Version 6 (IPv6) Packets

As noted for IPv4 packets, an IPv6 packet consists of a header field that carries control information that is used for addressing and routing and a data field that contains the payload data. The latter consists typically of transport layer messages, though ICMP and certain routing messages are at times also carried in a packet as data payloads. To simplify the format of IPv6 packets and reduce associated processing performed by routers, control information is included in a mandatory fixed header and in optional extension headers.

To avoid the processing overhead and ensuing transport latency involved in the fragmentation of packets and in the ensuing reassembly processing, routers do not fragment IPv6 packets that are longer than the maximum transmission unit (MTU) level. It is the responsibility of the source node to assure the creation of packets that can be carried by all routers located across the intended route. For this purpose, intermediate networks that a packet transits are required to use an MTU size that is equal to at least 1280 octets. To reduce end-to-end transport latency, source hosts can engage in Path MTU Discovery, which would enable them to route packets of lengths that are acceptable to all routers that are used along an underlying path.

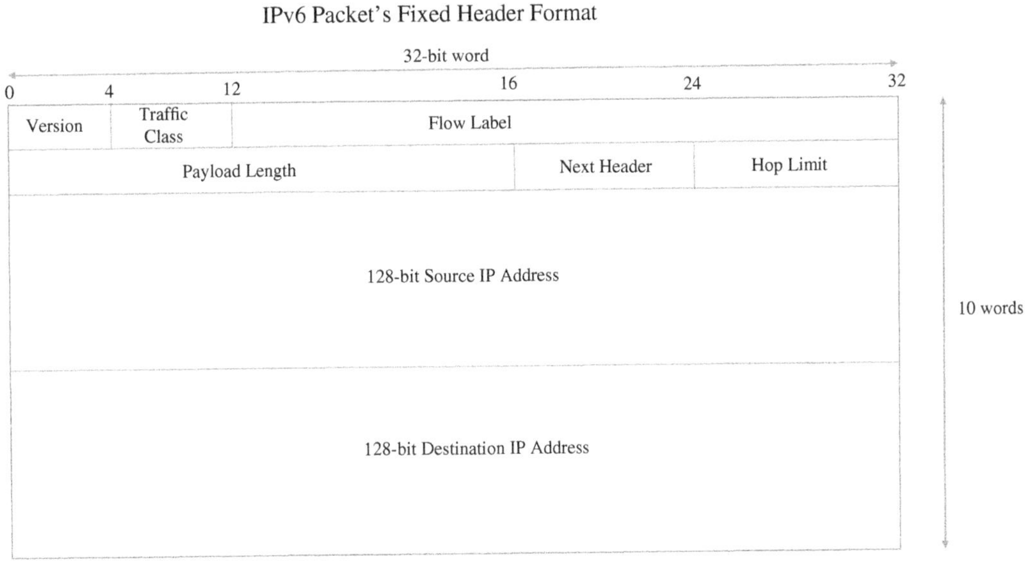

IPv6 Packet's Fixed Header Format

Figure 16.13 IP Version 6 (IPv6) Packet's Fixed Header Fields.

The format of the IPv6 packet's fixed header is shown in Fig. 16.13. The fixed header is placed at the start of the IPv6 packet. It contains 40 octets, or 320 bits, or 10 words (whereby each words is 32 bits long). The Traffic Class field is used to classify packets into traffic classes. The Flow Label field identifies packets as members of a flow whose packets are transported between the identified source and destination entities. The Payload Length field specifies (in octets) the size of the payload (it is set to 0 when a Hop-by-Hop extension header carries a Jumbo Payload option). The Next Header field specifies the type of the next header that is used. Typically, the transport layer protocol used by a packet's payload is specified here.

The Hop Limit field specifies the maximum number of residual hops that this packet is allowed to traverse. In IPv4, this role is assumed by the parameter specified in the TTL field. Its value is reduced by 1 at each forwarding node and the packet is discarded when its value is reduced to 0, though a destination node would process the packet normally even if received with a hop limit of 0.

The source address field specifies the unicast IPv6 address of the sending source node. The destination address field specifies the unicast or multicast IPv6 address of the destination node(s). No header checksum is included, as link-layer and transport layer protocols are generally configured to provide error protection means.

Extension headers are optionally included. Most of them are processed at the packet's destination, but Hop-by-Hop ones can be processed and modified by intermediate nodes. In the following, we discuss several extension headers.

The Hop-by-Hop Options extension header may be processed by nodes located on the packet's route, including its sending and receiving nodes. Its structure is shown in Fig. 16.14. The next header specifies the type of the next header. The header extension length specifies the length of the header in octets, not including the first eight octets. The Options fields contain various options while the padding is used to align options and to make the total header length equal to a multiple of eight octets. The Destination Options extension header is processed by the destination node(s) only.

The format of the Routing extension header is shown in Fig. 16.15. It is used to guide the packet to traverse one or several intermediate nodes before reaching its destination. Type specific data that

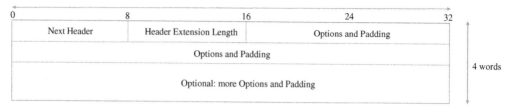

Figure 16.14 IP Version 6 (IPv6) Packet's Hop-by-Hop Options Extension Header.

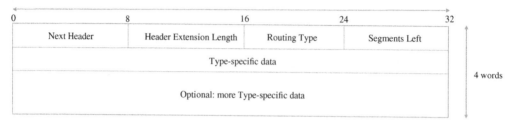

Figure 16.15 IP Version 6 (IPv6) Packet's Routing Extension Header.

relates to the function of this header can be included. The routing type is specified in a corresponding field. For example, Type-3 accounts for a Source Route Header, which is used to specify the successive nodes to be visited by the packet along the route. A type-4 specifies the Allowed Segment Routing data. The number of nodes that the packet still needs to visit before reaching its final destination is specified in the "Segments Left" field.

To send a packet that is longer than the path MTU, the sending node splits the packet into fragments. The Fragment Extension header carries the information necessary to identify the order of the fragments, including fragment offsets and an ID that binds together the fragments that belong to the same packet. This is required for the correct reassembling of the fragments into the original packet.

The Authentication Header and the Encapsulating Security Payload are part of IP Security (IPsec). IPsec provides end-to-end security over the IP layer (or OSI layer 3). It provides protocols for mutual authentication between agents at the start of a session and for negotiation of a session's cryptographic keys. Cryptographic security services are employed to protect communications over IP networks at the IP layer. Its services include network-level peer authentication, data origin authentication, data integrity, data confidentiality (encryption), and replay protection. Security services that are provided at other layers include Transport Layer Security (TLS) that operates above the Transport Layer and Secure Shell (SSH) that operates at the Application layer. Message transport Security methods are discussed in Chapter 22.

16.5 Transport Layer Protocols

16.5.1 Transmission Control Protocol (TCP)

In transporting packets across the Internet, the IP layer provides an unreliable network layer datagram service. While packet errors are detected, the IP protocol does not provide for the implementation of an Automatic Repeat Request (ARQ) error control scheme that enables hosts to acknowledge correctly received packets and to induce the possible retransmission of un-acknowledged packets. Such an error control service is of essential importance for the proper execution of many data transactions over the Internet.

The IP networking service does not provide for an orderly delivery of packet flows and streams. Packets in such flows must be reassembled at the destination host in a complete and proper order. Missing packets must therefore be recognized and their retransmission needs to be carried out.

Such services, contributing to the reliable and orderly end-to-end transfer of packet flows, are provided by the TCP. As a transport layer protocol, TCP provides its services to flows, including IP-based Internet flows, which are transacted between transport layer protocol entities that are located at end-user entities. To provide its services, TCP acts as an end-to-end **connection-oriented protocol**. A signaling protocol is defined and used to set up a TCP connection between the communicating transport layer protocol entities.

Key services provided by TCP include the following:

1) **Multiplexing** of flows between a source transport layer protocol entity and a destination transport layer protocol entity. Source and destination application-oriented identifiers, known as port numbers, are used to identify the originating and terminating application layer protocol entities. The transport layer protocol entity at the destination end user uses the destination port number to determine the application entity to which the received message should be delivered. Arriving TCP messages are identified as belonging to a specific TCP connection by their sockets, defined as the combination of source host address, source port, destination host address, and destination port. In this way, services are provided to data flows associated with distinct connections.

2) An **ARQ error-control** scheme is used to perform error detection and retransmission processes between the source and destination transport layer protocol entities. Sliding-window and select-and-repeat ARQ error-control schemes can be employed. Messages that are not positively acknowledged within a specified time threshold are scheduled for retransmission.

3) A **flow control** scheme is executed between the source and destination transport layer protocol entities. A sliding-window flow control, scheme that is similar to that used for error control purpose, is commonly employed. For each connection, the receiving transport layer entity dynamically issues an authorization for a specified window of data that it is willing at this time to receive from the corresponding sending transport layer entity.

4) **Congestion control** algorithms that are based on the TCP message transaction process can be enabled. An explicit congestion control scheme can make use of an ECN indicator embedded in a TCP message produced by a receiving TCP entity and aiming to inform the sending TCP entity about network congestion along its network route. More frequently employed is an implicit congestion control scheme under which the sending TCP entity regulates the rate at which it sends its messages across the network in accordance with its observations of the flow of data and ACK messages across its transport layer connection.

5) **Sequencing** of a flow's data segments that are transported between sending and receiving transport layer protocol entities is performed. Segments are reassembled at the destination transport layer entity in the proper order and in completion. Missing segments are detected and retransmitted.

In the following, in observing the format of a TCP message, we identify the parameters that are used by TCP to implement the services mentioned above.

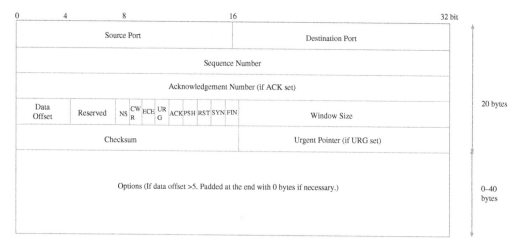

Figure 16.16 A Transmission Control Protocol (TCP) Message.

The format of a TCP message is shown in Fig. 16.16. The source port identifies the application entity that produced the message. The destination port identifies the port to be used by the destination TCP entity at the receiving end user as it delivers the payload of the received TCP segment to the destination application layer protocol entity.

TCP and UDP use port numbers to identify sending and receiving application end points on hosts. Each end point of a TCP connection has an associated 16-bit unsigned port number (0-65535) reserved by the application. An arriving TCP message is identified as belonging to a specific TCP connection by its socket. The latter is configured as the combination of source host address, source port, destination host address, and destination port. Using these IDs, a server computer can simultaneously provide services to several clients. A client that configures multiple simultaneous connections to a single destination port makes sure that the connections are initiated from different source ports.

The following well-known port numbers are used to identify applications that are frequently employed in formatting data messages that are transferred across the Internet by IP packets: FTP—ports 20, 21; Telnet—port 23; SMTP—port 25; Domain Name System (DNS)—port 53; DHCP—67 (server port), 68 (client port); HTTP used in WWW transfers—80; Network Time Protocol (NTP)—123; SNMP—161; Secure HTTP transfer (HTTPS) over TLS/SSL—443; DHCPv6—546,547.

The 32-bit **Sequence Number** field identifies the sequence number of this TCP segment. At the start of a TCP connection (when the SYN flag is set), it identifies the initial sequence number of the TCP segment issued by this TCP entity. Each TCP end point defines its own initial sequence number. This number represents the sequence number of the first data byte (i.e., octet) of the segment issued by the sending entity (at each end of a full duplex connection). An ACK that is issued by a receiving TCP entity is identified by the latter sequence number plus 1. If the underlying TCP data segment is part of an ongoing flow, past the establishment of the TCP connection, so that the SYN flag is clear (and set to 0), then the sequence number of the first data byte of this TCP segment is set equal to the accumulated sequence number for the ongoing session.

When the ACK flag is set, the 32-bit ACK number field contains the next sequence number that the receiver of a message expects. It is used to ACK the prior receipt of message octets. The first ACK sent by each TCP entity acknowledges solely the other entity's initial sequence number.

To illustrate, consider a TCP connection to have been established between TCP entity A and TCP entity B, located in their corresponding end-user hosts. Assume that TCP entity A wishes to send to TCP entity B two TCP messages, each of which is 1200 bytes long. In the first phase, a TCP connection is first established. Once it is done, the second phase consists of data transfer.

Considering the first phase, the following illustrates an initial settings of the corresponding sequence numbers for this TCP connection. The initial sequence number selected by entity A is, say, equal to 1500. It is ACKed by entity B by the latter specifying the ACK sequence number 1501. Similarly, an initial sequence number selected by entity B is, say, identified by sequence number 5600. Its ACK by entity A specifies the ACK sequence number 5601.

Consider next the second phase, during which data units flow across the configured connection. The first data segment issued by entity A is identified by TCP sequence number 1501. Assuming it to contain 1200 bytes and to be received correctly, the next segment sequence number expected to be received by entity B is equal to $1501 + 1200 = 2701$. Hence, the ACK sequence number set by entity B is equal to 2701. The sequence number of the next TCP data segment sent as part of this flow by entity A is thus set equal to 2701.

Assume now that the second TCP segment is corrupted by channel noise so that entity B determines it to contain errors. Under an employed ARQ scheme, entity B will then send an ACK message to entity A specifying an ACK sequence number that is still equal to 2701, as it still expects the next segment to be identified by the latter sequence number. This ACK serves as a Negative-ACK (NACK) message. Eventually, if and when its ARQ timeout timer expires, entity A will retransmit the latter TCP segment, identifying it by the same segment sequence number, 2701. Assuming it to now be received correctly at entity B, the ACK sequence number issued by entity B would be equal to $2701 + 1200 = 2901$. It informs entity A that the sequence number of a segment that it expects to receive next as part of this flow connection should be equal to 2901.

To increase the throughput efficiency of the employed TCP ARQ scheme, a selective ACK (SACK) scheme can also be used. Under this scheme, the receiver can ACK the receipt of discontinuous groups of TCP segments. Issued ACKs are set to identify the starting and ending sequence numbers of correctly received segment blocks by using a TCP SACK Option Header. Only the missing segments will be retransmitted. Selective acknowledgments have also been used under the Stream Control Transmission Protocol (SCTP).

The 16-bit Checksum field is employed for error checking of a composite number that consists of the TCP header, the payload and a pseudo-header. The pseudo-header consists of the source IP address, the destination IP address, the protocol number for the TCP protocol (6), and the length of the TCP header and payload (in bytes).

When an URG flag is set, the 16-bit Urgent Pointer field is used to indicate the last urgent data byte.

The CWR, ECE, and NS TCP flags are not commonly used. The CWR flag is used to indicate that the congestion window has been reduced, signaling that the sending host has received a TCP segment with the ECE flag set. The congestion window is an internal variable maintained by TCP to manage the size of the send window. The ECE flag indicates that a TCP peer is ECN capable during the TCP 3-way handshake (when the SYN flag is set) and (when the SYN flag is clear) to indicate that a TCP segment was received over the connection with the ECN field in its IP header set (to 11). This event signals to the TCP sender that a packet received at the receiving TCP entity has experienced ongoing (or impending) network congestion during its end-to-end transit. Such a procedure performs as a reactive congestion control process. It is discussed in Section 14.4.1.

The operation of a TCP implicit congestion control scheme is depicted in Fig. 14.6 and discussed in Section 14.4.1. In observing the rate at which data segments that are issued by a source TCP entity receive corresponding ACKs from the receiving TCP entity, the source estimates the throughput rate of its TCP connection. When it determines that the realized throughput rate has been reduced, it assumes it to be caused by congestion. It then proceeds to reduce, in an exponentially fast manner, the rate at which it feeds its TCP segments into the network. As it determines its connection's throughput rate's degradation to end, the source proceeds to gradually raise, in an additive manner, its segment feeding rate.

The Options field length varies between 0 to 320 bits. This field can be used to specify, for example, a Maximum Segment Size level. Other optional settings include the specification of values for a Window Scale (used to increase the specified receive window size by a defined scale parameter), and the setting of a Selective ACK scheme and its associated parameters.

The process employed by the TCP protocol scheme consists of three phases: (1) Connection establishment; (2) Data transfer; (3) Connection termination.

A TCP connection is managed by an operating system through a resource that represents the local end point for communications—the Internet socket. During the lifetime of a TCP connection, the local end point undergoes a series of state changes, in accordance with a specified protocol whose state process, as impacted by data inputs, is modeled as a Finite State Machine (FSM) type protocol.

The connection establishment phase includes the following transactions. It is noted that a segment is identified by a sequence number, while the sequence number of an ACK that may be carried by a packet identifies the sequence number of the next segment that the producer of this segment is expecting to receive.

1) **SYN**: A client sends a SYN message to a server. The client sets this segment's sequence number to a random value A.
2) **SYN-ACK**: In response, the server replies with a SYN-ACK message. Its ACK sequence number is set to A+1, indicating that the next segment that the server is expecting to receive is A+1, and in this manner ACKing the receipt of the SYN segment. The sequence number that the server chooses to identify the response (SYN-ACK) segment that it sends back to the client is set to a random number B.
3) **ACK**: Finally, the client sends an ACK segment back to the server. The sequence number of this segment is set to the received acknowledgment value; i.e., to A+1. The sequence number of this ACK segment is set to B+1.

   Upon the completion of these steps, when both client and server have received their expected acknowledgments, a full-duplex communication connection has been established.

The connection termination phase uses a four-way handshake, with each side of the connection terminating independently. When an end point wishes to stop its half of the connection, it transmits a FIN segment, which the other end acknowledges with an ACK. Therefore, a typical tear-down requires a pair of FIN and ACK segments from each TCP end point. After the side that sent the first FIN segment has responded with its final ACK segment, it waits for a timeout before finally closing the connection, during which time the local port is unavailable for new connections. This prevents possible confusion that can occur if delayed segments associated with a previous connection are delivered during a subsequent connection. It is also possible to terminate the connection by a 3-way handshake, when host A sends a FIN and host B replies with a FIN and an ACK and host A replies with an ACK.

Implementations tend to record an entry in a table that maps a session to a running operating system process. Because TCP segments do not include a session identifier, both end points identify the session by using the client's address and port. Whenever a segment is received, the TCP implementation must perform a lookup of this table to find the destination process. Each entry in the table is known as a Transmission Control Block or TCB. It contains information about the end points (IP and port), status of the connection, running data about the segment that are being exchanged and buffers for sending and receiving data.

The number of sessions in the server side is limited by memory and can grow as new connections arrive, but the client must allocate an ephemeral port before sending the first SYN to the server. This port remains allocated during the whole conversation and effectively limits the number of outgoing connections from each of the client's IP addresses. Both end points must also allocate space for unacknowledged segment and received (but unread) data.

A TCP data transfer process aims to produce orderly and reliable transport of TCP segments across an established TCP connection. Segment positions within a flow are identified by sequence numbers. Segments that are received in error are retransmitted. A sliding-window flow control scheme is used to throttle the rate used by a TCP sender to transfer TCP segments across its TCP connection. The receiver acts to pace the sending rate of TCP segments in a manner that depends on its current resource availability. As discussed in Section 14.4.1, a congestion control scheme is also employed, as lost segments trigger reductions in the data delivery rate.

16.5.2 User Datagram Protocol (UDP)

The **UDP** is a connectionless transport layer protocol. It provides an unreliable datagram service at the transport layer. An application entity that uses UDP to transport its messages across a network, such as the Internet, is not provided error control, flow control and sequencing services. The main service provided by UDP involves the specification of source and destination application layer SAPs, identified (as used by TCP) as source and destination port numbers. UDP provides a checksum that enables a UDP protocol layer entity to determine whether a received UDP datagram contains message errors, but not to correct them or to request for message retransmission.

While providing a simple yet unreliable transport layer protocol service, UDP is able to perform its functions in an efficient and prompt manner and requires less overhead than that used by TCP. This can reduce the time latency incurred in the transfer of messages. Such a performance feature is of interest for message flows that require rapid reactions and interactions. It is of interest to users that do not wish to use error-control-oriented retransmissions and flow control processes across the transport layer, or that engage in applications do not need them performed.

This is often the case for interactive realtime flows, such as voice and video streams that require low latency realtime transfer and are not sensitive to low message error and data loss rate levels. For example, for many real-time VoIP and video streams, the fidelity of the transport process is more sensitive to packet latency and delay jitters than to a modest level of packet loss. It is also of interest for the transport of certain network-management data packets that must be rapidly disseminated, and for the transport of status-update telemetry messaging. For such applications, the transmission of messages is often periodically repeated. Following the discard of a message by a receiving user, a more fresh status update message is expected to soon arrive, so that there is no need to request for the retransmission of an older message. Messages for certain such applications are short and are disseminated as individual protocol data units, so that they may not require the receiving entity to engage in a message reassembly process.

Induced by low latency performance requirements, and involving mostly individual (rather than stream based) message transactions, other Internet applications that use UDP include: Domain Name System (DNS); SNMP; Routing Information Protocol (RIP); DHCP.

As described in Section 16.5.3, QUIC is a connection-oriented transport layer protocol that provides multiplexing, error control and flow control services, and also includes the use of a secure

Figure 16.17 User Datagram Protocol (UDP) Datagram Header.

transfer protocol (such as TLS). It is used for low latency secure client–server interactions across the Internet, and it runs on top of UDP.

A UDP datagram consists of a header field and a data section. The data field that follows the header carries the data payload. The format of a UDP datagram header is shown in Fig. 16.17. It consists of 4 fields, each of which is 2 bytes (16 bits) long. As in TCP, the source and destination port numbers identify the corresponding application layer protocol entities at the SAPs of the interface between the transport and application protocol layers. Port numbers can also identify interfaces to other protocol entities that run over UDP (such as QUIC). Well-known port numbers are frequently employed, particularly when server computers are used. The use of a source port is optional in IPv6.

The checksum field is used for error checking of the header and data. This field is optional in IPv4, and mandatory in IPv6, noting that, over the Internet, UDP runs on top of IP.

The Length field specifies the length (in bytes) of the UDP header and the UDP data. The minimum length is 8 bytes, which is the length of the header. Since the size of the Length field is 16-bits, the maximum accommodated length is equal to 65,535 bytes, and thus 65,527 data bytes. If an IPv4 packet header of 20 bytes is included in the calculation, the resulting IPv4 allowable maximum packet size would be equal to 65,507 bytes. When using IPv6 jumbo-grams, longer UDP datagrams are permitted, and the Length field is then set to 0.

UDP is commonly employed as the transport layer protocol when streaming voice and video flows over the Internet, as well as over other IP networks. For the high-quality transport of voice and video streams in realtime, transport layer error and flow control regulation processes as those performed by TCP are not as essential as are low transport and processing delay performance levels.

To add message timeliness and sequencing features that are required by replay mechanisms executed at a receiving entity, higher layer message attributes as those included by **Realtime Transport Protocol (RTP)** are commonly used. As described in Sections 2.6.2.2 and 2.7.5, RTP is an application layer protocol that is used to attach time stamps and sequence numbers to voice or video messages. RTP messages are transported as payloads nested within UDP messages. Associated with RTP is the **Real-time Transport Control Protocol (RTCP)**. It supplies out-of-band statistics and control information for an RTP session. These statistics provide feedback on the quality of service (QoS) attained by a stream's transport process. An application at the source entity may use this data to make changes that enhance the transport quality of the stream.

RTP sessions are initiated, maintained, and terminated by using a *signaling protocol* such as the **Session Initiation Protocol (SIP)**. An SIP element can function as a user agent client (UAC) when it requests service and as a user agent server (UAS) when it responds to a request. A call established with SIP may consist of multiple media streams. Media type and associated parameter attributes are described by using a protocol such as the **Session Description Protocol (SDP)**. Its data units are transported as payloads of SIP messages. Often, SIP calls are managed through a *SIP* *server*, acting as a *SIP Proxy*. It takes requests from user agents to place, adjust and terminate calls. It routes calls to the proper destination and controls the application of call features. It negotiates session parameters using the SDP protocol, making adjustments during a session when requested to do so. Actual transmission and reception of stream messages are performed through the use of RTP.

16.5.3   QUIC: A Fast and Secure Transport Protocol

As noted in Section 16.2, HTTP/3 was developed in 2022 to provide faster, secure, and more robust client–server interactions. As shown in Fig. 16.5, HTTP/3 uses QUIC over UDP as its (layer-4) transport layer protocol.

A version of QUIC, which originally was an acronym for Quick UDP Internet Connections, was developed at Google around 2012. A different version has been developed by IETF. Its specification is described in the IETF RFC 9000 document [29] in association with RFC 8999, RFC 9001, and RFC 9002 [28].

QUIC is a connection-oriented transport layer protocol over UDP. It is used for supporting a stateful interactions between a client and a server. It provides transport layer services to an application such as HTTP, providing error, flow, and congestion control regulations of data streams that are multiplexed across an end-user to end-user connection.

QUIC services include the following features:

1) A handshake process is conducted to initiate a QUIC connection. In a single interaction, a shared secret is exchanged between client and server by using a QUIC-TLS cryptographic handshake protocol. This way, the QUIC protocol merges the TCP 3-way handshake and the TLS1.3 3-way handshake into a single 3-way packet exchange. As shown in Fig. 16.18, this shortens the duration of the initiation phase by a round-trip-time (RTT), which can enhance performance in a significant way.
2) In this manner, security is embedded within the QUIC connection process. Security data is exchanged during the initial handshake that is conducted between communicating end users. QUIC packets are encrypted individually and can therefore be individually decrypted, avoiding the need for a receiving end point to engage in resolving packet units within a continuous byte stream.
3) Multiple streams can be multiplexed across a single QUIC connection. This serves to speed-up the interaction process, avoiding the need for different streams to initiate and maintain separate transport layer connections.

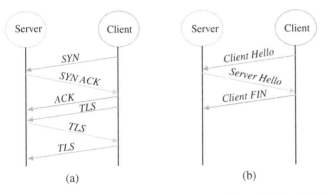

Figure 16.18   Handshake Transactions under TCP and QUIC: (a) TCP and TLS; (b) QUIC.

4) Under TCP, multiplexed messages form a *single serial flow of bytes* and are controlled accordingly. In contrast, **each QUIC stream is separately managed**. This functionality enhances the operation of error, flow, and congestion control schemes. Lacking such a separation, a loss of a packet that is embedded in a byte stream could block reception of another packet that belongs to another stream, even if the latter is received correctly. Such an occurrence is identified as *Head of the Line Blocking (HOLB)*, which would slow or disrupt the message transport process.

5) As they are transported across a network, TCP messages are often captured and processed by *middle-boxes* that may be placed at the boundaries of organizational domains, rather than processed solely by the targeted end-user entity, as intended. Such middle-box processors are able to observe TCP control data that are carried in *cleartext* and identified as *wire image* units. They are programmed at times to modify TCP control data, also identified as *meta data*. Such modifications may interfere with the transport process. Independently programmed manipulations by middle-boxes make it difficult to introduce protocol extensions and innovations. For example, certain middle-boxes may not be able to correctly recognize certain wire-image data that are used by an extended or a new protocols, and may therefore interfere with the proper operation of this protocol. This phenomena is identified as **protocol ossification**. Under QUIC, data and meta-data units are end-to-end encrypted, so that they cannot be manipulated by middle-boxes.

Based on the specifications provided in IETF RFC 9000 [29] and associated IETF documents, we outline in the following several key structural and operational elements of QUIC.

1) A **connection set up** involves an handshake between a client and a server. The QUIC handshake integrates the negotiation of transport and cryptographic (TLS) parameters. Hence, as illustrated in Fig. 16.18, rather than experiencing several RTTs in performing a cryptographic handshake and then establishing a transport connection, as done when configuring a TCP connection via a 3-way handshake process, application data can be sent as soon as possible (ASAP). Using a prior setting process, it is possible for clients to send, when ready, application data immediately, denoted as a 0-RTT interaction, rather than incurring an RTT delay.

2) Under a cryptographic handshake, an authenticated key exchange is carried out. The server and optionally the client are authenticated. Distinct keys are produced for every connection. Also authenticated are the transport parameter values set by the involved end points and the contents included in the negotiation process that involve the communicating application layer entities.

3) A QUIC connection is characterized by connection identifiers. Each end point configures one or several connection IDs, which are included by its peers in packets that they communicate. The QUIC connection is initiated by a client and is terminated at a server, each of which is identified as an *end point*. Using these identifiers, it is possible to transfer a connection from a current network path to a new network path. These identifiers also permit the maintenance of connections under changes in network topology or address mappings, such as those used under NAT.

4) A complete data unit used by QUIC is identified as a **QUIC packet**. QUIC packets are carried in **UDP datagrams**. One or several QUIC packets can be included in a single UDP datagram.

5) A Packet is identified by a 62-bit number. If data in a packet needs to be retransmitted, it is included in a newly formed packet, which is attached the sequentially next (per stream) packet number.

6) A QUIC packet is encrypted separately from other packets. Decryption delays induced by waiting for partially received packets are avoided. QUIC packets should thus not be fragmented. In comparison, under TCP, the decryption process is performed over a byte stream.

7) Packet types include initial, handshake, 0-RTT, and 1-RTT packets. Certain packet types are ACK-eliciting, requiring the sending end point to confirm delivery by receiving an acknowledgment from the receiving end point.

8) QUIC packets carry data by using multiple **frames** as payloads. Frames are classified into different frame types.

9) Application layer entities at end users interact with each other over a connection via (unidirectional or bidirectional) ordered flows of bytes, known as **streams**.

10) A QUIC connection can accommodate multiple simultaneous streams, whereby each stream consists of an ordered flow of bytes. A stream is identified by a unique stream ID, which is expressed as a 62-bit integer. It can be unidirectional or bi-directional, client initiated or server initiated.

11) A QUIC packet contains frames. Frames are of different types. Frames are used to open a stream, carry data, and close a stream.

12) Users interact across a QUIC end-to-end connection via **QUIC packets**, which carry application data and control information. Application data is encapsulated in frames. A stream ID and an offset identifier allow a QUIC end point to properly order the data that it receives. If required, it is responsible for delivering its received data to its application entity in correct order. Frames that are out of order, within a specified limit, are buffered.

13) Management of the transport protocol is transferred to an application layer entity. By transferring transport responsibility from a device kernel space to the user space, an application gains control independence and flexibility and the ability to hide control parameters from network entities.

14) Rule-based processes for connection tear-down can be specified.

15) **End-to-end error, flow, and congestion control** algorithms are employed.

16) For **flow and congestion control** purposes, the number of streams and the amount of data that can be exchanged are regulated through the use of a *credit management* scheme. Data transmission by an end point is flow controlled by its peer end point.

The multiplexing of multiple streams at an end point may be limited by existing resources. An application entity can specify a priority weight for a stream which would identify to QUIC the relative priority of allocating resources to an underlying stream.

To enable a rapid 0-RTT interaction, a client and server pair can use protocol parameters that were negotiated and stored under a previous connection. Also stored is information and parameters used for the application protocol and for the cryptographic handshake. These parameters are applied until the handshake is completed and the client then starts sending 1-RTT packets. The transport parameters configured under the latest handshake are employed. A 1-RTT packet includes a 1-RTT protected payload and a short header. The header includes identification of the destination connection ID and a packet number. A retry packet contains a token that the server can use to validate the client's address (confirming that the same token is included in the received client's response).

QUIC provides for address validation during connection establishment and during connection migration. If not validated, an end point could be maliciously used for packet amplification attacks. In this case, client A would send server B a message with a spoofed source address of client C. Client A could then generate and send long messages to server B and/or produce them at an excessively high rate. Such a scenario may flood client C with an excessively high rate of response messages. Prior to using the address validation method, an end point would protect itself by limiting its message send rate.

QUIC also provides for path validation. Either peer is able to verify at any time that a packet sent on a path to a peer actually reaches the targeted peer. Path-challenge and path-response frames are used for the validation process.

In performing error control, an end point informs its peer about application-level and transport-level errors that it detects. Application-level errors can be isolated to a specific stream. Connection errors can induce the closing of a connection. Stream errors may result in stream resetting or termination.

Data packets are protected by using the cryptographic keys obtained during the handshake phase. Various levels of protection are applied to control packets. Certain packets are not protected for confidentiality but only against accidental modification. Keys to use for decrypting such packets can be derived by using values that are visible on the wire.

For congestion control purposes, QUIC end points may detect and respond to congestion events by using an ECN approach. Upon detecting congestion, a QUIC end point can set an ECN-capable Transport (ECT) code point in the ECN field of an IP packet. An ECN end point reacts to such a setting by reducing its sending rate. The sending end point must first confirm that the path supports an ECN marking operation.

QUIC may include one or several packets within a UDP datagram. QUIC requires the largest UDP payload that can be sent across the network to be equal to at least 1200 bytes. It assumes a minimum IP packet size of 1280 bytes. Accounting for a UDP header size of 8 bytes and a minimum IP header size of 20 bytes under IPv4 and 40 bytes under IPv6, we conclude that the maximum datagram size would be equal to 1252 bytes under IPv4 and to 1232 bytes under IPv6. A Path Maximum Transmission Unit (PMTU) discovery process can be used to determine the maximum permissible size of a transmission unit for a flow that traverses a network path.

The following QUIC frames and associated parameters are noted:

1) A QUIC receiver acknowledges the highest (ACK eliciting) packet number received. If there are gaps in the received packet sequence, it also identifies the received contiguous blocks of lower-numbered packets. Up to 256 such number ranges can be identified in a single frame. In comparison, TCP selective ACK (SACK) is limited to 3 such ranges.
2) Retransmission of lost data is performed within the context of a QUIC stream. A lost numbered packet is not retransmitted. Rather, the lost payload data is included in a new packet that is associated with the same stream, and which is assigned a packet number that is compatible with the current sequence.
3) An ACK frame includes the following fields: the largest packet number being acknowledged; the locally incurred acknowledgment delay (in microseconds); the number of ACK range fields in the frame; first ACK range, which states the number of contiguous packets preceding the largest packet that is acknowledged; additional ranges of packets that are alternatively not acknowledged; ECN feedback, and report of the receipt of any QUIC packets with ECN code-points.
4) A stream frame offset identifier serves the same purpose as a TCP sequence number.
5) A crypto frame is used to transmit cryptographic handshake messages. It can be included in all packet types except 0-RTT.
6) A new token frame is sent by a server to a client so that the client would include it in the header of an initial packet for a future connection.

Algorithms that the QUIC protocol may use for error, congestion, and flow control operations are described in IETF RFC 9002 [28]. Relating to packet error control operations, the following is noted:

1) QUIC uses *separate packet number* spaces for each encryption level, except 0-RTT. All generations of 1-RTT keys use the same packet number space. Using separate packet number spaces ensures that the acknowledgment of packets sent with one level of encryption will not cause spurious retransmissions of packets sent with a different encryption level.

2) QUIC's packet number is strictly increasing within a packet number space and directly encodes transmission order. Delivery order is determined by the stream offset indicators included in stream frames. In comparison, under TCP, receiving order is linked with transmission order, which is generally signaled by associated sequence numbers.

3) Data frames that require reliable delivery are acknowledged. When they are declared lost, they are sent in new packets.

4) QUIC packets may contain multiple frames of different types. Packets that contain no ack-eliciting frames are only acknowledged along with ack-eliciting packets.

5) When a packet containing ack-eliciting frames is detected lost, QUIC includes it and its frames that require retransmission in a new packet that is identified by a new packet number. This process removes ambiguity about which packet is being acknowledged.

6) A loss epoch in QUIC starts when a packet is lost and ends when any subsequently sent packet is acknowledged. TCP would wait for sequence number gaps to be resolved. It may take several round trips for the loss epoch to resolve as a packet may be lost multiple times when consecutively transmitted. Under both protocols, the congestion window is reduced only once per epoch. QUIC would do it once for every round-trip that experiences loss.

7) As noted, QUIC ACK frames contain information similar to that used in TCP selective acknowledgments. QUIC supports the inclusion of many ACK ranges, which speeds recovery.

8) Various conditions are specified for declaring a transmitted packet to be lost. For example, an un-acknowledged packet that is sent earlier than an acknowledged packet, whereby the time difference is higher than a specified threshold level, would be declared lost.

9) When ack-eliciting packets are not acknowledged within an expected period of time, a Probe Timeout (PTO) signal is produced at the sender, inducing the sending of one or two probe datagrams.

10) A QUIC PTO value is calculated in a manner that is similar to that used by TCP in its computation of a retransmission timeout (RTO) level. PTO's computation takes into account a peer's measurement of the maximum expected local delay that is incurred when producing an acknowledgment.

11) A sender restarts its PTO timer every time an ack-eliciting packet is sent or acknowledged, so that the PTO is based on the latest estimate of the RTT incurred by correctly received packets. The duration of a PTO timer is increased when the timer expires.

12) Using monitored RTT samples, QUIC calculates an estimate of the time elapsed between the instant that a packet is sent to the time that it is acknowledged. As performed by TCP, an RTT calculation is based on using recently measured average and standard deviation delay values.

13) When a PTO timer expires, the sender transmits at least one or two ack-eliciting probe packets within the packet's number space. New data would be included in such packets. Previously sent data can be included if no new data.

Relating to flow and congestion control operations, we note the following [28]:

1) For flow control purposes, a QUIC end point receiver controls the maximum amount of data that it is willing to receive across a connection in the following ways. If a limit is exceeded, the connection may be blocked and terminated. When acceptable, the receiving end point may increase the limits.

a) Under **stream flow control**, it limits the amount of data that each stream is permitted to carry.

b) Under **connection flow control**, it limits the amount of data that can be sent by all streams sharing the connection.

c) Under **concurrency flow control**, an end point restricts the maximum number of streams that its peer can simultaneously open.

2) Separate **flow control** processes are performed for each stream. Flow control associated frames include:

a) A MAX_DATA frame is used to inform a peer of the maximum amount of data that can be sent on the connection.

b) A MAX_STREAM_DATA frame informs a peer of the maximum amount of data that can be sent on a stream.

c) A MAX_STREAMS frame informs the peer of the maximum cumulative number of streams of a given type that it is permitted to open.

3) A sender that is blocked due to connection or stream flow control, sends a Data_Blocked or a Stream_Data_Blocked frame.

4) RFC 9002 [28] describes a **congestion control algorithm** that is similar to a scheme that has been used for TCP congestion control. Congestion control is performed per path. A sender may choose to employ another congestion control algorithm.

5) Under the noted congestion control scheme, the congestion window is reduced exponentially fast in response to persistent congestion.

6) A sender should pace the sending of its packets by regulating the minimum spacing that is used between the times at which subsequent packets are sent. This is done in conjunction with the operation of the noted congestion control scheme. The duration of a burst period, which contains successive packet transmissions, is limited in accordance with the parameters set by the underlying congestion control algorithm.

In addition to supporting reliable streams, QUIC can also provide an unreliable (connectionless) datagram service. Data is secured via encryption and included in a datagram whose frames are not associated with a stream. Lost data is not retransmitted. Yet, if a packet containing such a datagram is acknowledged, the application entity can be notified of its successful delivery.

To support the proper operation of a transport layer process under QUIC, as it is being implemented in Internet systems for use by various service providers, network transport modifications are often required. Many Internet systems continue to use TCP to support connection-oriented reliable transport of data streams. UDP is being used to provide unreliable datagram services, such as DNS. Middle-boxes that regulate flows such as those that cross a boundary of an organizational domain have often been programmed to make use of (exposed) wire-images of TCP control data. They may use such data to enforce filtering, security, and traffic management operations, and to invoke desired flow control and server load balancing regulations. They may not be able to continue applying such operations when handling secure QUIC protocol-based flows.

16.6 Routing Over the Internet

16.6.1 Autonomous Systems as Routing Domains

As described in Section 15.3, to route packets in a scalable fashion over a large network such as the Internet, the network is divided into **routing domains**. A single routing domain consists of networked systems that employ routing protocols that are managed by a single administration.

Over the Internet, an **Autonomous System (AS)** serves as such a routing domain. Often, an AS is controlled by a single entity such as an ISP or by a large organization that consists of multiple interconnected network systems. Multiple network operators can provide management and control services for an AS while serving a single administrative entity.

As described in Chapter 15, over the Internet, intra-domain routing protocols are identified as **IGPs** while inter-domain routing protocols are presented as **EGPs**. Within each domain, messages are guided along desirable routes by domain routers. In turn, **boundary routers**, often identified as *Autonomous System Boundary Routers (ASBRs)*, are used as **border routers**, which serve to navigate messages across domains.

Routing of packets across multiple routing domains, or AS, is illustrated in Fig. 15.3. Each routing domain is identified by an Autonomous System Number (ASN). Commonly used IGP include RIP (Routing IP), OSPF (Open Shortest Path First) and CISCO's EIGRP. The Border Gateway Protocol (BGP) is commonly employed as an EGP, serving to route packets across ASs.

16.6.2   Intra-domain Routing: OSPF

Link state routing algorithms are capable of selecting good routes for the dissemination of packet flows within a routing domain. OSPF is such a commonly used routing algorithm. Under this algorithm, when the AS consists of a single OSPF area, each domain router obtains a complete topological map of the routing domain and uses it to find the best route to employ for navigating packets to their destinations.

To secure a full map of the domain, each router sends its link state vector to each other router in its domain by broadcasting periodically **Link State Advertisements (LSAs)**. Using this data, each router constructs a weighted graph model of its domain. Using an algorithm such as Dijkstra's Algorithm, also known as an OSPF algorithm, each router then proceeds to calculate the **Shortest Path Tree (SPT)** rooted at itself, which provides a shortest path from itself to any other router in its domain.

This algorithm and the ensuing routing process is described in Section 15.7. Under such an approach, it is possible to use specified link cost values to determine stable (noncyclic) high-performance routes through large and complex network routing domains. As link and path metrics or cost values change, the protocol is able to rapidly adapt.

Often, the link cost for a link connected to a specific interface of a router is related to the bit rate used at this interface. It is then possible to calculate a link cost as the ratio of a reference bit rate divided by the bit rate at this interface. For example, if a reference link bit rate of 1 Gbps = 1000 Mbps is selected, then the cost of a link which operates at a rate of 1 Gbps is equal to 1, and the costs of links whose bit rates are equal to 500 and 100 Mbps are set to 2 and 10, respectively. Link costs that are lower than 1 are often set to a value of 1.

Hosts or routers that reside in a single broadcast domain (such as one associated with an Ethernet LAN or an Ethernet Switch) have their packets routed to the domain's designated router (DR). An alternate designated router can also be specified and employed in case of failure of the primary DR.

A hello multicast process is used by each router to advertise the link costs of its interface links to its neighboring routers, forming its link state vector. Each router saves the link state advertisements that it receives in a **link-state database**, which it proceeds to advertise to its own neighbors. Each domain router multicasts its latest link-state database to its own neighbors. Eventually, the link state information advertised by each domain (AS) router is flooded throughout the network, leading to a state at which the link state databases at the routers are synchronized. Using this information, each router is able to construct a weighted graph model of the domain's network. Each

510 | 16 *The Internet*

router then calculates the least cost path to each other domain router, forming a SPT. Routers keep sending hello messages, so that topological and link state changes are rapidly propagated throughout the domain, leading to fast recalculations of optimal routes.

Accordingly, an OSPF router maintains the following tables: (1) A Neighbor Table, where it stores the identity of its neighbors. (2) A Topology Table, which describes the topological layout of the underlying network domain. (3) A Routing Table, which specifies the best route to use to reach identified entities.

When a packet is received at a router, the router seeks a match between the destination IP address specified at the header of the packet and an IP address prefix listed in its routing table. If multiple matches are detected, the longest address match is selected, as it specifies a route that is more specifically suitable to reach the desired destination. A default match forwarding entry is also provided, indicating the output port to use to forward a packet in case no match is detected. Such a configuration can reduce the size of the routing table when only packets with specific destination prefix addresses are forwarded to specified ports, while all other packets are switched to a specified default port.

To provide for a scalable calculation of routes for routing domains that contain a large number of routers, the OSPF network can be subdivided into hierarchically organized **routing areas**. An *OSPF backbone area* is defined, and denoted as area 0, or 0.0.0.0, using a 32-bit number as an area identifier. Each other area is connected by a router to the backbone area. Such a router is known as an *area border router (ABR)*. An ABR keeps a separate link-state database for each area it serves and maintains summarized routes for all areas in the network. A router whose all interfaces belong to the same area is identified as an *internal router*. A *backbone router* has an interface to the backbone area. An *autonomous system boundary router (ASBR)* is a router that uses more than one routing protocol. It exchanges routing information with other AS. ASBRs typically use static routes and/or run an exterior routing protocol such as BGP. An ASBR is used to distribute routes received from external ASs throughout its own AS. An ASBR creates External LSAs for external addresses and floods them to all areas via an ABR. Routers in other areas use ABRs as next hops to access external addresses. ABRs forward packets to an ASBR, which announces external addresses.

A router may run multiple OSPF processes. A router that is connected to several areas, and which receives routes from a BGP process connected to another AS, is both an area border router (ABR) and an ASBR.

In Fig. 16.19, we illustrate a layout of multiple routing areas under OSPF. Internal area routers are denoted as IRs, backbone routers as BRs, area border routers as ABRs, and an autonomous system boundary router is denoted as ASBR. In each area, hello packets are periodically sent by each router to maintain association with its neighboring routers, and detect changes in the status of its neighbors. Link-state advertisements (LSAs) are flooded by each internal router to any internal router, border router, and boundary router located in the underlying area (such as routers R1, R2, R4, and R5 for Area 0). Using disseminated link state data, each one of these routers constructs (e.g., by using Dijkstra's algorithm) the SPT rooted at itself, identifying the shortest path in the area from itself to each other area router. In this manner, each router in an area such as A0 is able to navigate a packet that it receives to its indicated destination, which could be a host computer attached to one of the area LANs, such as LAN0.1, LAN 0.2, LAN 0.3, and LAN 0.4. ABRs are used to route packets that flow between a non-backbone area and the backbone area, as well as between non-backbone areas. They disseminate summary-LSAs are used to inform other areas about inter-area routes and routers.

Packets that are transported between host computers that are located in non-backbone areas are routed in each non-backbone area to an ABR. The latter uses its backbone area routing table to

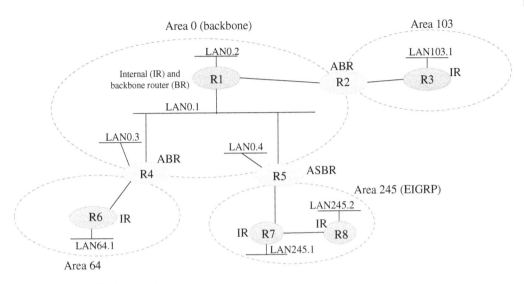

Figure 16.19 OSPF Routing Areas and Router Types.

route packets across the backbone area by using a shortest path to an ABR that connects to the area where the destination host is located.

Routing packets to and from an area that belongs to another AS (such as Area 245 that employs its own distinct routing algorithm) is performed by ASBR R5. An ASBR disseminates External-LSAs that contain information imported into OSPF from other routing processes. ASBR-Summary-LSAs are disseminated by ABRs, describing routes to AS boundary routers beyond its area. Jointly, ASBR-Summary and AS-External LSAs describe paths to use to reach an external router.

16.6.3 Inter-domain Routing: Border Gateway Protocol (BGP)

BGP is the commonly employed exterior gateway protocol used over the Internet for inter-domain (and thus inter-AS) routing. It is often referred to as a *path-vector routing protocol*. Decisions for routing packet flows across ASs are based on rules set by network administrators and network policies.

The processes used for routing packets across multiple ASs are illustrated in Fig. 15.3. An IGP is used to route packets within an AS, using a routing protocol such as RIP, OSPF, or EIGRP. To route packets across ASs, border routers (BRs) are used. A border router is located at the boundary of multiple ASs and is able to route packets across ASs. To determine an end-to-end inter-domain path to use for routing a packet whose destination router resides in a different AS, a border router employs an inter-domain routing protocol, identified as an EGP, such as BGP.

Considering the multidomain network system shown in Fig. 15.3, assume that source router R214, which is located in AS210, wishes to send packets to destination Router R101, located in AS100. The selected route would have to traverse several routing domains. The intra-domain protocol (IGP) employed by AS210 will be used to select a route that leads from R214 to border router BR211. This border router will then select a preferred inter-domain exterior path that traverses intermediate border routers, leading to a border router at the destination domain, AS100. One such inter-domain path traverses the three domains AS210-AS200-AS100, involving border routers BR211-BR202-BR201-BR102. The end-to-end path also includes interior routes located

512 | *16 The Internet*

in each involved domain, such as interior path R214-to-BR211 in domain AS210, interior path BR202-to-BR201 in domain AS200, and interior path BR102-to-R101 in domain AS100.

Another optional inter-domain path traverses the four domains AS210-AS200-AS300-AS100, transiting border routers BR211-BR202-BR201-BR301-BR302-BR102. The end-to-end path would include the interior intra-domain routes in each one of the traversed domains, selected in accordance with the IGP that is employed in each traversed AS.

In considering which exterior inter-domain path to select, a border router that is running BGP depends on path-vector advertisements that it receives from neighboring border routers. In its *reachability* advertisement message, a border router identifies the domains (ASs) that it can reach as well as the distance (or cost) metric value that characterizes each path that reaches each advertised destination domain. A commonly used such metric simply specifies the number of AS domains (also known as AS-hops) that the path traverses to reach the destination AS. For example, BR202 may advertise the above-mentioned two inter-domain routes that it can use to reach destination domain AS100, attaching to them the respective metric values of 2 and 3 AS-hops. BR211 may select the path that uses a smaller number of inter-domain hops, which in this case corresponds to selecting inter-domain path AS210-AS200-AS100.

BGP provides for the exchange of routing information among BGP routers that are located in different ASs. BGP uses the routing information to maintain a database of network layer reachability information (NLRI), which it exchanges with other BGP systems. BGP uses the network reachability information to construct a graph of AS connectivity, which enables BGP to remove routing loops and enforce policy decisions at the AS level.

Multiprotocol BGP (MBGP) extensions enable BGP to support IP version 6 (IPv6). NLRI update messages carry IPv6 address prefixes of feasible routes.

The general principle of operation of BGP is similar to that of a *distance-vector routing scheme*, whose operation is described in Section 15.6. As illustrated in Fig. 15.5, when applied to the operation of BGP, a BGP border router receives NLRI, including corresponding route metric values, from its neighboring BGP border routers with whom it is on speaking terms.

Neighboring BGP routers, called peers, interact over a TCP session that they manually establish (over port 179). A BGP speaker sends periodically keep-alive messages to maintain the connection. A border router that is located on the boundary of an AS exchanges BGP information with a peer router that is located on the border of another AS. Such routers tend to use direct connections. Routes learned by a BGP router from an external BGP (eBGP) peer are advertised to all its eBGP peers. During the peering session, BGP speakers can negotiate the parameters of the session.

A BGP router uses routing policies to choose among multiple paths to a destination and to control the redistribution of routing information. In doing so, the border router is able to calculate the end-to-end metric value of each candidate route by accounting for interior distance value components, such as those characterizing the cost of each interior route that connects it in its AS to each border router that terminates a candidate advertised inter-domain route.

When BGP runs between two peers in the same AS, it is referred to as Internal BGP (iBGP or Interior Border Gateway Protocol). When it runs between peers associated with different AS, it is called External BGP (eBGP or Exterior Border Gateway Protocol).

It is also feasible to run eBGP peering inside a VPN tunnel, allowing two remote sites to exchange routing information in a secure manner. eBGP peers are typically connected directly. In turn, iBGP peers can be interconnected through intermediate routers.

New routes learned from an eBGP peer are re-advertised to all iBGP and eBGP peers, while new routes learned from an iBGP peer are re-advertised to all eBGP peers only. Accordingly, iBGP peers inside an AS are typically interconnected in a full mesh by using iBGP sessions (though methods are available to achieve higher iBGP connection scalability by using confederation or reflector-based connectivity approaches).

The routing information that BGP systems exchange includes the complete route to each destination, as well as additional information about the route. Such a route is called an AS path. Additional route information is expressed as path attributes. BGP uses the AS path and path attributes to determine the network's topological layout. Once BGP determines the topology, it can detect and eliminate routing loops and select among groups of routes to enforce administrative preferences and to follow routing policy rules.

A BGP route is characterized by the IP address prefix of the destination node and the composition of the path leading to the destination. The latter includes the list of AS numbers that the underlying AS path traverses as well as path attributes.

BGP peers use update messages to advertise to each other the routes that each learns. A BGP router stores its routes in its **routing table**, which can include the following information: routing information learned from update messages received from peers; local routing information that BGP applies to routes because of local policies; information that BGP advertises to BGP peers in update messages.

For each prefix entry in the routing table, the routing process selects a single best path, called the active path. Unless you configure BGP to advertise multiple paths to the same destination, BGP advertises only the active path. A route advertised by a BGP is assigned one of the following values to identify its origin. During route selection, the lowest origin value is preferred: 0 – the router originally learned the route through an IGP (such as OSPF, IS-IS, or a static route); 1 – the router originally learned the route through an EGP (most likely BGP); 2 – the route's origin is unknown.

A multi-AS network layout is illustrated in Fig. 16.20. AS are shown to be linked by border routers. For example, AS500 and AS600 are linked by BR1.1 and BR2.1, which are directly connected to each other. Each pair of neighboring border routers establishes a TCP session across which they exchange BGP messages. Included are BGP update messages, which are used by the border routers to announce new routes that they learn as well as to indicate updates to prior advertised routes.

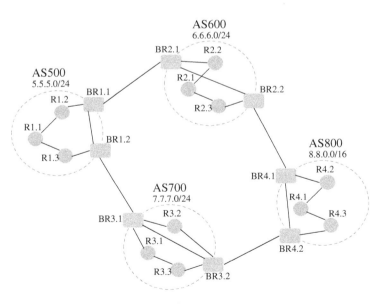

Figure 16.20 A Multi-AS Network Layout.

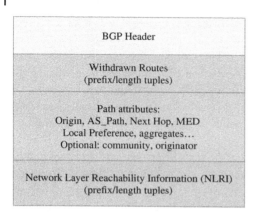

Figure 16.21 Contents of a BGP Update Message.

The contents of a BGP update message illustrated in Fig. 16.21. NLRI identifies paths that a border router advertises, identified by path prefix and path attributes.

Examining the network layout shown in Fig. 16.20, consider a packet flow that originates at router R1.1 that is located in AS500, which is routed to destination router R4.3, located in AS800. The border routers in AS500 learn about routes from the neighboring border routers with whom they exchange BGP messages.

To illustrate, note the following. Border routers in AS800 advertise to neighboring BGP border routers in AS600 and in AS700 about their reachability of routing prefix 8.8.0.0/16, where the destination router is located. A border router in AS600 which speaks to a border router in AS500, then advertises its reachability of the corresponding routing prefix by citing the AS-path AS600-AS800. Similarly, a border router in AS700 that communicated with a border router in AS500 would advertise the AS-path AS700-AS800.

Using this information, and their determination of other parameters, such as local preference parameters that characterize the desirability of using a specifically preferred interior path, one of the border routers in the source AS will elect itself, as the source border router, linked to an end-to-end AS path, and accordingly include a corresponding routing entry in its forwarding table. For example, assuming the latter router to be border router BR1.2, the end-to-end route will consist of the exterior path AS500-AS700-AS800, coupled with the selection of the corresponding internal paths in the source and destination ASs.

Key steps of the process used to update routes at a BGP border router are shown in Fig. 16.22. The physical implementation can vary among routers. Newly learned routes derived by advertisements received from adjacent (neighboring) routers are stored (per neighbor) in an Input Adjacent Routing Information Base (RIB), identified as Adj-RIB-In. Route selection policies will then be applied by the local BGP router. Applying such policies, which can vary among BGP router entities, preferred routes are selected and stored in a Local-RIB. Locally learned routes (including those derived through the application of an interior routing protocol such as OSPF or IS–IS) are incorporated in selecting and storing preferred routes in an IP Routing Table, as well as their inclusion in the Local-RIB.

The next hop to use to reach an underlying address prefix is then resolved and the outgoing interface is determined. This information is included as forwarding entries in a Forwarding Information Base (FIB). A forwarding entry specifies, for a destination address prefix, the next hop to use in routing an incoming packet. It also specifies the router's outgoing interface that is used to connect to the hop. Routes that are advertised to BGP neighbors are stored in the Adj-RIB-Out modules.

16.6 Routing Over the Internet | 515

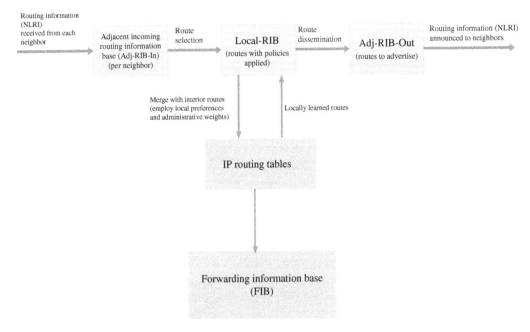

Figure 16.22 Route Update at a BGP Border Router.

In selecting the preferred routes, a BGP router is able to choose from a multitude of criteria. In case multiple routes yield similar preference weights, several tie-breaking rules are specified. A route learned from an exterior neighbor via eBGP is preferred to one learned from iBGP. Also preferred is a route that has lower interior cost to the next hop. BGP routers tend to be configured to prefer next hop entities that are reached via an established hardware connection.

After verifying that the next hop is reachable, if the route comes from an internal (i.e., iBGP) peer, the LOCAL_PREFERENCE attribute (related to the cost metric of the interior route) is examined. If there are several iBGP routes from a neighbor, the one with the highest LOCAL_PREFERENCE is selected unless there are several routes with the same LOCAL_PREFERENCE, in which case tie breakers are defined.

Following verification of next hop reachability, certain vendors consider a decision factor called WEIGHT which is local to the router (i.e., not transmitted by BGP). The route with the highest WEIGHT is then preferred. If the route was learned from an external peer, the per-neighbor BGP process computes a LOCAL_PREFERENCE value from local policy rules and then compares the LOCAL_PREFERENCE of all routes from the neighbor.

In selecting routes advertised by neighbors, it is generally preferable to choose the route with the **shortest AS_Path**. The distance metric value of such a path is equal to the number of AS domains that the route traverses. In case of ties, the first preference is to choose a route with the lowest value of ORIGIN attribute and then select a route with lowest MULTI_EXIT_DISC (multi-exit discriminator or MED) value. MED is an optional attribute that indicates to external neighbors about the preferred path into an AS that has multiple entry points. A lower MED value is preferred over a higher value. It is a valid component only when used between neighboring AS's.

BGP community identifiers are defined as (2 byte) attribute tags that can be applied to incoming or outgoing prefixes to achieve targeted route dissemination objectives. It is possible to configure in this manner restrictions on the advertisement of routes to selected domain communities. For example, if a No-Advertise community tag is attached to a route, the BGP speaker will not advertise the route to internal or external BGP peers. Under a No-Export community tag, the router does not advertise the route to external peers. Extended community attributes use 8-byte tags. The first 2 bytes specify the community type, while the last 6-bytes identify specific features relating to dissemination of routes for the underlying community. For example, as often used in Multi Protocol Label Switching-Virtual Private Network (MPLS-VPN), they can be used to identify routers that may receive a specific set of routes, or those that may inject into BGP a specific set of routes.

In comparing the **performance** of interior and exterior routing schemes, we note the following. Interior routing schemes generally handle smaller number of routes and prefix addresses, and therefore converge faster, though they may employ more complex routing algorithms. BGP, as an exterior routing scheme, must employ a very large number of candidate routes and network prefixes. Due to its relative simple algorithmic structure, it is able to operate in a highly scalable fashion and at reasonable timeliness, while allowing different AS organization to impose their own routing criteria or constraints.

It is of interest to characterize the **convergence times** for a routing algorithm based on the time that it takes the algorithm to reach its steady-state operation. In addition, it is of interest to evaluate the time that it would take a routing scheme to determine new routes following the failure of specified types of network node or link elements, or following changes in the network's topological layout. For example, convergence time measurements conducted on specific network layouts under various link failures and under the addition of various routers have shown that the convergence times for interior routing protocols such as EIGRP, IS–IS, and OSPF, were much shorter, typically of the order of several seconds. In turn, BGP tends to take much longer time to converge, often as long as a fraction of a minute to several minutes.

BGP serves a critical role in managing inter-domain routing across the Internet. BGP routers learn about a very large number of routes, store them in RIBs, select the preferred ones, compose forwarding tables, and advertise new and stale routes to their BGP neighbors. In assessing the challenges impacting the timely and stable operation of the BGP system, we discuss in the following two key processes: route update and packet forwarding.

The control message processing loading level that impacts a BGP router depends on the changes that take place while learning new routes and upon the deletion of older invalid routes. As a neighboring border router advertises to a BGP speaker about reachability to an IP address prefix, the BGP speaker compares this new information to stored routing data gained from previous advertisements. If the router determines that the new information yields a better path to a prefix, the speaker inserts this data, consisting of this prefix and the next-hop forwarding identifier to its local forwarding table. It then proceeds to inform its BGP neighbors of a new path to a prefix, which expresses itself as the next hop.

In turn, if a BGP speaker determines that it does not have an acceptable path to a specific prefix, it will advertise its withdrawal to its neighbors. Such a change may lead neighboring BGP routers to modify their preferred path to the latter prefix if an alternate path can be identified, and then inform their own neighbors about the change. Otherwise, a neighbor will disseminate withdrawal announcements to its own neighbors, indicating that it has no longer a viable path to reach the underlying prefix.

The stability of the routing system is impacted by the rates imposed on BGP routers by route update processes. When such a rate becomes higher than the native update capacity rate of a BGP router, unprocessed route update messages incur long queueing delays. Such occurrences can induce a significant slow-down in the route update convergence process. If such update rate conditions worsen and remain unresolved, BGP routers can get out of sync with current network topology, failing to learn about currently preferred routes. Routers may then attempt to forward packets across failed routes. Out-of-sync BGP routers can consequently advertise incorrect reachability information. Such conditions can lead to routing loops and the consequent saturation of transmission and route processing resources, leading to outages and excessive traffic loads.

Using data derived from online sources, we note in the following numerical features associated with the BGP operation over the Internet. The packet forwarding process involves finding a match between a destination address prefix embedded in an incoming packet and a corresponding forwarding entry in the forwarding table that specifies the best next hop and interface. Regarding BGP table size growth, the following is noted. As of 2014, many routers were limited to 512,000 IPv4 routing table entries. Consequently, in 2014, outages resulting from the impact of full tables at several organizations were noted. The number of advertised IPv4 routes reached a default limit. Routers attempted to compensate by using slow software routing, as opposed to fast hardware routing, which uses high-speed memory called Ternary Content Addressable Memory (TCAM). Also, route summarizations have been employed to reduce the number of routing entries. As of 2021, about 900,000 prefixes have been noted to be announced on the Internet, while about 70,000 AS domains have been observed.

It is noted that the speed required to perform timely packet forwarding processes increases as the attached communications links tend to employ higher data rates. For example, at a link data rate of 100 Gbps, the rate at which IP packets of size of 1000 bits arrive at the router becomes equal to 100 [MPackets/s]. Hence, a packet would arrive every 10 ns. A BGP router that may hold an order of 1 million entries would then have to find a match in a period of time that is of the order of a nanosecond, or a fraction of it (as several comparisons are generally required to determine the best match).

In BGP, there is no single authoritative view of the Internet's inter-domain routing pattern. Corresponding views are impacted by the perspective of each BGP speaker. Changes in routing are often induced not by the originator of the route but by interior transit preferences of individual organizations and by arrangements that they make to enhance their own cost and performance benefits, proceeding then accordingly to hide or expose groups of routes.

IPv4 BGP routing tables have been observed to grow from an order of 400,000 entries in 2012 to about 800,000 entries in 2021. Linear growth continues to be observed. Over this period, it has been noticed that the average BGP announcement size has dropped from about 7000 host addresses to about 3500 addresses. About 90% of all announced prefixes are noticed to be of size /20 or smaller. Thus, while the topology of the network has remained relatively consistent, the average AS path length has somewhat declined (from around 6 to about 5.5). The BGP network topology has thus been observed to have a higher density of peer BGP router connectivity.

A continued utilization of IPv6 addresses has been observed in the largest economies. The number of IPv6 BGP routing table entries over the 2012–2021 period has been noted to grow from about 10,000 entries to about 110,000 entries. Around January, 2021, about 105,000 prefixes and 21,400 AS domains were counted. A sustained growth in the use of IPv6 addresses has been forecasted.

BGP routers are configured through an application process. The distributed architecture of the BGP process depends on BGP routers accepting route advertisements from other BGP routers. While this architecture leads to a scalable and dynamically adaptive operation, it also entails major security risks. Improper advertisement of routes occurring due to accidental events or as result of actions by malicious actors, identified as BGP hijacking, can lead to major network connectivity blockages and route failures. Resolution approaches (including the use of cryptography to confirm the identity of BGP routers) are of prime importance.

518 | *16 The Internet*

Problems

16.1 Discuss the Internet Networking Architecture and describe the services provided by each layer.

16.2 Identify the layer interface points, corresponding to Service Access Points (SAPs) that are used by the TCP/IP protocol reference model and describe the function of each SAP.

16.3 Describe the services provided by the Logical Link Control (LLC) protocol.

16.4 Discuss and illustrate the formats of MAC addresses.

16.5 Define and describe the services provided by Address Resolution Protocol (ARP) and by Reverse Address Resolution Protocol (RARP).

16.6 Consider the IP Internetwork presented in Fig. 16.2. Describe the path and processes used in guiding the end-to-end flow of packets that are generated at source computer C1 and are destined to destination computer C7.

16.7 Describe the multitier organization of Internet Service Providers (ISPs).

16.8 Describe and discuss the operation of client–server communications over the Internet through the use of Hypertext Transfer Protocol (HTTP) and HTTPS.

16.9 Discuss the following fields of a URL used by HTTP or HTTPS:
a) Scheme.
b) Host.
c) Path.
d) Query string.

16.10 Describe the function of each one of the *methods* used by HTTP/1.1, HTTP/2, and HTTP/3.

16.11 Discuss how the use of the QUIC protocol enhances the performance of HTTP/3.

16.12 Describe the structure of IP Version 4 (IPv4) addresses and provide examples of such private and public addresses. Also, explain the following:
a) Routing prefix.
b) Host identifier.
c) CIDR (Classless Interdomain Routing) addressing system.

16.13 Discuss the structure of IP Version 6 (IPv6) addresses and provide several examples of such addresses.

16.14 Describe the operation of a Source Network Address Translation (S-NAT) process. Illustrate its operation by considering the network shown in Fig. 16.8.

16.15 Explain the IPv4 address assignments configured for the port entities shown in Fig. 16.9.

Problems | 519

16.16 Track and explain the process used in the network shown in Fig. 16.10 for a host that is attached to a private home LAN to communicate across the Internet with a web server.

16.17 Discuss the structure of IPv6 packet addresses.

16.18 Describe the operation of Transmission Control Protocol (TCP). Explain the following TCP services or terms:
a) Multiplexing service.
b) ARQ error control.
c) Flow control.
d) Congestion control.
e) Sequencing of data segments and sequence numbers.
f) Acknowledgment number.

16.19 Describe the handshake-based operation of the connection establishment process used by Transmission Control Protocol (TCP). Explain the role of the following messages:
a) SYN.
b) SYN-ACK.
c) ACK.

16.20 Describe the operation of User Datagram Protocol (UDP). Also, explain the following:
a) Source port.
b) Destination port.

16.21 Describe the operation of the QUIC Protocol. Discuss the following services as provided by QUIC:
a) Stream multiplexing and its avoidance of Head-of-the-Line (HOL) blocking.
b) Embedding of security protocol such as TLS.
c) Its ability to reduce the occurrence of protocol ossification.
d) Its ability to decrease the time that it takes for a client to connect with a server.
e) Its accommodation of error, flow, and congestion control algorithms.
f) Use of Probe Timeout (PTO) signals.

16.22 Describe the operation of OSPF (Open Shortest Path First) as an intra-domain routing protocol over the Internet.

16.23 Describe the operation of BGP (Border Gateway Protocol) as an inter-domain routing protocol over the Internet.

16.24 Describe and explain the communications and switching processes performed by network entities in routing packets that originate at Router R1.1 and that wish to reach destination router R4.3 across the multi-AS communications network shown in Fig. 16.20.

17

Local and Personal Area Wireless Networks

The Concept: *Wireless personal area networks (WPANs) provide low-power short distance communications over a wireless medium. A Bluetooth network is a WPAN that is used to replace cable connections between devices. A Zigbee network is used as short distance, low-power, and low data rate wireless network that is employed for monitoring and control purposes. Such applications are commonly used by wireless sensor network (WSN) systems. A wireless local area network (WLAN) that uses protocols based on IEEE 802.11 WiFi to Wi-Fi globally WiFi specifications provides wireless access to the Internet for local devices.*

The Methods: Member devices of a WPAN or a WLAN communicate their messages across a wireless medium. A medium access control (MAC) protocol is employed for sharing the medium. A Zigbee network provides communications over ranges of the order of 10–100 m. It is designed to require very low-energy resources, attaining a long battery activity lifespan. It accommodates a network that can consist of thousands of devices, such as sensor modules, operating at a data rate that does not exceeds 250 Kbps. A Carrier Sense Multiple Access with Collision Avoidance (CSMA/CA) based contention-oriented MAC protocol is employed. The network can be configured to assume a star or a mesh layout. For the later, ad hoc routing-based algorithmic approach is used to discover paths and carry out peer-to-peer communications. A Bluetooth network accommodates wireless communications between a small number of devices over a range of the order of 10 m under low power (2.5 mW) and a range of about 100 m under a higher transmit power level (such as 100 mW). It offers communications at data rates of the order of 1–3 Mbps. A star configuration is used. A session managing node allocates time slot resources to member devices so that the medium is shared in a contention-less manner. A WiFi WLAN provides for multiple access sharing of a wireless medium in offering communications between net devices and an access point (AP). It covers communications over a range of the order of 20 m indoors and 150 m outdoors, which can be expanded by using distance extension devices. In a hot-spot layout, local devices use the WiFi protocol to share a wireless medium for accessing the net's AP. The latter is attached to an IP router that is connected to an Internet Service Provider (ISP) node, which provides Internet access. Depending on the version and configuration used, data transmissions can be performed at 10 Mbps, 100 Mbps, and Gbps levels. The 802.11ax WiFi network offers data rates that span the range of 600 Mbps–9.6 Gbps. Using directional antenna configurations, extended ranges, higher data rates, and upgraded spectral utilization levels are attained. A star configuration is typically employed. Net member devices communicate through the AP. A contention oriented CSMA/CA MAC protocol is used as the basic algorithm for sharing the medium. Advanced WiFi modules employ medium reservation schemes and are able to dedicate wireless resources to specific users.

Principles of Data Transfer Through Communications Networks, the Internet, and Autonomous Mobiles, First Edition. Izhak Rubin.
© 2025 The Institute of Electrical and Electronics Engineers, Inc. Published 2025 by John Wiley & Sons, Inc.

17.1 Illustrative Personal Area and Local Area Wireless Networks

In this chapter, we review the principles of networking operations of wireless personal area networks (PANs) and wireless local area networks (WLANs).

Networking methods for wireless personal area networks (WPANs) are illustrated by discussing Bluetooth and Zigbee networks. A Bluetooth network serves to replace a short-range (of the order of meters) cable connection by using a wireless medium. It supports the sharing of a wireless channel among multiple devices, enabling a data rate of the order of 1–3 Mbps. A Zigbee network shares a wireless medium among low-power devices across a short range (of the order of meters) in carrying out monitoring and control functions, and in managing data transport for wireless sensor networks (WSNs). Communications is performed at a data rate that is typically not higher than 250 Kbps.

WLANs are widely employed for access to a backbone network, such as the Internet. A wireless medium is shared among multiple local devices. Such popular network systems are based on IEEE 802.11 standards and are identified as WiFi systems. We review the key features of such networks. They are capable of providing high speed data rates, ranging from Mbps to Gbps levels, covering distance ranges of 10's of meters and somewhat beyond. The principle of operation of key medium access control (MAC) schemes, algorithms, and protocols that are utilized by WiFi networks have been presented and discussed in Section 8.4.5.

In comparing the features offered by the above-mentioned networks, we note the following distinctions. In Fig. 17.1, we compare the extent of power consumption and complexity for versions of WiFi, Bluetooth, and Zigbee Systems, in relation to the data rate levels that they accommodate.

1) The battery lifespan of a Zigbee device can last for hundreds of days. In turn, WiFi and Bluetooth devices tend to experience battery lifespans of the order of hours/days and several days, respectively.
2) A Zigbee network can accommodate a much larger number of nodes. It can support an order of tens of thousands nodes. A Bluetooth net accommodates several (such as 7–10) nodes, and a WiFi net can normally accommodate several to tens of nodes.
3) As noted, the data rates offered by Zigbee, Bluetooth, and WiFi networks are of the order of 250 Kbps, 1–3 Mbps, and above 100 Mbps, respectively.
4) A Bluetooth net realizes a typical distance span of the order of 1–10 m, while the coverage span for WiFi is generally of the order of 1–100 m, and for Zigbee it is of the order of 1–75 m.

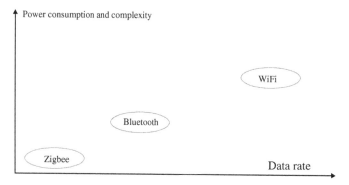

Figure 17.1 Comparison of Power Consumption and Complexity vs. Data Rate for Versions of WiFi, Bluetooth, and Zigbee Systems.

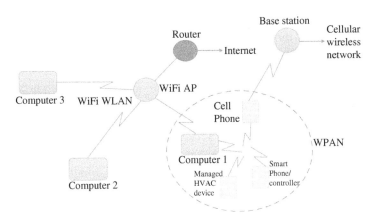

Figure 17.2 Illustrative Network Configuration Involving a Wireless Personal Area network (WPAN) and a Wireless Local Area Network (WLAN).

5) Topologically, WiFi and Bluetooth networks employ a star layout, under which a single node assumes a role as a central network manager. A Zigbee net can be formed to assume a star, tree or mesh layout.
6) The standby currents used by WiFi, Bluetooth, and Zigbee devices are of the order of 20 milli-amperes, 200 micro-amperes, and 3 micro-amperes, respectively.

An illustrative network configuration that employs a WPAN and a WLAN is shown in Fig. 17.2. The WPAN is shown to include multiple devices that are located at short distance from each other and that communicate over the net's wireless medium. Shown devices include a computer, a cell phone, a smart phone that is also used to control local devices and a managed Heating, Ventilation, and Air Conditioning (HVAC) device.

An Access Point (AP) of a WLAN, such as a WiFi network, serves to manage communications over the WLAN's wireless medium. The shown WLAN includes several computers as its user clients. The AP device is connected to an ISP router that provides connection with the Internet. A cell phone device is shown to have a wireless link that connects it to a Base Station (BS), which provides access to a cellular wireless network.

While devices that move are able to withdraw and join personal and local area networks, the associated processes are not designed to act rapidly in accommodating mobile entities and handle scenarios under which client memberships and network connectivity vary regularly. Principles of operation of mobile wireless networks are discussed in Chapters 18, 19, and 21.

A mobile wireless cellular network employs an infrastructure centered architecture. In turn, mobile ad hoc wireless networks (MANETs) use peer-to-peer message transport methods, avoiding the need to rely on a configured stationary backbone network. In Chapter 21, we discuss and illustrate the design of protocols and algorithms that are used to manage the networking operations of MANET and of hybrid infrastructure/MANET architectures for the autonomous highway.

17.2 WiFi: A Wireless Local Area Network (WLAN)

A WLAN is used for communications networking among devices (identified also as stations) that are located in close proximity to each other. Many such systems are based on IEEE 802.11 protocols, used in the implementation of different versions of WiFi WLAN systems.

WiFi stations that can communicate as members of a single net are said to form a **Basic Service Set (BSS)**. Communications within a BSS can be performed in a peer-to-peer manner, so that member stations can communicate directly with each other by sharing the BSS wireless medium. Such a BSS is identified as an Independent BSS (IBSS), as well as an ad hoc WLAN.

In contrast, an **infrastructure-based BSS** is designed to provide communications between user stations and a managing station identified as an **Access Point (AP)**. The AP shares the BSS net's wireless medium, which it uses to transmit data messages to its user station clients as well as to receive messages from its clients. Member stations of a BSS are able to communicate directly with the AP. The AP is often attached to a routing module, enabling it to route the messages that it receives from a client station to another client station. Included are source and destination user stations that are members of the BSS wireless cell.

The AP is typically attached to a backbone network, identified as an Infrastructure as well as a Distribution System (DS), such as the Internet, or a voice or data public or private network. Such a topological layout is shown in Fig. 18.1. Each wireless BSS is depicted as a cell that is managed by an AP station. The cell's wireless medium is shared by the cell's stations. It is used for frame transmissions from client (user) stations to the AP and for frame transmissions from the AP to client (user) stations.

The physical layer protocol data unit (PPDU) consists of preamble and data fields. The preamble field contains the *transmission vector (TXVECTOR)*, which includes frequency bandwidth and other transmission-associated parameters. The data field contains the user payload and higher layer headers, including *MAC* frame fields that are described in the following.

The primary MAC protocol used to share the WiFi wireless medium is based on the Carrier Sense Multiple Access with Collision Avoidance (CSMA/CA) algorithm described in Section 8.4.5. Advanced versions of the WiFi specification incorporate variations that lead to higher throughput efficiency. The features and operational principle of the basic CSMA/CA access protocol that is identified as Distributed Coordination Foundation (DCF) are described in Section 8.4.5.3. In Section 8.4.5.6, we review an expansion of the protocol to provide for priority-based access of frames, identified as Hybrid Coordination Function (HCF). In this section, we focus on the features of an advanced version, WiFi-6 (also known as IEEE 802.11ax) and its extension to WiFi-6E (also identified as IEEE 802.11ax-2021).

The basic structure of a WiFi frame as used for certain versions and frame types is shown in Fig. 17.3. The frame consists of fields that include: several MAC addresses; a frame control field; a duration field (which identifies the time duration that the medium will be occupied by the transmission of this frame and by a transmission of the subsequent ACK, when produced); a sequence control (SC) field, and Quality of Service (QoS) and High-Throughput (HT) control fields which are included in only certain frames such as QoS and certain management frames; a frame check (FCS) field, and the frame body that contains the data message payload carried by the frame. The FCS consists of a 32 bits CRC error detection code. The SC field consists of a sequence number (12 bits) and a fragment number (4 bits). A sequence number is used to filter duplicate

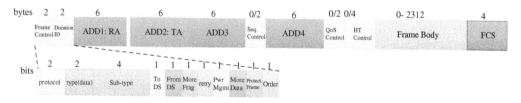

Figure 17.3 Format of a WiFi MAC Frame.

frames and the fragment number is used to order fragments. The payload of a WiFi frame is a higher layer PDU, such as an LLC frame that carries an IP packet as a payload.

The Type field specifies the function of the frame as being management, control or data. The Subtype field is used to specify association or beacon frames. The More Frag field indicates whether this frame is followed by other frames that are fragments of the same message. The Retransmission field indicates whether the frame is a retransmission of a previous one. The Power Management tag indicates whether the station operates in a power saving mode. The More Data tag indicates whether the sending station has more data to send. The WEP tag indicates whether a WiFi security is applied. The order tag indicates whether the frames must be processed in strict order.

The information included in the MAC addresses depends on the direction of the frame's flow. Address-1 designates the MAC address of the frame's receiving entity for a transmission performed across the BSS wireless channel, while address-2 specifies the MAC address of the frame's transmitting entity. For unicast data, the receiver sends its ACK frame to the transmitter identified by address-2.

The AP serves also as a gateway station. It is generally attached to a router that connects it to a DS. The WiFi protocol manages transmissions within the Wireless LAN's BSS. It does not specify the protocols and schemes used for message transport across the DS. The BSS Identifier (BSSID) represents an address associated with the AP.

As noted, the role of address components depends on the direction of the frame's flow, as identified by the To-DS /From-DS tags specified in the header's control field. When this pair is respectively set to 1/0, the transmitted frame is destined to the AP, so that address-1, representing the receiver's address (RA), is set to the BSSID. Address-2, which represents the transmitter's address (TA) is equal in this case to the source address (SA), TA=SA, which is identified by the MAC address of the source station that produced the payload PDU carried by the frame. Address-3 is used in this case to identify the destination address (DA) of the payload message.

When the To-DS /From-DS tag pair is set to 0/1, the frame is transmitted across the BSS medium by the AP. Hence, address-1 represents the receiver address (RA) as well as the destination address (DA), so that RA=DA. The BSSID is used as address-2, TA=BSSID, and address-3 is used then to identify the source address (SA).

A tag setting of 0/0 is used for an ad hoc net (identified as an Independent BSS). In this case, address-1 is set as RA=DA, address-2 is set as TA=SA, and address-3 identifies the BSS and thus states the BSSID.

The tag setting 1/1 is used when the BSS is employed as a wireless DS, in which case all four addresses are used. In this case, the 4 addresses are set (in order) as follows: (RA,TA,DA,SA).

The control field includes additional tags that are used to specify other MAC protocol parameters.

Key networking, operational and performance parameters associated with WiFi-4, WiFi-5, WiFi-6, and WiFi-6E are presented in Fig. 17.4. Prior to WiFi-6E, from its US start in 1985, Wi-Fi has been allocated a combined total of just 583 MHz of spectrum in the 2.4 and 5 GHz bands. Due its high utility and popularity, such a limited bandwidth allocation has proven to be highly insufficient. In 2020, a significant addition of 1200 MHz of spectrum between 5.925 GHz and 7.125 GHz was approved for unlicensed use; it is known as the 6 GHz band. It is utilized by Wi-Fi 6E, which is designed to accommodate Wi-Fi 6 and OFDMA-based structures only, rather than to stay compatible with less efficient earlier WiFi versions. WiFi 6E is able to provide a message transport service that uses a scheduled MAC scheme for offering certain designated stations and flows excellent delay-throughput performance, realizing lower interference, higher-throughput rate, and very low frame delay (of the order of 2 ms, or lower).

17 Local and Personal Area Wireless Networks

Attribute	WiFi 4 (802.11n)	WiFi 5 (802.11ac)	WiFi a6 and 6E (802.11ax, ax-2001)
Frequency bands	2.4 and 5 GHz	5 GHz	2.4 and 5 GHz and 6 GHz for 6E
Physical layer	High Throughput (HT)	Very High Throughput (VHT)	High Efficiency Wireless (HEW)
MU-MIMO	None	DL	DL and UL
Channel bandwidths	20, 40, 80 MHz	20, 40, 80, 160 MHz	20, 40, 80,160 MHz For 6E: total BW increase by 1200 MHz
Symbol Duration	3.2 micro-s	3.2 micro-s	12.8 micro-s
Guard intervals	800/400 ns	800/400 ns	800/1600/3200 ns
Spectral sharing	OFDM	OFDM	OFDM, OFDMA
High efficiency MCS	64 QAM	256 QAM	1024 QAM for MCS-10,11
Max speed	150 Mbps	433 Mbps (with 80 MHz and 1 SS); 6.933 Gbps with 160 MHz, 8 spatial streams (SS)	WiFi 6: 600.4 Mbps(with 80 MHz, 1 SS); 9.6078 Gbps (with 160 MHz, 8 SS)
Typical range	70 m (indoor), 250 meters (outdoor)	50–200 m; 80 m with 3 antennas	Longer range with beamforming
Power saving	STBC,U-APSD	STBC,U-APSD	STBC,U-APSD, Target Wake Time (TWT)
Spectral reuse			BSS Coloring
Release year	2009	2013	2020, 2021

Figure 17.4 Attributes of WiFi Versions.

The significant upgrade in delay-throughput efficiency attained by WiFi-6 (also known as IEEE 802.11ax) and WiFi-6E (also known as IEEE 802.11ax-2001) is driven by the inclusion of the following key mechanisms:

1) **Downlink and Uplink MU-MIMO**: To increase the spectral efficiency of the operation, directional antenna beams are formed so that multiple transmissions can take place. This way, the frequency band is reused. Multiple simultaneously transmitting and receiving antenna modules are configured and used by the corresponding stations.

   Such a configuration can be used for space-division multiplexing and for space-division multiple access (SDMA) operations. It is known as Multi-user Multiple-In Multiple-Out (MU-MIMO).

   The formation of the beams is managed by the AP. It is utilized to schedule spatial transmissions from the AP across the downlink (DL), as well as to direct transmissions from a client station to its AP across a specified uplink (UL). The operation allows a station to simultaneously configure packet transmissions to multiple spatially diverse users.

For this purpose, the AP calculates a channel gain matrix for each user and steers simultaneously active beams to different users. The AP can schedule the use of certain space segments, over specified frequency bands and time periods, for the transmission of frames to designated user stations. It can also designate the joint use of certain space segments and frequency/temporal resources for uplink transmission of frames by client stations.

2) **Orthogonal Frequency Division Multiple Access (OFDMA)**: While older WiFi versions (such as WiFi-4 and WiFi-5) have made use of Orthogonal Frequency Division Multiplexing (OFDM), which is a technique used by a station in transmitting signals across the wireless channel, WiFi-6 and 6E use, in addition, an Orthogonal Frequency Division Multiple Access (OFDMA) scheme. Under the latter, the AP schedules distinct frequency bands for exclusive uplink and/or downlink use by designated groups of stations. As a special case, a station can be designated a frequency band for its own exclusive use, providing its frames with deterministic-like delay-throughput performance. The AP integrates its ability to properly schedule uplink/downlink MU-MIMO configurations with its OFDMA based allocation of bandwidth resources. In this fashion, a WiFi-6 net operates in a manner that is similar to that used by an advanced cellular wireless network in sharing the resources of its Radio Access Network (RAN).

3) **Reduced Power Consumption—Target Wake Time (TWT)**: Under previous WiFi versions, a Power Save (PS) mode can be configured. A station can inform the AP that it enters a PS mode and then switch to a doze phase, functioning at much reduced operational power, and consequently save energy resources. The station wakes up at certain times to listen to a beacon message that is transmitted periodically by the AP. The beacon specifies whether the AP is storing a packet that is waiting for transmission to this station. If so, the station wakes up and prompts the AP to send it its packet. Following completion of its activity, the station returns to its doze phase. A station will also wake up when it has a frame that it wishes to transmit across the medium.

Under WiFi-6, a station can specify its TWT. This way, the client device can decide to stay in sleep mode for a longer period of time. Such an efficient power saving operation can be of critical importance in extending the operational lifetimes of energy limited Internet of Things (IoT) devices.

4) **Increased Spectral Efficiency Through Interference Management**: Multiple close-by WiFi cells are often simultaneously active, sharing a frequency band in providing access to a wide range of devices that may be located in a densely populated neighborhood. A transmission in a WiFi cell can then produce a signal that is well detected in another cell.

Under the CSMA/CA MAC scheme, a station with a frame to send would first listen to the channel to determine whether it is busy or idle. If the signal detected by a station exceeds a specified carrier-sensing (CS) threshold, the station determines the medium to be busy and will then avoid transmitting its own frame. This is a reasonable action if the detected signal originates at a station that belongs to the same cell (BSS) as that used by the underlying station. In turn, if this signal originates at a station that is a member of a neighboring cell, identified as an overlapping BSS (OBSS), it may not be necessary to allow it to prevent an underlying station from executing its pending transmission. Such two transmissions may be targeted for reception by stations that are located in distinct cells. The induced interference power levels at the latter targeted receiving stations may in the latter case be low, so that the simultaneous reception of the two frames issued by stations that are located in distinct cells could proceed successfully.

To accommodate such situations, and consequently upgrade the spectral utilization of the system, WiFi-6 provides stations with the ability to determine whether a detected message signal

528 | *17 Local and Personal Area Wireless Networks*

originates at a station belonging to the same cell or is issued by an OBSS. If the latter is the case, an increased carrier sensing (CS) threshold value is employed. The signal sensing station is then permitted to initiate its simultaneous transmission provided the detected CS signal power is lower than the raised threshold level.

5) **High Spectral Efficiency Through Adaptation of the Modulation/Coding Scheme (MCS)**: Jointly with the allocation of a frequency band and the configuration of MU-MIMO, a modulation/coding scheme (MCS) is selected in accordance with the expected SINR at the targeted receiver. The WiFi-6 system defines a wide range of available MCS settings, ranging from ones that attain lower spectral efficiency levels, such as one that employs a BPSK modulation and a rate 1/2 code, to ones that employ more spectrally efficient MCS settings, such as one that uses an 1024-QAM modulation and a rate 5/6 code. Over a frequency bandwidth of 160 MHz, a formerly employed MCS (identified as MCS code 0) attains a data rate that is equal 72 Mbps (under a guard interval that is equal to 800 ns, serving to provide spacing between consecutive transmissions to reduce inter-symbol interference effects), while a highly efficient MCS (identified as MCS code 11) realizes a transmission data rate that is equal to 1201 Mbps (under the same guard interval).

6) **Significant Increase in Spectral Resources Under WiFi-6E**: The significant increase in its allocation of frequency resources enables the WiFi-6E system to support many more users and user groups while allocating wider band channels. Resource allocations can be used to offer designated user stations predictable performance behavior. A wider range of high-performance message stream types can be accommodated. Such applications are highly useful in serving to sustain communications networking for autonomous highway vehicles, IoT devices, public safety systems, AI, and robotic-driven networked platforms.

17.3 Personal Area Networks (PANs) for Short-Range Wireless Communications

17.3.1 Personal Area networks (PANs)

PANs that use wireless communications channels provide for communications between devices that are at close range, usually 10 m or shorter. They allow users to communicate via their portable or mobile wireless smart devices with close-by entities, gaining connectivity to a wide range of services. System implementations employ simpler processing and communications networking protocols and are embedded in small chips. Operation requires low-energy resources, permitting long term use of devices that use tiny batteries. They provide essential communications connectivity for systems that use IoT devices and systems that make use of small sensor embedded modules.

In this section, we discuss such wireless networks by over-viewing aspects of the networking protocols that guide the operation of Bluetooth and Zigbee systems.

17.3.2 Short-Range Wireless Communications Using Bluetooth

Bluetooth is a Standards-based system that is used for short-range wireless communications between (stationary or mobile) devices over the ISM band, ranging between 2.402 and 2.48 GHz. It was originally standardized as IEEE 802.15.1 and has then transitioned to be managed by the Bluetooth Special Interest Group (SIG). Billions of Bluetooth chips and devices are installed annually.

17.3 Personal Area Networks (PANs) for Short-Range Wireless Communications | 529

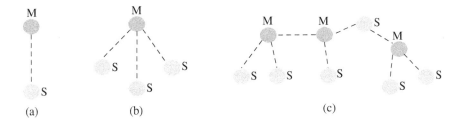

Figure 17.5 Bluetooth Net Configurations: (a) A Piconet Consisting of a Master (M) Device and a Slave (S) Device; (b) A single Master Device Communicating with Multiple Slave Devices; (c) A Scatternet Consisting of Several Interconnected Piconets.

Layout configurations of the Bluetooth system are illustrated in Fig. 17.5. The commonly employed layout shown in Fig. 17.5(a) represents the operation of a **piconet**. It consists of two Bluetooth devices communicating across a shared wireless channel. At a given point in time, one of the devices assumes the communications management role and is then identified as the Master (M) device, while the other one follows the networking control imposed by the master device and is identified as the slave (S) device. At other times, the roles of specific devices can be reversed.

As shown in Fig. 17.5(b), a master device can communicate with several slave devices. It is also possible for several piconets to interconnect in forming a configuration such as that shown in Fig. 17.5(c), known as a scatternet. In the latter, a (master or slave) member of a piconet is set to participate as a slave in a another piconet. This device can relay data between the piconets that it belongs to. The relating process would have to be managed by its host software, as such a function is not covered by the Bluetooth protocol.

As the roles of devices in acting as master or slave entities can change, the Bluetooth network is noted to operate on a *peer-to-peer* basis. Such a Bluetooth net is regarded as an *ad hoc network*. In contrast, an AP managed WiFi WLAN and Base-Station managed RAN in a cellular wireless network, operate as *infrastructure managed* architectures.

Data messages are segmented into packets. In the common layout of a piconet, the node that acts as the master can communicate with up to seven slaves. Piconet devices synchronize their clocks to that disseminated by the master. The master establishes time slots, which are recognized by all piconet members. The sharing of the wireless channel is based on using a TDM/TDMA time-division scheme. Under a commonly employed plan, the master transmits its packets on a TDM basis in even numbered slots, while slave devices transmit their packets on a TDMA basis in odd numbered slots. A packet may occupy a single slot or multiple slots. For a piconet that contains multiple slave devices, the master device manages the sharing of the channel. It allocates time slots for its own use and for the use of its clients. It can employ a round-robin medium access scheme, as it rotates the allocation of allocated client slots for use by different slave devices. A device is assigned time slots for its own transmission and a distinct set of time slots for its reception while sharing a single frequency band with other devices. Such a transmit/receive time slot arrangement is known as Time Division Duplex (TDD).

Development of the Bluetooth system started in 1999. The key Bluetooth system types developed over the period 1999–2010 are classified into Basic Rate (BR), Enhanced Data Rate (EDR), and High-Speed (HS) versions. These systems are also identified as **Bluetooth Classic**. They are used for short-range communication at relatively high-throughput rates, of the order of 1–4 Mbps. They continue to be used for a wide range of applications, such as access and data transfer systems that connect smartphones and other devices to headsets and to vehicular speakers.

17 Local and Personal Area Wireless Networks

Since 2010, starting with Bluetooth Version 4.0, Bluetooth system devices have been developed to provide low-energy operations at reduced data rates. These systems are not compatible with Bluetooth Classic systems and are known as **Bluetooth Low Energy (BLE)**. They are used for Ultra-Low-Power (ULP) and low bandwidth applications, as applicable to many IoT systems. Such systems have limited energy resources, using often coin-cell batteries. To reduce energy requirements, shorter packet sizes are used (such as a maximum packet payload length of 244 Bytes). They employ power saving (PS) schemes that turn-on radio modules infrequently and shut them down as soon as possible. Fast initiation and short connection times are imposed. The operation of a BLE module is designed to use a simple stateless process.

The wireless medium is divided into multiple designated Bluetooth channels. For Bluetooth Classic, 79 channels are specified and each channel has a bandwidth of 1 MHz. Under a basic rate (BR) mode, a data rate of 1 Mbps is attained. Under EDR modes, more efficient modulation schemes are used so that higher data rate values are realized, such as 2 and 3 Mbps under EDR2 and EDR3, respectively. A Bluetooth radio labeled as BR/EDR employs a combination of these modes. High Data Rate (HDR) implementations achieve data rate values of 4 and 8 Mbps. Forty channels are defined for use by **BLE** systems, where each channel has a bandwidth of 2 MHz.

At the physical layer, Bluetooth transmissions employ a frequency-hopping spread spectrum (FHSS) technique. For Bluetooth Classic, the hopping rate is 1600 hops per second. A pseudo-noise (PN) sequence is specified and used to define the pattern employed by a transmitter as it hops among channels in transmitting its data segments. This serves to control signal interference that may be caused by multiple Bluetooth nets that are simultaneously active in close proximity. A data message is divided into packet segments, and each segment is transmitted across one of the designated channels. Under BLE, a frequency hopping method is used as well though a different hopping technique is used.

Class 1 devices use a maximum power level of 100 mW, and can communicate over a range of about 100 m. Class 2 Bluetooth devices use a maximum transmit power level of 2.5 mW, providing for a communication range of the order of about 10 m. In turn, class 3 and 4 devices employ a maximum transmit power of only 1 and 0.5 mW, respectively, leading to respective communications ranges of the order of only 1 and 0.5 m. Communications between devices that are members of different classes is dominated by the features of the lower transmit power device.

When a Bluetooth device wishes to connect to another Bluetooth device, it starts a *paging* process, effectively calling the device that it wishes to connect to. This process consists of transmitting the Device Access Code (DAC) of the targeted device in the different channels that it is hopping across. The device initiating the page serves as the Master of the ensuing piconet, while the paged device becomes the Slave. The latter resides in Page Scan mode, so that it is listening for being called by some entity, and it wakes up at a specified frequency to detect paging transmissions that mention its ID. The Master listens between transmissions for the reception of a response from the Slave. In this manner, a connection is eventually configured.

To find the Bluetooth address of the targeted device and other local devices, the device enters an Inquiry phase during which it sends corresponding inquiries. A device is discoverable when it resides in Inquiry Scan state. It then detects an inquiry and responds to it. Using the Inquiry process, the inquiring device attains the Bluetooth addresses and clocks of the devices that respond, which are devices that reside in an Inquiry Scan state. Under Extended Inquiry Response (EIR), the response may also contain the local name and services of the responding discoverable device.

Once a connection is established and a piconet is configured, the data transmission process can take place. Data exchanges between Master and Slave entities can impose different transport requirements. For example, an audio flow may require realtime transport, entailing consistent

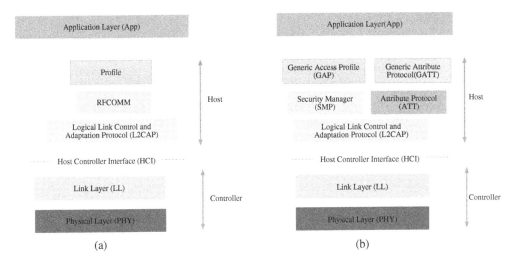

Figure 17.6 Bluetooth-Layered Protocol Stacks: (a) Bluetooth BR/EDR Classic; (b) Bluetooth Low Energy (BLE).

rate, low latency, low delay jitter, and low packet losses, while no packet retransmissions are invoked. Non-real-time data flows may tolerate longer message delay levels and would require the retransmission of erroneously received packets. Accordingly, several logical transport profiles are defined.

The layered architecture for *Bluetooth BR/EDR Classic* is shown in Fig. 17.6(a). The controller implements the physical layer and link-layer entities. The physical layer entity drives the BR/EDR radio. As noted, it operates in the 2.4 GHz frequency band and uses a Frequency Hopping technique. The radio hops in a pseudo-random way across 79 designated Bluetooth channels. Each Bluetooth channel has a bandwidth of 1 MHz. A TDD technique is used so that data transmissions occur in one direction at a time. Transmissions alternate in reverse directions. The link layer provides baseband packets to the physical layer for radio transmission. Message flows are arranged along one of the following logical links: 1. Synchronous connection-oriented (SCO), which supports real-time streaming, as that used for audio traffic; 2. Asynchronous connection-oriented logical Link (ACL), which supports data packet transmissions.

A Link Manager Protocol (LMP) is used for link setup and configuration, serving to connect Bluetooth devices. It negotiates and manages the Bluetooth connections between a Master and Slave, including logical transports and logical links. Each device contains a Link Manager (LM) that sends and receives LMP messages. LMP messages are exchanged over an ACL-C logical link. These messages have higher priority than other traffic on the channel. The LMP protocol is the main control mechanism used for linking tasks, including pairing, clock adjustment, security, adaptive frequency hopping (AFH, which serves to schedule hopping patters that avoid frequency bands that experience higher interference signals). Also managed are security processes that are used for authentication and encryption. Power modes and duty cycles of the Bluetooth radio module and the connection states of a Bluetooth device are also controlled.

The *Bluetooth Host* consists of the following layers. The *Logical Link Control and Adaptation Protocol (L2CAP)* layer is used to adapt higher layer protocols for operation over the link layer. Higher layer protocols are shielded from the detailed operation of lower layer protocols. L2CAP provides connection-oriented and connectionless services to higher-layer protocols, including multiplexing, segmentation, and reassembly operations.

532 | *17 Local and Personal Area Wireless Networks*

On top of L2CAP, as a service to the application layer, a service discovery protocol (SDP) is employed. It makes it possible for applications to query services and then follow up by establishing a connection.

The *RFCOMM* layer entity is used to emulate serial ports over L2CAP. It also provides multiple simultaneous connections to one device and enables connections to multiple devices. For telephony control service (TCS) purposes, a TCS-binary protocol defines call control signaling flows that are used to configure data and voice calls between Bluetooth devices. It resides on top of the L2CAP. It also executes mobility management procedures. In addition to the core protocols, the Bluetooth BR/EDR stack includes protocols adopted from other standard bodies, such as PPP, IP, UDP, TCP, and other.

The *Host Controller Interface (HCI)* provides a command interface to the BR/EDR radio and baseband controller, and the LM. It is a single standard interface for accessing the Bluetooth baseband capabilities, the hardware status, and control registers.

Bluetooth Classic Profile layers reside on top of the above noted Bluetooth layers. They define various profiles of Bluetooth message flows, such as: Advanced Audio Distribution Profile (A2DP) is used for streaming audio, as when one streams music from a smart phone to a vehicle (noting that it builds upon the Generic A/V Distribution Profile (GAVDP)); Hands-Free Profile (HFP) is used in Bluetooth headsets; Serial Port Profile (SPP) emulates serial ports over Bluetooth; Phone Book Access Profile (PBAP), which enables access to a Phone Book; Message Access Profile (MAP); A/V Remote Control Profile (AVRCP) is used to enable a Bluetooth device to act as a remote control device.

The **layered architecture for BLE** is shown in Fig. 17.6(b). As noted for the Classic module, the controller includes the BLE physical layer (PHY) module, a link layer (LL) module, and the controller-side host-controller-interface (HCI). As noted, the BLE radio operates at the 2.4 GHz ISM frequency band, over the range 2.4000 to 2.4835 GHz. The band is divided into 40 channels, whereby each is 2 MHz wide. Data packets are transmitted over the first 37 channels, while advertising packets are transmitted over the last 3 channels. Frequency hopping is used over the radio channel. The time between frequency hops can vary from 7.5 ms to 4 s and is set at the connection time for each slave. Transmission data rates include uncoded 1 Mbps and 2 Mbps rates and encoded symbols at 500 or 125 Kbps rates.

The *LL* provides MAC services. It manages the link state of the radio medium to define the role of a device as : Advertiser/Scanner (Initiator), Slave/Master, or Broadcaster/Observer.

The controller side of the HCI manages the interface between the host and the controller. It defines a set of commands and events for transmission and reception of packet data. When receiving packets from the controller, the HCI extracts raw data at the controller to send to the host, realizing logical end-to-end data communications.

The host module includes the host-side HCI, L2CAP, attribute protocol (ATT), generic attribute profile (GATT), security manager protocol (SMP), and generic access profile (GAP). It handles device discovery and connection-related services for the BLE device. The HCI defines a set of commands and events for transmission and reception of packet data. When transmitting data, HCI translates raw data into packets to send from the host to the controller. The L2CAP encapsulates data from BLE higher layers into standard BLE packet formats for transmission or extracts data from the standard BLE link layer packet on reception according to the link configuration specified by the ATT and SMP layer entities. The BLE L2CAP protocol entity provides link-layer protocol services, as performed by a Logical Link Control (LLC) protocol layer entity. It relies on lower layers for flow control and error control. It makes use of asynchronous (ACL) links and does

17.3 Personal Area Networks (PANs) for Short-Range Wireless Communications | 533

not use synchronous (SCO) links. Connectionless and connection-oriented service modes can be employed. A connectionless mode provides reliable datagram delivery service.

ATT is used by the BLE device to expose certain components of data or attributes. GATT defines the service framework, specifying sub-procedures to use ATT. Data communications between two BLE devices are handled through these sub-procedures. Applications and/or profiles use GATT directly. ATT transfers attribute data between clients and servers in accordance with GATT-based profiles.

ATT defines the roles of the client–server architecture. They typically correspond to the roles undertaken by the master and the slave entities, as defined in the link layer. In general, a device could be a client, a server, or both, irrespective of whether it is a master or a slave. ATT also performs data organization into attributes.

A client is a device that initiates GATT commands and requests and accepts responses, such as a smartphone. A server is a device that receives GATT requests and commands and returns responses, such as a temperature reading by a sensor.

BLE protocol layer entities interact with applications and profiles that reside at the application layer. Application inter-operability in the Bluetooth system is accomplished by making use of Bluetooth profiles. A profile defines the vertical interactions between the layers as well as the peer-to-peer interactions of specific layers between devices. A profile is composed of one or more services, addressing a particular use case. A service consists of its embedded characteristics or references to other services.

BLE devices are detected by broadcasting *advertising packets*. This is accomplished by using 3 separate frequency channels. The advertising device sends a packet on at least one of these three channels at a specified repetition period called the advertising interval. A random delay of up to 10 ms is added to each advertising interval to reduce the probability of collision between multiple such packet transmissions. The scanner listens to the channel for a specified time duration, called the scan window, periodically repeated every scan interval.

The mechanisms and protocols used by BLE aim to reduce energy consumption. In comparing operational parameters, we note that the wake latency from a non-connected state is typically of the order of 100 ms for the Classic system while it is of the order of 6 ms for BLE. Minimum total time for sending data for the two respective systems are 0.625 vs. 3 ms. Average power consumption is respectively of the order of 1 W vs. 0.01–0.50 W. During a connection state, a Bluetooth device can reside in one of the following states: Active state, when it is actively involved; Sniff state, when it is set to listen only on specified time slots for messages that are destined to it; Hold state, when it does not transmit for a long period of time; and a Park state, when it is not very active and is configured to consume very little energy resources.

17.3.3 Short-Range Low Data Rate Wireless Communications Using Zigbee

Zigbee is a short-range low data rate WPAN. It is used for monitoring and control-type applications applied for home or office applications. Applications include wireless light switches, home energy monitors, traffic management systems, and intermittent transmissions by sensor devices.

Communicating entities are located in close proximity to each other, forming a PAN. The wireless network sustains short-range low-rate data transfers at low-energy consumption levels. Small battery powered devices exhibit very long life spans.

Zigbee employs simple networking mechanisms that are lower cost to implement and operate than corresponding ones used by WiFi to Wi-Fi globally and Bluetooth systems. Transmission spans are typically of the order of 10–100 m, while the data rate is generally not higher than 250 Kbps.

17 Local and Personal Area Wireless Networks

To reach more distant Zigbee devices, a mesh Zigbee network can be configured by using Zigbee modules that act as routers. It was initiated in 1998, standardized in 2003, and has been undergoing later revisions. The name refers to the waggle dance of honey bees after their return to the beehive.

The Zigbee wireless radio module operates in the unlicensed ISM band, spanning 2.4 to 2.4835 GHz (worldwide), 902 to 928 MHz (Americas and Australia), and 868 to 868.6 MHz (Europe). In the 2.4 GHz band, 16 channels are allocated, 5 MHz apart. A bandwidth of 2 MHz is allocated for each channel. The radios use direct-sequence spread spectrum coding, so that each transmission is set to hop among multiple channels, whereby the hopping pattern is determined by a pseudo-noise (P-N) sequence. It helps in handling signal interference effects that are caused by simultaneous transmissions performed by several radios that are located in close proximity to each other. The over-the-air data rate is 250 Kbps per channel in the 2.4 GHz band, 40 Kbps per channel in the 915 MHz band, and 20 Kbps in the 868 MHz band. The output power of the radios is typically in the 1–100 mW range. At 2.4 GHz, a communications range of about 10–20 m is attained.

Typical Zigbee nets perform monitoring and control functions, such as the following ones: a. **Industrial Automation**: Monitoring parameters of equipment and serving to control and optimize operations. b. **Home Automation**: Remote control of home appliances such as lighting units, home appliances and systems, heating and cooling, surveillance and security. c. **Smart Metering**: Energy consumption response, cost management, and power systems. d. **Smart Grid monitoring**: Smart grid adaptive power management and status monitoring and control.

Zigbee devices are classified as follows:

1) A **Zigbee coordinator (ZC)** initiates the network and serves as a root of a tree layout as well as a bridge to other networks. It stores network data, including its security keys.
2) A **Zigbee router (ZR)** is capable of performing as a router, enabling the forwarding of data to other Zigbee devices and thereby extending the dissemination range of data.
3) A **Zigbee end device (ZED)** communicates with a parent node. The latter being either a coordinator or a router. It cannot relay data from other devices and is therefore able to switch into sleep mode for long periods of time, extending its battery lifespan. It uses lower memory and power resource levels.
4) A **ZigBee trust center (ZTC)** is a device that provides security management, security key distribution, and device authentication.
5) A **ZigBee gateway** is used to connect a ZigBee network to another network, such as a LAN, by performing protocol conversion.

Zigbee network topological layouts are shown in Fig. 17.7. The net initiator and key managing node the Zigbee coordinator (ZC). This device communicates with ZEDs across the wireless channel. Zigbee Routers (ZRs) are employed to extend the communications range. Depending upon the targeted destination of the packets that it receives, a Zigbee router forwards Zigbee packets to attached ZED devices or to other routers.

As shown in Fig. 17.7(a), under a **star topology**, the coordinator communicates directly across the wireless medium with end devices. Such a layout may lead to traffic overloading of the coordinator. Yet, the involved networking and other service protocols are simpler and each device is at a distance of two hops from every other device, leading devices to expend lower-energy resources.

Under a non-star **tree topology** such as that shown in Fig. 17.7(b), routers are used to extend the reach of the coordinator for communicating with end devices. An end device is identified as a child of a coordinator or of a router and the latter are identified as the parent of the end device. An end device can directly communicate only with its parent device. The network assumes a topological

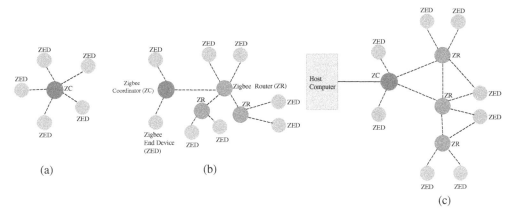

Figure 17.7 Zigbee Net Layouts: (a) Star Topology. (b) Tree Topology. (c) Mesh Topology.

layout of a tree graph rooted at central node, so that there exists only a single route between each end device and the coordinator. Under a tree topology, the coordinator maintains information about the network and uses this information to assign an address to each router. The routers, then assign addresses to their respective end devices (i.e., their children).

Under a **mesh topology** such as that shown in Fig. 17.7(c), packets may have a choice of (possibly multi hop) routes to take to reach their destinations. The operation is self-healing in that if a path fails, an alternate path may be taken. In a mesh topology, each router assigns a random address to its associated end devices. Any source device can communicate with any destination device across the network by using device to device routes. This operation is also referred to as a peer-to-peer networking scheme. It requires higher overhead and employs a more complex routing protocol than that used under a star topology.

The coordinator scans the wireless medium and selects a channel that incurs minimum interference. It allocates a network ID as well as an address to every device in its network. As noted above, messages issued by the coordinator reach directly end devices or reach a router that forwards them to either another router or a targeted end device. Small amount of data can be managed by end devices for communications with the coordinator. To transmit a data message to an end device, the message is sent to a parent node of the destination end device, such as a router. The latter holds the message until the targeted end device requests its transmission.

Zigbee protocols support beacon-enabled and non-beacon-enabled networks. When a **non-beacon-enabled network** is employed, a CSMA/CA channel access MAC protocol is used for sharing the wireless channel among the net's Zigbee devices. Devices are set to be continuously active to receive packets while transitioning to transmission mode at infrequent times. For example, a Zigbee device installed at a lamp is well powered so that it is able to stay continuously in receive mode and can act in a role of a ZC or ZR. The associated light switch, acting as a ZED, may be battery powered. It can be configured to reside in sleep mode until activated, then awaking and transmitting a control packet to the lamp device, and returning to sleep mode after receiving an ACK packet.

When operated as a **beacon-enabled network**, beacon messages are periodically transmitted by Zigbee routers to network nodes. For power saving, nodes may stay in sleep mode between times at which associated beacons are activated and stay awake during beacon transmitting time and corresponding periods when needed to engage in a transaction. Beacon interval duration levels are

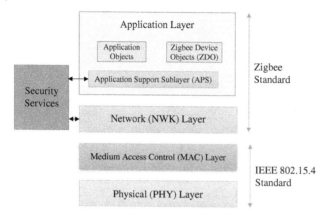

Figure 17.8 Protocol Layer Stack for a Zigbee System.

set in accordance with the planned data rate and energy consumption objectives. An interval period may range from 15.36 ms to 251.65824 seconds when operating at 250 Kbps.

The protocol layering for a Zigbee network is shown in Fig. 17.8. The physical and MAC layers are based on the IEEE 802.15.4 standard. The Zigbee standard defines the operation at the network, application, and security layers. The latter layers can also be synthesized in a customized manner by organizations that aim the system to operate in a specifically targeted manner. In the following, we overview key services performed by protocol layer entities.

The **Physical Layer** interfaces the wireless medium to provide signal transmission and reception services. Information bits received from the MAC layer are carried by physical layer PDUs. Modulation/Coding schemes (MCS) form channel signals. At receiving entity, demodulation/decoding schemes are used to process channel signals to extract payload data, which are then submitted to the associated MAC layer. The physical layer entity also performs channel selection, channel quality evaluation, and energy detection.

The 2.405 GHz channel band is employed worldwide, making use of channels 11–26, each offering a bandwidth of 2 MHz and a channel separation of 5 MHz, realizing a channel data rate of 250 Kbps. Also used in the USA is the frequency range of 902–928 MHz, which is divided into channels 1–10, operating at a channel data rate of 40 Kbps. In Europe, use has also been made of channel 0, operating at 868.3 MHz at a data rate of 20 Kbps. It spans a range of 10–100 m. In forming the network, a channel that is determined to carry lower interference signals is selected. A better channel may be selected at a later time if the current channel is determined to be degraded due to signal interference.

To protect a transmitted signal from signal interference by RF noise and nearby transmissions, a Direct Sequence Spread Spectrum (DSSS) technique is used. Each signal's time span is divided into short subintervals, called chips. A pseudo-noise (PN) sequence is used to encode the data being transmitted and to directly modulate the carrier signal so that it assumes different values at different chip periods. This operation spreads the frequency bandwidth occupied by the signal. The intended receiver correlates the received PN-encoded signal with the same employed PN sequence as it extracts the information destined for itself. Signals produced by transmitters that are members of another net encode their carriers by using a different (largely uncorrelated) PN code. Consequently, transmissions by a member of another net do not properly correlate at a receiver of this net and are effectively tuned out. This way, the impact of interference signals is much reduced. This operation provides for better coexistence with other active sources that use the same frequency band.

17.3 Personal Area Networks (PANs) for Short-Range Wireless Communications | 537

In comparison, we note that a Bluetooth network uses a FHSS method for signal transmissions. The signal's frequency hops quickly (at a rate of 1600 hops per second) among 79 frequency channels. In this manner, its signal will interfere with other RF systems, hitting common channels on a random occurrence basis over a very short period of time. In turn, as noted above, Zigbee uses a DSSS transmission method. For Zigbee operation over 2.4 GHz, a chip rate of the order of 2000 [Kchips/s] is used, having the signal spread over a channel whose bandwidth is equal to 2 MHz, attaining a bit rate of 250 Kbps. Interference by a Bluetooth signal occurs only when a Bluetooth device transmission occupies a frequency range that overlaps with the ZigBee net's channel spectral span, which occurs at a low rate of about once every 79 times. In this case, the ZigBee device would randomly back off while the Bluetooth signal continues to quickly hop to another frequency. Hence, a Bluetooth net does not induce much degradation to the operation of a Zigbee system. As a further reliability measure, the IEEE 802.15.4-2003 standard provides for the use of an acknowledge (ACK) frame, so that unacknowledged message can be retransmitted.

The **MAC layer** implements the protocols and algorithms used to share the wireless medium among the member nodes of the Zigbee network. It also implements mechanisms for the generation and management of Beacon messages, validation, and acknowledgment of data frames and interactions with higher layers. It serves also to define a network identifier (a PAN ID) and to perform network discovery through beacon requests. A MAC frame is formed by including the data packet payload and adding DA and transmit options.

The primary MAC algorithm used by Zigbee is CSMA/CA. It is similar to that used by a basic WiFi system. The transmitting radio module senses the medium and avoids transmission of its MAC frame if it determines the radio medium to be busy. It would then schedule another transmission attempt at a later time. At that time, if the medium is sensed to be idle, the radio module may be permitted to proceed with the transmission of its frame. Following transmission, it waits to receive a positive ACK. If an ACK message is not received prior to the expiration of an ACK threshold, the frame is scheduled for retransmission. To support the transmission of time critical frames, Zigbee also provides for the implementation of a Guaranteed Time Slot (GTS) management scheme, under which a specific station is allocated a time slot for its exclusive use in transmitting its frames.

While a Bluetooth network accommodates a relatively small number of devices, a Zigbee network can support up to about 64,000 devices. As noted, ZigBee employs a PAN coordinator, which may operate in a beacon-enabled mode or in a beaconless mode. In the beacon-enabled mode, the PAN coordinator defines a super-frame that starts and ends with beacon frames. These frames are used to also synchronize network nodes. In the contention access period (CAP) of the super-frame, a slotted CSMA/CA scheme is used. In a contention-free period (CFP) within the frame, a GTS access operation is enacted, where certain slots are configured for use by modules that have previously reserved them. In a beacon-less mode, no reservations are performed, and the operation is based on the basic CSMA/CA protocol. Full function devices (FFDs) are more capable and can be interconnected in a mesh topology.

The **Network layer** is responsible for the initiation of the Zigbee network, the assignment of node addresses, the configuration of new devices, the provision of transmission security and the routing of Zigbee packets. It is aware of the network layout, being configured as a star, tree, or mesh topology. The Zigbee routing scheme enables router nodes to forward packets to attached destination end-devices or to another router that is located along a desired route that leads to a destination end-device. As noted above, an end-device identifies an attached (over the wireless medium) parent node (such as a router or coordinator) which provides it with its packets.

538 | *17 Local and Personal Area Wireless Networks*

ZigBee modules follow the principle of operation of the *Ad hoc On-Demand Distance Vector (AODV)* protocol for routing packets across a mesh network, using a version of such a routing protocol identified as Z-AODV. Under an on-demand routing scheme, a source node discovers a route to a destination node only when it is needed. It issues then a discovery packet that is flooded across the network. It is retransmitted by all the nodes that it traverses until it reaches, when feasible, its desired destination node. The destination node then sends the first discovery packet that it receives along a reverse route back to the source node. In this way, a preferred route is identified. As part of this process, the corresponding forwarding entries are configured at each router that is located across the discovered route.

In large networks, the flooding of an excessively high rate of discovery packets may lead to high bandwidth and processing resource overuse. In this case, other routing approaches can be employed. One such alternative scheme that does not require the flooding of discovery packets by multiple nodes is a *many-to-one* source-routing scheme. A single data collector node, also known as a concentrator, such as the ZC, is selected. The collector node periodically issues a route discovery broadcast to all network nodes. Upon receipt of this broadcast, each router learns the next hop along a reverse route back to the collector. It stores this forwarding entry in its routing table.

To identify routes from the collector to each router in the network, the following process is carried out. Each router sends a unicast packet across the network to the collector. As this packet travels the network, the identity of the successive routers that it traverses are recorded, serving to identify the layout of a unicast route. By reversing the order of the flow, the collector learns a unicast route from itself to any other net router. It is then able to route packets that it receives from any source router to any destination router. In following a source routing method, the collector inserts the full descriptor of the route in the header of a packet that it wishes to send to another router, enabling its navigation to a targeted destination.

The principles of operation of on-demand routing protocols, using protocols such as AODV and source routing, are discussed in Chapter 19, where methods for the operation of MANETs are presented.

The **Application Layer** consists of the Application Support Sublayer (APS) and the Application Framework. The APS provides interface and control services. It works as a bridge between the network layer and the application layer. It filters and reassembles packets for end devices and checks for packet duplicates. It also performs automatic retries when the sender requests an ACK. It is involved in maintaining binding tables, which associate end points and nodes. It can be used to find devices for needed services. An address mapping process associates a 64-bit MAC address with a Zigbee 16-bit network address. The Application Framework depends on the vendor and on the specific application that is employed.

APS services are necessary for application objects (endpoints) and the Zigbee device object (ZDO) to interface with the network layer for data and management services. In aiding data transfer, APS produces request, confirm, and response messaging. The ZDO protocol provides device management, security keys, and policies. It defines the function of a device, such as its service as a coordinator or end device. It provides for the discovery of new devices on the network and for the identification of their offered services. It is involved in establishing secure links with external devices and for replies to binding requests.

Network objects communicate by using the services provided by APS, as supervised by ZDO interfaces. For example, an end point (EP) embedded in a Zigbee device that is used by a switch or by a control hub employs a corresponding application object to transmit (a unicast or multicast) control message across the Zigbee wireless medium to a Zigbee device (or a group of devices). These devices could be attached to modules such as light fixtures. They use their application objects to determine their reaction to the receipt of a control message, such as one that commands the turning on/off of a light fixture or the adjustment of temperature thresholds. In response, confirmation messages may be transmitted across the Zigbee medium. A node can contain multiple application objects.

Zigbee protocol definition has been subjected to modifications that offer extended functionalities. In 2007, ZigBee and ZigBee PRO extended protocols have been introduced. ZigBee PRO offers several significant improvements such as security and the capability to self-organize and self-heal the network. It is targeted for use in building automation and environmental and industrial applications that contain more than 30 nodes. ZigBee PRO is based on a mesh topology and is a beacon-less network. In ZigBee PRO, end devices are powered by batteries while router and coordinator nodes use main power. End devices can transition into sleep mode for power-saving purposes. When a device wakes up, it sends a message to a trust center to obtain an updated network key. ZigBee PRO offers two types of routing: many-to-one routing and multicast routing. Many-to-one routing makes use of a network concentrator node. For enhanced security, ZigBee PRO uses a trust center, which can be a router, coordinator, concentrator, or a specific device.

Problems

17.1 Describe the service and performance objectives of the following personal area and local area networks and discuss the differences between them:
a) Zigbee PAN.
b) Bluetooth PAN.
c) WiFi (802.11) WLAN.

17.2 Consider a WiFi WLAN. Define and explain the following:
a) Basic Service Set (BSS) and BSSID.
b) Access Point (AP).
c) CSMA/CA protocol.
d) The different addresses that can be included in a WiFi MAC frame.
e) Infrastructure vs. ad hoc modes of operation.
f) Slot (also known as mini-slot) duration. Discuss the impact of the slot length on the performance of the MAC scheme.
g) Clear Channel Assessment (CCA).
h) Virtual carrier sensing.
i) Purpose of the *Duration* field that is included in the header of WiFi MAC frame.
j) The method used by a WiFi WLAN for MAC layer ARQ error control.
k) The method used to resolve a *hidden terminal* problem by using an RTS/CTS process.
l) The method used to synchronize stations that are members of a WiFi net that is managed by an AP.

17.3 Summarize the key parameters, operational methods, and performance behavior features of the following versions of a WiFi net and identify the key performance-oriented differences between them:
a) WiFi 4 (802.11n).
b) WiFi 5 (802.11ac).
c) WiFi 6 and 6E (802.11ax, ax-2001).

540 | *17 Local and Personal Area Wireless Networks*

17.4 Describe the principle of operation of a Multiple-In Multiple-Out (MIMO) and of MU-MIMO antenna beam configurations as used by a WiFi medium access control (MAC) protocol.

17.5 Consider a WiFi network that provides priority-oriented access as described by the IEEE 802.11e protocol.
   a) Describe the operation of the employed priority-oriented access scheme used by EDCF, including the use of the following parameters: (CW_{min}, CW_{max}), *AIFS, TXOP Limit.*
   b) Describe the operation of a WiFi net that follows the Hybrid Coordination Function (HCF) mode, as polling and CSMA/CA access methods are jointly employed.

17.6 Describe the methods used by WiFi LANs to operate in a *power management* mode.

17.7 Illustrate and discuss the principle of operation of an advanced WiFi network that employs a DA/FDMA/TDMA/SDMA medium access control-type scheme.

17.8 Identify, describe, and illustrate the net configurations used by a Bluetooth network:
   a) A piconet that consists of a master device and a slave device.
   b) A configuration that consists of a single master device communicating with multiple slave devices.
   c) A scatternet that consists of interconnected piconets.

17.9 Identify the protocol layers and describe the function of each, in discussing the protocol architecture of Bluetooth Classic and of Bluetooth Low Energy.

17.10 Describe the operation of a Bluetooth Time Division Duplex technique.

17.11 Describe the objectives, principle of operation, parameters, and performance features of a Zigbee network.

17.12 Define and discuss the different classes of Zigbee devices.

17.13 Describe and discuss the different Zigbee net layouts: Star, tree, and mesh.

17.14 Describe the operation of a beacon-enabled Zigbee network.

17.15 Describe and discuss the layered architecture of a Zigbee system.

17.16 Compare the physical layer structures used by WiFi, Bluetooth, and Zigbee systems. Include the following elements: frequency spans, distance coverage ranges, transmit power levels, energy consumption levels, and network size.

18

Mobile Cellular Wireless Networks

The Objective: *Mobile wireless networks enable local and long distance communications that accommodate mobile endsers. They seamlessly support message flows produced and received by mobile users without interrupting the end-to-end integrity of transported data flows. Under a cellular architecture, an area of operations is divided into cells, where each cell is managed by a base station (BS). Mobile initiated and terminated data flows are carried out across a shared wireless medium, such as a Radio Access Network (RAN), as they connect with a mobile's associated BS. Under an infrastructure oriented architecture, a backbone network such as the Internet or a public telephone network is used to connect BSs.*

    The Methods: *To sustain session connectivity under mobility, a cellular network system employs a mobility control sub-system. A moving mobile associates with different BSs as it crosses cell boundaries. Communications resources that are available for allocation across the wireless access network are generally limited. Hence, their allocation impacts the networking performance of a cellular network. Efficient wireless medium sharing methods are used to enhance wireless resource utilization levels, while offering flows with transport performance levels that meet their desired quality of service (QoS) objectives. An infrastructure based cellular network system employs efficient multiplexing schemes for sharing downlink (BS to mobile) wireless communications channels. Demand-assigned multiple access schemes are used for allocating uplink (mobile to BS) wireless communications channels.*

18.1 Configurations of Mobile Wireless Networks

Transport of data flows between end-user stations across wireless communications channels is of prime utility. It is clearly essential for communications among user stations that are not tethered to wired media, as is the case for mobile users. It is also of key advantage in situations in which no wired infrastructure exists or can be readily deployed. This is the case when communicating in regions that do not offer a wired communications backbone facility. Wireless communications networks are widely used for the transport of multimedia message flows among mobile users as well as between mobile and stationary end-user entities.

    In examining the key architectural layouts of wireless networks, we observe the following configurations: infrastructure based, peer-to-peer oriented, and hybrid layouts.

    Under an **infrastructure based wireless network configuration**, a user (also identified as a user station or as a mobile) makes use of a **wireless access system**, such as a **Radio Access Network (RAN)**, to connect to an existing infrastructure system or a backbone network. A user connects across the RAN with an Access point (AP) station that is attached to a backbone

Principles of Data Transfer Through Communications Networks, the Internet, and Autonomous Mobiles, First Edition. Izhak Rubin.
© 2025 The Institute of Electrical and Electronics Engineers, Inc. Published 2025 by John Wiley & Sons, Inc.

18 Mobile Cellular Wireless Networks

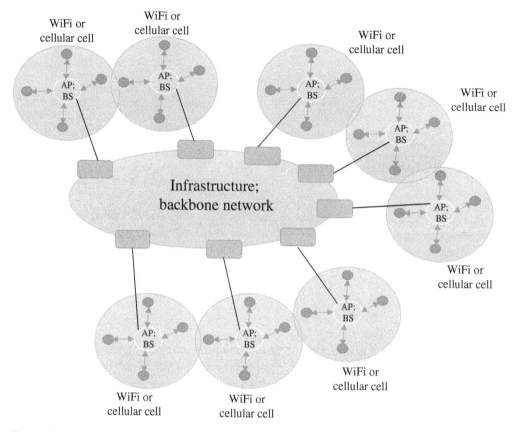

Figure 18.1 An Infrastructure-Based Wireless Network.

network. This is the case for (infrastructure based) WiFi wireless local area networks (WLANs), as well as for cellular wireless networks, whereby the AP is identified as a **Base Station (BS)**. Such a layout is illustrated in Fig. 18.1. End-users that associate and communicate with a central node (AP/BS, which is hereby identified in brief at times as either a BS or an AP) that is located in their vicinity, are said to become its client members. They are located in the **cell** area that it manages. To communicate a message, a station uses a wireless medium to have its message first transmitted across a RAN's **uplink wireless channel** to its associated AP or BS. The latter connects to the backbone network across which the message is routed to its targeted destination node.

A BS connects to a router that accesses the backbone network, such as the Internet. At the receiving end, if the destination end-user station is an entity (such as a mobile) that is a current member of a specific cell, the message is transported across the backbone network to a destination router that directs it to a BS that manages the cell in which the destination entity resides. The destination BS transmits the message across a **downlink wireless channel** to the targeted destination user. To handle mobile users, a mobility control management system is employed. It tracks the location of a destination user so that the cellular system can determine the specific cell with which the destination mobile is presently associated.

In certain situations, no infrastructure system is available to provide backbone transport services. In this case, **peer-to-peer** communications networking mechanisms are used to transport message

Figure 18.2 An Ad Hoc Wireless Network Employing Multi-hop Peer-to-Peer Communications.

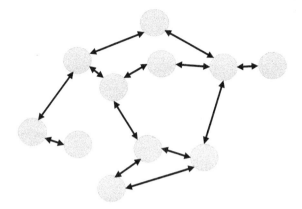

flows. *Infrastructure-less* networks are known as **ad hoc networks**. Such a scenario is illustrated in Fig. 18.2. The nodes shown in the figure represent users and their communications modules. They provide message transmission, routing, access control, and associated networking functions. Mobiles interconnect through the use of wireless communications channels. Two nodes that are able to directly communicate with each other are shown in the network's graph to be connected by a line (representing a link).

When nodes are mobile, a **mobile ad hoc network (MANET)** is formed. Autonomous vehicles moving across a highway may use MANET principles to disseminate their messages. Neighboring nodes communicate directly across wireless links. End-to-end communications between non-neighboring nodes are performed across multiple-hop paths. Under nodal mobility, the topological layout of the wireless network is continuously varying. The employed routing algorithm must be able to rapidly discover an advantageous path to use for reaching its targeted node(s). Methods for networking across mobile ad hoc wireless communications networks are discussed in Chapter 19.

A hybrid architecture is shown in Fig. 18.3. Both ad hoc and backbone networking methods are employed. A backbone network is used for communications among BS nodes. User nodes may also communicate with each other across the non-backbone segment of the network by using peer-to-peer ad hoc networking methods.

A source node may use a peer-to-peer networking scheme to reach a node that connects to a backbone network such as the Internet. The backbone network can then be used for wider span communications. Peer-to-peer ad hoc routes may be employed to disseminate messages among user nodes that are distributed away from the backbone network, as well as to transport messages from such users to or from the backbone.

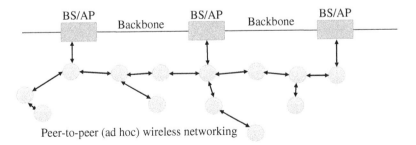

Figure 18.3 A Hybrid Infrastructure of a Wireless Network.

18.2 Architectural Elements of a Cellular Wireless Network

In the following, we outline key elements employed by a cellular wireless network system. Different generations of cellular network system implementations have been designed and employed, based on Standards recommendations. They have evolved over the years, progressively offering users improved support for communicating at higher data rates and lower message delay levels, while enabling higher concentrations of mobile users. The following outlined concepts, methods, and subsystems are essential ingredients of key mobile cellular wireless networks.

18.2.1 The Cellular Coverage

A cellular wireless network system serves its users by using an **infrastructure core**. Such a topology is illustrated in Fig. 18.1. The system divides its service area into access regions identified as **cells**. Each cell is managed by an entity identified as a **BS**. Mobiles that reside in a cell's region associate with their BS. Mobiles and their associated BS communicate across the cell's wireless media. Using radio signals, a **RAN** provides for communications between mobiles and their associated BS. An associated mobile receives messages that are transmitted to it by its BS across a specified **downlink wireless channel**. A mobile is assigned resources by its BS node for transmitting its own messages to the BS across an assigned **uplink wireless channel**. The wireless access network (such as a RAN) is a critical subsystem. It impacts in a significant way the performance behavior of a cellular wireless network. Its structure and the resource allocation algorithms that it employs for sharing its uplink and downlink wireless resources determine the utilization efficiency of these resources and the ensuing delay-throughput performance of message flows that use the network. Wireless access bandwidth resources are critical elements, induced by their limited availability and by the high cost of frequency bandwidth resources.

To provide seamless communications support of its users over its service area, a cellular system strives, when feasible, to fully cover mobile users located in its area of operations. An illustrative cellular coverage of a service area is shown in Fig. 18.4. For simplified representation, each cell is shown to assume an hexagonal coverage shape. A **fractional frequency reuse (FFR)** scheme is shown. In each cell, transmissions in the center of the cell (to/from the BS node) employ a frequency band that is commonly used in the interiors of all area cells. In turn, to reduce signal interference that occurs between transmissions that involve mobiles that are located around the boundary areas of neighboring cells, such mobiles are assigned distinct frequency bands, denoted as C1, C2, and C3. The latter are depicted as being associated with different colors. To improve the spectral efficiency of the operation, mobiles that reside in the boundary areas of non-neighboring cells, reuse frequency bands, or colors. As shown in the figure, under a reuse-3 pattern, 1/3-rd of the cells reuse the same frequency band at their boundary regions.

Cellular systems often employ a reuse index that is equal to 1, so that the same frequency resources are fully reused in every cell. Such systems act to mitigate signal interference effects by various means, such as by using proper directional antenna configurations and temporally coordinated transmission schedules. Cellular systems also employ methods for achieving signal directionality or localization through the use of other means. A common method involves the use of a coordinated configuration of multiple antennas that are employed by a single BS as it communicates with user stations (which may also use multiple antenna elements), identified as a Multiple-In Multiple-Out (MIMO) arrangement.

To enhance the simultaneous multicasting of a common data stream over a given area, a Coordinated Multi-Point (CoMP) process may be invoked. Multiple BS synchronize their signal

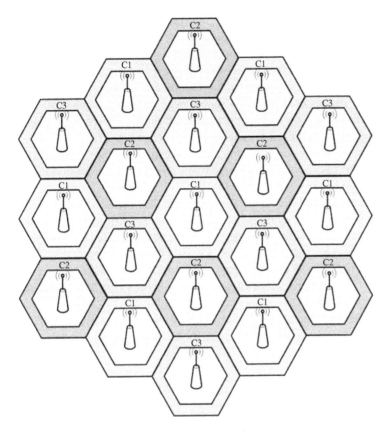

Figure 18.4 Spatial Reuse in a Cellular Network.

transmissions of a common stream that is multicasted over a specific region, so that a user may simultaneously receive the same multicasted data from multiple neighboring BSs, enhancing the quality of the signal received at intended user stations.

In an analogous manner, resource reuse advantages can be achieved by assigning, over a common frequency band, distinct time slots for operations in neighboring cells, while repeating time slot assignments for transmissions by entities that reside in non-neighboring cells. In general, effective resource reuse configurations can be jointly performed over time, frequency, and spatial dimensions.

Higher spatial reuse gains are attainable when **directional antennas** are employed. In many cellular wireless networks, BSs employ directional antennas, such as those that configure three 120-degrees, four 90-degrees, or six 60-degrees wide sectors. When each sector is associated with a separate transceiver, a BS is capable of simultaneously transmitting and receiving different data flows over different sectors of a cell. In Fig. 8.5(b), we depict such simultaneously carried uplink communications from mobiles located in four quadrant sectors of a cell to their BS node.

Each cell employs its own allocated wireless spectral resources. Hence, the total wireless bandwidth available for operations across a service area is proportional to the number of cells that are used to cover this service area. Yet, each cell requires the installation and operation of a BS, whose construction and maintenance costs are essential factors. The cellular concept provides for graceful growth as the number of users that are accommodated increases. Over busy regions, smaller cells are synthesized, so that each may share its bandwidth resources among a lower number of users.

546 | *18 Mobile Cellular Wireless Networks*

To accommodate the ever increasing multimedia throughput rates required by mobile users, next generation cellular systems transition to the use of higher carrier frequencies and frequency bands. Such bands provide the cellular network with much higher bandwidth resources for use across its wireless access systems. However, as the carrier frequency increases, signal propagation losses increase as well. Signals are then highly degraded by reflections and blockages caused by objects located across their paths. Hence, when used in advanced network systems such as 5G and 6G cellular systems, smaller cells are deployed. A larger number of small cells are used, where each provides ultrahigh-throughput rate and low message delay. At such a high-frequency range, antenna sizes are much smaller. A higher number of directional antenna arrays are used, leading to further increase in the spectral efficiency of the system, so that a significant increase in realized data throughput rate can be attained.

Small size cells are also used at times to accommodate special wireless access needs of users. For example, a home user that is located farther from a BS, or that experiences degraded BS signal reception, may be connected to an access point (or small BS) that manages a small cell, possibly also accommodating subscribing neighbors. The small cell's AP or BS may be connected through a high speed communications link to a large BS or to a backbone network. Small cells are categorized as follows in terms of their effective communications ranges: micro cell: lower than 2 km; pico cell: lower than 200 m; femto cell: around 10 m.

In this manner, the architecture of next generation cellular wireless network systems is evolving to contain sub-system segments that are capable of autonomously serve their clients, not having to rely on the traditional configuration of a BS-centric layout.

18.2.2 Cellular Networking Generations

Cellular network system technologies have been evolving over the years. It is customary to categorize them into distinct generations. **First-generation systems, identified as 1G**, were introduced around 1979 in Japan and in 1983 in the USA. The system was designed to provide mobile telephone service. Audio messages were encoded as analog radio signals, while signaling and control messages assumed the form of digitally encoded signals. The system has been identified in the USA as Advanced Mobile Phone System (AMPS). A Frequency Division Multiple Access (FDMA) technique was employed for sharing the wireless access medium. A mobile user that wished to engage in a telephone call, would send a signaling message request to the BS and be assigned two dedicated frequency bands (for a full duplex connection) to connect with the BS across the wireless access medium. 1G networks offered a voice channel bandwidth of 30 kHz and a data rate of 2.4 kbps. Based on a circuit switching approach, the BS initiated a process that resulted, in coordination with services provided by a regionally located Mobile Switching Center (MSC), in the establishment of an end-to-end voice connection over an allocated circuit. The sound quality was not high, security was low, and there was no encryption.

1G networks were replaced around 1991 by **second-generation (2G) cellular networks**. Analog signal communications and modulation methods were replaced by using digitally modulated voice signals, leading to significant enhancement in the system's capacity and security. Calls were encrypted and thus became more secure. In Europe, 2G systems were based on the **Global System for Mobile Communication (GSM)** standard, offering improved security and higher spectral efficiency and transport capacity. Channel bandwidths of 30–200 kHz were used. A **Time Division Multiple Access (TDMA)** method was used by mobiles to share the resources of the wireless medium of the RAN among multiple uplink voice circuit connections. The BS used a **Time Division Multiplexing (TDM)** method in allocating circuits for downlink transmissions.

In addition to providing improved mobile phone service, signaling channels were configured and made available to users to engage in low speed (up to 64 kbps) data transmissions. The latter included Short Message Service (SMS) text messaging, consisting of plain-text short message transmissions, as well as a Multimedia Message Service (MMS).

Evolutionary upgrades led to the introduction of 2.5G systems, which included packet switching based message networking service, providing data rates of up to 144 kbps. It was identified as General Packet Radio Service (GPRS), providing Enhanced Data rates for GSM Evolution (EDGE) 2G–3G systems. Applications of this technology included e-mail and web browsing communications.

Systems based on 2G North American Standards *IS-54 and IS-136*, deployed in 1993, have been known as **Digital AMPS**. Digital communication techniques were used for voice signal transmissions. The uplink medium was shared through the use of a TDMA scheme. Each 30 kHz channel was shared among multiple voice circuit connections. A TDM scheme was used to share the wireless resources of the downlink. A second approach used by certain US cellular service providers was based on the *IS-95 Standard*. Voice transmissions across the wireless channel were digitally encoded. The wireless access medium was shared by using a **Code Division Multiple Access (CDMA)** scheme. Design and implementation of this CDMA-based cellular system started around 1993–1995.

Third-Generation (3G) cellular systems, known as UMTS (standardized by 3GPP) in Europe and CDMA 2000 (standardized by 3GPP2) in the USA, emerged around the year 1998–2000. 3G mobile technology standards enabled the sending of voice, data, and signaling messages between mobile phones and BSs. The latter are also known as **cell sites**. Designed to be a backward compatible successor to the 2G cdmaOne (IS-95) system, it was used especially in North America, South Korea, China, Japan, Australia, and New Zealand.

3G technology provided information transfer rates that start at 144 kbps. Enhanced 3G releases, often denoted as 3.5G and 3.75G, provided mobile broadband access at several Mbps rates to smartphones and mobile modems. Such high rates enabled applications that included higher quality wireless voice telephony, mobile and fixed Internet access, mobile TV, video conferencing, music, video streaming, and video chat. **Smartphones** started to dominate as mobile user devices. Such devices included Blackberry units that were introduced in 2002 and iPhone devices that were introduced in 2007. iPhone devices generally replaced Blackberry units by around 2017. The introduction of smartphones accelerated the need for more capable and higher speed wireless networks.

The *3rd Generation Partnership Project (3GPP)* is an umbrella term for several standards organizations that have been developing protocols for mobile telecommunications. The 3GPP project was established in December 1998, aiming to develop a specification for a 3G mobile phone system that is based on the 2G GSM system. Its developments included 2G and 2.5G GSM, 3G UMTS, as well as protocols for 4G LTE, 5G New Radio (NR) and related 5G and subsequent standards.

Fourth-Generation (4G) cellular systems use **Long-Term Evolution (LTE)** technology to deliver download speeds that range between 10 Mbps and 1 Gbps. They have largely replaced earlier generation cellular systems. Commercial use started around the end of 2009. Using LTE technology, users have been able to transport data, video and voice flows at higher-throughput rates and at much lower message delays, producing upgraded message and stream transport quality. 4G systems are designed as **IP-based networks**. They can provide admitted data flows, voice, and video streams with prescribed quality of service (QoS) guarantees. Services include wireless broadband access, MMS, video chat, mobile TV, HDTV, and Digital Video Broadcasting (DVB). Popular cellphones that offer LTE include Apple's iPhone and Samsung's Galaxy smartphones.

548 | *18 Mobile Cellular Wireless Networks*

Fifth-Generation (5G) cellular systems aim to provide users with significantly higher data rates and very low message latency levels. Such enhanced performance measures are required for the support of many emerging applications, including augmented reality (AR), autonomous vehicles (AV), virtual reality (VR), and Internet of Things (IoT). Targeted performance metrics include throughput rates of up to 20 Gbps, which are higher by a factor of about 100 than those provided by 4G networks, and message latency levels of around 1–10 ms, in comparison with a packet delay of the order of 30–50 ms under 4G. A 5G system enables the implementation of near real-time response rates, while accommodating a user connection density of the order of 1000 devices per square kilometer (which is about 100 times higher than that offered by a typical 4G system). Such performance metrics are needed for realizing effective support of a large number of sensor nodes and IoT devices. 5G offers significantly wider frequency bandwidths, performing transmissions at higher-frequency carrier levels, spanning the 30–300 GHz range. 5G capabilities promise to enable a multitude of new services that require higher data rates and very low message latency values. Applicable applications include: Networked Artificial Intelligence (AI)-based technologies; remotely controlled robotics and tele-medicine; autonomous vehicle systems; advanced IoT and VR systems, distributed computing systems, and a multitude of newly emerging networked systems.

Sixth-Generation (6G) cellular systems aim to provide users with significantly higher data throughput rates at near real time and very low message latency levels. Such performance features are essential for the support of high-speed and high-performance mobile dynamic systems that require networking of information among widely distributed entities at high rates and at very low (of the order of 1 ms and lower) delays. Included are AI-oriented advanced networked autonomous transportation and other systems; mobile robotic systems; systems that jointly employ land, air and space networked platforms; eHealth systems; AR and VR systems; IoT; energy production and smart grid systems; security systems; urban, social, and agricultural production systems; education and training systems; and many other.

To achieve its objectives, 6G cellular wireless systems also make use of spectral resources in the THz domain, aiming to attain data rates of the order of 1 Tbps. Goals include offering a 10 times factor increase (in comparison with 5G) in spectral efficiency and a factor of about 100 increase in energy efficiency. In using higher carrier frequency levels, wavelengths are very low (of the order of 30 μm to 3 mm). As an antenna element size is of the order of a wavelength, it is possible under the 6G regime to place a very large number of antenna elements on a single small surface, enabling the production of a very compact intelligent antenna array system. It is applied to the implementation of a massive directional antenna beam array that is dynamically adaptable in forming directional communications links. Due to the high carrier frequency levels employed, short range, high capacity, line-of-sight (LOS) links are readily synthesized. Hub nodes that act as access points can be placed at key locations, serving as cluster-head points that sustain local autonomous mobility, and as access points that connect to a multi dimensional dynamically configured core network. Design schemes, operational techniques, and adaptation algorithms to be employed by 6G network systems would be supported by advanced *Artificial Intelligence and Machine learning (AI/ML)* techniques and would enhance the development of such methods.

Network architectures, functions and operations, and enabling technologies that are of prime importance for the development of next-generation mobile networks are discussed in Chapter 20. The following networking methods are noted [39]: Network Virtualization; Software Defined Networks (SDN); Network Functions Virtualization (NFV); Network Slicing; ML and AI-aided Automatic and Autonomous operations; Cloud Computing; Edge Computing; Open radio access network (O-RAN); access network convergence, employing a hybrid of terrestrial and nonterrestrial (airborne and space borne means, including UAV/Drone and satellite platforms and networks) systems; harmonization, in accommodating globally standardized protocols, interfaces, and assigned spectral bands; sustainable trust in providing secure services and message transport; employment of environmentally sustainable and energy efficient systems and telecommunications services.

18.2.3 Key Components of a Cellular Network Architecture

To illustrate the key architectural elements employed by a cellular network system, we consider in the following the **4G-LTE cellular system**. The **LTE high-level network architecture** consists, as shown in Fig. 18.5, of the following subsystems:

1) The end-user device, used to communicate over the cellular wireless network, is identified in LTE as the **User Equipment (UE)**. It includes the following modules: mobile termination (MT) that provides the communications functions; terminal equipment (TE) that terminates the data streams; Universal Integrated Circuit Card (UICC), known as the Subscriber Interface Module (SIM); also implemented as an embedded eSIM module. A SIM contains information that includes the user's phone number, home network identity, and security keys.
2) The **RAN** consisting of the wireless medium and the BS that communicates across the RAN with its users. A version of the RAN defined by LTE has been identified as the *Evolved UMTS Terrestrial Radio Access Network (E-UTRAN)*. The Universal Mobile Telecommunications System (UMTS) identifies the 3G mobile cellular system that was used by networks that are based on the GSM Standard. 3G cellular wireless systems used a WCDMA (wide-band CDMA) technique for sharing the access medium, while 4G-LTE and other later generation systems have been using a Demand Assigned joint TDMA/FDMA/SDMA approach.
3) The **core network**, which is identified in LTE as the *Evolved Packet Core (EPC)*. This system is responsible for interconnecting user flows with external networks, including *Packet Data Networks (PDNs)* such as the Internet. It also provides other essential services, such as *mobility management*.

As shown in Fig. 18.6, the LTE RAN, identified as E-UTRAN, provides radio communications between mobile users and the core system that is identified as the EPC. A RAN subsystem consists of BSs, identified in LTE as the evolved BS nodes (eNodeB or eNB). A mobile user communicates with a single BS at a time. A BS manages its client mobiles. It is responsible for controlling the sharing of its associated RAN's wireless media and for performing other functions that enable

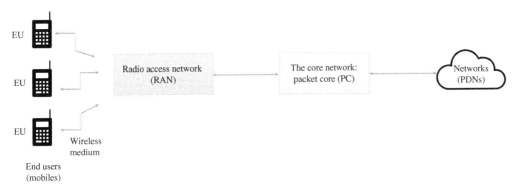

Figure 18.5 High-Level LTE Network Architecture.

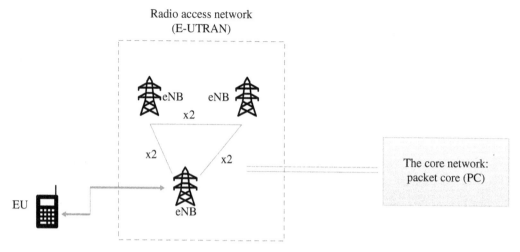

Figure 18.6 The LTE Radio Access Network Architecture.

communications to and from the mobiles that a BS serves. It allocates media resources for uplink and downlink data transmissions. It sends signaling messages that alert users about incoming flows that enable mobility management and that perform call and flow handover processes. As noted in Fig. 18.6, neighboring BSs interact with each other through the shown X2 interface. This is useful for accomplishing handover transactions, for scheduling wireless media resources, and for regulating signal interference effects that are caused by simultaneous transmissions that are conducted at near-by cells.

Key components of the LTE core network, known as the EPC, are shown in Fig. 18.7. The Home Subscriber Server (HSS) is a database that contains information about the network operator's subscribers. The **PDN Gateway (P-GW)** provides communications with external PDNs. The **Serving Gateway (S-GW)** performs a routing service as it forwards data between the underlying BS and the PDN gateway. Additional components, not shown in the figure, include the Policy Control and Charging Rules Function (PCRF) that is responsible for policy-based decision making and charging control. It is a function embedded in the Policy Control Enforcement Function (PCEF), which is contained in the P-GW. Also noted in the figure are interface identifiers between various components, noting that dashed lines represent signaling and control interactions and flows.

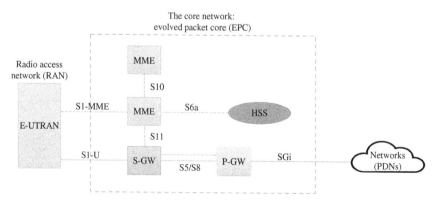

Figure 18.7 The LTE Evolved Packet Core (EPC).

A mobility management entity (MME) controls the tracking of users' locations as they move across different groups of cells. For this purpose, mobility related signaling messages are exchanged and user data stored in the HSS is used. A cellular wireless network operated by a specific service provider in one country (or certain regions) is known as a **Public Land Mobile Network (PLMN)**. Under a **roaming** process, a mobile user that moves outside her home network is able to use the networking services provided by a different organization, the visited-PLMN one. The roaming mobile is connected across the RAN, and the core's MME and S-GW modules of the visited network. It can be connected to an external PDN through the P-GW of the visited network or through the corresponding module located in its own home network.

18.3   Cellular Network Communications: The Process

Consider an end-user (EU) that travels in an area covered by one or several cellular service providers. Each cell is managed by a BS. The BS interacts with mobiles that reside in its cell, informing them about system parameters and about its allocations of networking resources. Control and management data is broadcast periodically by a BS across downlink broadcast channels. This data is included within specified fields that are nested in periodically recurring time frames, known as **beacons**. A beacon, as well as other messages that are of interest to all clients of a BS, are transmitted by the BS across a downlink broadcast signaling channel. Users listen to such messages to obtain system information that they require for their communications transport processes as well as to acquire system time synchronization.

As shown in Fig. 8.8, the wireless medium of the RAN is shared by multiple **uplink and downlink signaling and traffic channels**. Signaling channels are used for the transmission of signaling, control, and management messages, while traffic channels are used for the transmission of data bearing messages, which are formatted as MAC frames that carry data packets as their payloads. A beacon message that is periodically broadcast by a BS is also employed by mobile users to **synchronize their clocks** with that of the BS and thus gain common timing synchronization with other mobiles that share the same wireless medium. Accordingly, the boundaries of time slots assigned to a user by a BS can be well distinguished from those associated with neighboring time slots, so that message transmissions carried within distinct time slots do not temporally overlap. It enables a mobile user to unambiguously identify the proper timing of its allocated uplink and downlink time slots.

As a mobile user turns-on its cellular device, it listens to system information broadcast by nearby BSs. The mobile selects a BS to associate with. This is also known as an *attachment* process. Following the completion of a subsequent successful process of authentication and validation of its registration and record as a subscriber in good standing, the mobile user is accepted as a client of the selected cell's BS and its associated service provider. The underlying cell may be located in the subscriber's service provider's home region and operated by its home provider. This provider keeps its subscribers' features, locations and records in a **Home Location Register (HLR)**. As it moves, a mobile may visit and communicate in an area that is operated by another provider. The provider keeps information about visiting users in a **Visitor Location Register (VLR)**, enabling a roaming service for users that are located outside the area covered by their home network.

Consider the communications process involved in serving a mobile initiated (voice, data, or video) call. Having identified the resources used for configuring the cell's uplink and downlink signaling channels, the mobile user produces a call request packet that it transmits across an uplink signaling channel to its associated BS. The latter stores the call request packet in a queue and then processes it to determine whether it is able to allocate uplink and downlink wireless medium resources to accommodate the data needs of the requesting user's flow.

When the user requests the system to provide QoS levels in support of the transport of its messages across the wireless medium, the BS would have to determine whether it is able to currently allocate sufficient resources to grant the request. If so, the BS would reply to the requesting user by sending a resource allocation packet across a downlink signaling channel. This packet acknowledges the reception of the user's request as well as specifies uplink and downlink resources that the system is allocating for the transmission of messages associated with this flow.

Earlier cellular systems generations allocated resources for the full duration of a call, operating on a circuit-switched basis. Uplink and downlink resources were dedicated for use by a configured full duplex connection that is used to accommodate the associated two-way flow. One-way allocation can also be made. The latest cellular systems, such as 4G, 5G, and 6G use an IP-oriented packet-switching process. In these systems, uplink and downlink resources are allocated for the transmission of individual messages or message groups across the wireless medium.

Using its allocated wireless medium resources, the initiating mobile user is able to proceed in its communications with its associated BS, transmitting its packets across assigned uplink resources and receiving packets from the BS across assigned downlink resources. Resources are jointly specified across a subset of segments that span time, frequency, space, and possibly code dimensions. As described in further detail in Section 8.3.2, under a demand-assigned reservation **DA/FDMA/TDMA/SDMA** scheme, a station can be assigned a frequency band for use within specified time slots that belong to specified time frames. A BS or a user station may also be allocated a spatial segment(s) to use for orienting its transmissions through the use of directional antenna beams.

Upon receiving messages across an uplink traffic channel, the BS node proceeds to use core network resources to route these messages across a backbone network, such as the Internet or the public-switched telephone network (PSTN).

When the flow's packets are destined for reception by another mobile user, the following **mobile terminating call or flow communications** process takes place. The destination mobile may have moved since it has previously been tracked. Consequently, to reach a destination mobile, it is necessary to determine the cell in which it is currently located so that the packet destined to it can be routed across the backbone network to the BS node that manages the cell in which the destination mobile currently resides. For this purpose, a **location update process**, also known as a **mobility management** process, is performed.

To identify a mobile's location, cellular systems require the mobile to transmit **location update (LU) packets**. Such a packet identifies a specific region in which the mobile is currently located. Earlier cellular network generations have identified this region as a **Location Area (LA)**. It can consist of several cells (or even many smaller cells). BSs periodically transmit across a downlink broadcast channel the identity of the location area with which they are affiliated. As a mobile detects that it has crossed a location area boundary, it would generate a **location update message** and transmit it to its associated BS. This message will subsequently be conveyed to the network's management system. In this manner, the system knows in which location area a specific mobile is currently located, though it may not know the identity of the specific cell in which the mobile currently resides.

In later cellular network implementations, such as for a 4G-LTE network, the location area is identified as a **tracking area (TA)**. A mobile may be assigned a *list of such tracking areas* which when crossed would trigger the generation of a location update packet. Advanced systems can create the tracking list in a dynamic manner, adapting the rate at which location update packets

18.3 Cellular Network Communications: The Process 553

are generated to the underlying layout and to traffic loading characteristics. An area oriented update process is also used by 5G cellular wireless network systems.

To reach a destination mobile, BSs that reside in the area in which the mobile is currently roaming, would each broadcast (in its cell) a paging packet, informing the destination mobile of an impending message/call. Therefore, if a location area consists of 10 cells, so that it is managed by 10 corresponding BSs, 10 distinct paging packets would be broadcast by these BSs. Upon receiving a paging packet, a targeted destination mobile would respond by sending a message to its associated BS, informing it of its page message reception. Subsequently, the paging process is terminated, and the flow's packets would be routed across the backbone network to the identified destination BS. The latter will then configure resources across its wireless radio access network to enable, if feasible, downlink and uplink communications resources that will be used to communicate with the destination mobile user.

As it moves, a mobile user that is engaged in an ongoing flow may cross the boundary of a location or tracking area. The cellular system would then react by rapidly performing a **handover**, also known as **handoff** process, which directs the flow's packets to a new associated BS. Current cellular networks perform this handover process in a sufficiently rapid and seamless manner, without causing noticeable interruption in the integrity of the end-to-end message flow.

In 4G-LTE systems, three tracking area update (TAU) types are performed: (1) A Normal TAU is carried out when the tracking area to which the mobile user is attached is different from that associated with its previously registered cell. (2) Under a periodic TAU process, the mobile issues an update in a time periodic manner. (3) Under a combined tracking area update procedure, the mobile follows a tracking area update process that is specified and managed by the system. The PLMN system provides the user a list of tracking area codes (TACs). The mobile user generates a TAU message when it enters a tracking area that is not in the list of tracking areas that the mobile previously registered with the system's MME.

Mobility conditions can lead to the generation of a high rate of tracking area updates. The system may then modify those impacted tracking lists in aiming to balance the packet traffic loading rates that are produced across different areas.

The span of the tracking area impacts the produced update packet traffic rate. If the area is configured to cover a small geographical span, a higher tracking update packet traffic rate would be triggered, requiring the allocation of higher levels of critical wireless bandwidth resources for uplink transmission of location update packets and for downlink transmission of ensuing signaling and control packets. Such transmissions tend to use the system's frequency resources in an inefficient manner, as they tend to occur in a bursty fashion. High location update traffic rates would also increase the loading of MME service entities.

In turn, if the tracking area is set to cover a wider area, a higher rate of paging packets would be induced, as a larger number of BSs would then be required to broadcast paging packets, causing higher traffic loading of wireless resources. As noted above, to alert a destination mobile to an incoming flow, the BS system broadcasts paging messages throughput the tracking area span. An advanced mobility management process dynamically selects the tracking area list by taking into consideration multiple factors.

Other issues are also incorporated in the judicious setting of the parameters of the location update protocol, such as by taking into consideration the possibility that a mobile moving around a location area boundary may cross this boundary repeatedly via back-and-forth movements, triggering an excessively high traffic loading rate of location area update packets.

Mobility control schemes that are not handover based are also possible. A call redirection process may be used, though it would lead to a less smooth (and possibly somewhat disruptive) call transition process. Communications quality measurements of links that connect a mobile with its BS may be performed by the mobile and/or the BS and used to inform the selection of an alternate BS.

18.4 The 4G-LTE Protocol Architecture

To illustrate the protocol architecture of a cellular network system, we discuss in the following the 4G-LTE system. Our descriptions are based on definitions, outlines, and illustrations presented in 3GPP LTE Standards technical documents.

Transport of data messages is governed by **user plane**, also known as *data plane*, protocols and algorithms. An application protocol entity produces data messages that are processed by a transport protocol entity, typically TCP or UDP. At the network layer, the IP protocol is generally employed.

In the **control plane**, a radio resource control (RRC) protocol is employed for managing the allocation of wireless resources across the RAN. For this purpose, resource allocation signaling messages are transported between a BS and its mobile users.

In each plane, processing is performed by protocol entities that include: a packet data convergence protocol (PDCP) entity, radio link control (RLC) protocol entity and medium access control (MAC) protocol entity. The produced data is then passed to a physical layer entity for transmission across the wireless medium.

The LTE protocol architecture in the user plane is shown in Fig. 18.8(a) [2]. At the **physical layer (L1)**, the data frames produced by a MAC entity are transmitted over the wireless medium. The L1 entity performs the initial cell search, which is used for synchronization and handover purposes, implements adaptive modulation/coding (AMC), transmit power control and medium measurements. Signal to Interference and Noise Ratio (SINR)-oriented measurements are used by resource allocation schemes for assigning wireless resources over the RAN.

The **MAC** layer protocol entity implements the scheduling processes that are employed for sharing the RAN's wireless medium. In the downlink direction, protocol data units (PDUs) that arrive as service data units (SDUs) from a higher layer's logical channel are formed into MAC layer frames, which are identified as transport blocks (TBs). SDUs that arrive from several logical channels may be multiplexed within a TB. MAC frames are transferred to the physical layer entity.

TBs that are sent to a BS across an uplink channel are processed by the station's physical layer entity. Its product is delivered to the MAC layer protocol entity. They are de-multiplexed into their corresponding logical channels.

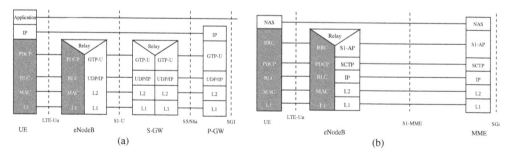

Figure 18.8 The LTE Protocol Architecture: (a) The User Plane; and (b) The Control Plane. Source: 3GPP LTE Standards, TS 23.401, version 9.14.0 release 9, authorized by Copyright License, ©2013-3GPP™.

The MAC layer entity is responsible for dynamic scheduling of the use of wireless medium resources. Demand-assigned multiplexing and multiple access schemes are employed. It manages the operation of QoS and priority-based scheduling schemes, data reporting that impacts the scheduling decision process, and performing error correction through the use of Hybrid ARQ (HARQ) schemes.

GPRS networks can connect to external networks, including the public Internet and a multitude of networks operated by partners and customers. GTP (GPRS Tunneling Protocol) is used between GPRS support nodes (GSNs). GTP is employed to establish a GTP tunnel between a Service Gateway (S-GW) and a PDN Gateway (P-GW), which serves as a GPRS Support Node (GGSN). The S-GW receives packets from end users and encapsulates them within a GTP header prior to forwarding them to the P-GW through the GTP tunnel. When the P-GW receives the packets, it decapsulates them and forwards them to the external host. Different tunneling protocols are used depending on the interface. The GTP is used on the S1 interface between the eNodeB and S-GW and on the S5/S8 interface between the S-GW and P-GW. GTP comprises a control plane and a user plane. Its data flows are often carried over UDP/IP.

The **RLC** layer provides three service mode types: Transparent Mode (TM), Unacknowledged Mode (UM), and Acknowledged Mode (AM). Its services include the following: transfer upper layer PDUs; perform ARQ error correction (under AM); execute concatenation, segmentation and reassembly of RLC SDUs (under UM and AM). It also is responsible for re-segmentation of RLC data PDUs (under AM), reordering of RLC data PDUs (under UM and AM), duplicate detection (under UM and AM), RLC SDU discard (under UM and AM), RLC re-establishment, and protocol error detection (under AM).

RRC layer services include: broadcasting of system information related to the access system, which is also known as the *Access Stratum (AS)*; broadcasting of system information related to the nonaccess system, known as the *Non-access Stratum (NAS)*; performing paging dissemination; maintenance and release of an RRC connection between the mobile user and the RAN; performing security functions; establishment, configuration, maintenance, and release of point-to-point Radio Bearers.

The **PDCP** layer is responsible for services that include: header compression and decompression of IP data; transfer of data (in the user plane or the control plane); maintenance of PDCP Sequence Numbers; in-sequence delivery of upper layer PDUs at the re-establishment of lower layers; duplicate elimination of lower layer SDUs at re-establishment of lower layers; ciphering and deciphering of user plane data and control plane data; integrity protection and verification of control plane data; timer based duplicate discarding.

The **protocol stack for the control plane** between a mobile (UE) and the MME entity is shown in Fig. 18.8(b), [2]. AS protocols are used for communications between a user, identified as equipment (UE), and a BS node (eNodeB). Lower layers perform the same functions as those noted for the user plane with the exception that there is no header compression function in the control plane. The NAS protocols support user mobility and perform session management procedures to establish and maintain IP connectivity between a mobile user and the PDN Gateway.

To provide end-to-end service involving connectivity between a mobile user and a destination peer user, a hierarchy of LTE bearers, as shown in Fig. 18.9 and presented in 3GPP LTE Standards [2, 19], is employed. A bearer is used to provide connectivity in the user plane between a mobile user and a PDN gateway. This bearer is identified as the Evolved Packet System (EPS) bearer. An initial EPS bearer is established when the UE registers with the network, using the Attach procedure. It is known as a default EPS bearer and it is used to provide always-on connectivity. Other EPS bearers, known as dedicated EPS bearers, can be configured to connect to other PDN gateways or

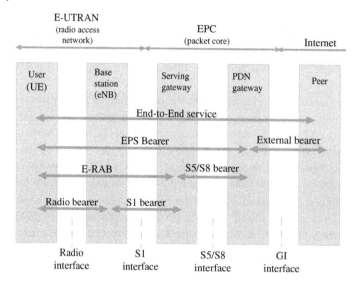

Figure 18.9 LTE Bearers.

to provide different LTE QoS support when connecting to the same PDN gateway. All user plane data transferred by using the same EPS bearer is granted the same QoS.

An EPS bearer is created as a concatenation of E-UTRAN radio access bearer (E-RAB) and S5/S8 bearer. The S1 bearer provides connectivity between a BS (eNB) and a home serving gateway. The S5 interface provides connectivity between a home serving gateway and a home PDN gateway. The S8 interface provides roaming connectivity between a visited serving gateway and a home PDN gateway. The connectivity over the RAN between a user (UE) and a BS (eNB) is identified as a radio bearer. An E-RAB is created when combining a radio bearer and S1 bearer.

The LTE RAN protocol layering stack is shown in Fig. 18.10, based on 3GPP LTE Standards [2]. Recall that a packet received by a layer entity is called a SDU while a packet produced by a layer entity (for examination by a peer layer entity) is identified as a PDU.

Consider user data traffic flowing in the downlink direction. IP (L3) PDUs are processed by the PDCP entity to produce PDCP PDUs. As noted above, a user data packet flow is associated with a hierarchy of bearers. PDCP PDUs are noted to be assigned a radio bearer as they are delivered as SDUs to the RLC layer entity. This bearer provides the flow's packets a default QoS performance behavior as they are transmitted across the radio access network. The processes and algorithms used by the RLC entity are driven by RLC control data that is produced by the RRC layer entity. The RLC PDUs produced by the RLC layer entity are delivered to the MAC layer entity across **Logical Channels**. These channels define *what type* of information is transmitted over the air, differentiating between traffic data, control/signaling, system broadcast channels, and other. Logical channel types include: Broadcast Control Channel (BCCH), Paging Control Channel (PCCH), Common Control Channel (CCCH), Dedicated Control Channel (DCCH), Multicast Control Channel (MCCH), Dedicated Traffic Channel (DTCH), and Multicast Traffic Channel (MTCH).

The MAC frame PDUs produced by the MAC layer entity are delivered to the physical layer entity across transport channels. A **transport channel** defines *how is* the underlying data unit transmitted over the air. It determines the encoding scheme and other mechanisms that are employed for the transmission of user and signaling data units. Transport channel types include: Broadcast Channel

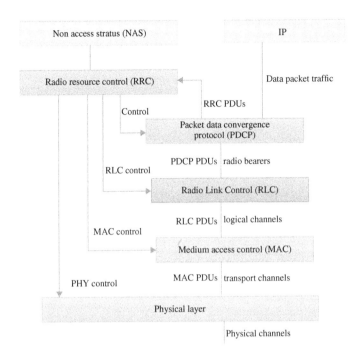

Figure 18.10 The LTE Protocol Stack.

(BCH), Downlink Shared Channel (DL-SCH), Paging Channel (PCH), Multicast Channel (MCH), Uplink Shared Channel (UL-SCH), and Random Access Channel (RACH).

The physical layer entity transmits the data and signaling MAC PDUs that it receives from the MAC layer entity across a **physical channel**. The latter identifies *where is* the transmission takes place, such as specifying the physical channel resources that are employed, spanning time, frequency and spatial dimensions. Downlink physical data channels include: Physical downlink shared channel (PDSCH), Physical broadcast channel (PBCH), and Physical multicast channel (PMCH). Uplink physical data channels include: Physical uplink shared channel (PUSCH) and Physical random access channel (PRACH). Physical control channels include the Physical downlink control channel (PDCCH) and the Physical uplink control channel (PUCCH).

Fig. 18.11, based on the 3GPP LTE Standards, shows an illustrative data flow across the RAN layer stack. A MAC PDU frame is noted to be sent by the MAC layer entity as a **TB** to a physical layer entity for transmission across the wireless medium. The illustration shows this block to be transmitted across the radio channel during a specified time period, known as a sub-frame, which consists of two time slots. The corresponding allocated wireless medium resource is also identified by an allocated frequency band, and at times also by a specified spatial segment (as is the case when using a multiple beam antenna array).

LTE uses an Orthogonal Frequency Division Multiplexing (OFDM) method for conducting its transmissions across the downlink. Assignment of wireless resource units in LTE is carried out over the time–frequency resource plane by allocating groups of *resource blocks* for physical layer transmission of TBs. Such a resource block is shown in Fig. 18.12 [3]. Each resource block occupies 180 kHz in the frequency domain and 0.5 ms in the time domain. A resource block consists of 12 sub-carriers. The frequency bandwidth of each subcarrier is equal to 15 kHz. Each 1 ms Transmission Time Interval (TTI) consists of two slots (T_{slot}). The figure shows each

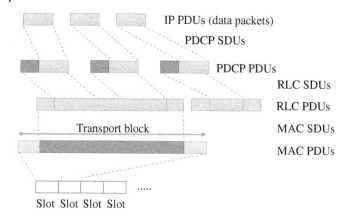

Figure 18.11 A Data Flow Across Radio Access Network (RAN) Layers.

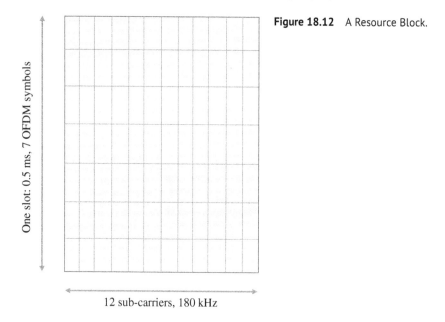

Figure 18.12 A Resource Block.

subcarrier of the resource block to carry 7 (channel signal modulation) symbols over a single time slot. The MAC scheduling scheme can allocate multiple resource blocks to accommodate the transmission of a TB.

18.4.1 Allocation of Wireless Access Resources

The wireless medium of the RAN is shared by mobiles and the BS for downlink and uplink transmissions of signaling and data packets. The sharing process is centrally managed by the cell's BS. The algorithms used for the dynamic allocation of downlink and uplink wireless resources, including the methods used to provide QoS performance guarantees, are not standardized. Different service providers may implement different QoS-oriented resource allocation schemes.

Prior to carrying out **downlink transmission of packets** by a BS to a mobile client user, the BS sends the user an allocation message across a downlink signaling channel. This message specifies the time/frequency/spatial resources that it allocates for this transmission. This information allows the user a brief time to get ready for the reception of the designated data. As the BS is often involved in the simultaneous transmission of data packets to multiple users, a dynamically adaptive statistical multiplexing scheme is employed. It is used to provide a bandwidth efficient allocation that meets, when requested and approved, a flow's performance requirements. Transmission resources are allocated for use during a flow's data activity bursts. During other times, resources are used for the accommodation of other active flows.

In aiming to meet QoS metric values, a downlink cross-layer adaptive multiplexing scheme is employed. For the transmission of data to a specific mobile user at a data rate that meets specified QoS objectives, including a targeted message error ratio at the receiving entity, the BS selects a proper modulation/coding scheme (MCS). For this purpose, the BS uses **Channel Quality Indicator (CQI)** reports that it receives from the destination mobile user. As described in Section 3.8.4.2 and demonstrated in Fig. 3.7(a), a higher CQI number indicates to the BS that its signal is received at the user at a higher *SINR* level. As demonstrated in Fig. 3.7(b), for a higher value for a CQI reported by a destination user, the BS is able to select a more throughput efficient MCS, achieving higher spectral efficiency. The frequency bandwidth allocated for the support of the underlying data transmission is accordingly determined.

In Chapter 6, we describe a multitude of multiplexing protocols and algorithms that can be employed by the BS for downlink transmission of data packets. Statistical multiplexing methods and related scheduling schemes are presented and discussed in Sections 6.3 and 6.4. Resources are allocated for the downlink transmission of a flow's packets in accordance with the performance metric values requested by the flow (or allocated to it by default).

As described in Chapter 5, messages impose distinct performance requirements that generally are dictated by the flow type. Such requirements include message delay, error rate, throughput rate, and message loss ratio. LTE-based metrics values are illustrated in Section 5.2.6. A **QoS Class Identifier (QCI)** is used to identify the performance treatment required by a flow. Such performance levels are shown in Fig. 5.4. Certain flow types are allocated resources on a **Guaranteed Bit Rate (GBR)** basis, where performance levels are guaranteed to meet or exceed minimal specified values. Other flows are designated to receive **non-GBR** treatment, so that they are not guaranteed performance objective values. Accordingly, no bandwidth resources may be reserved for the accommodation of non-GBR flows. A schematic display of the functional architecture of a multi-tier scheduling architecture is shown in Fig. 6.7 and discussed in Section 6.4.

As a BS would typically be involved in the downlink transmission of data packets to multiple mobile clients that are simultaneously active, it needs to allocate downlink wireless medium resources in a manner that is cognizant of the CQI value reported by each such user and of the QoS requirements associated with each active GBR-based flow. Multiplexing algorithms that take into consideration such conditions and requirements are described in Section 6.5. Also discussed there are algorithms that incorporate **fairness** considerations in the assignment of wireless media resources to multiple users. Included are the following schemes: Round Robin, Weighted Round Robin, Generalized Processor Sharing, Fair Queueing, Maximum Throughput, and **Proportional Fair Scheduling (PFS)**.

Dynamically adaptive multiple access schemes are used for the efficient allocation of **wireless uplink** RAN resources to active mobile users. Following a request for such an allocation that is sent by a mobile to its BS across an uplink signaling channel, the BS makes an allocation, if feasible, and uses a downlink signaling channel to inform the mobile of the time/frequency/spatial resources that are assigned for the uplink transmission of its data packets. The allocation can involve resources to be used for specific uplink packet transmissions, involving packets that are

560 | *18 Mobile Cellular Wireless Networks*

produced by a single flow or by multiple flows, in which case such packets are multiplexed by the user across its allocated uplink wireless resources.

Multiple access algorithms and protocols are described in Chapter 8. In particular, **Demand Assigned Multiple Access (DAMA)** and **Demand Assigned Reservation** schemes are presented and discussed in Sections 8.3 and 8.3.2. A mobile that wishes to send data to the BS and thus share the resources of the multiple access uplink medium, sends a reservation request to the BS across an uplink signaling channel. Under a **DA/FDMA/TDMA/SDMA** scheme, the requesting station is assigned a frequency band for use within specified time slots and time frames. Its transmission may also be allocated a spatial segment (in utilizing pre-set or synthesized antenna beams).

In assigning the corresponding resources to a mobile station, the BS takes into consideration the mobile's requested performance requirements (if any) that are linked to each flow. It also monitors the SINR level incurred by the user's signal at the BS's receiver. This information is used to determine the MCS to be used by the user's transmission module. The corresponding spectral efficiency level that is realized through the use of this MCS, determines the bandwidth resources that the BS needs to allocate to the user for realizing the throughput rate (and other performance objectives) that provides a GBR-oriented user its requested QoS level.

A **flow admission control** mechanism is employed to block the admission of flows to which the requested uplink resources cannot be allocated for a sufficiently high fraction of activity time. In determining the distribution of available uplink resources to active mobiles, algorithms that assure the fair allocation of resources to mobiles may be employed.

The implementation of a demand-assigned reservation scheme for sharing the resources of the uplink wireless medium requires a requesting mobile to use an uplink signaling channel for the transmission of its reservation request packet. The BS uses a downlink signaling channel for the transmission of its allocation packet. It specifies the uplink wireless resources that it assigns to the requesting mobile. The assignment accommodates the uplink transmission of a specified quantity of data units. Data units may correspond to individual message segments, a group of message segments, or segments that are transmitted over specified periods of time during which the user is active or expected to be active.

For data users that produce packet traffic in a highly bursty (and thus at a low duty cycle) and repetitive manner, reservation request and allocation packet traffic flows could lead to high signaling traffic loads. The latter can then occupy relatively high uplink and downlink bandwidth resources. Systems act to reduce such signaling loading rates by allocating uplink resources to such an active user for use over longer periods of time, reducing its request rate for resource allocation. For users that simultaneously engage in multiple flows, sufficient uplink resources may be allocated to enable such a user to statistically multiplex the data produced by its local flows.

18.5 Next-Generation 5G, 6G, and Millimeter-Wave Cellular Networks

Next generation, such as 5G, 6G, and beyond systems, aims to provide users with significantly higher uploading and downloading throughput rates, as well as transfer packets at very low latency levels. For this purpose, and to accommodate emerging applications that often require high bandwidth resources, these systems include operations that make use of higher carrier frequency levels. They offer much higher-frequency bandwidth spans.

5G systems use low-band, mid-band, or high-band frequency bands. Low-band systems use the 600–900 MHz frequency span, which is similar to that used by 4G systems. Data rates that range 30–250 Mbps are attained by low-band systems and by 4G systems. Mid-band systems use the frequency range 2.3–4.7 GHz (which is also used by 4G systems), attaining data rates of the order of 100–900 Mbps. A cell radius is of the order of several kilometers.

Applications of interest when using this technology include wireless communications used by autonomous vehicle systems, massive Machine-Type Communications (mMTC) that connect a large number of devices, and interconnections of a massive number of IoT embedded sensor systems.

A significant throughput-delay performance upgrade is attainable when employing a high-frequency band system, such as that uses the 24–57 GHz frequency range. Expected throughput rates can reach several Gbps levels. Future systems plan to use even higher-frequency bands, including frequency ranges that are higher than 60 GHz, which are in the upper **mmWave** band span.

While higher band and mmWave systems offer much higher throughput rates, signal propagation losses at such frequency levels are more significantly pronounced. Direct communications is more effective when entities are in LOS and at short range from each other. Obstacles encountered across the propagation path, including buildings, walls and other objects tend to cause significant signal attenuation. Weather conditions, including rain drops, can increase the attenuation level incurred by a mmWave signal. Indoor ranges are of the order of meters to tenths of meters, leading to the use of pico or femto cells. Outdoor direct communications ranges are of the order of 100–250 m, inducing the synthesis of micro and pico cells. Hence, it is preferred that at these bands, communicating mobiles travel at short distance and at LOS range from their associated BS nodes and for mobiles that they directly communicate with. When aiming to cover a wide range busy area, such as that characterizing an urban area, a very large number of BSs needs to be installed and interconnected by a core backbone network.

Millimeter wave antennas are much smaller, as the length of an antenna element is of the order of a wavelength. Consequently, higher dimensional MIMO systems can be effectively implemented, making it feasible for a BS to use a relatively smaller size antenna array that supports massive directional antenna beams. BSs make then use of spatial diversity to time-simultaneously communicate with many mobile users while reusing frequency resources.

System designs also involve the synthesis of a hybrid system that uses higher- and lower-frequency bands for the transport of different message types. For example, while higher-frequency bands are used to accommodate shorter range communications at ultrahigh-throughput rates that are required by multimedia data flows, lower-frequency bands can be used to disseminate signaling and management message flows over wider ranges. The later flows carry data that is used to manage and control the system, at longer time spans, such as those used for resource allocation and for coordinating operations among multiple subsystems. Such traffic tends to not require the higher-throughput rates demanded by many data flows, while requiring robust and reliable dissemination paths that span multiple subsystems across wider ranges.

Innovative approaches and methods have been developed and advanced ones are being planned for the design of future wireless network systems. Advanced networking techniques also make use of AI and ML-based methods. Dynamic system state processes are continuously monitored and analyzed. Ongoing statistical characterizations are performed, producing dynamic process models of user behavior, entity mobility, traffic loading processes, service and performance

18 Mobile Cellular Wireless Networks

demand profiles, utilization and availability levels of communications resource, and networking requirement patterns.

These processes are analyzed and used to autonomously adjust the network system's layout, and to perform adaptations of resource allocations, networking layouts, and control operations.

Hybrid networking architectures are synthesized. Infrastructure-based cellular and non-cellular architectures and networking techniques are employed in conjunction with the use of peer-to-peer ad hoc networking methods. Hybrids of multiple access wireless media and dynamically synthesized directional wireless links are used.

The topological and architectural symmetry of a current cellular system could be replaced at times by an amorphous structure that combines cellular components, using cells of varying sizes and shapes, with dynamically synthesized peer-to-peer multi-hop communications networking segments. The latter include infrastructure-less mobile ad hoc networking structures. Ad Hoc networking methods and protocols are presented and discussed in Chapter 19.

Network system architectures are also extended through the autonomous placement of multi-tier interconnected switching and relay nodes. The latter include mobile terrestrial relay platforms that are placed at strategic locations, coupled with the use of multi tier airborne and space-based interconnected switching nodes. Such networks may employ Unmanned Aerial Vehicle (UAV/drones) and low-altitude small satellite platforms. Interconnected multi attitude airborne systems are synthesized to form high-performance backbone networks that can be dynamically guided to provide communications coverage over a wide span area of operations. Unmanned ground vehicles (UGVs), acting as robotic platforms, are autonomously placed at best locations for the support of ground, air, and ground–air communications networking.

Dynamically adaptive architectures are designed to also provide robust support for rapid failure recovery. Mobile BSs and UGV/UAV modules are autonomously moved and placed at key locations that serve to aid communications in repairing a defective network layout that is caused by the failure of BSs and relays. Network system configuration, networking, and resource allocation operations are autonomously adapted and adjusted to best support the surviving network system.

Problems

18.1 Describe and discuss the architecture of an infrastructure based wireless network.

18.2 Consider the Radio Access Network managed by a Base Station (BS) in a mobile cellular wireless network. Define and explain the services provided by the following channels:
a) Uplink signaling channels.
b) Downlink signaling channels.
c) Uplink traffic (or data) channels.
d) Downlink traffic (or data) channels.

18.3 Consider the Radio Access Network (RAN) managed by a Base Station (BS) of a mobile cellular wireless network. Define and explain the functions of the following message types:
a) Uplink reservation message.
b) Downlink channel assignment message.
c) Downlink broadcast beacon message.
d) Uplink data message.
e) Downlink data message.

18.4 Consider the Radio Access Network (RAN) managed by a Base Station (BS) node of a mobile cellular wireless network. Define and explain the function and operation of the following message types:
a) Location update message.
b) Handoff (or handover) message.

18.5 Describe and discuss the spatial-reuse ("coloring") methods that are used to regulate signal interference that is caused by transmissions originated by sources that are located in neighboring or near-by cells. Include descriptions of the following interference mitigation techniques:
a) Frequency reuse and corresponding coloring-index.
b) Fractional Frequency Reuse (FFR).
c) Time slot reuse.
d) Joint frequency and time-slot reuse.
e) Spatial reuse using directional antenna beams (SDMA).
f) Joint FDMA/TDMA/SDMA.
g) Coordinated Multi-Point (CoMP) process for multicast message dissemination.

18.6 Describe the key principles of operation of different generations of cellular wireless networks and identify the main features that characterize each one.

18.7 Identify the key subsystem components of a cellular network system and describe the services provided by each one.

18.8 Describe the functions provided by each one of the following subsystems used in an LTE mobile cellular network and identify the services provided by each: RAN, core network, MME, S-GW, P-GW, and HSS.

18.9 Consider a mobile that travels to a location that is outside its home location. Describe the process used for this mobile to send a data message across a cellular wireless network to a destination mobile that is located in a different region.

18.10 Describe the process that is undertaken for a mobile to receive, a flow that is directed to it, across a cellular wireless network, and that originates at a fixed station or at a mobile.

18.11 Describe the contents stored in a Home Location Register (HLR) and in a Visitor Location register (VLR).

18.12 Describe the methods used by mobile cellular wireless networks to track the location of a mobile. Include in your descriptions the concepts of Location Area (LA) and of Tracking Area (TA).

18.13 Describe the layered protocol architecture of a 4G LTE cellular wireless network, as shown in Fig. 18.8:
a) Control plane interface between a user (UE) and its associated base-station (eNodeB).
b) Control plane interface between a base-station (eNodeB) and MME.
c) User plane interface between a user (UE) and its associated base-station (eNodeB).

18 *Mobile Cellular Wireless Networks*

    d) User plane interface between a base-station (eNodeB) and a service gateway (S-GW).

    e) User plane interface between a service gateway (S-GW) and a packet data network gateway (P-GW).

18.14   Describe the operations and services provided by the following entities in 4G LTE cellular wireless network:

    a) Medium Access Control (MAC) for uplink transmissions.

    b) Medium Access Control (MAC) for downlink transmissions.

    c) Radio Link Control (RLC).

    d) Packet Data Convergence Protocol (PDCP).

    e) Radio Resource Control (RRC).

18.15   Define the functions provided by LTE bearers in a 4G LTE cellular wireless network:

    a) Radio bearer.

    b) S1 bearer.

    c) EPS bearer.

    d) External bearer.

18.16   Define the functions provided by the following message types across RAN layers in a 4G LTE cellular wireless network:

    a) MAC PDU.

    b) MAC SDU.

    c) RLC PDU.

    d) RLC SDU.

    e) PDCP PDU.

    f) PDCP SDU.

    g) IP PDU.

18.17   Define the concept of an LTE Resource Block and describe its use in allocating RAN resources for message transmissions across a RAN.

18.18   Consider the determination process made by a base station of the resources to be allocated for downlink transmissions to a specific mobile station. Describe and discuss the following:

    a) The role played by the SINR level reported by a mobile to the base station, and the corresponding Channel Quality Indicator (CQI) descriptor.

    b) The selection of the associated Modulation/Coding scheme (MCS) and the determination of the ensuing spectral efficiency level.

    c) The process used by a base station to determine the size of the resource (such as over time/frequency dimensions) to be allocated for a downlink message flow that is destined to a specific mobile station.

18.19   Consider the determination process made by an LTE base station of the resources to be allocated for uplink transmissions from a specific mobile station. Describe ad discuss the following:

    a) The role played by the SINR level measured at the base station.

    b) The selection of the associated Modulation/Coding scheme (MCS) and the determination of the ensuing spectral efficiency level.

c) The setting of the size of the resource size (such as over time/frequency dimensions) to be allocated to a specific uplink message flow.

18.20 Describe the specification of quality-of-service (QoS) levels employed by an LTE network. Including:
a) Guaranteed Bit Rate (GBR) QoS metrics.
b) Non-GBR rating.

18.21 Describe the use of a Proportional Fair Scheduling (PFS) method employed by a BS for allocating resources to mobile stations across a Radio Access Network.

18.22 Describe key features of 5G and 6G cellular wireless networks.

18.23 Describe and discuss the key features of millimeter-wave cellular wireless networks.

18.24 Describe the characteristics of UAV (Unmanned Aerial Vehicle) and UGV (Unmanned Ground Vehicle) platforms as they are used to enhance the performance of a mobile wireless network system. Consider the use of system monitors and AI calculations to dynamically position such platforms in best locations, and/or altitudes and orbits, and to organize them into mobile swarms.

19

Mobile Ad Hoc Wireless Networks

The Objective: *A mobile ad hoc wireless network (MANET) does not make use of a backbone network. Rather, peer-to-peer communications methods are employed to discover and configure routes that connect mobiles across multihop wireless links.*

The Methods: *Under an On-Demand reactive approach, a user that wishes to transfer a message across a MANET proceeds to first discover a route to its destination entity. In turn, under a proactive approach, users configure and maintain routes to all other users that are located in their domain, updating them on a periodic basis and/or as needed, even when they are not interested in sending a message. Under either method, a message is transported across a configured route via peer-to-peer multihop communications. Users act as network nodes, as they employ their router modules, when members of a proper route, to relay messages that traverse the route. Routing, access control, and transmission operations dynamically adapt to fluctuations in the wireless network's topological layout, induced by nodal mobility.*

19.1 The Mobile Ad Hoc Wireless Networking Concept

An **ad hoc wireless network** provides for the transport of messages between end-users or nodes by not relying on the existence or use of a fixed infrastructure. When some or all of the network nodes are mobile, such a network is identified as a **mobile ad hoc wireless network (MANET)**. Configurations assumed by infrastructure-based networks and by ad hoc networks are illustrated in Section 18.1. A layout of a mobile ad hoc network is illustrated in Fig. 18.2. In a MANET system, nodes communicate in a *peer-to-peer* manner.

Neighboring nodes that are located a relative short range from each other are able to communicate directly with each other. In a MANET, a node uses a wireless link to directly communicate with a neighboring node. Such a direct link is identified as a **hop**. Control mechanisms are employed by each network node to continuously identify their neighboring nodes. Nodes periodically (or upon the occurrence of a triggering event) transmit **hello** messages, at a prescribed transmit power, over their neighborhood, to announce and identify themselves to their neighbors. As a node moves, the list of its detected neighbors dynamically changes. Omnidirectional, sector-oriented or directional antenna beams can be employed in forming links between neighboring nodes.

A medium access control scheme is employed by a node to regulate message transmissions across a wireless link. It spans the physical and link layers. The physical layer entity configures the modulation/coding scheme (MCS) and the transmit power level. The link layer consists of the lower Medium Access Control (MAC) sublayer and the upper Logical Link Control sublayer.

Principles of Data Transfer Through Communications Networks, the Internet, and Autonomous Mobiles, First Edition. Izhak Rubin.
© 2025 The Institute of Electrical and Electronics Engineers, Inc. Published 2025 by John Wiley & Sons, Inc.

19 Mobile Ad Hoc Wireless Networks

Frame transmissions that are conducted simultaneously in time by multiple neighboring nodes may induce signal interference at the corresponding receiving nodes, as such nodes may share the resources of a common wireless medium. Hence, the MAC layer protocol entity at a node implements a MAC scheduling scheme that regulates the sharing of a common wireless medium. Transmissions conducted by non-neighboring nodes can also induce signal interference at receiving nodes. Accordingly, the physical layer entity tends to employ an *Adaptive Modulation/Coding Scheme (AMCS)* that configures its schemes and parameters in a manner that assures reliable reception (aiming to achieve an acceptable SINR level) at a targeted neighboring node.

For example, the physical and MAC layer entities embedded in a Wi-Fi-based module could be employed. A link layer protocol is employed to handle link layer segmentation and reassembly functions and error control operations. MAC layer protocols, such as those used by a Wi-Fi module, can provide ARQ-based error control services, so that a transmitted frame that is not acknowledged by the expiration of a timeout threshold is scheduled for retransmission across the shared medium.

Communications between non-neighboring nodes is accomplished along a dynamically configured **multiple hop path**. For this purpose, each network node, which can be a mobile user, contains a **network layer protocol entity** that provides a routing service. Using an IP network layer protocol, it functions as an IP router. Impacted by nodal mobility, the *topological layout of the wireless network is continuously varying*. A **mobile ad hoc routing algorithm** is structured to rapidly discover and configure an advantageous path that a source node can use for transporting its messages to a destination node. Under a **cross-layer adaptive setting**, the network layer selection of a route is jointly performed with the setting of schemes and parameters at the link, MAC and physical layers, in aiming to best support the message transfer process.

To illustrate, consider node A that wishes to route a packet to a remotely located non-neighboring node D. Assume that node A's neighbors at this time are nodes B and C. Node B is at a closer range to node A and a frame transmission from A to B experiences a relatively high SINR level at node B's receiver. Node C is at a longer range from node A and may be subjected to similar interference signals. Hence, as the intended signal is received at lower power at node C, a frame transmission from A to C may then incur a relatively lower SINR level at node C's receiver. The routing scheme implemented by the node's network layer entity may decide to select an end-to-end path from A to D that includes link A–B, operating at a higher data rate. Alternatively, the routing scheme may prefer to use link A–C as it reduces the number of hops used to reach node D and therefore consumes lower bandwidth resources. Once the route has been selected at the network layer, the node's physical and MAC layer entities would act to execute reliable and robust transmissions across the route's links and physical media.

In the following, we describe several approaches that have been used for routing messages across MANET, noting the following ones:

1) Under a **Reactive**, also known as **On-Demand**, routing scheme, the source node proceeds to discover a route to the destination node at the time that it needs to communicate with this node. Commonly used routing algorithms that use this approach employ the **Ad hoc On-demand Distance Vector (AODV)** protocol or the **On-Demand Source Routing (DSR)** protocol.

2) Under a **Proactive** routing scheme, network routes are configured in advance, proactively, whether needed or not at a given time. The **Optimized Link State Routing (OLSR)** protocol uses such an algorithm.

3) Under a **hierarchical ad hoc routing** scheme, *hierarchically structured networking layers* are synthesized to aid the routing process and to enhance its efficiency and scalability. The **Mobile Backbone Network (MBN)** protocol implements such a scheme.

19.2 Ad Hoc On-Demand Distance Vector (AODV) Routing

The **AODV** protocol is used to discover, configure, and maintain routes in a MANET. As an on-demand protocol, it proceeds to discover a route between a source node and a designated destination node only when such a route is needed. While unicast or multicast routes can be configured, we focus our description on unicast flows. We describe the principle of operation of the protocol rather than provide a detailed specification of any of its versions.

AODV operations make use of the following processes. A **route discovery** process is used to find, select and configure a route to a designated destination node. For this purpose, as illustrated below, **Route Request (RREQ)** control packets are flooded across the network. When such a packet (or several packets) reach the destination node, a **Route Reply (RREP)** packet is formed by the destination. The latter disseminates it across a reverse unicast path, flowing from the destination node to the source node. Upon the successful completion of this process, routing entries are configured and stored in the routing tables of the routers located across the selected path.

Triggered by nodal mobility, communications links included in an active route may fail. The current route's ability to transport packets to the designated destination node is then disrupted. **Route maintenance** processes perform route repair or route rediscovery. For this purpose, upon a failure of a communications link, **Route Error (RERR)** messages are disseminated to affected nodes. Even if no failure occurs, a new route rediscovery process is triggered upon the expiration of a specified time period. To assure the freshness of routing entries, entries held by a node are purged if not used within a specified period of time.

Source node S that wishes to send packets to destination node D checks first its routing table to determine if it currently maintains a valid forwarding entry for the underlying destination node. Such a routing entry specifies the IP address of the destination node (D-IP-ADD) and the identity of the next hop to use in forwarding a packet that wishes to reach this destination node. The forwarding entry indicates the outgoing port that is attached to the communication link across which the packet is transmitted to reach the next node across the route. It identifies the IP address of the next hop's node (NH-IP-ADD).

Assume that the source node does not currently possess such a routing entry. In this case, node S initiates a **Route Discovery** process, aiming to find one or several routes (if any exists) that it can use to reach destination node D. This process involves the broadcasting across the network of a **RREQ** packet, denoted hereby RREQ or RREQ(S,D). For this purpose, a **flooding** protocol is used. Under such a protocol, the RREQ(S,D) packet is broadcast across the network so that it visits multiple network nodes. It inquires each node that it visits as to whether it possesses a currently valid routing entry that can be used to reach node D.

Under a flooding routing protocol, a node that receives a RREQ(S,D) packet and is not the destination node itself or it does not possess the requested forwarding entry, re-transmits the packet across each one of its links, avoiding its dissemination on links across which it has previously received this RREQ (S,D) packet.

570 | *19 Mobile Ad Hoc Wireless Networks*

To restrict the dissemination rate of these RREQ control packets, a node does not repeat the transmission of a packet that it has previously received, unless the new packet is identified by a more fresh timeliness identifier. RREQ control packets are uniquely identified by containing the involved source IP address (S-IP-ADD) and a source broadcast ID number (SB-N). The latter is set by the source node and is incremented each time that this node initiates a new RREQ packet.

As noted, an RREQ packet also contains the IP address of the destination node. In addition, to prevent routing loops and to maintain the freshness of routes (avoiding the use of routes and route components that are stale or that are impacted by failure events), source-sequence numbers (SSNs) and destination-sequence numbers (DSNs) are used. Every node maintains its own sequence numbers, increasing it monotonically whenever it detects a change in the topological layout of its neighborhood.

The designated destination node D does not forward received RREQ(S,D) packets, so that upon reaching node D, the route search process is terminated. Yet, RREQ packets may continue to be flooded across network routes that do not lead to node D.

To find a path to a destination node by using a flooding process, the source node first inquires its neighbors whether they know of such a path. If this is not the case, the process proceeds by each neighboring node then inquiring with each one of its own neighboring nodes. If there is a path that connects the source and destination nodes, the process of broadcasting RREQ(S,D) control packets around the network would result in the production of a RREP packet when either node D is reached or an intermediate node that possesses the requested forwarding data is encountered.

Such a flooding-based dissemination process can produce a very high rate of control message traffic across the network. In aiming to reduce this rate, an **Expanding Ring Search** scheme is employed. Under this scheme, the number of hops that a RREQ packet is set to traverse is limited by a specified Time to Live (TTL) parameter. For example, if TTL=2 is specified, the RREQ packet limits its search to a region that contains nodes that are within two hops from the source node. If the ensuing discovery process is not successful, a new discovery process is initiated, whereby a higher TTL value is specified (by assuming, e.g., a value that is double the previous one). The TTL limit is increased in this manner until a route is successfully discovered, or a configured hop limit is reached.

During the route discovery phase, a node may receive multiple RREQ(S,D) packets. They have been forwarded to it from several neighboring nodes. The node selects one of these RREQ packets, applying it for continued flooding, if it is a relay node, or for reply purposes, if it is a destination node. The selection can be based on various criteria. Often, the RREQ packet that is the earliest to be received at the node is the selected one. Such a selection is made when a higher weight is attached to the selection of lower delay paths. A node may delay its selection decision by waiting to receive multiple RREQ packets, each traversing a different route to reach it, and then employ a selection criterion that may incorporate a mix of parameters and associated weights.

Consider a node N that during the route search process selects a RREQ(S,D) packet that it has received from neighboring node M. Node N configures and saves a *reverse path forwarding (RPF)* entry that identifies its decision to use link (N,M) to forward packets that wish to reach node S. The protocol assumes the network's communications links to be capable of transporting packets in a bidirectional manner. Thus, a selected path that has been detected to be a good path for the transport of a packet from node S to node N is also deemed to be a good path to use to transport a packet from node N to node S. The corresponding RPF entry is saved temporarily by each such node N, as explained in the following.

We use the following network layout to illustrate and discuss the route discovery process and the ensuing route selection and maintenance processes. Consider the topological layout of an ad hoc

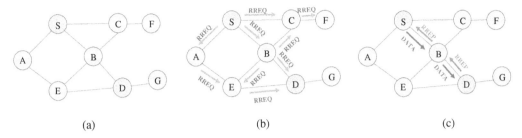

Figure 19.1 Illustrative AODV Protocol Messaging: (a) Topological Layout of an Ad Hoc Network; (b) Selective Route Request (RREQ) Packets; (c) Route Reply (RREP) Control Packet Flows and Data Packet Flows Across Selected Path.

network that is shown in Fig. 19.1(a). While this networking scenario involves a limited number of nodes and links, the same algorithmic principles apply to networks that include a higher number of nodes and links.

In the network's graph, two nodes are shown to be connected by a line if they can directly communicate (at the specified data rate and MCS settings) with each other. As noted, links are assumed to accommodate bidirectional communications. For example, node S is shown to have direct communications with neighboring nodes A, B, and C. Links may employ wireline or wireless media. If node S employs an omnidirectional antenna in transmitting its messages across a wireless channel, a single message transmission may be simultaneously received by nodes A, B, and C. A successful reception at a node will occur provided the signal is received at the node at an acceptable SINR (signal-to-interference plus noise ratio) level, so that the power level of the intended received signal is sufficiently high when compared with the power level induced by background noise plus that caused by transmissions that are performed by other nodes during overlapping periods of time. Under the displayed layout (as impacted also by current nodal geographical locations and the underlying terrain topology), a direct message transmission by node S cannot be (successfully) received at nodes E, D, F, and G. To reach the latter nodes, the message would have to be relayed by other nodes, acting as routers.

RREQ packet flows involved in the route discovery process are shown in Fig. 19.1(b). A source node S aims to discover a route to destination node D. The RREQ message initiated by node S, denoted as RREQ(S,D), is broadcast across the network by using a **flooding** protocol. The figure shows selective RREQ packet transmissions.

An illustrative dissemination scenario is shown. The original RREQ packet that is transmitted by node S is directly received by nodes A, B, and C, following its propagation across the corresponding 1-hop path (link) spans. Node A then proceeds to transmit this RREQ packet to its neighbors, not including the node from which it has received the packet. It is noted that if it had previously received this packet from another node and has already relayed it to its neighbors, it will not do it again. If, prior to proceeding with its relaying action, it is receiving multiple copies of the same RREQ packet from distinct neighbors, it will choose one of these packets, and record the identity of the corresponding neighbor, associating temporarily with it. It may choose to associate with the neighbor from which this RREQ packet was first to arrive.

In the illustrated scenario, node A proceeds to transmit its received RREQ packet to its neighboring node E. In this way, node E learns about the existence of a 2-hop path that leads from node S, passing through precursor node A. It also learns that a path from itself to node S passes through upstream neighboring node A.

19 Mobile Ad Hoc Wireless Networks

Similarly, node B is assumed to transmit the RREQ packet that it has received directly from node S to its neighboring nodes E, C, and D. Recalling that nodes that have already received the same RREQ packet from another node and have already forwarded it will delete the duplicate RREQ packet, the following is assumed. Node C has already received the RREQ packet directly from node S, which it has forwarded to neighboring node F. A duplicate copy of the RREQ packet that is received from node B, pointing to a 2-hop path from node S, will not be forwarded by C. A replica of the RREQ packet sent by node C that may be received at node B (a transmission that is not shown in the figure) will not be processed by node B since it has already processed a replica of the same RREQ packet that it has received from node S (pointing to a 1-hop path from S). Node E is shown to have received the flooded RREQ packet from nodes A and B and then to forward the selected RREQ packet to neighboring node D.

Node B selects a preferred received RREQ(S,D) packet. It then records a **RPF** entry that identifies the corresponding upstream neighboring node from which it has received this selected RREQ packet. This node is the next hop node that node B would use to reach node S. In this case, we assume that the first RREQ packet was received at node B across link (S,B). Consequently, the corresponding RPF entry recorded at node B points to reverse link (B,S). We denote it as RPF(B,S) = (B,S).

Similarly, node A is assumed to set RPF(A,S) = (A,S). Assume that node E has selected the RREQ packet that it has received from node A, so that it sets RPF(E,S) = (E,A). Thus, to reach node S, the RPF entry at node E points to link (E,A). Assuming that there is a route that connects node S with node D, the latter will eventually receive one or several RREQ packets. It does not forward these RREQ packets.

In the illustrated case, we show two RREQ packets to have reached destination node D. They have been received from its neighboring nodes B and E. Assume that the first RREQ packet to reach node D is the one received from node B and that this is the one that node D decides to select and to associate with. It accordingly forms a **RREP** packet, denoted as a RREP(S,D) packet. This packet is transported from node D to node S along a reverse path. This path consists of the RPF link selection made at node D and the subsequent RPF selections made at the upstream nodes that are situated along the reverse path, recalling that a single RPF link has been save by each upstream node.

In the illustrated scenario, node D transmits a RREP packet across link (D,B) to node B. Following the reception of a RREP(S,D) packet, a node forwards the packet toward node S by using its stored RFP (S,D) forwarding entry. In the underlying example, upon receiving this RREP packet, node B configures a data forwarding entry in its routing table that specifies the forwarding link to be used by forthcoming data packets that originate at S and wish to reach destination node D to an outgoing port that is attached to outgoing link (B,D). This forwarding entry is denoted as F(B,D) = (B,D).

Node B then examines its stored RPF entry for reaching node S, which is noted here to point to link (B,S). Accordingly, node B forwards across link (B,S) the RREP(S,D) packet that it has received.

The RREP packet is subsequently received at source node S, which then configures its data forwarding entry for reaching node D as F(S,D) = (S,B). At this time, the source node can start with its data packet transmission.

In this manner, as shown in Fig. 19.1(c), the reverse route used to send the RREP packet from Node D to node S is path (D, B, S). The selected forward data path that connects node S to node D is (S, B, D). Each node that is located on an active path keeps in its routing table also a list of nodal **precursors**, identifying the IP address of each of its (S,D) RFP upstream neighbor. In turn, reverse path entries that are being kept at non-selected nodes, which do not receive an associated RREP packet, are deleted after a specified time duration.

The network topological layout may change under nodal mobility and link failures. To react to link failures or layout variations, a **Route Maintenance** process is performed. This process aims to trigger the avoidance of failed routes and of routing cycles and induce a process that will provide for the configuration of alternate paths to replace failed paths.

If a node detects that an attached active link becomes inactive (e.g., by not receiving hello messages from an associated neighbor, or by not receiving MAC level ACK frames), it invalidates all the routes stored in its routing table that contain the failed link. It creates a **RERR** packet that lists, for routes whose packets it currently set to forward across the failed link, the corresponding destinations that have become unreachable by the underlying failure or change event.

RERR message notifications trigger adjustments in the routing entries used by impacted nodes. Under a common scenario, a RERR(S,D) packet would reach the impacted source node S. The latter then proceeds to initiate a new route discovery process, attaching a higher source sequence number to its new RREQ packet.

The control traffic loading rate induced by the generation of RERR messages is limited in scope since only impacted active nodes are informed. If certain RERR packets are lost, a new discovery process may lead to the generation of routing cycles. The frequency of occurrence of such loops can be resolved through the use of destination sequence numbers (DSNs). A node that possesses a forwarding entry that leads to a desired destination node specified in a received RREQ packet will not use its entry in forming a RREP that serves to create a new path when this entry is associated with a DSN that is lower than the corresponding number specified in the new RREQ message. The lower DSN indicates a possibly stale forwarding entry whose use may lead to the creation of routing loops.

For the underlying example, assume that link (B,D) has failed. Node B detects this failure, deletes its existing (S,D) route forwarding entry, and sends a RERR(S,D) packet across its reverse forwarding path, tagged with an incremented DSN. In this case, a RERR(S,D) packet would be sent across link (B,S). Node S would then initiate a new route discovery process, flooding across the network a new RREQ(S,D) packet which is tagged with an increased DSN. The subsequent flooding process may result in, for example, path (S,A,E,D) as the newly formed (S,D) route.

A local repair process may be undertaken. For example if node E stores a valid (E,D) route (which has not been impacted by an underlying failure), it may use this entry to respond to the new RREQ(S,D) packet received across path (S, A, E), forming a RREP(S,D) packet that will be forwarded to node S across the reverse (E, A, S) path.

19.3 Dynamic Source Routing (DSR)

Similar to the AODV protocol, the **DSR** protocol is an on-demand reactive routing protocol. It employs routines that discover, configure, and maintain routes in a mobile ad hoc network. As performed by AODV, a **route discovery** process is used by a source node to find a route to a designated destination node. For this purpose, **RREQ** packets are flooded across the network. However, rather than requiring a network node that receives and selects a RREQ packet to record a RPF entry, a RREQ packet is configured to carry in its header field the identity of the nodes that it has traversed as it travels to the destination node, as described in the following.

A source node S that wishes to find a route to destination node D forms a RREQ(S,D) packet, including a Request ID. A node that receives a RREQ(S,D) packet may select the first such packet to arrive to disseminate to its neighbors. Other criteria can be used for choosing which one of a received RREQ packets to select and forward. Examining a packet's ID, a node discards a RREQ packet that it has already processed. A node also discards a RREQ packet that specifies a route that contains the node's own address. In forwarding a selected RREQ(S,D) packet, an intermediate node adds its own nodal ID to the route's descriptor that is included in the packet, which lists the nodes that it has traversed.

Consider the network shown in Fig. 19.1(a) and the partial RREQ flows shown in Fig. 19.1(b). Upon receiving and selecting a RREQ packet received from node S, node B appends its own ID to form the source route descriptor (S,B), identifying nodes S and B as members of the upstream S-to-B route that it has selected. This descriptor is included in the RREQ packet forwarded by node B to its neighbors. Similarly, the source route descriptor included in the RREQ received by node E from node A specifies nodes (S,A) as its upstream path. Assuming node E to select the latter RREQ packet, it appends its own ID to form the source route descriptor (S,A,E), which it includes in the RREQ packet that it forwards to its neighbors. Destination node D would then receive two RREQ(S,D) packets: the first packet is received from neighboring node B, yielding the (S,D) path (S,B,D), while the second packet is received from neighboring node E, yielding the source route (S,A,E,D).

It is noted that when processing RREQ packets, an intermediate node may learn about network routes to various other nodes by examining the route descriptors included in the RREQ packets that it receives. It can then store this information in its cache of learned routes. For example, assuming bidirectional communications links, node E learns about a path to node S that consists of member nodes (E,A,S), as well as about the path (E,B,S). Such data can be employed to reduce future route discovery transmissions. It can also be used to identify alternate routes that may be invoked for rapid local route repair. An expiration time threshold is associated with each cached routing entry. The use of stale entries can lead to the configuration of ineffective routes.

Upon receiving RREQ(S,D) packets, node D selects one of this packets, such as the first one to arrive (or the one that uses the shortest or least congested path) as its preferred (S,D) route. It then forms a **RREP** packet and sends this packet across the network to source node S. In case the system contains asymmetrical communication links (whose communications features depend on the directionality of the flow across the link), the protocol provides for a separate discovery process to be initiated by the destination node for finding a route that leads from D to S. It then piggybacks the RREP packet on the latter RREQ(D,S) packet. In turn, when the network employs symmetrical bidirectional communication links, as is the case for many wireless links and as assumed to be the case when using AODV and we assume in the following, the reverse source path is utilized to send the RREP packet. Thus, the RREP packet traverses in reverse the same nodes that have been used to forward the RREQ packet that has been selected by node D. An intermediate node on this reverse path, say node B, uses the selected (S,B) source route to specify the IP address of its neighboring node to which it forwards the RREP packet. For example, for the network and scenario shown in Fig. 19.1(b), destination node D uses the RREQ(S,D) packet received from node B to form its RREP(S,D) packet. This packet is sent to node S along the reverse route, which consists of nodes (D,B,S).

When receiving the selected RREP(S,D) packet, source node S caches (i.e., stores) the ordered nodal composition of the cited (S,D) route. As long as this source route remains valid, node S proceeds to append the source route descriptor (specifying the IP addresses of the consecutively traversed nodes) to the header field of each one of the data packets that it sends to node D. An intermediate node located across the specified source route examines this header field and instantly determines the outgoing link across which it should forward the flow's data packets.

The source node maintains a send buffer in which it saves packets for which no source route has yet been discovered. After a timeout period, if no route is as yet determined, a rediscovery process is initiated. A packet is discarded if no route is discovered within a specified period of time.

Nodes can learn about network routes by monitoring route descriptors included in control packets, such as RREQ and RREP packets, as well as by monitoring the source route descriptors included in data packets. Considering a network that uses wireless links, nodes could learn network routes to various other nodes by overhearing control or data packet transmissions performed by neighboring nodes and examining their route descriptors.

Considering the example shown in Fig. 19.1, node C may overhear transmissions of (A,D) data messages by node B to learn and cache a route to node D and a route to node A that includes nodes (C,B,A). The multitude of learned routes stored at a nodal cache can be used by a node to form a RREP packet as a response to a received RREQ(S,D) packet when this node records in its cache a route to destination node D.

Another illustration of DSR protocol-based packet flows and route caches is also shown in Fig. 19.2. Source node A communicates with destination node D across source route (A,B,C,D). The packets that flow along this route include in their header the route's descriptor. Similarly, source node E communicates with destination node G across source route (E,F,G) whose descriptor is included in the headers of the corresponding flow's packets. Also noted in the figure is a list of routes cached by node C. It includes routes observed by packet flows across the (A,D) path, as well as routes that are deduced by overhearing packet flows across the (E,G) path, through examination of the source route descriptors included in the packets.

Under the DSR route maintenance process, a **RERR** packet is created by a node that detects an attached link to have failed. For example, considering the layout shown in Fig. 19.1, assume node B to have detected link (B,D) to have failed. Node B then forms a RERR packet, indicating the failure of link (B,D). The RERR packet is sent to impacted nodes to inform those nodes. In this example, RERR will be sent to node S across the route that consists of link (B,S). Node B may keep data concerning the operational status of link (B,D) for a relevant period of time and take it into consideration during route search and maintenance.

Upon overhearing across its wireless media the transmission of this RERR packet by node B, other nodes take into consideration this failure notice, deleting routes that they have cached that make use of the failed link. When creating a new RREQ packet, node S will include in it the RERR

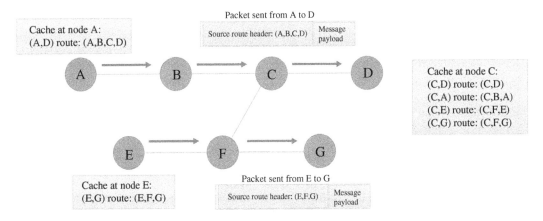

Figure 19.2 Illustrative DSR Protocol Packet Flows and Nodal Route Caches.

packet that it has received and triggered this rerouting process, making sure that other nodes participating in the induced route search process are aware of the underlying failure.

In assessing the performance and operational merits and disadvantages of the discussed on-demand ad hoc routing schemes, we note the following. By using the reactive nature of the on-demand routing operation, routes are searched and configured only when needed, leading to potential reduction in traffic loading and nodal processing rates. In turn, the flooding-based route search process can trigger very high control traffic rates. In particular, this will occur when the network carries a large number of flows and when a relatively high nodal mobility rate is experienced. Under such conditions, link failures can happen at a high rate, triggering a high occurrence rate of route repair and rediscovery processes. The throughput efficiency of the operation may then be reduced, becoming significantly degraded when the network contains a relatively large number of nodes. Route search operations may then induce the dissemination of control packets over many nodes over wide network spans.

Using the DSR scheme, nodes are able to use overhearing operations to learn about and cache a wider collection of primary and alternate routes. Such information must however be set to expire after proper time limits to avoid propagating stale topological data. The use of a source route descriptor that is included at the header of data packets can simplify and accelerate the nodal data and control packet forwarding process. However, it induces a higher packet overhead ratio and thus reduces the throughput efficiency of the network system, particularly when relatively long routes are involved.

In using wireless channels, the underlying operations conducted by the noted schemes can lead to packet collisions. Broadcast-oriented wireless transmissions, as performed by flooding-based route discovery processes, do not trigger packet retransmissions (as is the case for MAC layer operations under a Wi-Fi protocol). As a result, certain RREQ packets could collide and be lost, leading to the determination of less efficient routes. Such issues could be reduced when nodes are able to form directional antenna beams.

19.4   Optimized Link State Routing (OLSR): A Proactive Routing Algorithm

We consider a MANET network. A node may act as both an end-user and a router (and is henceforth referred to as a node or as a router). Communications media provide two-way (symmetric) links between neighboring nodes. Under a **proactive routing algorithm**, each node determines on an on-going basis the routes to use to reach other nodes in the network, which is assumed herein to constitute a single routing domain. Impacted by network layout variability, nodes act dynamically to update their views of network topology and consequently to update their routing tables. A nodal routing table contains forwarding entries that specify, for each destination node, the next node to which a packet should be forwarded. This way, when a node receives a packet that requires delivery to destination node, the corresponding routing entry is instantly found in its routing table and the packet is then forwarded to the designated outgoing port for transmission across the corresponding output link, avoiding the need to go through a preceding route discovery and configuration process.

A proactive routing algorithm that employs a **link-state routing** protocol is described in Section 15.7. Under a link-state routing algorithm, each node in a routing domain **broadcasts** its link-state message across the network, aiming it to reach all other nodes. A node's link-state message describes the status of each one of its attached links. It represents each link's operational condition (including its on/off state) and can also convey link performance metric values and other attributes.

Upon receiving link-state messages from each node in its domain, a node constructs a map of the domain's network's topological layout, which is represented as a weighted graph. Link-state messages are also identified as **Topology Control (TC)** messages. Using the network's weighted graph, a node calculates the shortest (also identified as "least cost") path from itself (as the source node) to every other node in the domain. These paths are also identified by examining a **Shortest Path Tree (SPT)** layout that is rooted at the source node. The routing table maintained at each node stores the corresponding forwarding entries.

The broadcasting of link-state messages in a MANET is often performed by using a **flooding** based routing scheme. The link-state (TC) messages that are repeatedly updated and reissued by each node are flooded across the network, each one of which is destined to reach each other node. Under a flooding routing protocol, a node retransmits each message that it receives to each one of its neighboring nodes, aiming it to be captured only by nodes that have not yet processed it. For this purpose, messages contain identifiers and sequence numbers. The flooding of TC messages may impose a low to moderate control traffic loading rate in a network that contains a relatively low number of nodes that are not densely situated, or whose nodes do not exhibit high mobility rates so that its topology and link metric values are mostly static.

In contrast, when considering a MANET that includes a relatively large number of densely situated nodes, whose topological layout is continuously varying, flooded control messages are likely to be produced at a high traffic rate. Such traffic may occupy a significant fraction of the network system's resources, including communications capacity, processing, storage, and energy resources, which can lead to low data packet throughput rates and high packet delay levels.

Smaller mobile user nodes tend to use omnidirectional antennas so that multiple nearby nodes may have their messages contend for accessing shared wireless media. To regulate the intensity of this sharing process, the transmission rate of control (or data) messages by neighboring (contending) nodes must be limited to sufficiently low levels. For example, when using a Wi-Fi-based MAC protocol to share a wireless channel, nodes must restrict their message transmission rates to avoid message reception collisions. Broadcast messages transmitted across a Wi-Fi network system do not involve the generation of ACK messages, so that colliding messages are not retransmitted and may be lost.

To increase the efficiency of the broadcasting process used by a proactive routing protocol in a wireless MANET, methods have been devised for the reduction of the produced control message rate, as well as for reducing the fraction of nodes that are actively involved in performing flooding and routing operations. In the following, we describe the principle of operation of such a protocol, identified as an **OLSR** protocol. Detailed description of a version of this protocol is included in IETF documents, such as in RFC 7181 for the OLSRv2 protocol, and in other related IETF documents.

The OLSR protocol realizes a proactive link-state routing algorithm for a MANET system. Rather than use all network nodes to relay control messages, it employs a scheme that is used to select a subset of the nodes to act as message relays, serving to flood and route control (such as link-state) messages. Its key element is the selection and configuration of certain nodes to act as such relays, identified as **Multipoint Relays (MPRs)**.

Under this process, each node selects a subset of its 1-hop neighboring nodes to act as its MPRs. The selection by a node, say node A, is performed in a manner that achieves the following objective for node A: the 1-hop neighbors of the nodes that are members of the MPR set of node A include all nodes that are 2-hops away from node A. Thus, while a message transmitted by node A reaches all its (1-hop) neighboring nodes, messages received and retransmitted by its MPRs, which are members of a set of nodes that is identified as MPR(A), reach all of node A's 2-hop neighbors.

Hence, message transmissions carried out by node A and by its MPRs cover the complete 2-hop neighborhood of node A (which consists of node A's 1-hop and 2-hop neighbors). As node A's neighbors that have not been selected as its MPR nodes do not retransmit (i.e., relay) its messages, this selection reduces the overall control message traffic rate.

To perform the MPR selection process, each node makes use of its knowledge of its 2-hop neighborhood, which is continuously monitored and updated. For this purpose, each node examines the **Hello messages** that it periodically receives from its neighbors. Such messages are used by nodes to discover their 1-hop and 2-hop neighbors. An Hello message production and dissemination protocol is described by IETF RFC 6130 and is identified as the MANET **Neighborhood Discovery Protocol (NHDP)**. In its Hello message, a node identifies itself and its status, and may include link metric values and other attributes. It also includes the names and attributes of its own neighbors, which it derives by monitoring the Hello messages that it receives from its own neighbors. In this manner, by collecting Hello message data received from its neighbors, *a node is able to identify its 1-hop and 2-hop neighbors and their inter-connectivity, and synthesize the topological layout of its 2-hop neighborhood.*

Hello messages are sent by nodes on a periodic basis, as well as upon the occurrence of special events, enabling nodes to update their 2-hop neighborhood graph and accordingly adjust their selection of MPR nodes. To reduce the network's control traffic rate, it is efficient for the system to have each node select its MPR set to include a low number of neighbors. This way, the number of employed relay nodes is reduced, leading to reduced network's control message flow rate. The derivation of a general algorithm that provides for the configuration of minimal MPR sets has been shown to be computationally demanding. Yet, computationally efficient heuristic MPR selection algorithms (not necessarily yielding the optimal solution) are available. Different network nodes may employ the same or different such algorithms.

To illustrate, consider the following greedy-type heuristic algorithm, as it is employed by node A in selecting its flooding MPR set of neighbors. Each neighbor of A that is selected to act as an MPR for node A is identified as MPR(A).

An algorithm for the selection by node A of its MPR set:

Algorithm 19.1

Step 1: Starting with its 2-hop neighborhood graph, node A first selects those neighbors that must be used to reach certain nodes that are 2-hops away (such as nodes that are the only ones that can reach these nodes).

Step 2: It then deletes from the neighborhood graph those selected MPR nodes as well as those nodes that are neighbors of selected MPR nodes and that are 2-hop neighbors of node A, producing a reduced neighborhood graph, identified as the residual graph.

Step 3: In the latest residual graph, node A selects as its MPR node a neighboring node that has the largest number of its own neighbors that are not 1-hop neighbors of node A and are thus 2-hop neighbors of node A.

Step 4: Applying the latest selection, node A deletes from the current graph the latest selected MPR node as well as its neighboring nodes, which are 2-hop neighbors of node A, hence producing the latest residual graph.

Successive performance of Steps 3 and 4: The process continues by repeating successively Steps 3 and 4 in each of the latest produced residual graph with node A selecting, at each step, the next MPR node as one of its neighboring nodes that has the largest number of its own neighbors in the latest residual graph, the latter being 2-hop neighbors of node A. It is followed by the corresponding nodal removal and the ensuing production of a new residual graph.

Termination: This iterative process continues until the produced residual graph is left with no nodes that are 2-hop neighbors of node A. At this point, all 2-hop neighbors of node A have been covered by the selected MPR nodes.

The set of selected MPR(A) nodes is said to *dominate* the 2-hop neighborhood graph of node A as every node in this graph is a neighbor of at least one node in the MPR(A) set.

A node may be selected as an MPR by several neighboring nodes. By denoting node B as MPR(A,C,D), we indicate that node B has been selected as an MPR node by its neighboring nodes A, C, and D. The set of the latter nodes, {A,C,D}, is identified as the **Selector Set** of MPR node B. When node B receives a message from either one of its neighboring nodes that is a member of its selector set, node B is tasked to relay this message. It will not relay messages that are received from a neighbor that is not a member of its selector set. A node that is not selected as an MPR node by any selecting neighbor, does not relay any of the messages that it may receive.

An MPR selection process is illustrated in Fig. 19.3. In Fig. 19.3(a), node A and its 2-hop neighborhood graph are shown. The graph's nodes include node A, its 1-hop neighbors and its 2-hop neighbors. The selection of node A's MPR set is performed by using the above-described heuristic algorithm. Node L is two hops away from node A and the only 1-hop node that covers it is node D. Hence, node D is selected as an MPR(A) node. The other 2-hop nodes that are covered by node D are nodes C and K. They are then deleted from the graph to yield the residual graph (which does not contain deleted nodes L,C,K).

In the residual graph, the degree values of node A's neighbors (which identifies for each neighboring node the number of its own neighbors that are not 1-hop neighbors of A and are thus 2-hop neighbors of A) are as follows: degree(B)=3, as its selection will cover 2-hop nodes J, I, and

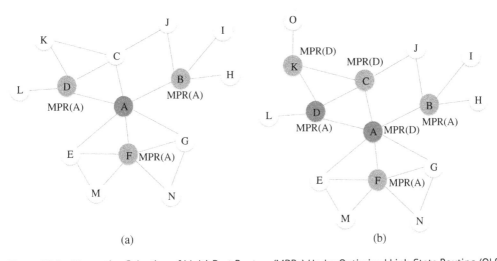

Figure 19.3 Illustrative Selection of Multi-Port Routers (MPRs) Under Optimized Link-State Routing (OLSR) Protocol in MANET: (a) 2-Hop Neighborhood of Node A and Its Selected MPR(A) Routers; (b) MPRs Selected by Node A and by Node D for the Underlying MANET Layout.

H; degree(G)=1; degree(E)=1; degree(F)=2 (i.e., it covers nodes N and M). Hence, node B, which exhibits the highest applicable degree, is then selected as another MPR(A) node. The next residual graph is produced by removing already covered nodes. In this latest graph, we note the residual neighboring nodes E, F, and G to assume the degree values noted above. Consequently, node F is selected as the third and final MPR(A) node. The MPR set for selector node A thus consists of nodes D, B, and F: MPR(A)={D, B, F}.

Consequently, a TC message that is created by node A for flooding over the network will be transmitted by node A to its 1-hop neighboring nodes D, C, B, G, F, and E. Among these six neighbors, only the MPR(A) nodes will relay these messages by transmitting them to their own direct neighbors. In this manner, the TC message is broadcast over the complete 2-hop neighborhood of node A by being relayed by only three neighboring (MPR) nodes, avoiding the need for the other three neighbors of A to relay its TC messages.

In Fig. 19.3(b), we show the 2-hop neighborhood of node A and its selected MPR(A) nodes, as shown in Fig. 19.3(a), and also display nodes that are members of the 2-hop neighborhood of node D. We illustrate the selection by node D of its MPR(D) nodes. They are used to relay the messages issued by node D. For the complete description of the 2-hop neighborhood of node D, node O is now included. This node is not a member of the 2-hop neighborhood of node A but it should be included for complete representation of the 2-hop neighborhood of node D.

By following the above-described heuristic algorithm, we note that node K must be selected as an MPR(D) node as it is the only one that can cover node O. The residual graph is obtained by removing nodes K and O. It is noted that node C is also a neighbor of node K but it is not a 2-hop neighbor of D as it is its 1-hop neighbor. The resulting residual 2-hop graph for node D includes nodes J, B, E, F, and G. The relevant degree of node A in the residual graph is equal to 4 (as it covers nodes B, E, F, and G, which are all 2-hop nodal neighbors of D) and it is then selected as an MPR(D) node. Subsequently, considering the next residual graph, it remains to cover node J, for which purpose node C is selected as an MPR(D) node.

It is noted that as an MPR(A) node, node D relays a message issued by and received from node A to its neighboring nodes, and in this way disseminates it to certain nodes that are 2-hops away from node A. As an MPR(D) node, node K relays messages received from its selectors to its neighbors, including to node O. Hence, a message issued by node D reaches node O, which is its 2-hop neighbor.

A TC message originated by node A is relayed by node D, which has been selected as an MPR(A) node. The latter is then disseminated within the 2-hop neighborhood of node D. This message will be relayed by node K, which has been selected as a MPR(D) node, reaching node O, which is three hops away from node A.

Under the IETF OLSRv2 protocol, which defines a version of the OLSR protocol, the following configurations and processes are included. Information is included in Hello and TC messages generated by a node to identify the status of the node's neighbors, its MPR selection, associated link metric values, content sequence number, and an MPR willingness indicator. The latter is a number that identifies the extent to which a node is willing to serve as an MPR. It serves as an indicator of the willingness of a node to serve as a flooding MPR and as a routing MPR. Based on energy or other resource constraints, certain nodes may restrict their willingness to serve in an MPR role. TC messages may also specify hop limit values to restrict the number of routers that they visit during their traversal across the network.

Each node selects two sets of MPRs: flooding MPRs and routing MPRs. Only nodes that have been selected as *flooding MPRs* act as relay nodes to flood messages. Messages are flooded across the network on a hop-by-hop forwarding basis. *A flooding MPR forwards control messages, such as TC (link-state) messages, only if they are received (for the first time, to eliminate duplicates) directly from a node that has selected it as its flooding MPR.* This serves to constrain the flooding loading rate of the network. The rate of contentions (and ensuing frame collisions) incurred in sharing wireless media is also reduced. To manage the occurrence of duplicate messages, a TC message includes its originator's address and a sequence number identifier. It also includes the addresses of its advertised neighbors.

Routing MPRs are used to reduce the data contents advertised by the link-state dissemination process while still leading to TC message dissemination across shortest paths. For this purpose, it is sufficient that routers announce their link-state information to only their MPR selectors. Hence, a TC message that is configured by an originating MPR that has been selected as an MPR router by certain nodes would advertise the states of only its links that connect to its MPR selectors. The TC message includes outgoing neighbor metric values, which are then used to calculate shortest paths. For such a reduced process to work, the selection of routing MPRs is performed in a manner that ensures the preservation of shortest paths over each 2-hop neighborhood of a router relative to the link metric being used, assuming the neighborhood under consideration to involve bidirectional links.

Only routers that are selected as routing MPRs are used to relay link-state messages, and they are configured to relay only the TC messages that are produced by nodes in their selector set. Routers that are not selected as routing MPRs by any router do not need to relay any link-state messages. This way, topology reduction is achieved, and the number and size of link-state messages that are disseminated across the network re reduced. Also, the quantity of TC information that each router needs to maintain is reduced. Routers would thus construct shortest multihop paths in which routing MPRs serve as intermediate routers.

As an example, consider the network layout and MPR selections illustrated in Fig. 19.3(b), assuming the selected MPRs to perform as both flooding and routing MPRs. Consider originating node A. Among its neighbors, its TC messages will be flooded by only routers D, B, and F, as they are its selected MPRs. Since router C has not been selected as a routing (or flooding) MPR by router A, router C will not declare the link-state status of link (C,A). Link-state data for links attached to node A will be declared for the links connecting node A with nodes D, B, and F, which are node A's selected routing MPR nodes. Router C is however selected as a flooding and routing MPR by node D, so that it will flood node D's messages as well as advertise the state of link (C,D).

Flooding and routing MPRs are selected separately. In selecting flooding MPRs, the use of link metrics is not required. When used, outgoing link metrics are employed. In selecting routing MPRs, the use of incoming neighbor metrics are required. Under properly employed link measures (such as when using a hop count metric), it is possible to select the same routers to serve as both flooding and routing MPRs.

The signaling and control messages distributed by the protocol are used by each router to construct a SPT. For this purpose, a shortest path construction algorithm, such as Dijkstra's algorithm, is employed. It determines the identity of a shortest path between the underlying router and any other router in the network domain. To keep the system's synthesized routing topology current, TC message dissemination and ensuing route calculations are repeated periodically. They may also be triggered by the occurrence of change events. Disseminated data and parameters are assigned expiration times. They need to be refreshed when their corresponding limit times expire. To provide system security, security measures and schemes that maintain and secure key exchange processes are often employed.

19.5 Mobile Backbone Networks (MBNs): Hierarchical Routing for Wireless Ad Hoc Networks

In this section, we illustrate the working of an hierarchical routing protocol for mobile wireless ad hoc networks (MANETs). We describe the operation of a version of a networking architecture that is identified as a **MBN** as developed and studied by Professor Izhak Rubin and his UCLA research group. This specific version employs a networking scheme that is identified as an **MBN-Protocol (MBNP)**.

We consider a MANET system that consists of nodes that reside or maneuver in a prescribed area of operations A. Each network node is capable of functioning as a layer-3 packet router. It may also represent a mobile or stationary end-user entity. A node is outfitted with a radio module whose operation is controlled by a physical layer protocol entity. This entity regulates its transmission process parameters by configuring the underlying MCS and transmit power levels. The sharing of wireless communications media by active nodes is regulated by an employed MAC protocol, such as a Wi-Fi scheme.

Consider a brief period of time during which network nodes are effectively stationary (or moving over a short distance), so that each is situated in a specific geographical segment that is located within the area of operations. Under the MBNP-networking concept, certain nodes are elected to act as **Backbone Nodes (BNs)** forming, whenever feasible, a **connected Backbone Network (Bnet)**. Each BN performs functions and provides services that are similar to those rendered by a *Base Station (BS)* in a cellular wireless network. Under MBNP, the following rules govern the configuration of the backbone network.

Network nodes are differentiated by the following classification. A node is identified as a **Backbone Capable Node (BCN)** if its current attributes qualify it to serve as a candidate for being elected as a BN. Noting that a BN is likely to have to carry and process higher message traffic rates and stay active for longer periods of time, a BCN is assume to have sufficient residual energy and other (such as transmit power, memory and processing) resources. In turn, a node that is currently not a candidate for serving as a BN is identified as a **Regular Node (RN)**. The latter may have limited communications and processing resources, as is the case at certain states for battery operated sensor nodes. This may also be the state of a node that has been operating as a BCN for a relatively long period of time and has exhausted its energy (or other) resources. RNs may also be configured to operate at a lower transmit power and data rate levels.

A layout of a MBN system is illustrated in Fig. 19.4. Under this configuration, the backbone network (Bnet) consists of four BCNs that have been elected to act as BNs. Each BN is able to connect to each other BN across the Bnet. The synthesized Bnet topology assures the 1-hop covering of all network BCNs. Thus, every BCN in the network that has not been elected at this time to serve as a BN is a neighbor of a BN. This is for example the case (among others) for BCN11 and BCN12, which are neighbors of BN1. A RN is provided access to the Bnet by either having a direct communication link to a BN, as is the case for RN11, by using a path that traverses a BCN, as is the case for RN21, or by using a path that leads to a BN that includes multiple RNs, as is the case for RN33.

The collection of nodes that include a BN and the BCNs and RNs that are provided paths to access the Bnet through this BN is said to form an **Access Net (Anet)**. The members of an Anet associate with the Anet's BN and are said to be clients of this BN. The latter serves as the managing node of its Anet and of its client members. The MBN illustrated in Fig. 19.4 uses a Bnet that consists of four BNs, each of which is configured to manage its own Anet. While the illustrated Bnet is structured as a tree graph (which provides a unique path between any pair of BNs), in general, as noted in the

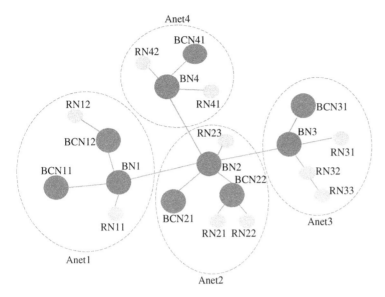

Figure 19.4 Illustrative Network Layout of a Mobile Backbone Network (MBN) Under the MBNP Protocol.

following, the MBNP topology synthesis algorithm may attempt to construct a Bnet that assumes a mesh layout, offering multiple routes for connecting BNs. In general, BCNs and RNs may also have multiple paths for communicating with their associated BN as well as for communicating with other BCNs and BNs.

In managing the operation of an MBN under MBNP, we observe the use of two key procedures:

1) An **MBN Topology Synthesis Algorithm (MBN-TSA)** is employed to select BCNs to act as BNs and to configure the underlying layout of the corresponding Bnet and Anet systems.
2) An **MBN Routing Protocol (MBN-RP)** is used to select routes that network nodes use to communicate with each other.

A multitude of BN election algorithms has been developed. Regarding each Anet to represent a cluster of nodes, BN-like access nodes have been identified at times as **cluster controllers**. Assuming all network nodes to be backbone capable, the network topological layout is modeled as a BCN graph. The nodes of this graph represent BCNs. A link is set to connect two nodes if they can directly communicate with each other. The selection of BNs would then correspond to the configuration of a **dominating nodal set**. A set of nodes is said to be dominating if every other node in the network is a neighbor of at least one node in this set. For operational efficiency and for reducing traffic rates, it is often of interest to select BNs to form a dominating set which contains a small number of BNs. The process used to find a minimal dominating set generally entails high computational complexity. However, efficient heuristic (and not necessarily optimal) algorithms are available. When employing a centrally managed scheme, it is assumed that the system employs a control station that continuously monitors the topological layout of the MANET system. This controller uses its knowledge of the complete network graph to synthesize a dominating set.

To synthesize a Bnet, it is of interest to select its BN members in such a manner that they are interconnected, forming a connected backbone. The corresponding objective requires the construction of a **connected dominating set**. The latter includes nodes that dominate the network (so that every other BCN is a neighbor of at least one BN member of this set) as well as nodes that assure the connectivity of the backbone, so that (when feasible) each BN can communicate with each other BN. Various available heuristic algorithms can be employed by a central controller for the purpose of synthesizing a connected Bnet.

However, in a typical MANET system, it is of interest to employ a **distributed**, rather than a centralized, scheme for the synthesis of a Bnet. Such a method is preferred as it can be employed to synthesize the Bnet in a rapid and robust manner, noting that network layout changes are regularly impacted by nodal mobility, communications media quality fluctuations, and nodal or link failures. A multitude of such algorithms has been developed. In the following, we describe the key principles employed by a version of the MBNP scheme. This scheme uses a distributed Topology Synthesis Algorithm (TSA).

Hello messages are exchanged between neighboring nodes. As used in other MANETs, an Hello message produced by a node specifies the node's attributes and status, as well as identifies the node's neighbors. In this manner, every node discovers the local connectivity of its 2-hop neighborhood involving its 1-hop and 2-hop neighbors. Under MBNP, BCN (and thus also BN) nodal attributes include an MBN-weight parameter, expressing the node's ability and willingness to currently serve as a BN. The MBN-weight value can be adjusted by the node based on its current resource availability state. Nodal attributes that affect the node's MBN-weight may include the node's current mobility speed, operational robustness and mobility steadiness, and the capacity of its attached communications links. To achieve a robust Bnet topology that is not subjected to excessively high fluctuation rate, it is preferable to select nodes to act as BNs if they move at lower speeds (or are even mostly stationary). Also preferred are nodes that are attached to higher-quality and steady communications links.

Due to nodal mobility and fluctuations in the quality of communications links, the topological layout of the network tends to fluctuate. It is however noted that the transmission rate of MBNP control packets that are used to adapt the backbone's layout is generally much higher than nodal speeds. To illustrate, consider a node that is moving at a speed of 100 [miles/h], or about 160 [km/h], or 44.44 [m/s]. It takes about 2.25 seconds for such a node to traverse a distance of 100 m. In comparison, the transmission of a control packet that contains 1000 bits at a rate of 1 Mbps takes only 1 ms. Hence, more than 2000 control packets can be transmitted during such a nodal traversal time. The MBN-TSA algorithm adapts rapidly to changes in the MANET's layout, enabling a stable fast converging adaptation process. The area of operations may include BCNs at a sufficiently high-density level so that a *connected Bnet* can be readily synthesized. When this is not the case, the algorithm will result in the production of a Bnet that covers only a subset of the nodes. It may then produce multiple disconnected backbones.

A BCN, whether newly arrived or an ongoing one, may observe to have several BCN neighbors. It may note that none of its neighbors is currently acting as a BN. It would then send an association request (which may be included in its Hello message) to a selected neighboring BCN, such as the one that displays the highest MBN-weight, requesting the latter to change its status to a BN. This is also the scenario that takes place at network initiation, or at a later time when a limited number of BNs are present. At other times, induced by topological changes, the Bnet layout may be impaired, causing certain BCNs to not be covered (i.e., not to be 1-hop neighbors of elected BNs). Association election requests and consequent actions would then be triggered. This is likely to result in the conversion of certain BCNs to BN state.

While such a selection process may result in full covering of network BCNs, the net layout may be such that certain BNs may not be connected to other BNs, so that the resulting topology does not serve as a connected Bnet. To achieve Bnet connectivity, additional BCNs may have to convert to BN state. Under MBNP, additional conversions to BN state may also be invoked for the purpose of increasing the level of connectivity of the synthesized Bnet, improving its **robust behavior** under

19.5 Mobile Backbone Networks (MBNs): Hierarchical Routing for Wireless Ad Hoc Networks | **585**

link failures. Such conversions lead to the construction of a connected backbone mesh that provides several alternate routes that can be used to connect certain BN pairs.

To adapt the synthesis of the Bnet to topological layout changes, the MBN-TSA employs the following process. During a Bnet synthesis initiation period, as well as in response to topology change events, a **BCN-to-BN conversion** process is performed. During this phase, identified as a Phase-I, certain BCN nodes convert their state to act as BN nodes. During a conversion phase, each node that is involved in the conversion process indicates its involvement in the Hello messages that it periodically sends, including a phase identifier. In this manner, neighboring nodes learn that the system is currently undergoing a specific conversion process. This way, conversions of a specific type that are performed by multiple neighborhood nodes are synchronized. Conversions performed by different nodes according to the rules specified at a given phase are performed during the same phase period. By the termination of their Phase-I operations, nodes that have changed their MBN state from BCN to BN indicate so in their Hello packet advertisements.

At the termination of the Phase I process, a connected Bnet would generally emerge, if such a layout is feasible. However, the constructed Bnet topology may contain an excessive number of BNs. To reduce networking complexity and traffic loading rates, selected nodes will then elect to convert their state from BN to BCN. Such a modification must be performed under the condition that it does not impair the desired attributes of the synthesized Bnet. Accordingly, at the termination of the Phase-I process, impacted nodes initiate a Phase-II conversion process. During the Phase-II period, certain BNs would switch from BN to BCN state. This phase is therefore identified as a **BN-to-BCN conversion** phase. By the termination of Phase-II, nodes that have changed their MBN state from BN to BCN indicate so in their Hello packet advertisements.

At the termination of a Phase-II process, a new Bnet layout is formed and advertised. A BCN (whether an existing one or a new one) that is not yet associated with a BN, and observes (using Hello messages that it receives) one or several BNs among its nodal neighbors, will associate with a BN neighbor that announces the highest MBN-weight (noting that nodal IDs, nodal addresses, or other attributes, can be used for tie-breaking purposes). Such a BCN then becomes a member of the Anet managed by its associated BN.

A 1-hop BCN-BN path is identified for each BCN. It is used for data communications between a BCN and its associated BN. Such associations are included in the Hello messages sent by each node.

A RN that learns about a BN neighbor proceeds to associate with this BN (selecting a preferred one if it has several BN neighbors). It includes this association descriptor in its Hello messages. Such an association identifies a 1-hop path between an RN and its associated BN.

For the illustrative network shown in Fig. 19.4, this is the case for regular nodes RN11, RN31, RN41, and RN42, each of which is directly attached to a BN. If the RN finds out that it has one or several BCN neighbors but no BN neighbor, it chooses a preferred BCN neighbor (such as the one that exhibits the highest MBN-weight or that is received at the highest signal power, or that satisfies a criterion that involves the latter two parameters) to connect to, selecting it as the *associated upstream node* in a path that leads to a BN. This path is used by the RN for associating with the Anet's BN and for registering itself as a member of this Anet. Such an association identifies a 2-hop path between the RN and the Anet's BN. For the illustrative network, this is the case for regular nodes RN12, RN21, and RN22. If an RN finds no BN or BCN neighbors, it is permitted to select an RN neighbor as its associated upstream node, provided the latter has announced itself as a covered member of the Anet (so that it has a path that leads to the Anet's BN). For the illustrative network, this is the case for regular node RN33, which selects RN32 as its upstream associated node, noting that RN32 is a neighbor of a BN and is therefore covered by the Bnet. Following a successful execution of any of these associations, the RN becomes a connected member of its Anet.

19 Mobile Ad Hoc Wireless Networks

While not including the detailed specification of the underlying Phase-I and Phase-II algorithms, the following description outlines key considerations and steps used by the MBN-TSA scheme that is employed by MBNP.

Algorithm 19.2   BCN to BN Conversion Algorithm: A BCN converts to BN state if it satisfies any of the following conditions:

1) **1-Hop Coverage**: This BCN has the highest MBN-weight among its BCN neighbors or it has received at least one association request.
2) **2-Hop BNet Connectivity**: This BCN observes at least one pair of its BN neighbors that do not connect to each other across a 1-hop or 2-hop Bnet path (i.e., a path that consists of BNs).
3) **3-Hop BNet Connectivity**: For this BCN, at least one of its BN neighbors and one of its BCN neighbors do not connect to each other across a 1-hop or 2-hop Bnet path.
4) **Optional Conversion Restriction Rules for Reducing Computational Complexity and Enhancing Convergence Time**: A BCN will not convert to BN state if the number of its BN neighbors exceeds a specified threshold and/or if this number increases during an Hello message transmission period.

BN to BCN Conversion Algorithm: A BN converts to BCN state if it satisfies all of the following conditions:

1) **Preserving Client Coverage Condition**: Each one of this BN's clients (i.e., associated nodes) has more than a single BN neighbor.
2) **Preserving 2-Hop BNet Connectivity**: Any two of this BN's neighboring BNs are either directly connected (and announce higher weights than this BN) or can be connected by a preferred (such as one that has higher weight) BN neighbor of these two BN neighbors.
3) **Preserving 3-Hop BNet Connectivity**: Any BN neighbor and BCN neighbor of this BN are either directly connected (and the BN neighbor has a higher weight than this BN) or can be connected by a preferred (such has one that has higher weight) BN neighbor of these neighboring BN and BCN.

The MBN-TSA scheme has been shown to display convergence time period, Bnet size (in terms of the number of its BNs) and control message lengths that are each of a constant order in term of assuming values that are independent of the number of network nodes. A basic lower bound on the number of BNs N_{BN} is expressed as follows. Consider the area of operations to be equal to $A\,[\text{m}^2]$ and the average communications range between any two BCN/BN nodes to be equal to $d\,[\text{m}]$. Assume each BN to cover an area of the order of $d^2\,[\text{m}^2]$. Assume the system to have a sufficient number of backbone capable nodes, and the latter to reside at locations that enable them to cover the full area of operations (otherwise, the conclusion still holds except that the synthesized backbone consists of several disconnected components). We then conclude that a lower bound on the number of BNs that are required to cover the area is of the order of $A/(d^2)$, which is noted to not depend on the number of backbone capable network nodes. For example, if $A = 400\,[\text{m}^2]$ and $d = 5\,[\text{m}]$, the corresponding lower bound indicates that the number of required BNs is equal to about 16, irrespective of whether the network contains 100 or 800 BCNs.

In Fig. 19.5, two instances of MBN network layouts are depicted. They were captured by running the MBNP-Simulator implemented by Professor Izhak Rubin and his development group at IRI Computer Communications Corporation; the contributions by Dr. Runhe Zhang and Professor Izhak Rubin in the development of such a preliminary simulation model of an earlier MBN version are also acknowledged. Each layout shows the structure of the underlying Bnet, which connects

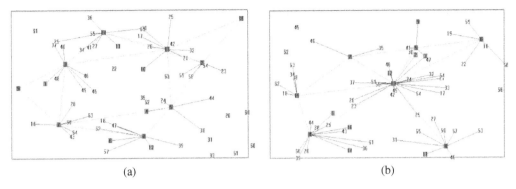

Figure 19.5 Instances of Mobile Backbone Network (MBN) Layouts.

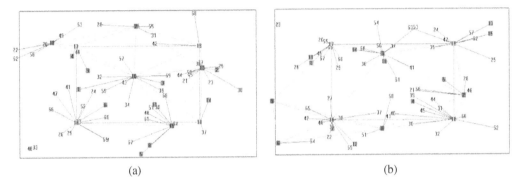

Figure 19.6 Two Instances of Unmanned Ground Vehicle (UGV) Aided Mobile Backbone Network (UGV-MBN) Synthesized by Using MBNP.

elected BNs. Also noted are the access nets (Anets), where each Anet consists of a BN and its BCN and RN clients. For the simulated system, BCNs (and thus also BNs) are assumed to be capable of setting higher transmission power levels, enabling them to communicate over longer links.

In Fig. 19.6, we depict a UGV-aided MBN topological layout (UGV-MBN) synthesized by using the MBNP scheme. Each **Unmanned Ground Vehicle (UGV)** is remotely controlled by an MBN managing controller, which places UGVs at desirable ground locations. A UGV stays at its location until the controller decides to move it to a more beneficial location, as induced by evolving nodal locations, mobility patterns and fluctuating traffic matrix profiles. A UGV acts as a **mobile relay or router**, functioning as a BCN, and thus possessing the capability to serve as a BN. Each one of the depicted UGV nodes can be assigned a sufficiently high MBN-weight so that it would be automatically elected as a BN by the MBN-TSA. In Fig. 19.6, four stationary UGVs are configured to serve as BN members of the Bnet. In addition, other BCNs are elected at certain times as BNs, when needed to enhance the Bnet's connectivity and coverage features.

In Fig. 19.7, we depict a UAV-aided MBN topological layout (UAV-MBN) synthesized by using the MBNP scheme. Each **Unmanned Aerial Vehicle (UAV)**, also known as a **drone** or as a **satellite**, is remotely controlled by an MBN managing controller, which places it at a desirable aerial altitude, orbit and location (relative to other UAV platforms and to the area of operations). It stays at this orbit and location until the controller moves it to a more beneficial location. A UAV acts as an **airborne mobile relay or router**, functioning as a BCN, and thus possessing the capability to serve as a BN. In the displayed system, each one of the depicted UAV nodes is assigned a sufficiently

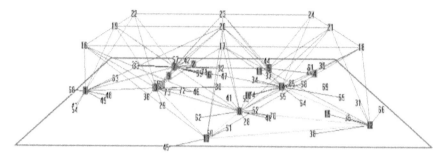

Figure 19.7 Unmanned Aerial Vehicle (UAV) Aided Mobile Backbone Network (UAV-MBN) Synthesized by Using MBNP.

high MBN-weight so that it is automatically (and autonomously) elected as a BN by the MBN-TSA. In Fig. 19.7, nine UAVs are shown to have been guided (or configured autonomously) to form an airborne backbone, which serves as a Bnet subnetwork.

Different groups of UAVs can have their UAV members placed at different orbits, at different altitudes, forming an hierarchical aerial backbone of UAV swarms. Supplemental ground nodes could be configured by the MBN-TSA to act as BNs to configure a desirable 3-dimensional Bnet layout. The use of a UAV, serving as a communications satellite, is advantageous in that it projects a wide footprint. It can provide communications connections between ground nodes that are located at longer distances from each other. In synthesizing a multidimensional backbone, orbit, and altitude position levels are dynamically set. A UAV covers a wider footprint when placed at a higher altitude but would then support a lower transmission data rate level for communicating across its ground–air link.

The **MBN routing scheme** that is used for transporting data packets across the MBN between source and destination nodes makes use of the synthesized Bnet and associated Anets. An approach used by the MBNP system, identified as an **MBN Routing Protocol (MBN-RP or MBNR)**, is described in the following. It makes use of two routines: 1. Intra-Anet routing; and 2. Intra-Bnet routing.

The **Intra-Anet** routing scheme is used to route messages across an Anet. Upon the formation of each Anet, its managing BN broadcasts across its Anet an *Anet registration message*. It is destined to all members of the Anet. The BN identifies itself as the Anet's manager, announces the Anet's ID, and includes relevant Anet attributes. Each BCN and RN in an Anet sends to its BN a *client registration and update message*, including its ID and attributes, and (when learned) its hop-distance from the BN. Such messages are sent by each Anet member at the time that it joins the Anet and are resent periodically. The BN records this information in a *client registration database*. As membership changes, this database is updated.

Intra-Anet routes are configured as follows. Downstream routes are used by the BN to communicate with a specific node, or with a group of nodes. An upstream route is used by a BCN or an RN to communicate with its BN. As each BCN is 1-hop away from its BN, the corresponding BN-to-BCN downstream path and the BCN-to-BN upstream path, would each consist of just the single link. Similarly, a single link path is used for communications with an RN that is a neighbor of its BN, as is the case for regular nodes RN11, RN31, RN41, or RN42 in the network shown in Fig. 19.4.

An RN that is not a neighbor of its BN but has a BCN as its neighbor is 2-hops away from its BN. A route for communicating between this RN and the BN is configured by having the BCN acting as an intermediate router. For example, for the network shown in Fig. 19.4, the downstream and upstream routes between BN1 and RN12 involve the use of BCN12 as an intermediate router. In examining disseminated Hello messages, BN1 learns that RN12 is a neighbor of BCN12 and RN12 learns that BN1 is a neighbor of BCN12. BCN12 learns that it can serve as a router that switches messages that flow between BN1 and RN12 by forwarding them across the corresponding output links. In case., RN12 has direct access across a wireless medium to both BCN12 and BCN11 (not explicitly shown in the figure), it may specify BCN12 as its upstream node (e.g., by comparing their MBN-weights) and BN1 may select BCN12 as its preferred downstream router for reaching RN12.

For the case that the Anet contains RNs that are multiple hops away from the BN, whereby the source RN of an upstream path to the BN (or the destination RN of a downstream path from the BN) has no neighboring BCNs, several other routing schemes may be used. A flooding scheme can be employed for this and the other scenarios to discover a route and then to set it up, as performed by on-demand MANET routing schemes. Alternatively, successive selection of upstream or downstream nodal associates can be used. For example, for the network shown in Fig. 19.4, for communications between RN33 and BN3, RN32 may be selected as the downstream (intermediate router) associate for BN3 and as the upstream associate for RN33. In this case, using hello message data, BN3 is aware that RN33 is a neighbor of RN32 and RN33 recognizes RN32 as a neighbor of BN3.

As another option, a proactive link state routing scheme may be employed. Since the Anet is managed by its BN, it is sufficient for each node to send periodically its link state vector only to the BN. The BN can then calculate the Anet's SPT rooted at itself. Using this tree, the BN identifies the shortest path from itself to each node in the Anet. It uses this information to place the corresponding downstream routing entries in each Anet node. This tree can also be used for multicasting (or broadcasting) messages downstream from the BN.

Furthermore, assuming *symmetric (bi-directional) links*, using the concept of RPF, the layout of the SPT that is calculated and flooded in the Anet by the BN can also be used by RNs for upstream routing to the BN. Each Anet node would then use the SPT link that is a member of the path from the BN to itself and that is attached to it, as the one to use to forward messages upstream to the BN. For example, if BN3 advertises its calculated SPT to be described by the tree topology shown for Anet3, RN32 will learn the downstream link from BN3 to be its (BN3, RN32) link. It will then forward upstream messages across this link. Similarly, RN33 would learn to forward upstream messages to RN32, as the SPT includes the link (RN32, RN33).

Intra-Anet message routing between Anet clients can be executed in several ways. Under one method, a star configuration with the BN serving as the focal hub is used. A source node sends its message upstream to the BN. The latter then routes it downstream to the destination node or nodes. This process can be reduced when the source and destination nodes have learned about a direct route that connects them. This would be the case when these nodes are 1-hop or 2-hop neighbors of each other, as learned by examining their Hello messages. Such routes can also be learned by observing periodic registration messages sent by the BN and by client nodes.

The primary MBNR routing approach is based on using the following method. A node that is a member of a certain Anet and that has a data packet to send proceeds first to attempt to send it across its Anet. If the destination node is an Anet node that is within 2-hops of the source node, a direct path can be used. Otherwise, the message is sent to the BN. In the case that the BN recognizes the destination node to be a member of its own Anet, it would relay the message to the destination node by sending it downstream across its Anet.

Under an extended operation, to reduce the Bnet's traffic loading rate, a source node that is a member of an Anet will execute a limited span search (setting a proper value for the maximum number of hops to be traversed by the search message) to discover (if not yet known) whether the destination node is located nearby, such as being a member of a neighboring Anet. If so, a peer-to-peer MANET routing protocol may be used. In this case, the BN will be informed to not use a Bnet aided routing process for this destination node.

In the case that the BN determines the destination node to not be its Anet's member, and no inter-Anet peer-to-peer routing scheme is used, the managing BN would proceed to route the data packet across the Bnet. For this purpose, the source BN first searches, across the Bnet, the destination BN that has saved the destination node's ID in its client registration data base, signaling that the destination node is a current member of its Anet. This data is cached (and aged after a specified period of time) by the source node (and possibly other nodes), serving to increase the efficiency of future discovery processes. An on-demand routing based scheme, using RREQ and RREP control packets, can be used for finding the identity of the targeted destination BN and establishing an intra-Bnet route, following the methods described for on-demand MANET routing protocols. Alternatively, after determining the identity of the destination BN, noting that at times only a limited number of nodes are used to form the Bnet, a proactive routing scheme, such as one that uses a link-state routing protocol, may be employed to configure preferred BN-to-BN paths. Once the data message reaches the destination BN, the latter would send it across its Anet to the destination node by using a downstream path or by employing a downstream flooding method.

For the network shown in Fig. 19.4, assume RN12 to be a source node that wishes to send data packets to destination node RN33. Assume that intra-Anet and inter-Anet (Bnet) routes have been configured. RN12 sends its flow request packet (or, the first data packet of its flow) upstream through BCN12, which forward it to BN1. At the same time, RN12 may also flood a RREQ packet in its neighborhood to find whether the destination node is located in a close-by Anet. Assume that this is not the case. As BN1 receives the packet, it examines its location register and determines that the destination node is not a member of its Anet. BN1 then uses a MANET-oriented routing scheme to discover the destination BN and establish a Bnet route to it. In this case, BN3 responds to the RREQ sent by BN1 producing a RREP that identifies itself as the desired destination BN since RN33 is a client of its Anet3. BN1 uses the configured Bnet path to route the data packets of this flow across the Bnet for reception by BN3. The latter then sends these packet across a downstream intra-Anet3 route to destination node RN33.

For regulating the performance behavior of data packet transport across the Bnet, a flow control (FC) scheme has been implemented, as described by the **MBNR-FC protocol**. Under this scheme, a new flow is not admitted by the source BN for transport across the Bnet if the source BN is unable to discover an acceptable Bnet route. When using an on-demand routing scheme, during the conduct of the flooding-based route discovery process, an intermediate BN that is currently experiencing a state of congestion would discard a new RREQ packet that it receives. Congestion state setting can be declared by a BN when observing its message queue of RREQ data packets waiting to be forwarded across the Bnet to exceed a specified threshold (whose value may depend on a performance metric cited by the RREQ packet). In this manner, a discovered route would not include intermediate BNs that have declared to be in state of congestion. If an acceptable route exists, the flooding-based route discovery process will successfully find a non-congested route that leads to the BN in whose Anet the destination node resides.

Alternatively, a proactive routing scheme can be employed. Each BN would send periodic advertisements to each other BN identifying itself and the state of its links, as well as disclosing its active client nodes. The source BN would prune from its constructed weighted graph that represents the Bnet's topology, those links that are reported to be unacceptably congested. The *pruned graph* is then used by the source BN to determine, when feasible, an acceptable end-to-end Bnet route. If such a route is determined to exist, the new flow is admitted and its packets are navigated across this route. Otherwise, the flow is blocked.

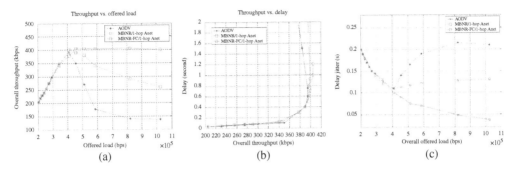

Figure 19.8 Comparative Performance Behavior of the MBNR, MBNR-FC and AODV ad hoc routing schemes: (a) Throughput vs. Offered Flow Rates; (b) Average Delay vs. Throughput Rate Performance; (c) Delay Jitter vs. Offered Traffic Load Rate. Source: X. Huang and I. Rubin [22], authorized by IEEE License.

Our studies have shown such a Bnet-oriented flow and admission control routing scheme to yield a much improved performance behavior. This has been confirmed in [22] and illustrated in Fig. 19.8, whereby the performance behavior of a MANET system that employs AODV, MBNR (without the use of the FC scheme), and MBNR-FC protocols is exhibited. In Fig. 19.8(a), we show the MBNR-FC scheme to yield a virtually ideal output traffic rate vs. offered traffic rate performance curve. As the offered traffic load rate increases, the MBNR-FC scheme induces the system to produce its highest feasible throughput rate and then keep the system's throughput rate at this maximum rate as the offered traffic rate further increases.

In contrast, as the traffic loading rates increases, the AODV scheme produces an exceedingly high control message traffic loading rate (as route discovery attempts are forced to be frequently unsuccessfully repeated), leading to a rapidly decreasing throughput rate.

Fig. 19.8(b) shows the average delay performance incurred by successfully transported data packets, for which a route has been successfully discovered. As the offered traffic load rate increases, the AODV scheme is often not able to discover an acceptable Bnet route, as flooded route discovery packets may not then be able to successfully reach their intended destinations. Lower throughput rates and higher delays are then experienced.

In turn, the MBNR-FC scheme acts to throttle the flow admission rate into the Bnet, serving to lower packet delays for admitted flows. In addition, as shown in Fig. 19.8(c), a data packet that is handled by the MBNR-FC scheme experiences much reduced packet delay variation, also identified as packet delay jitter. This is particularly the case under higher traffic loading rates, as then the realized throughput rate tends to fluctuate around its maximum feasible rate value, so that packets that belong to admitted flows experience similar queueing delay levels and thus incur lower packet delay variations.

Certain mobile nodes can be equipped with multiple radio modules, enabling the synthesis of a **multi-radio backbone mesh network**. Such nodes can configure multiple radios to simultaneously transmit messages across distinct frequency bands and/or schedule them for transmission at distinct time slots. To illustrate, we note the multi-radio operation studied in [31, 32].

These studies assume that BCNs are equipped with both higher-power and lower-power radio modules. A higher-power radio is capable of operating at a higher transmit power level, allowing it to communicate at higher data rates across longer distances. In turn, a lower-power radio module is used to communicate at lower data rates and use reduced energy resources, resulting in extended battery lifespan. A RN (such as a battery-driven smaller sensor node) is assumed to employ a single lower-power radio module.

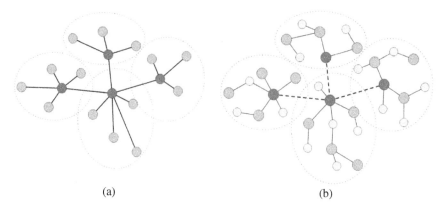

Figure 19.9 Connectivity Graphs for Multi-Radio MBN System: (a) when employing Higher-Power Radio Modules for Bnet and Anet Communications; (b) when using Lower Power Radio Modules for Intra-Anet Communications. Source: H-J Ju and I. Rubin [31, 32], authorized by IEEE License.

The synthesis and operation of a multi-radio-MBN system is set to use lower-power radios for RN-involved intra-Anet communications, while configuring BCNs to use higher-power radios to sustain longer distance and higher data rate communications links across the Bnet. Fig. 19.9(a) illustrates the topological layout of a synthesized Bnet, where higher-power radio modules are employed by BNs for Bnet communications. It also shows the use of higher-power radios for communications between BNs and BCNs. Higher-power radios can also be used for direct communications between Anet's BCNs. In Fig. 19.9(b), we show a layout of a network that employs intra-Anet communications links that are driven by the use of lower-power radio modules.

The overall performance behavior of a multi-radio MBN system has been shown to be much enhanced. It provides for the synthesis of a high-capacity Bnet while supporting energy constrained RNs such as sensor and IOT nodes. Capable BCNs can also be outfitted with programmable antenna arrays, enabling them to configure multiple simultaneously active directional communications beams and communicate with available UAV and UGV platforms, yielding enhanced performance behavior and spectral efficiency levels.

Problems

19.1 Describe the concept of operation of a mobile ad hoc wireless network (MANET).

19.2 Describe the principle of operation of the following MANET-based schemes:
a) On-demand (reactive) routing.
b) Proactive routing.
c) Hierarchical ad hoc routing.

19.3 Consider an On-Demand Distance Vector (AODV) routing protocol. Describe the functions provided by the following:
a) Concept of routing operation.
b) Route Request (RREQ) packets.
c) Route Reply (RREP) packets.
d) Route Error (RERR) packets.

19.4 Describe the operation of the *route discovery* process under an On-Demand Distance Vector (AODV) routing protocol. Discuss the illustrative operation shown in Fig. 19.1.

19.5 Describe and illustrate the operation of *Reverse Path Forwarding (RPF)* under the On-Demand Distance Vector (AODV) routing protocol.

19.6 Describe the principle of operation of the *Dynamic Source Routing (DSR)* protocol.

19.7 Illustrate the concept of route learning by network nodes that is performed under the DSR protocol in using the illustrative network layout presented in Fig. 19.2.

19.8 Compare the operation and performance features of the AODV and DSR ad hoc routing protocols.

19.9 Describe the principle of operation of the *Optimized Link State Routing (OLSR)* protocol.

19.10 Define and illustrate the concept of *Multipoint Relays (MPRs)* as used by the OLSR protocol. Discuss the performance advantages derived from their use.

19.11 Consider a MANET network that employs the OLSR protocol.
a) Describe an algorithm that can be used by a node to select its MPR set.
b) Draw a MANET network layout and illustrate the use of this algorithm in providing for the selection of an MPR set by network nodes.

19.12 Consider the version of the Mobile Backbone Network (MBN) protocol (MBNP) developed by Professor Izhak Rubin and his research group. Define and explain the following:
a) Backbone Capable Node (BCN).
b) Backbone Node (BN).
c) Regular Node (RN).
d) Backbone Network (Bnet).
e) MBN-Protocol (MBNP).

19.13 Describe and illustrate the operation of the MBN-Routing (MBNR) protocol.

19.14 For the MBN network layout shown in Fig. 19.4, describe the process used by node RN11 to route its packets to node RN33 by using an MBN-based routing protocol.

19.15 Explain the concept of MBN Topology Synthesis Algorithm (MBN-TSA) and illustrate its operation.

19.16 Explain the method of operation of an Unmanned Ground Vehicle (UGV)-aided Mobile Backbone Network (UGV-MBN), following the illustration presented in Fig. 19.6.

19.17 Explain the method of operation of an Unmanned Aerial Vehicle (UAV)-aided Mobile Backbone Network (UAV-MBN), following the illustration presented in Fig. 19.7.

594 | *19 Mobile Ad Hoc Wireless Networks*

19.18 Describe the principle of operation of the MBNR routing protocol when employing Flow Admission Control regulation (MBNR-FC) and discuss its performance enhancement capabilities.

19.19 Explain and compare the performance behavior characteristics of the MBNR, MBNR-FC, and AODV protocols as displayed in Fig. 19.8.

19.20 Describe an operation of a multi-radio MBN networking system under which BCNs are equipped with multiple radio modules. Discuss the performance enhancements that can be achieved when BCNs are also outfitted with enhanced capabilities to form directional antenna beams as well as to access optimally placed UGVs and UAVs (including groups of UAV swarming) platforms.

20

Next-Generation Networks: Enhancing Flexibility, Performance, and Scalability

The Objectives: *A network system that includes network nodes that employ proprietary means is difficult to manage. It is not readily reconfigured to accommodate new services and to rapidly adapt to support variations in traffic flows and service profiles. Methods and technologies that are being employed for enabling such regulation, adaptation, and support services that are provided by next-generation network systems include techniques that make use of: Network Virtualization; Network Slicing; Cloud and Edge Computing; Open Network systems and interfaces; Machine Learning (ML) and Artificial Intelligence (AI).*

The Methods: Key technologies that are employed to enable dynamic network service and flow operations, include SDN (Software-Defined Networking) and NFV (Network Functions Virtualization). Under SDN, separation is made between operations that are executed across the data plane and those that are performed at the control plane. Network control is carried out in a central manner by using software tools. Such controls are used to regulate data plane operations that are performed by network nodes. NFV uses software modules as applications that provide specific network functions (NFs), such as routing, mobility management and security, modeled as Virtualized Network Functions (VNFs). A Network Functions Virtualization infrastructure (NFVI) provides for the modeling of computation, storage, and networking infrastructure components, performed on a host computer system. These components are employed in the support of software that is used to execute networking applications. A Management, Automation, and Network Orchestration (MANO) module manages the NFV infrastructure. It also enables the provisioning of existing VNFs and for the configuration of new ones. In addition to SDN and NFV, the following methods are often employed: Network Slicing, providing for the sharing of system resources by multiple virtual networks; Cloud Computing, providing centralized network management and control; Edge Computing, including the placing of Point-of-Presence (PoP) entities closer to locations of served end-user groups; Open Networking and Interfaces, enabling the accommodation of multi-vendor systems and the continuing introduction of innovative networking techniques; Convergence of Technologies, involving the integrated use of terrestrial and nonterrestrial networking means, including airborne and spaceborne network systems, such as those that employ UAV (drone) and satellite platforms; ML and AI methods, in contributing to the provision of automatic and autonomous networking, service configurations and operations, threat detection and analyses, and network service and transport security.

20.1 Network Virtualization

As discussed and illustrated in Section 9.3.1, software-oriented control methods can be effectively used to configure the forwarding tables of network switches to perform in a manner that is

---

Principles of Data Transfer Through Communications Networks, the Internet, and Autonomous Mobiles, First Edition. Izhak Rubin.
© 2025 The Institute of Electrical and Electronics Engineers, Inc. Published 2025 by John Wiley & Sons, Inc.

responsive to Virtual LAN (VLAN) and Virtual Service Network (VSN)-based layouts. Dissemination of flows among end-user members of a Virtual Private Network (VPN) that are attached to distinct networks are enabled by using a control scheme to include VPN membership identifiers in messages and by configuring involved switches to associate specific ports with such VPN identifiers. In this manner, a control plane underlay is used to enhance the operational efficiency of a data plane overlay. The switching modules are responsible for performing transmission, switching and forwarding operations that are regulated by a higher layer control scheme. Such an operation is used to effectively provide for rapid, high performance, secure, and robust transport of data flows. It can regulate operations of a large number of networks in supporting multiple membership groups.

Also illustrated in Section 9.3.1 is the use of control plane schemes to program the operation of data plane modules in configuring the formation of VSNs. Authenticated message flows are switched and distributed across a fabric of interconnected networks, as they aim to reach service stations that have been configured as members of identified layer-2 or layer-3 VSNs. The formation of a multicast tree that is used to disseminate flows produced by VSN group members is illustrated in Fig. 9.11.

These methods demonstrate the use of software applications to configure the operation of data switching and routing modules. **Network Virtualization (NV)** enables the migration to software of network functions (NFs) that are performed by using specialized (and at times proprietary) hardware-based structures. *Using NV, a physical network system can be abstracted in software to support multiple separate virtual networks.*

Employing NV methods, networks are managed in software as they make use of an existing data packet forwarding infrastructure. Network provisioning, functions, services, and security operations can be automated and configured to rapidly adapt to changes in system traffic patterns, modifications in service prescriptions, and network topological variations. While physical network modules provide protocol layer 1–3 (L1–L3) data networking services, an NV is configured to provide virtualized (i.e., software imitated) L4–L7 services.

Adoption of NV-oriented methods, which as noted serve to separate operations at the data forwarding plane from those performed by the control plane, have been accelerated by the development of Software-Defined Networking (SDN) and Network Functions Virtualization (NFV) systems. In the following, we provide brief descriptions of key features of these technologies, while referring the reader to associated specifications, documents, and papers produced by corresponding Standards, vendor, manufacturing and service organizations.

Under **Cloud Computing**, a data center delivers services to its users over the Cloud (i.e., the Internet; though other backbone networks may also be used). Such services are often divided into the following types: Infrastructure-as-a-Service (IaaS), Platforms-as-a-Service (PaaS), and Software-as-a-Service (SaaS). A data center shares its resources, including storage, computing, database, streaming audio and video, computer programs, and software programs, among its active users. A resource is made available to a requesting user on-demand. A Cloud Computing vendor may distribute the placement of its resources across multiple data bases. A data center branch may be placed close to a concentrated location of clients. Such a data center is identified as a Point-of-Presence (PoP). It is a representative of the central cloud computing data center. Under such an architecture, users avoid the involved expense and need to maintain in-house expertise that is required to privately acquire, store, manage, and update such resources. The ensuing costs borne by the cloud vendor are efficiently shared among its active clients.

A **Virtual Machine (VM)** uses software to function as a virtual version of a computer. It is also regarded as a computer file, often called an image, that behaves as a computer. It resides on a physical host computer. It makes use of resource pools in its host computer, or of a remote server's memory, storage and CPU resources. It employs its own operating system (OS), operating independently of other VMs that may be running in the same host computer, independently of the operating system used by the host computer. A software entity that is called a *hypervisor* acts as a VM manager. It allocates host computer resources to individual VMs, enabling consequently the simultaneous running of multiple VMs on a single physical machine, where different VMs may use different operating systems. A VM can be moved to another hypervisor residing on a different machine. A cloud computing computer system may employ groups of hypervisors that are used for the support of multiple VMs.

20.2   Software-Defined Networking (SDN)

Under SDN, separation is made between operations that are executed across the data plane and those that are performed at the control plane. Network control functionalities are executed in a central manner by using software means. This way, network data plane operations and resources are software controlled. SDN controllers use monitored system states and service requirements to manage and automate NFs. System states can be derived by using data stored in the cloud. By employing cloud computing resources, collected data is used by resource allocation, scheduling, and other control and management schemes to determine an effective process to use to control network system operations, and consequently produce control signals that regulate the functioning of network elements (NEs). SDN controllers tend to make use of Open Standards and Application Programming Interfaces (APIs), enabling them to control and manage a wide multitude of network devices, including such that are manufactured by different companies.

In the following, we overview the architectural building blocks of an SDN system, following the corresponding descriptions provided in an SDN Architectural Overview document [40] published by the user-led Open Networking Foundation (ONF). ONF is involved in the development and promotion of the SDN **OpenFlow** standard. The OpenFlow protocol specifies open communications interfaces for SDN control and forwarding layers. Such interfaces support the use of SDN open source control software to interact with and control NEs, including proprietary switches and routers, either in a direct physical way or virtually, via an hypervisor.

The OpenFlow protocol is used to transport messages between SDN controllers and NEs in a secure manner, such as by using the Transport Layer Security (TLS) protocol which is run on top of a connection-oriented transport layer protocol (such as TCP).

The following description relating to the SDN architecture is included in [40]: "The SDN architecture specifies, at a high level, the reference points and open interfaces that enable the development of software that can control the connectivity provided by a set of network resources and the flow of network traffic through them, along with possible inspection and modification of traffic that may be performed in the network. Virtualization permits abstract views of network resources. These resources can be tailored to a particular client or application, and can be interrogated and manipulated by those clients or applications. A logically centralized, scalable control plane manages a wide range of data plane resources of possibly diverse data plane technologies. The SDN Architecture allows modeling of forwarding and processing behavior, supporting a variety of media and connectivity types; where processing includes any compute, storage or NFs. NFs and services may cover all Open Systems Interconnect (OSI) Layers and may be either physical or virtual."

The following basic principles of an SDN architecture have been noted [40]:

1) **Decoupling the SDN Control and Data Planes**: The SDN controller delegates functionality to NEs, while maintaining awareness of their states. Specifications are provided as to which functions should an SDN controller delegate to a NE.
2) **Logically Centralized Control**: The SDN controller is able to dynamically attain information relating to broad network system resources and states. A central controller (whether located physically within a single computer host or whether run in a logically central manner using distributed computers) makes use of broad network view to synthesize in a most advantageous manner network controls and subsequently produce control signals that enhance the system's performance behavior. To improve scalability, system state variables are expressed and disseminated at different precision levels to targeted SDN controllers that may be hierarchically stacked or functionally assigned to specific networking domains and trust regions.
3) **Exposure to External Applications**: Abstract network resources and states are exposed to external applications, at targeted granularity levels.

A high-level overview of the SDN architecture is illustrated in Fig. 20.1. As described in [40], it involves the following layers, interfaces, and services:

1) The **Data Plane** contains data NEs. NEs convey their resources and states to entities located in the controller plane across the data plane—controller plane interface (D-CPI).
2) In the **Controller Plane**, the SDN controller translates SDN application requirements to produce control signals that are conveyed to NEs. In the reverse direction, it transfers state information to SDN applications.
3) Agents embedded in NEs and the SDN controller support configuration, sharing or virtualizing schemes. Their settings determine the levels and boundaries of application of SDN control functionalities (such as which switch ports are exposed to SDN control). They also serve to isolate the services that are provided to different customers.
4) SDN applications reside in the **Application Plane**. Their network requirements are transferred to the Controller Plane via the A-CPI interface.

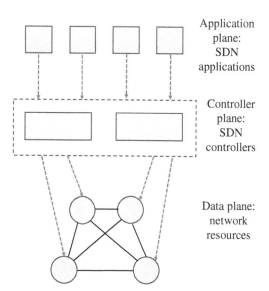

Figure 20.1 High-Level SDN Architecture.

5) Services are provided to applications by controller plane entities across the application-controller plane interface (A-CPI). System management and control and *resource allocation models* are employed. Their structures and parameters are impacted by specified policies and by dynamically received data that identify network system states.

6) *Management* entities serve to configure policies, operational scopes, and parameter levels specifying the operations conducted by layer entities, as dictated by Operations Support System (OSS) prescriptions. Such specifications are used by a telecommunications organization to control and manage the operation of its network system. The SDN controller is configured to provide **orchestration** services, acting to coordinate and manage the controls applied to multiple inter-related entities that may reside on inter-dependent platforms.

20.3 Network Functions Virtualization (NFV)

SDN and NFV are key technologies that are employed to enable dynamic service and network flow configurations and adaptations. NFV uses software applications that provide specific NFs, such as routing, mobility management, and security. They are identified as Virtualized Network Functions (VNFs). A Network Functions Virtualization infrastructure (NFVI) is used to model computation, storage and networking infrastructure components on a host computer system. These functions are employed for the support of software that is used to execute networking applications. Management, automation, and network orchestration (MANO) entities manage the NFV infrastructure. They also enable the setting and provisioning of VNFs.

Our description of the architecture of an NFV system is based on specifications prescribed in European Telecommunications Standards Institute (ETSI) based NFV documents, particularly [17] and [20].

Based on [20], high-level objectives of NFV include the following:

1) General purpose servers and storage devices are used and shared to provide NFs that are modeled by using software virtualization techniques. These NFs are referred as VNFs.
2) Flexibility in assigning VNFs to hardware. This aids scalability and largely decouples VNF functionality from location.
3) Rapid service innovation through software-based service deployment.
4) Improved operational efficiencies resulting from common automation and operating procedures.
5) Reduced power usage achieved by migrating workloads and powering down unused hardware.
6) Standardized and open interfaces between VNFs and the infrastructure and associated management entities, so that such decoupled elements can be provided by different vendors.

As shown in Fig. 20.2, a high-level NFV framework consists of the following domains:

1) VNFs domain, where each VNF is a software implementation of a NF that runs over the NFVI.
   a) VNFs represent, in software, NE entities such as routers, gateways, firewalls, management controllers, computing and storage systems, vendor services, and consumer service-oriented functions.
   b) The functional behavior and the external operational interfaces of a Physical Network Function (PNF) and a corresponding VNF are expected to be the same.
   c) A VNF can be implemented as one or multiple Virtual Machines (VMs) hosted on a single server or on multiple distributed servers. For example, control functions may be embedded in regional centers while data forwarding functions may be stored in edge cloud servers.

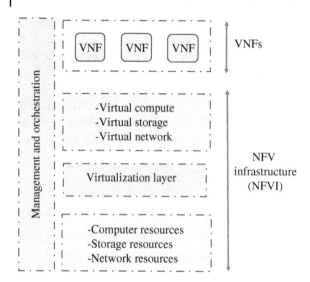

Figure 20.2 High-Level NFV Framework.

  d) A VNF implementation of a service may be embedded in multiple VMs, whereby each one may model a specific application.
  e) A VNF Forwarding Graph (VNF-FG) is used to model connectivity between VNFs. For example, it can model a chain of VNFs that are employed along a path that connects a client to a server, whereby VNF components may represent routers, firewalls, medium access controllers, and other modules. The layouts of VNF-FG paths that are traversed by traffic flows can be preconfigured or dynamically determined by signaling message flows.
2) The NFV Infrastructure (NFVI) supports the execution of the VNFs. It includes the hardware and software components that make up the network system in which the VNFs are deployed.
  a) The NFVI can span several locations, including places where NFVI-PoPs are located. The network system that is providing connectivity between these locations is regarded as part of the NFVI.
  b) From a VNF's perspective, the virtualization layer and hardware resources look like a single entity providing the VNF with desired virtualized resources.
  c) Hardware resources include computing, storage, and network entities that provide processing, storage, and connectivity to VNFs through the virtualization layer.
  d) Network resources include: (1) NFVI-PoP network, which interconnects the corresponding computing and storage resources contained in NFVI-PoP units. It also includes switching and routing devices, serving to provide external connectivity. (2) Transport network, which interconnects NFVI-PoPs to other networks and to other network appliances or terminals that are not contained within the NFVI-PoPs.
  e) The virtualization layer abstracts the hardware resources and decouples the VNF software from the underlying hardware.
  f) When virtualization is used in the network resource domain, network hardware is abstracted by the virtualization layer to realize virtualized network paths that provide connectivity between VMs of a VNF and/or between different VNFs instances. Techniques used for this purpose include network abstraction layers that isolate resources via virtual networks and network overlays. Examples include: Virtual Local Area Network (VLAN), Virtual Private LAN Service (VPLS), and Network Virtualization using Generic Routing Encapsulation (NVGRE).

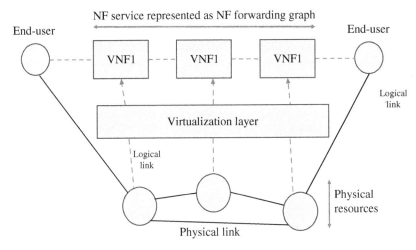

Figure 20.3 NFV-Based Network Service Represented as an NF Graph.

g) Other forms of virtualization of the transport network include centralizing the control plane of the transport network and separating it from the forwarding plane, as performed by SDN, or isolating the transport medium, as performed for example by using optical wavelengths.
3) NFV Management and Orchestration of physical and/or software resources that support the infrastructure, the virtualization process, and the management of VNFs.
4) The NFV Orchestrator is in charge of the orchestration and management of NFV infrastructure and of software resources and of realizing network services on NFVI.

An NF-Graph representation of a network service is illustrated in Fig. 20.3. The end points (shown as end users) and NFs of the network service are represented as nodes. End users are shown to be connected via logical interfaces that run across the NF modules (represented as dashed lines). Physical interconnections are provided by several (wireless or wired) network infrastructures and are depicted as solid lines.

An illustrative NFV-based network service that is modeled as a nested chain of VNFs is shown in Fig. 20.3. The NFVI-based virtualization layer abstracts the hardware resources of the NFV Infrastructure, serving to decouple the corresponding hardware and software NFs. The shown physical resources represent computation, storage and networking entities that are used by a network service provider. Dashed arrows represent the virtualization process. The VNF modules are produced as software abstractions of associated physical resources.

SDN and NFV processes are mutually beneficial. SDN provides mechanisms that are used for network control, enabling the provisioning of a network as a service. NFV provides management and orchestration methods and means for resource virtualization that are employed in the provisioning of NFs and their use in the synthesis of network services.

20.4 Network Slicing

Under the method of **Network Slicing**, multiple logical networks, each identified as a Network Slice, operate over the same physical network. Each slice is allocated (physically or logically) its own resources and is controlled in accordance with its stated service profiles and performance requirements, and its ensuing message traffic activity.

Network slicing is discussed in [37, 38], and [39] in connection with the multitude of services that need to be supported by **Next-Generation Mobile Networks (NGMNs)**, and particularly in sustaining the service provision and networking flexibility and scalability of the 5G network. Identified enabling technologies are also essential for the effective and efficient operations of other next-generation mobile and wired networks. In [38], network slicing is modeled as consisting of three layers: (1) The *service layer* contains the supported service instances. (2) The *network layer* consists of network slice instances. (3) The *resource layer* accounts for resources associated with the network infrastructure and for NFs, as well as for other entities that serve to support network slice instances.

A network slicing approach for 5G wireless networks is presented in [37]. Each network slice is configured to support its targeted connection types. Employed NFs can span all network domains. Different slices tend to require different NFs, involving different message flow dynamics and QoS requirements, and entailing the use of different network resources. NFs and resources can be shared by distinct slices, while each slice is set to make use of only the NFs and resources that it requires. Access for networking support within certain slices may be provided to organizations that specialize in offering services required by these network slices.

SDN and NFV are recognized as enabling technologies for the efficient, flexible, and salable implementation of network slicing. The use of NFV in supporting network slicing in a 5G network is illustrated in [10]. A 5G network example that supports the following network slicing types is discussed. The first slice corresponds to a network that supports typical traffic flows that are generated by smartphones. The NFV setup would then require the use of a wide range of NFs that run over broadband virtual links. The second slice is used to support the networking of safety messages produced by autonomous vehicles. NFs and resources are set to guarantee low message latency and high network transport reliability. The third slice is used to support a large Internet of Things (IoT) metering system. Support is required for network flows produced by embedded sensor nodes that do not require the use of mobility management oriented functions. Lower data flow rates are typically involved, so that utilized links are assigned lower data communications capacity levels.

SDN and NFV-aided network slicing synthesis, provisioning, and adaptation processes would be automated and optimized through the incorporation of Machine learning (ML) and Artificial Intelligence (AI) methods.

20.5 Edge Computing, Open Interfaces, Technology Convergence, Autonomous Operations

Based on [39], where the focus is on next generation 5G mobile networks, we note in the following several key technologies that serve a major role in enabling the design, control, management, operation and adaptation of networking services, and network operations for next-generation mobile systems.

1) **Cloud-Based Operation**: Mobile network operators (MNOs) aim to provide networking services and connectivity in a dynamic and scalable manner by employing VNFs within the NFVI framework for the separation between NFs that are software-based and NEs that are hardware based. A cloud infrastructure serves to provide interoperability in supporting VNFs that are used in a multi-vendor environment.

2) **Edge computing**: Due to the wide diversity of edge-user equipment, it is beneficial to locate resources at the *network edge*. In this way, the end user is able to locally customize its

service attributes to achieve optimized performance. Edge-based **Point of Presence (PoP)** server nodes run applications of interest to the user while geographically positioned closer to the end user. This way, they are able to efficiently off-download processing tasks from end-user equipment. An end-user can then be provided with network connectivity that is highly reliable and that offers very low latency in message transport, as well as effectively and rapidly provide service and networking adaptations across the user and data planes. An edge computing system can act to more readily sustain *precise location positioning* and *context aware* operations, including such that are required for industrial systems, IoT systems, and autonomous mobile systems.

3) **Open radio access network (O-RAN)**: The O-RAN Alliance has been working on a multi-vendor RAN that is open in its offering of interoperable interfaces. Legacy radio access network implementations have tended to use proprietary schemes, such as multiple access algorithms, multiplexing schemes, resource allocation protocols, interference mitigation processes, and others. An O-RAN system accommodates multi-vendor schemes and algorithms, resulting in the introduction and integration of innovative schemes that enhance operational and service performance, efficiency, and scalability.

4) **Autonomic management and control**: By using automatic and autonomic management and control schemes, and making use of cloud and edge computing-based architectures and of virtualization methods that are applied over network slices, a network system is able to dynamically and autonomously adapt to new or modified service scenarios, fluctuating traffic patterns, modified NFs, and to variations in the diversity of network applications. Cognitive, ML, and AI methods assist operations by aiding the calibration of effective techniques that learn the environment, track user and vendor behavior, and identify the dynamics of requirement profiles, as the system is induced to automatically and autonomously adapt to new and modified service and networking patterns.

5) **AI and ML**: AI and ML methods are employed for the detection, tracking, and analysis of ongoing data flows, network system behavioral states, and to the learning of environmental conditions and user quality of experience preferences. This information contributes to the dynamic synthesis of currently best service and networking processes and configurations. A wide range of performance objective functions are incorporated in performing the synthesis of the ensuing optimal resource allocation and networking processes.

6) **Access network convergence**: The use of diverse methods for regulating access to a wireless network is advantageous. Such a diversity yields a more robust access network, enhancing its ability to more effectively adapt to variations in network requirements and in service profiles. Such a system is capable to respond more effectively to system and subsystem failure conditions. Terrestrial and nonterrestrial access means are employed. Nonterrestrial access technologies use airborne and spaceborne vehicles. Included are drones, low and high earth orbit airborne satellite networks. High-altitude satellites that use a geosynchronous orbit offer long-range reachability, enabling connectivity among remote users across a single (or several) airborne hops. Networks that make use of interconnected low-altitude smaller satellite platforms offer airborne communications that can cover wide geographical spans at much lower message delay levels. A network layout that makes use of a hybrid of airborne and terrestrial hops can contribute to message transfers at upgraded delay-throughput performance levels. Nonterrestrial platforms can also be used to provide networking support for the access of users that are located in remote locations that do not offer terrestrial communications means. These platforms can also be employed to provide users with alternate routes when terrestrial paths are not available or have failed.

7) **Harmonization**: To provide mobiles with networking means on a global basis, it is essential to implement globally standardized access and communications protocols and means, and to globally harmonize access technologies and employed frequency bands. Harmonization would benefit from the standardization of device interfaces, including APIs. Harmonization enables the design of simpler, smaller, and more cost-effective intelligent user and vendor devices.

8) **Sustainable trust**: It is essential to assure users' trust in system services and operations, while maintaining system performance so that users are provided their desired quality of experience levels. High security and privacy are essential to gaining user trust and satisfaction. AI- and ML-aided cognitive modules enable the detection of service and transport anomalies, triggering actions that prevent or mitigate the impact of threat scenarios.

9) **Environmental Sustainability and energy consumption efficiency**: Employing technologies that contribute to environmental sustainability, including the maintenance of low environmental pollution rates. Of interest are methods that achieve low carbon footprints associated with user devices, network nodes, operator modules, and other entities. This is enabled by using techniques and technologies, including those mentioned above, that increase operational energy efficiency and such that make use of renewable energy sources.

Problems

20.1 Explain the concept of network virtualization.

20.2 Explain the and discuss the following:
a) Virtual Machine (VM).
b) Cloud Computing.
c) Virtual Private Network (VPN).

20.3 Explain and discuss the principle of operation of *Software-Defined Networking (SDN)*.

20.4 Explain the design and functioning of the SDN architecture shown in Fig. 20.1.

20.5 Explain and discuss the concept and principle of operation of *Network Functions Virtualization (NFV)*.

20.6 Describe and discuss the principles and functioning of the high-level NFV framework presented in Fig. 20.2.

20.7 Consider a Network Functions Virtualization (NFV) system. Define and explain the following:
a) Virtualized Network Function (VNF).
b) Network Function (NF).
c) Physical Network Function (PNF).
d) Virtualized Network Function Forwarding Graph (VNF-FG).
e) NFV Management and Orchestration.

20.8 Discuss the use of a network service that is modeled as a nested chain of VNFs as shown in Fig. 20.3.

20.9 Describe the principle of operation of Network Slicing. Provide 3 examples of network slices and discuss their uses.

20.10 Describe and discuss the following in relation to network system operations and message transfers across a network:
a) Cloud based operation.
b) Edge Computing using Point of Presence (PoP).
c) Open Radio Access Network (O-RAN).
d) Automatic Management and Control.
e) Access Network Convergence.
f) Artificial Intelligence (AI) and Machine Learning (ML)-aided networking.
g) Energy Efficiency and Environmental Sustainability.

21

Communications and Traffic Management for the Autonomous Highway

Objectives: *Vehicular wireless networks provide communications between vehicles. They enable a vehicle to disseminate messages to other vehicles moving along a highway segment, or in a specific region. As a primary objective, such transmissions are used to enhance driving safety. Vehicles gain up-to-date status of traffic and environmental conditions. They enable the implementation of traffic regulation schemes, including the dynamic synthesis of vehicular formations. Public safety networked systems employ Intelligent Transportation Systems (ITS) to disseminate control messages that regulate vehicular movements and speeds, serving to increase traffic safety and transit efficiency and to rapidly react to critical events. Vehicular wireless networks are essential enabling elements in the regulation and management of autonomous vehicle highway systems. Self-driving vehicles use the networks to interact with each other and with Road Side Units (RSUs), as they learn about system states and accordingly adjust their mobility patterns. Safety-centric applications make use of message communications that involve medical, fire, police, and first-responder vehicles. Vehicular networks also play key roles in enabling nonsafety applications such as those that involve the dissemination of business, commercial, entertainment, and educational message flows.*

Methods: *Key technologies employed for the operation of vehicular wireless networks include peer-to-peer mobile ad hoc networking systems, also known as vehicular ad hoc networks (VANETs), and infrastructure aided networking schemes. Under VANET, a mobile ad hoc network (MANET) system is synthesized and used for the dissemination of data messages from an active vehicle to neighboring and other targeted vehicles. Messages travel across multihop routes. Intermediate vehicles are dynamically selected to relay messages received from other vehicles. Particular VANET networking methods are used for vehicular wireless networking across highway lanes, where vehicle-to-vehicle (V2V) multihop paths are formed. Under an infrastructure-aided vehicular networking scheme, a backbone network (Bnet) aids in the dissemination of messages among vehicles. A wireless cellular network is often used for backbone communications. It employs base stations (BSs) that serve also as RSUs. These stations are connected to each other across the backbone by using wireline or wireless communications links. An active vehicle may be directed to transmit messages uplink to an associated RSU or BS, following a vehicle-to-infrastructure (V2I) or a vehicle-to-network (V2N) wireless networking protocol. Messages received at an RSU or a BS are transported across the Bnet to a destination RSU or BS. The latter transmits these messages across a wireless downlink to targeted vehicles. Under a hybrid networking scheme, both vehicle-to-vehicle (V2V) and backbone-based (V2I) transport processes are employed. Our recent development of such a hybrid network is based on the concept of a Vehicular Backbone Network (VBN). Bnet systems may also make use of 3-dimensional multitier array of unmanned aerial vehicles (UAVs), satellite systems, and Internet systems. Traffic management schemes, such as those based on our recent studies, are used to regulate the access of vehicles to the autonomous highway, to synthesize vehicle (platoon) formations, to set vehicle speeds,*

Principles of Data Transfer Through Communications Networks, the Internet, and Autonomous Mobiles, First Edition. Izhak Rubin.
© 2025 The Institute of Electrical and Electronics Engineers, Inc. Published 2025 by John Wiley & Sons, Inc.

and to regulate the routing of vehicles into highway lanes. They aim to achieve high vehicle flow rates while providing users with acceptable queueing delays and improved transit times.

21.1 Data Communications Services for Vehicular Wireless Networks

According to The US Department of Transportation (USDOT, www.its.dot.gov), "in 2013, there were 32,719 people killed in motor vehicle traffic crashes in the United States." It is also noted that "pedestrians, bicyclists, and other non-vehicle occupants made up 17% of these deaths." It was also reported that "there were an estimated 6.8 million police-reported crashes in 2019, resulting in 36,096 fatalities and an estimated 2.7 million people injured. **Connected vehicle communications** aim to address these issues, expecting the underlying technology to reduce crashes and accidents in a significant manner."

Wireless communications networking systems that enable highway vehicles to exchange data messages are essential for enhancing driving safety, for the implementation of safety and non-safety services, and for the coordinated safe functioning of autonomous and non-autonomous vehicular highway systems. A vehicle is equipped with one or several computer-based **Application Units (AUs)** that host traffic control and mobility management applications. An AU is connected in the vehicle to an **On Board Unit (OBU)**, which is used for exchanging data messages with other OBUs, and with Road Side Units (RSUs) or Access Points (APs), including Base Station (BS) nodes. Such applications also reside at RSUs. The OBU contains resources for storing and processing information and for a user interface. It includes networking and transmission modules that are used for wireless communications. The OBU provides networking, routing, and communications services to the AU. It serves to reliably route, transmit, and receive control and data packets that are sent to/from other vehicles and to/from road side stations. It is also responsible for implementing data security measures.

Using its AU, OBU, and embedded sensors, a vehicle collects and processes highway system status information. It uses this data to control its own mobility and to send related information to other vehicles and RSUs via wireless communications networks. RSUs communicate with other RSUs and with other widely distributed management and control stations, using the Internet and other backbone networks (Bnets). Communications paths are established for the dissemination of information that is collected by mobile units to traffic management centers. Status and control messages received from regional management and control stations are distributed across the Internet or other networks for dissemination to targeted highway vehicles.

Vehicle communications involve the following interactions:

1) **Vehicle to Vehicle (V2V)** communications are performed between vehicles, such as between vehicles that move along a highway and are located within a specified range from each other.
2) **Vehicle to Infrastructure (V2I)** communications take place between a vehicle and an associated roadside station. The latter may also be connected to Bnets, such as to a cellular wireless network and to the public (or a private) Internet. It may also be directed to designated traffic control and management stations.
3) **Vehicle to Pedestrian (V2P)** communications provide for interactions that serve to protect the safety of pedestrians and other non-vehicle entities.
4) **Vehicle to Network (V2N)** message communications connect a vehicle and a network. It is often used in referring to communications between a vehicle and a cellular network.

5) **Vehicle to Everything (V2X)** refers to communications between a vehicle and other entities. It includes the communications types mentioned above and other types, such as vehicle to device (V2D), vehicle to cloud (V2C), and vehicle to grid (V2G). The following systems are noted:

   a) **V2X in the context of VANET** involves direct communications between vehicles and other entities that are in communications range, including V2V and V2I communications. Ad hoc networking methods are employed so that vehicles can communicate with each other without using a communications infrastructure. Included are wireless transmissions and medium access control (MAC) techniques that are based on a wireless communications standard, such as IEEE 802.11p, and a fuller protocol stack standard known in the United States as Wireless Access in Vehicular Environments (WAVE) and in Europe as ITS-G5.

   b) **Cellular V2X (C-V2X)** communications involve the use of cellular networks. It has been defined initially as part of LTE in 3GPP Release 14 and then expanded to support 5G in 3GPP Release 15. C-V2X includes support of both direct communications between vehicles and cellular network-aided communications. It includes V2V, V2I, and V2N communications. Direct communications between a vehicle and other devices (V2V, V2I) use the so-called PC5 interface. V2N communications with a cellular network is performed through an air interface that is identified as the Uu interface. It connects the mobile user's equipment (UE) with the associated BS across an access medium. Multiple radios can be employed by a mobile to enable simultaneous V2V communications via the PC5 interface and to use V2N.C-V2X communications by using the Uu interface.

Safety messages are often repeated at a specified rate (e.g., at a rate of 1 message/s) and are expected to be disseminated over a targeted distance span within a specified maximum latency (such as 50–100 ms, or faster). Their dissemination must be performed at a high reliability, so that they reach, with high probability, a specified high fraction of their targeted destination vehicles. For example, a vehicle that slows down or is in the process of suddenly stopping would send a warning message to impacted vehicles, such as to all vehicles that follow it within a specified range.

Key safety services that are targeted for securing the operation of vehicular wireless networks, as noted in [13] and [48], are shown in Fig. 21.1. Category I services account for the production of vehicle status warning messages. Category II messages provide for vehicle-type warnings. Category III messages warn vehicles of traffic hazards. Category IV messages are used by vehicles to warn other vehicles about dynamically occurring events or maneuvers, such as lane changes and precrash event sensing.

V2V wireless data communications serve to significantly enhance the functioning and performance of crash avoidance systems. Such systems often employ radar, cameras, and various sensor devices, including such that are used to detect collision threats.

Category II safety services include the production of **Cooperative Awareness Messages (CAMs)**, also known as **Basic Safety Messages (BSMs)**. Such messages are periodically broadcast over a specified area or range. They can include information on vehicle's status, type, position, speed, resources, and other attributes that impact traffic safety conditions and identify the vehicle's parameters that affect its communications, data processing, storage, networking, and route planning. A key use of such messages is to convey road warning data. Other services manage the production of **Decentralized Environmental Notification (DEN)** messaging flows that notify other area vehicles of developing or current potential hazard conditions.

Illustrative message types that are produced by **non-safety-oriented services** are shown in Fig. 21.2. Included are message classes that are produced by traffic management and infotainment services.

Safety services category	User cases	Communication mode	Security/ reliability require-ments	Usage	Minimum frequency of periodic messages	Maximum latency
Category I: Vehicle status warning	Emergency electronic brake lights	Time limited periodic broadcast on event	High/high	Warn a sudden slowdown of the following vehicle	10 Hz	100 ms
	Abnormal condition warning	Time limited periodic broadcast on event	High/high	Warn the abnormal vehicle state	1 Hz	100 ms
Category II: Vehicle type warning	Emergency vehicle warning	Periodic triggered by vehicle mode	High/high	Reduce emergency vehicle's intervention time	10 Hz	100 ms
	Slow vehicle warning	Periodic triggered by vehicle mode	High/high	Improve the traffic fluidity	2 Hz	100 ms
	Motorcycle warning	V2X co-operative awareness	High/high	Collision avoidance	2 Hz	100 ms
	Vulnerable road user Warning	V2X co-operative awareness	High/high	Collision avoidance	1 Hz	100 ms
Category III: Traffic hazard warning	Wrong way driving warning	Time limited periodic broadcasting on event	High/high	Wrong way driving warning	10 Hz	100 ms
	Stationary vehicle warning	Time limited periodic broadcasting on event	High/high	Avoid succession of collisions	10 Hz	100 ms
	Traffic condition warning	Time limited periodic messages broadcasting/authoritative message triggered	High/high	Reduce the risk of longitudinal collision on traffic jam forming	1 Hz	100 ms
	Signal violation warning	Temporary messages broadcasting on event	High/high	Reduce the risk of a stop/traffic violation	10 Hz	100 ms
	Roadwork warning	Temporary messages broadcasting on event	High/high	Reduce the risk of accident at the level of roadwork	2 Hz	100 ms
	Decentralized floating car data	Time limited periodic broadcasting on event	High/high	Improve safety and traffic fluidity	10 Hz	100 ms
Category IV: Dynamic vehicle warning	Overtaking vehicle warning	V2X co-operative awareness	High/high	Reduce the risk of accident	10 Hz	100 ms
	Lane change assistance	V2X co-operative awareness	High/high	Active road safety	10 Hz	100 ms
	Pre-crash sensing warning	Broadcast of pre-crash state	High/high	Accident impact mitigation	10 Hz	50 ms
	Co-operative glare reduction	V2X co-operative awareness	Medium/ medium	Avoid the frontal collision	2 Hz	100 ms

Figure 21.1 Safety-Oriented Key Services and Message Types for Vehicular-Networked Systems. Sources: US Department of Transportation (DOT), National Highway Traffic Safety Administration (NHTSA) Report DOT HS 811 492A, September, 2011 [13]. Also Zheng et al. [48]. Authorized by IEEE License.

Other message types include roadside infrastructure information messages, Signal Phase and Timing (SPAT) messages, In-Vehicle Information (IVI) messages, and Service Request Messages (SRMs).

V2X performance requirements involving support for vehicular networks through the use of a 5G cellular wireless network are shown in Fig. 21.3 [4]. Noted are illustrative values involving latency, reliability, message length, and data rate requirements for vehicle platooning (entailing the synthesis of clusters of moving vehicles), advanced driving, extended sensors, and remote driving operations.

21.2 Configurations of Vehicular Data Communication Networks

In Fig. 21.4, we show several networking architectures that are used to provide for data communications among vehicles moving along a highway. The following key methods are noted. As shown in Fig. 21.4(a), the first method is based on forming a **Vehicular Ad hoc Network (VANET)** configuration, where vehicles communicate with each other through the formation of V2V wireless links. V2I wireless links are also configured and used for communications between vehicles and local RSUs. An RSU would often also connect across a wide area network including the Internet,

21.2 Configurations of Vehicular Data Communication Networks | 611

Non-safety services category	User cases	Communication mode	Security/ reliability requirements	Usage	Minimum frequency of periodic Messages	Maximum latency
Category I: Traffic management	Regulatory/contextual speed limits	Authoritative message triggered by traffic management entity	High/high	Enhance the traffic efficiency/reduce the vehicles' pollution	1 Hz	N/A
	Traffic light optimal speed advisory	Periodic, permanent messages broadcasting	High/high	Traffic regulation at an intersection	2 Hz	100 ms
	Intersection management	Periodic, permanent messages broadcasting	High/high	Road safety and traffic regulation at an intersection	1 Hz	100 ms
	Co-operative flexible lane change	Periodic broadcasting messages	High/high	Enhancement of mobility efficiency	1 Hz	500 ms
	Electronic toll collect	I2V broadcasting and unicast full duplex session	High/high	Traffic fluidity at the toll collect	1 Hz	500 ms
Category II: Infotainment	Point of interest notification	Periodic, permanent messages broadcasting	Medium/ medium	Driver and passengers comfort	1 Hz	500 ms
	Local electronic commerce	Duplex communication between RSU and vehicles	High/high	Vehicle driver/passenger comfort	1 Hz	500 ms
	Media download	User access to Internet for multimedia download	Medium/ medium	Passenger entertainment	1 Hz	500 ms
	Map download and update	Access to Internet for map download and update	Medium/ medium	Efficiency and comfort	1 Hz	500 ms

Figure 21.2 Non-safety-Oriented Key Services and Message Types for Vehicular-Networked Systems. Sources: US Department of Transportation (DOT), National Highway Traffic Safety Administration (NHTSA) Report DOT HS 811 492A, September, 2011 [13]. Also Zheng et al. [48]. Authorized by IEEE License.

Use case group	Payload (bytes)	Latency (ms)	Reliability (%)	Data rate (Mbps)
Vehicle platooning	50 – 6500	10 – 20	90 – 99.99	0.012 – 65
Advanced driving	300 – 12000	3 – 100	90 – 99.999	0.096 – 53
Extended sensors	1600	3 – 100	90 – 99.999	10 – 1000
Remote driving	—	5	99.999	UL: 25 DL:1

Figure 21.3 V2X Performance Requirements for Wireless Communications Systems Employed by Autonomous Vehicular Networked. Source: Adapted from 3GPP TS 22.186; v16.2.0.

sending and receiving system-wide data related to the state of the underlying highway transportation system. An RSU is often connected to local or regional traffic management and control stations. Such stations collect and disseminate transportation system and vehicular data. They may issue traffic management directives that inform vehicles about highway and environmental conditions that affect their mobility and routes. These stations employ traffic management tools that calculate vehicular mobility and communications parameters.

Another approach employs message communications networking services that are disseminated across a Bnet, in aiding communications between vehicles. Such a configuration is shown in Fig. 21.4(b). A cellular wireless network system (such as 4G LTE, 5G, and 6G), or another network system that supports high message throughput rates at low message transport latency levels, can be used as a Bnet. For example, data communications between vehicles is performed along a tandem chain that includes uplink and downlink wireless links. A source vehicle employs a wireless

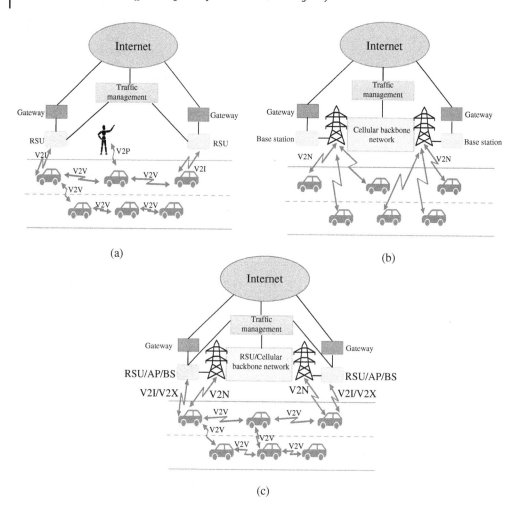

Figure 21.4 Vehicular Networking Configurations: (a) Vehicular Ad Hoc Network (VANET) Communications Using Vehicle-to-Vehicle (V2V) and Vehicle-to-Infrastructure (V2I) Links; (b) Backbone-Aided Vehicular Network Using Vehicle-to-Network (V2N) Communications; (c) Hybrid VANET and Backbone-Aided Vehicular Communications Network.

communications uplink channel across a cellular network's radio access network (RAN), using it to communicate its messages to its associated BS. The latter uses unicast or multicast/broadcast wireless downlink channels to communicate with a single vehicle or with multiple destination vehicles, respectively. An originating BS can use the Bnet to route data messages to other BSs whose locations enable them to send the messages over wireless downlink channels to targeted highway vehicles. BSs can also connect to other Bnets, such as to the Internet, and to local or regional traffic management and control centers.

A hybrid configuration is shown in Fig. 21.4(c). VANET-oriented V2V and V2I wireless communications links and networking schemes are combined with the use of a backbone networking system. Under such a configuration, a Bnet, such as a cellular network, is employed to aid in V2V communications. Direct peer-to-peer V2V communications using VANET methods or backbone-aided networking routes can also be used to disseminate message flows among vehicles and between vehicles and entities attached to (and reachable via) the infrastructure.

21.3 Vehicular Wireless Networking Methods

In the following, we describe a number of methods employed for V2V and V2I networking. We first review networking approaches that are based on the use of VANET methods, including such that employ modifications of techniques that have been used for WLAN and MANET systems. We then discuss infrastructure-aided methods.

21.3.1 VANET-Based Vehicle-to-Vehicle (V2V) Networking Protocols

A common VANET-based approach to vehicular networking uses a modified CSMA/CA MAC method. It is a modified version of the IEEE 802.11 protocol that has been used by Wi-Fi WLAN systems for sharing a wireless medium. It has been standardized around 2010 as IEEE 802.11p. This method is used for V2V and V2I wireless communications, as illustrated in Fig. 21.4(a). This protocol has been employed by a US Department of Transportation (DOT) project, using its *dedicated short-range communications (DSRC)* technology. It has also been used by ITS-G5, which has been standardized by the European Telecommunications Standards Institute (ETSI). The latter specifies VANET-type networking protocols for **Intelligent Transport Systems (ITS)**, using an IEEE 802.11p-oriented MAC protocol at the MAC layer and *GeoNetworking* (GN) schemes at the network layer.

In comparing the IEEE 802.11 and IEEE 802.11p protocol schemes, it is noted that the latter skips the process used by IEEE 802.11 systems that requires a mobile entity to associate with an AP as part of a transaction that establishes a basic service set (BSS). The communications link established between a mobile vehicle and an RSU, acting as its AP, is typically maintained for a relatively short period of time. Accordingly, message frames issued by IEEE 802.11p stations use the wildcard BSSID (consisting of all 1s) in their headers. A mobile can then start sending and receiving data frames as soon as it detects the communications medium. Higher layer protocols are used to provide authentication, confidentiality, and other security services. IEEE 802.11p RSUs broadcast management frames that advertise timing information. By using this information, network entities synchronize their clocks so that they are governed by a common time reference.

Under the IEEE 802.11p standard, a typically employed communications channel is assigned a 10 MHz bandwidth. In comparison, under IEEE 802.11, Wi-Fi systems can employ wider channel bands, such as the 20 MHz bandwidth employed by IEEE 802.11a. The difference is related to the higher signal interference, fading, and multipath effects that are caused in a vehicular communications environment, especially when operating in an urban setting, while requiring steady throughput behavior.

This technology is part of the WLAN IEEE 802.11 family of standards, known in the United States as WAVEs and in Europe as ITS-G5. These systems employ the IEEE 802.11p protocol at lower (Physical and MAC) layers. The WAVE protocol stack is shown in Fig. 21.5 [25].

In 1999, the US Federal Communications Commission (FCC) allocated 75 MHz in the spectral range of 5.850–5.925 GHz for ITS. In November 2020, the FCC reallocated the lower 45 MHz half of the DSRC spectrum to Wi-Fi and other unlicensed uses, noting that by that time only a limited number of vehicles were outfitted with this technology. The corresponding European system uses the 5.875–5.905 GHz frequency band for VANET communications that are used for ITS safety applications. Corresponding bands allocated for such systems in Japan and Australia are 5.770–5.850 GHz and 5.855–5.925 GHz, respectively.

IEEE 802.11p-based VANET systems provide for V2V and V2I communications by using a CSMA/CA MAC protocol. Priority-oriented access can be invoked, as defined by IEEE 802.11e.

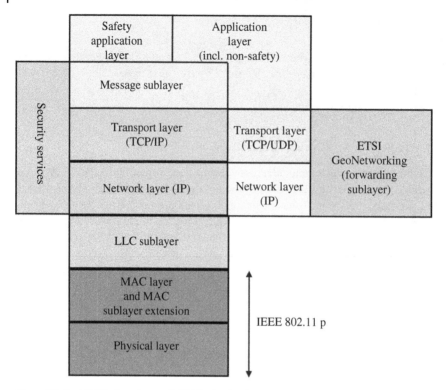

Figure 21.5 US Wireless Access in Vehicular Environments (WAVE) Protocol Stack and ETSI GeoNetworking layers.

Several **network layer forwarding and routing algorithms** have been considered and studied. To illustrate, we consider safety applications under which a vehicle aims to disseminate a safety message to all vehicles that reside within a specified range from itself and that travel along the same highway, or just along the same lane. For example, a vehicle may wish to broadcast multiple copies of an alert message to all vehicles that move along its lane within a distance of 300–500 m from itself, warning them of its immediate stopping or another maneuver. It is essential that such a critical safety message reaches, with high probability, all of its destined vehicles within a very brief period of time. For example, the specification may aim it to reach 99% of targeted vehicles within a delay level that is lower than 100 ms for 90% of the messages and also guarantee the dissemination delay to be not higher than 200 ms for 99% of the messages.

The realized V2V maximum message transmission range depends on several system parameters and conditions, including the transmission data rate and power level, signal interference conditions, effective SINR levels measured at targeted destination vehicular receivers, the employed modulation/coding scheme (MCS), and inter-vehicular distances. Parameters are calculated under the objective of meeting a specified packet error rate level, such as assuring the probability that a packet is received in error to not exceed 1%.

As assumed for the operation of MANET systems (in observing the periodic transmission of Hello messages), it is also the case for VANET and other vehicular networking systems that each vehicle broadcasts periodically to its neighborhood members a **Vehicle Parameters and Status** message, often identified as a **Beacon**. Such a status message may include data that cite the vehicle's mobility patterns (speed, direction, and routes), location, environmental conditions, communications means, relevant communications channel quality states, information about associated RSU stations and neighboring vehicles, lists of key events, and other attributes. The contents of such status messages and their periodic transmission rates can be system dependent and dynamically re-configured.

The following example illustrates a V2V forwarding process that may be used for the dissemination of a message between highway vehicles.

Example 21.1 Consider vehicles that move along a lane of a segment of a highway, as shown in Fig. 21.6. Assume that the maximum transmission range by a vehicle is equal to 100 m. Assume that the targeted dissemination range of a message (of a specific traffic class) issued by a vehicle is such that it reaches all vehicles that follow it along the same lane that are located within a range of 300 m from itself. Assume the current mobility pattern to be such that the inter-vehicular distance is equal to 30 m. Such a formation of vehicles is identified as a **platoon**. If not tightly regulated, the corresponding inter-vehicular spacing levels may fluctuate randomly. Using a vehicular wireless communications network, spacing distances between platoon vehicles may be properly regulated. In an autonomous highway system, platoon vehicles are often set to follow each other at fixed ranges.

Assume vehicles to follow each other at a specified distance level and also assume the highway's segment of interest to be fully occupied, as shown in Fig. 21.6(a). A message transmission by a source vehicle, denoted as Vehicle 1 (V1), which covers its maximum range of 100 m reaches a maximum of 3 downstream vehicles. This is the case since the third following vehicle (i.e., vehicle V4) is at a distance of 90 m from V1 while V5 is located at a distance of 120 m from V1 and will therefore be out of range. A message sent by V1 that aims to reach vehicles located within a range of 300 m needs to be successfully received by V11 as well as by all vehicles situated between V1 and V11.

A method that can be used to achieve the targeted dissemination span is to elect certain intermediate vehicles to act as **relays**, serving as **forwarding nodes**. For this purpose, each vehicle implements a **forwarding algorithm**, which determines whether it should act to forward/relay a received message, and then retransmit it to downstream vehicles. Each such forwarding operation would however induce a time-delay component. To limit the end-to-end message dissemination delay, it is desirable to reduce the total number of vehicles that are used to relay a message across its path. Under the forwarding configuration illustrated in Fig. 21.6(a), 3 vehicles are elected to act as forwarding nodes (in addition to V1 that acts as the forwarding source), so that 3 message retransmissions are executed, the lowest such feasible number for this scenario.

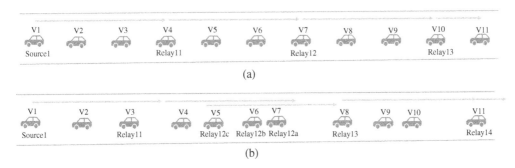

Figure 21.6 Vehicular Configurations and Relay Selections: (a) a Platoon of Vehicles Configured to follow each other at Fixed inter-vehicle distances; (b) Vehicles assuming randomly varying inter-vehicle distances.

The message created by an Application layer entity residing at V1 is transmitted across the shared wireless medium in accordance with the protocols and algorithms used by V1, such as those used in the WAVE system, which employs the physical and MAC protocols specified by IEEE 802.11p. In this case, MAC frames are transmitted across the medium in accordance with the employed CSMA/CA access scheme. This scheme, which is similar to that used by a Wi-Fi WLAN system, is contention based. Simultaneous message transmissions may take place and would result in a transmitted message experiencing either a successful reception or a collision (possibly leading to message loss). While a Wi-Fi-based unicast frame transmission requires the intended receiving station to produce, upon correct reception of a data frame, a positive acknowledgment (PACK) frame, so that a frame may be retransmitted when necessary, broadcast transmissions do not require receiving nodes to send PACKs. The MAC frame carries a network layer packet as its payload. To implement a packet routing algorithm, control data is included. Such data is processed by a *forwarding layer* entity, a sublayer of the network layer. The sender includes in its packet a sequence number (SN), which is used by a receiver to identify the reception of duplicate packets.

To aid in the dissemination of messages among vehicles, geographical (Geo) location information is often used. It identifies the location of a source vehicle and of a targeted destination vehicle or vehicles or, in case of a broadcast packet, the geographical region across which a packet should be disseminated. In making use of such information for networking purposes, it is assumed that each vehicle is capable of determining its current location as well as request and obtain the locations of destination and other vehicles. This can be accomplished by using GPS signal receptions, or by other means, such as by receiving location signals transmitted by nearby RSU stations that are situated at known positions, or by using inertial navigation methods or other techniques that do not require external reference signals. A location server may be queried by system vehicles. It tracks the locations of vehicles and keeps a data base of their locations. RSU-based signal broadcasts are used by a vehicle to also acquire precise system timing information, enabling the system to execute networking algorithms that require vehicular nodes to be time synchronized.

21.3.2 Selection of Relay Nodes

As discussed in Example 21.1, and as is typically the case for wireless data networking among vehicles that move across a highway lane, the selection of a packet's route is straightforward. As demonstrated by the configuration shown in Fig. 21.6(a), the linear layout of vehicles moving along a highway lane enables the realization of a unique end-to-end path. To traverse this path, a packet would be relayed by several intermediate vehicular nodes.

How would a VANET system proceed to select specific vehicles to serve as relay nodes? Several VANET forwarding and relay selection algorithms have been developed and studied. In comparing VANET and MANET routing, we note that the connectivity graph of a VANET-oriented network of vehicles that move along a highway lane is represented by a much simpler topology. Wireless connectivity patterns are constrained by the physical layout of the highway. In contrast, in a general MANET system, vehicular movements may follow spatially diverse stochastic patterns. Hence, the complexity of algorithms that are used for communications networking between highway vehicles can be much reduced. They can be designed to make use of nodal geographical locations and mobility constraints. This is the case when such schemes employ **GeoNetworking** algorithms.

In the following, we present and discuss the operation of several GN-based packet forwarding algorithms, as specified by ETSI and outlined in ETSI Standard documents EN-302-636-1 [16] and TS-102-636-4-2 [15]. We focus on a system that uses the ITS-G5 protocol as the employed medium access scheme. It uses the above-noted modified CSMA/CA protocol, as defined by IEEE 802.11p.

GN schemes are used for unicast and broadcast message dissemination. **GeoUnicasting** protocols are used for packet unicast routing, where a specific node is specified by a source vehicle as the packet's destination. **GeoBroadcasting** schemes are used for location aware packet broadcast routing. A source vehicle targets a packet for delivery to vehicles that reside in a specified geographical region. The destination region can be defined by specification of its geographical coordinates, or by the specification of parameters that define the area's shape (such as a circular area) and boundary (such as its radius).

Each GN-vehicle as well as an RSU acts as a network node that is able to operate as a router, called a GeoAdhoc router, and is identified by a unique GN address (GN_ADDR). This address is included in the *header of a GN packet (GN_Header)*. It contains identifiers that indicate whether this entity is a vehicle or an RSU, and a **MAC_ID (MID)**. For an IEEE 802.11p-based MAC scheme, the **IEEE 48-bit MAC address** is used as the MID. At the link layer, an ITS node is generally identified by its **Logical Link address (LL_Address)**.

Vehicles maintain geographical position data, keeping them as **Position Vectors (PVs)**, including their own positions and the positions of neighboring vehicles and other vehicles and entities of interest. This information is continuously refreshed. A vehicle's **neighboring vehicles** are those vehicles that can communicate with it across a single (link layer) hop at a specified transmission rate level (or range).

Each vehicle transmits periodically **beacon packets**. They are used to inform its nodal neighbors about its current attributes and parameters, including its location. Attribute data elements are kept by each node as *Location Table Entries (LocTE)*. Entries include: ITS station type (whether a vehicle or an RSU) and GN address; PV, which includes its geographical position and associated time stamp that identifies the time at which its position has been observed; speed; heading; and related accuracy parameters. Also maintained are various data elements of a node's neighbors and the sequence number (SN) of the last packet received from a specified source, provided it was identified by the node as not being a duplicate packet.

The GN packet's header (GN_header) consists of a common header that is supplemented by fields whose contents depend on the specific packet type. The common GN_header includes fields that identify the location of the source node, the Traffic Class (TC) associated with the packet (pointing to its desired performance behavior, which relates to the packet's payload type and its targeted reliability and latency objectives), and a hop limit (HL) value. For a GeoUnicast packet, additional fields provide the following information: a packet SN, which serves to eliminate reception of duplicate packets; a life-time (LT) parameter that specifies the maximum time latency that should be incurred by the packet until it reaches its destination; long PV for the source node (which is 28 octets long); and short PV for the destination node (set as 20 octets long).

Long PV data components for a node include its speed, heading, location coordinates (latitude, longitude, and altitude), associated time stamp, and accuracy level of data elements. A short PV includes a time stamp and longitude and altitude entries. These fields are used also for a GeoBroadcast packet, except that the field that includes the PV of the destination node is replaced by a field that identifies the geometric shape (and its associated parameters) of the destination area across which the packet should be broadcast.

A **Location Service (LS)** is maintained. It generates responses to requests made by nodes that wish to find the location of other specific nodes. The long PV of the requested node is provided. The execution of the LS process is transparent to protocol entities employed at higher layers. The function performed by the LS resides on top of the forwarding function and can therefore use any provided forwarding service.

The layer services provided by ETSI GN, in relation to the corresponding services provided by the WAVE system, are shown in Fig. 21.5. The noted ETSI services correspond to those provided by the WAVE's network and transport layers. The forwarding operation is shown to be embedded as a sublayer of the network layer. As for the WAVE system, the ETSI over ITS-G5 networking protocol entity receives link, MAC, and physical transmission services from the respective LLC, MAC and Physical layer protocol entities.

The protocol layering employed by the ETSI GN system is shown in Fig. 21.7 [15]. The payload carried by a GN packet is received at the GN Protocol layer from an upper layer. The payload can be an IP packet. Under IPv6, the packet is received through an GN-IPv6 Adaptation Sublayer entity. The latter entity submits a request to the GN protocol entity to transport the packet across the vehicular network so that it reaches the targeted destination nodes.

In the case of a **unicast packet**, the IP address of the destination node is provided. This address is resolved by a system utility that obtains the link-layer address, such as the MAC ID (MID) of the destination node. The GN protocol entity uses this information to construct a **GeoUnicast packet**. For this purpose, the underlying ITS LS is used to find the current location of the destination node. The GN protocol layer entity then uses the relevant **GeoForwarding Algorithm** to determine the link layer ID (such as the MAC address) of the relay node to which this packet should be forwarded. It is identified as the link-layer next-hop (NH) address (NH_LL_ADDR), citing a specific next-hop node or, when targeted for link-layer broadcasting, displaying a broadcast identifier. A next-hop node would then relay the packet to its determined next node. This process continues in this manner, as the GN packet is forwarded from the source vehicle to subsequently selected relay vehicles until it reaches the destination vehicle.

To transport the constructed GN packet across the wireless medium to its targeted next-hop node (whether a relay node or the destination node), the GN packet is moved down the protocol stack for handling by the link layer protocol entity. The latter forms a MAC frame. Its header identifies the destination MAC address of the next forwarding (or destination) node. The frame is then transmitted across the link's wireless medium in accordance with the underlying access scheme implemented by its MAC and physical layer entities, such as IEEE 802.11p. Upon the successful reception of a frame at the next forwarding (or destination) node, the frame's GN packet's payload is processed by the GN protocol entity. If it has not yet reached its destination node, this entity

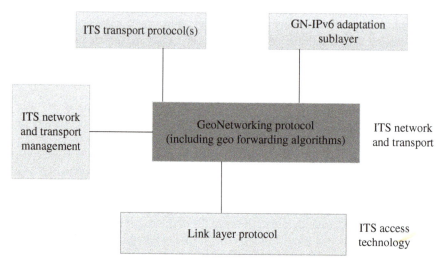

Figure 21.7 GeoNetworking (GN) Protocol Layers.

uses the forwarding sublayer entity to execute (for a unicast packet) the identified **GeoUnicast forwarding algorithm** to produce the link-layer identity of the next forwarding node. When the GN packet has reached its destination node, its payload is delivered up the protocol stack for processing by higher layer protocol entities.

As shown in Fig. 21.7, the transferred payload can include a packet that is produced by an ITS transport layer protocol entity. Such an ITS packet uses the services provided by the GN (including forwarding) and link-layer protocol entities for transfer to its destination node(s) as well as for the transfer of possible response packets.

A similar packet processing and forwarding process is used for the transport of a **GNBroadcasting packet**. At the forwarding sublayer, a **GeoBroadcast forwarding algorithm** is executed. It produces the link layer (MAC) address of the **next-hop node(s)** to which the packet should be forwarded. This address may identify a specific next-hop relay node. In turn, when aiming to broadcast the packet across its link, so that it is received by all its neighboring nodes, a link-layer next-hop BROADCAST identifier is used.

To simplify our descriptions, we assume in the following that vehicles use omni-directional antenna arrays. Under the above-discussed networking scheme, the wireless link configuration is assumed to possess a **broadcast property**. Signals produced by a message transmission that is issued by a vehicle moving across a highway would then propagate in the forward and reverse directions (and possibly sideways as well) relative to its movement. Such a transmission can be sensed by vehicles moving in either direction across the same lane, and possibly also by vehicles that travel across nearby lanes, provided the produced signals are received at a sufficiently high power level. While the packet's received power level may be too low for it to be processed, demodulated and decoded successfully by the receiver, it may be of sufficient intensity to trigger a Carrier Sensing (CS) signal indication at the receiver. Under the use of a CSMA/CA access scheme, while a node receives a CS signal, it avoids initiating a new frame transmission, as this signal indicates that the channel is occupied by other transmission(s). If at the CS reception time, the node has already initiated transmission of its own frame, a collision event may occur, which may lead to unsuccessful frame receptions by impacted receivers.

Assume message transmissions to be carried out at prescribed transmit power and data rate values, using an assigned frequency bandwidth level. Under a selected modulation/coding scheme (MCS), a targeted spectral efficiency ratio is attained. To achieve a desired bit error rate (or packet error ratio) level, it is necessary for the sender's signal to be detected at a targeted nodal receiver at a Signal-to-Interference-plus-Noise Ratio (SINR) level that is not lower than an intended value. Under observed noise and interference levels, this requirement induces a maximum distance range across which such a packet transmission would be successfully received. For instance, if the data transmission range is equal to 100 m while the CS range is equal to 200 m, a transmitting vehicle may be able to successfully transmit a packet to another vehicle that is located 100 m (or lower) away from it, while this transmission will induce CS signals at vehicles that are located at a distance range of 200 m (or lower) from it, preventing the latter vehicles from initiating at that time new packet transmissions.

The structure of a GN packet is as shown in Fig. 21.8 [19]. An extended header (not shown) is used when security is incorporated. The MAC header is composed of the MAC protocol entity, involving

MAC header	LLC header	GeoNetworking header	Payload (optional)

Figure 21.8 Structure of a GeoNetworking (GN) Packet.

the ITS-G5 access module. It is used to identify the next-hop address of the GN packet. The LLC header is the header of the Logical Link Control (LLC) PDU, following the IEEE 802.2 specification. It includes source and destination Service Access Points (SAPs) that identify the corresponding source and destination network-layer entities and thus enables multiplexing network layer PDUs over the link. It can also be used to provide link-layer error and flow control services. Such services tend to be generally employed at the MAC and transport layers. The GN-header is the header of the GN packet. For unicast GN packets, the location vectors of the source and destination nodes are included. For multicasting GN packets, it identifies the location vectors of the source node and of the targeted broadcast area. Also included are TC parameters and pointers to the next header type (such as pointing to an IP or ITS transport entity). The optional payload field contains user data that are created by upper protocol entities, such as an ITS transport entity that produces a T-SDU, or an IP entity whose product is formatted as a GN6-SDU. Certain GN packets, such as a Beacon Packet, do not carry a payload.

Under a CSMA/CA MAC scheme, the transmission of a unicast MAC frame is acknowledged by the MAC layer entity of the receiving node. In turn, no acknowledgments (ACKs) are produced in case of MAC frame broadcasting. Multiple nodes that are located within reception range, and are thus neighboring members of a broadcasting node, may be able to successfully receive a broadcast frame. Such *overhearing* by multiple nodes of a relayed transmission of a frame (with an identified SN) by node A also serves to inform these multiple nodes that node A has successfully received this frame and has subsequently proceeded to relay it to other nodes by transmitting it across the medium, aiming to forward it to other node (or nodes) that will be further forward it.

In simplifying the presentation of networking scenarios, the data units that are disseminated across a vehicular network are often referred to as frames or packets while the operation carried out by a relaying vehicle is referred to as transmission, retransmission, relaying, or forwarding. As described above, the data units transmitted across the wireless medium are MAC frames, while the nodes perform routing operations that involve MAC frames, LLC frames, forwarding layer-based protocol data units (PDUs) such as GN-packets, and data payloads. Operations at each node involve the corresponding protocol layer entities.

To illustrate, consider the layout shown in Fig. 21.6 and discussed in Example 21.1. Assume node V1 to broadcast a packet frame to its neighboring nodes V2, V3, and V4. It expects node V4 to successfully receive and retransmit this frame, forwarding it to its downstream neighboring vehicles. Assume that the transmission of this frame by node V1 is received successfully at nodes V2 and V3 but that its signal is degraded when received at node V4 and is consequently discarded by V4. Nodes V3 and V2 may wait for specified periods of time to overhear the retransmission of this frame by node V4. As this transmission is not forthcoming, upon the expiration of a prescribed timeout period, a forwarding algorithm may elect one of these nodes, say node V3, to then act as the forwarding node and retransmit the frame. Assume that node V3 is successful in retransmitting the frame. Then, node V2 would overhear this transmission, confirming that node V3 has acted as a forwarding node of the underlying frame. Node V2 could then delete a copy of this frame that it holds in its forwarding buffer.

In the following, we present several GN forwarding algorithms that are described by the underlying ETSI-defined GN system.

1) **GeoUnicast Forwarding Algorithms:** A Greedy Forwarding (GF) algorithm or a Contention-based Forwarding (CBF) algorithm may be employed.
   a) Under the **Greedy Forwarding (GF) algorithm**, starting with the source node, each forwarding node selects one of its (link layer) neighbors (which are vehicles that are within

transmission range of the underlying forwarding vehicle) that is located across the next-hop to act as the next forwarding node. In making this selection, a forwarding node examines the location of the destination node. This location is included in the GeoUnicast packet's header. To reduce the number of forwarding nodes used along a dissemination path, and thus improve the packet's transit time delay, it is advantageous for a forwarding vehicle to select a next forwarding vehicle whose location is closest to the location of the destination vehicle, while it is also within transmission range from the forwarding vehicle. Such a selection is based on what is known as the *Most Forward within Radius (MFR) policy*.

b) Under the **Contention-Based Forwarding (CBF) algorithm**, the selection of a forwarding node is performed by a receiver of a packet's broadcast transmission. In comparison, we note that under the GF algorithm the selection is made by a sending forwarding node. Under CBF, a node that receives a packet transmitted by another node, and is not the targeted destination, decides whether to act as a forwarder of this packet by examining the location of the destination node (as specified in the header of the GN packet). Starting with the source node, a forwarding node that is directed by the CBF algorithm to transmit a packet, marks **Broadcast** as its link-layer next-hop address. Its transmission across the wireless medium is thus intended for reception by all its link layer neighbors. Each node that receives this transmitted packet stores the packet in its CBF buffer and starts a **CBF timer**. The *expiration time or timeout* duration of this timer is set to a value that is proportional to the distance between this receiving node and the packet's destination node.

The CBF timer will expire after a shorter period of time at a receiving vehicle that is located closer to the destination vehicle. Prior to the expiration of the timer, the underlying node may receive a duplicate of this packet from a node whose CBF timeout duration is shorter (as it may be located closer to the destination node), by *overhearing* the transmission made by the latter node. In this case, if the same packet (identified by the same attributes and the same SN) still resides at the node's CBF buffer, it removes the packet from its buffer. It discards both packets and stops its timer so that redundant packets are deleted and are not rebroadcast. Otherwise, if no such overhearing transmission is received prior to the expiration time of its timer, the node proceeds to rebroadcast the packet at the time that its timer has expired.

2) **GeoBroadcast Forwarding Algorithms** include a Simple (S) algorithm and Advanced forwarding (AF) algorithms. We outline in the following the methods used for executing the Simple algorithm and for performing one of the AF algorithms (identified as Advanced GeoBroadcast Forwarding algorithm 1).

a) Under the **GeoBroadcast Simple (S) algorithm**, the packet is *flooded* in its target geographical area. Each node that receives the GeoBroadcast packet and is located inside or at the border of the targeted geographical area, rebroadcasts it once upon reception. If the node that receives a GeoBroadcast packet is located outside the target area, it uses the GF algorithm to forward the packet.

b) **GeoBroadcast Advanced Forwarding (AF) algorithm 1** combines the use of a Contention-Based Forwarding (CBF) algorithm with a Greedy Forwarding (GF) scheme. Thus, in addition to employing the CBF algorithm, the sender uses the GF algorithm to select one specific neighboring vehicle that is located within the targeted geographical area to serve as the next-hop forwarder. The destination location is now set as a point at the border of the geographical area that is located farthest away from the location of the source node. Under the GF scheme, upon correct reception of a packet by a designated forwarder, the latter acts to immediately forward it in accordance with the GF protocol. This combination is motivated by the following factors. The CBF forwarding mechanism handles situations of reception failures that may occur at intended forwarding nodes (including nodes selected by the GF scheme) by inducing other nodes to act as forwarders (as their timers may timeout when no transmission of the underlying packet by another node is overheard). However, under the CBF scheme, the occurrence of packet collisions and retransmissions may lead to higher packet dissemination latency. The GF component aims to reduce this latency by tasking a forwarding node to select by itself a specific next-hop node to forward the frame. At the same time, the use of a CBF access scheme is continued.

Example 21.2 Consider vehicles that move along a lane of a segment of a highway, as shown in Fig. 21.6 and discussed in Example 21.1. The distance span covered by a successfully received message transmission executed by a vehicle is assumed to be equal to 100 m. Consider a scenario under which vehicles move across a lane as a group, identified as a **platoon**, keeping fixed inter-vehicular distance of 30 m, as shown in Fig. 21.6(a). Assume vehicle V1 to have a **unicast** packet that it wishes to send to destination vehicle V11. A GeoUnicast forwarding algorithm is used for this purpose.

Under the **GeoUnicast Greedy Forwarding (GF) algorithm**, the following process takes place. The downstream neighbors of node V1 are those vehicles that are located within a range of 100 m from V1. These are vehicles V2, V3, and V4. Vehicle V1 selects the vehicle that is located closest to the destination node as its next-hop forwarding node. In this case, vehicle V4 is selected, noting that V4 is located 90 m away from V1, while V5 is located 120 m away from V1, which is outside V1's transmission range. Hence, V1 configures the MAC ID of V4 as the address of the next-hop node. V1 then submits the corresponding GN packet to its MAC layer entity, requesting the latter to send it across the wireless medium. This frame is then sent by the MAC entity of V1 across the wireless medium, following the access protocol specified by the IEEE 802.11p CSMA/CA scheme. Assume that no background signals or ongoing message transmissions cause reception interference at V1s neighboring nodes. In this case, the frame sent by V1 is successfully received by V1s neighboring nodes. The forwarding sublayer entity residing at node V4 examines the received GN packet, determining itself as the next-hop forwarding node and notes the identity and location of the node that is the final destination of the packet (i.e., V11). Vehicle V4 then initiates the forwarding of the received message. It executes the GF forwarding scheme as its unicast forwarding algorithm, determining the subsequent next-hop node to which the GN packet should be forwarded. It determines V7 as the next-hop forwarding node. Similarly, the forwarding sublayer entity at V7 selects V10 as the next-hop forwarding node. Subsequently, V10 transmits the packet across its wireless link, setting V11 as its ultimate destination. In this case, under assumed noninterference packet reception conditions, the GF unicast forwarding algorithm would perform well, selecting a route that consists of a low number of relaying vehicles.

Consider next a different scenario, involving the transport of the same unicast packet. Assume again that the nodes are configured to use the GF forwarding algorithm. Node V1 would then select V4 as its next-hop forwarding neighbor. However, the vehicular network is now assumed to be highly loaded by message traffic, including (unicast and broadcast) packets that share the same wireless medium. These packets may be sent by several vehicles and may be destined to various vehicles. Assume the following events to take place. The packet that has been transmitted by V1 is successfully received at V4. However, its transmission and retransmission by V4 have not been successfully received by V7. V4 may cancel a further retransmission attempt, failing to forward the packet to V7. The packet is prevented from being delivered to its ultimate destination node (V11).

Example 21.3 Consider next an illustrative forwarding process that may take place when the **CBF algorithm** is used. The forwarding layer entity at source node V1 submits the packet to its MAC layer entity, specifying BROADCAST as the next-hop address. Following its transmission across the shared medium in accordance with the access rules specified by the IEEE 802.11p CSMA/CA protocol, the frame is broadcast to V1's neighbors. Assume this MAC-frame transmission to be received successfully at all V1s neighboring nodes, consisting of nodes V2, V3, and V4. Each one of these nodes passes the received MAC frame to its forwarding layer entity, which then initiates a timer whose expiration time is set to a value that is proportional to the receiving node's distance to the location of destination node V11.

In accordance with the lane-oriented linear nodal layout, the timeout period is inversely proportional to the distance between the sending node and the receiving neighboring node. As V1s neighbor V4 is located closest to V11, its timer will expire first. Once expired, V4s forwarding layer entity submits the GN packet to its MAC layer entity for transmission across the shared wireless medium. This transmission is overheard by V4s neighbors, which consist of (assuming an omni-directional antenna to be used) downstream nodes V5, V6, and V7, and upstream nodes V1, V2, and V3. Assume this transmission to be correctly received at these neighboring nodes.

As a result of overhearing the successful transmission of this packet by forwarding node V4, upstream nodes V2 and V3 reset their timers and delete the copies of this packet that they hold in their CBF buffers. Such a reset and deletion can be executed successfully provided the forwarding layer entities at nodes V2 and V3 find out about the re-relaying performed by V4 before their timers have expired. Assume that this is the case under the scenario discussed here, so that the contention and transmission times incurred at V4 turn out to be relatively short, avoiding too early timer expiration events to occur at V2 and V3. It is however noted that when the shared medium is subjected to higher packet traffic loading rates, this may not be the case at some or all of the nodes that hold packets at their CBF buffers. Once a node's timer expires, it submits its packet to its MAC layer entity for transmission across the shared medium. Once the node has initiated the latter request, this request cannot be canceled at a later time, at which time the forwarding layer entity may receive feedback from the MAC layer entity that it has overhead a transmission of the same packet, inducing consequently a possible occurrence of a MAC layer collision event.

If all MAC layer transmission and reception events for this illustrative scenario are performed successfully, the outcome of the forwarding process enacted by using the CBF algorithm leads to forwarding of the packet by subsequent nodes V7 and V10, as intended. In turn, upon the occurrence of failed packet receptions at certain nodes, other nodes may assume forwarding roles, acting as relay vehicles. To illustrate, consider the following scenario. Assume that the frame transmission performed by V1 to its neighbors is not successfully received at nodes V4 and V3. Subsequently, failing to overhear forwarding transmissions executed by V4 and V3, the forwarding layer entity at V2 (whose timer is assumed to not yet expire) waits for its timer to expire. It then broadcasts the underlying packet to its neighbors, which are nodes V3, V4, and V5. The forwarding transport process would thus involve a transmission by node V2. As illustrated, we note that the CBF forwarding algorithm may induce longer end-to-end packet transit delays. In turn, the use of autonomous selections by receiving nodes to act as packet forwarders when needed may induce a more reliable packet dissemination process, yielding potentially higher packet delivery rates.

Consider next a scenario under which, as shown in Fig. 21.6(b), inter-vehicular distances assume variable values. In this case, vehicles that are located closer to each other will set their timers to timeout values that are relatively close to each other. Under relatively high traffic loading conditions, the wireless medium may be busy over longer time periods, resulting in multiple close-by nodes simultaneously incurring timer expiration events, leading to multiple frame access contentions that are more likely to induce frame collisions. Such scenarios would tend to cause a significant degradation in the realized frame throughput rate, leading to longer frame transit delay times. The illustrative scenario shown in Fig. 21.6(b) displays an event under which a frame that is forwarded by V4 to its neighbors is successfully received by V5, V6, and V7. However, the close-by locations of V6 and V7 induce frame forwarding failures, resulting in V5 eventually acting as the forwarding vehicle.

The following example illustrates the operation when forwarding a broadcast packet.

Example 21.4 Consider the same vehicular layout as that described in Example 21.2 and depicted in Fig. 21.6. Assume now that vehicle V1 has produced a safety packet that it wishes to **broadcast** to all other vehicles that travel along the underlying segment of the highway, aiming it to reach vehicles V2–V11. A **GeoBroadcast forwarding algorithm** is used for this purpose.

Under the **GeoBroadcast Simple (S) algorithm**, the packet is flooded in its targeted geographical area. Each node that receives the GeoBroadcast packet and is located inside or at the border of the target geographical area, would rebroadcast it once upon reception. The target area consists of the segment of the highway in which the underlying vehicles travel. To illustrate the ensuing transmission and forwarding process, consider the following scenarios. Under the first scenario, we assume the wireless medium to be lightly loaded. The transmission by V1 of its frame is assumed to be successfully received by each one of its neighbors, V2, V3, and V4. The forwarding layer at each of the latter nodes would then submit the received packet to its MAC layer entity for transmission as a broadcast packet to its neighboring nodes. Assuming the MAC layer entities of the latter nodes to select distinct CSMA/CA-based access minislots for their targeted initiation of frame transmission, the node that selects the earlier minislot will capture the medium. Assuming this transmission to induce CS signal at the other nodes, the MAC layer entities of the other nodes will defer their frame transmissions, scheduling them to possibly take place at a later time. To illustrate the process, assume the winning node to be node V3. Each one of its neighboring nodes (including nodes V1, V2, V4, V5, and V6) may overhear V3's forwarding transmission and will then pass the received GN packet (assuming it to be successfully received) to its higher layers for processing.

If at this time, the frame layer entities at nodes V2 and V4 have already requested their MAC layer entities to send this packet over the wireless channel, it would be too late for them to cancel this request. These vehicles may then attempt to send their frames at later times. Such transmissions may be successful, but the ensuing received frames will be deleted as duplicates by certain nodes (including those that have already successfully received these frames or those that have noted them to have been forwarded by other nodes). In turn, the later transmissions may cause packet collisions. Collided broadcast packets are not retransmitted by their senders. It is noted that if random transmission time delays were configured at the forwarding layers, the probability of occurrence of collisions may be lowered at the cost of increasing frame transit latency. The transmission and forwarding process progresses in this manner across the targeted highway's segment. For the frame's broadcasting process to be successfully completed, attaining a frame delivery ratio of 100%, at least a single copy of each frame needs to be successfully received by each vehicle. To meet this objective, the loading rate of the shared wireless medium should be properly regulated.

Example 21.5 Consider next the use of the **GeoBroadcast Advanced Forwarding algorithm 1 (AF1)**. A CBF algorithm is combined with a Greedy Forwarding (GF) scheme. The execution of the CBF algorithm for the underlying example proceeds in a manner that is similar to that described above for the CBF-based GeoUnicasting process. The destination location is set as the location of V11, which for this scenario is the area's border location. The CBF component of the algorithm would induce each receiving node to set its timer expiration time to a level whose value is proportional to its distance from V11.

To illustrate, consider the following scenario, which can take place under the configurations shown in Fig. 21.6(a) or in Fig. 21.6(b). Source vehicle V1 may select V4 as its GF node. Assume the frame sent by its MAC layer entity to be successfully received at vehicles V2 and V3 but not at vehicle V4. Consequently, not overhearing a successful forwarding transmission of this packet by V4, the CBF-based timer used by the forwarding layer entity at V3 would expire, causing it to submit the received frame to its MAC layer entity for broadcast transmission to its neighbors. The forwarding entity at V3 would select, in accordance with its GF scheme, the farthest node to serve as a candidate forwarder. Assume this node to be vehicle V6. Assume also that the frame transmission performed by V3 is received correctly by V4, V5, and V6. Vehicle V6, as the GF-selected forwarder, would immediately schedule the forwarding of this frame. Assume that the frame transmission executed by V6 collides with other frame transmissions so that it is not received correctly at any neighboring node (including the node selected by its GF scheme). In this case, not overhearing a successful transmission, the CBF scheme may cause the timer to expire at a node that holds this frame in its CBF buffer, such as vehicle V5. The latter would then select itself to act as the frame's forwarder. The forwarding process then continues in the same manner, employing a combined execution of the CBF and GF schemes.

21.3.3   Flow and Congestion Controls

Examining the scenarios and packet forwarding and transmission processes illustrated above, it is noted that if the wireless medium is subjected to low-level traffic loading conditions, the simpler schemes may yield low message transit delays, while the integration of a CBF procedure into the forwarding scheme provides a more robust outcome, leading to improved packet delivery ratio. Under medium to high traffic loading rate conditions, frame transmissions may frequently not be successfully received by targeted nodes, leading to high packet delay levels and low packet delivery ratios.

Several mechanisms can be employed to handle such high traffic loading scenarios. Critical safety packets that are broadcast by a vehicle to all vehicles that are located in a target area can be marked as higher priority packets. The IEEE 802.11p CSMA/CA MAC scheme supports the granting of higher access priority to specified TCs (as described for the IEEE 802.11e priority oriented EDCA scheme).

To increase the delivery ratio of a critical packet, a source node may repeat the sending of the packet over a specified period of time. Multiple copies of the packet can be sent by the source node. They would be forwarded by nodes that are selected (or self-selected) in accordance with the employed forwarding algorithm. Such an operation could indeed increase the packet delivery ratio. However, the ensuing higher traffic rate could lead to higher signal interference power, causing lower SINR levels to be detected at nodal receivers, leading to higher rate of packet reception failures.

To avoid high signal interference and packet collision rates, it is essential to prevent excessive congestion of the wireless medium. For this purpose, flow control schemes are employed. They regulate the packet send rate. To illustrate, consider a flow regulation process used at a vehicle that transmits a flow's burst of packets. To throttle the loading of the medium, the node is regulating the **inter-packet temporal spacing** levels, **pacing** the transmission of its packets. For example, consider the layout shown in Fig. 21.6(a). Source node V1 waits for a period of time that is equal to the expected time that it takes its first packet to be received and transmitted by its second (or third) downstream forwarding node before it transmits its second packet. This way, at the time that the second packet is transmitted by V1, the first packet would be transmitted by V4.

This scheme enables a successful operation that provides for *simultaneous packet transmissions to be carried out across sufficiently distant spatial segments of the highway*. Such an operation can increase the vehicular system's packet throughput rate due to its **spatial reuse** sharing of medium resources. However, in considering the latest illustrative scenario, the following may occur. While receiving the second frame, V4s receiver may be detecting an interference signal that is caused by the transmission of the first frame by V7. Such signal interference can cause the second frame to not be successfully received at V4. As will be noted later, under a properly configured cross-layer operation, the packet's transmission data rate across the medium, and the employed modulation/coding scheme (MCS), would be set to proper values to assure correct packet reception (under a specified packet error ratio) under expected interference scenarios, adapting to the underlying signal interference (and ensuing SINR) levels.

To assure acceptable networking performance across the vehicular system, it is essential that traffic loading rates be properly regulated. For this purpose, an access-layer-oriented **Distributed Congestion Control (DCC)** (denoted as DCC_Access) scheme is employed. To reduce the rate a node's access scheme triggers a node's packet transmission rate, it is essential for higher layer forwarding, networking, and transport layer entities to properly throttle the rate used to submit packets for transmission across the medium. Such a rate reduction process may however lead to back-pressure on higher layer processes. For messages processed by a higher layer protocol entity to consequently not experience a state of high congestion, which may lead to packet delays and losses, a higher-layer flow control mechanism should be applied. The use of a cross-layer DCC operation should be planned to adapt traffic regulation parameters that are used by a layer protocol entity in accordance with the needs imposed by dependent lower layer entities. Such a DCC architecture, as described in [19], is shown in Fig. 21.9.

The DCC mechanism described for the underlying GN ETSI system regulates wireless channel congestion by ensuring that the channel is not overloaded by induced transmissions over a certain geographical span. As specified in ETSI document [18] and shown in the reference architecture displayed in Fig. 21.9, the DCC functionality is distributed across multiple layers. DCC regulation is undertaken by protocol entities at the access layer (DCC_ACC), network & transport layer (DCC_NET), facilities (and application) layers (DCC_FAC), and by a cross-layer entity (DCC_CROSS). The DCC_NET layer entity periodically calculates a *Global Channel Busy Ratio*

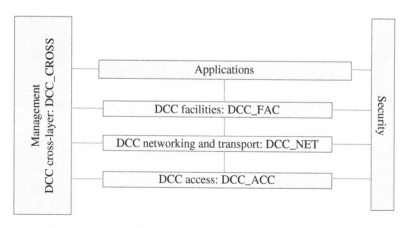

Figure 21.9 Cross-Layer Distributed Congestion Control (DCC).

(CBR_G) parameter, which identifies (as explained below) the current fraction of time that the wireless channel is kept busy. This data is passed to the DCC_CROSS entity, which makes it available to DCC entities at other layers. DCC-related information is also exchanged by GN nodes.

For each radio channel, a node collects data about the fraction of time that the channel is busy. Such local channel busy ratio (CBR_Local) measurements involve a node's local channel and its 1-hop and 2-hop away channels. Nodes periodically broadcast their local CBR measurements and other DCC parameters to neighboring nodes that share the same channel. The global channel busy ratio (CBR_G) for a given channel is calculated as the maximum of the latter CBR local measures. Using these and other metrics (including the number of neighboring nodes), a node calculates the current level of occupancy of wireless medium resources, determining accordingly the current level of resource usage. The result is used to classify the channel's present congestion level and calculate the level of flow control that should be applied.

The DCC flow control scheme retrieves messages from DCC queues in accordance with their priorities and transfers them to the radio module for transmission. The DCC scheme is also used to assure a priority-oriented fair allocation of resources. Channel access resources are left available for the transmission of critical safety messages that may show up in a stochastic manner as a result of occurrence of high priority events. Priority-oriented buffers are reserved in the access module to temporarily store transmission requests. Stored messages form separate, per priority level, DCC queues. They are based on a message's EDCA access category (as noted for IEEE 802.11e).

DCC schemes regulate congestion by using a cross-layer operation of mechanisms that adjust flow rates, as well as forwarding and transmission oriented parameters. Such mechanisms include the following: Transmit Power Control (TPC), Transmit Rate Control (TRC), and Transmit Data rate Control (TDC). Under a reactive scheme, the observed current flow rate is used to trigger a mix of these control mechanisms. Under an adaptive scheme, a transmit rate control is used, whereby the time spacing between successive packets submitted to the access layer is regulated.

DCC_NET may provide DCC-related information to the GN forwarding algorithm. Using transmit rate control, the packet flow rate loading of the MAC layer entity is controlled by adapting the time spacing between consecutive packets that are fed to the access layer. This spacing is increased as CBR increases. Such an operation can lead to longer packet queueing times in the transmit queues.

Consider the implication of such high queueing delays on the operation of the CBF forwarding scheme. Under high traffic conditions, the queueing delay times incurred by packets waiting in a CBF buffer could exceed the maximum expiration times of the underlying CBF timers. In this case, the CBF overhearing process will fail. Packets will be passed to the MAC layer entity for retransmission and may then wait for a long time for their transmission or re-transmission. While a packet is waiting at a transmit DCC queue, the network layer entity may receive a replica of the same packet from a transmission that has been performed successfully by another node. Without the use of a cross layer operation, the MAC layer entity cannot be notified of such a duplicate reception and will proceed to transmit a duplicate packet. Such scenarios cause the wireless channel to be flooded with redundant packets, leading to delay–throughput performance degradation.

A mechanism that can be used to resolve such a degradation factor, which is described in [19], makes use of the cross-layer operation of the DCC scheme. A DCC flow control entity embedded as a sublayer of the access module (DCC_ACC) interacts across layers through DCC_CROSS with the DCC_NET protocol entity that is embedded in the Networking & Transport layer. When a duplicate packet is received at a node, the Network & Transport layer's protocol entity marks a corresponding entry in a Duplicate Packet List (DPL). The DCC flow control protocol entity employed at the access module examines this duplicate list to review the SNs of received duplicate packets. When the DCC flow control entity selects (in a priority oriented manner) a packet from a transmit DCC queue (of the access module), it checks whether this packet has been marked as a duplicate. If so, it discards this packet, avoiding the transmission of duplicate messages by the radio module, and consequently reducing the congestion state of the wireless channel. Otherwise, it passes this packet to the radio module for transmission across the medium.

21.3.4 Vehicular Backbone Networks (VBNs): Hierarchical Networking Using Cluster Formations

In addition to the VANET techniques discussed in Sections 21.1, 21.2, 21.3.1 and 21.3.2, a wide range of other V2V wireless networking methods have been investigated. Under a *clustering oriented approach*, vehicles are dynamically organized into groups based on their location proximity and other possible attributes, such as lane association, and parameters that may involve their capabilities related to communications, processing, energy resources, and speed levels. A specific vehicle, often identified as a *cluster-head*, is selected to serve as a manager and as a gateway for its cluster group. Acting as a gateway, it is used to relay messages to and from members of its group as well as relay messages to gateway nodes that manage other groups. In this manner, a hierarchical networking structure is formed, separating between intracluster communications and intercluster networking.

In this section, we illustrate such an architecture, identified as a **VBN**. It specializes the concept of a *Mobile Backbone Networks (MBNs)* to vehicular highway networks. Both MBN and VBN architectures were developed by Professor Izhak Rubin and his research group. Increased efficiency of networking operations is achieved by VBN systems in comparison to MBN systems, as the configured hierarchical networking algorithms that are used for VBN systems take advantage of linear arrangements of vehicles along highway lanes.

The architecture of a VBN is illustrated in Fig. 21.10. V2V data flows are transported by using a network architecture that consists of two hierarchical levels: a Bnet and access nets (Anets). The **BNet** consists of **Backbone Nodes (BNs)**, also identified as **Relay Nodes (RNs)**, and the communications links that they employ for their BN-to-BN communications. An **Access Network (Anet)** consists of a Backbone Node (BN) and close-by vehicular **client nodes**, which are vehicles that are associated with this BN. An Anet is also identified as a **VBN cell** or as an **Anet cell**. An Anet cell is managed by its BN, and its user nodes or mobiles are recognized as the BN's clients. Each client associates with its managing BN. The VBN configuration shown in Fig. 21.10 identifies the Bnet to include RN_1, RN_2, and RN_3 as its BNs, and the backbone communications links that connect these backbone nodes.

A vehicle that is an $Anet_i$ member associates with BN RN_i. Such a vehicle can communicate with other $Anet_i$ vehicles either by broadcasting its data packets across $Anet_i$ or by first sending a data packet to its associated BN and then have the latter broadcast the packet in its Anet, as performed by an AP of a Wi-Fi wireless LAN (WLAN). In turn, to send its data packets to vehicles that are further located, such as those that roam outside its Anet, a source vehicle makes use of the Bnet to efficiently traverse the vehicular system.

For example, consider a vehicle that is located in $Anet_1$ and that is the source of a multicast packet flow whose data packets are targeted for broadcast to vehicles that are members of $Anet_1$, $Anet_2$ and $Anet_3$. This vehicle sends its packets to its associated BN, RN_1. The later then broadcasts the packets across its Anet as well as directs them across a BNet link to RN_2, which serves as the backbone node

21.3 Vehicular Wireless Networking Methods

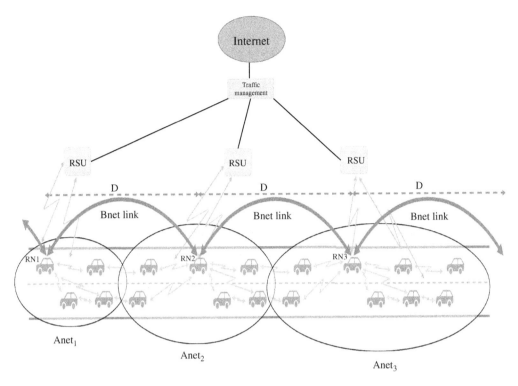

Figure 21.10 Architecture of a Vehicular Backbone Network (VBN).

of $Anet_2$. RN_2 then broadcasts the packets across its own Anet as well as directs them across a BNet link to RN_3, which serves as the backbone node of $Anet_3$. RN_3 then broadcasts the packets across its Anet.

Vehicles can also communicate with RSUs. An RSU communicates with near-by highway vehicles through the use of V2I communications links. The latter can involve the use of Wi-Fi, cellular wireless networks, or other radio access technologies. RSU-based communications platforms, as well as advanced vehicular communications systems, may employ antenna arrays that are configured for broad area coverage, as well as such that form single or multiple directional antenna beams.

The sharing of the wireless media across and within each hierarchical level can be managed by using different multiple access and multiplexing methods. For example, FDMA-based multiple frequency bands, and/or TDMA-based distinct time slots, can be allocated for packet transmissions that are carried out across the Bnet, within each Anet, and for V2I and I2V communications. The following methods are noted:

1) **Intra-Anet Communications, accounting for communications between mobiles, (also identified as vehicles or user nodes) that are members of the same Anet**:
    a) Uplink mobile-to-BN communications within each Anet. For example, Anet medium can be shared by using a multiple access protocol, such as a Wi-Fi-type CSMA/CA or DA/FDMA/TDMA/SDMA schemes. Under the use of DA/TDMA/FDMA/SDMA schemes, control channels are employed by vehicles to communicate signaling packets that are used by entities to request communications resources. They are also employed by managing nodes to allocate system resources.

b) Downlink BN-to-mobile communications within an Anet can be performed by setting the BN to broadcast packets across its Anet. Media-sharing protocols can include Wi-Fi-type CSMA/CA, statistical multiplexing schemes such as DA/FDM/TDM/SDM. Directional communications can be employed for unicast networking purposes.

c) Peer-to-peer methods, as employed by using a CSMA/CA multiple access protocol, can be used for intra-Anet communications as well as for communications with vehicles that travel in neighboring Anets.

d) Distinct **priority-oriented Anet service channels** may be configured for intra-Anet data packet communications. Critical safety data packets are provided higher priority access to wireless media resources.

e) Distinct **Anet control and management channels** are configured to provide for the transmission of management, control and status packets in the Anet. Such channels are used by vehicles to disseminate packets that convey status vectors. They are used to provide status updates to Anet vehicles and to associated backbone nodes. A status vector includes a node's regional and neighborhood observables, mobility and communications states and requirements, and other attributes.

Signaling channels are used for broadcasting of control, management, and status packets across the Anet. Status data messages may be collected and reissued in summarized format by the Anet's BN. They may be forwarded by an Anet's Gateway to a local management station via an associated RSU. Status control packets may be used by the BN, a management center, and by the vehicles themselves. Machine learning (ML) and artificial intelligence (AI)-oriented techniques may be used to configure communications networking and mobility parameters and to specify and manage ongoing operations.

f) **Anet-Spatial-Reuse** settings are configured to provide for the spatial reuse of frequency bands and time slots that are allocated for intra-Anet communications in distinct Anet cells. Under Anet-Spatial-Reuse-Level-k, whereby $k \geq 1$, a frequency band and/or time slots are reused across distinct Anets that are k-neighbors of each other (i.e., they are separated by $k - 1$ intermediate Anet cells).

To illustrate, consider the network layout shown in Fig. 21.10. Under Anet-Spatial-Reuse level that is equal to $k = 1$, a broadcast packet transmission performed over $Anet_2$ by RN_2 may trigger signal interference at receivers of vehicles located in $Anet_1$ which are targeted for reception of data packets transmitted by RN_1. Such signal interference effects may degrade the ensuing SINR levels measured at vehicular receivers located in $Anet_1$. The employed MCS would then be adapted to enable the successful reception of packets, while adjusting the underlying transmission data rate.

When setting the Anet-Spatial-Reuse level to $k = 2$, data transmissions conducted over neighboring Anets are assigned distinct frequency bands or time slots. For the illustrative system, assuming an FDMA-based operation, so that $Anet_1$ and $Anet_3$ are assigned the same frequency bands, while $Anet_2$ is assigned a different frequency band. Such a higher spatial reuse level, also known as a *coloring factor*, produces lower intercell signal interference, leading to the recording of higher SINR levels at Anet nodal receivers. Anet nodal transmitters can then adapt their modulation/coding schemes to achieve higher spectral efficiency levels, enabling (under a prescribed bandwidth assignment) packet transmissions that are performed at higher data rates.

In turn, since (under $k = 2$) neighboring Anet cells use distinct frequency bands, each cell is now allocated a bandwidth level that is reduced by a factor of 2 (assuming similar traffic activity levels across the cells and a prescribed global bandwidth resource level).

The overall effect may result in either a net increase or a net decrease in the realized Anet's throughput rate. Depending on the underlying parameters of the communications channel, as well as on the underlying network's connectivity and traffic activity patterns, an optimal Anet-Spatial-Reuse factor, denoted as k_{opt}, can be determined. Based on ongoing observations of system behavior and on instituting a learning system that tracks the system's activity patterns and derives associated modeling parameters, vehicles (possibly aided by their BNs or management stations) could employ dynamic algorithms that are used to calculate the best values to set for **cross-layer control process parameters**, leading to enhanced delay–throughput performance behavior.

2) **V2I Communications**:

a) V2I uplink communications is used by a vehicle to reach an associated RSU, including an AP or a BS node, that connects with a roadside backbone infrastructure such as a wireless cellular network. Such a transmission can be performed by the vehicle across a direct communications link that connects it to an RSU. A vehicle that does not employ such a direct link can use intra-Anet communications to reach another vehicle within the Anet that has been assigned to act as a *V2I-gateway node*. An Anet's backbone node may be assigned to act as a V2I-Gateway. The gateway forwards packets to the RSU by using a V2I communications link, which may be sharing a multiple access medium on an FDMA or TDMA basis.

b) V2I downlink communications from an RSU to Anet vehicles may use a wide-beam antenna at the RSU. A broadcast channel is formed, providing for packet multicasting to Anet clients. An RSU may also configure an antenna array that produces multiple steerable narrow beams. These beams are used for direct transmissions to targeted Anet vehicles. Another option is for the system to use a directional downlink beam for an RSU to communicate with an Anet's V2I-gateway node. The gateway node would then forward the packets that it receives from the RSU to designated Anet clients through intra-Anet communications.

c) **Priority oriented V2I service channels** are configured for the support of essential V2I packet communications. Critical safety packets are provided higher-priority access to communications resources. They may be sharing communications resources that are allocated for their exclusive use.

d) **Control and management V2I channels** are configured to provide for V2I transmission of management, control, and status messages. Such messages are periodically issued by a local management station and are disseminated to highway vehicles. They are sent by an RSU station to associated vehicles across V2I downlink channels. Status messages are also periodically compiled and sent by vehicles to other vehicles and to management and control stations via intra-Anet and inter-Anet V2V and V2I communications.

3) **Backbone (Bnet) Synthesis and Communications**:

a) **Bnet Synthesis**: A Bnet topology synthesis algorithm (BTSA) is used by vehicular nodes to form the Bnet. The algorithm uses system state observations, as collected by each vehicle by exchanging status packets with area vehicles and by receiving such system status data through V2I downlink transmissions issued by local management stations and associated RSUs.

b) The **BTSA** is used to determine the best cross-layer settings of the following: (a) the preferred distance $D = D_{opt}$ between neighboring backbone nodes (BNs). (b) The formation of each Bnet link, which connects neighboring BNs, and the configuration of priority-oriented communications channels across each Bnet link. Frequency bands, time slots, and spatial resources can be allocated to form a Bnet link. (c) The spatial reuse factor M to be employed across the Bnet, determining the spatial distance between Bnet links that employ the same frequency bands or time slots. (d) Use of (intended and interference) signal power-related measurements to calculate the SINR values detected at targeted receivers, the consequent setting of the MCS, the corresponding determination of the realized spectral efficiency levels, and the transmission data rate values to be used across each Bnet link.

c) A BTSA protocol is used to select specific nodes to act as BNs. When feasible, inter-BN distances are set to assume values that are close to the currently calculated $D = D_{opt}$ level. Autonomous algorithms may be employed, so that vehicles self-elect themselves to start acting as BNs and to relinquish this role when it is better served by other vehicles. Also, infrastructure-aided Bnet synthesis protocols can be employed. Roadside management stations track the status and mobility patterns of vehicles that flow across highway lanes and select vehicles that are at preferred positions and at desired distance ranges from each other as BNs. Selections adapt to changes in vehicular speeds, locations, resource availability and mobility patterns.

d) Different methods have been considered for the implementation of BN election algorithms, including the following approaches and considerations.

   i) Under a *road markup* approach, specific geographic locations are identified and announced so that vehicles that are situated close to these locations elect themselves as BNs. A node's role as a BN is updated when it deviates beyond a specified range from a desired marked location.

   ii) Under a relative location approach, vehicles that travel as a platoon or in a group formation (possibly across multiple lanes) mutually engage in the election of BNs through data interactions.

   iii) To enhance the topological stability of the Bnet and improve data communications rates, preference is provided to the election of vehicles that exhibit preferential features to serve as BNs. For example, such vehicles may be stationary (e.g., vehicles parked along the side of the road), move at lower speeds, report higher-quality communications links, and possess higher-energy resources. When feasible, stations that are associated with *highway-embedded* or RSUs, including such that are attached to highway poles or traffic control fixtures, may be selected to manage Anet cells and thus serve as BNs.

e) A BN uses the transport services provided by the Bnet to disseminate packets across multiple Anet cells. It stores incoming packets in priority oriented queues that it forms at its Bnet transmit buffers. Dynamic multiplexing techniques are employed by the BN in sharing its Bnet communication links. The parameters and service policies employed by these schemes are configured at each BN, aiming to best serve the system's packet flow processes in accordance with the quality-of-service (QoS) objectives associated with each packet flow's TC.

f) Flow control and security regulations are applied to protect Bnet (and Anet) resources, enabling the provision of QoS performance guarantees.

g) Parameter settings for system transmission and networking schemes and for the allocation of bandwidth resources are performed on a system-wide basis. Recognizing the ongoing statistics of traffic flow rates and paths, resource allocations aim to avoid the emergence of bottleneck conditions, whereby one segment of a flow's end-to-end route is lacking sufficient wireless communications resources while another segment is allocated excess resources.

21.3.5 Vehicular Backbone Networks (VBNs): Backbone Network Synthesis

To illustrate the VBN concept and the synthesis of a Bnet, we examine the network configuration shown in Fig. 21.11 (as presented in [43]). The VBN networking methods described in the following are based on the developments and studies carried out by Professor Izhak Rubin and by his research collaborators at UCLA and at the Sapienza University of Rome. References to several of these works include: [43, 45], and [12].

As an illustrative example, we consider a Bnet system that supports the following traffic flow scenario. A RSU station, identified as Relay-Node 1, and denoted as RN_1, is the source of a flow of high-priority data packets that it wishes to broadcast to vehicles that move across a segment of a highway. A frequency band has been dedicated across the Bnet for the broadcasting of packets that belong to this flow.

The figure shows the location, at a given point in time, of vehicles moving across a two-lane highway segment. Assuming high traffic loading conditions, a flow control mechanism is employed at the source RSU node. It keeps the allocated bandwidth resource loaded while minimizing the queueing delays that are incurred by disseminated packets at backbone nodes. Using the location of the RSU station as a reference position, vehicles that are located at a targeted distance of about D [m] from each other are set (or self-elected) to act as backbone nodes (BNs), also identified as relay nodes (RNs). Neighboring BNs are interconnected by Bnet links, which utilize the shared Bnet spectral resource.

The network layer routing and forwarding processes that are used to disseminate RSU packets across the Bnet perform as follows. Vehicles are assumed to each use an omnidirectional antenna system. Assume the highway segment to be divided into successive spatial intervals, or segments, where each interval is about D [m] long. Packet forwarding transmissions are scheduled in the following manner. A packet transmitted by RN_1 is set to be successfully received by BN vehicle RN_2. The same transmission is also intended for reception by other vehicles that are located in the first spatial interval. Upon receiving a data packet, forwarding BN vehicle RN_2 schedules its broadcast retransmission. This retransmission is received by nearby vehicles, including nodes that are currently located nearby in the first and second intervals, as well as by forwarding BN vehicle RN_3. Vehicles located in the first interval that have already correctly received this packet would proceed to delete received duplicate copies of the same packet. RN_3 then retransmits the received packet to nodes located nearby in the second and third interval as well as forwarding it to BN vehicle RN_1 that is located in $Anet3$. Continuing this process, packets are routed across the designated stretch of the highway by using the BNs to act as forwarding nodes.

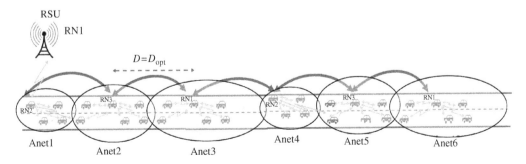

Figure 21.11 A Vehicular Backbone Network (VBN) Configuration in Support of Broadcast of a Data Packet Flow Originating at RSU Node RN1.

As stated above, Bnet links share an frequency band that has been allocated for the exclusive dissemination of the packets that belong to the flow that originates at the RSU node. The scheduling of the flow's packet transmissions across the Bnet links is assumed (for this illustrative network system) to be governed by a **spatial-reuse TDMA** MAC scheme. A realization of this scheme is depicted in Fig. 21.11. It is assumed to employ a *spatial reuse factor* that is equal to $M = 3$. Backbone nodes are time synchronized with respect to the temporal structure of the corresponding TDMA frame, which consists of three slots. The slot duration, T_S [sec], is sized so that it accommodates the transmission of a data packet at a designated data rate, R_D [bps]. Each relay node (acting as a BN) is assigned a specific time slot position in each time frame to use for the transmission of a packet. Under this reuse-3 TDMA scheme, a relay vehicle that is identified as node RN_i is assigned slot-i in each frame for its use. Thus, RSU node $RN1$ is able to transmit a data packet in slot-1 of each time frame. The relay node located in *Anet3* is also denoted in the figure as node $RN1$ and is thus also assigned to transmit a data packet in slot-1 of each time frame. Node $RN2$ is scheduled to transmit a data packet that it holds in its Bnet transmit queue, in slot-2 of each time frame. The latter data packet includes packet-1, the first packet that it has received from node $RN1$, and subsequently received packets. The relay node located in *Anet4* is also denoted as node $RN2$. It is assigned to transmit a data packet that is stored in its own Bnet transmit queue in slot-2 of every time frame. Relay nodes that are labeled as $RN3$ are each scheduled to transmit their packets in slot-3 of each time frame. This way, multiple simultaneous data packet transmissions are conducted across the Bnet, while sharing a common frequency band.

To properly regulate the congestion level that may be incurred by packets transferred across the Bnet, and to assure low packet queueing delays at the BNs that they traverse, a **flow control** mechanism is employed at source node $RN1$. The time intervals imposed between the transmissions by $RN1$ of successive data packets are kept longer than a threshold level. For the dissemination process illustrated in Fig. 21.11, the corresponding time-spacing levels are dictated by the structure of the employed reuse-3 ($M = 3$) TDMA scheme. The maximum flow rate of data packets that are fed across the Bnet circuit is limited to 1 packet per time frame, realizing a normalized throughput rate of $s = 1/3$, or 1 packet per every 3 slots. For example, at the same time that the *third packet of the flow* is transmitted in slot-1 of a frame by the RSU, acting as $RN1$, over *Anet1*, the second packet of the flow is transmitted by backbone node $RN1$ that is located in *Anet3*. The latter transmission is forwarded over the Bnet to RN_2 that is located in *Anet4*. Such transmissions are also received by the Anet clients of the corresponding forwarding BNs. As noted below, the employed modulation/coding schemes and transmission data rates are set to values that enable the successful conduct of these simultaneous transmissions. Due to the pacing method used to set the intervals between transmissions performed by the source RSU, when a packet is received at a BN, it finds the transmit queue to effectively be empty, so that it does not incur queueing delays. In this manner, this flow's data packets incur minimal queueing delays at the relay nodes that they traverse while being disseminated across the Bnet, essentially experiencing just transmission time delays across the Bnet links.

Under such a Bnet scheduling scheme, no packet collisions are incurred and a system-wide high throughput rate is achieved. It is however important to select the transmission data rate level by properly configuring the MCS to assure the ability of intended BN receivers to successfully decode their received packets. For this purpose, the SINR levels incurred at these receivers are used.

To illustrate, consider the simultaneously executed Bnet transmissions performed in slot-2 of every frame by two neighboring RN_2 nodes, such as those illustrated in Fig. 21.11. Busy $RN2$ nodes simultaneously transmit their packets in their associated slot-2 slots. Such packet transmissions across the Bnet are intended for reception by traversed Anet clients as well as for dissemination

21.3 *Vehicular Wireless Networking Methods* | **635**

across the Bnet and reception by their neighboring $RN3$ nodes. The SINR values measured at the latter backbone nodes are calculated by determining the ratio of the received power of the intended signal, denoted as P_r, and the power level measured at this receiver that is equal to the sum of the noise power, denoted as P_N, and of the interference signal power, denoted as P_I. To illustrate, consider a single exponent Physical-layer Loss Model (PLM), so that the signal power is attenuated over distance D in proportion to $\frac{1}{D^\alpha}$, where α is an attenuation exponent, whose typical value lies between 2 and 4. The power of the intended signal measured at a receiving backbone node, such as at node $RN3$ that is located in the third interval, is equal to $P_r = C\frac{P_t}{D^\alpha}$, where C is an attenuation parameter, and P_t is the transmit power level.

In presenting a simplified expression for the evaluation of the interference power, assume that the main component of the interfering signal is contributed by a simultaneously transmitting backbone node that is situated closest to the intended receiving node. For the illustrative case, consider a packet transmission by RN_2 across a backbone link whose intended recipient is backbone node RN_3. The closest interferer impacting this reception is another transmitting RN_2 backbone node that is situated at a distance of about 2D away from the intended receiver. In general, if a reuse-M spatial-TDMA scheme is used, the closest interferer is located at a distance that is equal to about $(M - 1)D$, where the distance between neighboring BNs located across the Bnet is assume to be equal to D. It is noted that the next key interferer could be located on the upstream side of an intended receiver at distance $(M + 1)D$. Accounting only for the closest interferer, the corresponding approximate value of the SINR level measured at a typical BN receiver is calculated by using the following formula:

$$SINR = \frac{\frac{CP_t}{D^\alpha}}{P_N + \frac{CP_t}{((M-1)D)^\alpha}}. \tag{21.1}$$

For systems that commonly employ inter-BN distances (D) that are relative short, the signal power measured at a receiving vehicle is dominated by interference power (i.e., by the message signals that are produced by simultaneously transmitting backbone nodes). The transmission operation is then said to reside in an *interference-dominated mode*. In this case, neglecting the impact of the background noise component, P_N, and assuming each BN to use the same fixed transmit power level, we note the SINR value to be of the order of $(M - 1)^\alpha$, so that it is not sensitive to the value of D. Under $P_N > 0$, the SINR is noted to be monotonically decreasing with D. As the inter-vehicular distance D increases to a level under which the backbone communications link's transmission process resides in a *noise-dominated mode*, the contribution of background noise becomes critical, and a higher SINR level is then achieved by reducing the inter-vehicular D level, as it leads to a corresponding increase of the received intended signal power level.

To achieve a performance efficient operation, a **cross-layer** method is used to set the parameters of the VBN system. Key system performance metrics include the network's throughput rate (f_{TH}), whose maximum value is identified for the underlying illustrative broadcast operation as the system's **broadcast capacity (C_B)**, and the packet's dissemination time delay (D_P [sec]) across the identified segment. Under a reuse-M spatial-TDMA scheme, the maximum attainable throughput rate is equal to $f_{TH} = 1/M$ [packets/slot], as the RSU source node is regulated to drive the system at a rate of 1 packet every time frame (which is M slots long). Assuming a packet length of L_P [bits/packet], the packet transmission time (and slot duration) across a Bnet link is equal to $T_S = L_P/R_D$, as the slot duration is set as the time that it takes to transmit a packet at the configured transmission data rate, which is set to R_D [bits/s]. Hence, the broadcast throughput capacity rate is expressed as $f_{TH} = C_B = L_P/(MT_S) = R_D/M$ [bits/s].

For the underlying system, the time latency incurred in the dissemination of a data packet across the Bnet is proportional to the number of backbone nodes that it traverses. For a dissemination backbone path that spans a distance of L [m], the number of backbone relay nodes is of the order of $N_R = L/D$. Since each relay node is set to transmit a single packet during each time frame, the packet's dissemination latency is of the order of $D_P = N_R M T_S = L M T_S/D$.

The backbone system's **Transport Throughput (C_T)** metric is used to account for both throughput and delay metrics. It is defined as a measure that is proportional to the ratio of the system's throughput rate and the incurred packet's time delay. Thus, it is calculated as the ratio $C_T = C_B/D_P$. Using the relationships noted above, we express the transport throughput metric as $C_T = D(C_B)^2/(LL_P)$. Its assumed values are thus proportional to $D(C_B)^2$. For the system under consideration, as the inter-BN distance (D) level is decreased the system's broadcast throughput rate (C_B) is noted to increase, while the packet's dissemination delay increases as well.

As noted above, the attained backbone system's broadcast throughput rate (C_B) is proportional to R_D/M. Under a prescribed bandwidth level, the highest attainable Bnet link data rate R_D is determined by the SINR levels measured at the intended BN receivers (and the prescribed packet error rate). The realized SINR value depends on the setting of the system's inter-BN distance D and the Bnet's reuse level (M). For a single exponent PLM, we can note that higher broadcast throughput rates are achieved when lower D levels are configured. Using a more detailed two-exponent PLM, the results shown in Fig. 21.12(a) demonstrate the optimal (D_{opt}) inter-BN distance values to be set under several M levels. See also [43, 45], and [12].

When considering the performance behavior of the transport throughput rate (C_T), it is noted that it is not advantageous to set a too low D level, as the latter would induce higher packet dissemination latency across the Bnet. This is confirmed by observing the performance curves for the transport throughput metric that are exhibited in Fig. 21.12(a).

It is further noticed that the involved cross-layer parameters of the Bnet system, which include the spatial reuse index (M), the inter-BN distance (D), the MCS, and the associated data rate (R_D), must be jointly configured.

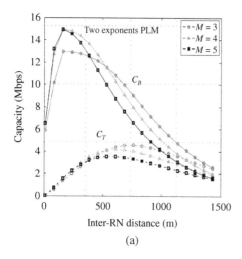
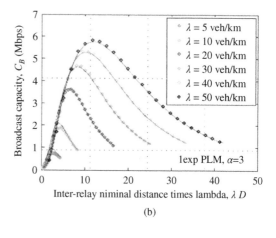

Figure 21.12 Performance Behavior of a Backbone Network Synthesized for a Vehicular Backbone Network (VBN): (a) Broadcast Throughput Capacity (C_B) and Transport Throughput Capacity (C_T) Performance Curves vs. the Inter-BN Distance (D) Under Several Reuse-M Levels; (b) Broadcast Throughput Capacity Curves vs. the Product of the Vehicular Traffic Density Rate and the Inter-BN Distance (λD).

21.3 Vehicular Wireless Networking Methods | **637**

The performance curves presented in Fig. 21.12(a) illustrate the merit of properly setting the cross-layer configuration. Reuse level M values that are set to 4 or 5 yield a higher peak throughput rate, in comparison to a setting of $M = 3$, when jointly configuring the proper levels for the D and R_D parameters, and accordingly the adjusted MCS parameters. As noted above, the setting of a higher M value serves to throttle the data flow's packet input rate and reduce its realized throughput rate, whose level is proportional to $1/M$. In turn, a higher reuse-M level causes simultaneously transmitting backbone nodes to be more spatially distant, inducing lower interference signal power at targeted BN receivers. By configuring the proper cross-layer settings, one can affect the backbone system's performance in a significant way. For example, a value of $M = 5$ yields a high-throughput rate when D is equal to about 100 m, while for higher or lower-values of D, such as 50 or 800 m, a much lower throughput rate is realized. Under $D = 800$–1200 m, a setting of $M = 3$ yields a higher-throughput rate. For such longer distance (D) levels, the $1/M$ throughput factor becomes a more dominant component, as the signal interference factor is then not as critical.

Observing the results presented in Fig. 21.12(a), it is noted that the lower reuse level, set as $M = 3$, yields higher transport throughput values. The choice of such a lower M level enhances the throughput in proportion to $1/M$ while increasing signal interference and consequently lowering the Bnet link's data rate. We have noted that the packet's delay level is impacted by a factor that depends on the M/D ratio. By applying Eq. 21.1, we observe that when the Bnet link's channel resides in interference dominated mode, as is the case under lower D values, the SINR level at a targeted receiver is not very sensitive to the inter-vehicular D value.

Under longer D values, a lower interference power is detected at an intended receiver. The ensuing reduced number of relay nodes (N_R) leads to a lower number of dissemination hops. The system operates in a noise dominated mode, so that the SINR level recorded at the intended receiver is reduced as D increases, leading to lower-throughput rate (C_B). A higher D value leads to reduced packet delay, yielding transport throughput rate (C_T) that is not a sensitive function of D as that noted for the throughput rate metric. Also, the setting of a lower M level (such as $M = 3$) reduces the transit time per hop (which consists of M slots), leading to improved C_T levels.

The results presented in Fig. 21.12(a) have been derived under the assumption that the vehicular traffic density, denoted as λ [vehicles/m], along a lane of the underlying segment of the highway, is sufficiently high. In this case, the system can always find vehicles to serve as backbone nodes such that they are located at (or close to) the desired inter-BN distance (D). In this case, nearly fixed inter-BN distance values that are equal to the optimal levels can be configured. In turn, when vehicles are not regulated, or their moving positions are set autonomously, it may not be possible to guide them to travel in formations that meet targeted inter-vehicle spacing.

Assume therefore vehicles to incur random inter-vehicle distances that follow a probability distribution, with an average inter-vehicular spacing of $1/\lambda$ [m]. In this case, it will still be feasible to elect BN vehicles that are at a spacing of approximately D [m] provided the product λD is sufficiently high. This product expresses the average number of vehicles that travel along a lane in a spatial interval of length D. For lower λD product values, the distances incurred between neighboring backbone nodes will exhibit higher random variations, degrading the realized broadcast throughput rate.

For the underlying system, the achievable broadcast throughput capacity rate (C_B) is calculated as follows. Assume the system to be able to select BNs so that neighboring BNs are at fixed distance D from each other. Let the Bnet's link-i (denoted as BL_i) identify the ith communications link connecting neighboring BNs. Let the throughput capacity rate attained by packets communicating across link-i be denoted as C_i. It represents the maximum data rate at which packets can be transmitted across the ith wireless backbone link, when this link is made available for such transmissions on a full time basis. As noted above, a link's maximum throughput rate is calculated by determining the SINR value recorded at its receiving BN, as illustrated by the calculation presented in Eq. 21.1. For example, using *Shannon's Channel Capacity Formula*, an upper bound on the channel's capacity level is expressed as:

$$C_i = (1/W)[\log_2(1 + SINR_i/\gamma)], \tag{21.2}$$

where W is the backbone link's channel bandwidth that is allocated to the circuit that is used to support the underlying flow, $SINR_i$ is the SINR value measured at the receiving backbone node that terminates the ith backbone link, and $\gamma > 1$ is an implementation inefficiency factor.

When inter-BN distances randomly fluctuate, it is noted that across different backbone links, BN receivers will be measuring different SINR levels, inducing variable backbone link capacity data rates. Noting that the underlying Bnet is configured as a *tandem chain of backbone links*, we conclude that the flow's throughput capacity rate is limited by the throughput capacity rate of the link (or links) that attain the lowest such level. It is denoted as C_{min}, and a link that exhibits such a low level is identified as a **bottleneck backbone link**. Under a reuse-M TDMA scheme, the source node is regulated so that it feeds into the network a single packet during each (M slot) time frame. Hence, the attained system's broadcast throughput capacity rate C_B is given by the following expression:

$$C_B = (1/M)\min_i C_i = (1/M)C_{min}. \tag{21.3}$$

The corresponding performance behavior of such a Bnet system is exhibited in Fig. 21.12(b), where a single exponent physical channel loss model (PLM) is assumed, and vehicles are situated across a single lane at locations that are governed by the statistics of a spatial **Poisson Point Process**, at a vehicular traffic density of λ. Under a higher λD product value, the highway's vehicular traffic density is relatively high, so that typically a pair of neighboring BNs could be elected to be at a distance from each other that is quite close to the desired D level, which yields high-performance behavior. In turn, under lower highway vehicular density levels, and thus lower λD product values, the distance between elected neighboring backbone nodes would be more variable, leading to performance degradation.

The importance of properly setting the parameters of the cross-layer scheme is noted again. The magnitude of the fluctuating distances between neighboring vehicles moving along a lane impacts the variability of inter-BN link spans, resulting in a more frequent occurrence of more restrictive bottleneck backbone links, which leads to lower-throughput capacity rate (C_{min}).

21.3.6 Infrastructure-Aided Vehicle-to-Vehicle (V2V) Networking

V2V and V2I communications can be enhanced by using the data transport services provided by an infrastructure-based backbone communications network. In the architecture illustrated in Fig. 21.13, vehicles moving along the highway are dynamically grouped to form multiple cells. As described under the VBN architecture, these clusters form **Anets**. An Anet may include the following relay nodes: a **V2V Backbone Node (V2V_BN)** and a **V2I Gateway Node (V2I_GN)**. As employed in a VBN system, V2V backbone nodes are interconnected by wireless links to form a VBN Bnet (a V2V_Bnet) that is used for longer-range V2V wireless communications. A V2I gateway node provides for (typically two-way) communications with an infrastructure access station, such as an RSU, an AP, or a BS of a cellular wireless network.

In Fig. 21.13, the V2I gateway node associated with $Anet_i$ is denoted as GN_i. GN_i may be configured as the manager of $Anet_i$, as well as act as a VBN-type backbone node (V2V_BN). V2V_BN

21.3 Vehicular Wireless Networking Methods

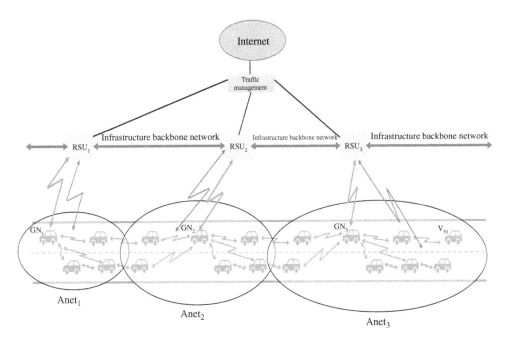

Figure 21.13 A VBN-Based V2V Network That Is Aided by the Use of an Infrastructure Core Network.

and V2I_GN stations may generally be associated with distinct cells. It is desirable to dynamically select vehicles that are located in positions that enable high-quality V2I communications to act as V2I gateway nodes. For vehicle communications over a longer distance, it is preferred to use long-ranging backbone paths. While long backbone paths are stable, as they use stationary links, long V2V paths are liable to experience link disconnects that are induced by movements of intermediate vehicles.

Vehicles that reside in the same Anet may communicate with each other by using VANET based V2V networking methods, without making use of the services provided by the core Bnet. A V2V multihop communications networking protocol can be used for a vehicle to communicate with a close-by vehicle such as one that is located in a neighboring cell. In turn, to communicate with a vehicle that is located farther away, it is preferable for a vehicle to use the transport service provided by the Infrastructure Backbone Network (IBN), or by a V2V_BN, or a hybrid of both.

For the network system shown in Fig. 21.13, we assume each Anet to include a gateway node that offers high-quality V2I communications. A vehicle located in an $Anet_i$ cell communicates its packets by using a V2V path across its Anet to reach its GN_i gateway. The latter transmits the packets to its associated backbone access station, denoted as RSU_i. In the case that such a gateway node cannot be configured for a certain Anet, as there may not be a near-by RSU node or the communications paths to the IBN may be degraded or blocked, a source vehicle may use an intra-Anet V2V wireless path to send its packets to a local V2V_BN. The latter may then use a VBN-oriented V2V_Bnet to send the packets to an Anet cell that includes a capable gateway node. The use of VBN-based Bnet communications can be avoided for implementations that assure high-quality V2I communications links.

Following the path taken by a data packet that travels a longer distance, we note the following. Upon reception at a gateway node, the latter transmits the packet to the IBN via its V2I link. A unicast packet is routed across the IBN to its destination RSU. The later transmits the packet via a V2I downlink to a gateway node that is located in the destination Anet's cell. The gateway node then uses an intra-Anet V2V wireless networking path to reach the destination vehicle. If the destination cell does not contain a V2I gateway, the packet is sent to a reachable near-by cell and is then directed to the destination cell via V2V means.

A packet that is targeted for **multicasting** over a specified region, such as a safety-oriented message that is multicasted to all vehicles located over a specified range from the source vehicle, is broadcast across the downlink by each RSU that it traverses across the IBN. An RSU uses for this purpose a wide-angle antenna beam, so that a transmission of a single packet can be simultaneously received by all targeted Anet client nodes. Alternatively, it can be directed to corresponding cell gateways for V2V broadcasting across targeted cells.

An advantageous implementation over a busy metropolitan highway is expected to employ a large number of closely located access stations. In this case, communications flows can be disseminated mostly through the IBN, supplemented when needed by the use of V2V paths. When most mobiles are able to directly access the IBN, so that each is acting as its own V2I gateway node, as targeted for use in advanced CV2X systems, the occurrence rate of V2V wireless flows will be much reduced. Yet, V2V flows are expected to still be essential under a variety of system architectural, environmental and layout scenarios, such as those that involve country-side roads. They also play an essential role in providing alternate paths for systems that have been affected by infrastructure failures and jams.

Using vehicular connectivity means, it is advantageous to have vehicles organize themselves into **vehicular platoons**. A platoon consists of a group of vehicles that move along the highway as a coherent unit, at specified speed and fixed inter-vehicular spans. Illustrative platoon formations are shown in Fig. 21.14. In this system, vehicles that move across a lane have been organized into platoons. Platoon teams that consist of vehicles that move cohesively along multiple lanes can also be formed. As will be noted when discussing autonomous vehicular highway systems, platoon formations are useful in managing and regulating autonomous vehicular flows.

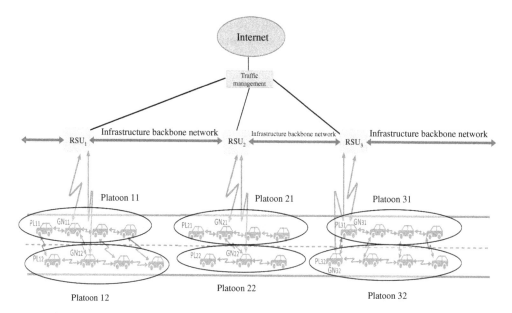

Figure 21.14 Infrastructure-Aided Vehicular Network Using Platoon Formations.

Effective vehicular networking methods can be designed by taking advantage of platoon formations. As illustrated in Fig. 21.14, specific vehicles within a platoon can be assigned to perform specific functions. A **Platoon Manager (PM)** is responsible for organizing and managing its platoon formation, as well as for disseminating data communications networking parameters to its members. The vehicle that moves at the head of a platoon is identified as its **Platoon Leader (PL)**. A platoon vehicle that serves as a **Platoon Gateway Node (GN)** provides for V2I communications, serving to access an infrastructure access node (IAN). The latter can serve as, or attach to, a cellular BS, a Wi-Fi AP node, or another station. A regional manager collects status data concerning the vehicular system, its traffic flows, safety and security conditions and parameters, and environmental conditions and disseminates this information regionally to highway vehicles either by using direct V2I means or by forwarding it through gateway nodes. A vehicle can be selected to act in multiple such roles.

As illustrated in Fig. 21.14, vehicular highway networking methods can advantageously capitalize on the formation of vehicular platoons. Communications between vehicles that are members of the same platoon, as well as with vehicles that are members of a nearby platoon (such as between vehicle members of Platoons 11 and 12), can be carried out by using V2V methods. Associated forwarding algorithms and protocols can be managed by PMs, RSUs, and/or autonomously by the vehicles themselves. For longer distance communications, the gateway node of the source platoon is used to relay packets to an IAN, which then disseminates the packets across the IBN. The disseminated packets are transmitted between IANs. An IAN can broadcast the packets that it receives across a downlink communication channel, aiming them to reach targeted client vehicles. It can also use unicast transmissions to deliver them to a Gateway Node, which will then disseminate them locally to targeted vehicles. Gateway nodes and/or direct V2V links can be used to disseminate packet flows across platoons that use distinct highway lanes. A platoon can employ multiple gateway nodes, reducing V2V traffic loading rates.

It is noted that while sub-6 GHz frequency bands have been commonly employed, advanced and future developments are expected to make more use of mmWave frequency bands, such as using spectral resources around 60 GHz. Signal transmissions at such a high frequency are subjected to much higher-power attenuation effects. They are mostly effective when used across a line-of-sight link. Such mmWave-based systems offer significantly broader frequency bandwidths, which is highly important when ultrahigh data rates are involved, such as those required when sharing detailed maps or environmental views that are used for navigating vehicles in an urban environment. Enhanced spectral efficiency levels are attained by forming directional antenna beams. In reviewing the layout shown in Fig. 21.14, it is noted that such directional beams are effectively used for ultra high speed communications between neighboring vehicles that belong to a single platoon (and possibly also neighboring platoons) as well as for V2I communications between a vehicle and an associated RSU. While directional intra-platoon links can be maintained over longer periods of time, for as long as platoon vehicles move as a cohesive team, antenna beam directionality must be periodically readjusted for V2I links, at a rate that depends on the width of the antenna beam and the speed of the vehicle. Infrastructure-aided vehicle network systems that make use of a cellular wireless network infrastructure, identified as CV2X systems, are discussed in Section 21.3.7.

21.3.7 Cellular Vehicle-to-Everything (CV2X) Networking

Methods for providing V2X data communications through the use of cellular wireless networks, known as **Cellular Vehicle-to-Everything (CV2X)**, have been developed by the

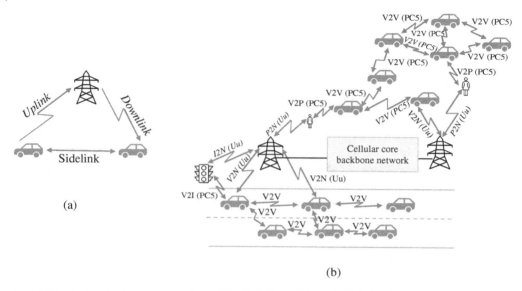

Figure 21.15 Cellular V2X Configurations: (a) Uplink, Downlink, and Sidelink; (b) A CV2X Network Configuration Showing Infrastructure Access (Using the Uu Interface) I2N Links and Sidelink Based (Using the PC5 Interface) V2V, V2I, and V2P Links.

Third-Generation Partnership Project (3GPP). 3GPP Release-16 specifications include V2X services, identified also as **5G NR V2X**, employing the 5G New Radio (NR) air interface. Included are performance requirements and mechanisms that are useful for connected and automated driving systems. In addition to employing V2I communications links that connect vehicles with the cellular infrastructure network by using corresponding uplinks and downlinks, the concept of V2V communications through the use of sidelinks is employed. As shown in Fig. 21.15(a), a **Sidelink (SL)** provides for direct communications between vehicles, which are represented as User Equipment (UE) entities, without going through the Bnet, and thus not using uplink and downlink communications. Ongoing enhanced V2X (eV2X) developments provide enhanced support for the implementation of connected and automated driving systems. The following overview of CV2X networking methods includes (nonexhaustive) descriptions that are provided in related 3GPP Standards documents and in related tutorials and papers, such as [8].

Spatially, CV2X systems make use of the following networking paths: (1) An uplink path carries message transmissions from a mobile user to its associated BS. Multiple access algorithms and protocols are used to dynamically share uplink resources of a RAN. (2) A downlink path carries message transmissions from a BS to a mobile user that resides in its cell. Multiplexing algorithms and protocols are used by the **BS** to dynamically share the downlink resources of a RAN. (3) A sidelink is used to provide for direct communications between mobiles. Algorithms and protocols are employed to regulate the sharing of wireless resources that mobiles use for sidelink transport, without using the communications resources offered by the system's uplinks and downlinks and without making use of transport resources maintained by the cellular backbone infrastructure. A depiction of such paths is shown in Fig. 21.15(a) and (b).

Transmissions across the uplink and downlink RAN of the cellular wireless system are based on a reservation scheme, under which wireless media resources are dynamically allocated by the BS over the joint frequency/time/spatial dimensions. A busy mobile sends a request for resource allocation to its associated BS across an uplink signaling channel. When available, the BS would then send the requesting mobile, across a downlink signaling channel, an allocation message that specifies the resources assigned to the requesting user, aiming to meet, when specified, the requesting mobile's QoS performance requirements.

V2X communications were introduced under 3GPP Releases 14 and 15, using the LTE air interface. LTE V2X was set in various markets to use the 5.9 GHz frequency band reserved for ITS services. LTE V2X services include basic cooperative active traffic safety, traffic management, and telematics applications. Typically, such message flows require the broadcast transmission of short awareness packets, such as Cooperative Awareness Messages (CAMs) or Basic Safety Messages (BSMs). These messages include information about a vehicle's location, direction, and speed. LTE V2X systems implement new PHY (Physical) and MAC layers for V2X applications. They reuse the upper V2X layers and protocols specified by Standards organizations such as ETSI (European Telecommunications Standardization Institute), IEEE (Institute of Electrical and Electronic Engineers), and SAE (Society of Automotive Engineers).

For V2X Sidelink (SL) communications, LTE V2X defines two new resource allocation modes, mode 3 and mode 4. Under mode 3, communications is managed by the BS (identified as eNB), as it assigns time/frequency sub-channel resources. Using a DA/TDMA/FDMA-based **dynamic scheduling** scheme, vehicles send reservation packets to their associated BSs, requesting the assignment of sub-channels for the transmission of each Transport Block (TB). To improve performance efficiency, a **semi-persistent scheduling (SPS)** algorithm can be employed. Under SPS, the BS assigns sub-channels to a vehicle for use by the vehicle for the transmission of several TBs. The BS specifies the periodicity of each sub-channel, which determines the number of time slots allocated to the user during a certain number of time periods, over a specified frequency band. The BS informs the details of this specification to the vehicle by sending it over a *Physical Downlink Control Channel*.

Under mode 4, vehicles can be out of cellular network coverage. They autonomously configure and manage the sub-channels that they will use. A sensing-based SPS scheme was specified in Rel. 14/15. A vehicle uses this scheme to select sub-channels, announcing the selection of transmission resources that it will use during the next TB time frame. A vehicle that selects sub-channel(s) at subframe t, informs neighboring vehicles that they are reserved for its use within subframe $(t + RRI)$, where RRI denotes a specified Resource Reservation Interval. Through such an advanced reservation announcement, the vehicle aims to avoid simultaneous selections of the same time/frequency/space resource by several vehicles. RRI values are selected from a specified list, such as one that contains the values 0, 20, 50, 100 ms, or any multiple of 100 ms up to a maximum value of 1000 ms. A vehicle sets the RRI equal to 0 ms to announce to its neighboring vehicles that it is not reserving the same sub-channels for its next TB time frame.

A Reselection Counter is maintained and is decremented by one after transmitting a TB. A vehicle must select new sub-channels with probability $(1 - p)$ when the Reselection Counter is equal to zero, where $0 \le p \le 0.8$. New sub-channels must also be selected if a new TB does not fit in the previously reserved sub-channels or if the current reservation cannot satisfy the latency deadline of a new TB. In selecting new sub-channel(s), vehicles use a sensing-based SPS algorithm. Under this protocol, the vehicle excludes resources that are being used as well as candidate resources that it estimates will be used by other vehicles, yielding a first list of candidate resources. The vehicle also creates a second list, which consists of a subset of the candidate resources included in the first list, by selecting those resources that experience the lowest average Received Signal Strength Indicator (RSSI) values measured over a sensing window. The vehicle randomly selects one of the candidate resources included in the second list. This way, the selection is expected to induce a lower probability of simultaneous selection of a resource by several vehicles. The selected candidate resource is used by the vehicle to transmit its new TB. The vehicle maintains the selection for its next (Reselection-Counter)-1 transmission.

Though no specific congestion control scheme is specified, it is essential to regulate the transmission rate by active vehicles in a manner that will reduce the probability of packet collisions. For this purpose, an active vehicle keeps track of system activity indicators, including *Channel Busy Ratio (CBR)* and *Channel Occupancy Ratio (CR)*. The CBR metric is defined (in considering sub-channels in a resource pool) as the ratio of the number of sub-channels that experience a Received Signal Strength Indicator (RSSI) higher than a preset threshold to the total number of sub-channels used in the previous 100 sub-frames. The CR is a metric that represents a ratio of the total number of sub-channels that are used (over a specified number of sub-frames) by a specific transmitting vehicle. A vehicle regulates its transmissions to not surpass a specified maximum CR value.

LTE V2X supports basic active safety and traffic management use cases, while 5G NR V2X supports advanced use cases and higher automation levels. NR-V2X has been designed to coexist and support interworking with LTE-V2X. The 5G NR V2X 3GPP use cases are classified into the following key groups:

1) **Vehicles Platooning**: Mechanisms for the dynamic formation of groups of vehicles into platoons and for platoon management. Vehicles in a platoon exchange data periodically to ensure the correct functioning of the platoon. The inter-vehicle distance between vehicles in a platoon may be determined in relation to specified QoS objectives.
2) **Advanced Driving**: Enabling semi-automated or fully-automated driving. Vehicles share data obtained from their local sensors with area vehicles. Vehicles also share their driving intentions, serving to coordinate routes and driving patterns and maneuvers.
3) **Extended Sensing**: Sensor data flows are exchanged among vehicles, RSUs, pedestrian devices, and V2X application servers.
4) **Remote Driving**: Enabling a remote driver or a V2X application to operate a vehicle. Aiding of safe automatic driving through complex terrains or environments.

The 5G Automotive Association (5GAA) connects telecom industry and vehicle manufacturers in aiming to develop end-to-end solutions for future mobility and transportation services. A 5GAA Working Group responsible for specifying Use Cases and Technical Requirements has defined a classification that includes the following performance objective attributes: Safety; Vehicle operations management; Convenience; Autonomous driving; Platooning; Traffic efficiency and environmental friendliness; Society and community.

Performance requirements specified by 3GPP per each one of the following operations have been noted as:

1) **Vehicle Platooning**: Data throughput rates are not high, data packet transport latency should be low and the packet delivery rate should be high. For example, end-to-end packet latency should be of the order of 10–25 ms, and dissemination reliability should be of the order of 90–99.99%. Message length is of the order of 50–6000 Bytes and the flow rate is of the order of 2–50 [packets/s]. The required data rate is typically not higher than 65 Mbps. The communications range is of the order of 80–250 m.
2) **Advanced Driving**: End-to-end packet latency should be of the order of 10–100 ms; dissemination reliability should be of the order of 90–99.999%. Message length is of the order of 1600 Bytes and the flow rate is of the order of 10–100 [packets/s] across the SL and about 50 [packets/s] across the uplink. The required data rate is around 10–50 Mbps for the SL, 0.25–10 Mbps for the uplink, and 50 Mbps for the downlink. The communications range is of the order of 360–700 m.

3) **Extended Sensing**: End-to-end packet latency should be of the order of 3–100 ms; dissemination reliability should be of the order of 90–99.999%. Message length is of the order of 1600 Bytes and the flow rate is of the order of 10 [packets/s]. The required data rate is around 10–1000 Mbps and the communications range is of the order of 50–1000 m.

4) **Remote Driving**: End-to-end packet latency should be of the order of 3–100 ms; dissemination reliability should be of the order of 90–99.999%. Message length is of the order of 1600–41,700 Bytes and the flow rate is of the order of 5 [packets/s]. The required data rate is around 25 Mbps for the uplink and around 1 Mbps for the downlink. The required communications range is 1000 m or higher.

The 5G system architecture supports V2X communication over the PC5 interface and over the Uu interface. The *PC5 interface* supports SL V2X communications for New Radio (NR) and LTE. V2X communications over *interface Uu for uplink (UL) and downlink (DL)* transmissions can be carried out under NR Non-Standalone (NSA) and Standalone (SA) deployments. The 5G System (5GS) consists of the Next-Generation Radio Access Network (NG-RAN) and the 5G Core network (5GC) domains. The 5GC provides several Network Functions (NF) such as the Access and Mobility Management Function (AMF), Policy Control Function (PCF), Network Data Analytics Function (NWDAF), Network Repository Function (NRF), Network Exposure Function (NEF), Unified Data Repository (UDR), Unified Data Management (UDM), Session Management Function (SMF), and User Plane Function (UPF), among others. V2X applications over the end user (UE) interface are included in the Application Server (AS). Application layer information and configuration parameters that are useful for the UE to configure its V2X communications are exchanged over Application layer interfaces.

At the Physical Layer, the following *frequency bands* are utilized. Rel. 16 NR V2X sidelink operates (as for Rel. 15 NR Uu) over the following frequency ranges: 2. Frequency range 1 (FR1): 410 MHz–7.125 GHz; and Frequency range 2 (FR2): 24.25–52.6 GHz. Transmissions use the orthogonal frequency division multiplexing (OFDM) waveform. Resources are allocated from the available time–frequency joint space. The SL channel is organized in frames, each with a duration of 10 ms. A radio frame is divided into 10 subframes, each with a duration of 1 ms. The number of slots per subframe and the subcarrier spacing (SCS) for the OFDM waveform are flexible. Only certain slots are set to accommodate SL transmissions. A SL transmission can use one or multiple sub-channels, selected from allocated resource pools (RPs). A resource pool (RP) can be shared by several users (UEs) for their SL transmissions. A UE can be configured with multiple RPs that it can use for its transmissions and with multiple RPs for use in reception. A UE can receive data on resource pools used for SL transmissions by other UEs, while it can transmit across the SL by using its own transmit resource pools.

The following physical layer channels are defined: (1) Physical Sidelink Control Channel (PSCCH), which is used to carry control information in the sidelink. (2) Physical Sidelink Shared Channel (PSSCH) carries data payload in the sidelink and additional control information. (3) Physical Sidelink Broadcast Channel (PSBCH) carries information for supporting synchronization in the sidelink. PSBCH is sent within a sidelink synchronization signal block (S-SSB). (4) Physical Sidelink Feedback Channel (PSFCH) carries feedback data related to the successful or failed reception outcome of a sidelink transmission. While LTE V2X supports only broadcast transmissions in the sidelink, NR V2X provides physical layer support for unicast, groupcast (which is similar to network-layer based multicast except that it refers to application layer based groups), and broadcast transmissions in the SL.

The LTE QoS model that is based on the use of QoS class identifiers (QCIs) is employed for specifying performance metrics for LTE-based V2X communications. QoS of SL communications for LTE V2X functions is managed on a per-packet basis. The Application Layer protocol entity associates V2X packets with a priority level values, denoted as PPPP in identifying Proximity service Per-Packet Priority, and optionally reliability, denoted as PPPR, identifying Proximity service Per-Packet Reliability. These values are used by the implementer to configure the related SL logical channels.

QoS performance requirements are specified for NR V2X communications. Configured parameters include *transmission rate, latency, reliability, data rate, communication range and priority level.* QoS management in NR V2X SL is configured by the network and is based on the 5G QoS model, where QoS requirements are associated with QoS flows. A *QoS flow is attached a specified QoS Profile that is based on its V2X application.* Procedures are defined for the support of **service continuity** in reaction to QoS variations. As noted in [5], a **QoS Flow** is defined by 3GPP as "the finest granularity for QoS forwarding treatment in the 5G System. All traffic flows mapped to the same 5G QoS Flow receive the same forwarding treatment (e.g., scheduling policy, queue management policy, rate shaping policy, Radio Link Control (RLC) configuration). Providing different QoS forwarding treatment requires separate 5G QoS Flow." Packet flows are mapped to QoS Flows at the V2X layer based on the QoS requirements associated with the underlying application.

Each QoS Flow is identified by a *PC5 QoS Flow ID (PFI).* QoS flows are mapped to **Sidelink Radio Bearers (SLRBs)** using rules provided to the end user (UE) by the network. The radio bearer provides a layer-2 service for the transfer of data. Each rule specifies the associated configuration and parameters of layer-1 and layer-2 operations. The mapping from QoS Flows to radio bearers is performed at the Service Data Adaptation Protocol (SDAP) layer. For unicast data networking, the SLRB is established by a UE with its peer node, configuring all parameters of the Access Stratum layer, including those involving Packet Data Convergence Protocol (PDCP), Radio Link Control (RLC), MAC, and PHY layers. Procedures are also defined for the allocation of resources for the configuration of SLRBs that are used for groupcast and broadcast communications.

SL QoS flows are characterized by QoS parameters and features that are referred to as QoS Profiles. For each cast type (i.e., unicast, broadcast, and groupcast), the UE maintains the mappings of PFIs (PC5 QoS Flow ID) to the PC5 QoS parameters and PC5 QoS Rules per destination. Under Rel. 16, the following PC5 QoS parameters that are associated with PC5 QoS Flows are defined. A PC5 5G NR Standardized QoS Identifier (PQI) is used for referring to specific PC5 QoS characteristics associated with V2X services. A flow bitrate parameter set is specified for a QoS Flow with guaranteed bit rate (GBR). Included are parameter values for Guaranteed Flow Bit Rate (GFBR) and Maximum Flow Bit Rate (MFBR). For a non-GBR link, an aggregate bit rate level is specified, accounting for all non-GBR flows using the link. For groupcast communications, a distance range parameter is specified, representing the minimum range covered between the transmitting and receiving UEs, for which the underlying QoS parameters are guaranteed.

The following PC5 QoS characteristics are identified by the PQI: (1) Resource Type, classified as GBR, Delay-critical GBR, or non-GBR. GBR flows require the dedicated allocation of network resources. Delay-critical GBR flows require specific Packet Delay Budget and Packet Error Rate specifications. Also specified is the Maximum Data Burst Volume (MDBV). (2) Priority Level. (3) Packet Delay Budget (PDB), which sets an upper bound on the packet delay. For GBR QoS flows, 98% of the packets are required to not experience a delay over the PDB value. For such flows, a packet that is delayed for longer than the specified PDB is counted as lost and included in the PER metric value (unless the GFBR or MDBV is exceeded). Services using Non-GBR QoS flows can experience packet drops and delays during network congestion states. In un-congested situations, 98% of the packets should not experience a delay over PDB. (4) Packet Error Rate (PER) serves as an upper bound on the error rate of packets processed by the RLC layer of a transmitting user. Packets that are not successfully delivered by the receiving user to its PDCP layer are noted. (5) An Averaging Window (for GBR and Delay-critical GBR resource types only) specifies the duration over which the GFBR and MFBR metrics are calculated. (6) Maximum Data Burst Volume (for Delay-critical GBR resource type only), MDBV, denotes the largest amount of data that the PC5 link is required to accommodate, within a specified PDB period, for QoS flows that are served under Delay-critical GBR.

PC5-RRC (Radio Resource Control) in LTE V2X is used to exchange synchronization information between UEs on the Sidelink Broadcast Control Channel (SBCCH). PC5-RRC comprises RRC protocol and signaling message units that are transported over the protocol stack of the Access Stratum (AS) PC5 control plane. It is composed of the RRC, PDCP, RLC, MAC, and PHY sublayers. PC5-RRC is defined and used in NR V2X to provide functionalities for the support of SL unicast communications. It includes provisions for exchanging AS (Access Stratum)-level information that transmitting and receiving end users require for supporting unicast communications.

Release-16 also supports *service continuity in NR V2X communications over Uu*. If the initial QoS profile is not supported, the use of a specified alternative QoS profile is provided. The related application traffic is able to adapt to a transport service that meets specified alternative QoS metrics. While QoS profiles over PC5 are denoted as PQI, equivalent profiles for 5G communications over Uu are referred to as 5QI. The alternative QoS profile contains one or more QoS reference parameters in a prioritized order. This enables the system to select alternative service profiles in case of QoS changes. The NG-RAN (the Next Generation Radio Access Network, as used by 5G wireless networks) notifies the 5GC (Core) that an alternative QoS profile can be supported and then 5GC can provide this notification to the V2X AS.

In Release-16, mechanisms are defined for a 5G System (5GS) for monitoring, collecting, and reporting information about experienced QoS features. Advance notifications of estimated changes in QoS can be sent to affected V2X applications. This procedure is referred to in 3GPP standards as QoS Sustainability Analytics. It helps the V2X application to decide in a proactive manner if there is a need for an application change when the QoS degrades. A V2X AS may request notifications regarding QoS sustainability analytics for indicated geographic area and time period. The V2X AS then provides the network with location information in the form of a path of interest or a geographical area where to receive notifications of potential changes in QoS. The V2X AS also sends to the 5GS the QoS parameters that should be monitored, as well as the corresponding QoS thresholds to be used for signifying efficient and safe operation of an application. The 5GS system compares QoS thresholds with predicted values of QoS metric values to decide if it should notify the V2X AS of an expected QoS change.

Release-16 NR V2X supports V2N communications over the Uu interface. It introduces enhancements to Release-15 NR Uu and LTE Uu interfaces, in aiming to meet requirements imposed by advanced eV2X services. It does not directly support multicast V2N communications. Among these enhancements, we note the following. (1) Multiple uplink flows are enabled, whereby each flow can be allocated a semi-persistent resource assignment. (2) End users can report to their BSs information that includes Uu and SL traffic characteristics, such as periodicity, packet latency requirements, and maximum TB size. BSs use this information for dynamic adjustment of resource allocations. This data may be used by BSs to enhance V2N communications. It is employed to best schedule the sharing of wireless resources among SL and Uu transmissions, to satisfy performance requirements and to control signal interference. It can also be employed to plan for enhanced time/frequency/space resource reuse processes. Users that are sufficiently far from each other may be assigned the same resources. Resource pools can be shared based on data that identify

648 | 21 *Communications and Traffic Management for the Autonomous Highway*

vehicular geolocations and flow patterns. (3) Enhanced handovers (HOs) between cells are defined over the Uu interface (and can be used for remote driving). Under its Dual Active Protocol Stack (DAPS), UEs can have two active links during an handover process, one with the source cell and the other one with the target cell. Such a configuration increases the probability that the handover process will be performed successfully. Settings are provided to specify target values for handover delay and for the associated maximum ensuing interrupt time. (4) A conditional handover (CHO) process is defined and used to improve the reliability of the handover process. For this purpose, RRC messages are exchanged between the source cell and the UE. The UE constantly compares HO measurements with the conditions specified by the CHO settings. When CHO conditions are met, the mobile would automatically execute an handover process, without having to wait for issuing an handover command.

21.3.8 Networking Automated and Autonomous Vehicles

An autonomous vehicle (AV) is equipped with sensing, communications, and mobility systems that enable it to monitor its environment and communicate with other vehicles, with RSUs, and with management stations. It uses this information to derive mobility parameters and to execute its maneuvering plan. Such a vehicle is also identified as a self-driving car or a robotic car. Its advanced sensing systems include a wide range of devices, such as cameras, radar, lidar, sonar, GPS, and a host of other environmental measurement and sensing systems. Sensor data obtained by onboard devices and through communications with other vehicles and RSU nodes are fused (using often background maps as reference images) to provide the vehicle with precise image and status of the geographical regions that it is traveling through and that it plans to traverse.

Generally, vehicles share the roads that they move across. By communicating their travel plans to each other, vehicles are able to better utilize available transportation means. They may adjust their travel routes and schedules to accommodate each other and assure safety while striving to reach their destinations at targeted times.

Autonomy in vehicles has been categorized by SAE International into six levels:

1) **SAE Automation Level 0**: No automation is embedded, so that the vehicle is managed by a human driver on a full time basis.
2) **SAE Automation Level 1**: A driver assistance system is provided. Information about the environment is provided to the driver, assisting the driver in dynamically maneuvering the vehicle.
3) **SAE Automation Level 2**: A partial automation system is used. A driver assistance system involving both steering and acceleration/deceleration is provided. A human is required to monitor the driving environment and to be available to take over.
4) **SAE Automation Level 3**: A conditionally automated driving system that monitors the driving environment is employed. A human driver is expected to intervene at times.
5) **SAE Automation Level 4**: An automated driving system that monitors the driving environment and dynamically controls the driving function is employed under many driving modes. A human driver may respond to a request to intervene, but the vehicle can perform safely if such a response is not forthcoming.
6) **SAE Automation Level 5**: A fully automated driving system that monitors the driving environment and dynamically controls the driving function is employed under all driving modes and environmental conditions that can be managed by a human driver.

Data communications between vehicles (V2V) and between vehicles and road side units (V2I) are essential for the operation and regulation of automated and AVs and systems (which henceforth are both referred to as AVs and systems). As noted above, the formation of **vehicular platoons** is used to simplify and enhance V2V and V2I data communications. As discussed in Section 21.3.9, platoon formations can be utilized to enable the effective control of autonomous mobility and to manage vehicle traffic flows over a highway system. Such controls include the regulation of vehicle admissions into the highway system, the setting of inter-platoon and intra-platoon spacing levels and vehicle speeds. Their configuration aims to achieve high vehicle flow rates while meeting high safety requirements and lowering vehicle queueing and transit delays.

The implementation of high-performance autonomous and semi-autonomous vehicular flows impose high data throughput and low packet delay requirements on implemented V2V and V2I network systems. The vehicular network must accommodate the transport of critical safety messages, producing relatively low-throughput rates and strictly limited message communications delays, while also provide for the transmission of other data flow types that demand higher-throughput rates. The latter include raw and processed sensor and vehicular status flows and maneuvering data plans. Automated driving systems also make use in realtime of high definition maps, assuring the safe maneuvering of vehicles.

To meet such demanding performance requirements, the use of communications media that offer higher-frequency bandwidth levels is highly beneficial. Frequency bands centered around 5, 30, and 60 GHz have been considered. The use of **mmWave frequency bands** can enable communications at higher data rates while yielding low packet latency levels.

At these bands, smaller antenna arrays become feasible. The employment of electronically synthesized multiple directional antenna beams offers higher antenna gains and upgraded spatial and frequency reuse levels. In turn, operation at such high frequencies induces higher communications propagation losses. The presence of obstacles in a line-of-sight path of a transmitted signal may effectively block its reception.

The transmission of a V2I signal between a vehicle and an RSU or a BS could make use of highly directional antenna beams. A rather precise beam alignment process would then be required. As a vehicle moves, its line-of-sight connectivity that points toward its targeted RSU station would fluctuate, requiring reactivation of the antenna alignment process. To enhance the performance behavior of this process, ML schemes can be employed, improving the efficiency of the alignment process by making use of location, speed, and topographical sensing data, coupled with the use of mobility-oriented environmentally dependent prediction calculations.

The IEEE 802.11ad standard includes the specification of an advanced wireless LAN system that operates at 60 GHz. It provides a data rate of up to 6.75 Gbps over a communications channel that is allocated a 2.16 GHz bandwidth. At least 7 GHz spectrum has been allocated in the United States in the 57–71 GHz band. ETSI has specified 63–64 GHz for mmWave-based V2X applications. Future NR-based V2X systems, including Sidelink communications, are expected to employ mmWave bands in addition to using the 5.9 GHz band.

The hybrid V2V/V2I/V2N/V2X networking configurations illustrated in Fig. 21.14, and in Fig. 21.15, would be enhanced to offer high-performance operations when accommodating the use of mmWave communication channels. Communications between neighboring vehicles that are platoon members involve signal propagation over short distance ranges. Such relatively steady V2V (unobstructed) line-of-sight links can well accommodate mmWave transmissions. Communications between close-by vehicles that move in a synchronous manner in neighboring lanes can also be well executed by using mmWave bands. V2I signals between gateway vehicles and close-by RSUs would benefit as well by employing directional antenna beams.

To increase the availability of unobstructed V2I and V2N communications pathways, a system designer can make use of RSU and BS nodes that are mounted at higher altitude, such as on

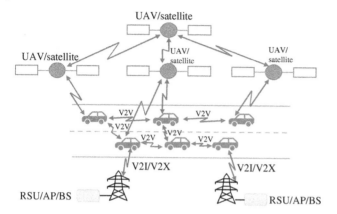

Figure 21.16 Vehicular Network System That Employs V2V, V2I and C-V2X Communications Links Supplemented by the Use of UAVs and/or Satellites.

side-road poles or traffic lights. Unmanned Airborne Vehicles (UAV drones) and low-altitude satellite network systems can be beneficially employed. Transmissions at lower (such as sub-6 GHz) frequency bands from such higher altitude platforms can be utilized to *broadcast* control and status update data to widely distributed vehicles. The latter would be covered by the wider transmission footprints realized by using UAV or satellite platforms. They could serve to accommodate the dissemination of status update data issued by active vehicles and management stations over wide geographical areas. End user vehicles may be outfitted with micro satellite/UAV communications access means. High data rate and ultralow latency wide-span communications links could be enabled. Such a configuration is illustrated in Fig. 21.16.

21.3.9 Traffic Management of Autonomous Highway Systems

21.3.9.1 Achieving the Highest Vehicle Flow Rate

The regulation of vehicular access to a highway coupled with the synthesis of vehicle formations impacts the highway's vehicular congestion state and its achievable vehicle throughput capacity rate. These settings affect the time delay levels experienced by vehicles in waiting to access the highway and those incurred in their transit along the highway. In this section, we present key features of a traffic management process. Our presentation follows the models developed by Professor Izhak Rubin at UCLA and Professor Andrea Baiocchi at the Sapienza University of Rome, as principal researchers, and their associate researchers, including models presented in [44].

Consider groups of vehicle platoons that flow across a segment of the highway, as shown in Fig. 21.17. Platoon-oriented formation and speed parameters are shown in Fig. 21.18. To simplify, we use in the following discussions and analyses the same parameter values to characterize

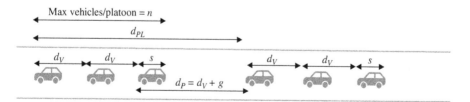

Figure 21.17 Flow of Vehicle Platoons Across a Highway Link.

Symbol	Definition
d_V	Minimum distance between vehicle members of the same platoon [m]
s	Mean length of a vehicle [m = meters]
v	Mean speed of a vehicle [m/s]
n	Maximum number of vehicles forming a platoon
g	Excess gap range $= \psi d_V$ [m], $\psi \geq 1$
d_P	Minimum gap distance between neighboring platoons $= d_V + g = (1+\psi)d_V$
d_{PL}	Minimum distance between leading heads of neighboring platoons $= nd_V + g = (n + \psi)d_V$
u	Nominal deceleration of a vehicle [m/s^2]
ζ	Fractional deviation from nominal deceleration, indicative of $\Delta u / u$

Figure 21.18 Platoon-Oriented Parameters.

different vehicles and different vehicle platoon formations. A platoon is assumed to consist of a maximum of n vehicles, each moving at a speed of v [m/s]. Within a platoon, the inter-vehicle spacing is denoted as d_V [m], as measured between the front of a vehicle and the front of the neighboring vehicle that belongs to the same platoon and travels along the same lane. The length of a vehicle is denoted as s [m]. The minimal space between a platoon and the following platoon is denoted as d_P [m]. It is set as $d_P = d_V + g$, where g [m] is a configured excess gap distance. We set $g = \psi d_V$ [m], where $\psi \geq 1$. Thus, for $\psi = 2$, we have $g = 2d_V$, so that the minimal gap between neighboring platoons is then set equal to $d_P = 3d_V$. These gaps are used to accommodate vehicles that change lanes, such as those that cross a lane to join vehicle flows that use a faster lane, those that change lanes for the purpose of exiting the road or those that switch lanes to access another road. Assuming henceforth a loaded highway under which platoons have been placed at minimal inter-platoon distances. The distance between the leading heads of neighboring platoons is then equal to $d_{PL} = nd_V + g = (n + \psi)d_V$. It is noted that wider gaps, and thus higher ψ values, may be configured across slower lanes, leaving wider open spaces for lane crossings.

To illustrate the calculation of the minimum spacing to be maintained between vehicles, we assume here that this distance is dictated by the minimum braking range required by a vehicle that follows a vehicle that has triggered a braking action. We assume the vehicular communications network system to reliably and promptly produce and communicate the underlying braking warning message issued by the braking vehicle, delivering it to impacted vehicles that follow the braking vehicle. The message transfer time is very short (within an order of milliseconds or less). Upon receiving this message, a following vehicle reacts by triggering its braking mechanism, decelerating to reduce its speed until it reaches a standstill position, as it aims to avoid collision. The *deceleration rate* of a vehicle is assumed to have a nominal value of u [m/sec^2]. Since a braking vehicle would still move forward for a period of time until it completely stops, it is clear that if the reaction and braking mechanisms used by all vehicles exhibit exactly that same response time and deceleration rate values, the headway spacing between neighboring vehicles can be much reduced. However, realistically, vehicles exhibit variable reaction and deceleration rates, leading to a maximum deviation from the nominal value u that is denoted as Δu, inducing a deceleration rate variation over the interval $(u - \Delta u, u + \Delta u)$. As noted in Fig. 21.18, we set the factor ζ, where typically $0 < \zeta \leq 1$, to

652 | *21 Communications and Traffic Management for the Autonomous Highway*

express the maximum fractional deviation from the nominal deceleration level, assuming a value that is indicative of $2\Delta u/u$. For notation convenience, we set $b = \zeta/(2u)$.

A vehicle that moves at speed v [m/s] and decelerates at rate u [m/s^2] would reach a stopping point after covering a braking distance d_b that is equal to:

$$d_b = \frac{v^2}{2u}. \tag{21.4}$$

To motivate the formula given by this expression, we note the following. As a decelerating vehicle slows its speed from an initial value of v [m/s] to a final value of 0 [m/s], it proceeds to reduce its speed by a rate of u [m/s] per every second. Hence, it will require a braking time period of $t_b = v/u$ [seconds] to reach stand still. During this period, its average speed level is equal to $v/2$ [m/s]. Hence, the braking vehicle will span a braking distance, denoted as d_b [m], that is equal to $d_b = (v/2)t_b = v^2/(2u)$ [m].

To calculate a lower bound for the required minimum distance between neighboring vehicles, assume that the deceleration rates of a leading vehicle and a vehicle that follows it differ so that the leading vehicle decelerates at a higher rate that equal to $(u + \Delta u)$, while the following vehicle decelerates at a lower rate that equal to $(u - \Delta u)$. The following vehicle would then get closer and closer to the leading vehicle before it is able to reach a full stop. We assume that $\Delta u \ll u$. Hence, we set the desired minimum spacing distance as equal to: $d_s = \frac{v^2}{2u}\frac{2\Delta u}{u} = \zeta\frac{v^2}{2u}$.

The braking distance limits the spacing distance between the rear end of a vehicle and the front end of the following vehicle. As noted in Fig. 21.17, the corresponding minimum value for the *headway spacing*, denoted as d_V, is calculated as $d_V = s + d_s$. Hence, the minimum spacing between platoon vehicles must satisfy: $d_{min} = d_V - s \geq \zeta d_b$, or $d_{min} = d_s$, yielding the following expression for the minimal value of the headway spacing:

$$d_V \geq s + \zeta\frac{v^2}{2u}. \tag{21.5}$$

Observing the distance components shown in Fig. 21.17, we conclude the minimum distance between the leading heads of neighboring platoons, denoted as d_{PL}, to be given as:

$$d_{PL} = (n-1)d_V + d_P = nd_V + g = ns + n\zeta\frac{v^2}{2u} + g. \tag{21.6}$$

Considering a highway system, the key performance metrics associated with a vehicle traffic flow process involve its **flow, speed, and density** levels. Averaged over the spatial dimensions of the underlying system (such as over a specific lane or over multiple lanes in a segment of a highway, or over a network of roads spanning a geographical region) these metrics are defined as follows:

flow (f) = Average number of vehicles that transit a highway system per unit time [vehicles/s].

$$\tag{21.7}$$

speed (v) = Average velocity of a vehicle that moves along a highway system [m/s]. $\quad$ (21.8)

density (δ) = Average number of vehicles traveling a highway per unit length [vehicles/m].

$$\tag{21.9}$$

The flow rate metric, denoted as f, represents the average number of vehicles that transit a road or a road system per unit time. This can involve a single lane road, a multiple lane road, or a highway system that consists of multiple interconnected roads or highways. Under equilibrium (steady-state) conditions, the vehicle departing rate measured at the exit from such an open system is equal to the rate of vehicles admitted into the system. It is identified as the *vehicle traffic throughput rate.* For example, when the flow metric is measured to be $f = 1.5$ [vehicles/s], we observe the system to sustain an average flow of 3 vehicles every 2 seconds, or equivalently an average flow of 5400 vehicles/h.

The vehicle density metric, denoted as δ, measures the average number of vehicles occupying a road system per unit distance. For example, under a density metric value of $\delta = 0.1$ [vehicles/m/lane], we observe an average of a single vehicle to occupy a road's lane per stretch that is 10 m long.

The vehicle speed metric expresses the average velocity at which a vehicle moves over the system under consideration. For example, under a speed metric value of $v = 15$ [m/s], which is also equal to 54 [km/h], or about 33.75 [miles/h], a vehicle would cover a distance of 15 m during each second, or 54 km during an hour. The speed can be computed by averaging the values assumed over different roads or lanes. Speed levels may vary over lanes. Vehicles are able to move at higher speeds across lanes that are occupied at lower density.

It is readily noted that the value assumed by a vehicle flow rate (f) is equal to the product of the values assumed by the corresponding density (δ) and speed (v) metrics, referring to the same underlying road or highway system. We thus state the following formula, often identified as the *fundamental equation of traffic flow*:

Fundamental Equation of Traffic Flow:

$$f = v\delta. \tag{21.10}$$

In the following, we illustrate the calculation of these metrics and observe general trends in the behavior of the vehicle traffic flow rate metric as a function of key system parameters. To simplify the illustration, we consider a single lane road segment across which moving vehicles are configured into platoon formations. We assume this road segment to be highly loaded so that it is fully occupied by vehicle platoons. Each platoon consists of n vehicles. Platoon formations are characterized by the distance parameters shown in Fig. 21.17. A multilane highway system will be considered in Section 21.3.9.2.

Using the underlying model, consider vehicles that move across a lane. The distance between platoons is calculated by using Eq. 21.6. We assume that vehicles are directed to keep as short a spacing as safety allows to best utilize the road's spatial resources and thus reach the highest density that provides for safe mobility (accounting for safe braking distances). Setting $g = \psi d_V$, and letting d_V assume the lower bound value expressed by Eq. 21.5, we obtain the following expression for the density function:

$$\delta = \frac{1}{1 + \frac{\psi}{n}} \frac{1}{bv^2 + s}, \tag{21.11}$$

where $b = \zeta/(2u)$.

Using Eq. 21.10, we conclude the corresponding vehicle flow rate $f = f(v)$ across the lane to be expressed as:

$$f = v\delta = \frac{1}{1 + \frac{\psi}{n}} \frac{v}{bv^2 + s}, \tag{21.12}$$

where $b = \zeta/(2u)$.

Using the formulas expressed by Eqs. 21.11 and 21.12, we examine the variation of the vehicle density and flow rate metrics across the lane as the vehicle's speed value v is increased. When vehicles move at a relatively low speed, the safe inter-vehicle spacing level (in each platoon and between platoons) assumes a relatively low value, so that vehicles can be set to move at a close distance from each other. The corresponding density (δ) level then assumes a high value.

654 | *21 Communications and Traffic Management for the Autonomous Highway*

However, since the speed is low, the ensuing product, which yields the flow rate, yields a low vehicle throughput flow rate. The highway system is noted to then operate in a congested state.

In turn, under an uncongested (or less congested) state, the road accommodates a lower number of vehicles, so that longer inter-vehicle distances can be maintained, leading to lower vehicle density, hence permitting an increase in the vehicle's speed level v. The ensuing product of density and speed may then result in a higher vehicle flow rate. As the vehicle's speed is further increased, inter-vehicle ranges must be increased (to maintain a safe breaking spacing) so that the ensuing vehicle density level is reduced. Such a reduction can be maintained by applying flow control, so that the number of vehicles admitted into the underlying segment of the highway is properly limited. Since the required inter-vehicle spacing distance is proportional to v^2, the density level rapidly decreases as the vehicle speed level v is further increased, leading to reduced vehicle flow rates.

Based on these observations, we conclude that the vehicle speed level v can be set to a level that would cause the vehicle flow rate across the underlying segment of the highway to reach its maximum feasible level. Using Eq. 21.12, we calculate the highest flow rate level that is achievable by adjusting the vehicle speed level. It is denoted as f_{max}. The speed level that achieves this peak throughput rate is denoted as v_0. We calculate the corresponding peak flow rate by taking the derivative of the formula for f with respect to the speed variable v and setting it equal to 0. The ensuing equation is solved to yield v_0. The ensuing results are presented as follows. The derived optimal speed level, v_0, is given as:

The optimal speed level:

$$v_0 = \sqrt{s/b}. \tag{21.13}$$

The optimal flow rate achieved by configuring the vehicle speed level to v_0 and regulating the vehicle admission rate accordingly (assuring that the corresponding minimum required safety spacing is maintained) is then expressed as follows:

The optimal flow rate level:

$$f_{max} = \frac{1}{1 + \frac{\psi}{n}} \frac{1}{2\sqrt{bs}}, \tag{21.14}$$

where $b = \zeta/(2u)$.

In the following, we illustrate the performance behavior of an autonomous highway as the vehicle speed level v is varied while maintaining safe inter-vehicle spacing. For this purpose, we use the formulas presented above, in which we set the following parameter values: $n = 10$, so that each platoon consists of 10 vehicles (noting above that the system's performance behavior is not highly sensitive to the value of n as long as it is not too small); $\psi = 2$, so that $g = 2d_V$; $s = 6$ [m]; $u = 3$ [m/s^2]; $\zeta = 0.2$, so that the deviation of the deceleration level among highway vehicles is assumed to be within 20%.

The variation of the flow rate f as a function of the vehicle speed v is shown in Fig. 21.19(a). It is noted that when the loading of the highway is relatively light, so that the road is said to reside in an uncongested or free-flow state, vehicles can be configured to move at higher speed levels. In contrast, when the highway experiences higher traffic loading rates so that it resides in a congested state, vehicles must reduce their speed levels to allow for shorter inter-vehicle ranges, accommodating the higher vehicle density level incurred along the highway.

Using the formula given by Eq. 21.13, we calculate the speed at which the highest flow level is attained, for the underlying scenario and parameters, to be equal to $v_0 = 13.4$ [m/s], or 48.3 [km/h]. The corresponding value for the highest vehicle flow rate across this road is obtained by Eq. 21.14 to be equal to $f_{max} = 0.932$ [vehicles/s], or $f_{max} \approx 3354$ [vehicles/h].

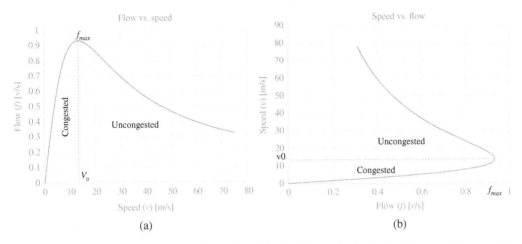

Figure 21.19 Illustrative Performance Curves for Vehicular Platoon Traffic Moving Along an Autonomous Highway: (a) Flow vs. Speed and (b) Speed vs. Flow.

As noted in the figure, when the road resides in congested mode, whereby $v \leq v_0$, increasing the vehicle speed yields a distinctive noticeable increase in the flow rate. In turn, when in uncongested mode, whereby $v > v_0$, increasing the vehicle speed results in a (moderate) decrease in the vehicle throughput flow rate. The inverse of the latter performance curve, exhibiting the variation of the vehicle speed level v as a function of the attained flow rate f (with the speed as the varying parameter) is shown in Fig. 21.19(b). One notes the change in the system's behavior as it transitions between congested and free-flow (uncongested) states.

As the vehicle speed v value is varied, the ensuing vehicle density is calculated by using Eq. 21.11. For the underlying example, the variation of the flow rate f as a function of the vehicle density is shown in Fig. 21.20(a). When the highway resides in the uncongested (free-flow) state, a proper increase in the vehicle density level yields a higher flow rate. As the highway experiences higher loading and the corresponding road density level exceeds the corresponding density value $\delta_0 = f_{max}/v_0$, yielding for this illustrative case the density level $\delta_0 = 0.0746$ [vehicles/m], or an average of 7.46 vehicles across each stretch of 100 m, the highway transitions into congested state. When in a congested state, an increase in the vehicle density level leads to lower flow rate.

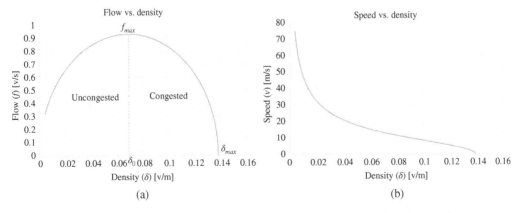

Figure 21.20 Illustrative Performance Curves for Vehicular Platoon Traffic Moving Along an Autonomous Highway: (a) Flow vs. Density and (b) Speed vs. Density.

The variation of the speed level vs. the road density level, as the speed level is varied, is shown in Fig. 21.20(b). It is noted that when the system resides in a free-flow (uncongested) state (while operating at lower density and higher speed levels), a distinctively large decrease in the speed level induces only a modest increase in the density level. In turn, while operating at lower-speed levels in a congested mode, a modest decrease in the speed level induces a high increase of the density level.

21.3.9.2 Traffic Management Under Queueing and Transit Delay Limits

To achieve the highest feasible vehicle flow rate across a lane of a highway that is equal to the value that was calculated above, vehicles must be arranged in platoon formations that are configured by using the above noted parameter and speed levels.

However, to assure a vehicle with high performance experience, it is of interest to consider the time delay incurred by a vehicle. Key delay components consist of a vehicle's on-ramp queueing delay, representing the time incurred while waiting to be admitted into the highway, and its transit delay, representing the time that it takes a vehicle to travel the highway from its entry point to its exit point.

To assure a vehicle that is admitted to the highway with acceptable access and transit time delay levels, it is necessary to limit the traffic rate of vehicles that flow in the system, setting it to a level that is lower than the one that was calculated above, which yields a peak flow rate. The traffic management scheme needs then to employ a flow admission control scheme that limits the vehicle traffic loading of the road. In the following, we illustrate the design and performance features of such a scheme.

We start by presenting a traffic and analysis model for a single-lane single-link road system. We then outline our approach to traffic management for 3 other highway road configurations: single-link multiple-lane segment, multiple-link single-lane interconnected road network, and multiple-link multiple-lane interconnected road network. Analysis details are presented in [44].

For the **single-link single-lane road** system, we consider a segment of the highway that spans a road that stretches between two consecutive terminations. Such a segment is shown in Fig. 21.21. Vehicles moving along this link include vehicles that arrive from neighboring segments as well as vehicles that arrive from outside the highway system. External vehicles, when admitted into the highway, enter the road across an input ramp. For this first case model, we assume that all vehicles moving along the link access it at its input ramp and then exit the road at its outgoing ramp. Transit vehicles that travel across other links are not directed to this link, or do not impact the evaluated

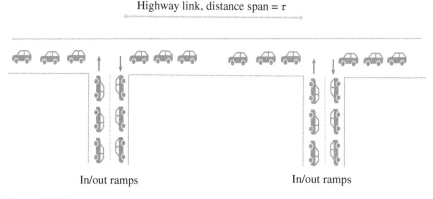

Figure 21.21 A Link Span of a Highway.

traffic flowing across this link. At each link termination point, incoming and outgoing ramps are provided. Vehicles exit the highway by using an output ramp. Vehicles that arrive to the highway from outside the highway (identified henceforth as outside vehicles) use incoming ramps to access the highway. Vehicles form a queueing line as they wait across an incoming ramp until they are permitted to enter the highway.

As noted above, the formation and speeds of vehicles moving across a lane of a road segment can be configured to yield the highest feasible vehicle flow rate, serving to enhance the traffic capacity of the road, as the highest feasible number of vehicles per unit time is accommodated. To assure the traffic loading of the road to be performed in accordance with desired optimal road density conditions, when the outside rate of vehicles is relatively high, a traffic admission flow control regulation scheme is employed. It is used to regulate the admission rate of vehicles into the lane. For the illustrative system presented above, the maximum achievable flow rate across the link is equal to $f_{max} = 0.932$ [vehicles/s]. Hence, the admitted vehicle flow rate into the link, denoted as λ [vehicles/s], must be limited to a value that is not higher than the flow capacity of the highway link. For the underlying example, assuming that admitted vehicles are set to move at the optimal speed ($v = v_0$), every 10 seconds, no more than an average of about 9 vehicles should be admitted for travel across the lane.

Vehicles waiting in line along an input ramp for admission into the road form a queue (i.e., a wait line) whose behavior is governed by the dynamics of a **queueing system**. Vehicles are regarded as customers that are provided service by being admitted for transit along the road. The *input traffic* process that feeds this queueing system consists of vehicles that arrive at the input ramp from the *outside*. We denote the arrival rate of such external vehicles as λ [vehicles/s]. The service rate of this queuing system is denoted as μ [vehicles/s]. It represents the rate at which the system is set to admit waiting vehicles into the road. The service rate is thus set as $\mu = f$ for a road that is configured to achieve a flow rate of f [vehicles/s]. When considering a road along which vehicles are configured to move at speed v, we denote the ensuing flow rate as $f(v)$. The corresponding service rate is then denoted as $\mu(v) = f(v)$. For example, considering the above noted illustrative system, when the road system is optimized to yield the highest feasible flow rate f_{max}, the corresponding speed of admitted vehicles is set to v_0, and the service rate of the on-ramp queueing system attains the peak value $\mu = \mu(v_0) = f_{max}$.

To assure a vehicle that is eventually admitted for transit along the link with a limited (and not excessively high) waiting time at its input ramp's queue, it is necessary to regulate the input traffic that loads this queue so that it is lower than the link's service rate. This requirement is expressed as follows:

$$\lambda < \mu. \tag{21.15}$$

Movement of vehicles at the above noted optimal speed level (v_0) may however lead to unacceptably long transit time delay across the road. To reduce this delay, the traffic manager may increase the speed levels used by admitted vehicles. In doing so, the system's parameter settings would deviate from the settings that lead to a maximal flow rate, resulting in a lower realized flow rate, implying consequently a lower value for the service rate $\mu = \mu(v)$. To satisfy Eq. 21.15, the admission rate of outside vehicle traffic (λ) may have to be reduced so that the condition $\lambda < \mu(v)$ is maintained.

To calculate the statistical behavior of the time delay incurred by vehicles that wait in an on-ramp queue, we model the on-ramp waiting line system as an M/M/1 queueing system. It is a single server queueing system that is loaded by vehicles (which are the system's customers) that arrive at a rate of λ [vehicles/s] and are served at a rate of $\mu(v)$ [vehicles/s].

658 | *21 Communications and Traffic Management for the Autonomous Highway*

Under this queueing system model, assuming external arrivals to be regulated so that $\lambda < \mu$, the (steady state) average delay incurred by a waiting vehicle, denoted as D_{ramp}, is calculated by using the formula expressed by Eq. 7.23, yielding the following result:

On-ramp average vehicle delay:

$$D_{ramp} = \frac{1}{\mu(v) - \lambda}, \quad \text{where} \quad \lambda < \mu(v). \tag{21.16}$$

An admitted vehicle incurs an overall average time delay that consists of two components: an on-ramp average delay that is calculated as D_{ramp} and a transit time delay as it moves along the road. The latter represents the time that it takes the vehicle to reach its destination. It is identified as its *transit delay* and its average level is denoted as $D_{transit}$. In moving, from entry to exit, across a road of length l [m], a vehicle's transit delay is calculated as $D_{transit} = l/v$, assuming it to move at speed level v.

In the following, we illustrate the synthesis of a traffic management scheme that aims to operate the highway at the highest feasible vehicle flow rate while at the same time imposing a limit on the average cumulative delay incurred by an admitted vehicle.

This average cumulative delay limit is denoted as D_0. It represents an upper bound on an average of the vehicle's aggregate weighted sum of its incurred on-ramp queueing delay and its transit time. It is expressed as: $D_0 = D_{0q} + \frac{l}{v_{tg}}$ where D_{0q} is an upper bound that is imposed on the average time delay (D_{ramp}) incurred by a vehicle in waiting on-ramp to access its assigned lane, and $D_T = \frac{l}{v_{tg}}$ is an upper bound term that accounts for the transit time incurred by a vehicle that moves a distance l along its lane. The average aggregate time spent by a vehicle in waiting and in transit is thus upper bounded as follows:

$$\frac{1}{\mu(v) - \lambda} + \frac{l}{v} \leq D_0 = D_{0q} + \frac{l}{v_{tg}}, \quad \text{where} \quad \lambda < \mu(v). \tag{21.17}$$

It is noted that the upper bound's delay component contributed by the vehicle's transit along its lane is expressed as $\frac{l}{v_{tg}}$ in terms of a speed parameter identified as v_{tg}. The expression for the delay limit is calculated as the sum of its two limit components, rather than by imposing a separate limit on each component. In this context, the target speed parameter, v_{tg}, is used to signify the relative weight applied to the transit delay component rather than a targeted speed objective. The higher the value attached to the target speed parameter, the higher is the weight attached to the time delay incurred in-transit along the road, as compared to the delay experienced by the vehicle in its on-ramp queue.

Using the inequality given by Eq. 21.17, we derive the following inequality on the outside vehicle arrival rate, where we use "D" as an indication that the associated expressions relate to the delay constrained model:

$$\lambda \leq f(v, D) = \frac{1}{1 + \frac{\psi}{n}} \frac{v}{bv^2 + s} - \frac{1}{D_{0q} + l/v_{tg} - l/v}, \tag{21.18}$$

when system and performance objective parameters assume values that yield a realizable positive value for $f(v, D)$.

The highest feasible level of the external vehicle arrival rate that should be admissible into the input ramp and then into the road, while assuring an average upper bound D_0 on the cumulative vehicle delay, is denoted as $\lambda_{max}(D)$. Under the condition that (for at least a single value of v) the configured system parameters yield feasible values for $f(v, D)$, which is calculated by using Eq. 21.18, we calculate the highest vehicle admissible flow rate, denoted as $f_{max}(D) = \lambda_{max}(D)$, by finding the optimal speed $v = v_{opt} = v_0$ that yields the highest value attainable by the upper bound given by Eq. 21.18:

$$\lambda_{max}(D) = f_{max}(D) = \max_{v \geq 0} \left[\frac{1}{1 + \frac{\psi}{n}} \frac{v}{bv^2 + s} - \frac{1}{D_{0q} + l/v_{tg} - l/v} \right], \quad (21.19)$$

when system and performance objective parameters are assumed to take values in a range that yields such a realizable maximal value. The imposition of too strict delay objectives coupled with the setting of certain platoon formation and mobility conditions may not yield a feasible solution.

The delay-constrained performance of a highway link of length l that accommodates the mobility of vehicle platoons is illustrated in Fig. 21.22 [44]. For this system, the following parameters are assumed: $\psi = 2, s = 6$ [m]; $u = 3 [m/s^2]; \zeta = 0.2; D_{0q} = 3$ [minutes], $v_{tg} = 108$ [km/h]. The number of vehicles included in a platoon, n, is varied from $n = 2$ to $n = 10$. The optimal value of the flow rate across a lane (in units of [vehicles/h]) vs. the link length (in [km]) is shown in Fig. 21.22(b), while the corresponding settings of the optimal speed levels are shown in Fig. 21.22(a). As expected, longer platoons yield higher flow rates, as it serves to moderate the throughput reduction impact of inter-platoon gaps. The corresponding improvement is however noted to saturate as the number (n) of platoon members increases. The membership size of the platoon does not have much impact on the setting of the optimal speed level. Over longer links, higher speed levels are employed to reduce the ensuing transit delay.

The maximal attainable flow rates are reduced when longer links are used, though only moderately. For example, the maximal flow rate (which is equal to the corresponding admitted vehicle arrival rate) attains values that are equal to about 3300 and 2940 [vehicles/h] as the length of the link increases from 2 to 10 km, respectively. Using the calculated optimal speed levels, the transit delay across the road is noted to be equal to about 2.4 and 8 minutes under road lengths of 2 and 10 km, respectively. The specified upper bound on the vehicle average on-ramp queueing delay (D_{0q}) has been set to 3 minutes, in each illustrative case. In comparison, using the calculated optimal speed levels, the corresponding transit delay limit values (i.e., l/v_{tg}) are equal to about 1.1 [minutes] and 5.5 [minutes] under link spans of 2 and 10 km, respectively. It is thus noted that the transit delay component tends to become a dominant factor for longer links, provided the realized service rate is sufficiently high to not cause a too high vehicle wait time at the on-ramp queueing system.

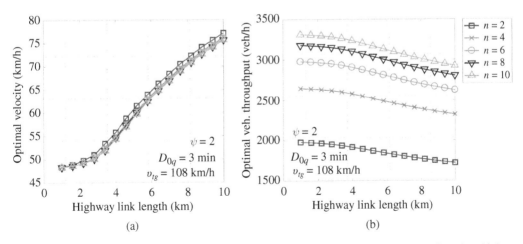

Figure 21.22 Delay Constrained Performance of Platoon Flows Along a Lane: (a) Optimal Speed vs. Link Length; (b) Optimal Vehicle Flow Rate vs. Link Length. Source: Rubin et al. [44]/with permission of Elsevier.

When comparing the formulas presented by Eqs. 21.19 and 21.12, it is noted that higher vehicle throughput rate reductions are incurred as more strict delay limits are imposed. The delay limited setting of the optimal speed level deviates from the optimal speed value (i.e., v_0) configured when no such limits are imposed, leading to lower-throughput flow rates. To illustrate such a comparison with the flow rates attained under a corresponding system that imposes no delay limits, we note that for the underlying example, the delay constrained flow rate achieved for an $l = 2$ km link is only moderately lower than that achieved when no delay constraints are imposed. In turn, for travel along a $l = 10$ km link, the attained delay constrained vehicle throughput rate of about 2940 [vehicles/h] is lower than that achieved under no delay limits, which is equal to about 3350 [vehicles/h]. Lower-throughput rates are attained when the system is designed to meet stricter vehicle delay requirements. Higher flow rate and lower delay levels are attainable when more efficient system parameter values, such as deceleration rates, are employed.

We next consider a **single-link multiple-lane road system**. A single road link that accommodates vehicles traveling along multiple lanes is assumed. A total of l lanes, $l \geq 2$, is assumed, where lane-1 is the fastest while lane-l is the slowest. To synthesize the employed traffic management mechanism, we use an extension of the system model presented above for the single-link single-lane case, modified by the introduction of corresponding lane-based parameters. Platoons are arranged across each lane, following the formation structures introduced above. Parameters that are associated with lane-i are identified by subscript i. Thus, the speed of a vehicle moving along lane-i is denoted as v_i. The inter-platoon gap parameter for lane-i is denoted as ψ_i, noting that along slower lanes, longer gaps may be configured to allow for the higher rate of vehicles that are expected to cross the lane as they enter or exit the highway and as they switch lanes. For the calculation of the delay constraint D_0, we assume that the transit delay speed parameter for lane-i is denoted as $v_{tg,i}$. Faster lanes may be configured to allow vehicles to move at higher speeds. The on-ramp waiting time limit parameter is set to D_{0q}, independently of the identity of the associated lane.

A 2-lane system is illustrated in Fig. 21.23. The external arrival rate of vehicles that are admitted for on-ramp waiting and are assigned to move into lane-i is denoted as λ_i, $i = 1, 2$. The serving rate of lane-i vehicles that form the lane-i on-ramp queueing system is shown to be denoted as $\mu_i = \mu_i(v_i)$. It expresses, the rate at which waiting vehicles are served by (i.e., admitted into) lane-i. Under this scenario, all vehicles travel along a single link, so that the service rate is equal to the maximum flow rate that can be accommodated along lane-i. The flow rate across lane-i is denoted as f_i.

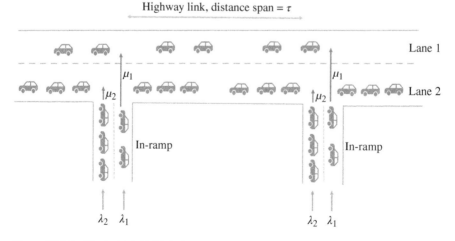

Figure 21.23 Single-Link Multiple Lane Configuration.

Arriving vehicles that are assigned by the traffic manager to move across lane-i join the lane-i on-ramp queue's waiting line, so that separate per-lane on-ramp queues are formed. The fraction of admitted arriving vehicles that are assigned for travel along lane-i is denoted as r_i, whereby $\sum_i r_i = 1$. Under steady-state conditions, the rate of vehicles admitted into the on-ramp queue that is associated with lane-i is equal to the rate of vehicles that are admitted into (and flow along) lane-i, so that we have: $f_i = \lambda_i$, for lane-i, for each i. The aggregate arrival and flow rates are, respectively, equal to $\lambda = \sum_i \lambda_i, f = \sum_i f_i$. Following the analysis presented above for the single-link single-lane case, the following expression is used for the calculation of the delay-limited average cumulative (i.e., queueing plus transit) delay incurred by a vehicle that is admitted into lane-i, for each i:

$$\frac{1}{\mu_i(v_i) - r_i\lambda} + \frac{l}{v_i} \leq D_0 = D_{0q} + \frac{l}{v_{tg_i}}, \quad \text{where} \quad r_i\lambda < \mu_i(v_i). \tag{21.20}$$

Using this expression, we conclude the following upper bound on the flow rate occupying lane-i:

$$r_i\lambda \leq f_i(v_i, D) = \mu_i(v_i) - \frac{1}{D_{0q} + l/v_{tg_i} - l/v_i}, \tag{21.21}$$

when system and performance objective parameters assume values that yield a realizable positive value for $f_i(v_i, D)$.

To synthesize the system so that, across each lane, the highest feasible flow rate is attained, while meeting the underlying delay objectives, we proceed to calculate, separately for each lane, the maximum value achieved by the upper bound given in Eq. 21.21, by varying for each lane the underlying vehicle speed parameter (v_i). The vehicle speed along lane-i is thus set to a value that yields this upper bound and is denoted as $v_{i,opt}$. Applying this value, we use Eq. 21.21 to calculate the optimal flow rate across lane-i, $f_{i,opt}(D) = f_i(v_{i,opt}, D)$, noting that the corresponding optimal outside arrival rate that is assigned to lane-i is equal to $\lambda_{i,opt}(D) = f_i(v_{i,opt}, D)$. The summation of the corresponding terms over the lanes yields the overall outside arrival rate $\lambda_{opt}(D) = \sum_i \lambda_{i,opt}(D)$ and aggregate (over all lanes) flow rate $f_{opt}(D) = \sum_i f_i(v_{i,opt}, D)$. The ensuing aggregate vehicle admitted arrival rate is denoted as λ_{opt}. We have:

$$\lambda \leq \lambda_{opt} = \sum_{i=1}^{l} f_{i,opt}(D). \tag{21.22}$$

To illustrate a possible configuration of system flow parameters, we assume a design objective that aims to maximize the aggregate throughput rate λ level, under the prescribed per-lane delay constraints. For this purpose, the fraction of admitted vehicle arrival rate that is directed to each lane should be set as follows:

$$r_i = \frac{f_{i,opt}(v_{i,opt}, D)}{\sum_{i=1}^{l} f_{i,opt}(v_{i,opt}, D)}, \quad \text{for} \quad i = 1, 2, \dots, l. \tag{21.23}$$

Thus, under this single-link multilane model, the per-lane optimal vehicle speed is calculated first by maximizing (over v_i) the upper bound given by Eq. 21.21. The routing fractions are then calculated in accordance with Eq. 21.23. The highest vehicle aggregate throughput rate that is accommodated under the prescribed delay limits is then equal to $\lambda = \sum_{i=1}^{l} f_{i,opt}(D)$.

We next consider a **multiple-link road system that employs single-lane links.** Model parameters are similar to those used above, except that this system represents a *transportation network*. The transportation network's topology is modeled as a graph, which consists of multiple road links and nodes. Assume the network graph to consist of m links and N nodes. A node represents an in/out ramp entity and/or a switching junction. Vehicles traverse the network by moving along routes that may consist of multiple links.

As performed when representing a Jackson-type queueing network model, a random routing scheme is used. Thus, upon completing its movement across link l_h, a vehicle will transition (i.e., switch across a corresponding junction node) to moving across link l_k with probability p_{hk}. An $N+1_{st}$ virtual node, denoted as n_0, and a corresponding virtual link l_0 are introduced into the model to represent the outside system. Admitted vehicles that arrive from the outside and join a queue of vehicles that are waiting to access the highway at link l_k are modeled to be governed by the statistics of a Poisson process whose rate is equal to v_k. The aggregate admitted arrival rate is equal to $\lambda = \sum_k v_k$. It is noted that upon completing its movement over link h, a vehicle will exit the network (i.e., switch to virtual link l_0 or to node 0) with probability p_{h0}. It is also noted that $p_{hk} = 0$ when link h is not connected to link k. The consequent average vehicle flow rate along link l_k is denoted as ϕ_k. Noting that the latter flow consists of vehicles that arrive from the outside as well as of vehicles that arrive from neighboring links, we conclude the following set of equations:

$$\phi_k = v_k + \sum_{h \in L} \phi_h p_{hk}, \quad \text{for } k \in L. \tag{21.24}$$

We assume henceforth that set of Eq. 21.24 yields a unique solution, which expresses the vehicle flow rates across the network links.

Such an illustrative transportation network is shown in Fig. 21.24. The network is shown to be loaded by two outside vehicle flows. Vehicles that are members of the first class arrive to an on-ramp entry at admitted rate v_1, forming the on-ramp queue Q_1. These vehicles aim to access the network at junction N_1 and then start transit across the network by flowing along link l_1. 50% of the vehicles in the class-1 flow exit the highway at junction node N_3 (following transit through links l_1 and l_2), while the other vehicles in this class continue along link l_3 and then exit the highway at node N_4. Class-2 flow consists of vehicles that access the network at node N_2, move along links l_2 and l_3 and then exit the network at node N_4. Link flow values, ϕ_1, ϕ_2, and ϕ_3, are noted.

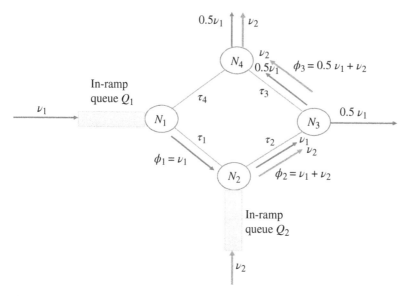

Figure 21.24   Illustrative Transportation Network.

To illustrate the synthesis of a traffic management scheme for this network model, we assume that the designer's aim is to the maximize the value attained by an objective function that is structured as a weighted sum of the flow rates, under a prescribed average vehicle delay metric level. The latter is composed by prescribing an average delay limit for each link. For this purpose, we express the average aggregate number of vehicles associated with link l_k to consist of the average number of vehicles moving across this link joined by the average number of vehicles that are waiting along the input ramp for access to this link. For this calculation, we apply Little's Formula. As described in Property 7.11, the average number of vehicles resident (under equilibrium conditions) at a service system is related to the average delay incurred by a vehicle that is served by the system. Accordingly, considering link-k, the average number of vehicles associated with this link, denoted as $E(X_k)$, is equal to the product of the applicable flow rate along the link and the ensuing vehicle delay. For the underlying model, the average number of vehicles residing at an on-ramp entry (for link-k) queue, noting each such vehicle to be delayed for a period of time that is equal to D_{Q_k}, is equal to $v_k D_{Q_k}$, where $D_{Q_k} = \frac{1}{\mu(v_k)-\phi_k}$, for each link-$k$.

In examining the latter expression, it is noted that for an M/M/1 queueing system, as expressed by Eq. 7.23, the average delay is given by $\frac{1}{\mu_{MM1}-\lambda_{MM1}}$, for $\lambda_{MM1} < \mu_{MM1}$. Let $\lambda(itransit)_k$ represent the rate of vehicles that arrive from internal network links and transit along link-k. Assume the system to employ a flow control scheme under which such vehicles essentially experience no queueing delays while transitioning into link-k. For this purpose, the employed flow control policy provides preferential access to a link to vehicles that arrive from other internal links (while the access of on-ramp queued vehicles is deferred while such internal vehicles, if any, are accommodated). Accordingly, the effective service rate of the on-ramp queue is expressed as $\mu_{MM1} = \mu(v_k) - \lambda(itransit)_k$.

The rate of outside vehicles loading the link-k on-ramp queue is equal to $\lambda_{MM1} = v_k$. It is noted that under steady-state conditions (whereby the admission rate of on-ramp queued vehicles into link-k is equal to their outside arrival rate), the flow rate across link-k is calculated as the sum of the two types of streams that feed traffic into the link, so that $\phi_k = \lambda(itransit)_k + v_k$.

The M/M/1 formula for the average number of vehicles resident in the on-ramp queue, as expressed by Eq. 7.20, is therefore written as $\frac{v_k}{\mu_{MM1}-\lambda_{MM1}} = \frac{v_k}{\mu(v_k)-\phi_k}$.

The average number of in-transit vehicles that reside at the same time along link-k is equal to $\phi_k l_k / v_k$, as the flow rate across the link is equal to ϕ_k and each vehicle spends an average transit time along the link that is equal to l_k / v_k. Hence, the average number of vehicles that are, at a given time, being served by link-k (whether queued or in-transit) is given as:

$$\frac{v_k}{\mu(v_k)-\phi_k} + \phi_k \frac{l_k}{v_k}, \quad \text{for} \quad \phi_k < \mu(v_k). \tag{21.25}$$

To illustrate the calculation of incurred vehicle delays, we examine the network shown in Fig. 21.24. The following parameter assumptions are made. The flow capacity rate across link-k is assumed to be configured to $\mu(v_k) = 2000$ [vehicles/h], for each k. The outside vehicle arrival rates are equal to $v_1 = 1000$ [vehicles/h] and $v_2 = 900$ [vehicles/h]. The ensuing flow rates across the three loaded links are then calculated to be equal to: $\phi_1 = 1000, \phi_2 = 1900, \phi_3 = 1400$ [vehicles/h]. The on-ramp queueing system that feeds vehicles to link-1 produces a traffic flow across this link whose rate is equal to 1000 [vehicles/h]. The corresponding on-ramp queueing system experiences a traffic intensity (expressing the ratio of its loading rate in to its service rate) that is equal to $\rho_1 = 50\%$. The average delay incurred by a waiting message at this queue is calculated to be: $\frac{1}{\mu_1-\phi_1}$ $= \frac{1}{2000-1000}$ [h/vehicle] = 3.6 [s/vehicle]. In turn, the on-ramp queueing system that feeds vehicles to link-2 produces a traffic flow across this link whose rate is equal to 1900 [vehicles/h], so that

21 Communications and Traffic Management for the Autonomous Highway

the corresponding on-ramp queueing system experiences a traffic intensity index that is equal to $\rho_2 = 95\%$. The average delay incurred by a message waiting at this queue is calculated to be: $\frac{1}{\mu_2 - \phi_2} = \frac{1}{2000 - 1900}$ [h/vehicle] = 36 [s/vehicle].

The delay limit per link is set as the sum of a term that accounts for the on-ramp delay, D_{0q}, and a term that accounts for the transit time incurred by a vehicle that moves along the link. We invoke Little's Formula in converting the ensuing mean delays to corresponding mean system size (i.e., number of vehicles) levels, yielding the following inequality:

$$\frac{v_k}{\mu(v_k) - \phi_k} + \phi_k \frac{l_k}{v_k} \leq v_k D_{0q} + \phi_k \frac{l_k}{v_{tg}}, \quad \text{for } \phi_k < \mu(v_k), \text{at each link } l_k. \tag{21.26}$$

By setting $\chi_k \equiv \phi_k / v_k$, we obtain the following:

$$\frac{1}{\mu(v_k) - \phi_k} + \frac{\chi_k l_k}{v_k} \leq D_{0q} + \frac{\chi_k l_k}{v_{tg}}, \quad \text{for } \phi_k < \mu(v_k), \text{at each link } l_k. \tag{21.27}$$

This inequality is the same as that presented for the corresponding single-link single-lane case with the following exceptions for each link-k: flow rate ϕ_k replaces arrival rate λ; the link-k length is modified to $\chi_k l_k$. Rearranging terms in Eq. 21.24, we derive the following set of equations, where we set $\alpha_k = v_k / \lambda$, for each link-k, to denote the fraction of the overall arrival rate that feeds link l_k:

$$\alpha_k \chi_k = \alpha_k + \sum_{h \in L} \alpha_h \chi_h p_{hk}, \quad \text{for each } k \in L. \tag{21.28}$$

Using this set of equations, we calculate the coefficients $\{\chi_k\}$. These coefficients depend only on the distribution of vehicular traffic across the input ramps and on the routing scheme. Given these entities, these coefficients do not depend on the underlying traffic intensity values.

To illustrate the use of the model to the design of a traffic management scheme, we consider an objective function O that is expressed as a weighted sum of the network's link flows: $O = \sum_{k \in L} w_k \phi_k$, where $\{w_k, k \in L\}$ are specified link weights. The weight w_k may express the net gain (e.g., profit minus cost) earned per unit-flow directed along link l_k. The traffic management scheme to be synthesized is then derived by solving the following constrained optimization problem, yielding an optimal set of link flows $\phi = \{\phi_k, k \in L\}$:

$$\max_{\phi} \sum_{k \in L} w_k \phi_k, \tag{21.29}$$

such that

$$\phi_k \leq \mu(v_k) - \frac{1}{D_{0q} + \chi_k l_k (1/v_{tg} - 1/v_k)} = f(v_k, x_k l_k, D), \tag{21.30}$$

for each $k \in L$.

Next, consider a traffic management model for a transportation network that consists of **multiple-links and multiple lanes per link**. It can be modeled, evaluated, and optimized by using a combination of the models and methods noted above. The routing pattern can again be based on using a random routing model with routing parameters $\{p_{hk}\}$. Under the model studied in the referenced paper [44], a vehicle admitted to link k was assigned to lane i with probability r_{ik}, independently of other assignments. Using the indices (i, k) to represent lane-i of link-k, an optimization problem that combines the elements that are described above has been presented. It serves to calculate the optimal speeds $\{v_{ik}\}$ of vehicles moving along each lane-i of link-k, the optimal lane assignment ratios $\{r_{ik}\}$, and the optimal flow rates $\{\phi_{ik}\}$, for each network link and lane.

Problems | 665

Problems

21.1 Describe the following communications methods as used in a vehicular network system:
a) Vehicle to Vehicle (V2V).
b) Vehicle to Infrastructure (V2I).
c) Vehicle to Network (V2N).
d) Vehicle to Everything (V2X).
e) Cellular – V2X (C-V2X).

21.2 Provide examples of safety messages that are used in an autonomous vehicular network system.

21.3 Provide examples of nonsafety messages that are used in an autonomous vehicular network system.

21.4 Describe the vehicular networking configurations displayed in Fig. 21.4.

21.5 Describe and discuss the US WAVE architecture and the ETSI GeoNetworking approach and their corresponding communications layers.

21.6 Describe the V2V networking method used under IEEE 802.11p.

21.7 Consider the ETSI-based GeoNetworking methods. Define and discuss the following:
a) GeoUnicasting.
b) GeoBroadcasting.
c) MAC_ ID (MID).
d) LL Address.
e) Position Vector (PV).
f) Location Service.
g) Beacon packets.
h) GN_Header.

21.8 Describe the services provided by the protocol layer entities employed by the ETSI GeoNetworking (GN) system as shown in Fig. 21.7.

21.9 Describe and explain the structure of an ETSI GeoNetworking packet.

21.10 Describe the function and operation of forwarding algorithms in an V2V system.

21.11 Describe and illustrate the operation of the following ETSI GeoUnicast forwarding algorithms:
a) Greedy Forwarding (GF) algorithm.
b) Contention-Based Forwarding (CBF) algorithm.

21.12 Describe and illustrate the operation of the following ETSI GeoBroadcast forwarding algorithms:
a) GeoBroadcast Simple (S) algorithm.
b) GeoBroadcast Advanced Forwarding (AF) algorithm 1.

666 | *21 Communications and Traffic Management for the Autonomous Highway*

21.13   Describe the operation of Distributed Congestion Control (DCC) schemes that are employed to regulate message flows in an ETSI V2V system.

21.14   Describe the operation of a V2V Vehicular Backbone Network (VBN), including the following entities:
a) Backbone Nodes (BNs).
b) Relay Nodes (RNs).
c) Backbone Network (Bnet).
d) Access Network (Anet); Anet cell.
e) VBN architecture.

21.15   Describe methods that are used under the VBN architecture for performing the following communications networking functions:
a) Intra-Anet communications; including the use of methods to attain Anet spatial reuse.
b) V2I communications; use of V2I gateways.
c) Backbone Network (Bnet) synthesis and Bnet communications networking methods.

21.16   Explain the principle of operation of the VBN network system illustrated in Fig. 21.11, under the use of a spatial-TDMA medium access control scheme.

21.17   Define and explain the calculation of the following VBN performance metrics:
a) Broadcast Capacity.
b) Transport Throughput.

21.18   Explain and discuss the performance results shown in Fig. 21.12 relating to the optimal setting of inter-BN distance values and the impact of spatial-reuse (M) and vehicle density level parameters. Also describe the method used to determine the best joint setting of the MCS and the system's inter-BN distance levels.

21.19   Describe the configuration and operation of a VBN-based V2V network that is aided by an Infrastructure core network, as illustrated in Fig. 21.13.

21.20   Describe the configuration and operation of a VBN-based V2V network that is aided by an Infrastructure core network and makes use of vehicular platoons, as illustrated in Fig. 21.14.

21.21   Describe and explain the definition, use and operation of the different types of CV2X communications links and elements, including the following:
a) Uplink, downlink, and sidelink links.
b) Infrastructure access interfaces, including Uu interface.
c) I2N links and Sidelinks using PC5 interface in providing V2V, V2I, and V2P communications.
d) Modes 3 and 4 resource allocation methods for V2X sidelink operations.
e) Channel Burst Ratio (CBR) and Channel Occupancy Ratio (CR).

21.22   Describe the 5G NR V2X 3GPP use cases that are used for performing the following functions:

a) Vehicle platooning.
b) Advanced driving.
c) Extended sensing.
d) Remote driving.

21.23 Describe and explain the performance requirements specified by 3GPP for the following operations:
a) Vehicle Platooning.
b) Advanced Driving.
c) Extended sensing.
d) Remote Driving.

21.24 Describe and explain the following QoS performance parameters and terms:
a) PC5 QoS Flow ID (PFI).
b) Guaranteed Flow Bit Rate (GFBR) and Maximum Flow Bit Rate (MFBR).
c) Delay Critical GBR.
d) Maximum Data Burst Volume (MDBV).
e) Packet Delay Budget (PDB).
f) V2N communications over the Uu interface.
g) Dual Active Protocol Stack (DAPS).

21.25 Define the various levels of Vehicle Automation and describe the processes employed at each level.

21.26 Describe the concept of *vehicle platoons* and explain the advantages gained by synthesizing platoon configurations for supporting and sustaining traffic management and data communications for autonomous vehicular highway flows.

21.27 Discuss the benefits gained through the use of satellites and UAV platforms in supporting the operations of connected autonomous vehicle systems.

21.28 Explain the Fundamental Equation of Traffic Flow, as presented in Eq. 21.10. Define and illustrate its flow rate, velocity and density metric components.

21.29 Derive and explain the relationship presented by Eq. 21.12. Assume values for the corresponding parameters and calculate and draw ensuing figures that illustrate the variation of the flow rate as a function of the vehicular speed.

21.30 Consider the model described in the chapter for autonomous vehicular flows when no delay constraints are imposed. Assume two different sets of system parameters and use Eqs. 21.13 and 21.14, to calculate the following:
a) The optimal speed level.
b) The ensuing optimal flow rate level.

21.31 Explain and discuss the flow-rate vs. speed and speed vs. flow-rate performance functions displayed in Fig. 21.19, identifying the characteristics of the operations in uncongested and congested modes.

668 | *21 Communications and Traffic Management for the Autonomous Highway*

21.32 Explain and discuss the flow-rate vs. density performance function displayed in Fig. 21.20(a), identifying the characteristics of the operations in uncongested and congested modes.

21.33 Consider the *Single-link Single-lane* model presented in the chapter for the traffic management of autonomous highway vehicles under queueing and transit delay limits.
   a) Derive and explain the formula presented by Eq. 21.19 for the calculation of an upper bound on the delay constrained maximum vehicular flow rate (and admission rate) level.
   b) Explain the performance results presented in Fig. 21.22.
   c) Assume two different sets of system parameters and obtain performance curves corresponding to those presented in Fig. 21.22.

21.34 Consider the *Single-link Multiple-lane* model presented in the chapter for the traffic management of autonomous highway vehicles under queueing and transit delay limits and explain the derived performance formulas.

21.35 Consider the *Multiple-link Single-lane* model presented in the chapter for the traffic management of autonomous highway vehicles under queueing and transit delay limits.
   a) Explain the method used to carry out the system's performance analysis and design.
   b) Explain the performance model and analysis carried out for the example presented in Fig. 21.24.
   c) Assume a different illustrative layout of a highway network and its loading by multiple vehicular flows. Derive formulas for calculating the ensuing highest feasible delay-constrained flow throughput levels.

21.36 Consider the *Multiple-link Multiple-lane* model presented in the chapter for the traffic management of autonomous highway vehicles under queueing and transit delay limits. Explain the approach noted for carrying out its performance analysis and design.

21.37 Consider a highway segment (A,B) connecting nodal (ramp, access and exit point, intersection) locations A and B. Assume this link to be 20 km long and to support two lanes: lane 1 (the fast lane) and lane 2. It is followed (linearly) by a two-lane highway segment (B,C) of length 40 km. Class-0 vehicles arrive at intersection node A from another highway at a rate of $\lambda_0 = 600$ [vehicles/h]. These vehicles are directed to travel along lane-2 of highway segment (A,B).

Vehicles arrive (in accordance with a Poisson process) at node A also from a local on-ramp at a rate λ_A [vehicles/h]. These vehicles are categorized into two classes, identified as Class-1 and Class-2. They are admitted into the highway in accordance with the following considerations. Class-2 vehicles that are admitted at on-ramp A join Class-0 vehicles and are together (being reorganized into platoons, in accordance with Lane-2 platoon and inter-platoon parameters) guided to travel along Lane-2 of link (A,B). Note: Class-0 vehicles are given priority for admission into link (A,B). Hence, we assume that the service rate available to admit and accommodate Class-2 vehicles is reduced by the Class-0 arrival rate.

60% of the vehicles traveling in Lane-2 of (A,B), whether of Class-0 or of Class-2, depart the highway at node B. The remaining 40% continue their travel along Lane-2 of (B,C).

Problems | 669

Class-1 vehicles that have been admitted at on-ramp A are guided to travel along Lane-1 of link (A,B) (organized into platoons in accordance with Lane-1 parameters). They continue their travel on Lane-1 of highway segment (B,C), employing then the same speed and platoon structure as used across segment (A,B).

The system traffic manager organizes vehicle flows across the highway's segments in accordance with the following principles:

a) All highway vehicles are arranged to travel in platoons, whereby each platoon (across either lane) is assumed to be of size $N = 12$ (i.e., containing 12 vehicle members).

b) The flow of vehicles across Lane-2 of highway segment (A,B) is organized in such a manner that the flow rate of the vehicles in this segment achieves its highest feasible level under imposed on-ramp delay level D_{0q} and targeted speed parameter of $V_{tg}(2)$. Note: We assume that separate on-ramp access queues are formed for Lane-1 and Lane-2 vehicles.

c) The vehicular flow across Lane-1 traverses segments (A,B) and (B,C) and is also denoted as Lane-1 of link (A,C). It is organized in such a manner that the flow rate of vehicles along this combined segment stretch achieves its highest feasible level under required on-ramp delay level (at A) D_{0q} and a targeted speed parameter value along (A,C) of $V_{tg}(1)$.

Using the model developed by Professor Izhak Rubin et al. and presented in the chapter, calculate the following, assuming that the system parameters that are not subjected to determination by your calculations use the following values:

$u = 4\ m/(s^2); \xi = \Delta = 0.2; D_{0q} = 4\ \text{minutes}; V_{tg}^{(1)} = 120\ \text{km/h}; V_{tg}^{(2)} = 80\ \text{km/h}; s = 5\ \text{m}; \psi^{(1)} = 1.4; \psi^{(2)} = 2; P = \text{Packet length (including overhead)} = 2400\ \text{bits}.$

a) Design the layout of vehicle flows along Lane-2 of segment (A,B). Accordingly, determine the following:

   i) The maximal level of admitted (from the on-ramp) rate at A of vehicles that are routed to Lane-2 of (A,B), $\lambda_A^{(2)}$ [vehicles/h].

   ii) Calculate the speed assigned for the travel of vehicles along Lane-2 of (A,B) based on the above-derived design.

   iii) Calculate the density of vehicles traveling along Lane-2 of (A,B) based on the above-derived design.

b) Design the vehicular formation that is used for the flow of vehicles traveling along Lane-1 of segment (A,C). Accordingly, determine the following:

   i) The maximal admitted (from the on-ramp) rate level of class-1 vehicles that are routed to Lane-1 of (A,C), $\lambda_A^{(1)}$ [vehicles/h].

   ii) The speed of vehicles traveling along Lane-1 of (A,C) based on the above-derived design.

   iii) The density of vehicles traveling along Lane-1 of (A,C) based on the above-derived design.

c) Determine the best speed to use for vehicles traveling along Lane-2 of (B,C), $v(B, C)^{(2)}$, by assuming that the traffic flow design performed along this lane aims to minimize the travel time of vehicles moving along this lane and segment. Assume the platoon size to be set at $N = 12$.

22

Networking Security

The Objective: *Provide secure message transport across a communications network such as the Internet. Protection of the confidentiality and integrity of messages as they are transported across the network and as they are exchanged between end users, clients, and servers. Protection from unauthorized access into and modification of resources, algorithms, and protocols of a communications network, including operations at the data, control, and management planes. Secure message transfer and networking while preserving network performance and availability.*

    The Methods: *Under a **symmetric-key encryption** system, communicating users exchange an encryption key, which is used as their shared secret. This key is employed to encrypt sent messages and to decrypt received messages. Such secure operations are generally conducted in an efficient and timely manner. A secure key exchange process is efficiently implemented by using a **public-key encryption** system. Under such a system, each user configures a public-key and a secret private-key. The public-key is distributed openly to other users. The private-key is kept as a secret by its user. The source user encrypts a message by using the public-key of the destination user. The destination user decrypts the received message by using its secret private-key. Procedures are also available for securely transporting a digital signature that enables a targeted destination entity to confirm the identity of the author of a received message and to determine its integrity. Schemes are employed to preserve message confidentiality, integrity, and identity as it is transported across a network during a client–server session.*

22.1 Network Security Architecture and Cybersecurity Frameworks

End users that communicate with remote users, servers, and service organizations across a network system have been facing a wide range of security issues. Even more severe security threats are presented when users communicate across public networks such as the public Internet. End users expect a network system to provide them with reliable and robust connections, rendering high-quality secure message transport experience. The communications of critical data across a network generally requires the use of a transport mechanism that provides confidentiality and integrity. A recipient of data transported across a network would often wish to confirm the validity of its claimed source, as well as its integrity, in checking that its data contents were not corrupted by unauthorized intruders during its transport across the network. Network system message transfer protection measures must assure the secure, effective, and efficient operation of the network.

    Digital attacks can be targeted to harm the hardware and software entities of a network system, affecting the mechanisms that are used by the system to function, and damage or steal information records stored at user facilities, at data centers, or directly transferred by users. They may attempt to steal-sensitive information embedded in messages that are transported by

---

Principles of Data Transfer Through Communications Networks, the Internet, and Autonomous Mobiles, First Edition. Izhak Rubin.
© 2025 The Institute of Electrical and Electronics Engineers, Inc. Published 2025 by John Wiley & Sons, Inc.

22 Networking Security

users across a network such as the Internet. Such intrusions may capture confidential credit card numbers, financial data, personal identity data, healthcare information, critical business information, and public infrastructure data. They may interrupt or subvert the secure and orderly activities of business, education, healthcare, governmental (including military and defense units), and e-commerce organizations, and damage the orderly functioning of social sites that serve many individuals and organizations. They may aim to harm computer and network systems that are used to operate and control critical urban infrastructure systems, such as those run by water/electric/gas utilities and degrade the functioning of transportation companies. They may incapacitate networks that are critical to the operation of public safety systems that are designed to prevent, resolve, or mitigate the impact of such digital attacks.

Examples of common cybersecurity-oriented attacks include the following: 1) Phishing— whereby users receive fraudulent emails that are similar to those that they receive from valid sources, which are used to steal confidential data and connectivity information. 2) Malware— including software that penetrates user devices, aiming to damage them. 3) Ransomware—software schemes that are used to extort money by blocking a user's access to his files or computer system. 4) Social engineering oriented attacks that are designed to trick a user to divulge sensitive information.

A communications network is designed to provide its users with high quality of experience (QoE) by instituting protocols and algorithms that implement quality of performance (QoP) measures. To meet this aim, it is essential to protect the network's means that are used to provide users with measures of **confidentiality, integrity, and availability (CIA)**, also known as the CIA triad. *Confidentiality* measures assure a user that its critical messages are not intercepted and read by unauthorized entities. In preserving the *integrity* of message transport, a user is assured that a message that it receives has not been altered by an intercepting intruder. Measures are instituted so that destination users are able to trust the identity of the source cited in a received message. In protecting the *availability* of the network system, we assure users that the network is operational and capable of performing its transport mission during a prescribed period of time. Network software and hardware resources, algorithms, protocols, and transport processes are secured in a manner that protects them from being degraded by unauthorized intruders.

The Open Security Architecture (OSA) organization defines *Information Technology (IT) security architecture* as "the design artifacts that describe how the security controls (security countermeasures) are positioned, and how they relate to the overall information technology architecture. These controls serve the purpose to maintain the system's quality attributes: confidentiality, integrity, availability, accountability and assurance services." Also known as **cybersecurity**, such measures serve to protect networked systems and associated applications from digital attacks.

To protect a network system from threats, it is necessary to institute security measures that involve all critical elements of the system. Included are mechanisms that provide security measures for the execution of applications, networking, data storage, data processing, and information transport. Critical data and its associated data processing, storage, and networking processes must be protected from unauthorized access, modification, and theft.

Mobile security is of key importance for the protection of mobile platforms, including systems that use wireless access networks. Such measures support users and other entities that provide networking mechanisms for mobile platforms, including autonomous robotics and vehicle systems.

A security system architecture implements means for assuring the effective and efficient functioning of a targeted system, in protecting data and networking operations even when the security system itself is under attack. To be effective, system-wide protection means are implemented.

They encompass: end-users, their facilities and associated servers, and service resources; the boundaries between end users and the network; the access network; and the core network. For mobile users, such domains and their boundaries must be dynamically identified and protected.

Security mechanisms provide protections for software and hardware entities. Across the network system, each protocol layer is secured. Means are employed for protecting user sessions, and corresponding end-to-end data message transactions. Secure operations are implemented at the network's data, control, and management domains.

Security is enhanced by using protections that are based on segmentation of the network into distinct physical and logical domains, and into subsystems and subnetworks. Segmentation can be based on organizational, functional, service category, traffic type, or other attributes. Such a segmentation enhances the effectiveness of processes that are used to control access and isolate infections. They aid in the management and operation of mechanisms that protect identified and combined critical system elements.

Security mechanisms are enacted at the edge of a network to protect end-user entities and organizations. They protect the access of a user system to the transport network system. Protection mechanisms that are used for this purpose include the employment of **firewall**-oriented filtering processes. Interconnected LAN systems are often logically segmented into multiple Virtual Private Networks (VPNs), whereby a traffic flow associated with a specific user group is authorized access to only specifically defined resources and network systems. Access-control filtering processes regulate the access of flows into and out of identified subnetworks and network domains and user-network edges. Such mechanisms often employ artificial intelligence (AI) and machine learning (ML)-based methods, dynamically learning the statistical patterns of authorized traffic processes and of unauthorized intrusion processes. Traffic signatures are identified and used to detect and mitigate threats.

For further protection, end-users often institute a **DMZ (de-militarized zone) network**, which is placed between a private network and a public network such as the Internet. To help secure its private network, a user organization would then place services and resources that interface the Internet (such as web servers, and DNS, FTP, VoIP, and other assets) in the DMZ. These services are provided only limited access into a private network, isolating the latter from the (less secure) public network.

The *NIST Cybersecurity Framework*, as described in [36], includes the following key functions (see Fig. 22.1):

1) **Identify**:
   a) Identification of critical processes and assets in the organization.
   b) Documentation of information flows.

Figure 22.1 National Institute of Standards and Technology (NIST) Cybersecurity Framework.

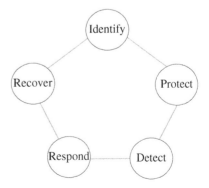

674 | 22 *Networking Security*

    c) Maintaining hardware and software inventory.

    d) Establishment of policies for cybersecurity, including roles and responsibilities; risk management.

    e) Identification of threats, vulnerabilities, and risks to assets.

2) **Protect**:

    a) Access control: managing access to assets and information.

    b) Data security: protection of sensitive data; information protection processes and procedures.

    c) Conducting regular backups.

    d) Protection of devices.

    e) Management of device vulnerability; maintenance.

    f) Training of users.

3) **Detect**:

    a) Characterization of anomalies and threat events, development of detection processes, and use of continuous security monitoring to detect threats.

    b) Testing and updating detection processes.

    c) Knowledge of the expected data flows of the enterprise.

    d) Maintenance and monitoring of logs.

    e) Understanding the impact of cybersecurity events.

4) **Respond**:

    a) Development of response plans and mitigation processes.

    b) Testing and analyzing response plans.

    c) Updating and improving response plans.

    d) Coordination with internal and external stakeholders, including service providers.

5) **Recover**:

    a) Development of recovery plans.

    b) Communications with internal and external stakeholders.

    c) Updating recovery plans.

    d) Management of public relations.

Other network system security architectures have been defined and presented by commercial and governmental outfits and by Standards organizations. The latter include the International Organization for Standardization (ISO) and the International Electrotechnical Commission (IEC). They have produced the ISO/IEC 27000 family of standards, which include an information security management system (ISMS) standard. Cybersecurity standards for road vehicles have been under development by ISO and SAE working groups, including ISO / SAE (Society of Automotive Engineers) Standard 21434, entitled "Road vehicles–Cybersecurity engineering." Cybersecurity standards have also been developed by the European Union and other standards committees in the USA and throughput the globe. For example, the ETSI EN 303-645 standard provides a set of requirements for security in Internet of Things (IoT) devices.

In the following sections, we focus on outlining methods for *secure end-to-end message transfer across a network system*. Security objectives for such transfers include: **confidentiality**, so that only the intended destination is capable of reading its message; **integrity**, assuring the destination party that the message has not been altered in transit; **authenticity**, via the use of **digital signature**, certifying to the receiving party the identity of the party that has produced and signed the message.

22.2 Message Confidentiality: Symmetric Encryption

Interacting end users often establish an end-to-end session that is protected through the use of passwords. Messages and passwords may be transported across a public communication network, such as the Internet. Unauthorized parties may be able to monitor and access user and/or network hardware facilities or software tools and programs and intercept these messages, gaining access to their sensitive contents. A commonly used approach to secure **message confidentiality** is to **encrypt** sensitive data in messages that are transferred across a network.

Under an encryption process, the original text of a message, known as **plaintext**, is processed to produce encoded text, known as **ciphertext**, through the use of an **encryption algorithm**. The encryption process uses a *secret code*, known as a **key**, or a **private key**. To preserve confidentiality, it is essential that the decoding of the message, identified as a **decryption** process, which maps the received ciphertext back to its original plaintext, can only be performed by an authorized destination user. An unauthorized party, which does not have access to the private key, should not be able to decipher the message and would thus not be able to gain access to its data contents. **Symmetric Key Encryption (SKE)** and **Public Key Encryption (PKE)** are commonly used encryption methods. We discuss the former one in this section and the latter one in Section 22.3.

Under **Symmetric Key Encryption (SKE)**, a **symmetric key** is employed. It is used to both encrypt the message and to decrypt it, employing an **encryption/decryption algorithm**, also identified henceforth in brief as an **encryption algorithm**, which the communicating parties agree upon. Communicating entities regard the **private symmetric key** as their **shared secret**. Unauthorized entities do not possess the symmetric key and are therefore not able to decrypt encrypted messages that are exchanged between the parties, even when the employed encryption algorithm (but not the shared key) is publicly known.

This process, known as **symmetric cryptography**, is illustrated in Fig. 22.2. The illustrated scenario shows the sending of an encrypted message between two end-users, identified here as Sammy and Jonny. Sammy uses the shared private key to encrypt her plaintext message by using an encryption algorithm. She then sends the encrypted ciphertext message across a network, such as the Internet, to Jonny's facility (or computer). Upon reception of the ciphertext message at Jonny's facility, the message is decrypted by using the shared secret key, which is locally stored, and the same encryption/decryption algorithm that was used by Sammy's system. The original plaintext message is then recovered.

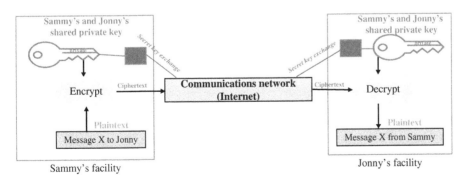

Figure 22.2 Symmetric Key Encryption.

The use of symmetric key cryptography carries several key advantages. It is computationally practically infeasible to decipher a well-encrypted message by a party that does not possess the shared private key. This has been demonstrated to be the case for variety of widely used encryption schemes. Encryption algorithms often operate by encrypting the message stream on a block by block basis. For example, a message may be divided into blocks where each block may contain 128 bits of data, using key sizes that are 128 bits to 256 bits long. Symmetric-key encryption/decryption algorithms perform their operations in a timely and computationally effective manner.

The symmetric cryptographic process also provides a measure of **authentication** since the decrypting party must employ the same symmetric-key used by the encrypting user. Hence, as long as the symmetric key is kept secret by the parties, the ability by a receiving user to decrypt the message assures this user that the message was produced by an authorized party (the one that keeps their shared secret). Thus, Jonny's ability to decrypt the message sent by Sammy assures him that the message was produced by Sammy, as nobody else is in possession of their shared secret key.

For the symmetric key cryptographic process to retain its strong measures of confidentiality and authentication, it is critical that the shared key is kept secret. Hence, the employed **key exchange process** is a critical element of this operation. If not performed properly and securely, the effectiveness of the scheme would be degraded. Consider a situation under which an unauthorized intruding entity gains access to the secret key. In this case, the intruding party may use the stolen private key to decrypt messages that are transported across the network. The intruder can then modify the contents of a captured and decrypted message, replacing its contents with newly composed data, and then encrypt (assuming the encryption scheme to be known) it and send it across the network to the proper destination party. The latter would be able to decrypt it but may not be able to confirm its authenticity.

Thus, it is critical that the secret key is exchanged between authorized communicating parties in a secret manner. It is often performed through a process that takes place prior to the start of the message transport phase, whose data must not leak to unauthorized parties. Earlier systems employed a physically secure channel for passing a secret key between communicating parties. More recent systems utilize a *PKE* process for the distribution of a symmetric secret key, in a manner that is described in Section 22.3. A new secret key is often produced for each new session and when new session participants are involved.

A pseudo-random key generator is typically used to produce the employed symmetric session key. To preserve high security, it is essential that these generators synthesize streams that offer a very high measure of randomness, so that an eavesdropper would not be able to exploit any embedded statistical biases.

A multitude of symmetric encryption algorithms that offer ultrahigh measures of privacy have been available and employed. The **Advanced Encryption Standard (AES)** has been adopted by the US government in 2002 as a US standard. It superseded the *Data Encryption Standard (DES)* that was published in 1977. Under AES, the following process is currently employed. Data stream encryption is performed on a block by clock basis, using a fixed block size of 128 bits, and a key size of 128, 192, or 256 bits. Multiple rounds of substitution–permutation operations, which depend upon the selected encryption key, are applied over a 4 x 4 array of bytes. The number of transformations is set to 10 rounds for 128-bit keys, 12 rounds for 192-bit keys, and 14 rounds for 256-bit keys. For increased security, a plaintext is modified by first randomizing it and then encrypting it. Two ciphertexts that correspond to two identical original plaintexts would not be the same.

22.3 Public Key Encryption (PKE)

Under **Public Key Cryptography**, a user maintains a pair of keys, consisting of a **public key** and a **private key**. The user makes the public key publicly known, so that it is openly available to users that wish to communicate with it. In turn, its private key is privately held, keeping it secret. Messages sent by user-A to user-B are encrypted by user-A by using the public key that user-B has made publicly known. Upon receipt of the encrypted message, user-B will proceed to decrypt it by using the paired private key that it holds and keeps secret.

The public-key cryptographic process is illustrated in Fig. 22.3. Sammy is shown to produce a plaintext message that she wishes to send to Jonny. It is essential for her that the message is transferred in a confidential manner, so that its contents would be available only to Jonny. Other parties who may intercept the message should be unable to decrypt it. Sammy encrypts the message by using Jonny's public key. The resulting ciphertext is then transported across a network, such as the Internet, to Jonny's facility. At the latter, the received message is decrypted by using Jonny's private key, which is secretly held by Jonny. The outcome produces the intended plaintext message.

Processes used by cryptographic algorithms to generate such key pairs are based on the mathematics of problems addressing **one-way functions**. To illustrate, consider the problem of factoring a large number. In one direction, one can readily compute such a large number by multiplying two large prime numbers. In the other direction, it is computationally very difficult to find the factors of the latter large number.

The concept of public-key cryptography was published in 1976 by Whitfield Diffie and Martin Hellman [14]. An implementation of such a function and a corresponding algorithm for producing private and public keys for a public key cryptosystem was described in 1977 by Ron Rivest et al. [41], referred to as **RSA**. Its security derives from the difficulty of factoring a number that is computed as the product of two large prime numbers. A brief overview of their approach is noted in the following, without exploring its mathematical and algorithmic details. Its involved computations make it a relatively slow algorithm. Hence, it is often not used to directly encrypt data but rather to securely exchange shared keys that are then used for the implementation of a symmetric-key cryptographical system.

The RSA algorithm involves the use of the following very large positive integers: a **public key exponent** e_p, a **private secret key exponent** e_s, and a **modulus** integer n. The principle of operation is based on the following properties:

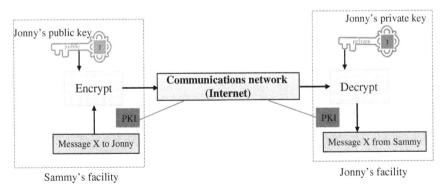

Figure 22.3 An Illustrative Message Transfer Under Public Key Cryptography.

678 | *22 Networking Security*

1) It is practical to find such integers that meet the identity:

$$(m^{e_p})^{e_s} \equiv m \,(\text{mod}\, n), \tag{22.1}$$

for all integers m such that $0 \leq m < n$.

2) Given the integers e_p and n, and even m, it is very difficult to determine the integer e_s.

The *modulo* (denoted as *mod*) operation produces the remainder of a division of one integer by another. For example, $17\,(\text{mod}\, 5) = 2$, and $127\,(\text{mod}\, 5) = 2$ as well, so that we can write that $127 \equiv 17\,(\text{mod}\, 5)$. For certain operations, it is convenient to have the order of operations reversed to have:

$$(m^{e_s})^{e_p} \equiv m \,(\text{mod}\, n), \tag{22.2}$$

for all integers m such that $0 \leq m < n$.

Under RSA cryptography, the public key consists of the modulus n and the public key exponent e_p. The private key consists of the secret key exponent e_s, and the same modulus n. The integer m is derived from the message. User-A that wishes to send a plaintext message M to user-B first maps the message into an integer m by using an agreed upon reversible mapping (so that m is readily inverted to yield M) known as a *padding protocol*, such that $0 \leq m < n$. User-A then produces the ciphertext c by employing user's-B's public key exponent, e_p to encrypt its message m, performing:

$$c \equiv m^{e_p} \,(\text{mod}\, n). \tag{22.3}$$

Upon receiving the encrypted message, user-B uses her secret private key exponent e_s to recover the plaintext message m, by performing:

$$c^{e_s} \equiv (m^{e_p})^{e_s} \equiv m \,(\text{mod}\, n). \tag{22.4}$$

The original plaintext message M is then recovered by using the inverse mapping specified by the padding function.

Under RSA, the process followed by a user for selecting its public and private codes requires the user to compute the modulus n as a product of two large distinct prime numbers, p and q. They are chosen as random numbers of similar magnitude and different lengths. The user keeps the numbers p and q secret, while n is publicly released. The length of n is the key length. Using the selected values for p and q, the user follows various computations that produce its public encryption and private encryption exponents, e_p and e_s, respectively. For this purpose, the following process has been described. The number $\lambda(n) = lcm(p - 1, q - 1)$ is computed, whereby $lcm(a, b)$ denotes the least common multiple of a and b, so that it represents the lowest positive integer that is divisible by both numbers. Then, e_p is chosen as an integer that satisfies $1 < e_p < \lambda(n)$, so that e_p and $\lambda(n)$ are co-prime (so that one is not an integral factor of the other). Then, e_s is computed as a solution to the modular identity $e_p e_s \equiv 1 \,(\text{mod}\, \lambda(n))$. It is noted that an intruder that is able to factor the modulus number n into its p and q factors would then be able solve the latter identity, obtaining the secret encryption exponent e_s. No polynomial-time (i.e., time effective) method for factoring such a large integer by using a current computer is presently known.

How do we know that an advertised public key for an organization is one that is associated with this organization rather than announced by an intruder that intends to steal the public key identity of the targeted organization? It is clearly essential that a user is able to trust the authenticity of the public key that is used to encrypt the messages that are sent to a targeted entity. If intruder C were to announce its own public key, $e_p(C)$, and claim that it represents the public key of organization B, messages sent by user A to user B may be encrypted by using $e_p(C)$. Intruder C would then be able to use its own private key to decrypt the messages sent by user A to user B.

To resolve this problem, a **Public Key Infrastructure (PKI)** is maintained. It produces, distributes, and manages public-key cryptographic systems. In particular, it is responsible for creating, distributing, storing, validating, revoking, and managing a PKE system. It issues certificates that bind public keys with the identities of their respective (personal or organizational) owners. Through this process, it is possible to trust the ability of a PKE system to provide for the secure distribution of messages to their intended destinations. A certificate binding operation can be carried out by using an automated process over a secure network connection, or via human interactions. For this purpose, *registration authority (RA)* entities are configured. Under PKI, certification authority (CA) domains are defined and domain based CA entities are identified. A validation authority (VA) can act as a third party that provides this information on behalf of a CA. The X.509 standard defines widely used CA formats.

22.4 Digital Signature

Secure **digital signature** schemes are used to provide an entity that receives a message with a trusted authentication of the source that created the message. Several digital signature schemes have been developed and widely employed. In the following, we describe the elements of a digital signature scheme that is based on the above-described RSA public-key cryptosystem [41].

The concept is illustrated in Fig. 22.4, considering a digitally signed message that is sent over the Internet from Sammy to Jonny. Under this algorithm, the process undertaken at the sender's facility computes a *hash of the message*. A hash function maps the source digital message stream into a relatively short (fixed length) number that is identified as a **message digest**, or a message hash. A hash function maps distinct digital streams to distinct hash (also known as digest) values. The computation of the hash is performed at very high speed. The mapping is done in a one-way manner, so that given the hash value, one cannot determine the original stream that has produced it. Two digital streams that produce the same hash can be effectively assured to be the same. Various such hash functions have been widely employed.

The message digest is then encrypted by using the source's private key, yielding an integer that represents the message digital signature. For added security, the original message is often padded by the addition of a random sequence of bytes.

The data message and the encrypted digital signature message are sent by Sammy to Jonny across a network such as the Internet. Upon reception of this combined message at Jonny's facility, a *verify algorithm* is performed to authenticate the sender of the message. The message hash is computed first, producing the original hash plaintext (as computed at the source). Note that if the

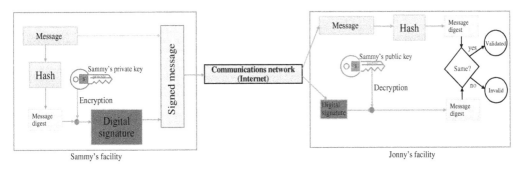

Figure 22.4 A Digital Signature Scheme Employing Public Key Cryptography.

combined message was encrypted by Sammy to provide for confidentiality in its transport across the network, then it would be first decrypted at Jonny's facility, yielding the source data message stream's plaintext and the encrypted message hash (i.e., the digital signature). The received encrypted message hash is then decrypted by using Sammy's public key, yielding the original hash plaintext. The latter is then compared with the plaintext hash of the data message. If they agree, the receiving party is assured that the signature is valid and that this message was signed by Sammy.

Other digital signature schemes include elliptic curve-based signature algorithms. They produce shorter key lengths and have been shown to offer improved performance.

22.5   Secure Exchange of Cryptographic Keys

It is more computationally efficient to use a symmetric-key cryptographic system for implementing confidential message transfer over an un-secure network such as the Internet. However, this system requires the source and destination entities to hold a shared secret key. For this purpose, the parties must be provided a mechanism for exchanging cryptographic keys in a secure manner. The exchanged keys must be securely encrypted, preventing an entity that intercepts their transmission from being able to decrypt them.

Processes, protocols, and associated algorithms have been developed to enable the secure exchange of encryption keys over an un-secure network such as the Internet. Such an interchange allows the communicating parties to configure their shared secret key. The latter is then used by the communicating participants as the encryption/decryption key.

In the following, we illustrate the operation undertaken by such a secure key exchange protocol. It is based on the concept of public-key cryptography, as described in a paper published in 1976 by Diffie and Hellman [14]. It is known as a Diffie–Hellman (DH) key exchange scheme, or as the Diffie–Hellman–Merkle key exchange scheme, in recognition of contributions to public-key cryptography made by Ralph Merkle.

A simple implementation version of the DH algorithm includes a prime integer n that is used as the **modulus** and a prime integer g that is used as the **base**. Both parameters are kept public and may thus be known to an eavesdropper. The base g is selected as an integer that is a *primitive root modulo n*. This means that for any positive integer m that is co-prime to n, there is some integer k for which $g^k = m \pmod{n}$. To illustrate, assume $n = 5$ and $g = 2$. One readily finds that $g^k \pmod{n}$ yields the periodic sequence of numbers [2, 4, 3, 1, 2, 4, 3, 1, 2, …], so that the desired presentation holds for every possible positive integer m that is co-prime to 5, which must be a member of the set of all integers whose values vary from 1 to $n - 1 = 4$. It is also noted that the set $g^k \pmod{n}, k \geq 1$, forms a cyclic sequence of positive integers whose cycle length is equal to $n - 1$.

The DH algorithm for the production of a shared key between user-A and user-B then proceeds as follows. The parties *publicly* mutually agree on proper values for the modulus n and the base g. For security purposes, the modulus is selected as a large prime integer (generally, at least hundreds of digits long) while the base g can be selected as a small positive integer that is a primitive root modulo n. As illustrated in Fig. 22.5, the key exchange process proceeds as follows:

1) **User-A's Formation of a Private Key**: User-A selects a secret integer, exponent e_{sA}, which serves to form its **private key**. It is randomly chosen in the range $1 < e_{sA} < n$.
2) **User-A's Formation of a Public Key**: User-A derives its public key by calculating $e_{pA} = g^{e_{sA}} \pmod{n}$, which is known as *modular exponentiation* and can be rapidly executed at low computational complexity even for large numbers.

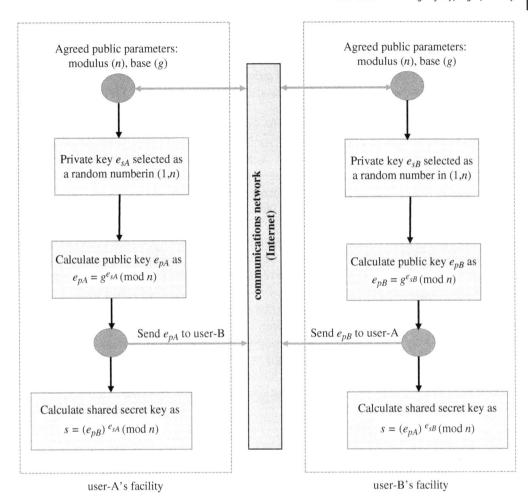

Figure 22.5 The Diffie–Hellman (DH) Secure Key Exchange Algorithm.

3) **Public Transport Across the Network by User-A**: User-A openly sends its public key, exponent e_{pA}, across the network to user-B.
4) **User-B's Formation of a Private Key**: User-B selects a secret integer, exponent e_{sB}, which serves to form its **private key**. It is randomly chosen in the range $1 < e_{sB} < n$.
5) **User-B's Formation of a Public Key**: User-B derives its public key by calculating $e_{pB} = g^{e_{sB}} \pmod{n}$.
6) **Public Transport Across the Network by User-B**: User-B openly sends its public key, exponent e_{pB}, across the network to user-A.
7) **The shared secret key** s is computed by user-A by calculating it as $s = (e_{pB})^{e_{sA}} \pmod{n}$.
8) **The shared secret key** s is computed by user-B by calculating it as $s = (e_{pA})^{e_{sB}} \pmod{n}$.

It is noted that by following this process, user-A and user-B derive the same shared secret key s as:

$$(e_{pB})^{e_{sA}} \pmod{n} = g^{e_{sA} e_{sB}} \pmod{n} = (e_{pA})^{e_{sB}} \pmod{n} = s. \qquad (22.5)$$

The secrecy of the described key exchange scheme is rooted in the observation that it is very difficult (and computationally practically infeasible by using current computing resources) to

682 | *22 Networking Security*

calculate $g^{e_{sA}e_{sB}} \pmod n = g^{e_{sB}e_{sA}} \pmod n$, by knowing n, g, e_{pA}, and e_{pB}. It is also difficult for user-A to solve for user-B's secret key and for user-B to solve for user-A's secret key based on known data and parameters. Similarly, an intruding interceptor is not able to solve for the shared key. It is also noted that an interceptor is able to monitor the publicly available public keys, but by computing their product, the interceptor obtains $g^{e_{sA}+e_{sB}} \pmod n$, which does not yield the desired shared secret key.

The DH algorithm by itself does not provide for user authentication. It is necessary to provide user authentication (or use other variants of the algorithm) to prevent a *man-in-the-middle* attack. Under such an attack, an intruder can separately form a mutual secret with user-A (masquerading as user-B) and with user-B (masquerading as user-A), and then decrypt and re-encrypt (and gain access to the contents of) messages that are exchanged between these users.

This is resolved by algorithms that incorporate secure user authentication into the key exchange process. For example, the Station-to-Station (STS) protocol incorporates into the DH algorithm a user authentication process. Such a protocol is identified as an authenticated key agreement with key confirmation (AKC) protocol. The process includes the following steps (performed by using the respective modular computations). User-A sends to user-B the underlying exponential $g^{e_{sA}}$. Using the DH scheme, user-B then computes the shared secret key. User B then sends to user A his exponential $g^{e_{sB}}$ as well as a **signed** message that contains the exponentials $(g^{e_{sB}}, g^{e_{sA}})$, which is encrypted by using the shared key. Using the received data, user-A computes the shared key, and then proceeds to decrypt and verify user-B's signature by using user-B's asymmetric public-key. User-A then creates a message that contains the corresponding exponentials, signs it with her asymmetric private-key and encrypts the signature by using the shared key. This message is then sent to user-B, which would then decrypt the message and verify user-A's signature by using her asymmetric public key.

22.6 Secure Client–Server Message Transport Over the Network

In this section, we overview the operation of the **Transport Layer Security (TLS)** protocol. It is widely used for message communications between clients and servers over networks, including the Internet. The protocol employs cryptographic means to secure such communications, providing confidentiality, integrity, and authenticity. It is executed at the application layer, so that a TLS application layer protocol entity at the client communicates with a corresponding entity at the server. It has replaced the earlier Secure Sockets Layer (SSL) protocol. Defined as a proposed Internet Engineering Task Force (IETF) standard in 1999, the TLS 1.3 version was defined in 2018.

At the start of a session, the client requests the server to set up a TLS connection. While transport layer port 80 is used for un-encrypted HTTP traffic, port 443 is often used for establishing a transport layer connection that is used for encrypted HTTPS message flows. Subsequently, a stateful handshaking process is followed by the communicating server and its client, as the following process is used:

1) During a preliminary **handshaking communications phase**, secure session parameter settings are configured, including the configuration of a session-specific **shared encryption key**. For this purpose, an asymmetric cryptological process is used, typically employing the following steps:

    a) The client connects to the server, presenting a list of supported ciphers and hash functions.

b) The server selects from the list a cipher and a hash function that it supports and informs the client of its selection.

c) The server provides the client a digital certificate that identifies itself and its **public encryption key**. It also cites the *certificate authority (CA)* that can be used to verify the authenticity of the certificate.

d) The client processes the received information and confirms the validity of the certificate.

e) A unique shared session key is generated, serving to encrypt and decrypt the data that is subsequently sent during the session. Different methods are supported for exchanging keys, encrypting data, and authenticating message integrity. For example, methods that can be used for the creation of a shared session secret key include:

   i) The client uses the server's public key to encrypt a random number and send the outcome to the server. The server decrypts it by using its private key. This random number serves as the shared secret key.

   ii) Alternatively, a Diffie–Hellman key exchange process can be used to create a session shared secret key. This method also provides *forward secrecy*, as if a private key is compromised and becomes known in the future, an intruder is still not able to use it to decrypt past data exchanged during the session.

2) During the subsequent **data communications phase**, exchanged data messages use a symmetric-key encryption process. For this purpose, the session-specific shared secret key that has been configured during the handshaking phase is employed. Such a key, which is only valid during the lifetime of a communications session, is also known as an *ephemeral key*.

Over time, TLS and other secure message transport protocol implementations are being updated and extended as security weaknesses are being discovered.

As noted before, steps can be taken to assure the integrity of a message and provide for non-repudiation of its authorship. **Message integrity** assures the destination entity that the message has not been altered by an intruder that intercepts it during its transport across the network. For this purpose, the following is a commonly employed scheme. The source uses a hash function to calculate a message digest that is added to the encrypted data that is sent across the network. When received by the destination user, the data is decrypted, a message digest is recalculated (using the mutually agreed hash function) and compared to the sent message digest. An agreement of the latter two values attests to the integrity of the received message. Since the message is encrypted, a hacker is not able to gain access to the message data, modify it, and recalculate and insert a corresponding modified digest. To provide **non-repudiation**, a **digital signature** message would be attached. A message author cannot then dispute the authorship or the validity of the associated document.

Problems

22.1 Explain the functions provided by the following (CIA triad) measures and present methods that realize these measures:

   a) Confidentiality.

   b) Integrity.

   c) Availability.

22.2 Describe and explain the functions and measures included in the NIST Cybersecurity Framework.

684 | *22 Networking Security*

22.3 Describe the principle of operation of Symmetric Encryption in aiming to provide confidentiality. Include the definitions and functions of the following elements:
a) Plaintext.
b) Ciphertext.
c) Secret Code.
d) Private Key.
e) Encryption algorithm.

22.4 Describe the principle of operation of Public Key Encryption (PKE). Include the definitions and functions of the following elements:
a) Public Key.
b) Private Key.
c) PKE algorithm.
d) Padding protocol.
e) RSA cryptography.
f) Public Key Infrastructure (PKI).
g) Registration Authority.
h) Certification Authority (CA) domains.

22.5 Describe the purpose of using Digital Signature. Include explanations of the following:
a) A digital signature scheme that employs Public Key Cryptography.
b) The role of a message hash.

22.6 Explain the operation of the Diffie–Hellman (DH) (also known as the Diffie–Hellman–Merkle) key exchange scheme. Include illustration of the exchange process via reference to Fig. 22.5.

22.7 Describe and explain the operation of Transport Layer Security (TLS).

22.8 Describe methods that are used to enhance TLS by providing the following measures:
a) Message Integrity.
b) Non-repudiation.

References

1 3GPP. Evolved universal terrestrial radio access (E-UTRA) physical layer procedure. *3GPP TS 36.213; v8.3.0; Release 8*, 2008.

2 3GPP. Evolved universal terrestrial radio access network (E-UTRAN) GPRS enhancements. *3GPP TS 23.401; v9.14.0; Release 9*, 2013.

3 3GPP. LTE evolved universal terrestrial radio access (E-UTRA) physical channels and modulation. *3GPP TS 36.211; v15.4.0; Release 15*, May 2019.

4 3GPP. Service requirements for enhanced V2X scenarios. *3GPP TS 22.186; v16.2.0; Release 16*, November 2020.

5 3GPP. System architecture for the 5G system. *3GPP TS 23.501; v16.4.0; Release 16*, March 2020.

6 3GPP. Policy and charging control architecture. *3GPP TS 23.203; Release 17*, pages 53–56, 2021.

7 N. Abramson. The throughput of packet broadcasting channels. *IEEE Transactions on Communications*, 25(1):117–128, January 1977.

8 M.H. Castaneda-Garcia, A. Molina-Galan, M. Boban, J. Gozalvez, B. Coll-Perales, T. Sahin, and A. Kousaridas. A tutorial on 5G NR V2X communications. *IEEE Communications Surveys & Tutorials*, 23(3):1972–2026, 2021.

9 H.-B. Chang and I. Rubin. Optimal downlink and uplink fractional frequency reuse in cellular wireless networks. *IEEE Journal on Vehicular Technology*, 65:2295–2308, April 2016.

10 B. Chatras, S.T. Kwong U, and N. Bihannic. NFV enabling network slicing for 5G. *Proceedings of 2017 20th Conference on Innovations in Clouds, Internet and Networks (ICIN)*, March 2017.

11 J.W. Cohen. *The Single Server Queue*. North-Holland, 1982.

12 F. Cuomo, I. Rubin, A. Baiocchi, and P. Salvo. Enhanced VANET broadcast throughput capacity via a dynamic backbone architecture. *Ad Hoc Networks Journal*, 21:42–59, 2014.

13 Department of Transportation. Vehicle safety communications applications. *National Highway Traffic Safety Administration (NHTSA) Report DOT HS 811 492A*, September 2011.

14 W. Diffie and M. E. Hellman. New directions in cryptography. *IEEE Transactions on Information Theory*, 22(6):644–654, November 1976.

15 ETSI. Intelligent transport systems (ITS) vehicular communications GeoNetworking. *Technical Specification ETSI TS 102 636-4-1 V1.1.1 Part 4 Sub-part 1*, June 2011.

16 ETSI. Intelligent transport systems (ITS) vehicular communications GeoNetworking. *Technical Specification ETSI EN 302 636-1 V1.2.1 Part 1*, April 2014.

17 ETSI. Network functions virtualisation (NFV) architectural framework. *Technical Specification ETSI GS NFV 002 V1.2.1*, 2014.

18 ETSI. Intelligent transport systems (ITS) cross layer DCC management entity for operation in the its G5A and its G5B medium. *Technical Specification ETSI TS 103 175 V1.1.1*, June 2015.

Principles of Data Transfer Through Communications Networks, the Internet, and Autonomous Mobiles, First Edition. Izhak Rubin.
© 2025 The Institute of Electrical and Electronics Engineers, Inc. Published 2025 by John Wiley & Sons, Inc.

686 | *References*

19 ETSI. Intelligent transport systems (ITS) vehicular communications GeoNetworking. *Technical Specification ETSI TS 102 636-4-2 V1.4.1 Part 4*, 2021.

20 ETSI. Network functions virtualisation (NFV) management and orchestration. *Technical Specification ETSI GS NFV 006 V4.4.1 Release 4*, 2022.

21 D. Gross, J.F. Shortle, J.M. Thompson, and C.M. Harris. *Fundamentals of Queueing Theory.* Wiley-Interscience, 2008.

22 X. Huang and I. Rubin. A mobile backbone network routing protocol with flow control. *Proceedings of IEEE MILCOM 2004 Conference*, 2004.

23 IEEE. Wireless LAN medium access control (MAC) and physical layer (PHY) specifications. *Technical Specification IEEE Standard 802.11-2007 Part 11*, June 2007.

24 IEEE. Wireless LAN medium access control (MAC) and physical layer (PHY) specifications. *Technical Specification IEEE Standard 802.11-2009 Part 11*, October 2009.

25 IEEE. IEEE guide for wireless access in vehicular environments (WAVE) architecture. *Technical Specification IEEE Standard 1609.0-2019*, February 2019.

26 ITU-T. End-user multimedia QoS categories. *ITU-T G.1010*, November 2001.

27 ITU-T. Quality of experience requirements for IPTV services. *ITU-T G.1080*, December 2008.

28 J. Iyengar and I. Swett. QUIC loss detection and congestion control. *Internet Engineering Task Force (IETF) RFC 9002*, May 2021.

29 J. Iyengar and M. Thomson. QUIC: A UDP-based multiplexed and secure transport. *Internet Engineering Task Force (IETF) RFC 9000*, May 2021.

30 J.R. Jackson. Networks of waiting lines. *Operations Research*, 5(4):518–521, 1957.

31 H.-J. Ju and I. Rubin. Backbone topology synthesis for multi-radio meshed wireless LANs. *Proceedings of IEEE INFOCOM 2006 Conference*, April 2006.

32 H.-J. Ju and I. Rubin. Backbone topology synthesis for multiradio mesh networks. *IEEE Journal on Selected Areas in Communications*, 24(11):2116–2126, November 2006.

33 L. Kleinrock. *Queueing Systems I: Theory.* Wiley-Interscience, 1975.

34 L. Kleinrock. *Queueing Systems II: Computer Applications.* Wiley-Interscience, 1976.

35 L. Kleinrock and F. Tobagi. Packet switching in radio channels: Part I-carrier sense multiple-access modes and their throughput-delay characteristics. *IEEE Transactions on Communications*, 23(12):1400–1416, December 1975.

36 National Institute of Standards and Technology. NIST cybersecurity framework. *NIST Document 1.1 and NIST Publication 1271*, 2021.

37 NGMN. Next generation mobile networks (NGMN) 5G white paper 1. *Next Generation Mobile Networks (NGMN) Alliance*, February 2015.

38 NGMN. Description of network slicing concept. *Next Generation Mobile Networks (NGMN) Alliance 5G White Paper*, January 2016.

39 NGMN. Next generation mobile networks (NGMN) 5G white paper 2. *Next Generation Mobile Networks (NGMN) Alliance*, July 2020.

40 Open Networking Foundation. SDN architecture overview. *Technical Report ONF TR-504 Version 1.1*, November 2014.

41 R. Rivest, A. Shamir, and L. Adleman. A method for obtaining digital signatures and public-key cryptosystems. *Communications of the ACM*, 21(2):120–126, February 1978.

42 I. Rubin. Message path delays in packet-switching communication networks. *IEEE Transactions on Communications*, 23(2):186–192, February 1975.

43 I. Rubin, A. Baiocchi, F. Cuomo, and P. Salvo. GPS aided inter-vehicular wireless networking. *Proceedings Information Theory and Applications workshop (ITA)*, San Diego, CA, USA, pages 125–148, February 2013.

44 I. Rubin, A. Baiocchi, Y. Sunyoto, and I. Turcanu. Traffic management and networking for autonomous vehicular networks. *Ad Hoc Networks Journal*, 83:125–148, 2019.

45 I. Rubin, Y.-Y. Lin, A. Baiocchi, F. Cuomo, and P. Salvo. Micro base station aided vehicular ad hoc networking. *Proceedings ICNC 2014 Conference*, February 2014.

46 I. Rubin and R. Zhang. Robust throughput and routing for mobile ad hoc wireless networks. *Ad Hoc Networks Journal*, 7(2):265–280, March 2009.

47 J.D. Spragins, J.L. Hammond, and K. Pawlikowski. *Telecommunications Protocols and Design*. Addison Wesley, Reading, MA, 1994.

48 K. Zheng, Q. Zheng, P. Chatzimisios, W. Xiang, and Y. Zhou. Heterogeneous vehicular networking: a survey on architecture challenges and solutions. *IEEE Communications Surveys & Tutorials*, 17(4):2377–2396, 2015.

Index

a

access link 3
analog signals 54
analog to digital discretization 55
audio streaming 62, 66
autonomous vehicular highway
  automation levels 648
  autonomy 650
  Cellular Vehicle-to-Everything (CV2X)
    Networking 641–648
  data networking services 608–610
  flow and congestion control 625–628
  GeoNetworking 616
  Infrastructure Aided V2V Networking
    638–641
  networking configurations 610–612
  networking methods 613
  selection of relay nodes 616–625
  traffic management 650–664
  VBN backbone synthesis 633–635
  VBN performance 635–638
  Vehicular Backbone Networks (VBNs)
    628–632
autonomous vehicular networks 43

b

bandwidth 58
broadcast flow 5
broadcast link 4
broadcasting 4
bursty data flow 90

c

cellular wireless networks
  architecture 544–546

Channel Quality Indicator (CQI) 559
components 549
core network 555
4G LTE system 554–560
handoff 553
location update 552, 554
network generations 546–549, 551
next generation networks 560–562
paging 553
QoS Class Identifier (QCI) 559
radio access network (RAN) resource reuse
    544, 546
radio link control (RLC) 555
radio resource control (RRC) 555
the process 551–554
transport blocks 554
channel capacity 104–106
circuit 7
circuit switching 7, 345–367
  cross-connect networking 359
  Grade of Service (GOS) 358, 359
  performance 356–359
  signaling 354–356
  switching fabric 351–354
communications channel 98–102
communications link 3
communications system 97, 98
connection oriented networking 7
connection oriented packet switching
  method 372–373
  Asynchronous Transfer Mode (ATM) networks
    376
  performance 379–384
  processes 373–374
  technologies 376–379

Principles of Data Transfer Through Communications Networks, the Internet, and Autonomous Mobiles, First Edition. Izhak Rubin.
© 2025 The Institute of Electrical and Electronics Engineers, Inc. Published 2025 by John Wiley & Sons, Inc.

690 | *Index*

d

datagram packet switching
the method   387–390
performance   392–395
the router   390–392
demand assigned multiple access (DAMA)   258
digital image   79
digital modulations   110–115
digital signals   54
discrete cosine transform (DCT)   74
discrete-event simulations   229
duplex link   3

e

end-node   5
end-to-end (ETE) flow   3
end users   2, 47
error control
Automatic Repeat Request (ARQ)   406–407
Error Detection Coding   404–406
forward error correction (FEC)   400–404
Go-Back-N sliding window ARQ   415–420
Hybrid ARQ (HARQ)   423–428
methods   397–399
parameters   399–400
selective-repeat ARQ   420–423
Stop-and-Wait ARQ   407–415

f

fixed multiple access
Code Division Multiple Access (CDMA)
253–258
fixed multiple access scheme   244
Frequency Division Multiple Access (FDMA)
246–249
Space Division Multiple Access (SDMA)
249–253
Time Division Multiple Access (TDMA)
244–246
flow and congestion control
closed-loop flow control   434–436
configurations   431–434
leaky-bucket algorithm   443
open-loop flow control   436–444
proactive congestion control   448–451
reactive congestion control   444–448
token-bucket algorithm   439–443

g

frequency spectra   59, 62
full-duplex (FDX) link   26

g

Generalized Multi-Protocol Label Switching
(GMPLS)   367
Geocast flow   5
geometric point process   130

h

half-duplex (HDX) link   26

i

the Internet
HTTP   482–484
inter-domain routing using BGP   511–517
Internet Protocol (IP)   478
intra-domain routing using OSPF   509–511
IP addresses   485–492
IP packets   492–496
ISP hierarchy   480, 482
networking architecture   476–482
QUIC protocol   503–508
routing   508–517
TCP/IP model   478
Transmission Control Protocol (TCP)
496–501
User Datagram Protocol (UDP)   501–503
Internet of Things (IOT)   44

l

label switching   366
layer-2 switching
bridging   321–341
control plane   334–337
equal cost path routing   337
flooding protocol   324–325
IS-IS protocol   334
local area network (LAN)   321
Shortest Path Bridging (SPB)   321, 330–334
spanning tree protocol   325, 330
Virtual LAN (VLAN)   321
LTE Channel Quality Indicator (CQI)   115

m

message flow(s)   5, 48, 51
mobile ad hoc wireless networks (MANETs)

AODV protocol   569–573
DSR protocol   573–576
MBN performance   587
MBN routing   588, 591
methods   567, 569
Mobile Backbone Networks (MBNs)   582
multi-radio MBN   591
OLSR protocol   576–581
UAV aided MBN   587
UGV aided MBN   587
mobile wireless networks
  configurations   541–543
modulation/coding schemes (MCS)   26, 107–110
Monte Carlo computer simulation   228
Monte Carlo Simulation(s)   227–230
multi protocol label switching (MPLS)   366
multi-level traffic model   119–122
multicast flow   5
Multiple Access
  ALOHA random access   275–284
  Carrier Sense Multiple Access (CSMA)
    284–291
  Carrier Sense Multiple Access with Collision
    Avoidance (CSMA/CA) and WiFi WLAN
    297–306
  Carrier Sense Multiple Access with Collision
    Detection (CSMA/CD)   291–295
  demand assigned reservation   258–262
  Ethernet Local Area Network (LAN)   295, 297
  FDDI token ring network   268
  hub polling   263
  implicit polling   269
  multiple access scheme   242
  Performance of ALOHA algorithms   277–284
  performance of polling schemes   269–272
  polling   262–269
  random access schemes   272
  resource reuse and graph coloring   249, 253
  Slotted ALOHA algorithm   276–277
  token-passing polling   266
  token ring protocol   267
  Unslotted ALOHA algorithm   275
multiple access link   4
Multiple-In Multiple-Out (MIMO) antenna
    configuration   252, 253
multiplexing   162, 164
  Code Division Multiplexing (CDM)   170–171

fixed multiplexing   164–165
Frequency Division Multiplexing (FDM)
    169–170
One-to-Many Media   183–186
Proportional Fair Scheduling (PFS)   185, 186
scheduling algorithms   173–183
Space Division Multiplexing (SDM)   171
statistical multiplexing   171–173
Time Division Multiplexing (TDM)
    166–169
Wavelength Division Multiplexing (WDM)
    170
multipoint-to-point link   4

n

network architecture   9
  control plane   27, 30–32
  data plane   27, 29–30
  management plane   28, 32–34
network graph   5
  connected graph   5
network switch   4
network topology   5
networking security
  digital signature   679–680
  framework   671–674
  key exchange   680–682
  public key encryption (PKE)   677–679
  symmetric encryption   675–676
  transport layer security (TLS)   682, 683
Next Generation Networking
  Access Network Convergence   603
  Artificial Intelligence (AI) and Machine
    Learning (ML)   603
  Autonomic Management and Control   603
  Cloud based operation   602
  Cloud Computing   596
  Edge Computing   602
  Network Functions Virtualization (NFV)
    599–601
  Network Slicing   601–602
  Network Virtualization   597
  Open Radio Access Network   603
  Software Defined Networking (SDN)
    597–599
  Virtual Machine (VM)   596
nodal degree   5

692 | *Index*

node
  end-user node 4
  network node 4

p

packet switching 8
peer-to-peer 88
performance measures 87, 90
  quality of experience (QoE) 88
  quality of service (QoS) 88
personal area networks (PANs) 528–539
  Bluetooth 528–533, 539
  Zigbee 533–539
point-to-multipoint link 4
point-to-point link 4
Poisson process
  Poisson counting process 128
  Poisson point process 126
protocol entity 14
Protocols
  Application Layer 19, 20
  Link Layer Control (LLC) protocol 25
  Link Layer Protocols 25, 26
  Logical Link Control (LLC) protocol 25
  Medium Access Control (MAC) protocol 26
  Network Layer Protocols 24, 25
  networking protocols 9
  OSI Reference Model 18, 20, 26
  Physical Layer Protocol 26
  Presentation Layer 20
  protocol layer entity 15
  protocol layers 14, 17
  Session Layer 20
  Transport Layer Protocols 20, 24
proxy server 84
Public Switched Telephone Network (PSTN) 67

q

quality of experience (QoE) 157, 159
quality of service (QoS) 144, 157
  availability 151
  cellular wireless network 155
  error rate 150
  message delay 148
  reliability 152
  security 153
  throughput 144

queueing
  basic model 189
  blocking probability 196
  busy cycle 195
  busy period 195
  delay time 194
  Discrete-Event MM1 Simulation 230, 234
  idle period 195
  Little's Formula 197
  M/G/1 system 213, 218
  M/M/1 system 201, 207
  M/M/1/N system 207, 208
  M/M/m, M/M/m/N systems 208, 213
  Markovian systems 199
  number served messages 196
  offered rate 196
  priority queueing 218, 221
  processes 192, 196
  properties 196, 199
  queue size 194
  queueing network 222, 227
  service size 194
  system size 194
  throughput 196
  unfinished load 195
  utilization 196
  virtual waiting time 195
  waiting time 194
QUIC 21

r

renewal point process 125, 129
routing
  alternate routing 8
  autonomous system routing domains 457–460
  centralized routing 461
  Dijkstra's Algorithm 472
  distance vector routing 465–470
  distributed routing 461
  dynamic routing 461
  label switching 463
  link state routing 470, 473
  on-demand routing 463
  path selection 453, 455
  proactive routing 463
  reactive routing 463

route metrics 455, 457
router 5
routing 5
routing algorithm 5, 6, 8
routing table 5
Shortest Path Tree (SPT) 464, 465
source routing 462
static routing 461

s

satellite communications networks 41, 42
service classes 51, 53
service rate 6
Session Initiation Protocol (SIP) 89
Signal-to-Noise (SNR) Ratio 102
Signal-to-Noise-plus-Interference Ratio (SINR) 102
signaling 7
  signaling phase 7
simplex link 3, 26
spectral efficiency 104
streaming 50
switching
  autonomous vehicle highway 320
  circuit switching 315
  datagram packet switching 316
  on-board processing satellite 318, 320
  packet switching 315, 316
  reflecting repeating hub 318
  repeaters 317, 320
  switching nodes 309, 315
  virtual circuit switching 316

t

TCP 21
TCP/IP model 26, 27
time domain and frequency domain 56

time-driven simulations 229
traffic
  burst level rates 134
  carried load rate 134
  Erlang loading 132
  message level rates 134–136
  offered load rate 133
  terminal flow rates 136
  traffic matrix 137
  traffic rates 132, 136
traffic intensity 6
traffic process
  message arrivals and lengths 123, 129
tunneling 366

u

unicast flow 5
User Datagram Protocol (UDP) 21
utilization 6

v

Vehicular Ad Hoc Network (VANET) 43
video compression 81–83
video streams 77, 81
video stream transport 83–86
virtual circuit switching (VCS) 7
Voice over IP (VoIP) 68–75
  quality metrics 75–77

w

wireless communications media 95–97
wireless local area network (WLAN)
  WiFi 523–528
wireline communications media 94–95
  coaxial cable 95
  copper cable 94
  fiberoptic cable 95